Three hundred years of gravitation

PHILOSOPHIÆ

NATURALIS

PRINCIPIA

MATHEMATICA·

Autore *IS. NEWTON*, *Trin. Coll. Cantab. Soc.* Matheseos
Professore *Lucasiano*, & Societatis Regalis Sodali.

IMPRIMATUR·
S. PEPYS, *Reg. Soc.* PRÆSES.
Julii 5. 1686.

LONDINI,

Jussu *Societatis Regiæ* ac Typis *Josephi Streater*. Prostat apud
plures Bibliopolas. *Anno* MDCLXXXVII.

Three hundred years of gravitation

EDITED BY

S.W.HAWKING

Lucasian Professor of Mathematics, University of Cambridge

W.ISRAEL

*Professor of Physics, University of Alberta, and
Senior Fellow, Canadian Institute for Advanced Research*

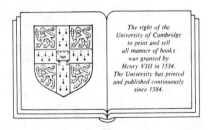

The right of the
University of Cambridge
to print and sell
all manner of books
was granted by
Henry VIII in 1534.
The University has printed
and published continuously
since 1584.

CAMBRIDGE UNIVERSITY PRESS

Cambridge

New York   New Rochelle   Melbourne   Sydney

Published by the Press Syndicate of the University of Cambridge
The Pitt Building, Trumpington Street, Cambridge CB2 1RP
32 East 57th Street, New York, NY 10022, USA
10 Stamford Road, Oakleigh, Melbourne 3166, Australia

First published 1987

Printed and bound in Great Britain by
Redwood Burn Limited, Trowbridge, Wiltshire

British Library cataloguing in publication data
Three hundred years of gravitation.
1. Gravitation
I. Hawking, S. W.   II. Israel, W.
531'.14   QC178

Library of Congress cataloguing in publication data
Three hundred years of gravitation.
Includes bibliographies.
1. Gravitation.   2. Cosmology.   3. Astrophysics.
I. Hawking, S. W. (Stephen W.)   II. Israel, W.
III. Title: 300 years of gravitation.
QC178.T47   1987   531'.14   87-10364

ISBN 0 521 34312 7

MP

Contents

Contributors

R. Blandford
Department of Theoretical Astrophysics, California Institute of Technology, Pasadena, California 91125, USA

S. K. Blau
Department of Physics, University of Texas, Austin, Texas 78712, USA

A. H. Cook
The Master's Lodge, Selwyn College, Cambridge

C. Crnković
Department of Physics, Joseph Henry Laboratory, Princeton, New Jersey 08544, USA

T. Damour
Department of Fundamental Astrophysics, Observatoire de Paris, 5 Place Janssen, 92195 Meudon, France

A. Guth
Department of Physics, Massachusetts Institute of Technology, Cambridge, Massachusetts 02139, USA

S. W. Hawking
Department of Applied Mathematics and Theoretical Physics, Silver Street, Cambridge CB3 9EW, UK

W. Israel
Avadh Bhatia Physics Laboratory, University of Alberta, Edmonton T6G 2J1, Canada

A. Linde
P. N. Lebedev Institute of Physics, Academy of Sciences of USSR, Leninski Prospect 53, 117924 Moscow, USSR

R. Penrose
Mathematical Institute, 24–29 St Giles, Oxford OX1 3LB, UK

M. J. Rees
Institute of Astronomy, Madingley Road, Cambridge CB3 0HA, UK

J. H. Schwarz
Department of Theoretical Astrophysics, California Institute of Technology, Pasadena, California 91125, USA

K. S. Thorne
Department of Theoretical Astrophysics, California Institute of Technology, Pasadena, California 91125, USA

A. Vilenkin
Department of Physics and Astronomy, Tufts University, Medford, Massachusetts 02155, USA

S. Weinberg
Department of Physics, University of Texas at Austin, Texas 78712, USA

C. M. Will
McDonnell Centre for Space Science, Department of Physics, Washington University, St Louis, Missouri 63130, USA

E. Witten
Department of Physics, Joseph Henry Laboratory, Princeton, New Jersey 08544, USA

Preface

On 5 July 1687 Edmond Halley wrote from London to Isaac Newton in Cambridge concerning his book, the *Philosophiae Naturalis Principia Mathematica*, which, in view of the Royal Society's straitened circumstances at the time, Halley had undertaken to publish at his own expense:

Honoured Sr

I have at length brought your Book to an end, and hope it will please you. The last errata came just in time to be inserted. I will present from you the books you desire to the R. Society, Mr Boyle, Mr Pagit, Mr Flamsteed and if there be any elce in town that you design to gratifie that way; and I have sent you to bestow on your friends in the University 20 Copies, which I entreat you to accept.

Three centuries later, the heavens themselves appear to have joined in the commemoration of this historic moment. At the time of writing, neutrino, X-ray and optical astronomers are sorting out the first messages from Supernova 1987a, at once a tribute to the birth of gravitational theory and a signpost to its future.

We had the same dual purpose in mind when soliciting contributions to this volume. We wanted to mark the 300th anniversary of this important occasion with suitably reflective essays. But gravitational theory is very far from being a closed chapter of scientific history, and most of the articles review current developments and future prospects. They are self-contained, but designed also to supplement and update the articles in a book we edited previously to mark the centenary of Albert Einstein's birth.†

---

† *General Relativity: An Einstein Centenary Survey*, ed. S. W. Hawking and W. Israel. Cambridge University Press: Cambridge 1979.

The intervening eight years have witnessed dramatic progress on both the observational and theoretical fronts. The earlier volume was already in press when Joseph Taylor announced measurements of the orbital speed-up of the binary pulsar which confirmed the value predicted from Einstein's formula for the loss of energy by gravitational radiation. In 1979, this spectacular vindication could be noted only as a last-minute tabular entry in the proof stages of Clifford Will's article. In Chapter 5 Will carries the story of this and other experimental tests of Einstein's theory up to the present.

The flood of data from the binary pulsar has spurred a vigorous theoretical effort finally to bring under control the two-body problem in general relativity, long embroiled in controversy and confusion. Thibault Damour, who has been in the vanguard of these advances, reviews their current status in Chapter 6.

It is now a realistic prospect that, before the turn of the century, direct observation of gravitational waves using laser interferometry, will have become possible and even routine. In Chapter 9, Kip Thorne presents a comprehensive review of the relevant theoretical and experimental aspects.

Cygnus X-1, the only firm black hole candidate known in the 1970s, has since been joined by other binary X-ray sources in our galaxy and in the Large Magellanic Cloud. Meanwhile circumstantial evidence for supermassive holes in the nuclei of galaxies (including our own) has continued to build up steadily. Roger Blandford critically reviews the current observational evidence for black holes in Chapter 8.

Of the recent theoretical developments, one of the most publicised and most exciting has been superstring theory (reviewed in Chapter 15 by John Schwarz, a leading pioneer in the field), which for the first time offers the hope of including the gravitational force within a unified and finite theory. Another is the great upsurge of interest in the very early universe, sparked by Alan Guth's formulation of inflationary cosmology in 1981. Guth's idea was, in part, inspired by a brilliantly imaginative chapter in the Einstein centenary volume co-authored by Robert Dicke and Jim Peebles, which spotlighted the horizon problem of the standard big-bang cosmology as well as 'the remarkable balance between mass density and expansion rate'. Chapters 12–14 by Blau and Guth, Linde and Hawking survey current ideas on inflationary and quantum cosmology.

The inflationary hypothesis makes it difficult to escape the conclusion that perhaps 99% of the matter in the universe must be in some invisible form. Martin Rees in Chapter 10 discusses various possibilities for the nature of this hypothetical dark matter and its influence on the manner in which galaxies form. The mechanism of galaxy formation is being actively studied, and this branch of cosmology has developed into a mature science in which detailed theoretical scenarios are matched to sophisticated observational data on the clustering of galaxies, the isotropy of the microwave background and the abundance of the light elements.

These brief remarks must serve as a very selective and inadequate introduction to the many notable recent advances and to the contents of this volume.

The book is being published in association with a Newton Tercentenary Conference (29 June to 4 July 1987) at Trinity College, Cambridge – Newton's home from 1661 until 1696. We thank the contributors for their excellent work and the staff of Cambridge University Press for performing their customary miracle. Our secretaries, Cheryl Billington and Mary Yiu, as usual have been a tower of strength. We gratefully acknowledge the financial support of Trinity College, Cambridge, the Ralph Smith Foundation, the Canadian Institute for Advanced Research and the Natural Sciences and Engineering Research Council of Canada.

S. W. Hawking
W. Israel

1

Newton's Principia

S.W.HAWKING

The *Philosophiae Naturalis Principia Mathematica* by Isaac Newton, first published in Latin in 1687, is probably the most important single work ever published in the physical sciences. Its significance is equalled in the biological sciences only by *The Origin of Species* by Charles Darwin. The original impulse which caused Newton to write the *Principia* was a question from Edmond Halley as to whether the elliptical orbits of the planets could be accounted for on the hypothesis of an inverse square force directed towards the Sun. This was something that Newton had worked out some years earlier but had not published, like most of his work on mathematics and physics. However, Halley's challenge, and the desire to refute the suggestions of others such as Hooke and Descartes, spurred Newton to try to write a proper account of this result. In order to do so he had first to develop a Theory of Mechanics and the mathematical techniques needed to analyse this theory. The old Aristotelian idea had been that rest was the natural state of a body and that a body moved only if driven by some cause. However, Newton substituted for this his First Law: 'Every body continues in its state of rest, or of uniform motion in a straight line, unless it is acted upon by a force.' Newton's Second Law stated that the acceleration, or rate of change of velocity, was proportional to the force.

These laws formalised the Theory of Mechanics that Galileo had implicitly assumed though not explicitly stated. In the case of a force that was constant in magnitude and direction, like the gravitational force near the surface of the Earth, Galileo had shown that they implied that the distance moved by a body initially at rest would be proportional to the square of the time elapsed. However, the question of the motion of the planets was much more difficult because the force on them varied both in magnitude and direction. Newton seems to have solved the problem originally using what we would now call differential calculus, which he had invented along with integral calculus. However, for some reason he did not think that calculus, or 'the theory of fluxions', as he called it, was suitable for publication in the *Principia*. He therefore developed a new method to show that a body that was attracted towards a fixed point by a force proportional to the inverse square of the distance would move on an elliptical orbit and would obey the laws that Kepler had discovered by observation. Newton's failure to publish his theory of fluxions led Leibnitz to discover calculus independently and to a bitter dispute about priority.

Not only did Newton explain the elliptical orbits of the planets, but he also showed that the same force of gravity was responsible for the orbits of the satellites of Jupiter and Saturn, the orbit of the Moon about the Earth, the tides in the sea and the fall of an apple off a tree. In other words, Gravity is Universal: every body in the universe attracts every other body with a force that is proportional to its mass and inversely proportional to the square of the distance between them. By Newton's Third Law, which states that action and reaction are equal and opposite, every body is similarly attracted towards every other body. It was natural to suppose that this universality might extend to light which was first shown to travel at a finite velocity by Roemer in 1676. Roemer observed that there were apparent discrepancies between the timing of the orbits of the moons of Jupiter and the predictions of Kepler's theory. These discrepancies disappeared if he assumed that light travels at a finite velocity and therefore takes a variable amount of time to cross the variable distance between Jupiter and the Earth. If light travels at a finite velocity and if it is affected by gravity in the same way as planets or cannonballs, then one can imagine a situation in which one has a body which is so massive and compact that light cannot escape from it but gets dragged back by gravity. The possible existence of such objects was pointed out by Michell in 1783 and by Laplace in 1799 but not much notice was taken at the time. Indeed, Laplace removed the suggestion from the later editions of his book *Exposition du Système du Monde*. However, we now believe that these objects, which we call black holes, are the final stage of evolution of massive stars and that matter falling into them provides the power for compact X-ray objects and the radio and optical emission that we observe from quasars, very compact and luminous sources that we can see even at the other side of the universe.

Another consequence of the universality of gravity that was not properly appreciated until much later was that it was inconsistent with the prevailing idea that the universe was infinite in extent and filled with a distribution of stars with a density that was more or less constant in space and time. If the stars were at rest with respect to each other at one time, the gravitational attraction of each star for the others would cause them to start to fall together and the density would go up with time. Alternatively, the stars could be moving away from each other. In that case the density would decrease with time. Gravity would act to slow down the recession of stars from one another and might eventually stop the expansion of the universe and convert it into a contraction. Newton was aware of this problem but he argued that, in an infinite distribution of stars, the force on a star caused by the attraction of other stars to one side of it (say, the right side) would be almost exactly balanced by the force of attraction of the stars to the left of the star. The net force on the star would therefore be small and the system of stars could continue to exist at more or less constant distances from each other. However, there is a flaw in this argument. The force of attraction produced by the stars to the right of the given star would be infinite because, although the force produced by each star to the right gets smaller the further away the star is from the given star, the number of stars at a given distance gets larger the further one is from the given star. Similarly, the force of attraction produced by the stars to the left would be infinite. When one subtracts infinity from infinity, it is well known that one can get any answer one wants. We now know that the universal attractive nature of gravity is inconsistent with a static infinite universe: the universe has to be either expanding or contracting. This was not realised, however, until in the 1920s observations of distant galaxies revealed that the universe is expanding. It was a great missed opportunity for theoretical physics: Newton could have predicted the expansion of the universe.

In the *Principia* Newton introduced the concepts of Absolute Time and Absolute Space. By Absolute Time he meant that there was a quantity called time which would be measured by all properly constructed clocks, regardless of their motion. This was an idea that nobody questioned until the advent of the Theory of Relativity at the beginning of this century. Much more controversial was Newton's advocacy of Absolute Space. Newton argued that in the case of circular motion there was a preferred state of rest in which there were no centrifugal forces. He therefore claimed that by analogy there ought to be a preferred state of Absolute Rest with respect to linear motion, though he admitted that it would be difficult to determine this state of rest by observations. He suggested, as a hypothesis, that this state of rest coincided with the centre of gravity of the solar system. In fact Newton's

Laws of Motion are the same in all frames of reference moving uniformly in straight lines though they are not the same in rotating frames. Thus the laws do not determine a state of rest and so do not support the concept of Absolute Space, though they are not inconsistent with it. They do, however, require the concept of Absolute Time. In other words, in Newton's theory one can determine whether two events at different points in space occur at the same time but one cannot tell whether events at different times occur at the same point of space: that depends on the frame of reference and would be different in two frames that were moving relative to each other.

The laws presented in the *Principia* remained the accepted theories of mechanics and gravity for more than two hundred years. Even today they are the basis of nearly all practical calculations. It is only in very extreme situations that one has to take into account the modifications introduced by the Special and General Theories of Relativity formulated by Albert Einstein in 1905 and 1915 respectively. In the Theory of Relativity one has to abandon the concept of Absolute Time as well as that of Absolute Space. The time measured by a clock now depends on its velocity and one can no longer determine whether events at different points of space occur at the same time: again this depends on the motion of the frame of reference. The philosophical implications of this change from Absolute Space and Time to Relative Space and Time have been profound but have often not been properly appreciated. Lenin wrote a pamphlet attacking Relativity because he thought it was a threat to the absolute system of Hegel and Marx. It is only in the last thirty years that the study of relativity has become respectable in the Soviet Union. Other people in the West reacted in a similar way. Einstein was accused of undermining moral standards by suggesting that everything was relative. However, this attack and that of Lenin were based on a misunderstanding of the Theory of Relativity: it is a mathematical model of space and time. It makes no statement about how human affairs should be conducted or organised.

Einstein is the only figure in the physical sciences with a stature that can be compared with Newton. Newton is reported to have said: 'If I have seen further than other men, it is because I have stood on the shoulders of giants.' This remark is even more true of Einstein who stood on the shoulders of Newton. Both Newton and Einstein put forward a theory of mechanics and a theory of gravity but Einstein was able to base General Relativity on the mathematical theory of curved spaces that had been constructed by Riemann while Newton had to develop his own mathematical machinery. It is therefore appropriate to acclaim Newton as the greatest figure in mathematical physics and the *Principia* is his greatest achievement.

2

Newtonianism and today's physics

STEVEN WEINBERG

The written version of my talk at the Newton tercentenary celebration in Cambridge was not ready in time to appear in this book. Instead, I have submitted this talk, which was originally presented at a similar celebration at the University of Texas at Dallas, on 16 April 1986. I am grateful to the editors for allowing me to make this substitution.

Newton was a complicated man, who lived in complicated times. It would take the talents of a historian to say anything new and useful about Newton himself. I am not a historian, and know better than to try. I am a physicist, and will talk here not about Newton, but Newtonianism; that is, about Newton's work, his achievement, as an example which over the last three hundred years has guided the evolution of the science in which I now work; Newton, in other words, to use a much over-used word, as paradigm.

  The central part of the Newtonian achievement I suppose is his theory of the solar system, which had three parts, each one of which would by itself guarantee his immortality. First, there is Newton's theory of gravity, which explains how the distribution of all the matter in the universe determines the gravitational force on any one particle, whether that particle is a particle here on earth or a particle on the sun or the moon or the stars. Second, there is a law of motion that describes how, given the forces acting on a particle, one can calculate its motion. Third, there is a method of calculation, the mathematical tool now known as calculus. (I gather that Newton was apparently not at ease about calculus. He often presents his derivations in terms of historically sanctified geometrical methods. However, calculus has become the machinery by which subsequent generations of physicists have learned to think about these problems.)

The first question that I as a physicist should address is: how well does it work? How does Newtonianism stand up after three hundred years of experimental testing? It stands up very well. We do now understand that there are small corrections to the Newtonian picture of gravity and dynamics. The most interesting and important corrections are those provided by Einstein's general theory of relativity, but they're awfully small. For example, in the motion of the earth around the sun, the corrections due to general relativity are of the order of one part in a hundred million, so small in fact that they've never been detected. We detect the effects of relativity only for more rapidly moving bodies like the planet Mercury, and for light itself.

However, whether the corrections are large or small doesn't seem to me to be the point. In 1919, when Einstein's theory began to be popular, *The Times* of London proclaimed that Einstein had disproved Newton. That's very far from the truth. Einstein's theory reduces to Newton's theory in the limit of slowly moving bodies at large distances from each other, which is certainly true of the outer part of the solar system, and indeed of most of the universe.

In fact, it is fair to say that, not only does Einstein's theory not supplant Newton's theory, it explains Newton's theory. (I think this is a point that's not often appreciated.) In Newton's work the inverse square law appears as a means of accounting for the observations of the solar system, particularly Kepler's interpretation of Brahe's observations in terms of a relationship between periods and radii of orbits. For Newton it was just a fact, learned from experiment, that the gravitational force fell off as the inverse square of the distance. As far as Newton was concerned, if the force had fallen off as the inverse cube of the distance or the inverse 2.1 power of the distance that would have been perfectly fine. There was no explanation in Newton's theory of why it had to be an inverse square law. This explanation was finally provided in 1915 and 1916, by Einstein's general theory of relativity. If you adopt Einstein's insight that the force of gravity is due to a curvature in spacetime, and follow that insight to where it leads you, you find that you cannot without standing on your head make a theory of gravity in which the force at large distances is anything but an inverse square law. (What the force is at short distances is a very complicated question to which I will return.)

Newton's theory then works very well, and it now has a rationale that it lacked in Newton's time, provided by general relativity. We use it to follow the motion of objects in the solar system, some of them objects that we put there ourselves. I think nothing was more dramatic as an example of our faith in the Newtonian theory than the Apollo project. In its early stages, remember that astronauts were sent out in spacecraft which circled the moon, orbited several times, were given a little boost from a rocket motor, and then came back to the earth with essentially all the fuel gone; just relying on the validity of Newton's laws. It was somehow very appropriate that Major William Anders on the return from the first circumlunar voyage in December 1968, made the remark, 'I think Isaac Newton is doing most of the driving right now.' (I'm sure that other talks and articles about Newton this year will provide you with wonderful quotes about Newton, quotes from Pope and Halley and Blake; I just wanted to drag in one quote that they probably hadn't used.)

So Newton's theory is not challenged as a theory of the solar system. What about Newton's theory as a model for physical theory in general?

I think the qualitative feature of Newton's theory that has come under the most intense and controversial discussion in this century has been its determinism. (Though Newton himself may not have been so much of a determinist – not as much, say, as Leibniz.) Look at the picture that Newtonianism provides: At every instant each particle in the universe experiences a force determined by the positions of all the other particles and their masses. Knowing that force at that instant you can calculate the acceleration; knowing the acceleration and velocity at one instant you can then calculate the velocity that the particle will have in the next instant. Also, knowing the velocity it has, you can calculate the position it will have in the next instant. So, the solar system, and in fact the whole universe, goes grinding along, with its configuration in any one instant determining what its configuration will be in the next instant. This is a picture that is deeply satisfying to some and frightening to others.

In this century the determinism of Newton's theory came under profound challenge from the greatest revolution in physics since the time of Newton, the advent of quantum mechanics in the 1920s. Quantum mechanics was a new way of looking at nature, that was particularly appropriate for matter on the scale of atoms and, below the scale of atoms, for atomic nuclei and elementary particles. Quantum mechanics is the basic tool we use to understand physics in the small. Quantum mechanics, as I think almost everyone has heard, describes nature not by means of a language of particles that have definite positions and velocities and move along clearly defined trajectories, but instead in what sounds like a much cloudier language, a language of probabilities. We are not allowed to ask where an electron is inside a hydrogen atom at any one time, and if we try to find out, the experiment that we do breaks up the atom, and we are unable to answer the question that we set out to answer. Because in quantum mechanics one talks in terms of probabilities, there has grown up especially in the last decade or so an idea, fostered by some popular books, that quantum mechanics is somehow closer to a gentler, more mystical, view of nature than the hard, brutal deterministic view of Newton. Here Newton is being cast as a kind of cosmic Scrooge who denies particles any volition, whereas the developers of quantum mechanics in the 1920s are seen as gentle, soft-spoken flower children who are going to return physics to a universal mysticism.

Nothing could be farther from the truth. Quantum mechanics in fact provides a completely deterministic view of the evolution of physical states. In quantum mechanics, if you know the state at one instant, and of course if you know the machinery of the system, you can then calculate with no lack of precision what the state is in any subsequent instant. The indeterminacy of quantum mechanics comes in when human beings try to make measurements, because in making measurements we inevitably disturb the system in ways that can't be predicted, but the system itself evolves in a completely deterministic way.

In a sense a more interesting challenge to the determinism of Newton's theory came more recently than quantum mechanics, in the last decade or so, within Newtonian mechanics itself, with the discovery of the importance of what is called chaos. Chaos has become a technical term, referring to the practical unpredictability of systems that are often very simple. It has always been of course understood that if you have to deal with a very complicated system with many individual parts, such as the stock market or the weather, it will be impossible, just because the system is so complicated, to predict what will happen. What has been realized in recent years is that even very simple systems exhibit behaviour which can best be called chaotic, because it is for all practical purposes unpredictable. Let me give you an example, and, since this is the year of Halley's comet, let me give you an example involving three bodies: a comet, a planet (let's say Jupiter), and the sun. Imagine them all going around in their orbits, and suppose for simplicity (what is not true for Halley's comet) that they're all on the same plane, the plane of the ecliptic. Also for the sake of simplicity forget their finite size, so that we don't have to worry about whether the comet is going to run into the sun or Jupiter is going to hit the comet. The comet goes way out very far from the sun, and then comes back on a very eccentric orbit. Jupiter goes chugging around the sun in its nearly circular orbit. Every once in a while the two bodies come close together. When that happens the comet is perturbed by the gravitational field of Jupiter and goes off in a slightly different direction. Now, there is no question that if we know with absolute mathematical precision the motion of the comet at one instant (and if we had an infinitely capable computer) we would be able to predict it at all future instants. But that's never the way the world works. In this situation I think you can see that if we only know the motion of the comet to, say, a tenth of a per cent before its close encounter with Jupiter then, since what happens when it makes the close encounter with Jupiter depends so sensitively on exactly at what angle it approaches the planet, its motion after its encounter with Jupiter will be more uncertain, let's say by one per cent. (Just take these numbers as an example.) Then on the next encounter the uncertainty with which the orbit of the comet is known increases by another factor of ten, and so on. Clearly no matter what kind of accuracy you have when you first observe the comet, whether you observe the elements of its orbits to one part in a million or one part in a hundred million or what you will, after a hundred or so encounters with Jupiter the comet's motion will be completely unknowable, because the system puts such stringent demands on the accuracy of the initial data (and on the accuracy of your computations) that as time passes eventually no conceivable precision would allow you to predict what would happen. The system is chaotic; determinism has lost its point.

This is a rather obvious example, one of many that have been known for generations. What has been realized in the last few decades is that chaos is itself interesting; there are universal rules of chaos. As long as you don't try to follow a precise trajectory, but think statistically of families of trajectories, you can describe chaotic behavior in terms that turn out to have common features for the orbits of comets, for the concentrations of various chemicals in chaotic chemical reactions, for convection in fluids, and so on.

Now there are international conferences on chaos, and journals devoted to chaos, but it took a while to realize how interesting chaos is. One of the reasons it took so long is that our solar system is not particularly chaotic. The orbits of the planets are quite regular (a good thing for us!). Just in the last year evidence has come in that one of the moons of Saturn, Hyperion, a moon a few hundred miles in diameter shaped somewhat like a large hamburger, is tumbling in what appears to be a chaotic manner. Here Newtonian mechanics is dealing with what is after all a fairly simple system, the moons of Saturn, a dozen or so moons going around the planet, the planet going around the sun. Yet Newtonian mechanics is for all practical purposes incapable of calculating what will happen next in the rotation of the moon, Hyperion. However, such chaos is very rare in the solar system. Apparently the reason that there is so little visible chaos in the solar system is because of a principle I suppose could be called 'survival of the stablest'. All those bodies which happened to be formed in orbits where they would exhibit chaotic motion have by now received kicks which have knocked them into other orbits or out of the solar system. If you look, for example, at a photo of the rings of Saturn, you will see beautiful bright rings, separated by dark lanes where there are no little moonlets. These dark lanes are the regions within the rings where the orbits would be chaotic. You don't see chaotic orbits in the solar system because if they were chaotic they wouldn't last very long.

This discussion of determinism has pointed up a moral. Great ideas like Newtonian determinism are not easily overthrown, but they do get refined as time passes. Both quantum mechanics and the theory of chaos have forced us to sharpen our understanding of determinism, in different ways. Neither atomic physics nor celestial mechanics are entirely deterministic, but nevertheless there is a sense in which underlying these phenomena there operates a completely deterministic mechanics, whether it's a Newtonian mechanics or quantum mechanics.

I would not say that either the broad picture of a deterministic mechanical universe, or the particular details of Newton's law of gravity, or Newton's equation of motion, or even the invention of the calculus constitute the heart of the Newtonian achievement. In preparing this, I did some soul searching about why Newtonianism seems to me to be such an incredible watershed in the history of our species. I think it is because with Newton's work mankind for the first time saw the glimpse of a possibility of a comprehensive quantitative understanding of all of nature.

Now of course Newton was not the first person to think about nature in quantitative terms. The Hellenistic Greeks, such as Eratosthenes, Ptolemy, and Archimedes, and then in the century before Newton physicists like Huygens and Galileo, were masters at bringing quantitative methods to bear on physical reality. Also Newton was not the first person who tried to think about nature in comprehensive terms, who tried to make a unified theory of everything. That goes back even further than the Hellenistic Greeks, to Thales and Anaximenes, and then, closer to Newton's time, Descartes. But the two styles had never come together before. Those who thought comprehensively about nature from Thales to Descartes had never confronted the necessity of carrying their comprehensive world picture through to a quantitative understanding of nature in all its aspects. And those who thought quantitatively about nature had never seen the beginning of a hope of formulating laws that would describe everything.

With Newton, for the first time one saw the possibility of a quantitative understanding of everything, not just the motion of the planets around the sun, but of all physical phenomena. Newton showed us what could be done. In this sense, the most important part of Newton's opus was not his theory of planetary motion, but his theory of the motion of the moon, in book three of the *Principia*. There Newton observes that on earth the force of gravity makes objects fall with a certain acceleration, which can be described by saying that a body allowed to drop from rest in the first second will fall sixteen feet. The surface of the earth where we do our experiments is a certain distance from the center of the earth, and Newton knew that you could think of the earth as if all its mass was concentrated at the center. (This was in fact the hardest part of the whole argument to demonstrate mathematically.) The moon is sixty times farther from the center of the earth than we are here on the earth's surface, so from the inverse square law the moon, if allowed to drop from rest, in the first second should fall not sixteen feet, but an amount less than that by a factor of sixty squared, or 3600. You can easily work out that sixteen feet divided by 3600 is about a twentieth of an inch. So if you let the moon drop from rest, it should fall toward the earth a twentieth of an inch in the first second. Now of course the moon isn't at rest, it's going around in a circular orbit, but it's really the same thing. If the moon were not attracted to the earth it would go off in a straight line, a tangent to wherever it was in its orbit at that moment. Instead of going in a straight line it curves toward the earth; that's what a circle does. And any one second it curves toward the earth an amount which brings it closer to the earth than if it had gone in a straight line in that second by an amount which is in fact just a twentieth of an inch. The beautiful thing about this argument is that Newton was not just relating celestial phenomena to each other. He was relating a celestial phenomenon, the moon's orbit, to something here on earth, an apple falling on his head in Cambridgeshire. (No, I don't know if the apple ever fell on his head.) Newton broke down the barrier between the celestial and terrestrial styles in physics, and in so doing he not only demystified the heavens, he opened up the possibility of an understanding which would embrace the heavens and the earth, all in one synthesis.

I don't think there have been many things in the history of man more important. It changed the mentality of human beings, the way they look at the universe and at their position in the universe. The historian

Trevor-Roper has credited Newton and the ideas that flowed from Newton with the disappearance of the witch craze in Europe in the eighteenth century. Historians have argued to what extent Newtonianism, the physics of Newton and his followers, was responsible for the industrial revolution a century later, and I don't have an opinion of my own. The general view seems to be that the industrial revolution was made by men like Watt and Stephenson and Edison who were not very learned in the science of their time (or even of Newton's time), but who operated within a scientifically oriented style that had been created by the scientific revolution of the seventeenth century. It's not surprising that the British historian, Herbert Butterfield, a historian both of science and of the politics of the seventeenth century, made the remark, 'Since the rise of Christianity there is no landmark in history that is worthy to be compared with this'.

But Newton of course only provided us with a glimpse of a really comprehensive system of nature. Newton knew there were other forces in nature besides gravitation, and he knew that he didn't know what they were, but he hoped that the same kind of mathematical reasoning, the same clarity of vision that had revealed the nature of the force of gravitation through its role in the solar system, would reveal the nature of the other forces, and their role governing all the phenomena of nature. In the preface to the first edition of the *Principia*, which Newton wrote at Trinity College on 8 May 1686, he said, 'I wish we could derive the rest of the phenomena of nature by the same kind of reasoning for mechanical principles, for I am induced by many reasons to suspect that they may all depend on certain forces.' But he didn't know what those forces were, and it took a long time to understand what they were. There was the understanding in the nineteenth century by organic chemists that the chemicals of living things were not subject to separate chemical laws, but were subject to the same chemical laws as ordinary inorganic chemicals. There was the understanding by Darwin and Wallace that the growth of various species of living things on earth did not have to be accounted for by some extra biological laws separate from the ordinary laws of physics and chemistry, but could be accounted for by the operations of chance on inheritable variations among organisms. There was the realization by Maxwell that light was not something separate from the rest of nature, but was a manifestation of oscillating electric and magnetic fields.

And so on. These were all great steps in the development of a unified view of nature, but by the beginning of the twentieth century we were still a long way from understanding even the possibility of a really unified view.

Remember the famous remark, made in 1903 by the American physicist

A. A. Michelson. In his book, *Light Waves and Their Uses* (as quoted by R. S. Andersen), he said that 'The more important fundamental laws and facts of physical science have all been discovered, and these are now so firmly established that the possibility of their ever being supplemented in consequence of new discoveries is exceedingly remote.' Physicists have laughed at this ever since. Well, of course Michelson was wrong, but I think most of those who laugh at this remark may miss the point about where he went wrong. A certain kind of physics was indeed coming to an end in 1903. It was the physics of the macroscopic motions of fluids and solid bodies. Michelson could not possibly have thought that the physicists of his time had succeeded in explaining chemical forces, for example. To Michelson, chemistry and physics were two different sciences; it was not the responsibility of physics to explain chemistry. Michelson's mistake I would guess was in having too unambitious a view of what would ultimately be explained by the methods of physics. Indeed it was not until the discovery of direct evidence that nature is composed of atoms, and the beginning of the task of pinning down their properties – measuring the charge of the electron, measuring the mass of the hydrogen atom – in the first few decades of the twentieth century, that physicists really began to see that the properties of matter – chemical forces, frictional forces, all the familiar things that happen when you bang things together, could be accounted for in terms of physical forces, just the way Newton had hoped.

By 1918, Einstein had concluded that we were well along the path to a unified view of nature, and in a speech he gave in that year he said, 'The supreme task of the physicist is to arrive at those universal elementary laws from which the cosmos can be built up by pure deduction' – just Newton's hope.

Now there's a bad name for this sort of hope: it's sometimes called reductionism. In fact there's no question that a naive kind of reductionism can do a great deal of harm. It is certainly not true that physics is going to replace the other sciences. The work of the chemists will continue to be done in terms of chemical phenomena, and chemists will not reduce all of their efforts to solving the Schrödinger equation of quantum mechanics. It's even more true of the work of biologists and sociologists and economists. They will all operate within their own disciplines, because at a certain level of complexity nature begins to exhibit phenomena that cannot be usefully reduced to the motion of elementary particles.

We even see this within physics itself. Thermodynamics, the science of heat, is a separate branch of physics. We do not study the behavior of hot bodies when they cool by going back at every point to the motions of the elementary particles within them. It isn't useful; in fact if you could follow the motion of every elementary particle in a glass of water as the water boils, you would have an incredible amount of information about the trajectory of every particle, and nowhere in the mountain of computer tape would you get the impression that water was boiling.

So reductionism has its limits. Another reductionist fallacy that has to be avoided is seeing physics as a model for the other sciences, or for human thought in general. I think a great deal of harm has been done by those from Herbert Spencer on who tried to see the social sciences for example as being based on physics as a model. I think physics is a terrible model for the social sciences. In fact it's probably a terrible model for everything except physics itself.

Nevertheless, despite all the caveats that I've presented you with about the dangers of reductionism, still there is a sense that among all the holes in our understanding of nature, there is a special importance to the hole at the bottom, that is, in our understanding of the forces which govern the particles, out of which the atoms, out of which ordinary matter is made.

Of course we are still struggling to complete this understanding. We now understand the electroweak forces which are responsible for electricity and magnetism and for the weak nuclear interactions. We understand the strong nuclear forces which hold together the quarks inside the particles inside the nucleus of the atom, and we understand some aspects of gravitation. Oddly enough, of all the forces that we know about, the one that we've studied longest, gravitation, is the one we understand least. We have a good mathematical theory of the strong nuclear forces, known as quantum chromodynamics, developed by about a dozen physicists in the early 1970s. Quantum chromodynamics is a very satisfactory mathematical theory, in which there apparently are no mathematical inconsistencies, and which accounts as far as we can tell for all the phenomena having to do with the strong nuclear forces, the forces that hold the nucleus of the atom together. The electroweak forces are also well understood though with certain large gaps in our understanding, which we hope will be closed by the next generation of accelerators. These theories make mathematical sense, they've been reasonably well tested by experiment, and they even have a kind of compelling quality. They haven't been carefully adjusted to fit the data; they are what they are, because you can't think of anything else.

Gravity is in a very different position. We have a theory of gravity, Einstein's theory of general relativity, which as I said reduces to Newton's theory at large distances and small velocities. This theory of gravity works very well on the scale of the solar system or the galaxy or here on the scale of everyday life on the surface of the earth, but it is a theory which when pushed to very short distances and high energies begins to give mathematical nonsense. If you ask questions like: What is the contribution to the force between two particles due to the emission and reabsorption of quanta of gravitational radiation of arbitrarily short wavelength?, the answer that the theory gives you is that the force is infinite. It's nonsense – it's clear the theory is simply breaking down for very short wavelengths. How then do we make sense of these two great revolutions of the twentieth century? How do we combine our understanding of gravity with the quantum mechanics which is supposed to govern nature in the very small?

In recent years there has been the development of a remarkable theory called superstring theory, which perhaps now for the first time provides a mathematically consistent theory of gravity. In superstring theory the fundamental constituents of nature are seen to be not particles or waves, but little strings, either opened or closed, continually joining and breaking apart. Each string can vibrate in many modes and the vibrations of these strings are supposed to stand in one-to-one correspondence with the various species of elementary particles. This is a theory that apparently is free of all the mathematical inconsistencies of all previous theories of gravity and, perhaps even more intriguing, this is a theory that not only allows a mathematically consistent theory of gravity, but can't help it: You can't in this theory imagine that gravity didn't exist. So we have taken three big steps. First, Newton described gravity successfully in terms of an inverse square law, but he could always have imagined that it wasn't an inverse square law – it was just the data that forced it to be an inverse square law. Then Einstein in general relativity developed a theory of gravity which included Newton's as a limiting case, and which explained why in that limiting case it had to be an inverse square law. But Einstein's general theory of relativity still left open the question: Well why is there any gravity at all? Why does space curve? Why isn't space just perfectly flat and there is no such thing as gravity? Who asked for that? In the superstring theory we have for the first time a theory in which gravity cannot be left out.

The history of this theory is rather amusing. It was developed as an attempt to explain strong nuclear forces, at a time before the advent of quantum chromodynamics. The mathematical physicists who developed it at the end of the 1960s were terribly embarrassed at the fact that the theory predicted the existence of a particle which looked like the quantum of gravitational radiation, the so-called graviton, and yet it was supposed to be a theory of the strong nuclear forces, and so they couldn't understand what this graviton was doing in their theory. Quantum chromodynamics came along in the early 1970s, and put the string theorists out of business for a while. Some of them continued to beaver away in obscurity, not getting promoted. (I speak here not with bitterness, but with embarrassment, because I was not one of them. In fact I was one of the people who helped to ignore them.) Just in the last few years it's been widely realized that superstring theories offer a good chance of a unified theory of all particles and forces.

These theories have not been tested experimentally, and we are desperately hoping that society will continue to fund the efforts of particle physicists to pin down our rather speculative ideas. The only way they can be pinned down is by building large accelerators which will allow us to test our ideas, and refine our ideas. After all, Newton didn't invent the theory of gravity without knowing about the existence of the solar system.

So it goes. Newton's hope, 'I wish we could derive the rest of the phenomena of nature . . .', has not been fulfilled, but we are working on it, very much in the Newtonian tradition: the formulation of increasingly comprehensive quantitative laws. From this high viewpoint, all that has happened since 1687 is a gloss on the *Principia*.

Acknowledgement

Research supported by Robert A. Welch Foundation and NSF grant PHY 8605978. I wish to thank Gerald Holton and Victor Szebehely for their valuable help in the preparation of the published version of this talk.

3

Newton, quantum theory and reality

ROGER PENROSE

3.1 Newton's corpuscular–undulatory view of light

The supreme stature of Newton as a scientist cannot be doubted. As an experimental physicist, he had superb natural skill, and was profoundly ingenious, as well as being exceptionally careful in the construction and execution of his experiments. As a mathematician he possessed extraordinary power, and, indeed, had few mathematical peers over the whole of history. In fact, it was very necessary for him, in developing his scientific theories, to explore and innovate in mathematical ways, and he uncovered many deep mathematical truths – the fundamental theorem of calculus being but one. The *Principia* itself, whose three-hundredth anniversary I am most honoured to be able to celebrate with this essay, contained among its many propositions, multitudes of mathematical gems and numerous profound mathematical ideas. In addition to these scientific qualities, and most importantly for the development of his science, Newton also possessed superlative instincts into the workings of Nature – instincts which often fell hardly short of miraculous.

In one particular instance, concerning Newton's preference for a corpuscular theory of light to a wave theory, the extent of his insight is most intriguing – and perhaps somewhat debatable. The experimental evidence in Newton's day was inconclusive, yet it would seem that the balance of that evidence should have been in support of a wave theory of light rather than a corpuscular one. Ironically, it was Newton's own discovery and painstaking analysis of an optical interference phenomenon – Newton's rings – which should have provided the strongest evidence in favour of the opposing wave theory! (Other types of interference had been observed earlier by Grimaldi.) Yet Newton stuck doggedly to a corpuscular view, and was led to a picture of light quanta possessing both particle-like and wave-like properties. Can this apparent anticipation of modern quantum theory – a theory which was not formulated until over two hundred years later – be other than the merest fluke, resulting perhaps from Newton's obstinate refusal to give in to his adversaries and accept their wave picture of light?

In his preface to the 1952 Dover edition to Newton's *Opticks*, I. Bernard Cohen rightly warns us not to make too much of Newton's apparent anticipation of our present quantum-mechanical concepts. Yet, one cannot but be struck by the fact that this master among scientists found himself driven to a view of the physical nature of light that was not only seemingly almost absurd, with its curious intermixture of corpuscular and undulatory properties, but absurd in ways which seem close to the absurdities of modern quantum theory! Is not this an indication that Newton possessed insights of such power that they even gave him hints of revolutionary developments not to occur until over two centuries later? On the other hand, perhaps the similarity with quantum mechanics is indeed illusory. Is there really much detailed correspondence between Newton's undulatory 'fits of easy transmission and reflection' when his corpuscles of light interact with the material of a refractive medium, and the wave-like aspects of photons in their interactions with matter, as we understand them today? I would accept that these analogies should not be pressed too far, and that the correspondence between Newton's picture of light and that presented by modern quantum mechanics would not stand up to much detailed analysis. I can believe, also, that Newton possessed the mathematical wherewithal, so that in principle he could have pressed forward with an entirely wave picture of light, had he wished to do so, which would have been well in accord with the observations of his day. Yet I think that it is likely that Newton did indeed have powerful reasons for rejecting a purely wave picture of light; and there was evidence available even at that time, vaguely discernible to a scientist of Newton's sensitivity, for a view of reality in which undulatory and corpuscular ingredients of reality co-exist at a fundamental level. However, such reasons would have to have been ones that Newton could not easily have articulated – or perhaps would not have wished to do so, owing to his avowed dislike of purely 'hypothetical' contentions.

3.2 Stated reasons for rejection of wave theory

Let us first briefly examine Newton's stated reasons for rejecting the wave theory. The main one seems to have been the rectilinear propagation of light, and the fact that light does not bend into the shadow of an obscuring body. As we now know, from the phenomenon of destructive interference discovered by Thomas Young a century later, an effective rectilinear propagation does indeed occur for wave propagation when the wavelength is short, such as that which occurs with visible light – and light does indeed bend slightly into the shadow. We do not blame Newton and his contemporaries for not knowing this. Yet I suspect that if Newton had preferred the wave theory for other reasons, it might not have been beyond him to suggest something of the nature of destructive interference. (The wavelengths of visible light were, in effect, known to him from his 'Newton's rings' observations.) However, his preferences were for a corpuscular theory, which rectilinear propagation seemed to support, so he did not find himself compelled to examine this matter more deeply.

Newton's other stated reason for rejecting the wave theory of light was that light has 'sides' – which was Newton's term for linear polarization. The observations of Erasmus Bartholin and Christiaan Huygens concerning the phenomenon of double refraction in calcite, or Iceland spar, had led Newton to conclude that a beam of light consists of components which are not rotationally symmetric (i.e., they are polarized). This is, indeed, quite incompatible with the type of longitudinal compression wave that Huygens had suggested for light. However, a transverse oscillation, as was later suggested by Young, would fit the bill. Apparently it occurred to neither Huygens nor Newton that light might be a transverse oscillation. One can again speculate that if Newton had had other strong reasons for preferring a wave theory of light, he might perhaps have been driven to think of the idea of transverse oscillations as the explanation of 'sides'. However, he had not and he was not.

These two arguments for rejecting the wave theory of light, though undoubtedly impressive at the time, must be counted as 'wrong'; so if this was actually all that Newton had to go on, we should, indeed, have to consider that it was a 'fluke' that he found himself driven to a combined corpuscular–undulatory theory of light. However, it seems to me to be likely that Newton may well have had various other reasons for disliking an entirely undulatory theory of light. One can only surmise what some of these might have been. One possibility that has often been suggested is that Newton favoured an atomistic view of Nature. Perhaps there is some force to such an argument. If we are to believe that light is a wave motion, then surely it must be a wave in some medium. On the atomistic view, this medium must itself be composed of particles. This gives light a character secondary to that of the particles of the medium. Yet, there is perhaps something primitive about light that argues against it having such a secondary nature. If new particles are to be invoked in order to compose such a hypothetical all-pervasive medium, is it not much more economical to use such new particles as the constituents merely of the light itself?

Newton seems to have had some difficulties in coming to terms with such an all-pervasive medium, and he discussed, at some length, the strange hypothetical properties that such a medium would need to have (in *Queries 18–22* of *Book Three* of his *Opticks*). Even when contemplating such a medium, he preferred to have light constituted separately as particles of its own. (Influences in the medium were imagined somehow to be able to overtake the light corpuscles.) He also had some idea that an explanation of motion under *gravity* might be found in terms of 'refraction' by such an all-pervasive medium, and he presented a picture which even has something in common with the ideas of general relativity! (Perhaps, instead, Newton's ideas were foreshadowing De Broglie's realization that matter itself possessed wave-like qualities. Take your pick!) Newton appeared to have no qualms about suggesting such a view of the nature of gravity, while at the same time holding to his picture of an inverse square law of force holding between all pairs of particles in the universe. No doubt he regarded all these pictures as highly provisional. He admitted to not knowing what the nature of such a medium might be (*Query* 21). The medium would need to have some strange and powerful properties; and I suspect that Newton's view was that the medium, if it existed, could not be material in any ordinary sense, but would have to be of some mysterious substance quite different from that of matter. Perhaps this was the 'absolute space' that Newton was really groping for. Perhaps, indeed, if the modern ideas could have been explained to him, Newton might not have been averse to viewing this medium in terms of Einstein's picture of a 'flexible' spacetime continuum!

None of this implies that light must be corpuscular. Ordinary atomism might require merely that the medium, if it exists, be corpuscular. It is just economy, or perhaps aesthetic requirements, which, together with atomism, would imply that light must be corpuscular. I wish to suggest that there may, instead, have been another quite different motivation underlying Newton's preference for a corpuscular view of light. This is the motivation from Galilean relativity.

3.3 Newton and relativity

The reader may well object that there is not much evidence that Newton was at heart a dynamical 'relativist'. He was certainly well aware that his dynamical laws are invariant under change to a uniformly moving frame (as well as under shift of origin and under rotation). This much he made clear in his *Principia*. But what evidence is there that Newton at any time shared Einstein's *conviction* that physics *must* be invariant under change to uniform motion? The common view seems to be that, on the contrary, 'Newton believed in absolute space' – with its decidedly non-relativistic connotations. If space (as opposed to spacetime) is indeed absolute, so that it is meaningful to talk about the same point in space at two different times, then there is an absolutely preferred state of rest. It then becomes a 'fluke' that Newton's dynamical laws are Galilei-invariant. There is no need to believe that this fluke should continue to hold when additional ingredients of physics are added to the basic dynamical laws. One such additional ingredient is the velocity of the medium in which light travels, if it indeed be the case that light merely consists of wave motion in some all-pervading medium. If we insist that light is merely a wave motion, then we must give up Galilean relativity.

Of course, Lorentz, Einstein and Poincaré have now taught us that there is a way out of this. Light can still consist of a wave motion – this time, the waves of oscillating electric and magnetic fields of the Maxwell theory – and invariance under change to a uniformly moving frame can be maintained. However, this requires a drastic revision of dynamical laws (for speeds approaching that of light) and also an overturning of one's very notions of space and time. It is thus, as we now know, that Galilean relativity can be given up and replaced by *another* relativity – the relativity of Einstein and Poincaré.

I am certainly not attempting to suggest that Newton could have had the absurd foresight to sense the need, or even the possibility of such a revision of basic principles. However, if we do accept that this route was *not* effectively open to him, then we see that Newton was presented with a clear choice: give up Galilean relativity or stick to a corpuscular theory of light. Now what evidence is there that Newton believed 'in his bones' that some kind of dynamical relativity must hold? I contend that, despite what appears to be a common belief to the contrary, such evidence does exist. I refer the reader to Richard Westfall's fine biography of Newton (*Never at Rest*, 1980). On p. 147 Westfall describes how, in January 1665, recorded in private notes to himself in a work referred to as his *Waste Book*, the youthful Newton derived the laws of impact for two perfectly elastic bodies. (Huygen's slightly earlier treatment of this problem had not yet been published.) Newton's procedure was to consider the two bodies as a single system, and he came to the conclusion that their common centre of gravity moves inertially whether or not the bodies impinge upon one another. Realizing that every impact can be viewed from the special frame of reference of the common centre of gravity of the two bodies, Newton appears to have used the impact behaviour in *that* frame – in which the behaviour may be regarded as 'obvious' – and then assumed *Galilean relativity* to derive the elastic impact law in the general case. This was more than twenty years before the publication of *Principia*, but Westfall remarks, concerning the consequent uniform motion of the centre of gravity, that 'Newton never forgot this conclusion. It contained the first adumbration of his third law of motion, . . .' – and it seems to have formed one of the central guiding ideas behind Newton's dynamics, despite the fact that in *Principia* Newton chose to derive his dynamics from different axiomatic principles.

In his account, Westfall does not draw particular attention to the fact that Newton was implicitly relying on Galilean invariance to derive his impact law,† although Newton's procedure cannot be used without such an assumption. Without it, one has no right to 'pass to the special frame of reference of the common centre of gravity' and to assume that the same dynamical laws must hold in that frame as in the original frame. It seems to me to be highly likely that by the age of 23, when Newton produced this argument, some strong intuition concerning the relativity of motion was already an essential ingredient of his thinking. It is, after all, a necessary ingredient of the Copernican–Galilean mode of thought that the dynamics that we employ at the surface of the Earth should be insensitive to the Earth's uniform motion about the Sun. Newton was well aware of this, of course. (It appears that, at least by 1666, Newton had read Galileo's *Dialogues*, which appeared in translation in England in 1665.) It would have

---

† Towards the end of his life, Galileo had claimed a solution of the problem of the dynamics of impulses. However, no account of this appears to have survived (cf. Galilei, 1638; Dover edn, p. 293). It is intriguing to speculate whether Galileo might have been guided along similar lines of reasoning.

Although Huygens's treatment of this problem seems to have been a little different from that of Newton, he also used Galilean invariance (cf. Mach, 1972). It is striking to note that Huygens did *not* believe that this Galilean invariance extended to light, in the sense that I mean it here, since he strongly promoted the wave theory! Huygens's view seems to have been that the individual particles of the medium might be subject to a Galilei-invariant dynamics even though the medium as a whole might define a rest-frame. However, this viewpoint attaches an 'accidental' rather than a 'fundamental' role to this medium – and therefore to light itself. It also presupposes that the Galilei-invariance of the local dynamics of *macroscopic* bodies is insensitive to the presence of the medium – another 'accidental' (and, moreover, highly improbable) feature. While the example of Huygens shows that it was possible for a great scientist of the time to hold to the wave theory and also to be a professed relativist, this is not totally being a relativist in the 'deep' sense that I mean here.

been natural for him to suppose that other physical properties (such as the behaviour of light) should also be governed by the same beautiful invariance principles. Perhaps, at that time, he had not explicitly formulated, even to himself, what such 'invariance principles' might entail, but my guess would be that he knew pretty well within himself what he was about. His treatment of the problem of impulses seems to me clearly to indicate that he instinctively felt that a relativity principle must be intrinsic to Nature's ways.

At roughly the same age as he studied the impulse problem, he was also becoming profoundly interested in the nature of light. On Newton's own account, he 'had the Theory of Colours' in January 1666 (Westfall, 1980, p. 156). He may even have had some inkling (as had Galileo before him – cf. Drake, 1957, p. 278, concerning a remark made in *The Assayer*) that light has some profound role to play in governing the minute structure of matter. One can readily imagine that Newton's great intuition guided him to believe strongly in some kind of relativity *principle* encompassing both the dynamics of matter and the behaviour of light. This intuition would have guided him surely to a corpuscular rather than to a wave picture of light. That would have been in the years when his profound views on the nature of light were first being formed. These years must have set the stage for all his later thinking. When at a later time in his life he became fully confronted by the experimental facts of optical interference phenomena, he would have been reluctant to give up the intuitions that he had built up during this formative time. I suggest that in his 'bones' he *knew* that light must be corpuscular, even though in later years the good reasons for these early intuitions may well have receded from him. And unlike the stated reasons that Newton later put forward in his *Opticks* in favour of a corpuscular picture of light – which, technically, were 'incorrect' – this unstated one, that a relativity principle must hold, was, I am claiming, profoundly 'correct'!

When Newton came to write his *Principia*, some twenty years after these intuitions must have been formed, he needed to place his logical development on a clear footing. To state his basic laws in a definite mathematical way, he needed the framework of 'absolute space' so as to be able to formulate things precisely. Even with the abstract mathematical machinery of today, it is not so easy to give a totally Galilean-relativistic treatment of space, time and dynamics right from the start. It can be done, but one needs an absolute spacetime, rather than an absolute space. (See Arnol'd, 1978, for an excellent modern such treatment.) Newton did take pains to be completely clear that Galilean invariance was indeed a feature of his laws (cf. Scholium 5, following Definition VIII), but this invariance was not a necessary principle. It might even be that the technicalities involved in writing his *Principia* guided Newton away from what I believe must have been his early relativistic intuitions. Once Galilean relativity has become a *consequence* of the basic laws, rather than a basic principle, there appears to be nothing against adding to those laws something – such as an all-pervasive medium – which does not share in that Galilean invariance. Accepting all the profound influence, power and comprehensiveness of *Principia*, and accepting all its magic in deriving from such simple axioms so much of the dynamical workings of the world, one can still contemplate the case that its very strength of mathematical structure could have obscured a youthful insight – an insight that had to wait nearly two and one half centuries to be profoundly reinstated in a new form by Einstein in 1905: that one should take relativity as a *principle*, rather than as a seemingly accidental consequence of other laws. (See note added in proof, p. 49.)

3.4 Newton's route to an undulatory–corpuscular picture?

I am conjecturing that, despite Newton's speculations in his *Opticks*, concerning the possibility of an all-pervasive medium, and despite the seeming implications, in *Principia*, that he might be happy with an absolute space, Newton carried forward from his youth a deep insight into the relativistic nature of physics. Indeed, he well recognized that his space and his medium, if they 'existed' in any ordinary sense, would have to have profoundly peculiar properties. However, it would be far too much to expect of him that the realities eventually revealed by Maxwell concerning electromagnetism and light, and particularly those revealed by Minkowski and Einstein concerning the essentially four-dimensional structure of spacetime, could have been discernible by him. That would have been a quite unbelievable route for him to follow. Instead, he chose the alternative Galilean-relativistic, but, for the time, equally valid, route. His early relativistic insights, I am suggesting, may well have been largely responsible for an intuitive belief in a corpuscular picture of light. Once he was firmly down that road, and having to come to terms with optical interference phenomena, he was led almost inevitably (and, in a sense, 'correctly') to a picture of light in which corpuscular and undulatory features had somehow to coexist, leading to an eerie foreshadowing of quantum wave–particle duality over two hundred years before its time.

It may well be that there were other 'valid' hints that a combined undulatory and corpuscular picture had to be entertained. Newton's interest in the colours of actual objects, and his alchemical work concerning the transmutation of substances, must have convinced him of a deep interrelation between light and matter. He certainly speculated that matter and light might be convertible into one another (*Opticks, Query* 30). One is reminded of Einstein's early struggles concerning the absorption and emission of light (see Pais, 1980). It was not only Planck's explanation of the black-body spectrum which disturbed Einstein. The whole idea that individual atoms could discretely pick up energy from what, at that scale of things, would have been a barely oscillating and nebulous medium seemed almost preposterous. Einstein's intuitions and insights guided him to the view that, despite the fact that light is waves, light must also be particles: indeed Nature *is* preposterous! It is not inconceivable to me that a few of the many worries that so haunted Einstein in the early twentieth century might already have been vaguely discernible to Newton over two hundred years earlier!

3.5 Quantum mechanics

For the rest of this essay, I shall depart from such speculative considerations of pseudo-history and concentrate on this very preposterous nature of reality that the observational facts of modern quantum physics have presented us with. The questions of quantum reality are closely related to some of those we have just been discussing. We shall find that these questions will lead us into some speculative considerations of a different kind!

Let us remind ourselves of one of the very basic rules of quantum theory, namely *linearity*. If $|\psi\rangle$ and $|\phi\rangle$ are two vectors representing quantum states, then any quantity of the form

$$\lambda|\psi\rangle + \mu|\phi\rangle,$$

where λ and μ are complex numbers not both zero, also represents an allowable quantum state. Recall that multiplying a quantum state-vector by a non-zero complex number yields a state-vector representing physically the same quantum state as before, while all other state-vectors describe physically distinct quantum states. Thus, taking all the different possible ratios λ/μ (including ∞), we are provided with an array of physically allowable alternative states, obtainable by quantum linearity from the original two states. The abstract space of such complex ratios has the topology of a spherical surface – the *Riemann sphere*.

This sphere is implicitly present in any two-state quantum system. Sometimes it has a very clear geometrical role to play; sometimes this role is rather subtle. It is clearest for the case of the spin-state space of a spin one-half massive particle, say an electron. Here each physically distinct state of spin corresponds to a definite direction in space. This direction is the one with the property that if the spin of the electron is measured in that direction, then the answer to the measurement is: *yes*, with *certainty*. In the case of polarization states of a photon, the geometrical role of this sphere is slightly less direct. Suppose that the photon is moving upwards. Then the north and south poles correspond, respectively, to the states of right-handed and left-handed circular polarization. The points around the equator correspond to the states of linear polarization (Newton's 'sides'), but since a linear polarization state corresponds to an *un*directed line element, the two opposite directions of this line element must correspond to the same point on the equator. More directly, it is the 'square root' Riemann sphere for $(\lambda/\mu)^{1/2}$ which has a clear-cut geometrical correspondence with polarization directions (Stokes vectors).

Sometimes the geometry is not at all direct, such as with the various linear superposition states of two different positions of a particle. Suppose a particle is passing through a pair of slits A and B. If the particle is at A, then we represent this by the north pole of our Riemann sphere. If it is at B, then by the south pole. The various other points of the sphere correspond to states in which the particle is partly at A and partly at B. As we know, it is not a question of mere probabilities of the particle being at A or B, but of *amplitudes* of it being at A or B – whatever that might mean! The space of different possible ways that the particle can be located as it passes through the pair of slits has the structure of a *sphere* – not a line segment, as would be the case with probabilities (the range of probabilities being the real-line segment $[0, 1]$).

3.6 Physical reality

We are here beginning to be confronted by the odd picture of reality that quantum mechanics presents us with. Do we really *believe* that the particle can be in two places at once, with complex weighting factors? Our intuitions are not so stretched in the case of the spin states of the electron. In that case we can visualize the electron as 'spinning' about an axis which has some definite direction in space. The photon, too, can be thought of as having some definite geometrical polarization state. Some people feel uneasy with such clear-cut geometrical pictures of quantum states. They say that we should not try to form a picture of reality when applying the rules of quantum mechanics: just follow the rules of the quantum mechanical formalism, do not try to form pictures and do not ask questions about reality! This seems to me to be wholly unreasonable. Physics, after all, constitutes our best way of groping for the true nature of the real world in which we find ourselves. Some would have us believe that physics is to be regarded as merely providing a means of effective prediction about the future. It is true that one of the most powerful shifts in attitude in science came about with the methodology of Galileo and Newton: form a mathematical theory from which predictions can be made and test these predictions against observation. Do not ask *what* it is that constitutes the substance of things, nor *why* things behave as they do. Just ask *how* they behave and try to form an elegant mathematical structure that mimics that behaviour as closely as possible. ('*Hypotheses non fingo*'!)

However, the power of such methodology should not obscure the fact that we are still, nevertheless, groping for the realities of the physical world. The methodology is powerful because it enables us to clear from our minds a good many preconceived and incorrect notions concerning the nature of this reality. It does not, and it should not, lead us to believe that we can dispense with the notion of physical reality altogether – as some would seem to have us believe! What, after all, is the point of making 'predictions about the future' unless we are allowed to assume that there is some reality about this future whose state we are proposing to predict? Despite the operationalistic nature of the methodology of Galileo and Newton, their dynamical theory provided a picture of 'reality' which was very much clearer than that which had gone before. In fact, this picture has become so 'clear' to us that there is now a great reluctance, even after we have been forced to accept the physical existence of quantum phenomena, to believe that 'reality' can take any other form.

It is clear from Newton's *Queries* in his *Opticks*, that he was, indeed, well prepared to speculate on the nature of physical reality and that, in the case of gravity, the existence of a wonderful mathematical theory did not stop him from speculating on 'causes' (cf. his refractive all-pervasive medium referred to above). '*Hypotheses non fingo*' did not apply to *Opticks* (nor, indeed, did it apply to Newton's earlier thinking)! It seems likely, also, that Newton's picture of reality was actually *not* very close to the 'clear' picture of reality that we have subsequently built up from a familiarity with Newtonian dynamics.

3.7 Reality of the state vector

Let us return to quantum mechanics and to the strange pictures that it presents us with. We have seen that, with electron spin, our geometrical picture of the quantum state-vector is not at all an unreasonable one. However, even here, many would say that we are deluding ourselves, and to form such a picture is misleading. Somehow the electron has only *two* ways in which its spin can point, not a whole spherical continuum of ways. Any experiment that we may perform on the electron to measure its spin can give us but *one bit* of information: the spin can turn out to be one way, or it can turn out to be the other, but it can be nothing in between. That is the way quantum mechanics works.

However, when there are just two such 'alternatives', the space of states is indeed a whole Riemann sphere. We cannot say which the alternatives are. They might be up/down or north/south or east/west or anything in between. There is nothing to choose between any of these pairs of alternatives. But given any *particular* state of spin of the electron, there will be precisely one direction in space for which the state gives certainty that the spin is in that direction. Given the state, there is a 'reality' about the direction in which the state points. An observation *might* be performed measuring the spin in that direction and, if so, the state has to be prepared to say 'yes', with certainty, to that measurement. Somehow the state has to 'know' that direction, even though there is no experiment which can be performed to determine which direction in space it actually is. So long as we are sticking to standard quantum mechanics, the different possible 'realities' for the states of spin of an electron are indeed the points of the Riemann sphere. This is for a system with just *two* alternative states. In a sense, the whole Riemann sphere 'counts' as just two!

For massive particles of higher spin there is also a corresponding geometrical picture for the states. Take the spin to be $n/2$, then in place of the Riemann sphere we have a complex n-dimensional projective space of different possible physical states. But we can still use the Riemann sphere to describe the individual states. Each physical state of spin $\dot{n}/2$ corresponds uniquely to an unordered set of n points on the sphere of directions.† A characterization of these points is that if the spin is measured in that direction then there is zero probability that the spin turns out to be totally in the opposite direction. Again, the state is uniquely characterized by the fact

---

† I am told that this result is due originally to Majorana, but I do not know of a reference. The result is not hard to see using 2-spinors (cf. Penrose and Rindler, 1984, p. 162). If $\psi_{AB\ldots N}$ is an n-index symmetric spinor representing the state, then it has a canonical decomposition $\psi_{AB\ldots N} = \alpha_{(A}\beta_B \ldots \nu_{N)}$ (round brackets denoting symmetrization), and the n directions are those represented by the spinors $\alpha_A, \beta_B, \ldots, \nu_N$.

that if measured in any one of these n directions, then with certainty it cannot spin totally the other way. The state has to 'know' that it must respond this way in the event that any one of these measurements is carried out, so, up to proportionality, it is again objectively characterized.

The same objectivity holds, in principle, for any quantum state whatever – so long as the rules of quantum mechanics are presumed to hold rigorously. Suppose the state-vector is $|\psi\rangle$. Then we can consider making an observation corresponding to the observable

$$Q = |\psi\rangle\langle\psi|.$$

The state $|\psi\rangle$ is the only one, up to proportionality, for which the observable Q yields the result 1 with certainty. The state must 'know' that it has to produce this result in the event that the observation Q is actually performed. This is a completely objective property, so the fact that the state is (proportional to) $|\psi\rangle$ is an objective property of the world.

I am expressing a point of view here which, for some reason, is not often maintained. One frequently hears the opposing view that 'the state-vector merely expresses our state of knowledge about a system', or 'the state-vector is expressing a property of ensembles of systems rather than of a single system'. However, the point of view that I am expressing is that the state-vector is clearly defining an objective property of a *single* system. It is not a 'testable' property of that system in the sense that we can perform an experiment on it to determine what its state $|\psi\rangle$ actually is, but it is an 'objective' property of that individual system in the sense that the state of that one system is characterized by the results of experiments that one *might* perform on it.

3.8 Quantum non-locality

Why is this objectivity not something that is more frequently expressed by theorists? The answer to this question is not really clear to me, but I suspect the reason has to do with the strange picture of objective reality that it presents us with. Take the situation considered above in which we had a particle passing through two slits A and B. The objective reality of the system, I claim, actually consists of the particle being partly at A and partly at B, both together in some complex linear combination (since for each choice of complex linear combination of being at A and being at B there is an experiment that could be performed which gives the answer 'yes' with certainty *only* for that particular combination). However, we like to think of particles as tiny point-like objects, not as things that can be in two places at once. Moreover, we may choose to measure whether the particle is at A, or to measure whether it is at B, and if we find it at one place we cannot find it at the other. Somehow, if we perform such a measurement, we find that at the next moment the particle *is* either at A or B, so we like to think that, a moment before, the particle really *was* in one place or the other, but because of some kind of funny statistics which holds at the quantum level the alternatives must be weighted with complex rather than real coefficients. If we take the realistic view of the state-vector that I maintain, in view of the above discussion, is really almost forced upon us, then we must accept a degree of non-locality: finding the particle at A instantly forces it to be *not* found at B and, conversely, *not* finding it at A instantly forces it to *be* at B. This latter type of observation is an instance of the particularly confusing situation referred to as a 'null measurement'. The mere fact that a detector somewhere has *not* registered the presence of some particle forces the state of that particle to be different from what it was before in some region perhaps distant from the detector itself. It is especially puzzling that the *absence* of a physical interaction can affect the state of a particle in some distant region!

This non-locality is perhaps even more striking when there is more than one particle involved. The state-vector may consist of linear superpositions of different arrangements of the particles. Observing the state of one of the particles may instantly force the others into some state correlated with it, even though they may be well separated from it. As an instance of this sort of thing, let us consider a variant of Bohm's version of the Einstein–Podolsky–Rosen thought experiment. Suppose that we have a particle in a spin-zero state which decays (non-relativistically) into two particles each of spin one-half. Suppose also that there is no orbital contribution to the angular momentum. We can express the various spins in terms of eigenstates of spin in the vertical direction. Then we can think of the initial spin state $|0\rangle$ to be decomposed, after decay, into the difference between two states, one state having one of the particles spinning up and the other down, and the other state having the spins of the two particles both reversed:

$$|0\rangle = |{\uparrow}\rangle|{\downarrow}\rangle - |{\downarrow}\rangle|{\uparrow}\rangle.$$

However, there is nothing special about the choice of the vertical direction here. We could equally well have chosen left/right:

$$|0\rangle = |{\leftarrow}\rangle|{\rightarrow}\rangle - |{\rightarrow}\rangle|{\leftarrow}\rangle,$$

or, of course, any other pair of opposite directions. Neither of the resulting particles has, by itself, a state. Only the pair considered as a whole has. If we observe the spin of one of these particles to have some particular direction, then the other instantly acquires an 'objective spin state', namely the state corresponding to spin with certainty in the opposite direction. This applies equally to any direction that we may choose. Moreover, there is no requirement that the two particles be at all close to one another. In principle they could be hundreds of miles apart. By observing the spin of one of the particles, we instantly put the other particle into a spin state whose direction is fixed by our *direction of observation* on the first particle! (We note that no message can be sent from one particle to the other by this means.) This is a decidedly non-local picture of reality.

I am regarding the total state on the right-hand side of these equations as being 'real'. But how else are we to regard it? We are always at liberty to choose to measure the spins of neither particle but, instead, to reflect both of them – carefully, so as not to disturb their spins – back to their original position so that they can recombine as a spin-zero state. The spin of the recombined state must indeed be zero with certainty, and the possibility of observation of this final spin assigns a 'reality' to the non-local spin state of the combined system. Other situations of this same general nature exhibit a similar non-locality. This is particularly striking when the two emitted particles are photons since the effect of observing one of them would, if it were a signal in the ordinary sense, have to travel faster than light in order to reach the other one. The experiments of Freedman and Clauser (1972) and Aspect (1976) and coworkers, etc., have shown that a non-locality of this type (over a distance of some twelve metres, in the latter case) is an actual feature of the world we live in and not just a theoretical fiction.

3.9 Quantum mechanics and macroscopic physics

We see that this quantum view of reality is very different from the one that we have become accustomed to from classical physics, where particles can be only in one place at a time, where the physics is local (except for action at a distance) and where each particle is a separate individual object which, when it is in free flight, can be considered in isolation from any other particle. All these classical conceptions must be overturned once we accept the reality of the state-vector. It is perhaps little wonder that most people are reluctant to do this. However, the quantum view of reality is more worrisome even than this. For the linear superpositions between different particle positions can get coupled in with macroscopically differing states of *macroscopic* objects. In that case, according to the strict rules of quantum mechanics (and assuming that no 'observation' is deemed to have been made) we must carry over the complex linear superpositions to the macroscopic objects themselves. Thus we can be presented with a situation like that of Schrödinger's cat, where we seem to have to consider states representing linear combinations of a dead cat and a live cat, with complex coefficients. There is a very reasonable reluctance to accept such a state as actually describing reality! Surely, at that level the cat *is* either dead or alive. However, there still remains the possibility, in principle, that one might perform the observation described by $Q = |\psi\rangle\langle\psi|$, as suggested above, that requires objectivity for the state $|\psi\rangle$ *whatever* it is. In this instance, where $|\psi\rangle$ represents some complex linear combination of a dead cat and a live one, it might well be objected that the observation described by Q is quite out of the question practically. But quantum theory, as it stands, has nothing to say about what is 'practical' or 'impractical'. To place some restriction on which operators Q are allowed as observables, and which are not, would be to introduce a *change* into the theory.

Some people take the view that a quantum linear superposition at that level is still perfectly in order. A conscious human observer could also be in such a superposition but, because of some quirk of the nature of conscious perception, his consciousness would 'split', so that each state of awareness would perceive just one of the macroscopically distinct states under superposition, and would not be aware of the actual superposition. This (or something like it) is the many-worlds view of Everett and his followers. Personally, I should be most reluctant to go to such lengths to save the superposition principle for macroscopic bodies. Not least, it is worrisome to have to rely on a yet undiscovered theory of consciousness in order to find agreement between the picture of things presented by theory and what is observed in practice. It seems to me that something must go wrong with the superposition principle a good deal before this. Macroscopic physics *is* like classical physics, where particles occupy single localized places and behave as individual separate entities.

3.10 Linearity and time-evolution

So far, I have not discussed the time-evolution of quantum states. As we know, there appear to be two completely different modes of evolution in quantum theory – and this decidedly odd fact has, no doubt, added to the reluctance that many people may feel about taking the state-vector as actually representing reality. On the one hand, we have unitary evolution (Schrödinger or Heisenberg, according to one's description). This is completely deterministic, as in classical physics, and it preserves quantum linearity. On the other hand, we have 'state-vector reduction', which is probabilistic and does not preserve quantum linearity. If, as I am proposing here, one takes a realistic view of the state-vector (and supposing that one rejects the many-worlds view – or even if one does not), then one must come to terms with these two different modes of evolution.

Some would take the line that state-vector reduction is an illusion which appears to arise because one is ignoring the complicated interconnections between the system under consideration and the outside environment. The claim is, on this view, that unitary evolution *would* hold if we could take everything into account, but, since we are ignoring the outside, we are led to an approximation which appears like a reduction process. I have to say that I do not find this viewpoint at all plausible. I really do not see how the linear superposition of Schrödinger's dead and alive cat can get resolved to one alternative or the other, simply by taking the environment into account. The best that one can hope to do with this type of procedure is to show that somehow the cat can be treated *in practice* as some probability-weighted mixture of alternatives. It cannot explain how the reality of the situation can shift from a complex linear superposition to alternatives. Unless we change the theoretical structure of quantum mechanics at some stage we still have the possibility *in principle* of performing the measurement Q referred to above, so 'reality' must remain the complex linear superposition!

Another suggestion is what is known as the 'Wigner view' (Wigner, 1961), according to which unitary evolution holds until consciousness gets involved in the physical system. At that stage, some non-unitary evolution takes over and state-vector reduction objectively takes place. Again I am worried that a theory of consciousness seems to be required in order that the theory can be used. But a stronger objection, to my mind, is that the view seems to assign a privileged role to those corners of the universe where consciousness resides – and they may be very rare indeed. The objective evolution of the state-vector would be quite different in those corners from elsewhere. The effect might not be 'observable' but the asymmetry in the pattern of time-evolution presents a disturbing view of reality.

My own picture is different from any of these, though it shares with the Wigner view the idea that unitary evolution is to be supplanted, under certain physical circumstances, by an objective evolution which closely approximates state-vector reduction. However, I am *not* suggesting that consciousness is needed for this. The reduction process would normally be completed long before the results of the reduction are perceived by any conscious observer. Indeed, an actual conscious observer would normally be playing no role whatever in the reduction process. Instead, I am suggesting that, at a certain level of scale, a failure of quantum linear superposition becomes important, and unitary evolution is taken over by some specific non-unitary procedure. It might be that this procedure can be described by some non-linear (and non-local) differential equation, non-linear instabilities arising at a certain scale, so that the system then 'flops' into one or the other of two macroscopically distinguishable states. On the other hand, the procedure might have to be described in some quite new way. Non-linearity is essential, since the *conventional* reduction procedure does not preserve quantum linearity; but it is not clear to me that a differential equation is the right thing to look for. Also I do not wish to prejudice the issue as to whether the reduction procedure should be genuinely stochastic or whether there should be some underlying determinism which just has the *appearance* of randomness. The linearity of quantum mechanics would, on this view, be merely an approximation, albeit an excellent one at the scales where quantum mechanics has been tested.

Many people feel uncomfortable about dispensing with quantum linearity, rightly pointing out that complex linearity is the source of so much of the extraordinary mathematical elegance of conventional quantum theory. To this I can only respond that we have a precedent. The remarkable mathematical elegance of Newton's theory of universal gravitation, as so magnificently expounded in *Principia*, also owes much to the fact that the total gravitational force at any point is a *linear* function of the sources. Yet when this theory was finally superseded by a more comprehensive one, namely Einstein's general theory of relativity, it was by an astonishing *non-linear* theory of even greater mathematical elegance! I would anticipate that the appropriate theory which replaces our present quantum ideas must, when it comes, exceed even quantum mechanics in its mathematical beauty, economy and grandeur. Unfortunately, I am in no position to make clear suggestions as to what form such a theory might take. However, I *am* making a definite proposal as to the kind of physics which should, in my view, be relevant. This proposal leads us to a fairly concrete suggestion, enabling us to give, in principle, very rough estimates as to when linear evolution should fail and be taken over by some non-linear procedure resembling state-vector reduction.

3.11 Quantum gravity and time-asymmetry

The new physical ingredient which must be added to quantum theory is, I claim, *general relativity*. The new physical theory we are seeking is thus a form of *quantum gravity*. As of now, no satisfactory theory of quantum gravity exists. Ideas abound and vast calculations continue to be performed, but there is still no adequate theory. My own ideas concerning the form that this theory should take – and even concerning the very purpose of such a theory – seem to differ from those of most workers in the field. Most people seem not to be concerned with trying to alter the structure of quantum mechanics. On the contrary, the impulsive reaction has generally been to modify the structure of general relativity instead (to supergravity, more complicated Lagrangians, higher dimensions, or strings), as soon as the encounter between these two great theories runs into difficulties!

Why do I believe that quantum theory is likely to require revision when unified with general relativity? Aspects of my views on this question have appeared in other articles (Penrose, 1976, 1983, 1986); also, I refer the reader to other related viewpoints which had been put forward earlier (Károlyházy, 1966; Károlyházy, Frenkel and Lukás, 1986; Komar, 1969). Part of the reason for believing that the true quantum gravity theory may entail a change in the structure of quantum mechanics is to be found in the very difficulty that one encounters in trying to 'quantize' general relativity in a standard way. Often these difficulties seem to confront one with foundational problems concerning quantum observations. I do not propose to go into a discussion of these here. Many of these arguments are familiar in the literature (cf. Wheeler, 1964; Kuchař, 1981). In my opinion there is a considerable force behind these arguments. Attempts to build up an *S*-matrix theory of gravity from perturbation series starting from flat spacetime, for example, simply do not face up to many of these long-standing foundational problems.

I wish to stress, on the other hand, a point of a different nature – one which might even be called an *observational fact* of quantum gravity! One of the few situations in nature where it is generally accepted that quantum gravity effects should be important is in governing the physics at *spacetime singularities* (or, at whatever quantum-geometrical regions are to replace such classical singularities). From the assumption that cosmic censorship holds, we can deduce that these singularities can be divided into two classes: past singularities (out of which particles can emerge) and future singularities (into which particles can disappear). (See Penrose, 1978.) The one past singularity that we know of is the *big bang*. Future singularities are expected to arise inside black holes and perhaps in the big crunch (if the universe eventually recollapses to a singularity). Now the big bang was very far from being a 'generic' singularity. We know this both from the observed uniformity of the early universe and from the very existence of a second law of thermodynamics (Penrose, 1979). (This argument is not affected by current considerations of 'inflation'.) If we imagine the big bang time-reversed, we see more clearly how special it really is. In a generic big crunch, we expect many black holes congealing finally into one very complicated final singularity. The geometry close to the end would be extremely irregular and not at all close to the highly uniform and ordered singularity that is observed to be the case with the big bang. In the final big crunch (if it occurs – or, in any case, as a theoretical consideration) we have an irregular, extremely high entropy singularity. To a somewhat lesser (but still very considerable) degree, the situation is similar for the singularities inside black holes; each such singularity would be of a very high entropy type. However, the big bang was a singularity of a uniform and very low entropy type. The lowness of this entropy lay in the fact that the gravitational degrees of freedom were somehow initially not 'thermalized' with the matter and, moreover, have not much become so because of the weakness of the gravitational coupling. As time progressed, the matter in the universe has gradually been able to take advantage of this imbalance and clump gravitationally. Our Sun, itself, is a result of such gravitational clumping and has become the hot spot in the sky that all life on Earth ultimately relies upon as its source of low entropy.

Using the Bekenstein–Hawking formula for black-hole entropy, we can estimate the entropy that a final big crunch would have, and hence deduce how extraordinarily special the big bang actually was. The answer is remarkable: for a closed universe with, say, 10^{80} baryons, we find that the big bang was special to the degree roughly

$$\text{one part in } 10^{10^{123}}$$

where this figure refers to the region of phase-space volume that corresponds to a big-bang singularity of the uniformity that is observed in the universe as we know it. For a larger number of baryons, this figure would be even larger. Allowing for inflation makes virtually no difference to the argument.

The specialness of the big bang can be accounted for if we assume that for some reason – presumably as a consequence of the quantum gravity theory that we are striving for – the Weyl curvature is necessarily zero (or at least very small compared with the Ricci curvature) at any past singularity. The same constraint cannot apply to future singularities, for, if it did, we should not have a second law of thermodynamics. Moreover, there would then be all sorts of teleological behaviour preventing the formation of black holes. We see that this one 'observed' consequence of quantum gravity is grossly time-asymmetrical! From this we can deduce the important moral: *the true quantum gravity must be a time-asymmetrical theory.*

3.12 Time-asymmetry of state-vector reduction

Notice that this is not at all the expected conclusion if we think of quantum gravity as arising as the result of the application of some standard quantum-mechanical procedure (which would be time-symmetrical) to standard general relativity (which *is* time-symmetrical). What is needed is some new procedure which allows for some time-asymmetry.

Now, state-vector reduction – or the more complete process that underlies it – is, I claim, a time-*a*symmetrical process. In the first place, the realistic view of the state-vector that I have been promoting itself entails a time-asymmetry that is not necessarily present in other attitudes to quantum mechanics. For we note that immediately *after* an observation is made, the state-vector is in an eigenstate of the relevant Hermitian operator, but this does not generally apply to the state immediately *before* an observation. (Time-symmetry for quantum evolution can be partially reinstated at the expense of the realistic view, where one regards transition probabilities as providing the only observational aspects of quantum theory; see, for example, Penrose, 1979, p. 584. However, I am now disputing my earlier acquiescence in the common contention that transition probabilities are time-reversible – as we shall see in a moment!) Note that this picture is tied up with the fact that our realistic view seemed forced upon us because of our presumed freedom, in principle, to carry out a quantum-mechanical measurement corresponding to the Hermitian operator of our choice, here the operator $Q = |\psi\rangle\langle\psi|$. The state has to be 'prepared for' the possibility that this measurement might 'at any moment' be performed on it. The use of such an argument in the past direction of time seems to carry much less weight, since one does not seem to have the freedom to carry out 'unexpected' experiments in the past! One thinks intuitively that the state 'already knew' what was done to it in the past, but 'cannot know' what we might unexpectedly do to it in the future. Of course these feelings may be entirely inappropriate and illusory. For very minute intervals of time, I would myself not give them much credence. Yet, for a state which has persisted for a while, I cannot help feeling that there is something in the 'common sense' attitude that the state is being governed by what was done to it in the past rather than by what is going to be done to it in the future. (This use of 'governed' has not really much to do with the determinacy/indeterminacy question. I am referring, here, to the requirement that the state be an eigenstate of the past measurement immediately after it was performed, but not an eigenstate of the future measurement immediately before if will be performed.) For the moment, at least, I am allowing this use of 'common sense' to instruct us that our objective state-vector must indeed behave in a time-asymmetrical way under the reduction process.

This time-asymmetry seems to be bound up in some way with our 'freedom of choice' to perform measurements. It is probably also, therefore, bound up with the second law of thermodynamics – although the precise relation between this question of 'choice' and the second law is, to my mind, somewhat obscure. However, there is a second aspect of time-asymmetry in the 'measurement process' which, in a certain way, seems also to be connected with the second law. Imagine a two-state system, with an orthogonal basis for states that I symbolically write as $|\uparrow\rangle$, $|\downarrow\rangle$. Let us assume that the system is initially prepared in the state $|\uparrow\rangle$. Suppose that a measurement is then performed on the system, the eigenstates of the measurement being $|\rightarrow\rangle$ and $|\leftarrow\rangle$. (This need have nothing to do with spin. I just mean these to be orthogonal states different from the original pair.) There will be some resulting probability that the outcome is $|\rightarrow\rangle$ and some probability that it is $|\leftarrow\rangle$. Two possible universe histories that must be allowed by the reduction procedure are therefore:

$$\text{first } |\uparrow\rangle \quad \text{and then } |\rightarrow\rangle$$

and

$$\text{first } |\uparrow\rangle \quad \text{and then } |\leftarrow\rangle.$$

Now if this procedure were time-symmetric, it would have also to allow both

$$\text{first } |\uparrow\rangle \quad \text{and then } |\rightarrow\rangle$$

and

$$\text{first } |\downarrow\rangle \quad \text{and then } |\rightarrow\rangle,$$

if we assume that the final state is given as $|\rightarrow\rangle$, and also both

$$\text{first } |\uparrow\rangle \quad \text{and then } |\leftarrow\rangle$$

and

$$\text{first } |\downarrow\rangle \quad \text{and then } |\leftarrow\rangle$$

for a given final state $|\leftarrow\rangle$ (where the words 'first' and 'then' are being used in their normal temporal sense), the same probabilities arising as before (in reverse order), where now it is $|\rightarrow\rangle$, or $|\leftarrow\rangle$, that is, 'given' in the time-reversed reduction procedure.

Now, in practically any plausible experimental set-up, the input $|\uparrow\rangle$, and the two alternative outcomes $|\rightarrow\rangle$ and $|\leftarrow\rangle$, are all perfectly reasonable things to occur. However, $|\downarrow\rangle$ will normally be 'absurd' – at least in the sense that it represents an extreme improbability which violates the second law – this probability being nothing like what would be predicted by the standard quantum-mechanical computation. For example, one might have a Stern–Gerlach apparatus preparing spin one-half atoms in a spin-up state ($|\uparrow\rangle$). Interposed in the beam we insert a second Stern–Gerlach apparatus oriented at right angles so as to measure the spins in the left/right direction $(|\rightarrow\rangle, |\leftarrow\rangle)$, and the beam duly splits as required. However, if we try to imagine the alternative input, corresponding to $|\downarrow\rangle$, we find that, instead of being the alternative beam of the original Stern–Gerlach apparatus, it is yet another beam, coming from the laboratory wall or some irrelevant part of the apparatus! Such a behaviour would normally be thought of as extremely improbable, if not totally impossible – and not *at all* that given by the quantum probabilities applied in the time-reversed sense.

An even more elementary situation of this type occurs if we consider a photon, coming from a laboratory source (state $|\uparrow\rangle$), and simply reflect it off a half-silvered mirror, where we set up photo-cells in both the reflected and transmitted positions (states $|\rightarrow\rangle$ and $|\leftarrow\rangle$, respectively). The state $|\downarrow\rangle$ represents the photon coming from such a direction that if reflected it would reach the transmitted position and if transmitted it would reach the reflected position. There would normally be no source in that direction. The photon, in state $|\downarrow\rangle$, like the atoms above, would have to be just 'absurdly' ejected from the laboratory wall!

I believe this to be a key issue; and, since it is easily misunderstood, let me try to be somewhat clearer about the sense in which these probabilities are to be interpreted. It is necessary to adopt a viewpoint on this which allows conditional probabilities to be understood in a way that is totally unbiassed with respect to the direction in which time is taken to be running. Let me simplify the above experimental set-up even further so that we have just one photo-cell and one 'lamp' (the photon source). Half-way between the photo-cell and lamp we have a half-reflecting mirror whose plane is perpendicular to the line joining photo-cell to lamp. If desired, we can add an ellipsoidal mirror (a prolate ellipsoid of revolution) surrounding all three, the photo-cell **P** being at one focus and the lamp **L** being at the other, so that we need not worry about the direction in which the photon enters or leaves the photo-cell or lamp. (See Fig. 3.1.) There are four possible ways that the photon can go. It can start at **L** and end at **P**, or it can start at **L** and end at **L**. (These are the 'normal' cases, of everyday experience.) Moreover, it can start at **P** and end at **P**, or it can start at **P** and end at **L**. (These cases are also

needed in order to complete the space of quantum-mechanical transitions.)
The respective spacetime descriptions of these four possibilities are given in
Fig. 3.2.

Now suppose that this 'experiment' is performed a very large number of
times at various different locations throughout spacetime (where, to be on
the safe side, I am restricting these locations to be in our own galaxy in the
present era). To see what the various conditional probabilities are, we
examine the spacetime (Fig. 3.3) and select those cases in which the in-state
is the one desired ($|..\rangle$) and count the various alternative out-states ($\langle..|$)
(normal use of conditional probabilities), or else we select those cases in
which the out-state is the one desired ($\langle..|$) and count the various
alternative in-states ($|..\rangle$) (time-reversed use of conditional probabilities).
All the quantum-mechanical matrix elements $\langle P|L\rangle, \langle L|L\rangle, \langle P|P\rangle, \langle L|P\rangle$
have squared modulus equal to one-half. These give the *correct* probabilities
if we select for the in-state $|L\rangle$ and count how many times the photon ends
up in $\langle P|$ and how many times it ends up in $\langle L|$ (closely 50 per cent of each).
This is using the quantum-mechanical probabilities in the normal direction
in time. However, if we select for the out-state $\langle L|$ and count how many
times the photon started out in $|P\rangle$ and how many times it started out in $|L\rangle$
(closely 0 per cent and 100 per cent, respectively), we find that the reversed-
direction quantum-mechanical calculation gives completely the *wrong*
answer (50 per cent for each)!

It is no good trying to retrieve the situation by bringing in other factors
that we do or do not know about the universe (e.g. the second law of
thermodynamics). We do not ask that other factors be taken into
consideration when calculating quantum-mechanical probabilities in the
normal direction in time. Quantum-mechanical probabilities are supposed

Fig. 3.1. Is state-vector reduction time-symmetric? In this simple
experimental set-up a photo-cell **P** and a lamp **L** are separated by a half-
silvered mirror. A photon travels between them.

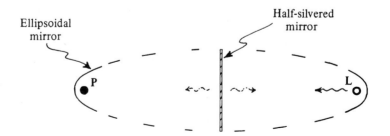

Ellipsoidal
mirror

Half-silvered
mirror

to be entirely stochastic, and not influenced by any other such factors. The standard rules of quantum mechanics were obtained by observing (in our own galaxy in the present era!) the way in which the probabilities behave in the normal direction in time. These particular quantum-mechanical rules for calculating probabilities simply *do not work* when used in the reverse direction in time.

I am proposing to disallow these 'absurd' states (like the in-state $|\mathbf{P}\rangle$ above), and simply *accept* that the objective physical process underlying state-vector reduction must be time-asymmetrical. (Most such 'absurd' states, like the in-state $|\mathbf{P}\rangle$, would, when traced backwards in time, ultimately becomes incompatible with the hypothesis of vanishing Weyl

Fig. 3.2. Spacetime diagrams of the four different ways that the photon can travel. The top two diagrams depict the 'normal' situation, with the photon coming from the lamp. The two lower possibilities are needed in order to complete the list of possible transitions between the quantum states.

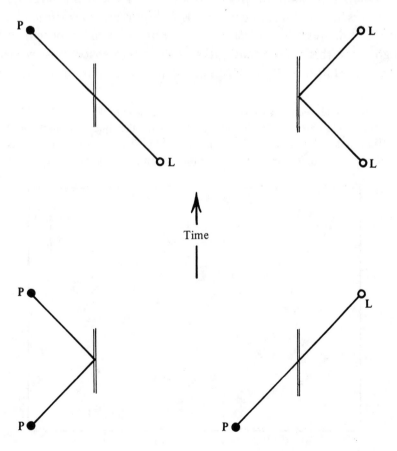

curvature at initial singularities, since they are in contradiction with the second law of thermodynamics.) Thus, one input can have several alternative possible outcomes, but there would not generally be two different inputs leading to the same outcome. This would mean that if we follow a path in the phase space of an isolated system, then at certain stages, namely where the state-vector reduction process objectively takes place, this path would split into several paths, but different paths would not generally unite together in this process. Thus, the evolution in phase space that is implied by such a state-vector reduction process will be time-asymmetric: paths can sometimes bifurcate but they do not rejoin.

3.13 Reduction and the (longitudinal) graviton count

Apart from merely this fact of time-asymmetry that I am claiming for the reduction process, is there any more direct evidence to link this process with gravity? I have argued elsewhere that there may well be (Penrose, 1979, 1981, 1986). The quantum gravity effect that I claim must restrict past singularities to have vanishing (or small) Weyl curvature also *disallows white holes* (the time-reverses of black holes). For an isolated system of large enough mass, this entails that paths in the phase space must sometimes become *united* into a single path. This occurs because information can get

Fig. 3.3. The entire space-time is examined. If the *initial* state is selected for, the final state ratios are compatible with the quantum-mechanical calculation, but if the *final* state is selected for, the same calculation gives quite the *wrong* answer for the initial state ratios.

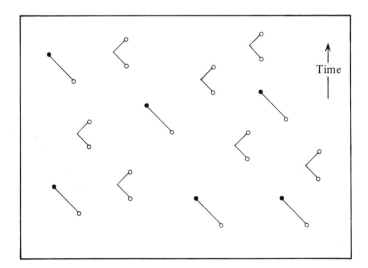

lost at the singularities, all of which are of the *future* type (into which matter can disappear: black holes) rather than of the past type (from which matter could have emerged: white holes), the latter having to be effectively excluded owing to the above-mentioned claimed quantum gravity effect. As I have suggested elsewhere (Penrose, 1981, 1986), it is very plausible to suggest that the uniting of phase-space paths that result from the above singularity discussion is exactly balanced by the splitting of phase-space paths which I argued earlier should result from the proposed objective state-vector reduction procedure. I am not, of course, arguing for any direct connection between specific spacetime singularities and specific instances of the reduction process. Clearly one does not need a black hole in the laboratory in order to perform a Stern–Gerlach experiment! I am speaking merely of an overall balance between phase-space loss from spacetime singularities and phase-space gain arising from state-vector reduction. The argument is that if the first is a quantum gravity effect, then so also must be the second.

In earlier accounts I have phrased things in terms of some (undefined) concept of 'gravitational entropy'. My present opinion is that this concept is both inappropriate and unnecessary for the discussion. (I am grateful to Boris Zel'dovich and Bob Wald for some very pertinent criticisms of my earlier view.) Instead, I believe what is required is some measure of (longitudinal) graviton number. The idea is that when two states in linear superposition become differently coupled to the gravitational field, to the extent that the difference between the fields is of the general order of *one* (*longitudinal*) *graviton*, then failure of linear superposition becomes significant, and some non-linear procedure (instability?) forces the state out of the superposition and into one alternative or the other. This procedure, if it could be found, would provide our sought-for objective state-vector reduction.

The immediate question that arises is whether the orders of magnitude involved make this seem at all plausible. While I have not been able to explore this question in a great deal of detail, it seems to me that the signs are not altogether discouraging. Let me give an example, which I shall try to describe according to the kind of picture that I have in mind. Suppose we have an atom surrounded, at some distance, by a particle detector – say a cloud chamber or a bubble chamber. The atom decays, emitting a charged particle. As the particle moves away from the atom, its *reality* consists of a spherical wave moving outwards and centred on the atom. Quantum linearity holds excellently at this level, and the wave can, if preferred, be viewed as a linear superposition of straight particle tracks directed outwards from the atom. As these tracks enter the detector, they each cause streaks of ionization which in turn cause trails of droplets, or bubbles, to be formed. The reality of the situation is still a complex linear combination of all these possible different trails together, so long as the movements of energy are sufficiently small that the coupling to gravity is significantly below the level of one graviton. Eventually, however, when the droplets, or bubbles, grow to a sufficient size, the differences between the gravitational fields produced by the different linearly superposed trails will grow to become quantum-mechanically significant – by which I mean that the (longitudinal) graviton count for the differences between the various gravitational fields of the individual trails reaches order unity. When this happens, non-linear effects become important and the system 'flops' into a state in which the spacetime geometry is (at the level of one graviton or so) well defined. Just one of the complex-linearly superposed spacetime geometries is singled out, and so just one of the possible trails reaches realization. The idea is that, in a sense, Nature abhors complex linear combinations of differing spacetime geometries!

One must do a calculation to see if this is in any way plausible. My own crude attempts at this have been superseded by a calculation performed by Abhay Ashtekar. Let us, for simplicity, assume that we have just a single droplet. (The case of a bubble is essentially similar: we simply count its mass as negative – which occurs in the formula *squared*. This mass is the difference between the mass of the bubble and that of the ambient medium.) Let the droplet's mass be m and its radius be r. Let the radius of the material from which the droplet has condensed be R. Take everything to be spherically symmetrical, and suppose, crudely, that there is a region of vacuum formed between the radii R and r. Adapting, to the case of (linearized) gravity, Zel'dovich's (1966) procedure for estimating the number of photons in a classical free electromagnetic field and applying the expression to the situation of a static field with source (where the procedure does not strictly apply – but what else can one do?), and adopting a particular gauge condition, Ashtekar obtains the following expression for the expectation value of the number of (longitudinal) gravitons arising in the region between radii r and R (this being the region where the linearized Weyl curvature is non-zero):

$$768\pi^3(m/m_{\mathrm{P}})^2 \log(R/r).$$

Here m_{P} is the Planck mass ($\sim 10^{-5}$ g). The coefficient $768\pi^3$ is about twenty-four thousand, so effects at the 'one-graviton' level could show up when *m* is around 10^{-7} g or so. The hope is that the corresponding formula for a whole string of correlated droplets (or bubbles), and other cases without spherical symmetry, would lead to a somewhat smaller mass giving a one-graviton effect.

The situations with other types of measurement, such as (developed?) photographic plates, spark chambers, computer memories and, most importantly, bundles of nerve signals (since the retina can be used as a quantum detector for small numbers of photons – and because the only time we *know* that reduction has taken place is when we personally experience it!) remain to be examined. My impression is that it would be more encouraging if plausible theoretical reasons could be found to suggest that the non-linear reduction process should take effect at a somewhat smaller scale than is estimated at present. Perhaps taking into account asymmetries, or detailed granular structure, might lead to such a conclusion. Alternatively, it might be that one needs a slightly different criterion for the onset of reduction, still involving a 'graviton' criterion, but where there would be significant probabilities for reduction over short timescales even well below the 'one-graviton' level. I feel that things are not too discouraging, in view of the primitive nature of all these considerations, so far.

3.14 Non-locality in quantum geometry

From the discussion given above, any physical reduction process must have decidedly non-local and peculiar features. In this respect, it is encouraging that the expression for graviton number in a linearized gravitational field is indeed a non-local one. However, the reader should be warned, in my opinion, that any successful objective physical theory of state-vector reduction must be non-local in a way that fundamentally affects even the very fabric of spacetime. I do not expect, except perhaps at some temporary provisional level, that we shall be able to get away with a theory which describes an objective reality taking place *within* some ambient spacetime. The spacetime must itself become subject to this non-local description.

One reason for believing this stems from consideration of the kind of non-locality that arises in situations of the Einstein–Podolsky–Rosen type as exhibited by the experiments of Freedman and Clauser, and Aspect and his coworkers. We can obtain a consistent description of the photon pairs using the conventional quantum-mechanical reduction procedure, but this must be applied *simultaneously* over a region of space encompassing both photons: a partial measurement is made on the system when one photon is observed, and this instantly affects the state of the other photon, putting that one into a state which becomes subject to the observation on *it* which immediately follows. However, this picture is not at all Lorentz invariant. If we allow a different Lorentz frame to describe the situation, we can arrange that it is the observation on this second photon which occurs first and causes the partial reduction, the result of which 'then' becomes subject to the original observation on the first photon. In this second Lorentz frame our picture of 'reality' is completely different from the one in the first frame – where I am taking the conventional description in terms of evolving and reducing state-vectors to represent 'reality'.

At this point we could simply abandon relativity for our picture of reality, if we wish. There would be no actual conflict with the results of experiment if we do so. But there is conflict with the *spirit* of relativity in such a procedure (cf. Bell, 1987). Once we adopt a non-relativistic view for our picture of reality, we find it to be a 'fluke' that relativity actually holds. To adopt such a view would be to ignore the insight that Einstein impressed upon us in 1905 – and also the closely related insight that I am suggesting guided the youthful Newton to an important aspect of his views on the nature of light. This is the insight that relativity should be taken as a *principle* rather than as an accidental feature of the precise form of the physical laws. It may be argued that cosmology (and, indeed, the big-bang singularity) now supplies us with a rest-frame, so the case is certainly debatable. Nevertheless, my own money, for what it is worth, would go on a description of reality which, at this level, should indeed be relativistically invariant. The best suggestion that I can make at this stage would be for a picture involving some sort of partially formed, partially bifurcating spacetime, where the nature of the spacetime has not been adequately resolved until the second photon observation has taken place. I have been suggesting, after all, that the reduction process should be an intimate feature of the interrelation between quantum mechanics and general relativity. General relativity describes the structure of spacetime, so, when quantum mechanics becomes intimately involved, we must expect some radically different pre-spacetime notions to enter the picture.

Another puzzle concerning the reduction procedure is the one mentioned above concerning 'null measurements'. *Not* observing a particle in a detector can effect a partial reduction of the state-vector. On the view that I am putting forward, this is not unreasonable. The spacetime geometry which gets coupled to the situation resulting from the particle entering the detector differs from the geometry resulting from it not entering the detector, and their difference might well reach the one-graviton level. At this point Nature would have to choose between (partial?) spacetime geometries. She might well choose the geometry in which the particle has not entered the detector – and that choice would non-locally affect the state elsewhere.

How is one to develop such a non-local quantum spacetime theory? I have my own ideas about what might be the fruitful line to pursue, but it would not be appropriate for me to try to enter into these here since the ideas, even after very many years of labour, are still much too ill-formulated to suggest any reliable insights. It will surprise no-one if I say that it is still my opinion that the formalism of twistor theory, with its essentially non-local description of spacetime ideas, ought to provide an important input to such a quantum spacetime theory (cf. Penrose and Rindler, 1986). Also, I still have some hankering after ideas that I entertained many years ago in evolving a theory of spin-networks (Penrose, 1971, 1972). Thought experiments of the Bohm–Einstein–Podolsky–Rosen type had an important motivational role to play in that theory, and the idea was to build up the concept of space as a limiting implicit structure when large numbers of particles are involved. However, neither twistor theory nor spin-network theory have any time-asymmetrical ingredients, as things stand. It is clear to me that some essential new input is needed.

Whatever the future may hold for the development of physical theory, it is clear that quantum theory, relativity, and even Maxwell's electromagnetic theory, have led us – or should have led us, if we have examined the facts – to a view of reality very different from that presented by classical Newtonian particle mechanics. From his writings, it is equally clear that Isaac Newton was not, in the strict sense, a 'classical Newtonian'. His view of Nature contained very much more mystery than that which the standard picture of Newtonian mechanics now conjures up. Perhaps his readiness to put forward such a strange corpuscular yet wave-like conception of light is indicative of his belief in the depths of Nature's mysteries. If so, he was certainly right!

Acknowledgements

I am grateful to Abhay Ashtekar and Dipankar Home for valuable suggestions, and to Karel Kuchař for very helpful criticisms of the manuscript. I thank, also, the Institute for Theoretical Physics at the University of California in Santa Barbara for its hospitality and stimulation, where discussions with D. Boulware, I. Bialynicki-Birula, E. T. Newman, W. G. Unruh and R. Wald were most valuable. This research was supported in part by the National Science Foundation under Grant No. PHY82-17853, supplemented by funds from the National Aeronautics and Space Administration, at the University of California at Santa Barbara and in part by the U.S. Air Force Office of Scientific Research.

References

Arnol'd, V. I. (1978). *Mathematical Methods of Classical Mechanics*. Springer: New York.

Aspect, A. (1976). *Phys. Rev.* **D14**, 1944.

Bell, J. (1987). Beables for Quantum Field Theory in *Quantum Implications: essays in honour of D. Bohm* (eds. B. J. Hiley & S. D. Peat, Routledge & Kegan Paul) p. 227

Drake, S. (1957). *Discoveries and Opinions of Galileo*. Doubleday: New York.

Freedman, S. J. and Clauser, J. F. (1972). *Phys. Rev. Lett.*, **28**, 938.

Galilei, G. (1638). *Dialogues Concerning Two New Sciences* (Macmillan edn, 1914). Dover: New York.

Károlyházy, F. (1966). *Nuovo Cim.*, **A42**, 390.

Károlyházy, F., Frenkel, A. and Lukás, B. (1986). On the possible role of gravity in the reduction of the wave function. In *Quantum Concepts in Space and Time*, ed. C. J. Isham and R. Penrose. Oxford University Press: Oxford.

Komar, A. B. (1969). *Int. J. Theor. Phys.*, **2**, 257.

Kuchař, K. (1981). Canonical Methods of Quantization in *Quantum Gravity 2: a second Oxford Symposium* (ed. C. J. Isham, R. Penrose & D. W. Sciama, Clarendon Press, Oxford) p. 329

Mache, E. (1972). *The Science of Mechanics*. Open Court: La Salle, Ill.

Newton, I. (1730). *Opticks*. (1952) Dover: New York.

Newton, I. (1687). *Principia*.

Pais, A. (1980). *Subtle is the Lord*. Oxford University Press: Oxford.

Penrose, R. (1971). Angular momentum: an approach to combinatorial space-time. In *Quantum Theory and Beyond*, ed. T. Bastin. Cambridge University Press: Cambridge.

Penrose, R. (1972). On the nature of quantum gravity. In *Magic Without Magic*, ed. J. R. Klauder. Freeman: San Francisco.

Penrose, R. (1976). *Gen. Rel. Grav.*, 7, 31–52.

Penrose, R. (1978). Singularities of space-time. In *Theoretical Principles in Astrophysics and Relativity*, ed. N. R. Leibowitz, W. H. Reid and P. O. Vandervoort. Chicago University Press: Chicago.

Penrose, R. (1979). Singularities and time-asymmetry. In *General Relativity: An Einstein Centenary*, ed. S. W. Hawking and W. Israel. Cambridge University Press, Cambridge.

Penrose, R. (1981). Time-asymmetry and quantum gravity. In *Quantum Gravity 2*, ed. C. J. Isham, R. Penrose and D. W. Sciama. Oxford University Press: Oxford.

Penrose, R. (1983). Donaldson's moduli space: a 'model' for quantum gravity? In *Quantum Theory of Gravity*, ed. S. M. Christensen. Adam Hilger: Bristol.

Penrose, R. (1986). Gravity and state-vector reduction. In *Quantum Concepts in Space and Time*, ed. C. J. Isham and R. Penrose. Oxford University Press: Oxford.

Penrose, R. and Rindler, W. (1984). *Spinors and Space-Time*, Volume 1. Cambridge University Press: Cambridge.

Penrose, R. and Rindler, W. (1986). *Spinors and Space-Time*, Volume 2. Cambridge University Press: Cambridge.

Westfall, R. S. (1980). *Never at Rest*. Cambridge University Press: Cambridge.

Wheeler, J. A. (1964). In *Relativity, Groups and Topology: the 1963 Les Houches Lectures*, ed. B. S. De Witt and C. M. De Witt. Gordon and Breach: New York.

Wheeler, J. A. (1964). *Geometrodynamics and the Issue of the Final State in Relativite, Groupes et Topologie* (eds. C. deWitt & B. deWitt, Gordon and Breach, New York & London)

Wigner, E. P. (1961). Remarks on the mind–body question. In *The Scientist Speculates*, ed. I. J. Good. Heinemann: London.

Zel'dovich, Ya. B. (1966). *Sov. Phys. Dokl.*, **10**, 771–2.

Note added in proof

It has been brought to my attention, by N. M. J. Woodhouse, that in Newton's manuscript fragment *De motu corporum in mediis regulariter cedentibus* (a precursor of *Principia* written in 1684) Newton had originally proposed to base his mechanics on five (or six) fundamental laws rather than the three that have come down to us through *Principia*. Law 4 was actually a clear statement of the (Galilean) relativity principle! Newton was well aware that these laws were not independent of one another, and for *Principia* he settled on the three that are now familiar to us. What is remarkable – apparently supporting the viewpoint tentatively put forward in this article – is that Newton had at one time, indeed, seriously contemplated using the relativity principle *as a fundamental principle* (despite what his views on 'absolute space' may or may not have been)!

4

Experiments on gravitation

A.H.COOK

We at least know something in knowing what qualities gravitation does not possess and when the time shall come for explanation, all these laborious and, at first sight, useless experiments will take their place in the foundation on which that explanation will be built.

J. H. Poynting (1900)

4.1 Introduction

Newton's law of gravitation has long intrigued and baffled experimenters and theorists alike by its simplicity and generality and by the failure of all attempts to establish any departure from it or any dependence on extraneous circumstances. The only known deviation is that consequent upon general relativity in the neighbourhood of a massive body which leads, for example, to the anomalous precession of the perihelion of Mercury. The situation is in contrast to that of the force between electrically charged particles where the inverse square law attraction between isolated stationary particles is modified by the presence of other material bodies and by the velocities of particles. Those deviations from the simple Coulomb law have enabled the nature of the electromagnetic force and the electrical structure of materials to be studied by experiment and, in the absence of similar effects in gravitation, the gravitational force resists experimental investigation.

It may be that gravitation truly is nothing but a manifestation of the geometrical structure of nature as set out in general relativity, but it may be that the reason that no other effects have so far been detected is that they are very small and that experiments are insufficiently sensitive. As will appear later on, experiments on gravitation are indeed much more difficult than those on electromagnetism, for two reasons: first, the gravitational force is only 10^{-40} of the electrostatic force between baryons, so that disturbing forces are much more troublesome; and, secondly, the techniques available for mechanical measurements are far less delicate than those for electrical measurements, in which very sensitive electronic devices can be used. Thus, for example, the inverse square law variation of gravitation has been shown to be correct to within about 1 part in 10^4 at distances of 0.1 m, whereas the limits set by experiment on the inverse square law of electrostatic attraction are many orders of magnitude less. Again, the constant of gravitation is known to little better than 1 part in 10^4 whereas almost all other fundamental constants of physics are known to 1 part in 10^6 and some very much better. Experiments on gravitation are therefore a challenge to the technique and imagination of the experimenter and so continue to attract skilled and ingenious physicists.

Experiments on gravitation are complementary to analyses of celestial mechanics. It is well known that despite the most thorough attempts (including some with notorious errors in dynamical theory) no departure that can unambiguously be attributed to a deviation from the inverse square law has ever been established, with the sole exception of the consequences of general relativity. In particular, the inverse square law itself is firmly established in celestial mechanics and the force of gravitation shows no dependence on the direction of the line joining masses. These two properties can indeed be investigated in the laboratory at short distances, but with very much greater uncertainties. On the other hand, celestial mechanics tells nothing about the dependence of the force on the constitution of the masses (nor indeed that it is just proportional to the product of the masses) whereas that is something that can be studied with great precision in the laboratory.

Something can also be learned from geophysics, at distances intermediate between those of the laboratory and celestial mechanics, and limits may be placed on directional effects and on shielding. Celestial mechanics, geophysics and laboratory investigations must therefore all be taken into account in assessing the validity of Newton's law, but this article is concerned almost entirely with experiment, that is to say, with observations which can be made in a laboratory where conditions are under the control of the experimenter. Nothing is therefore said about the laser ranging observations of the Moon, which belong to celestial mechanics. Recently it has become possible to think of making experiments in space craft and that has somewhat extended the range of effects that may be studied – however, no such experiment has yet been carried out.

Experiments on gravitation have been made from the beginning. Newton himself, no mean experimenter, used pendulums of different materials to see if there were any dependence on constitution and was able to set a limit of about 1 part in 10^3 (Newton, 1687). He also made some estimates of the gravitational attraction of a mountain on a plumb bob and of the time it would take two masses to come together under their mutual gravitation, both with the idea of determining the constant of gravitation. He concluded that the effects were too small to observe but the first was employed by Bouguer and by Maskelyne (1775) (at Schiehallion) while the second was the principle of an experiment attempted some twenty years ago (Luther *et al.*, 1976). The most important advance in experiments on gravitation and other delicate measurements was the introduction of the torsion balance by Michell and its use by Cavendish (1798) in gravitation and electrostatics. It has been the basis of all the most significant experiments on gravitation ever since. It is also a measure of the inherent difficulty of gravitational experiments that, despite many refinements of technique and design, the accuracy of Cavendish's determination of G has been little improved upon until very recently. By contrast, the study of possible differences between materials has been carried to great refinement, with limits now about 1 part in 10^{12} instead of the 1 in 10^3 that Newton attained. These and other experiments of historical interest will be described to some extent in the appropriate sections.

4.2 Theoretical framework

This section is in no sense a technical account of theory. There are indeed two approaches that may be taken to the theoretical grounds for experiments. One is to explore the inferences to be drawn from a consistent theory of gravitation and to see what experiments may be conducted to check them. Will (1981) has shown that many theories may be treated on a uniform basis in the parametrised post-Newtonian (PPN) formulation and it is convenient to follow him in setting out the implications of particular experiments. On the other hand, one could, in a rather simple-minded way, just take Newton's law and ask what features of it might be tested experimentally: is force just proportional to the product of masses, are non-linear or shielding effects absent? is the force inverse square? does it depend on direction? on the constitution of the masses? on time? on velocities? It is convenient to combine the two approaches – the type of experiment that can be done is best classified according to the second scheme, the interpretation is conveniently given in PPN terms.

The inverse square law itself is somewhat special because of a direct connection with the inverse square law for electrostatics, the validity of which has been established experimentally to very high precision.

The conditions for the inverse square law to be followed by a force are that the geometry of space should be locally Euclidean and that the mass of the force field should be zero. The first condition is equivalent to the area of the sphere drawn about a point source being proportional to the square of the radius, or to the proper distance from the source being independent of source strength, for Gauss's theorem shows that in a massless field the source strength is equal to the integral of the force over a sphere drawn around the source and, if the inverse square law is to apply, the area of the sphere must be proportional to the square of the radius. In the PPN formulation, the proper distance from a source of potential U is $r(1+\gamma U)$ where γ is one of the PPN parameters (see below); if γ is zero, the proper distance is independent of U and then the area of a sphere is proportional to the square of the radius.

The condition that the field should be massless is not satisfied by, for example, a Yukawa potential; effectively the total energy stored within a sphere now depends on the size of the sphere, even with Euclidean geometry. The electromagnetic field is massless as is shown by the fact that electromagnetic waves in vacuum have no dispersion. The electrostatic law of force is also inverse square to very close limits, as shown experimentally by the absence of any detectable field outside a closed conductor. It may therefore be inferred that the geometry is locally Euclidean, at least for the values of masses that can be placed in a terrestrial laboratory.

If then, it were found that the inverse square law did not hold for gravitation, it would follow that the gravitational field carried mass, and the analysis of experiments on the inverse square law is then naturally made in terms of a Yukawa potential (Fujii, 1971, 1972)

$$-\frac{k}{r}(1+\alpha\,e^{-\mu r}).$$

Experiments on the inverse square law seek to establish limits on α and μ.

The general structure of the PPN formulation of metric theories of gravitation has been described by Will (1981). The theories are metric because the motion of a particle is completely determined by the metric of the geometry, a metric which is specified by ten parameters.

The elements of the metric form are
$$g_{00} = -1 + 2U - 2\beta U^2 - 2\xi\Phi_W + (2\gamma + 2 + \alpha_3 + \zeta_1 - 2\xi)\Phi_1$$
$$+ 2(3\gamma - 2\beta + 1 + \zeta_2 + \xi)\Phi_2 + 2(1 + \zeta_3)\Phi_3 + 2(3\gamma + 3\zeta_4 - 2\xi)\Phi_4$$
$$- (\zeta_1 - 2\xi)\mathcal{A} - (\alpha_1 - \alpha_2 - \alpha_3)w^2 U - \alpha_2 w^i w^j U_{ij} + (2\alpha_3 - \alpha_1)w^i V_i;$$
$$g_{0i} = -\tfrac{1}{2}(4\gamma + 3 + \alpha_1 - \alpha_2 - 2\xi)V_i - \tfrac{1}{2}(1 + \alpha_2 - \zeta_1 + 2\xi)W_i$$
$$- \tfrac{1}{2}(\alpha_1 - 2\alpha_2)w^i U - \alpha_2 w^j U_{ij};$$
$$g_{ij} = (1 + 2\gamma U)\delta_{ij}.$$

where U is the Newtonian scalar potential, Φ_W is a potential introduced by Whitehead, Φ_1, Φ_2, Φ_3, Φ_4 and $\mathscr{A}$ are other scalars, V_i and W_i are vector potentials and U_{ij} is a tensor potential. In the absence of sources, the metric coform is that of special relativity, diag $(-1, 1, 1, 1)$. w^i is the velocity of the local frame relative to some absolute frame.

In general relativity, $\beta = \gamma = 1$ and all other parameters are zero, and so the metric is

$$g_{00} = -1 + 2U - 2U^2 + 2(2\Phi_1 + 4\Phi_2 + \Phi_3 + 3\Phi_4);$$
$$g_{0i} = -\tfrac{7}{2}V_i - \tfrac{1}{2}W_i;$$
$$g_{ij} = (1 + 2U)\delta_{ij},$$

where Φ_1, Φ_2 and Φ_3 are scalar potentials of which the sources are kinetic, potential and internal energy not included in the rest masses. Most of the experiments that are considered below will be related to the PPN parameters.

A crucial postulate of general relativity is the *weak equivalence principle*, according to which the passive gravitational mass of a test particle bears always the same ratio to its inertial mass, whatever its material constitution. The motions of all test particles in a gravitational field are therefore all the same, in fact along geodesics. Thus all bodies in vacuum fall towards the Earth with the same velocity. If this were not so in celestial mechanics, Kepler's law,

$$GM = a^3 n^2,$$

relating the (active) mass M of a primary to the semi-major axis, a, and mean motion, n, of a secondary would not be followed, for it assumes that the inertial mass, m_i, and (passive) gravitational mass, m_g, of the secondary are the same. Without that assumption,

$$GM = a^3 n^2 (m_g/m_i)$$

and if secondaries differed in the ratio m_g/m_i (as, say, between the Earth and Jupiter) there might be detectable differences.

The weak equivalence principle does not hold in all PPN theories, for the ratios of passive and active gravitational masses (m_P and m_A) to inertial mass, m_i, are given by (Will, 1981)

$$m_P/m_i = 1 + (4\beta - \gamma - 3 - \tfrac{10}{3}\xi - \alpha_1 + \tfrac{2}{3}\alpha_2 - \tfrac{2}{3}\zeta_1 - \tfrac{1}{3}\zeta_2)\Omega/m$$

and

$$m_A/m_i = 1 + (4\beta - \gamma - 3 - \tfrac{10}{3}\xi - \tfrac{1}{3}\alpha_3 - \tfrac{1}{3}\zeta_1 - 2\zeta_2)\Omega/m$$
$$+ \zeta_3 E'/m' - (\tfrac{3}{2}\alpha_3 + \zeta_1 - 3\zeta_4)P'/m',$$

where m_i is the total mass energy of the body (rest energy, internal kinetic, potential and gravitational energy), Ω is the internal self-energy of gravitation, E' is the integral of internal energy and P' that of pressure for the *attracted* body of mass m'. In general relativity $4\beta = \gamma + 3$ and all other parameters are zero, so $m_A = m_P = m_i$. Experimental studies of the weak equivalence principle are crucial to gravitational physics.

The crucial observations that first established the validity of general relativity were those on the deflexion of light in the gravitational field of the Sun and the anomalous advance of the perihelion of Mercury, corresponding to a term in the potential of the Sun proportional to $1/r^2$. The deflexion of light may be seen as the consequence of a reduction of the speed of light near a massive body; the reduction, which is proportional to $\frac{1}{2}(1+\gamma)$, may be observed either through the shift of the apparent direction of a source or through an increase in the time of passage of light. Both observations can now be made rather precisely by radio means, long-base line interferometry for direction (Fomalont and Sramek, 1977), and radar measurements (Shapiro *et al.*, 1972) and radio transponders (Anderson *et al.*, 1975; Reasenberg *et al.*, 1979) for time delays. The time delay measurements are the most precise and show that γ is 1 to within 0.1 per cent (Will, 1981) in accordance with general relativity. The overall precession of the perihelion of Mercury is obtained both from long series of optical observations (to within 1 per cent) and from radar observations of the distance of Mercury from the Earth (to within 0.5 per cent) (Shapiro *et al.*, 1972; Shapiro *et al.*, 1976). In Newtonian theory, the attractions of the other planets contribute to the precession, as does the quadrupole moment of the Sun and there is an effect of the general precession of the equinox. The relativistic effect is the residue. The planetary and equinoctial terms are well known, but the value of the quadrupole moment of the Sun has been questioned. The relativistic precession is proportional to $\frac{1}{3}(2+2\gamma-\beta)$ and with the best radar results for the overall precession and discounting doubts about the internal constitution of the Sun, β is found to be 0.991 ± 0.015, again in accordance with general relativity.

It thus appears that those PPN parameters, β and γ, which are non-zero in general relativity, have the expected values of 1. There is also nothing in the motions of planets or satellites which requires there to be a preferred frame of reference for the solar system. In the PPN formulation, the equations of motion contain terms proportional to the velocity relative to the preferred frame, multiplied by various linear combinations of the parameters α_1, α_2 and α_3. The absence of preferred frame effects means that α_1, α_2 and α_3 are all zero. Evidence from Earth tides (Section 4.5) also sets limits on the αs.

4.3 The inverse square law

In the hands of Cavendish, the torsion balance was used to establish the inverse square law for electrostatic forces. Subsequently, the inverse square law in electrostatics has been demonstrated to the highest precision by showing that no field can be detected outside a closed conductor containing charge (the Faraday cage). Experiments on gravitation are far more difficult and it is only with the refinement of the torsion balance in recent years that useful results have been obtained, Cavendish (1798), in his initial determination of the constant of gravitation, mentions that he has made a check of the inverse square law but gives no details, and in subsequent determinations (for example, Baily, 1842; Cornu and Baille, 1873, 1878) no check on the law seems to have been attempted, but the overall consistency of results from experiments in which the distances between masses were different, implies a rough agreement with the inverse square law. Mackenzie (1895) in his experiments on the attractions of crystals, placed the masses at separations of 3.6, 5.5 and 7.4 cm and claimed that he had confirmed the inverse square law to 1/500 between 3.5 and 7.3 cm.

C. V. Boys (1895) greatly increased the sensitivity and stability of the torsion balance when he made very fine quartz fibres and, subsequently, tungsten wires properly treated (Braginski and Manukin, 1977; Chen, Cook and Metherell, 1984) have proved as satisfactory. It is not sufficient that a fibre should have a small restoring couple; more important, its rest point should be stable. The other improvement has been in the means of detecting the angular position of the beam of a torsion balance; the optical lever used by Boys has been replaced by optical interferometers, capacitance bridges or position-sensitive photo-detectors, each of which is more convenient and capable of much greater sensitivity than the traditional optical lever, to such an extent that it is not necessary to use the highest sensitivity of which a torsion balance is capable, allowing more attention to be paid to stability.

In recent years three experiments on the inverse square law have been completed, in two of which forces between masses at different distances from a test mass were compared while, in the third, a gravitational equivalent of a Faraday cage was used. The interpretation of all such experiments depends on being able to calculate the forces corresponding to a specific non-Newtonian potential, in particular one in the form

$$-(k/r)(1 + \alpha\, e^{-\mu r}).$$

The form

$$G = G_0[1 + \varepsilon \ln(r/r_0)]$$

has also been used for the non-Newtonian force; the two are equivalent at short distances if

$$\varepsilon r_0^{-1} = \alpha\mu.$$

The necessary calculations have been thoroughly investigated by Chen (1982) (also Cook and Chen, 1982) particularly for the cylindrical masses that he used in his own experiment (Chen, Cook and Metherell, 1984).

   Long (1976) appears to have been the first after Mackenzie to have investigated the inverse square law at small distances. He placed two objects of different mass in turn opposite a test mass on one end of the beam of a torsion balance, at such distances that the forces on the test mass should be equal for Newtonian attractions. His attracting objects were rings of different radii. If the radius of a ring is r and the mass of the ring is m, then the axial attraction at a distance z from the plane of the ring is

$$\frac{G2\pi mz}{(r^2 + z^2)^{3/2}}.$$

The attraction is a maximum when z^2 equals $\frac{1}{2}r^2$, when it has the value

$$4\pi Gm/3^{3/2}r^2.$$

If the rings are placed so that the attractions they exert upon the test masses are maximum, it is not necessary to measure the (possibly ill-defined) distances from the test mass. Long claimed that his results showed significant deviations from the inverse square law (see below).

   The principle of the experiment of Newman and his colleagues (Spero *et al.*, 1980) was different: the field at the centre of an infinitely long hollow cylinder would be constant if the inverse square law applied. Such a cylinder is the gravitational equivalent of the Faraday cage and a test mass inside it would experience no change of force as the cylinder was moved from side to side. The test mass must be suspended from the arm of a torsion balance used to detect changes of force, and so the cylinder cannot be infinitely long and must be open at the ends. There will then be end effects such that the mass will experience some residual variable force under the inverse square law, but it is sufficient to calculate that from the Newtonian law. Spero *et al.* (1980) found no significant deviation from the inverse square law but Long (1980) criticised the principle of their experiment on the grounds that it did not exclude all possible non-Newtonian effects and Chen (Cook, Chen and Metherell, 1984) has performed an experiment the design of which meets Long's criticism and which has set slightly different limits to deviations from the inverse square law from those of Spero *et al.* (1980).

Like Long, Chen compared the effects of different masses on a test mass, but his masses were quite large cylinders and he used three; one was placed in a fixed position relative to the test mass and its attraction was balanced either by a small cylinder close to the test mass or a larger one further away. Thus the test mass was always observed in the same position, or nearly so. In consequence, it is not necessary to find the position of the test mass relative to the cylinders. Let m_1, m_2, m_3 be the masses of the three cylinders, of which 1 is kept fixed. Likewise, let r_1, r_2 and r_3 be the distances from the test mass. Then it is difficult to measure those distances from the mobile mass but straightforward to measure $(r_1 + r_2)$, $(r_1 + r_3)$ and $(r_2 - r_3)$. With the Newtonian law,

$$r_2 = (r_1 + r_2)m_2^{1/2}/(m_1^{1/2} + m_2^{1/2})$$

and

$$r_3 = (r_1 + r_3)m_3^{1/2}/(m_1^{1/2} + m_3^{1/2}),$$

whence $(r_2 - r_3)$ as calculated from the masses and the measured sums $(r_1 + r_2)$ and $(r_1 + r_3)$, may be compared with the observed $(r_2 - r_3)$ for the respective conditions of balance for masses 2 or 3 with 1. Chen placed his torsion balance in an evacuated enclosure and observed changes of position by reflecting a beam of laser light from a mirror on the beam to a position-sensitive photo-detector.

The centres of mass of large objects used in experiments on gravitation cannot be found exactly because the density may not be uniform but the uncertainty may be eliminated by rotating a mass such as a cylinder about its geometrical axis.

Chen in fact carried out two groups of experiment, in the one the net force on the test mass was very close to zero, while in the other there was a significant net force. The purpose was to check the existence of a vacuum mass polarisation in the presence of a net field.

The results of experiments on the inverse square law may be expressed as permitted ranges of the parameters α and μ. According to Fujii (1971, 1972), α should be $\frac{1}{3}$ in which case the following limits are placed on μ:

Long (1976):   (non-null) $1.8 > \mu^{-1} > 2.2$ m (Chen, 1982)
Spero *et al.* (1980):   (null) $\mu^{-1} > 3.4$ m (Chen, 1982)
Chen, Cook and Metherell (1984): (non-null) $\mu^{-1} > 3.4$ m
                                          (null) $\mu^{-1} > 4.9$ m

Alternatively, the results may be expressed as values of ε in the law $G = G_0(1 + \varepsilon \ln(r/r_0))$

Long (1976): $(20 \pm 4) \times 10^{-4}$

Spero *et al.* (1980): $(1\pm)\times 10^{-5}$

Chen, Cook and Metherell (1984): (non-null) $(-4.3\pm5.1)\times 10^{-4}$

                                         (null) $(2.0\pm6.6)\times 10^{-5}$

Spero *et al.* (1980) have shown how the result of an inverse square law experiment may be displayed in a diagram of allowed combinations of α and μ and the diagrams for the Long, Spero and Chen experiments are reproduced in Fig. 4.1. A summary of the results is that the inverse square law is followed to within 1 part in 10^{-4} at distances of about 0.1 m and that the characteristic length, μ^{-1} in a Fujii-type potential is at least 3.5 m (see also, Gibbons and Whiting, 1981).

Other laboratory experiments have been performed in recent years but

Fig. 4.1. Allowed values of α and μ in a Spero diagram (after Chen, Cook and Metherell, 1984).

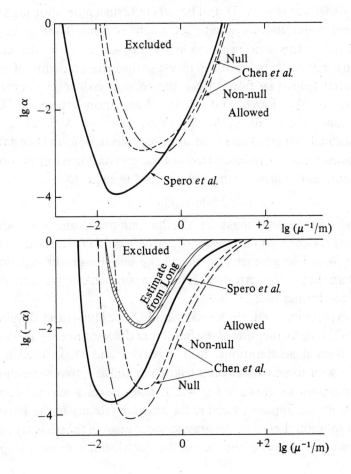

they are less precise than the three just described (Panov and Frontov, 1979; Ogawa *et al.*, 1982; Chan *et al.*, 1982). Experiments at greater distances have also been attempted (Yu *et al.*, 1979) while Stacey and Tuck (1981) have argued from geophysical measurements that the corresponding value of G is some 0.5–1.0 per cent greater than laboratory determinations. However, the precision of these results also is poor (see also next section).

4.4 The weak equivalence principle

As already mentioned, Newton himself did some experiments with pendulums of different materials. He made two pendulums with equal wooden containers hung from wires 11 ft long, and compared the periods when one container was filled with wood and the other with gold, silver, lead, glass, sand, salt or water. He found the periods to be the same to about 1 part in 1000 (Newton, 1687), and Bessel (1832) in a more thorough study of 12 different materials was able to set a similar limit. The big advance in experiments in this field was made by Eötvös who, in the course of developing a torsion balance for investigating the gradients of the local gravitational field, saw how to use the torsion balance to compare the attractions of the Earth and of the Sun upon masses of different composition. Experiments of this sort in fact test for differences in the ratio of gravitational to inertial mass, for some way has to be found of establishing when masses are equal, independently of the gravitational forces upon them. In pendulum experiments, where the period is equal to

$$2\pi(m_g gh/m_i r^2)^{1/2}$$

(m_g is the gravitational mass, m_i the inertial mass, and r the radius of gyration) it is clear that differences of period would correspond to different ratios (m_g/m_i). The geometry of the Dicke and Braginski experiments is clearer than that of the terrestrial Eötvös experiment, so it is convenient to consider the former first.

In the experiments of Roll, Krotkov and Dicke (1964) and Braginski and Panov (1972) the attractions of the Sun upon different masses are compared with the inertial accelerations. Roll, Krotkov and Dicke used a torsion pendulum with three masses on a triangular frame – two were aluminium and one was gold, as shown in Fig. 4.2. Let $M_\odot$ be the mass of the Sun, R its distance from the apparatus and ω the angular velocity of the Earth about its axis of rotation. Let m_i be the gravitational mass of the ith body and m_i' its inertial mass. Let p_i be the perpendicular distance from the point of suspension to the line from the Sun to the *i*th body, projected upon the plane of the triangle of mass.

The gravitational torque about the suspension fibre is then

$$\frac{GM_\odot}{R^2}(m_1p_1 + m_2p_2 + m_3p_3)$$

while the inertial acceleration is

$$R\omega^2(m_1'p_i + m_2'p_2 + m_3'p_3).$$

For the Earth as a whole R, the orbital distance from the Sun to the centre of mass of the Earth, adjusts itself so that $GM_\odot/R^2$ is equal to $R\omega^2$ (Kepler's law).

If the ratios of gravitational to inertial mass are the same for each of the three masses on the pendulum then the ratio

$$(m_1p_1 + m_2p_2 + m_3p_3)/(m_1'p_1 + m_2'p_2 + m_3'p_3)$$

will be unchanged when the p_i vary as the direction of the Sun changes with the rotation of the Earth, but if the ratios do differ between materials then

Fig. 4.2. The Dicke experiment on the weak equivalence principle.

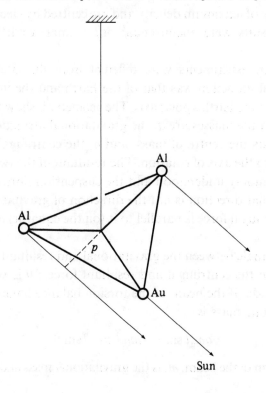

the ratio of torques will not be constant, but will vary with a period of 24 h. Roll, Krotkov and Dicke combined a very sensitive detector of the angular position of the triangular frame with electrodes fed with a voltage to apply a restoring force, so that the angular position of the frame was kept constant as the Earth rotated; a record was made of the voltage that had to be applied.

One of the main problems that had to be faced was that many disturbing effects depend on the position of the Sun, whether directly through the diurnal variation of the temperature or otherwise, so that very great care had to be taken to isolate the apparatus (in a special vault) to reduce extraneous effects of diurnal period so far as possible. The limit put upon any difference of the ratio of gravitational to inertial mass between gold and aluminium was 1 part in 10^{11}, but an unexplained feature of the observations is that there was a variation of torque with a semi-diurnal period. Subsequently Braginski and Panov (1972) carried out a similar experiment but with improved discrimination, so that they were able to set a limit of 1 part in 10^{12} on any difference between materials.

Another experiment was started about 1979 (Keiser and Faller, 1979) in which the test masses are not hung from a fibre but are floated on water at the temperature of maximum density and are centred by electrostatic forces. Preliminary results were encouraging but a final result has still to be announced.

Eötvös's main experiments were different from the later ones, in that the gravitational attraction was that of the Earth and the inertial force that of rotation about the Earth's polar axis. The geometry is shown in Fig. 4.3. The forces acting on the masses are g, the gravitational attraction of the Earth which is towards the centre of mass, and a, the centrifugal force which is perpendicular to the axis of rotation. The resultant of the two is the force of gravity as commonly understood and the suspension fibre is tangential to the resultant; that direction is not the direction of gravitational attraction unless the centrifugal force is parallel to it (on the equator) or is zero (at the poles).

Let α be the angle between the gravitational and resultant forces and let θ be that between the centrifugal and resultant forces; θ is very nearly 180° minus the latitude. If the beam of the torsion balance hangs east–west, the net torque on one mass is

$$gm_g l \sin \alpha - am_i l \cos \theta \sin \theta,$$

where l is the arm of the beam, m_g is the gravitational mass and m_i is the inertial mass. With two masses, one at each end of the beam, the net torque is

$$g(m_{1g}l_1 - m_{2g}l_2)\sin\alpha - \tfrac{1}{2}a(m_{1i}l_1 - m_{2i}l_2)\sin 2\theta.$$

If the apparatus is turned round through $180°$ the signs of the differences $(m_{1g}l_1 - m_{2g}l_2)$ and $(m_{1i}l_1 - m_{2i}l_2)$ are reversed. If the two factors are equal there is no change of torque on reversing the balance, but if they differ because the ratios $(m_g/m_i)_1$ and $(m_g/m_i)_2$ are unequal, there will be a net change of torque.

Eötvös, Pekar and Fekete (1922) examined a number of pairs of different substances and found no difference of gravitational mass relative to inertial mass to exceed 1 part in 10^9 and their results were confirmed in a later set of experiments by Renner (1935).

Recently, the results have been reexamined by Fishbach *et al.* (1986).

Fig. 4.3. The Eötvös experiment on the weak equivalence principle.

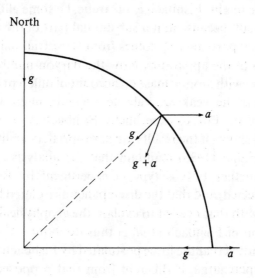

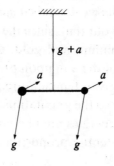

They point out that if there is an additional non-Newtonian potential of the form

$$-\frac{Gm_1 m_2}{r}\, \alpha\, e^{-\mu r},$$

then the constants in that expression might depend on the material constitution and in particular on the baryon numbers of the substances.

Fishbach *et al.* also appeal to certain geophysical evidence which may indicate that the effective constant of gravitation at distances of some hundreds of metres is greater than that for laboratory distances (Stacey and Tuck, 1981; Holding and Tuck, 1984) and would seem to entail a value of α of $(-7.2 \pm 3.6) \times 10^{-3}$ and of μ^{-1} of (200 ± 50) m. Such values could not be detected by the inverse square law experiments of Section 4.3. They would also not affect the experiment of Roll, Krotkov and Dicke (1964), nor of Braginski and Panov (1972) for there r is the distance to the Sun and $e^{-\mu r}$ is negligible. There might, Fishbach *et al.* argue, be some effect in the original Eötvös experiment, because in it a substantial part of the attraction on the test masses, a few parts in 10^3, comes from terrestrial matter within a few hundred metres of the apparatus. Now the baryon number of material of given mass varies with atomic mass on account of different binding energies, and so a failure of the weak equivalence principle might be expected in the circumstances of the Eötvös experiment. Fishbach *et al.* argue that such a failure is indeed shown if the results are appropriately analysed, but Keyser, Niebauer and Faller (1986) maintain that the analysis is faulty, that the results of the further Eötvös type of experiment by Renner (1935) are incorrectly neglected, and that the discrepancy as isolated by Fishbach *et al.* is inconsistent with that needed to explain the geophysical data. The status of the suggestion of Fishbach *et al.* is thus doubtful.

It should be noted that the force postulated by Fishbach *et al.*, involving a new type of hypercharge, is different from that proposed by Fujii (1971, 1972) which is purely gravitational.

The implications of the weak equivalence results are rather far-reaching. Thus Dicke has pointed out that, since the relative numbers of protons and neutrons differ in aluminium and gold, there is no detectable difference between the attraction of the Sun upon protons and neutrons. Further, the internal velocities at a given temperature are different because the atomic masses are different and so the gravitational force cannot depend on atomic velocities. (The kinetic energies are of course the same, so the experiments say nothing about a potential proportional to kinetic energy.)

4.5 Preferred frames and locations

In the PPN formulation the velocity of a local frame relative to some preferred celestial frame enters the metric through the following terms:

in g_{00}: $-(\alpha_1 - \alpha_2 - \alpha_3)w^2 U - \alpha_2 w^i w^j U_{ij} + (2\alpha_3 - \alpha_1)w^i V_i$,

in g_{0i}: $-\frac{1}{2}(\alpha_1 - 2\alpha_2)w^i U - \alpha_2 w^j U_{ij}$,

and as a result (Will, 1981) the measured constant of gravitation, as found from the acceleration of a test mass towards a source mass, contains the term

$$\tfrac{1}{2}G\alpha_2(1 - 3I/mr^2)(\mathbf{w} \cdot \mathbf{n})^2 - \tfrac{1}{2}G[\alpha_1 - \alpha_3 - \alpha_2(1 - I/mr^2)]w^2,$$

where $\mathbf{n}$ is the unit vector along the line joining the two bodies and m, r and I are the rest mass, radius and moment of inertia of the source mass, supposed spherical.

If the source density is uniform, $I = \tfrac{2}{5}mr^2$ and the first term is

$$-\tfrac{1}{10}G\alpha_2(\mathbf{w} \cdot \mathbf{n})^2$$

(w is in units of the velocity of light), while the second is

$$-\tfrac{1}{2}G(\alpha_1 - \alpha_3 - \tfrac{3}{5}\alpha_2)w^2.$$

Suppose an experiment is performed in a laboratory on the Earth's surface, or in an artificial satellite, having a velocity $\mathbf{v}$ relative to the centre of the Earth, itself having velocity $\mathbf{w}_0$ in the preferred frame. Then

$$\mathbf{w} = \mathbf{w}_0 + \mathbf{v},$$
$$w^2 = w_0^2 + v^2 + 2\mathbf{w}_0 \cdot \mathbf{v}$$

and

$$(\mathbf{w} \cdot \mathbf{n})^2 = [(\mathbf{w}_0 + \mathbf{v}) \cdot \mathbf{n}]^2$$
$$= (\mathbf{w}_0 \cdot \mathbf{n})^2 + (\mathbf{v} \cdot \mathbf{n})^2 + 2(\mathbf{w}_0 \cdot \mathbf{n})(\mathbf{v} \cdot \mathbf{n}).$$

If $\mathbf{n}$ is kept fixed in the laboratory frame, $\mathbf{v} \cdot \mathbf{n}$ will be constant as will w^2 and v^2, but $\mathbf{w}_0 \cdot \mathbf{n}$ and $\mathbf{w}_0 \cdot \mathbf{v}$ vary with $\mathbf{v}$ and $\mathbf{n}$.

This indicates that there might be two ways of doing experiments on preferred frame effects – one would be to observe changes in G as the laboratory goes round with the Earth or space craft, the other would be to vary $\mathbf{n}$ periodically by turning the experiment in the laboratory.

As yet, no such experiments have been performed. If the difficulties of setting up an experiment in a space craft could be overcome, that would be the most promising scheme, for if the space craft were close to the Earth, the direction of $\mathbf{v}$ would change with a period of about 100 minutes (50 minutes if the experiment were symmetrical with respect to the masses) instead of 24 h for a terrestrial experiment, and the environment would probably be much quieter.

The alternative, for a terrestrial experiment, is to change $\mathbf{n}$. Chen (private communication) has designed an experiment in which the position of a source mass attracting the test mass of a torsion pendulum is moved periodically, so changing $\mathbf{n}$; $\mathbf{v}$ is the velocity of the laboratory about the centre of the Earth, so $\mathbf{w}_0 \cdot \mathbf{v}$ has a daily period, while $\mathbf{v} \cdot \mathbf{n}$ has the period with which $\mathbf{n}$ is changed and $\mathbf{w}_0 \cdot \mathbf{n}$ has periods equal to the sum and difference of the two other periods, all thus readily identified in the observations.

No such experiment has yet been performed and the utility of them would depend on whether they could attain appreciably better limits on the PPN parameters than are set by observations on the Earth tides. As the Earth rotates, the potential at any point on the surface varies because of the changing positions of the Sun and Moon relative to the point, and because a point on the surface does not coincide with the centre of mass of the Earth, where the gravitational attraction of the Sun (or Moon) balances the inertial acceleration in the orbit. The changing potential at the surface is the tidal potential and can be calculated very accurately on Newtonian principles. The tidal potential is about 1 part in 10^7 of the Earth's own attraction, and gives rise to a variation in the magnitude of the local gravitational acceleration of about 1 part in 10^7, again accurately calculable. It is, however, not the only effect. The Earth, being elastic, yields under the tidal stresses, the direct consequence of which is that the radial distance of the surface from the centre of mass changes, and hence also the gravitational acceleration due to the mass of the Earth. The indirect consequence is that the moment of inertia (and with it higher multipole moments) changes with the tidal potential, giving rise to a further change in gravity at the surface. The elastic yielding of the Earth was at one time somewhat uncertain, but the present knowledge of the distribution of density and elastic moduli within the Earth (see Bullen and Bolt, 1985) is now such that the gravitational effects of tidal yielding can be calculated quite well. A third effect is less certain. The largest motions produced by the tidal potential are of course the tides in the oceans, and they in turn have a direct and indirect effect upon the surface gravity. The gravitational attraction at any point includes a contribution from the oceans, and that changes with the tides. In addition, the pressure exerted by the water upon the surface of the solid Earth has a tidal variation which leads to an additional distortion of the solid Earth, and so to a further change in surface gravity. Finally, the tidal displacement of the sea bed gives rise to yet further components of the ocean tides. All these effects are again of the order of 1 part in 10^7 of surface gravity, but the ocean tides and their effects cannot be calculated so well as those of the solid Earth assuming the oceans absent. The ocean basins have complicated shapes, and elaborate numerical integrations of the equations of motion are required for meaningful results. It has also, until recently, been difficult to check the solutions. Tides around the ocean basins are naturally well known, but those far from land, which contribute most to the various gravitational effects, could not be observed until fairly recently.

If there were a preferred frame of reference or a preferred location, the local surface acceleration due to the gravitational attraction of the Earth as a whole would vary as the Earth rotated and altered the alignment of the direction of the centre of the Earth, and the position of the surface point relative to the preferred position. Will (1981) gives a thorough analysis of the respective changes in gravity. The variations would mimic part of the Earth tides and because the Earth tides are about 1 part in 10^7 of gravity, it can be said that any preferred frame or location effects must be less than 1 part in 10^7, and in fact less by at least one order of magnitude since to within an order of magnitude the observed tidal effects agree with the calculations on the best geophysical models.

The measurement of variations of 1 part in 10^8 in local gravity is nowadays not so difficult, and the measurements themselves are not the limitation in setting bounds on the PPN parameters. Spring gravity meters as are commonly used in geophysical surveys are sufficiently sensitive, but need to be provided with automatic recorders and isolated from mechanical and thermal disturbance. With such precautions the drift of the gravity meter can be kept low, as is necessary for good tidal observations. Some special instruments have been constructed for tidal observations, in particular one in which a ball with a superconducting coating is suspended electromagnetically in the field of superconducting magnet. The sensitivity $(10^{-11}\,\mathrm{g})$ and stability are much superior to spring instruments. Analysing a run of measurements at Piñon Flat in Southern California, Warburton and Goodkind (1976) were able to set a limit of 10^{-3} on α_2.

4.6 Additional deviations from general relativity

Besides the main effects discussed in the previous sections, there are others that have been the subject of experiment or proposed experiment. First, it was seen above that the general expression for the gravitational mass when a body is considered as a source (active) is different from that when it is considered as a test mass (passive). An experiment by Kreuzer (1968) sets a limit to such a difference. A cylinder of polytetrafluoroethylene (Teflon) of which fluorine comprises about 75 per cent by weight was placed in a tank containing a mixture of trichlorethylene and dibromomethane adjusted in composition (74 per cent of bromine by weight) so that it had the same density as the Teflon cylinder which floated in it with neutral buoyancy. Consequently, the passive gravitational masses in the gravity field of the Earth of the Teflon and the liquid it displaced, were identical.

The Teflon cylinder was moved from side to side and the torsion balance was used to detect any change of attraction on a test mass. At the limits of the experimental sensitivity, it was found that the relative difference of the ratio, active to passive gravitational mass, for fluorine and bromine was less than 10^{-5}. Will (1981) shows that the result entails for the PPN parameter ζ_3,

$$|\zeta_3| < 6 \times 10^{-2}.$$

A possible dependence of gravitational force on temperature was examined by Poynting and Phillips (1905). A mass was heated to the temperature of steam and cooled to that of liquid air and its weight (that is, its gravitational attraction to the Earth) was found to change by less than 1 part in 10^{10} per degree Centigrade, whether it was heated or cooled.

Gravitational shielding or non-linear superposition has also been investigated. In Newtonian physics and general relativity there is no shielding or polarising effect, and so, when a third body is interposed between two others, it should not change the attraction between them.

It might also be that the gravitational attraction between two bodies would depend on preferred directions (such as an optic axis) in the crystalline structure. Mackenzie (1895) used a torsion balance in direct observations of attractions between anisotropic and isotropic bodies, but detected no difference; he also confirmed that, to within his accuracy, the attraction was proportional to the product of the attracting masses. Poynting and Gray (1899) performed a further experiment in which they attempted to excite rotation of a cylinder of crystalline quartz by rotating a second cylinder close to it. They concluded that any directional effect dependent on the crystal structure was less than 6×10^{-5} at 5.9 cm.

A laboratory experiment has been performed by Austin and Thwring (1897) who employed a torsion balance of similar design to that of Boys (1895) and found that when screens of lead, zinc, mercury, water or alcohol were introduced, any reduction of the attraction was less than 1 part in 500.

Again, when the Moon comes between the Sun and the Earth at a total eclipse of the Sun, the combined attraction of the Sun and the Moon at a point on the Earth should be just the sum of the two separately. A total eclipse has been seen near the sunset to the west of Trieste where the variation of the horizontal attraction could be measured with very sensitive horizontal pendulums of long period (Marussi in Cook, 1971). Nothing inconsistent with the straight sum of the attractions before and after eclipse was seen during eclipse. The test is not perhaps very rigorous for with a similar geometrical disposition, electrostatic or magnetostatic screening would be slight; for substantial screening, the Moon would need to subtend more nearly 90°.

A proposed test of the precession of a gyroscope in a space craft may be considered as an *experiment*, rather than as an *observation*, for the apparatus is under an experimenter's control. The experiment was first suggested by Schiff (1960a, b) and has been in preparation for more than a decade, but still awaits a suitable space flight. The rate of change of the spin vector **S** is (Will, 1981)

$$dS/d\tau = \mathbf{\Omega} \wedge \mathbf{S}$$

where $\mathbf{\Omega}$ is an angular velocity:

$$\mathbf{\Omega} = -\tfrac{1}{2}\mathbf{v} \wedge \mathbf{a} - \tfrac{1}{2}\mathbf{V} \wedge \mathbf{g} + (\gamma + \tfrac{1}{2})\mathbf{v} \wedge \nabla U$$

(U is the gravitational potential, **g** is a vector related to the metric tensor: $\mathbf{g} = g_{0j}\mathbf{e}_j$ (where $\mathbf{e}_j$ is a unit vector in the *j*-direction) and **a** is the spatial component of the 4-acceleration and is zero in free fall, as in a satellite in orbit about the Earth).

Thus if **V** is the centre of mass 3-velocity,

$$\mathbf{\Omega} = \tfrac{1}{4}(4 + 4\gamma + \alpha_1)\mathbf{V} \wedge \mathbf{V} - \tfrac{1}{2}\alpha_1\mathbf{w} \wedge \nabla U + (\gamma + \tfrac{1}{2})\mathbf{v} \wedge \nabla U,$$

where **w** is the velocity of the coordinate system relative to a preferred frame.

The observed precession is the resultant of a number of terms in the equation for $dS/d\tau$. First, there is the geodetic precession, resulting from the curvature of space near massive bodies. For a gyroscope in a circular orbit of radius *a* about the Earth (of mass m_E) the change of the direction of the spin axis in one orbit is

$$\delta S = -2\pi(\gamma + \tfrac{1}{2})(m_E/a)(\mathbf{S} \wedge \mathbf{h}),$$

where **h** is the unit vector perpendicular to the plane of the orbit. The rate is approximately 7 arc seconds per year for a close satellite; and is much the largest contribution to the precession.

The Lense–Thirring precession would be of the order of 0.01 arc seconds per year, while a periodic precession arising from the velocity of a preferred frame would have an amplitude of some $10^{-3}\alpha_1$ seconds.

The practical difficulties are very great. The prime requirement is that the gyroscope should be subject to no torques other than those arising from the effects sought, and that means that it must be uniform in shape and homogeneous in density to about 1 part in a million. Secondly, the recognition of the spin axis must involve no significant anisotropy or torque and, lastly, the direction of spin has to be referred to a reference frame constituted by distant stars, the directions of which should be established to 10^{-3} arc seconds. All these requirements present severe problems (Lippa and Everitt, 1978; Cabrera and van Kann, 1978).

Lastly, in this section, the possible dependence of the constant of gravitation upon time is mentioned, if only to remark that at present it seems most unlikely that any dependence could be examined experimentally. General relativity treats G as a fundamental constant, part of the system of definition of the theory, and so not susceptible to experimental test (the velocity of light in special relativity occupies a similar position) but other theories admit variation, with the implication that G is not an independent constant of nature, but a quantity related in some way to other constants of nature so that it may meaningfully be compared with them to establish whether or not it changes. Dirac's hypothesis of large numbers leads to a rate of change of an order of magnitude in the lifetime of the universe, an amount that could only be checked by reference to geophysical or astronomical observations. Analyses of, for example, palaeomagnetic data about possible changes in the radius of the Earth have been inconclusive. Within the laboratory, an experiment on the time dependence of G would need to establish the value of G in relation to the standards of length, time and mass (in actual current practice, the velocity of light, the standard of frequency and the standard kilogram) with sufficient precision at different times to enable a change of 10^{-10} per year to be detected. Since, as will be seen, the value of G is known at best to rather better than 1 part in 10^4, and since the standard kilogram is established to no better than 1 part in 10^9, it will be clear that there is no prospect of any realistic experimental check upon a possible variation of G in time.

4.7 The measurement of the constant of gravitation

The measurement of the constant of gravitation has a curious place in physics. On the one hand, the value of G is relatively poorly known compared with other natural constants, such as Planck's constant, or the mass of the electron, most of which are known to 1 part in 10^6 or better. The reasons for this state of affairs have already been alluded to: gravitational forces are very small compared with the electromagnetic forces and so experiments are very susceptible to extraneous disturbances, while the reliable and sensitive electromagnetic techniques that can be applied to other constants are not available for gravitation. Yet, while the measured value of G is so uncertain the relevance of G to the rest of physics is slight. The other principal constants of physics form an interconnected set and a good knowledge of their values has consequences in both fundamental theory (relativistic quantum mechanics, for example) and in practical measurement of high precision (establishment of standards for electrical measurements as an instance). Almost no such requirements or implications apply to knowledge of the value of G. It is, so far as is known or postulated (but see the previous section) independent of all the other constants and is not required in any practical system of measurement. It enters into the conversion of orbital parameters in celestial mechanics into masses (in kilograms) of celestial bodies, but no theory of the constitution of such bodies is so well established in detail as to require the masses to be known in terms of the terrestrial standard to better than a few per cent. The only possible exception to that statement has to do with the internal constitution of the Earth. The variations of density and elastic moduli within the Earth are nowadays established to about 1 part in a thousand throughout most of the Earth from analysis of numerous seismological observations. In essence those analyses give the value of the ratios K/ρ and μ/ρ, where K is the bulk modulus, μ the shear modulus and ρ the density. In those ratios the standard of mass enters both the numerators and denominators and so appears in neither of the ratios. At the same time, the absolute value of density is set by the condition that the density must integrate up to the total mass of the Earth, which is fixed by the attraction of gravity of the Earth; the conversion of density to units of the terrestrial standards of mass and length therefore requires a knowledge of the constant of gravitation. The point of such a conversion is to make comparisons with laboratory determinations of equations of state of possible constituents of the body of the Earth, in which the density is measured against the usual terrestrial standards and is independent of the value of G.

Five methods have been used for the laboratory determination of the constant of gravitation. The earliest uses the deflexion of a torsion balance to measure the force between two masses. In the second method, a torsion balance is used as an oscillator, the gravitational potential of attracting masses being used to alter the restoring force and hence the period of oscillation. In a variant of those two methods, the attracting masses were moved back and forth around the torsional pendulum in resonance with the period of oscillation of the pendulum and the steady amplitude of the oscillations was measured. The force exerted by an attracting mass on a test mass can also be measured by balancing it against the gravitational attraction of the Earth (with a known acceleration due to gravity) on a sensitive equal arm chemical balance. Lastly, a method first suggested in principle by Newton has been attempted. Newton (see Section 4.1) calculated the time it would take two masses moving freely to come together from an initial separation; the experiment of Luther *et al.* (1976) is somewhat different, in that the separation of two masses was kept constant by a servo-system which moved the one away from the other in opposition to the gravitational acceleration. One mass (the controlled one) was on a rotating table, the other was suspended from a torsion balance and the servo-control of the first mass was adjusted to maintain deflexion of the torsion balance constant, when the acceleration of the masses towards each other was equal to the acceleration of the controlled mass.

In the eighteenth century, estimates of the constant of gravitation were also made from the gravitational attraction of some isolated mountain (in Scotland, Schiehallion and Arthur's Seat) which was found from the amount by which a plumb bob was deflected towards the mountain – the tangent of that angle is the ratio of the attraction of the mountain to that of the whole Earth. The radical weakness of the method is the determination of the attraction of the mountain, for the internal distribution of density can never be known with any exactness and there is no such thing as an isolated mountain with a clearly defined boundary. The interest of such observations lies in what they tell us about sub-surface distributions of mass associated with mountains and not in the values of G that have been derived.

All experimental determinations of the constant of gravitation suffer from the same problems – the measurement of the very small gravitational forces in the face of noise from other causes, and the location of the centres of mass of the attracting and test masses and the measurement of their separations. In the torsion balance, the force on the test mass is balanced by the torque corresponding to the twist of the fibre of the balance, and that is so whether a steady value of the twist is measured, as in the original work of Cavendish, or whether the torque is the restoring force in free oscillations of the pendulum, as in the determinations of Braun and others following him. The essential advantage of finding the torque from the period of free oscillations is that, if the pendulum is allowed to oscillate for many periods, a period can be found much more precisely than a steady deflexion. In either case, the torsional constant of the pendulum has to be found, usually from the free period in the absence of attracting masses. The other problem common to all experimental determinations is that the distribution of density within masses may not be uniform and therefore the position of the centre of attraction cannot be found from the dimensions of the exterior surface. That problem can to a large extent be overcome by turning the attracting masses so that the centre of attraction rotates about some known axis, and taking observations in different positions. The test mass is usually made small enough that the density can be assumed to be uniform. There is also the problem of measuring from a fixed attracting mass to a mobile test mass, but it usually is possible to arrange the observations so that the critical distances are those between different positions of the attracting masses (compare the experiment of Chen on the inverse square law – Section 4.3).

In all measurements with the torsion balance, the stability of the position of rest of the balance beam is critical. The measurement of a period is much less subject to instability of the rest position, though not entirely so. The cause of the instability is the movement of dislocations of the material of the fibre under stress and a considerable improvement in technique was made when Boys developed fused silica fibres which not only can be made much finer than metal or other fibres, and so attain a greater sensitivity, but also lack the dislocations of polycrystalline metals although as glasses they undergo viscous drift. Thus, since the experiments of Boys, fibres of fused silica have generally been used for the most delicate balances, although in recent years Braginski and Manukin (1977), followed by Chen (Chen, Cook and Metherell, 1984), have shown how to remove dislocations from tungsten wires by a combination of thermal and mechanical stress, and so have produced very stable torsion balances.

All the earlier determinations of the constant of gravitation depended on the deflexion of a torsion balance (Cavendish, 1798; Baily, 1842; Cornu and Baille, 1873, 1878) and perforce were performed in an enclosure with air at atmospheric pressure; they were greatly disturbed by convection currents. Two important advances were made by Boys (1895) and Braun (1897). Boys introduced quartz fibres for the torsion suspension and showed that it was an advantage to reduce the size of a deflexion experiment. Braun, who worked in relative isolation in his Jesuit house, was the first to evacuate the enclosure in which the torsion balance hung and also performed the first experiment using the period of the balance instead of its deflexion – he used the deflexion method as well.

Braun's use of the period of the torsion balance has been followed by Heyl

(1930), Heyl and Chrzanowski (1942) and Luther and Towler (1982). If attracting masses are placed in line with the beam of the torsion balance, the torque they exert is in the same sense as that of the suspension and increases the frequency of oscillation, whereas if they are placed at right angles to the beam, the attraction decreases the frequency. The method seems to have given the best results so far, by far the most precise result being, it seems, that of Luther and Towler (Table 4.1).

A variant of the method of periods was introduced by Zahradniček (1933). Two coaxial torsion balances were set up, a light one with masses of about 9 g and a heavy one with masses of about 11 kg. The periods were nearly but not quite the same. The heavy balance was set into free oscillations driving the light balance through the gravitational attraction; the value of G is found from the amplitude of the forced oscillation of the light balance.

A very sensitive equal arm common balance was used by von Jolly (1881) and by Poynting (1891). Von Jolly placed a container which could be filled with mercury below one pan, while in Poynting's determination a large block of lead was placed below the mass on one pan of the balance and the increase in weight of the mass was found by simple weighing. Poynting devised a very sensitive means of measuring the deflexion of the balance and hence the small increase in weight.

The accuracy of his result is comparable with that of other determinations of that time (Table 4.1). It is clear that if the sensitivity of the balance were to

Table 4.1. *Determinations of the constant of gravitation*

Author	Method	Result $(10^{-11}\,\mathrm{Nm^2/kg^2})$
Cavendish, 1798	Torsion balance, deflexion	6.754
Poynting, 1891	Common balance	6.698
Boys, 1895	Torsion balance, deflexion	6.658
Braun, 1897	Torsion balance, period and deflexion	6.658
Heyl, 1930	Torsion balance, period:	
	gold	6.678
	platinum	6.664
	glass	6.674
Zahradniček, 1933	Torsion balance, resonance	6.659
Heyl and Chrzanowski, 1942	Torsion balance, period:	
	hard drawn wire	6.676
	annealed wire	6.668
Luther and Towler, 1982	Torsion balance, period	6.6726

be improved, the measurement of the distance between the centres of mass of the weight on the pan and the heavy mass placed below it would be a limitation. Deflexions of the apparatus as a result of putting the heavy attracting mass in position and taking it away may also be a severe problem.

Lastly, Luther *et al.* (1976) developed the method first thought of by Newton to measure directly the acceleration of two nearby masses. Whereas Newton estimated the time it would take for two masses to come together, Luther and his colleagues devised a servo-system that kept the two masses always at the same distance from each other, and measured the acceleration of two nearby masses. A final result seems not to have been published.

4.8 Conclusion

The experiments discussed in this article are those that have been made in a laboratory on Earth or that could be made in one in a space craft. Thus nothing has been said of attempts to observe gravitational radiation, for, although they use some of the most advanced techniques of experimental physics, they depend upon radiation from sources which are clearly not under the control of the experimenter; the experiments considered here involve for the most part sources within the laboratory and at the experimenter's disposition. In addition the techniques are all similar, involving delicate mechanical measurements of very small forces. Tests of general relativity or other theories of gravitation by means of analyses of celestial mechanics and geophysics are also not considered.

The limitation to experimental tests means that the range of distances over which theories can be checked is of the order of 1 m or somewhat less. Taken with geophysical and astronomical observations (for the most part in the solar system), laboratory experiments test the validity of theories of gravitation from a few centimetres out to the extent of the solar system, so that there are ranges of both larger and small distances for which there is no experimental test.

Newton's law of gravitation implies that the attraction between two bodies is proportional to mass and independent of material constitution, that the law is the inverse square law and that G is a constant independent of time, or any preferred coordinate frame or any preferred position. Experimental results confirm the first implication at least to the extent that the ratio of passive gravitational mass to inertial mass is the same for all materials to within 1 part in 10^{12}, and they confirm that the law is the inverse square to within 1 part in 10^4. Geophysical evidence from Earth tides indicates that any effects of a preferred frame or location are small and nothing useful can be said about variation in time. There is some evidence against screening of a gravitational source by other material.

Looked at from the point of view of PPN theory, observations on the scale of the solar system show that the parameters γ and β must have values close to unity as predicted by general relativity, but the only other conclusions to be drawn from experiment or observation is that the parameter α_2 is zero and that certain combinations of other parameters are also zero. All parameters other than β and γ are zero in general relativity.

In a number of instances, the results of experiments show unexplained systematic effects, of which three have already been mentioned. The results of the experiment by Roll, Krotkov and Dicke (1964) showed a 12 h period in the recorded signal; that cannot arise from the attraction of the Sun on the masses and its origin has not been accounted for. Heyl (1930), in his determination of the gravitational constant, used different materials for the masses in various groups of observations and found different results for G (Table 4.1), which is inconsistent with the Eötvös group of experiments. Similarly, there appear to be some systematic differences between materials within the Eötvös experiment itself, even though the interpretation proposed by Fishbach *et al.* (1986) may be invalid. Again, Heyl and Chrzanowski (1942) found that whether the tungsten torsion fibre was hard drawn or annealed gave different values of G (Table 4.1). The significance of these examples is not so much the physical interpretation that may be put upon them, rather it is that it is difficult to attain an adequate understanding of experiments at the limit of available techniques.

In view of the rather limited nature of experiments so far performed, and in view of the difficulties in understanding the outcomes of these experiments, there must be considerable interest in devising new experiments, both better ones to reexamine existing results and new ones to investigate other aspects of gravitation. Braginski *et al.* (1977) have proposed three new techniques of detection which they argue should be very sensitive and have suggested a number of experiments of which they might form the basis. None of the experiments has, however, yet been performed.

The naive experimenter may be allowed to conclude that general relativity remains the best description of gravitation, but he is also well aware that much ingenuity, care and imagination will be required before his experiments have the delicacy that will enable them to contribute, as Poynting hoped, to the foundations of the explanation of gravitation.

References

Anderson, J. D., Esposito, P. B., Martin, W., Thornton, C. L. and Muhleman, D. O. (1975). Experimental test of general relativity using time-delay data from *Mariner* 6 and *Mariner* 7. *Astrophys. J.*, **200**, 221–33.

Austin, L. W. and Thwring, C. B. (1897). An experimental research on gravitational permeability. *Phys. Rev.*, **5**, 294–300.

Baily, F. (1842). An account of some experiments with the torsion rod for determining the mean density of the Earth. *Mon. Not. R. Astron. Soc.*, **5**, 197–206.

Bessel, F. W. (1832). Versuche über die Kraft mit welcher die Erde Körper von verschiedene Beschaffenheit anzieht. *Ann. Phys. Chem. (Poggendorf)*, **25**, 401–17.

Boys, C. V. (1895). On the Newtonian Constant of Gravitation. *Phil. Trans. Roy. Soc.*, **A182**, 1–72.

Braginski, V. B., Caves, C. M. and Thorne, Kip. S. (1977). Laboratory experiments to test relativistic gravity. *Phys. Rev.*, **D15**, 2047–68.

Braginski, V. B. and Manukin, A. B. (1977). In *Measurements of Weak Forces in Physics Experiments*, ed. D. H. Douglass. Chicago University Press: Chicago.

Braginski, V. B. and Panov, V. I. (1972). Verification of the equivalence of inertial and gravitational mass. *Sov. Phys. J.E.T.P.*, **34**, 463–6.

Braun, C. (1897). Die Gravitationskonstante, die Masse und Mittlere Dichte der Erde. *Denschr. Akad. Wiss.* (Wien), *Math. naturwiss. Kl.*, **64**, 187–258.

Bullen, K. E. and Bolt, B. A. (1985). *An Introduction to the Theory of Seismology.* Cambridge University Press: Cambridge.

Cabrera, B. and van Kann, F. J. (1978). Ultra-low magnetic field apparatus for a cryogenic gyroscope. *Acta Astronautica*, **5**, 125–30.

Cavendish, H. (1798). Experiments to determine the density of the Earth. *Phil. Trans. Roy. Soc.*, **88**, 469–526.

Chan, H. A., Moody, M. V. and Paik, H. J. (1982). Null test of the gravitational inverse square law. *Phys. Rev. Lett.*, **49**, 1745–8.

Chen, Y. T. (1982). The gravitational field inside a long hollow cylinder of infinite length. *Proc. Roy. Soc.*, **A382**, 75–82.

Chen, Y. T., Cook, A. H. and Metherell, A. J. F. (1984). An experimental test of inverse square law of gravitation. *Proc. Roy. Soc.*, **A394**, 47–68.

Cook, A. H. (1971). The experimental determination of the constant of gravitation. *Nat. Bur. Stds. Sp. Publ.*, **343**, 475–83.

Cook, A. H. and Chen, Y. T. (1982). On the significance of the radial gravitational force of the finite cylinder. *J. Phys.*, **A15**, 1591–7.

Cornu, A. and Baille, J. (1873). Determination nouvelle de la constante de l'attraction et de la densité moyenne de la Terre. *C.R. Acad. Sci. (Paris)*, **76**, 954–5.

Cornu, A. and Baille, J. (1878). Sur la mesure de la densité moyenne de la Terre. *C.R. Acad. Sci. (Paris)*, **86**, 699–702.

Eötvös, R. V., Pekar, V. and Fekete, E. (1922). Beitrag zum Gesetze der Proportionalität von Trägheit und Gravität. *Ann. Phys.*, **68**, 11–66.

Fishbach, E., Sadarsk, E., Szafer, A., Talmadge, C. and Aronson, S. H. (1986). Reanalysis of the Eötvös experiment. *Phys. Rev. Lett.*, **56**, 3–6.

Formalont, E. B. and Sramak, R. A. (1977). The deflection of radio waves by the Sun. *Comm. Astrophys.*, **7**, 19–33.

Fujii, Y. (1971). Dilation and possible non-Newtonian gravity. *Nature Phys. Sci.*, **234**, 5–7.

Fujii, Y. (1972). Scale invariance and gravity of hadrons. *Ann. Phys.*, **69**, 494–521.

Gibbons, G. W. and Whiting, B. F. (1981). Newtonian gravity measurements impose constraints on unification theories. *Nature (London)*, **291**, 636–8.

Heyl, P. R. (1930). A redetermination of the constant of gravitation. *J. Res. Nat. Bur. Stds.*, **5**, 1243–90.

Heyl, P. R. and Chrzanowski, P. (1942). A new determination of the constant of gravitation. *J. Res. Nat. Bur. Stds.*, **29**, 1–31.

Holding, S. C. and Tuck, G. J. (1984). A new mine determination of the Newtonian gravitational constant. *Nature (London)*, **307**, 714–16.

Keiser, G. M. and Faller, J. E. (1979). A new approach to the Eötvös experiment. *Bull. Amer. Phys. Soc.*, **24**, 579.

Keyser, P. T., Niebauer, T. and Faller, J. E. (1986). Comment on 'Reanalysis of the Eötvös experiment'. *Phys. Rev. Lett.*, **56**, 2425.

Kreuzer, L. B. (1968). Experimental measurement of the equivalence of active and passive gravitational mass. *Phys. Rev.*, **169**, 1007–12.

Lippa, J. A. and Everitt, C. W. F. (1978). The role of cryogenics in the gyroscope experiment. *Acta Astron.*, **5**, 119–123.

Long, D. R. (1976). Experimental examination of the gravitational inverse square law. *Nature (London)*, **260**, 417–18.

Long, D. R. (1980). *Nuovo Cim.*, **B55**, 252.

Luther, G. G. and Towler, W. R. (1982). Redetermination of the Newtonian gravitational constant, G. *Phys. Rev. Lett.*, **48**, 121–3.

Luther, G. G., Towler, W. R., Deslattes, R. D., Lowry, R. and Beams, J. (1976). *Int. Conf. on Atomic Masses and Fundamental Constants – 5*, ed. J. H. Sanders and A. H. Wapstra, p. 592. Plenum: New York.

Mackenzie, A. S. (1895). On the attraction of crystalline and isotropic masses at small distances. *Phys. Rev.*, **2**, 321–43.

Maskelyne, N. (1775). A proposal for measuring the attractions of some hill in the kingdom by astronomical observations. *Phil. Trans. Roy. Soc.*, **65**, 495–9.

Newton, I. (1687). *Philosophiae Naturalis Principia Mathematica*, III. Prop. VI, Th. VI.

Ogawa, Y., Tsubona, K. and Hirakawa, H. (1982). Experimental test of the law of gravitation. *Phys. Rev.* **D26**, 729–34.

Panov, V. I. and Frontov, V. N. (1979). The Cavendish experiment at large distances. *Soviet Physics, J.E.T.P.*, **50**, 852–6.

Poynting, J. H. (1891). On the determination of the mean density of the Earth and the gravitation constant by means of the common balance. *Phil. Trans. Roy. Soc.*, **A182**, 565–656.

Poynting, J. H. (1900). Recent studies in gravitation. *Proc. Roy. Inst. G.B.*, **16**, 278–94.

Poynting, J. H. and Gray, P. L. (1899). An experiment in search of a directive action of one quartz crystal by another. *Phil. Trans. Roy. Soc.*, **A192**, 245–56.

Poynting, J. H. and Phillips, P. (1905). An experiment with the balance to find if change of temperature has any effect on weight. *Proc. Roy. Soc.*, **A76**, 445–57.

Reasenberg, R. D., Shapiro, I. I., MacNeil, P. E., Goldstein, R. B., Breidenthal, J. C., Brenkle, J. P., Cain, D. L., Kaufman, T. M., Komarek, T. A. and Zygielbaum, A. I. (1979). Viking relativity experiment: Verification of signal retardation by solar gravity. *Astrophys. J.*, **234**, 219–21.

Renner, J. (1935). *Hung. Acad. Sci.*, **53** (II), 542–70.

Roll, P. G., Krotkov, R. and Dicke, R. H. (1964). The equivalence of inertial and passive gravitational mass. *Ann. Phys. (New York)*, **26**, 442–517.

Schiff, L. I. (1960a). Motion of a gyroscope according to Einstein's general theory of relativity. *Proc. Nat. Acad. Sci. (USA)*, **46**, 871–82.

Schiff, L. I. (1960b). Possible new test of general relativity theory. *Phys. Rev. Lett.*, **4**, 215–17.

Shapiro, I. I., Counselman, C. C. III and King, R. W. (1976). Verification of the principle of equivalence for massive bodies. *Phys. Rev. Lett.*, **36**, 555–8.

Shapiro, I. I., Pettengill, G. H., Ash, M. E., Ingills, R. P., Campbell, D. B. and Dyce, R. B. (1972). Mercury's perihelion advance – determination by radar. *Phys. Rev. Lett.*, **28**, 1594–7.

Spero, R. E., Hoskins, J. K., Newman, R., Pellam, J. and Schultz, J. (1980). Tests of the gravitational inverse square law at laboratory distances. *Phys. Rev. Lett.*, **44**, 1645–8.

Stacey, F. D. and Tuck, G. J. (1981). Geophysical evidence for non-Newtonian gravity. *Nature (London)*, **292**, 230–2.

von Jolly, Ph. (1881). Die Anwendung der Wage auf Probleme der Gravitation. *Ann. der Physik u. Chemie (Wiedener)*, N.F., **14**, 331–55.

Warburton, R. J. and Goodkind, J. M. (1976). Search for evidence for a preferred reference frame. *Astrophys. J.*, **208**, 881–6.

Will, C. M. (1981). *Theory and Experiment in Gravitational Physics*. Cambridge University Press: Cambridge.

Yu, H-T., Ni, W-T., Hu, C-G., Liu, F-H., Yang, C-H. and Lin, W-N. (1979). Experimental determination of the gravitational forces at separations around 10 meters. *Phys. Rev.*, **D20**, 1813–15.

Zahradniček, J. (1933). Resonanz methode für die Messüng der Gravitations konstante mittels der Drehwaage. *Phys. Zeit.*, **34**, 126–33.

5

Experimental gravitation from Newton's Principia *to Einstein's general relativity*

CLIFFORD M.WILL

5.1 Introduction

The tercentenary of the publication of Newton's *Principia* has occurred at a propitious time in the subject of gravitation physics. We find ourselves in the midst of a renaissance for general relativity, the theory of gravitation that superseded Newtonian gravity, and that is now one of the most active and exciting branches of physics. General relativity has become an important theoretical tool for the astrophysicist, as well as a fundamental ingredient in the quest for unification of the theories of the basic interactions.

Like any other branch of physics, gravitation has a strong experimental component as well. The confrontation between theory and experiment was a central element in the *Principia*; there Newton reported results of his own experiments to verify the principle of equivalence, and made detailed comparisons of the predictions of his theory of gravity with astronomical observations. In the two centuries following publication of the *Principia*, Newtonian gravitation was put to the test in a variety of ways, and passed every test but one with flying colors. Nor did Einstein shy away from experimental confrontation, for he proposed three important tests of general relativity. Although two of Einstein's three tests were confirmed immediately or shortly after publication of his theory in 1916, further experimental progress during the next 45 years was very slow, largely because of a lack of experimental technology of sufficient accuracy to measure the extremely small predicted effects.

However, during the two decades 1960–80, there was a rebirth in the subject of experimental gravitation. This rebirth coincided with the renaissance of general relativity as a whole, and was propelled by astronomical discoveries that indicated a role for relativistic gravity in astrophysics, by new theoretical insights into the observable consequences of general relativity, and by advances in the technology of laboratory and space experimentation that provided high-precision tools to carry out gravity tests.

The period 1960–80 became the 'decades for testing general relativity'. The systematic high-precision testing of gravitational theory emerged as an active and challenging field, with improved versions of Einstein's tests such as the deflection of light and the gravitational redshift, and brand new tests, such as the Shapiro time delay and the Nordtvedt effect in lunar motion. Several sophisticated theoretical frameworks were developed to compare the relative merits of various experiments, to help discover new tests and to classify and compare alternative theories of gravitation.

As did Newtonian theory in earlier years, so too did general relativity pass virtually every experimental test with equally flying colors. Although the entire story of experimental gravitation is far from over, it is reasonable to say that the second chapter of the story has been written. The first chapter consisted of the verification of Newtonian gravity, between 1687 and the end of the nineteenth century. The second chapter consisted of the verification of the principal predictions of general relativity, with the peak of activity during the decades 1960–80. The next chapter of the story puts us at the frontiers of experimental gravitation, where the observable effects are very small and the experiments very difficult. The effects to be detected correspond in some sense to the finer predictions of general relativity, such as the dragging of inertial frames and its effect on orbiting gyroscopes, or the presence of higher-order effects in phenomena such as the deflection of light. We may enter the next century before this chapter is concluded.

This is therefore an appropriate occasion for a summary and consolidation of the experimental evidence that forms the empirical foundation both for Newtonian gravitation as well as for general relativity. We shall begin this summary by placing the two respective theories in the proper context. Newton's and Einstein's conceptions of gravity had a number of common elements, such as the equivalence principle, yet differed radically in other elements, such as the nature of the spacetime underlying gravity. Section 5.2 reviews the commonalities and contrasts between the two theories. In Section 5.3, we discuss four fundamentals of gravitation that, in some ways, are common to both theories: the principle of equivalence, the inverse square law, the equality of action and reaction, and the constancy of the gravitational constant, and review the experimental evidence supporting them. The testable differences between the theories show up in applications where general relativity mainly predicts corrections to the Newtonian behavior. In Section 5.4, we describe these applications, specifically to the motions of planets, the motion of the Moon, the motion of light, and the motion of binary star systems, and describe the extent to which Newtonian gravity and general relativity have been verified. Section 5.5 presents concluding remarks. In an Appendix, we summarize the parametrized post-Newtonian (PPN) formalism, which is a useful framework for discussing experimental gravitation in the solar system.

Our goal is not to give an exhaustive, completely up-to-date technical review of this subject with a complete bibliography, rather it is to honor Newton on this occasion by demonstrating the remarkable success of gravitational theory, both the Newtonian and the Einsteinian, in the confrontation with experiment. As a result, we will not provide complete references to work done in this field, but instead will refer the reader to the appropriate technical review articles and monographs, specifically to *Theory and Experiment in Gravitational Physics* (Will, 1981), hereafter referred to as TEGP, and to 'The Confrontation between General Relativity and Experiment: an Update' (Will, 1984), hereafter referred to as UPDATE. Other references will be confined to the key original papers, and to important recent papers that are not included in TEGP or UPDATE. A popular account of this subject, with historical sidelights and anecdotes, can be found in *Was Einstein Right?* (Will, 1986).

5.2 Newtonian gravitation and general relativity: commonalities and contrasts

Newtonian gravitational theory and general relativity are radically different theories of the gravitational interaction. Newtonian theory is founded upon the notion of absolute space and absolute time, while general relativity is based upon spacetime and the compatibility with special relativity. In Newtonian physics, the permissible coordinate transformations are those of the Galilei group, while in general relativity, arbitrary coordinate transformations are allowed. The gravitational potential in Newtonian theory is a field that is 'pasted on' the absolute space and time underlying the theory, while in general relativity, the gravitational potential is the spacetime metric itself, and the effects of gravity are consequences of curved spacetime. Since the gravitational interaction in Newtonian theory is assumed to be instantaneous (although Newton himself never made this assumption) gravitational waves do not exist, but, more than this, the actual causal structure of the theory is very different from that of general relativity, whose causal structure is based upon null cones or light cones that reflect the finite propagation speed of all interactions, including gravitation. Newtonian gravitation is a linear theory, whereas general relativity is highly non-linear. These differences are illustrated by the kinds of topics in general relativity that are discussed in other articles in this volume: black holes, gravitational radiation, cosmological models, geometry of spacetime, and so on, for which the predictions of Newtonian theory are very different from those of general relativity, or are even non-existent.

On the other hand, there are many ways in which the two theories are similar. Both theories are based on a principle of equivalence whose main element is the concept that bodies fall with the same acceleration regardless of their structure or composition. Both theories embody the law of 'action equals reaction', Newtonian theory by explicit assumption, general relativity by being based on an invariant action principle, one of whose consequences is the conservation of momentum or the equality of action and reaction. In both theories, the fundamental constant that determines the strength of gravity, the gravitational constant G, is a true constant of nature.

Even the inverse square law, one of the main tenets of Newtonian gravity, can be seen to be a common feature of the two theories, if viewed properly. To be sure, in general relativity, the inverse square law for the gravitational force is valid only as a first approximation in situations of weak gravitational fields, but there is a sense in which the inverse square law is an exact parallel between the two theories. In Newtonian theory, the inverse square law for the gravitational acceleration implies that the divergence of the acceleration $(\vec{\nabla} \cdot \vec{g})$, which is the same as the Laplacian of the potential $(\nabla^2 U)$, must vanish in vacuum. The same statement holds in general relativity. There, in a local freely falling frame, the measured gradients of the acceleration experienced by a collection of particles at rest are the 'electric' components of the Riemann tensor (R_{0i0j}); the divergence of the acceleration is thus the trace of R_{0i0j}, which is the Ricci tensor component R_{00}, which vanishes in vacuum, according to Einstein's field equations.

Yet there is one area in which the two theories are very similar though slightly different, and this is the area of prime importance for experimental gravitation. This is the weak-field, slow-motion limit of general relativity, the limit appropriate for discussing motions in the solar system and in other astronomical arenas. In this limit, the Einstein field equations can be solved by successive approximations. At the first approximation, the field equations and equations of motion for matter are equivalent to those of Newtonian gravity. This is the 'Newtonian limit' of general relativity. At the next level of approximation, known as the 'post-Newtonian approximation', correction terms occur that lead to the important, experimentally observable effects such as the perihelion advance of Mercury, and the deflection of light. It may soon be possible to devise experiments to look for effects that enter at the next, or 'post-post-Newtonian' level of approximation.

One feature of the post-Newtonian approximation that has proved to be very important in the discussion of experimental gravitation is the fact that most alternative spacetime metric theories of gravitation to general relativity, such as the Brans–Dicke scalar–tensor theory, have post-Newtonian limits with the same general structure, the only difference from one theory to the next being the values of a set of numerical coefficients that appear in the post-Newtonian metric. This circumstance has made it possible to develop a very useful formalism, called the 'parametrized post-Newtonian' formalism for treating experimental tests and for analysing alternative theories of gravity. The main features of this framework are summarized in an Appendix.

The formal link between general relativity and Newtonian gravity has received renewed attention recently. The issue is, can Newtonian gravity, as formulated for example in the geometric, coordinate-free language of Cartan, be obtained formally from general relativity as a limit of a sequence of spacetimes? The limiting equations of motion are not at issue, rather it is the limiting global structure of the spacetime that is the question. In particular, how can one go from the causal structure of general relativity with null cones and gravitational radiation to that of Newtonian gravity with neither null cones nor gravitational radiation, along a smooth sequence of spacetimes? Understanding this issue may contribute to the rigorous foundation of equations of motion and of formulae for gravitational radiation in the weak-field slow-motion limit of general relativity (Winicour, 1983; Futamase and Schutz, 1985).

One important commonality between Newtonian gravity and general relativity: both have been subjected to strenuous experimental testing, and it is to these tests that we now turn.

5.3. Testing the fundamentals of gravitation theory

5.3.1 The principle of equivalence

The principle of equivalence has historically played an important role in the development of gravitation theory. Newton regarded this principle as such a cornerstone of mechanics that he devoted the opening paragraph of the

Principia to it, stating in Definition I, 'this quantity that I mean hereafter everywhere under the name of . . . mass . . . is known by the weight . . for it is proportional to the weight, as I have found by experiments on pendulums, very accurately made . . .'.† In 1907, Einstein used the principle as a basic element of general relativity. We now regard the principle of equivalence as the foundation, not of Newtonian gravity or of general relativity, but of the broader idea that spacetime is curved.

One elementary equivalence principle is the kind Newton had in mind when he stated that the property of a body called 'mass' is proportional to the 'weight', and is known as the weak equivalence principle (WEP). An alternative statement of WEP is that the trajectory of a freely falling body (one not acted upon by such forces as electromagnetism and too small to be affected by tidal gravitational forces) is independent of its internal structure and composition. In the simplest case of dropping two different bodies in a gravitational field, the weak equivalence principle states that the bodies fall with the same acceleration.

A much more powerful and far-reaching equivalence principle is known as the Einstein equivalence principle (EEP). It states (i) WEP is valid, (ii) the outcome of any local non-gravitational experiment is independent of the velocity of the freely falling reference frame in which it is performed, and (iii) the outcome of any local non-gravitational experiment is independent of where and when in the universe it is performed. The second piece of EEP is called local Lorentz invariance (LLI), and the third piece is called local position invariance (LPI). For example, a measurement of the electric force between two charged bodies is a local non-gravitational experiment; a measurement of the gravitational force between two bodies (Cavendish experiment) is not.

More so than the weak principle, EEP is the heart and soul of gravitational theory, for it is possible to argue convincingly that if EEP is valid, then gravitation must be a 'curved spacetime' phenomenon, in other words, the effects of gravity must be equivalent to the effects of living in a curved spacetime. As a consequence of this argument, the only theories of gravity that can embody EEP are those that satisfy the postulates of 'metric theories of gravity', which are (i) spacetime is endowed with a symmetric metric, (ii) the trajectories of freely falling bodies are geodesics of that metric, and (iii) in local freely falling reference frames, the non-gravitational laws of

---

† This and other quotations from the *Principia* are taken from the so-called 'Cajori' edition (Cajori, 1934).

physics are those written in the language of special relativity. The argument that leads to this conclusion simply notes that, if EEP is valid, then in local freely falling frames, the laws governing experiments must be independent of the velocity of the frame (LLI), with constant values for the various atomic constants (in order to be independent of location). The only laws we know of that fulfil this are those that are compatible with special relativity, such as Maxwell's equations of electromagnetism. Furthermore, in local freely falling frames, test bodies appear to be unaccelerated, in other words, they move on straight lines, but such 'locally straight' lines simply correspond to 'geodesics' in a curved spacetime (TEGP, section 2.3).

General relativity is a metric theory of gravity, but then so are many others, including the Brans–Dicke theory. Because Newtonian gravitation is not compatible with LLI, it is not a metric theory (although it can be expressed in geometrical, coordinate-free language). So the notion of curved spacetime is a very general and fundamental one, and therefore it is important to test the various aspects of EEP thoroughly.

A direct test of WEP is the comparison of the acceleration of two laboratory-sized bodies of different composition in an external gravitational field. Such tests of WEP predate Newton, including Philiponos (5th or 6th C.), Stevin (1586) and Galileo (c. 1590). If the principle were violated, then the accelerations of different bodies would differ. The simplest way to quantify such possible violations of WEP in a form suitable for comparison with experiment is to suppose that for a body with inertial mass m_I, the passive gravitational mass m_P is no longer equal to m_I, so that in a gravitational field g, the acceleration is given by $m_I a = m_P g$. Now the inertial mass of a typical laboratory body is made up of several types of mass-energy: rest energy, electromagnetic energy, weak-interaction energy, and so on. If one of these forms of energy contributes to m_P differently than it does to m_I, a violation of WEP would result. One could then write

$$m_P = m_I + \sum_A \eta^A E^A / c^2, \tag{1}$$

where E^A is the internal energy of the body generated by interaction A, and η^A is a dimensionless parameter that measures the strength of the violation of WEP induced by that interaction, and c is the speed of light. A measurement or limit on the fractional difference in acceleration between two bodies then yields a quantity called the 'Eötvös ratio' given by

$$\eta = \frac{2|a_1 - a_2|}{|a_1 + a_2|} = \sum_A \eta^A \left(\frac{E_1^A}{m_1 c^2} - \frac{E_2^A}{m_2 c^2} \right). \tag{2}$$

Thus, experimental limits on η place limits on the WEP-violation parameters η^A.

Many high-precision Eötvös-type experiments have been performed, from the pendulum experiments of Newton, Bessel and Potter, to the classic torsion balance measurements of Eötvös, Dicke, Braginsky and their collaborators. Newton, for example, suspended identical boxes of wood from wires 11 feet long. He filled one box with wood and the other with an equal weight of gold and set the pair of pendula in motion, observing that they kept in step to high accuracy. Repetitions of the experiment using other materials led to the conclusion that the accelerations of different materials are the same to about one part in a thousand. In the modern torsion balance experiments, two objects of different composition are connected by a rod and suspended in a horizontal orientation by a fine wire. If the gravitational acceleration of the bodies differs, there will be a torque induced on the suspension wire, related to the angle between the wire and the direction of the gravitational acceleration g. If the entire apparatus is rotated about some direction with angular velocity ω, the torque will be modulated with period $2\pi/\omega$. In the experiments of Eötvös and his collaborators, the wire and g were not quite parallel because of the centripetal acceleration on the apparatus due to the Earth's rotation; the apparatus was rotated about the direction of the wire. In the Princeton (Dicke *et al.*) and Moscow (Braginsky *et al.*) experiments, g was that of the Sun, and the rotation of the Earth provided the modulation of the torque at a period of 24 h. The modulated torque was determined either by measuring the torsional motions of the rod or by measuring the force required to counteract the torque and keep the rod in place. The resulting upper limits on measurable torques yielded limits on η given by $|\eta| < 1 \times 10^{-11}$ (Princeton, 1964) and $|\eta| < 1 \times 10^{-12}$ (Moscow, 1972), where the limits are 1σ formal standard deviations (see TEGP, section 2.4(*a*), for further discussion and references). The primary sources of error in these experiments are seismic noise and coupling of the torsion balance to gradients in the external gravitational field. Attempts to improve these results have centered on different forms of suspension of the masses, including magnetic levitation, flotation on liquids, and potential use of orbiting platforms. The results of various experiments are summarized in Table 5.1.

In order to determine the limits placed on individual parameters η^A by these experiments, it is necessary to estimate the coefficients E^A/mc^2 for different interactions and different materials. For laboratory-sized bodies, the dominant contribution comes from the nucleus. Table 5.2 summarizes these estimates and limits (based on the Moscow experiments) for various interactions (see TEGP, section 2.4(a), for details).

Recently, there has been a renewal of interest in Eötvös's original version of the experiment as a consequence of a reanalysis of the Eötvös data by Fischbach *et al.* (1986a). One of the goals of that reanalysis was to search for the effects of a hypothetical short-range (≈ 100 m) force, known as the 'fifth'

Table 5.1. *Tests of the weak equivalence principle[a]*

Experiment	Name	Method	Substances tested	Limit on $|\eta|$
Newton	Newton	Pendula	Various	10^{-3}
Bessel	Bessel	Pendula	Various	5×10^{-5}
Eötvös	Eötvös, Pekár and Fekete	Torsion balance	Various	5×10^{-9}
Potter	Potter	Pendula	Various	2×10^{-5}
Renner	Renner	Torsion balance	Various	2×10^{-9}
Princeton	Roll, Krotkov and Dicke	Torsion balance	Aluminum and gold	10^{-11}
Moscow	Braginsky and Panov	Torsion balance	Aluminum and platinum	10^{-12}
Munich	Koester	Free fall	Neutrons	3×10^{-4}
Stanford	Worden	Magnetic suspension	Niobium, Earth	10^{-4}
Boulder	Keiser and Faller	Flotation on water	Copper, tungsten	4×10^{-11}

[a] For references see TEGP, section 2.4(a).

Table 5.2. *Experimental limits on equivalence principle violations for various interactions[a]*

Interaction	Limit on η^A from Moscow Eötvös experiment
Strong	5×10^{-10}
Electromagnetic	
Electrostatic	4×10^{-10}
Magnetostatic	6×10^{-6}
Hyperfine	2×10^{-7}
Weak	10^{-2}
Gravitational	No limit

[a] For discussion and references, see TEGP, section 2.4(a).

force, that could couple to hypercharge or baryon number. Such a force would cause a difference in acceleration between bodies whose baryon number to mass ratio differed, and would yield a measurable Eötvös ratio given by

$$\eta = \eta^Y \left(\frac{B_1}{\mu_1} - \frac{B_2}{\mu_2} \right), \tag{3}$$

where μ_i is the mass of the ith body measured in atomic mass units, B_i is its total baryon number, and η^Y is a parameter that depends on the strength and range of the putative force, and on the detailed distribution of local matter within that range. Fischbach *et al.* claimed that Eötvös's data showed a significant dependence of η on the baryon-number-per-unit-mass difference, and they quoted a value $\eta^Y = (5.65 \pm 0.71) \times 10^{-6}$. This result, they argued, was qualitatively in accord with measured deviations from the inverse square law of gravity using gravimeter data from deep mines (Holding *et al.*, 1986), and with anomalous energy dependences in the fundamental parameters that characterize the behavior of the $K^0 - \bar{K}^0$ mesons, effects that would also be consequences of such a fifth force. The more precise Princeton and Moscow experiments would not be sensitive to this effect because the relevant source of gravity in those cases was the Sun, and the effect of a 100 m short-range force would therefore be negligible.

A number of authors subsequently took issue with the results of this reanalysis. It was argued by some that the evidence for the behavior indicated by (3) was much weaker than claimed by Fischbach *et al.* because of a number of factors which were inadequately taken into account in the reanalysis, including a sign error in the interpretation of one of Eötvös's conventions, uncertainties in the isotopic composition of each element and the chemical composition of each compound used in the experiment, uncertainties in the individual masses of the substances used and of the containers in which they were placed, and unknown systematic errors that might have affected the outcomes of Eötvös's experiments, given that three different methods were used by Eötvös and his colleagues. Reanalyses by others of Eötvös's data, and of the data from a series of 1935 experiments using the same apparatus by Renner failed to support the Fischbach *et al.* value of η^Y. Other authors pointed out that, even if one accepts the dependence of η on baryon number claimed by Fischbach *et al.*, it is virtually impossible to infer anything about the nature of a short-range fifth force, because its putative effect on Eötvös's experiments is extremely sensitive to the details of the nearby mass distribution (such as the mass of the building next to Eötvös's Budapest laboratory), which are unknowable nearly a century after the fact.† At present the situation is inconclusive. However, the Eötvös reanalysis has stimulated extensive interest in new experiments to test this idea, and several are in progress.

The second ingredient of EEP, LLI, can be said to be tested every time that special relativity is confirmed in the laboratory. However, many such experiments, especially in high-energy physics, are not 'clean' tests, because in many cases it is unlikely that a violation of Lorentz invariance could be distinguished from effects due to the complicated strong and weak interactions.

However, there is one class of experiments that have been interpreted as 'clean' tests of local Lorentz invariance, and with ultra-high precision at that. This is the class of 'mass anisotropy' experiments: the classic version is the Hughes–Drever experiment, performed in the period 1959–60 independently by Hughes and collaborators at Yale University, and by Drever at Glasgow University (TEGP, section 2.4(b)); improved versions have been carried out recently by Prestage et al. (1985) and by Lamoreaux et al. (1986). A simple and useful way of interpreting these experiments is to suppose that the electromagnetic interactions suffer a slight violation of Lorentz invariance, through a change in the speed of light relative to the limiting speed of test particles, in other words, $c_{light} \neq c_0$ (TEGP, section 2.6; see also Haugan (1986b)). Such a violation necessarily selects a preferred universal rest-frame, presumably that of the cosmic background radiation, through which we are moving at about 300 km s^{-1}. Such a Lorentz-non-invariant electromagnetic interaction would cause shifts in the energy levels of atoms and nuclei that depend on the orientation of the quantization axis of the state relative to our universal velocity vector, and on the quantum numbers of the state. The presence or absence of such energy shifts can be examined by measuring the energy of one such state relative to another state that is either unaffected or is affected differently by the supposed violation. One way is to look for a change in the energy levels of states that are ordinarily equally spaced, such as the four $J = 3/2$ ground states of the lithium-7 nucleus in a magnetic field (Drever experiment); another is to compare the levels of a complex nucleus with the atomic

---

† Most of the papers commenting on the Fischbach et al. analysis were contained in the 2 June 1986 issue of *Physical Review Letters*: they include Neufield (1986), Thieberger (1986), Nussinov (1986), Thodberg (1986), Keyser et al. (1986), together with responses by Fischbach et al. (1986b). Also listed by the editors are the authors of similar papers which were not published for reasons of space; see also De Rujula (1986).

hyperfine levels of a hydrogen-maser clock (Prestage *et al.* experiment), and another is to monitor the Larmor precession of mercury-201 nuclei in a constant magnetic field (Lamoreaux experiment). These experiments have all yielded extremely accurate results, quoted as limits on the parameter $[(c_{light})^2/(c_0)^2 - 1]$ in Table 5.3.

The principle of LPI, the third part of EEP, can be tested by the gravitational redshift experiment, the first experimental test of gravitation proposed by Einstein. Despite the fact that Einstein regarded this as a crucial test of general relativity, we now realize that it does not distinguish between general relativity and any other metric theory of gravity, instead it is a test only of EEP. A typical gravitational redshift experiment measures the frequency or wavelength shift $Z \equiv \Delta v/v = -\Delta\lambda/\lambda$ between two identical frequency standards (clocks) placed at rest at different heights in a static gravitational field. If the frequency of a given type of atomic clock is the same when measured in a local, momentarily comoving freely falling frame (Lorentz frame) independent of the location or velocity of that frame, then the comparison of frequencies of two clocks at rest at different locations boils down to a comparison of the velocities of two local Lorentz frames, one at rest with respect to one clock at the moment of emission of its signal, the other at rest with respect to the other clock at the moment of reception of the signal. The frequency shift is then a consequence of the first-order Doppler shift between the frames. The structure of the clock plays no role whatsoever. The result is a shift

$$Z = \Delta U/c^2, \tag{4}$$

where ΔU is the difference in the Newtonian gravitational potential between the receiver and the emitter. If LPI is not valid, then it turns out that the shift can be written

$$Z = (1 + \alpha)\,\Delta U/c^2, \tag{5}$$

Table 5.3. *Mass anisotropy tests of local Lorentz invariance*[a]

Experiment	Substances tested	Limit on $\lvert(c_{light})^2/(c_0)^2 - 1\rvert$
Hughes–Drever	Lithium-7	5×10^{-16}
Prestage *et al.*	Beryllum-9	10^{-18}
Lamoreaux *et al.*	Mercury-201	10^{-20}

[a] For discussion see TEGP, sections 2.4(*b*), 2.6(*e*), and Haugan (1986*b*).

where the parameter α may depend upon the nature of the clock whose shift is being measured (see TEGP, section 2.4(b), for details).

Although there were several attempts following the publication of the general theory of relativity to measure the gravitational redshift of spectral lines from white dwarf stars, the results were inconclusive (see Bertotti *et al.* (1962) for a review). The first successful, high-precision redshift measurement was the series of Pound–Rebka–Snider experiments of 1960–5, that measured the frequency shift of γ-ray photons from iron-57 as they ascended or descended the Jefferson Physical Laboratory tower at Harvard University. The high accuracy achieved – 1 per cent – was obtained by making use of the Mössbauer effect to produce a narrow resonance line whose shift could be accurately determined. Other experiments since 1960 measured the shift of spectral lines in the Sun's gravitational field and the change in rate of atomic clocks transported aloft on aircraft, rockets and satellites. Table 5.4 summarizes the important redshift experiments that have been performed since 1960.

The most recent experiments have taken advantage of the development of frequency standards of ultra-high stability – parts in 10^{15}–10^{16} over averaging times of 10–100 seconds and longer, such as hydrogen-maser clocks, and superconducting-cavity stabilized oscillator (SCSO) clocks.

Table 5.4. *Gravitational redshift experiments*[a]

Experiment	Name	Method	Limit on $|\alpha|$
Pound–Rebka–Snider	Pound and Rebka, Pound and Snider	Fall of photons from Mössbauer emitters	10^{-2}
Brault	Brault	Solar spectral lines	5×10^{-2}
Jenkins	Jenkins	Crystal oscillator clocks on GEOS-1 Satellite	9×10^{-2}
Snider	Snider	Solar spectral lines	6×10^{-2}
Jet-lagged clocks (A)	Hafele and Keating	Cesium beam clocks on jet aircraft	10^{-1}
Jet-lagged clocks (B)	Alley	Rubidium clocks on jet aircraft	2×10^{-2}
Vessot–Levine rocket redshift experiment	Vessot *et al.*	Hydrogen maser on rocket	2×10^{-4}
Null redshift experiment	Turneaure *et al.*	Hydrogen maser vs. SCSO	2×10^{-2}

[a] For references see TEGP, section 2.4(c).

The first such experiment (and the most precise to date) was the Vessot–Levine rocket redshift experiment that took place in June 1976 (in NASA circles, the experiment was denoted Gravity Probe A or GPA). A hydrogen-maser clock was flown on a rocket to an altitude of about 10 000 km and its frequency compared with a similar clock on the ground. The experiment took advantage of the masers' frequency stability by monitoring the frequency shift as a function of altitude. A sophisticated data acquisition scheme accurately eliminated all effects of the first-order Doppler shift due to the rocket's motion, while tracking data were used to determine the payload's location and velocity (to evaluate the potential difference ΔU, and the time dilation). Analysis of the data yielded a limit $|\alpha| < 2 \times 10^{-4}$ (Vessot *et al.*, 1980). Improvement of this limit may be possible by placing such clocks on Earth-orbiting satellites or on a spacecraft in a very eccentric solar orbit.

Advances in stable clocks have also made possible a redshift experiment that is a more direct test of LPI: a 'null' gravitational redshift experiment that compares two different types of clocks, side by side in the same laboratory. If LPI is violated, then not only is the proper ticking rate of an atomic clock dependent upon position, but the position dependence must itself depend on the structure and composition of the clock, otherwise all clocks would vary with position in a universal way and there would be no operational way to detect the effect (since one clock must be selected as a standard and ratios taken relative to that clock). From (5) it is easy to see that a comparison of two different clocks A and B at the same location would measure variations in their frequency ratio that depend upon gravitational potential according to

$$v_A/v_B = (v_A/v_B)_0 [1 + (\alpha^A - \alpha^B)U/c^2]. \tag{6}$$

where α^A and α^B are α-parameters for each clock type, and $(v_A/v_B)_0$ is the constant frequency ratio at some fiducial spacetime location from which U is measured.

A null redshift experiment of this type was performed in April 1978 at Stanford University. The rates of two hydrogen-maser clocks and of an ensemble of three SCSO clocks were compared over a 10-day period. During this period, the solar potential U/c^2 changed sinusoidally with a 24-hour period by 3×10^{-13} because of the Earth's rotation, and changed linearly at 3×10^{-12} per day because the Earth is 90° from perihelion in April. However, analysis of the data revealed no variations of either type within experimental errors, leading to a limit on the LPI violation parameter $|\alpha^H - \alpha^{SCSO}| < 2 \times 10^{-2}$ (Turneaure *et al.*, 1983).

The art of atomic timekeeping has advanced to such a state that it is now routine to take redshift and time-dilation corrections into account in making comparisons between timekeeping installations at different altitudes and latitudes, and in navigation systems using Earth-orbiting atomic clocks. For other tests of LPI, including the constancy of the fundamental non-gravitational constants, see TEGP, section 2.4(*c*), and UPDATE, section 2.3.4.

Newton's WEP has stood the test of more than 300 years of experimental scrutiny, without any confirmed indication of a violation. When augmented by the well-tested principles of LLI and LPI to form EEP, it becomes the basis for the modern view of gravity as a consequence of curved spacetime. The overwhelming empirical evidence supporting these fundamental principles has convinced theorists that only metric theories of gravity have a hope of being viable. As a consequence, most of the discussion of experimental gravitation centers on metric theories. However, before we deal with those tests, we must consider several more fundamental concepts that are common to Newtonian gravity and general relativity.

5.3.2 *The inverse square law*

Although the inverse square law is one of the centerpieces of Newtonian gravitation, the *Principia* does not contain a general statement of the law, analogous, say, to the statement of the three laws of motion that appear in the section headed 'Axioms, or Laws of Motion'. The closest Newton comes to a 'law' of gravity occurs in Book III, in the General Scholium to Proposition XLII, where he states that gravity operates 'according to the quantity of the solid matter which they [the sun and planets] contain, and propagates its virtue on all sides to immense distances, decreasing always as the inverse square of the distances'. Apart from the fact that this law accounted for the motions of the planets, Newton gave a direct demonstration of its validity by showing, in Book III, Proposition IV, that the force required to keep the Moon in its orbit is related to the force of gravity on the surface of the Earth by the inverse square of the Moon's orbital radius in units of the Earth's radius.

As we shall see in Section 5 4, up to the accuracy at which post-Newtonian corrections play a role, the inverse square law of Newtonian gravity has been verified to high accuracy in planetary and lunar motion. Observations of planetary, solar, and stellar structure support the law as applied to bulk matter. Further agreement has been found for determinations of the Earth's gravitational field via comparisons of surface, rocket, and satellite measurements, and via comparison of laser ranging to the Moon and to the LAGEOS satellite. These observations confirm the inverse square law for distances typically from 1 km to several astronomical units.

However, at shorter distances, the inverse square law has recently come under more intense theoretical and experimental scrutiny. One reason for this interest is the possibility that examination of the inverse square law could reveal information about theories of the strong, weak, electromagnetic, and gravitational interactions. In some models, exchange of various types of particles of small but non-zero mass (axions, hyperphotons, etc.) can lead to a Yukawa-type contribution to the interaction potential between bodies that could be of sufficiently long range to be observable. In some models, the resulting force can be combined with the Newtonian gravitational force, and can be written

$$F = -G(r)m_1 m_2/r^2. \tag{7}$$

where m_1 and m_2 are the masses of the bodies, r is their separation, and $G(r)$ is an 'effective' gravitational constant, given by

$$G(r) = G_\infty [1 + \alpha(1 + r/\lambda) e^{-r/\lambda}], \tag{8}$$

where α characterizes the strength of the coupling of the massive particle, and $\lambda = \hbar/mc$ is its Compton wavelength.

Despite an early report of a deviation from the inverse square law in the range of 5–30 cm, later experiments ruled this out. The methods in these laboratory inverse square experiments range from variants of the Cavendish experiment using various massive sources (rings, cylinders, oil tanks, etc.) to direct measurements of the divergence of $\vec{g}$, which vanishes if the inverse square law is valid (see Section 5.2), using an array of orthogonal gravity gradiometers. (For reviews and references see Newman (1983), UPDATE, section 2.1(iv), Hoskins *et al.* (1985), and Stacey *et al.* (1986).) However, in the range 1–1000 m, measurements of the gravity profile in deep mines have yielded values of the gravitational constant at depth that appear to differ from the laboratory values obtained from Cavendish-type experiments. In these experiments, measurements of the local acceleration g and its gradient at different depths in mines are combined with determinations of the local matter density from core samples, in order to take into account deviations in density from that of an idealized Earth model. The resulting values of G are between 0.5 and 1 per cent higher than the laboratory values. One source of systematic error in such experiments is the possibility of unsampled density anomalies. Other measurements of this type involve determining the gravitational acceleration at various depths in large bodies of water, such as the oceans or reservoirs (Holding *et al.*, 1986; Stacey *et al.*, 1986). Fig. 5.1 shows the current limits on the parameters α and λ in (8) that are imposed by the best current experiments.

5.3.3 *Does action equal reaction?*

Perhaps the most original and far reaching of Newton's laws of motion is the third law: 'to every action there is always opposed an equal reaction: or the mutual actions of two bodies upon each other are always equal, and directed to contrary parts'. (The first two laws are the law of inertia and the force law.) This law allowed Newton to free himself from the constraint of treating gravitation as strictly a one-body problem, and to deal with two or more

Fig. 5.1. Limits on coupling strength α and range λ of a hypothetical short-range component of gravitational force. Hatched regions are disallowed at 1σ level. Constraint in the range 10^{-3}–10 m comes from laboratory experiments (Hoskins *et al.*, 1985); constraint in the range $> 10^2$ m comes from satellite data. Shaded region in $\alpha < 0$ portion of figure shows values of α and λ implied by mine measurements (Stacey *et al.*, 1986).

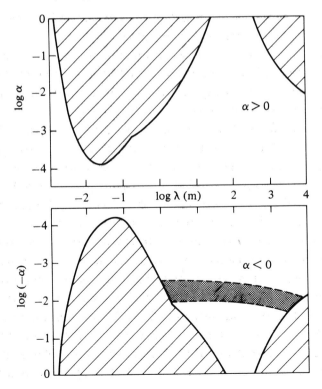

bodies in mutual gravitational attraction moving about their common center of mass. But this law goes even deeper, for it embodies the concept of the conservation of momentum for isolated gravitating systems: thus, as Newton states in Corollary IV to the Laws of Motion, 'the common centre of gravity of all bodies acting upon each other (excluding external actions and impediments) is either at rest, or moves uniformly in a right line.'

Conservation of momentum is of course one of the cornerstones of modern physical theory; it is normally built into the structure of a given theory by basing the theory on an action principle. Then, according to Noether's theorem, conservation laws arise naturally in such theories when the action embodies a symmetry, in the case of momentum conservation, a translational symmetry. Newtonian gravitation theory can be derived from an action, as can general relativity and many other theories of gravitation, so conservation of momentum is a natural feature. It is remarkable that, without the benefit of modern concepts of action and symmetry, Newton still grasped the meaning of conservation of momentum (it is implicit in the first law of motion) and built it into gravitation through the third law of motion.

Although there is ample experimental evidence for conservation of momentum in atomic and high-energy physics, there is rather less direct evidence of it in specifically gravitational situations. An alternative statement of Newton's third law for gravitating systems is that the 'active gravitational mass', that is, the mass that determines the gravitational potential exhibited by a body, should equal the 'passive gravitational mass', the mass that determines the force on a body in a gravitational field. Such an equality guarantees the equality of action and reaction and of conservation of momentum, at least in the Newtonian limit.

A classic test of Newton's third law for gravitating systems was carried out by Kreuzer (1968). Kreuzer's experiment used a Cavendish balance to compare the Newtonian gravitational force generated by a cylinder of Teflon (76 per cent fluorine by weight) with the force generated by that amount of a liquid mixture of trichloroethylene and dibromomethane (74 per cent bromine by weight) that had the same passive gravitational mass as the cylinder (namely, the amount of liquid displaced by the cylinder at neutral buoyancy). Kreuzer's conclusion was that the forces were the same, and hence that the ratio of active to passive mass for fluorine and bromine were the same to 5 parts in 10^5, that is,

$$\left| \frac{(m_A/m_P)_{Fl} - (m_A/m_P)_{Br}}{(m_A/m_P)_{Br}} \right| < 5 \times 10^{-5}. \tag{9}$$

A remarkable planetary test of Newton's third law was recently reported by Bartlett and van Buren (1986). They noted that current understanding of the structure of the Moon involves an iron-rich, aluminum-poor mantle whose center of mass is offset about 10 km from the center of mass of an aluminum-rich, iron-poor crust. The direction of offset is toward the Earth, about 14° to the east of the Earth–Moon line. Such a model accounts for the basaltic maria which face the Earth, and the aluminum-rich highlands on the Moon's far side, and for a 2 km offset between the observed center of mass and center of figure for the Moon. Because of this asymmetry, a violation of Newton's third law for aluminum and iron would result in a momentum non-conserving self-force on the Moon, whose component along the orbital direction would contribute to the secular acceleration of the lunar orbit. Improved knowledge of the lunar orbit through lunar laser ranging, and a better understanding of tidal effects in the Earth–Moon system (which also contribute to the secular acceleration) through LAGEOS satellite data, severely limit any anomalous secular acceleration, with the resulting limit

$$\left| \frac{(m_A/m_P)_{Al} - (m_A/m_P)_{Fe}}{(m_A/m_P)_{Fe}} \right| < 4 \times 10^{-12}. \tag{10}$$

According to the PPN formalism, in a theory of gravity that violates conservation of momentum, the electrostatic binding energy E_e of an atomic nucleus could make a contribution to the ratio of active to passive mass of the form

$$m_A = m_P + \tfrac{1}{2}\zeta_3 E_e/c^2, \tag{11}$$

where ζ_3 is a PPN 'conservation law' parameter, whose value is zero in any conservative theory. The resulting limits on ζ_3 from the Kreuzer and lunar experiments are $|\zeta_3| < 6 \times 10^{-2}$ and $|\zeta_3| < 1 \times 10^{-8}$, respectively. Other tests of conservation laws in post-Newtonian gravity are described in TEGP, section 9.3.

5.3.4 Is Newton's gravitational constant constant?

It is interesting to note that the term 'gravitational constant' never occurs in the *Principia*. In fact, it seems that the universal constant of proportionality that we now call G, that relates the gravitational force between two bodies to the product of their masses and the inverse square of their separations, does not make an appearance until well into the eighteenth century, in Laplace's *Mécanique Céleste*, for example, and later in Poisson's *Traité de Mécanique*. The reason is that, in applying Newtonian theory to planetary motions, the absolute value of G is irrelevant, since it is only the product of G with the unknown solar or planetary mass M that is important, thus the entire discussion of celestial mechanics can be carried out in terms of ratios referred to the Sun or the Earth, as it is in the *Principia*. Even when Bouguer in the 1730s carried out laboratory-scale gravitational experiments using pendula and plumb lines to measure the variation of gravity with altitude and the gravitational effect of mountains, the goal was not to measure G directly, but to determine the mean density of the Earth. Similarly, the classic torsion-pendulum experiments performed by Cavendish around 1797 make no mention of G; they too were designed to measure the Earth's mean density. Even though G could in principle be calculated from the density, the local gravitational acceleration g, and the radius of the Earth, it was 1890 before experimentalists such as Poynting, Boys, and others began to quote absolute values for G, and to regard it as a fundamental constant, for which a standard value could be determined.†

Whether the gravitational constant was discussed explicitly or not, one theme that ran as a thread through all gravitational studies up to the twentieth century was the idea that gravity is unchanging with time, in other words, that G is constant in time. Such a notion meshed with the prevalent world view that held that the universe outside the solar system was static, therefore there was no mechanism for G to vary, no natural timescale for a variation.

However, the discovery that the universe is expanding generated numerous proposals for a gravitational constant that could vary as the universe evolves, ranging from the 'large numbers hypothesis' of Dirac in the 1930s to the Machian viewpoint of Dicke and the scalar–tensor theory in the 1960s (see Wesson (1978) for a review). In most of these models, the value of G was viewed as being modulated by a coupling between gravity and some additional field of cosmological origin. If G does change with cosmic evolution, its rate of variation should be of the order of the expansion rate of the universe, i.e., $\dot{G}/G = \sigma H_0$, where H_0 is the Hubble expansion parameter whose value we shall adopt to be $H_0 \approx 55 \text{ km s}^{-1} \text{ Mpc}^{-1} \approx (2 \times 10^{10} \text{ yr})^{-1}$, and σ is a dimensionless parameter whose value depends upon the theory of gravity under study and upon the detailed cosmological model. In Newtonian gravity and general relativity, of course, G is precisely constant $(\sigma = 0)$.

Several observational constraints have been placed on $\dot{G}/G$, using

† For an informative history of the Bouguer, Cavendish, and other geophysical measurements up to 1900, see Mackenzie (1900).

methods that include studies of the evolution of the Sun, observations of
lunar occultations (including analyses of ancient eclipse data), planetary
radar-ranging measurements, and lunar laser-ranging measurements. The
present status of these experiments in summarized in Table 5.5 (for reviews
of some of these methods see Halpern (1978), Gillies (1982) and Reasenberg
(1983)). Some authors have claimed that the non-zero results for σ shown in
Table 5.5 are significant and support the hypothesis of a varying
gravitational constant, while others have argued that unavoidable errors in
the models and in the different data sets used in the numerical estimation of
parameters such as $\dot{G}/G$ may seriously degrade such estimates. The most
promising limit or estimate for $\dot{G}/G$ may come from ranging measurements
to Viking. The combination of three factors: (i) extremely accurate range
measurements made possible by anchoring of the landers and orbiters, (ii)
the unexpectedly long lifetime of the spacecraft (Lander 2 finally died in
November 1982), and (iii) the ability to combine Viking data consistently
with other data sets such as Mercury and Venus passive radar, Mariner 9
radar and lunar laser-ranging data, has made it possible to look for $\dot{G}/G$ at
levels below 10^{-11} yr^{-1}. Current results give upper limits ranging from 3 to
0.6 parts in 10^{11} yr^{-1}. The major factors limiting the accuracy of these
estimates (and responsible in part for the difference between the two

Table 5.5. *Tests of the constancy of the gravitational constant*[a]

Method	$\sigma = (\dot{G})$ $\times (2 \times 10^{10}$ yr)	Name
Solar evolution	$|\sigma| < 2$	Chin and Stothers
Lunar occultations and eclipses	$|\sigma| < 0.8$	Morrison
	$\sigma = -(0.6 \pm 0.3)$	Van Flandern
	$\sigma = -(0.5 \pm 0.3)$	Muller
	$\sigma = -(2.5 \pm 0.7)$	Newton
	$\sigma = -(0.3 \pm 0.1)$	Van Flandern
Planetary and spacecraft radar	$|\sigma| < 8$	Shapiro *et al.*
	$|\sigma| < 3$	Reasenberg and Shapiro
		Anderson *et al.*
Lunar laser ranging	$|\sigma| < 0.6$	Williams *et al.*
Viking radar	$|\sigma| < 0.6$	Reasenberg
	$\sigma = 0.04 \pm 0.08$	Hellings *et al.*

[a] For references see UPDATE, section 5.3.

estimates, despite being based upon similar data sets) are the uncertainty in the masses and distributions of the asteroids, and the level of correlations among the many parameters to be estimated in the models. Some authors have suggested that radar observations of a Mercury orbiter over a two-year mission (30 cm accuracy in range) could yield $\Delta(\dot{G}/G) \sim 3 \times 10^{-13} \, \text{yr}^{-1}$ (see UPDATE, section 5.3).

5.4 Testing the applications of gravitation theory

5.4.1 Motion of the planets

In Proposition XIII of Book III of the *Principia*, Newton states 'the planets move in ellipses which have their common focus in the centre of the Sun; and by radii drawn to that centre, they describe areas proportional to the times of description'. This statement was simultaneously a summary of two of Kepler's three empirical laws of planetary motion, and a statement of the predictions of the law of gravity. Elsewhere in the *Principia*, Newton similarly deals with Kepler's third law, that the square of the period of the orbit of a planet is proportional to the cube of its semimajor axis. Moreover, the universality of the proposed law of gravitation meant that the same statements should apply to the Moon, to the satellites of Jupiter and Saturn, to the comets, in fact, to any massive bodies. Thus on the broad level, Newtonian gravity led to a complete understanding of the principles of planetary motion. One of the first successes of Newton's theory of gravitation was its use by Halley to determine that the orbits of the comets that had appeared in 1531, 1607 and 1682 were actually the same orbit, and that the comet should reappear approximately every 75 years. The reappearance on schedule of Halley's comet in 1986 was a fitting celestial herald of the tercentenary year.

In reality, of course, the detailed motions of the planets were more complicated than the simple Keplerian picture, but these could be explained as being due to the perturbations of a given orbit by the gravitational forces due to other planets. Thus, for example, Proposition XIV, Book III, states 'the aphelions and nodes of the orbits of the planets are fixed'; nevertheless, Newton recognized that this is an approximation, and that the inner planets (Mercury, Venus, Earth and Mars) will experience motions of their aphelia because of the perturbing effect of Jupiter and Saturn.

Newton could only provide estimates of the effects of perturbations. During the eighteenth century, such figures as D'Alembert, Clairaut, Euler, Lagrange, and Laplace brought increasing mathematical sophistication to bear on the problem of celestial mechanics, so that, by 1758, Clairaut was able to make an accurate prediction of the return of Halley's comet (confirmed to within one month), Euler and Lagrange were able to demonstrate that the observed changes in the period of Jupiter and Saturn were not secular, but were periodic over timescales of centuries, and Laplace's *Mécanique Céleste* gave a complete mathematical accounting of planetary motions.†

The success of Newtonian gravity applied to the planets reached its zenith in the middle of the nineteenth century, when Adams and Le Verrier independently predicted the approximate location of a hitherto unknown planet that had to be present in order to explain anomalies in the motion of Uranus. Astronomical searches of the predicted location in 1846 revealed such a planet: Neptune.

But a cloud soon appeared on the Newtonian horizon with the announcement by Leverrier in 1859 that, after the perturbing effects of the planets on Mercury's orbit had been accounted for, and after the effect of the precession of the equinoxes on the astronomical coordinate system had been subtracted off, there remained in the data an unexplained advance in the perihelion of Mercury. The modern value for this discrepancy is 43 arc seconds per century. A number of *ad hoc* proposals were made in an attempt to account for this excess, including, among others, the existence of a new planet Vulcan near the Sun, a ring of planetoids, a solar quadrupole moment and a deviation from the inverse square law of gravitation (for a review, see Roseveare, 1982). Although these proposals could account for the perihelion advance of Mercury, they either involved objects that were detectable by direct optical observation yet were not observed, or predicted perturbations on the other planets (for example, regressions of nodes, changes in orbital inclinations) that were inconsistent with observations. Thus they were doomed to failure. General relativity accounted for the anomalous shift in a natural way without disturbing the agreement with other planetary observations.

However, since 1967, there has been periodic controversy over whether the perihelion shift is a confirmation or a refutation of general relativity because of the apparent existence of a solar quadrupole moment that could contribute a portion of the observed perihelion shift. To date, the question of the size of the solar quadrupole moment remains an open one.

The predicted advance, $\Delta\tilde\omega$, per orbit, including both the dominant

† For the early history of celestial mechanics, see Moulton (1914).

relativistic PPN contributions and the Newtonian contribution of the modified gravitational potential due to a possible oblateness, is given by

$$\Delta\tilde{\omega} = (6\pi Gm/pc^2)[\tfrac{1}{3}(2+2\gamma-\beta)+J_2(R^2c^2/2Gmp)], \qquad (12)$$

where $m \equiv m_1 + m_2$ is the total mass of the two-body system; $p = a(1-e^2)$ is the semilatus rectum of the orbit, with a the semimajor axis and e the eccentricity; R is the mean radius of the oblate body; and J_2 is a dimensionless measure of its quadrupole moment, given by $J_2 = (C-A)/m_1 R^2$, where C and A are the moments of inertia about the body's rotation and equatorial axes, respectively (for details of the derivation see TEGP, section 7.3).

The first term in (12) is the classical perihelion shift, which depends upon the PPN parameters γ and β. (For a summary of the PPN formalism, see the Appendix.) In general relativity, $\gamma = \beta = 1$. The second term depends upon the solar quadrupole moment J_2. For a Sun that rotates uniformly with its observed surface angular velocity, so that the quadrupole moment is produced by centrifugal flattening, one may estimate J_2 to be $\sim 1 \times 10^{-7}$. Normalizing J_2 by this value and substituting standard orbital elements and physical constants for Mercury and the Sun we obtain the rate of perihelion shift $\dot{\tilde{\omega}}$, in seconds of arc per century,

$$\dot{\tilde{\omega}} = 42''98 \, \lambda_p \, c^{-1},$$

$$\lambda_p = [\tfrac{1}{3}(2+2\gamma-\beta)+3 \times 10^{-4}(J_2/10^{-7})]. \qquad (13)$$

The measured perihelion shift is accurately known: after the effects of the general precession of the equinoxes ($5000'' \, c^{-1}$) and the perturbing effects of the other planets ($280'' \, c^{-1}$ from Venus, $150'' \, c^{-1}$ from Jupiter, $100'' \, c^{-1}$ from the rest) have been accounted for, the remaining perihelion shift is known (i) to a precision of about 1 per cent from optical observations of Mercury during the past three centuries, and (ii) to about 0.5 per cent from radar observations during 1966–76. Unfortunately, measurements of the orbit of Mercury alone are incapable at present of separating the effects of relativistic gravity and of solar quadrupole moment in the determination of λ_p. Thus, in two analyses of radar distance measurements to Mercury, J_2 was *assumed* to have a value corresponding to uniform rotation (effect on λ_p negligible), and the PPN parameter combination was estimated. The results were

$$\tfrac{1}{3}(2+2\gamma-\beta) = \begin{cases} 1.005 \pm 0.020 \ (1966\text{--}71 \ \text{data}) \\ 1.003 \pm 0.005 \ (1966\text{--}76 \ \text{data}) \end{cases}, \qquad (14)$$

(see TEGP, section 7.3, for references) where the quoted errors are 1σ

estimates of the realistic error (taking into account possible systematic errors).

The origin of the uncertainty that has clouded the interpretation of perihelion-shift measurements is a series of experiments performed in 1966 by Dicke and Goldenberg. Those experiments measured the visual oblateness or flattening of the Sun's disk and found a difference between the apparent polar and equatorial angular radii that was interpreted as corresponding to a value $J_2 = (2.47 \pm 0.23) \times 10^{-5}$. A value of J_2 this large would have contributed about $3'' \, c^{-1}$ to Mercury's perihelion shift, and thus would have put general relativity in serious disagreement with the observations, while on the other hand supporting Brans–Dicke theory with a value $\omega \simeq 7$, whose post-Newtonian contribution to the perihelion shift would thus have been $40'' \, c^{-1}$.

These results generated considerable controversy within the relativity and solar physics communities, and a mammoth number of papers was produced, some in support of solar oblateness, some in opposition. The controversy abated somewhat in 1973, when Hill and his collaborators performed a similar visual oblateness measurement that yielded $J_2 = (0.10 \pm 0.43) \times 10^{-5}$, an upper limit five times smaller than Dicke's value (TEGP, section 7.3).

In 1982 the controversy was revived by reported estimates for J_2 as large as Hill et al.'s previous upper limit, obtained from analyses of the global five-minute oscillations of the Sun. The discovery of these oscillations around 1976 provided a possible new probe of the internal structure of the Sun, in particular of its internal rotation. In the absence of rotation, the $2l + 1$ normal modes of non-radial oscillation corresponding to the spherical harmonic indices (l, m) are degenerate, i.e., they all have the same eigenfrequency. However, rotation splits the modes by amounts that depend on the mode number, on (l, m) and on certain weighted averages of the angular velocity distribution of the Sun over its interior. Analyses of the observed splittings by Hill et al. (1982, 1986), Gough (1982), Campbell et al. (1983), and Duvall et al. (1984) produced estimates for the angular velocity distribution in the interior of the Sun which were then used in solar models to obtain values for J_2. The results ranged between 10^{-7} and 6×10^{-6}. Meanwhile, new visual oblateness measurements by Dicke and colleagues in 1983 gave a reported value of about 8×10^{-6} (Dicke et al., 1985).

The present situation is illustrated in Fig. 5.2: plotted are measured values of J_2 as a function of time. Also shown are the range of values of J_2 expected in a uniformly rotating solar model and the maximum value of J_2 that would be compatible with general relativity within the 1σ and 2σ upper bounds on λ_p obtained from radar observations of Mercury.

One of the major difficulties in relating visual solar oblateness results and solar oscillation data to J_2 is that a considerable amount of complex solar physics theory must be employed. There is, however, a way of determining J_2 unambiguously, namely by probing the solar gravity field at different distances from the Sun, thereby separating the effects of J_2 from those of relativistic gravitation through their different radial dependences. One method would compare the perihelion shifts of different planets. Another method would take advantage of Mercury's orbital eccentricity ($e \sim 0.2$) and

Fig. 5.2. The problem of J_2. Open circles represent visual oblateness measurements, filled circles represent solar oscillation inferences. 'General relativity 1σ (2σ)' lines represent maximum values of J_2 that would be compatible with general relativity within 1σ (2σ) errors in radar determinations of Mercury's perihelion shift. Shaded area represents values of J_2 that would be expected from a conventional, uniformly rotating solar model.

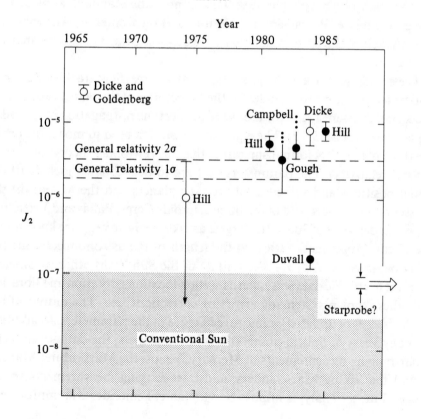

search for the different *periodic* orbital perturbations induced by J_2 and by relativistic gravity. The accuracy required for such measurements would necessitate tracking of a spacecraft in orbit around Mercury, but some studies have shown that J_2 could be determined to within a few parts in 10^7. A direct measurement of J_2 could be provided by a mission that was at one time under study by NASA. Known as Starprobe, it is a spacecraft whose orbit would take it to within four solar radii of the Sun. Feasibility studies indicated that J_2 could be measured to an accuracy of 10 per cent of its conventional value of 10^{-7}. Unfortunately, this mission is not a part of NASA's current plans.

From the time that mathematical techniques for celestial mechanics became sophisticated enough to carry out detailed perturbation calculations, the determination of ephemerides of the planets has been carried out using Newtonian gravity. However, in the past 20 years, the tremendous improvements in the accuracy of planetary and spacecraft tracking, and in the ability of theorists to model their motions, especially using computers, has made it possible to turn from Newtonian ephemerides, to post-Newtonian ephemerides. This is now the standard approach at places like the Jet Propulsion Laboratory and the Center for Astrophysics, where much of the current analysis of relativistic solar-system dynamics is carried out.

These analyses use a multiparameter least-squares fit of tracking data and optical observations to a model for the trajectories of the planets and of any spacecraft of interest, and for the curved-spacetime propagation of the radar signals (see Section 5.4.3). The equations of motion used to model the orbits are those of the PPN formalism, rather than of Newtonian theory (or even of general relativity). The parameters that are to be estimated include (i) the initial positions and velocities of the nine planets and the Moon; (ii) the masses of the planets, and of the three asteroids Ceres, Pallas and Vesta; (iii) the mean densities of 200 of the largest asteroids whose radii are known; (iv) the Earth–Moon mass ratio; (v) the length of the astronomical unit; (vi) PPN parameters, γ, β, α_1, ... ; (vii) J_2 of the Sun; (viii) other parameters relevant to specific data sets, such as station locations, rotation and libration of bodies, known systematic errors or corrections, etc. The output of the model is a set of updated or improved values of the parameters. In addition to many years of optical observations of the planets, the data set includes radar-ranging measurements to Mercury, Venus, the Mars orbiter Mariner 9, and the Viking Mars landers, and laser-ranging measurements to the Moon. Further analyses of post-Newtonian ephemerides will improve our knowledge of PPN parameters, of J_2, and of the dynamics of the solar system. (For further details, see Hellings (1984), Reasenberg (1983), and UPDATE, section 4.4.)

5.4.2 Motion of the Moon

The Moon plays a central role in Newtonian gravitation and in the *Principia*. It gave Newton the first test of the validity of the inverse square law, in his demonstration, carried out at least 20 years before the publication of the *Principia*, that the force required to keep the Moon in its orbit, and the force of gravity at the surface of the Earth were related by the ratio of the inverse square of the respective distances from the center of the Earth. Newton withheld announcement of this result until around 1685, when he was fully satisfied with his proof that the field outside a spherical body was the same as that of a mass point (Cajori, 1928).

The motion of the lunar perigee, caused by the Sun's gravitational perturbation, was also treated by Newton, incorrectly as it turns out. Although it is believed that Newton at some point carried out a correct treatment of the problem, he never incorporated it into any editions of the *Principia*, instead the derivation he gave in the *Principia* leads to a value that is about half the observed value (Book I, Proposition XLV, Corollary I). In Book III, he alludes to a correct derivation, but without any calculational details. The correct derivation is contained in a set of Newton's unpublished papers now called the Portsmouth collection, but astronomers of the day were unaware of it. This celebrated discrepancy induced the English government and several scientific societies across Europe to offer prizes for a correct calculation of the effect. The problem was finally resolved in 1749 by Clairaut, and the result was another triumph for Newtonian theory.

Another difficulty in the motion of the Moon was the 'secular acceleration', an observed systematic increase in the lunar orbital period. This problem was attacked by the great mathematicians of the nineteenth century, such as Euler, Lagrange, and Laplace. These efforts were only partially successful in accounting for observations, not because of a defect of Newtonian theory, but because the strong solar perturbation of the Earth–Moon system made it difficult to find a reasonably convergent approximation method, and because of the complicating dissipative effect of ocean tides on the Earth. It was late in the nineteenth century before an accurate and reliable model, the Hill–Brown theory, was devised. The problem of the secular acceleration was further complicated by the fact that part of the observed effect is due to the slowing down of the rotation of the

Earth. Although Kant was the first to suggest, in 1754, that the Earth must be slowing down (arguing that since the Moon has spun down until it presents the same face to the Earth, presumably from tidal effects, the Earth must suffer the same slowing down), he later retracted the hypothesis, and it was not until the work of Delauney in 1866 and Spencer-Jones in 1939 that the importance of this effect was appreciated (for a brief review, see the article by Muller in Halpern (1978)).

Today, the agreement of the theory of the Moon's motion with observations carried out using such modern tools as lunar laser ranging represents another triumph for Newtonian gravity. Although some workers have incorporated general relativistic effects into the theory, most of them are too small to be detectable with current technology (TEGP, section 8.3).

The effect of tidal dissipation in the Earth on the Moon's secular acceleration remains an important uncertainty in any attempts to place tight limits on a cosmological variation in G (see Section 5.3.4), but these uncertainties may soon be reduced significantly through laser ranging to the Earth-orbiting LAGEOS satellite, from which improved determinations of the effects of ocean tides can be obtained.

However, there is one way in which the Moon has provided an important test of relativistic gravity, and it hearkens back to Newton's WEP. According to Newton, the equivalence principle should apply to all bodies, laboratory-sized bodies as well as planets. Indeed, since Newtonian gravity is a linear theory, it is straightforward to show that the acceleration of a gravitating body such as a planet in an external gravitational field is the same as that of a body of negligible size (ignoring tidal effects). But in most relativistic theories of gravity, including general relativity, gravity is non-linear, so that the internal gravitational field of a massive body can interact with the external gravitational field, in a manner that could differ from the interaction of the matter comprising the body. The result would be a violation of WEP for massive, self-gravitating bodies.

In a pioneering calculation using an early form of the PPN formalism, Nordtvedt (1968) showed that many metric theories predicted just such a violation, in other words, they predicted that bodies could fall with different accelerations depending on their gravitational self-energy. For a spherically symmetric body, the acceleration from rest in an external gravitational potential U has the form

$$\mathbf{a} = (m_\mathrm{P}/m_\mathrm{I})\,\nabla U,$$

$$m_\mathrm{P} = m_\mathrm{I} - \eta E_\mathrm{g}/c^2,$$

$$\eta = 4\beta - \gamma - 3 - \tfrac{10}{3}\xi - \alpha_1 + \tfrac{2}{3}\alpha_2 - \tfrac{2}{3}\zeta_1 - \tfrac{1}{3}\zeta_2, \tag{15}$$

where $\gamma, \beta, \xi, \alpha_1, \alpha_2, \zeta_1$, and ζ_2 are PPN parameters (see Appendix), and E_g is the absolute value of the gravitational self-energy of the body. This violation of the massive-body equivalence principle is known as the 'Nordtvedt effect'. The effect is absent in general relativity ($\eta = 0$), but present in Brans–Dicke theory ($\eta = 1/(2 + \omega)$).

The fact that the Nordtvedt effect is absent in general relativity is part of what has come to be called the 'strong equivalence principle' (TEGP, section 3.3). It has also been shown to be valid for the acceleration of highly relativistic bodies in general relativity, such as neutron stars and black holes. In other words, *all* bodies in general relativity fall with the same acceleration.

The existence of the Nordtvedt effect does not violate the results of laboratory Eötvös experiments, since for laboratory-sized objects, $E_g/m_I \sim 10^{-27}$, far below the sensitivity of current or future experiments. However, for astronomical bodies, E_g/m_I may be significant (10^{-5} for the Sun, 10^{-8} for Jupiter, 4.6×10^{-10} for the Earth, 0.2×10^{-10} for the Moon), and in fact it is an 'Eötvös experiment' using solar-system bodies that has yielded a significant test of the Nordtvedt effect. If the Nordtvedt effect is present ($\eta \neq 0$) then the Earth should fall toward the Sun with a slightly different acceleration than the Moon. This perturbation in the Earth–Moon orbit leads to a polarization of the orbit that is directed toward the Sun as it moves around the Earth–Moon system, as seen from Earth, resulting in a perturbation in the Earth–Moon distance of the form $\delta r = 9.2\,\eta \cos(\omega_0 - \omega_s)t$ m, where ω_0 and ω_s are the angular frequencies of the orbits of the Moon and Sun around the Earth (see TEGP, section 8.1, for detailed derivations).

Since August 1969, when the first successful acquisition was made of a laser signal reflected from the Apollo 11 retroreflector on the Moon, the lunar laser-ranging experiment has made regular measurements of the round-trip travel times of laser pulses between McDonald Observatory in Texas and the lunar retroreflectors, with accuracies in range of 15–30 cm. These measurements were fit using the method of least squares to a theoretical model for the lunar motion that took into account perturbations due to the other planets, tidal interactions, and post-Newtonian gravitational effects. The predicted round-trip travel times between retroreflector and telescope also took into account the librations of the Moon, the orientation of the Earth, the location of the observatory and atmospheric effects on the signal propagation. The 'Nordtvedt' parameter η along with several other important parameters of the model were then estimated in the least-squares method.

An important issue in this analysis is whether other perturbations of the Earth–Moon orbit could mask the Nordtvedt effect. Most perturbations produce effects in δr, which, when decomposed into sinusoidal components occur at frequencies different than that of the Nordtvedt term (e.g., at angular frequencies ω_0, $2(\omega_0 - \omega_s)$), and thus can be separated cleanly from it using the multiyear span of data. However, there is one perturbation, due to the tidal part of the solar potential, that does have a component at the frequency $(\omega_0 - \omega_s)$, with an amplitude of 110 km.

Although this term is ten thousand times larger than the nominal amplitude of the Nordtvedt effect, it turns out, fortunately, that the parameters that determine its numerical value, such as the Earth–Moon and Earth–Sun mass ratios, the astronomical unit and the mean Earth–Moon distance, are known with sufficient accuracy, partly from previous measurements and partly from the laser-ranging data itself, that this term can be accounted for to a precision of about 2 cm.

Two independent analyses of the data taken between acquisition and 1975 were carried out, both finding no evidence, within experimental uncertainty, for the Nordtvedt effect. Their results for η were $\eta = 0.00 \pm 0.03$ (Williams et al., 1976) and $\eta = 0.001 \pm 0.015$ (Shapiro et al., 1976), where the quoted errors are 1σ, obtained by estimating the sensitivity of η to possible systematic errors in the data or in the theoretical model. The formal statistical errors that emerged from the data analysis were typically much smaller, of order $\sigma(\eta)_{\text{FORMAL}} \sim \pm 0.004$. This represents a limit on a possible violation of WEP for massive bodies of 7 parts in 10^{12} (compare Table 5.1).

Improvements in the measurement accuracy and in the theoretical analysis of the lunar motion may tighten this limit by an order of magnitude, while a comparable test of the Nordtvedt effect may be possible using the Sun–Mars–Jupiter system (see TEGP, section 8.1, and UPDATE, section 5.1, for details and references).

5.4.3 Motion of light

The behavior of light in a gravitational field was not a subject that originated with Einstein. In fact Newton raised the question in Query I of the *Opticks*: 'Do not bodies act upon light at a distance, and by their action bend its rays; and is not this action (*coeteris paribus*) strongest at the least distance?' In the late eighteenth century, Michell and Laplace independently considered the nature of a body for which the escape velocity would equal that of light (what we now call a black hole). Around the same time, Cavendish evidently calculated the deflection of a light ray that grazes a body, using Newtonian gravity and the corpuscular theory of light, as we know from an unpublished scrap of paper found among his manuscripts (Thorpe, 1921). The first published calculation of the Newtonian deflection of light was that of Soldner in 1803 (see Jaki (1978) for a translation and commentary on Soldner's paper). In 1911, unaware of Soldner's work, Einstein used the principle of equivalence to derive the deflection of light, obtaining the same value.

In 1915, with the full equations of general relativity in hand, Einstein obtained double the Newtonian deflection. The reason is as follows: the classic derivations that use only the principle of equivalence or Newtonian theory yield the deflection of light relative to local straight lines, as defined, for example, by rigid rods. However, general relativity predicts that space is curved around the Sun, so local straight lines are bent relative to asymptotic straight lines far from the Sun by just enough to give double the deflection.

The first contribution to the deflection is universal: it is the same in any theory compatible with the equivalence principle; the second contribution, due to space curvature, varies from theory to theory. Within the PPN formalism, this variation is expressed by the PPN parameter γ. Specifically, a light ray which grazes the Sun is deflected by an amount

$$\delta\theta = \tfrac{1}{2}(1+\gamma)(4GM_\odot/dc^2)$$
$$= \tfrac{1}{2}(1+\gamma)1''.7.5, \tag{16}$$

where $M_\odot$ is the mass of the Sun, and d is the solar radius (see TEGP, section 7.1, for derivation). Thus measurements of the deflection of light can be viewed as measurements of the coefficient $\tfrac{1}{2}(1+\gamma)$.

The prediction of the full bending of light by the Sun was one of the great successes of Einstein's general relativity. Eddington's confirmation of the bending of optical starlight observed during a solar eclipse in the first days following World War I helped make Einstein famous. However, the experiments of Eddington and his coworkers had only 30 per cent accuracy, and succeeding experiments were not much better: the results were scattered between one half and twice the Einstein value, and the accuracies were low (for a review, see Bertotti *et al.*, 1962). The most recent optical measurement, during the solar eclipse of 30 June 1973, yielded a value $\tfrac{1}{2}(1+\gamma) = 0.95 \pm 0.11$.

However, the development of long-baseline radio interferometry has produced greatly improved determinations of the deflection of light. Long-baseline and very-long-baseline (VLBI) interferometric techniques have the capability in principle of measuring angular separations and changes in angles as small as 3×10^{-4} seconds of arc. Coupled with this technological advance is a series of heavenly coincidences: each year, groups of strong quasi stellar radio sources pass very close to the Sun (as seen from the Earth), including the group 3C273, 3C279, and 3C48, and the group 0111+02, 0119+11 and 0116+08. As the Earth moves in orbit, changing the lines of sight of the quasars relative to the Sun, the angular separation between pairs of quasars varies, and the variation can be measured accurately by the radio interferometer. A number of measurements of this kind over the period 1969–75 yielded an accurate determination of the coefficient $\frac{1}{2}(1+\gamma)$, which has the value unity in general relativity. Their results are shown in Fig. 5.3.

A recent series of transcontinental and intercontinental interferometric quasar observations made primarily to monitor the Earth's rotation was

Fig. 5.3. Results of radio-wave deflection measurements 1969–84 (for references, see TEGP). Horizontal scale at the top represents values of the PPN parameter combination $\frac{1}{2}(1+\gamma)$, whose value in general relativity is unity. Bottom scale gives the value of this combination for the corresponding value of the Brans–Dicke coupling constant ω. In the limit $\omega \to \infty$, Brans–Dicke theory becomes indistinguishable from general relativity.

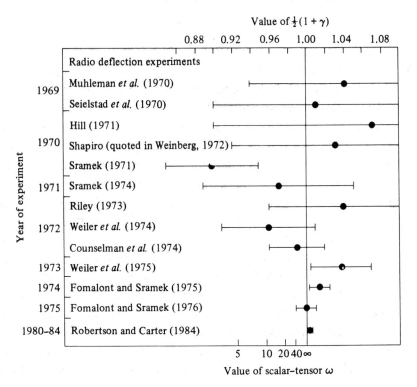

sensitive to the deflection of light over almost the entire celestial sphere (at 90° from the Sun, the deflection is still 4 milliarcseconds). The data yielded a value $\frac{1}{2}(1+\gamma)=1.004\pm0.003$, where the error is a formal standard error (Robertson and Carter, 1984).

One of the major sources of error in these experiments is the solar corona, which bends radio waves much more strongly than it bent the visible light rays which Eddington observed. Improvements in dual frequency techniques improved accuracies by allowing the coronal bending, which depends on the frequency of the wave, to be measured separately from the gravitational bending, which does not (for a review, see Fomalont and Sramek, 1977).

Measurements of optical light deflection using telescopes in space have recently come under serious study. Two proposals along this line are (i) Hipparcos, an astrometry satellite planned for launch in the late 1980s by the European Space Agency, designed to measure positions of 10^5 stars to 5 milliarcsecond accuracy (anticipated accuracy in γ of 10^{-3}); and (ii) POINTS (precision optical interferometry in space), a proposed orbiting optical interferometer with microarcsecond accuracy, capable in principle of detecting second-order or post-post-Newtonian effects in light deflection. For further details concerning the deflection of light, see TEGP, section 7.1, and UPDATE, section 4.1.

The discovery in 1979 of the 'double' quasar Q0957+561 and its subsequent interpretation as a multiple image of a single quasar caused by the gravitational lensing effect of an intervening galaxy made the deflection of light a useful tool in astrophysics and cosmology. The number and characteristics of images in such lensed systems can be used as a probe of the mass distribution of the lensing galaxy or cluster of galaxies.

Another important effect of gravity on the propagation of light was considered neither by Newton nor by Einstein. Instead it was discovered by Shapiro, in 1964. It is now called the Shapiro time delay, a retardation of light signals that pass near a massive body, such as the Sun (Shapiro, 1964). For instance, a radar signal sent across the solar system past the Sun to a planet or satellite and returned to the Earth suffers an additional non-Newtonian delay in its round-trip travel time, given by

$$\delta t \simeq \frac{1}{2}(1+\gamma)[240-20\ln(d^2/r)] \ \mu s, \qquad (17)$$

where d is the distance of closest approach of the ray in solar radii, and r is the distance of the planet or satellite from the Sun, in astronomical units (see TEGP, section 7.2, for detailed derivation and references).

In the two decades following Shapiro's discovery of this effect, several high-precision measurements were made using radar ranging to targets passing through superior conjunction. Since one does not have access to a 'Newtonian' signal against which to compare the round-trip travel time of the observed signal, it is necessary to do a differential measurement of the variations in round-trip travel times as the target passes through superior conjunction, and to look for the logarithmic behavior of (17). In order to do this accurately, however, one must take into account the variations in round-trip travel time due to the orbital motion of the target relative to the Earth. This is done by using radar-ranging (and possibly other) data on the target taken when it is far from superior conjunction (i.e., when the time-delay term is negligible) to determine an accurate ephemeris for the target, using the ephemeris to predict the PPN coordinate trajectory near superior conjunction, then combining that trajectory with the trajectory of the Earth to determine the Newtonian round-trip time and the logarithmic term in (17). The resulting predicted round-trip travel times in terms of the unknown coefficient $\frac{1}{2}(1+\gamma)$ are then fit to the measured travel times using the method of least squares, and an estimate obtained for $\frac{1}{2}(1+\gamma)$. The time delay is built into current implementations of post-Newtonian ephemerides (Section 5.4.1), and is solved for automatically.

Three types of target were used. The first type is a planet, such as Mercury or Venus, used as a passive reflector of the radar signals ('passive radar'). One of the major difficulties with this method is that the largely unknown planetary topography can introduce errors in round-trip travel times as much as $5\,\mu s$ (i.e., the subradar point could be a mountain-top or a valley) that thus introduce errors in both the planetary ephemeris and, more importantly, in the round-trip travel times at superior conjunction. Several sophisticated attempts were made to overcome this problem.

The second type of target is an artificial satellite, such as the Mariners 6 and 7, used as active retransmitters of the radar signals ('active radar'). Here topography is not an issue, and the on-board transponders permit accurate determination of the true range to the spacecraft. Unfortunately, spacecraft suffer random perturbing accelerations from a variety of sources, including random fluctuations in the forces caused by solar wind and solar radiation pressure, and random forces from on-board attitude-control devices. These random accelerations can cause the trajectory of the spacecraft near superior conjunction to differ by as much as $50\,m$ or $0.1\,\mu s$ from the predicted trajectory in an essentially unknown way. Special methods of analyzing the ranging data were devised to alleviate this problem.

The third target is the result of an attempt to combine the transponding capabilities of spacecraft with the imperturbable motions of planets by anchoring satellites to planets. Examples are the Mariner 9 Mars orbiter and the Viking Mars landers and orbiters.

In all of these cases, as in the radio-wave deflection measurements, the solar corona causes uncertainties because of its slowing down of the radar signal; again, dual frequency ranging helps reduce these errors. Still, the corona problem was a major factor limiting the accuracy of the most recent Viking time-delay measurements, despite the use of a dual-frequency downlink from the orbiters.

The results for the coefficient $\frac{1}{2}(1+\gamma)$ of all radar time-delay measurements performed to 1980 are shown in Fig. 5.4. Analyses of Viking data resulted in a 0.1 per cent measurement (Reasenberg *et al.*, 1979).

5.4.4 *Motion of stars in binary systems*

Newtonian gravitational theory is a standard tool in the analysis and understanding of binary and multiple star systems, yet there is no mention in the *Principia* of such systems. Newton restricted his attention to the solar

Fig. 5.4. Results of radar time-delay measurements 1968–79 (for references see TEGP).

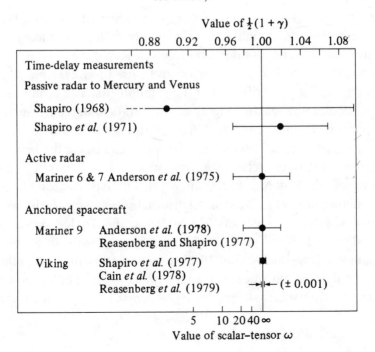

Value of $\frac{1}{2}(1+\gamma)$

Value of scalar–tensor ω

system. The reason is that the first 'double star' (conventionally referring to a pair of stars seen to be less than a few arcseconds apart) was discovered only around 1650 and, until around 1779, such occurrences were viewed merely as curiosities caused by the chance juxtaposition of stellar images. In 1767, Michell pointed out that the probability of such a random juxtaposition was sufficiently low for the closest pairs, that they may 'really consist of stars placed near together, and under the influence of some general law'. But in 1779, Mayer suggested the serious possibility of small suns revolving around larger ones, and Herschel began a systematic search for double stars. By 1803, Herschel had demonstrated conclusively that the changes in relative positions of some of the double star systems that he had followed over a 25 year period could only be accounted for if the two stars were 'intimately held together by the bonds of mutual attraction'. The motions were seen to correspond to elliptical orbits projected onto the plane of the sky, a confirmation of the Newtonian inverse square law. Between 1827 and 1832, the problem of deriving the orbit from observational data was solved by Savary, Encke, and Herschel. In 1889, the first of a new class of binary star systems, the spectroscopic binaries, was discovered. These are binary systems in which the two stars cannot be resolved telescopically, instead the varying Doppler shifts in one or both of the spectra are measured.†

Analysis of binary-system orbits provides information about the masses of the components, which are basic quantities for the theory of stellar structure and evolution. Observations of non-Keplerian apsidal advances in close binaries with tidal interactions can be used to infer something about the interior density distribution of the components. The evolution of close binaries containing massive highly-evolved stars undergoing mass transfer is important in understanding binary X-ray sources containing neutron stars and black holes. Newtonian celestial mechanics is the key tool in all these orbital analyses. Relativistic gravity has typically played a minor role.

However, there is one system in which relativistic celestial mechanics must be used instead of Newtonian celestial mechanics: the binary pulsar. Discovered in the summer of 1974 by Hulse and Taylor, it is a pulsar of nominal period 59 ms in a close binary system with an as yet unseen companion (Hulse and Taylor, 1975). From detailed analyses of the arrival times of pulses (which amount to an integrated version of the Doppler-shift methods used in spectroscopic binary systems) extremely accurate orbital

† For the history of binary stars, see Aitken (1963).

and physical parameters for the system have been obtained (see Table 5.6). Because the orbit is so close ($\approx 1R_\odot$) and because there is no evidence of an eclipse of the pulsar signal or of mass transfer from the companion, it is generally believed that the companion is compact: a white dwarf, a black hole, or (most likely) a second neutron star. Thus the orbital motion is very clean, free from tidal or other complicating effects. Furthermore, the data acquisition is 'clean' in the sense that the observers can keep track of the pulsar phase with an accuracy of $20\,\mu s$, despite gaps of up to six months between observing sessions. The pulsar has shown no evidence of 'glitches' in its pulse period.

Three factors make this system an arena where relativistic celestial mechanics must be used: the relatively large size of relativistic effects $[v_{orbit}/c \approx (GM/c^2R)^{1/2} \approx 10^{-3}]$; the short orbital period (8 hours), allowing secular effects to build up rapidly; and the cleanliness of the system, allowing

Table 5.6. *Measured parameters of the binary pulsar*[a]

Parameter	Symbol (units)	Value (data to Feb. 1985)
(i) 'Physical' parameters		
Right ascension	α	$19^h 13^m 12\overset{s}{.}468 \pm 0\overset{s}{.}001$
Declination	δ	$16°01'08''.16 \pm 0''.0.2$
Pulsar period	P_p (s)	$0.059\,029\,995\,2709 \pm 20$
Derivative of period	$\dot{P}_p$ (ss^{-1})	$(8.63 \pm 0.02) \times 10^{-18}$
(ii) 'Classical' parameters		
Projected semimajor axis	$a_p \sin i$ (light-sec)	$2.341\,85 \pm 0.000\,12$
Eccentricity	e	$0.617\,127 \pm 0.000\,003$
Orbital period	P_b (s)	$27\,906.981\,63 \pm 0.000\,02$
Longitude of periastron	ω_0 (deg)	178.8643 ± 0.0009
Julian ephemeris date of periastron and reference time for P_b and ω_0	T_0	$2\,442\,321.433\,2084 \pm 0.000\,0012$
(iii) 'Relativistic' parameters		
Mean rate of periastron advance	$\langle \dot{\omega} \rangle$ (deg yr^{-1})	4.2263 ± 0.0003
Gravitational redshift and time dilation	γ (s)	$0.004\,38 \pm 0.000\,12$
Orbital period derivative	$\dot{P}_b$ (ss^{-1})	$(-2.40 \pm 0.09) \times 10^{-12}$
Orbital inclination	$\sin i$	0.72 ± 0.05

[a] Data from Weisberg and Taylor (1984) and Haugan (1986a).

accurate determinations of small effects. Just as Newtonian gravity is used as a tool for measuring astrophysical parameters of ordinary binary systems, so general relativity is used as a tool for measuring astrophysical parameters in the binary pulsar.

The observational parameters that are obtained from a least-squares solution of the arrival time data fall into three groups: (i) non-orbital parameters, such as the pulsar period and its rate of change, and the position of the pulsar on the sky; (ii) five 'classical' parameters, most closely related to those appropriate for standard Newtonian systems, such as the eccentricity and the orbital period; and (iii) four 'relativistic' parameters. The four relativistic parameters are $\langle \dot{\omega} \rangle$, the average rate of periastron advance; γ, the amplitude of delays in arrival of pulses caused by the varying effects of the gravitational redshift and time dilation as the pulsar moves in its elliptical orbit at varying distances from the companion and with varying speeds; $\dot{P}_b$, the rate of change of orbital period; and $\sin i$, where i is the angle of inclination of the orbit relative to the plane of the sky.

In general relativity, these parameters can be related to the masses of the two bodies and to measured 'classical' parameters by the equations

$$\langle \dot{\omega} \rangle = 3G^{2/3}c^{-2}(P_b/2\pi)^{-5/3}(1-e^2)(m_p+m_c)^{2/3} \tag{18}$$

$$\gamma = G^{2/3}c^{-2}e(P_b/2\pi)^{1/3}m_c(m_p+2m_c)(m_p+m_c)^{-4/3}, \tag{19}$$

$$\dot{P}_b = -\frac{192\pi G^{5/3}}{5c^5}\left(\frac{P_b}{2\pi}\right)^{-5/3}(1-e^2)^{-7/2}(1+\tfrac{73}{24}e^2+\tfrac{37}{96}e^4)m_p m_c(m_p+m_c)^{-1/3}, \tag{20}$$

$$\sin i = G^{-1/3}c(a_p \sin i/m_c)(P_b/2\pi)^{-2/3}(m_p+m_c)^{2/3}, \tag{21}$$

where m_p and m_c denote the pulsar and companion masses, respectively. The formula for $\langle \dot{\omega} \rangle$ ignores possible non-relativistic contributions to the periastron shift, such as tidally or rotationally induced effects caused by the companion (for discussion of these effects, see TEGP, section 12.1(c)). The formula for $\dot{P}_b$ represents the effect of energy loss through the emission of gravitational radiation, and makes use of the 'quadrupole formula' of general relativity (for a survey of the quadrupole approximation for gravitational radiation, see Schutz and Will (1987)); it ignores other sources of energy loss, such as tidal dissipation (see TEGP, section 12.1(f)). The parameter $\sin i$ represents the effect of the Shapiro time delay of the pulsar signal as it passes the companion.

The current timing model that contains these parameters was developed by Haugan from earlier treatments by Blandford and Teukolsky and by

Epstein (see Haugan, 1985, for review and references). An alternative timing model, using a slightly different set of relativistic parameters, devised by Damour and Deruelle (1986), has also been used in the data analysis, with equivalent results. The values shown in Table 5.6 are from data taken through February 1985 (Haugan, 1986a); results through August 1983 were presented by Weisberg and Taylor (1984).

The most convenient way to display these results is to plot the constraints they imply for the two masses m_p and m_c. These are shown in Fig. 5.5. From $\langle \dot{\omega} \rangle$ and γ we obtain the values $m_p = (1.42 \pm 0.03)m_\odot$ and $m_c = (1.40 \pm 0.03)m_\odot$. Equations (20) and (21) then predict values for the remaining parameters $\dot{P}_b = -2.403 \pm 0.002 \times 10^{-12}$ and $\sin i = 0.72 \pm 0.03$, in complete agreement with the measured values in Table 5.6. This consistency is also displayed in Fig. 5.5, in which the regions allowed by the four measured constraints have a single common overlap. This consistency provides a test of the assumption that the two bodies behave as 'point' masses, without complicated tidal effects.

Fig. 5.5. Curves showing constraints on the mass of pulsar and its companion provided by measured values and estimated errors of the parameters $\langle \dot{\omega} \rangle$, γ, $\dot{P}_b$, and $\sin i$. Uncertainty in $\langle \dot{\omega} \rangle$ is less than width of sloping straight line. All four constraints overlap in the region near 1.4 solar masses for each body.

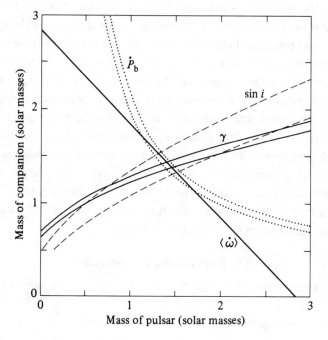

One of the most important results of the analysis of the binary pulsar is the confirmation of the existence of gravitational radiation through the loss of orbital energy. Although it is normally stated that Newtonian gravity does not admit gravitational radiation because the interaction is instantaneous, this was not necessarily Newton's intent. In the gravitational calculations in the *Principia* the interaction was assumed to be instantaneous as a necessary approximation, while in other writings Newton made it clear that he had in mind a propagating action, in which the ether between two moving, gravitating bodies might 'begin to rarify'.† In an unsuccessful attempt to explain the secular acceleration of the Moon, Laplace assumed that the speed of gravitational propagation was seven million times that of light. Nevertheless, the simplest mathematical formulation of Newtonian gravity assumes instantaneous gravitational interactions. On the other hand, gravitational radiation is a necessary outcome of general relativity, since the theory is compatible with special relativity, and with a limiting speed for all interactions.

By the same token, alternative metric theories of gravity, because of their compatibility with Lorentz invariance at some level, also predict gravitational radiation. However, most such theories predict, in addition to a contribution from 'quadrupole' gravitational radiation analogous to that of general relativity, a contribution of 'dipole' gravitational radiation. In such cases, the amount of orbital period decay is expected to be in significant disagreement with the constraints imposed on the masses by the other three relativistic parameters. This difficulty dealt a mortal blow to the 'Rosen bimetric theory', for example (see TEGP, section 12.3, for discussion).

The binary pulsar has yielded a remarkably rich test of gravitation theory. confirming Newtonian and general relativistic celestial mechanics, and the existence of gravitational radiation. It has also demonstrated a role for general relativity in the determination of astrophysical parameters, such as the mass of a neutron star. The first announcement of the observation of the orbital period decrease came in late 1978, just at the beginning of the Einstein centenary year, and toward the end of the 'decades for testing general relativity'. It is now appropriate to summarize and to attempt to look to the future of experimental gravitation.

5.5 Beyond the tercentenary

On the occasion of the tercentenary of Newton's *Principia*, we find Einstein's

---

† Letter from Newton to Boyle, quoted by Cajori in Appendix to the *Principia*.

general relativity in a situation not unlike that of Newtonian theory around the sesquicentenary. In 1837, Newtonian gravity was one of the most successful theories of nature ever formulated. It passed every test with flying colors, explaining the detailed motions of the planets and comets, including a host of perturbing effects. It accounted for the motion of the Moon in detail; any unresolved discrepancies simply awaited the advent of more sophisticated approximation methods. Its realm of validity was extended far beyond the solar system, with its successful application to binary star systems. Within a decade, its crowning achievement, the discovery of Neptune, would be accomplished.

However, a change was soon in the air. In 1843, the quest for higher observational and calculational accuracy led Le Verrier to discover the problem of Mercury's perihelion, although he did not announce it publicly for another 16 years. In addition, a revolution was underway in another area of physics. The new understanding of electromagnetism brought about by Faraday and Maxwell between 1830 and 1870 led to the new world view of special relativity, and Einstein's quest for the unification of gravity with special relativity led to general relativity and to the overthrow of Newtonian gravity.

In 1987, general relativity is also viewed as one of the most successful theories of all time. It, too, has passed every experimental test. In a similar quest for higher accuracy, theorists and experimentalists are contemplating and designing the next generation of experiments to look for possible deviations from general relativity, although we see no indication of serious deviations on the horizon. A similar revolution is underway in elementary particle physics, and the new understanding of gauge theories of the non-gravitational interactions has spawned the current quest for a unification of all the interactions.

Will these modern developments lead to an overthrow of general relativity and to new gravitational experiments? Most observers believe that the current attempts at unification, if successful, will not lead to observable effects at the level of macroscopic physics or at the energies accessible for the foreseeable future. This superunification, unlike that of Einstein, will not overthrow the standard, macroscopic, or 'low energy' version of general relativity. Instead, any modifications are expected to occur at the 'Planck energy' appropriate to the very early universe, or at singularities inside black holes. Consequently, gravitation experimenters of the coming decades must content themselves with testing standard general relativity, in new regimes and to higher orders perhaps, but with little reason to anticipate a failure of Einstein's theory.

On the other hand, although physics is normally concerned with making predictions about the behavior of systems, physicists are notoriously poor at predicting the future of their own subject. So even after 300 years of experimental gravitation, who is to say that there will not be surprises around the corner?

Acknowledgement

This work was supported in part by the National Science Foundation [PHY85-13953].

Appendix: The parametrized post-Newtonian formalism

The comparison of metric theories of gravity with each other and with experiment becomes particularly simple when one takes the slow-motion, weak-field limit. This approximation, known as the post-Newtonian limit, is sufficiently accurate to encompass most solar-system tests that can be performed in the foreseeable future. It turns out that, in this limit, the spacetime metric $g_{\mu\nu}$ predicted by nearly every metric theory of gravity has the same structure. It can be written as an expansion about the Minkowski metric $[\eta_{\mu\nu} = \text{diag}(-1, 1, 1, 1)]$ in terms of dimensionless gravitational potentials of varying degrees of smallness. These potentials are constructed from the matter variables (Table 5.7) and from the Newtonian gravitational potential

$$U(\mathbf{x}, t) \equiv \int \rho(x', t)|\mathbf{x} - \mathbf{x}'|^{-1} \, \mathrm{d}^3 x'.$$

The 'order of smallness' is determined according to the rules $U \sim v^2 \sim \Pi \sim p/\rho \sim O(2)$, $v^i \sim |\mathrm{d}/\mathrm{d}t|/|\mathrm{d}/\mathrm{d}x| \sim O(1)$, and so on. A consistent post-Newtonian limit requires determination of g_{00} correct through $O(4)$, g_{0i} through $O(3)$ and g_{ij} through $O(2)$ (for details see TEGP, section 4.1). The only way that one metric theory differs from another is in the numerical values of the coefficients that appear in front of the metric potentials. The parametrized post-Newtonian (PPN) formalism inserts parameters in place of these coefficients, parameters whose values depend on the theory under study. In the standard version of the PPN formalism, summarized in Table 5.7, ten parameters are used, chosen in such a manner that they measure or indicate general properties of metric theories of gravity (Table 5.8). The parameters γ and β are the usual Eddington–Robertson–Schiff parameters used to

Table 5.7. *The parametrized post-Newtonian formalism*

A. Coordinate system:

The framework uses a nearly globally Lorentz coordinate system in which the coordinates are (t, x^1, x^2, x^3). Three-dimensional, Euclidean vector notation is used throughout. All coordinate arbitrariness ('gauge freedom') has been removed by specialization of the coordinates to the standard PPN gauge (TEGP, section 4.2). Units are chosen so that $G = c = 1$.

B. Matter variables:

1. ρ = density of rest mass as measured in a local freely falling frame momentarily comoving with the gravitating matter.

2. $v^i = (dx^i/dt)$ = coordinate velocity of the matter.

3. w^i = coordinate velocity of PPN coordinate system relative to the mean rest-frame of the universe.

4. p = pressure as measured in a local freely falling frame momentarily comoving with the matter.

5. Π = internal energy per unit rest mass. It includes all forms of non-rest-mass, non-gravitational energy – e.g., energy of compression and thermal energy.

C. PPN parameters:

$\gamma, \beta, \xi, \alpha_1, \alpha_2, \alpha_3, \zeta_1, \zeta_2, \zeta_3, \zeta_4$.

D. Metric:

$$g_{00} = -1 + 2U - 2\beta U^2 - 2\xi\Phi_w + (2\gamma + 2 + \alpha_3 + \zeta_1 - 2\xi)\Phi_1 +$$
$$2(3\gamma - 2\beta + 1 + \zeta_2 + \xi)\Phi_2 + 2(1+\zeta_3)\Phi_3 + 2(3\gamma + 3\zeta_4 - 2\xi)\Phi_4 - (\zeta_1 - 2\xi)A -$$
$$(\alpha_1 - \alpha_2 - \alpha_3)w^2 U - \alpha_2 w^i w^j U_{ij} + (2\alpha_3 - \alpha_1)w^i V_i$$
$$g_{0i} = -\tfrac{1}{2}(4\gamma + 3 + \alpha_1 - \alpha_2 + \zeta_1 - 2\xi)V_i - \tfrac{1}{2}(1 + \alpha_2 - \zeta_1 + 2\xi)W_i - \tfrac{1}{2}(\alpha_1 - 2\alpha_2)w^i U -$$
$$\alpha_2 w^j U_{ij}$$
$$g_{ij} = (1 + 2\gamma U)\delta_{ij}$$

E. Metric potentials:

$$U = \int \frac{\rho'}{|\mathbf{x} - \mathbf{x}'|} d^3x', \qquad U_{ij} = \int \frac{\rho'(x - x')_i(x - x')_j}{|\mathbf{x} - \mathbf{x}'|^3} d^3x'$$

$$\Phi_w = \int \frac{\rho'\rho''(\mathbf{x} - \mathbf{x}')}{|\mathbf{x} - \mathbf{x}'|^3} \cdot \left(\frac{\mathbf{x}' - \mathbf{x}''}{|\mathbf{x} - \mathbf{x}''|} - \frac{\mathbf{x} - \mathbf{x}''}{|\mathbf{x}' - \mathbf{x}''|} \right) d^3x' d^3x''$$

$$A = \int \frac{\rho'[\mathbf{v}' \cdot (\mathbf{x} - \mathbf{x}')]^2}{|\mathbf{x} - \mathbf{x}'|^3} d^3x'$$

$$\Phi_1 = \int \frac{\rho'v'^2}{|\mathbf{x} - \mathbf{x}'|} d^3x', \qquad \Phi_2 = \int \frac{\rho'U'}{|\mathbf{x} - \mathbf{x}'|} d^3x'$$

continued

Table 5.7 (cont.)

E. Metric potentials:

$$\Phi_3 = \int \frac{\rho' \Pi'}{|\mathbf{x} - \mathbf{x}'|} \, d^3x', \qquad \Phi_4 = \int \frac{p'}{|\mathbf{x} - \mathbf{x}'|} \, d^3x'$$

$$V_i = \int \frac{\rho' v_i'}{|\mathbf{x} - \mathbf{x}'|} \, d^3x', \qquad W_i = \int \frac{\rho'[\mathbf{v}' \cdot (\mathbf{x} - \mathbf{x}')](x - x')_i}{|\mathbf{x} - \mathbf{x}'|^3} \, d^3x'$$

F. Stress-energy tensor (perfect fluid)

$$T^{00} = \rho(1 + \Pi + v^2 + 2U)$$
$$T^{0i} = \rho(1 + \Pi + v^2 + 2U + p/\rho)v^i$$
$$T_{ij} = \rho v^i v^j(1 + \Pi + v^2 + 2U + p/\rho) + p\delta^{ij}(1 - 2\gamma U)$$

G. Equations of motion

1. Stressed matter, $T^{\mu\nu}_{\ ;\nu} = 0$
2. Test bodies, $d^2 x^\mu / d\lambda^2 + \Gamma^\mu_{\ \nu\lambda}(dx^\nu/d\lambda)(dx^\lambda/d\lambda) = 0$
3. Maxwell's equations, $F^{\mu\nu}_{\ ;\nu} = 4\pi J^\mu, \quad F_{\mu\nu} = A_{\nu;\mu} - A_{\mu;\nu}$

Table 5.8. The PPN parameters and their significance[a]

Parameter	What it measures relative to general relativity	Value in general relativity	Value in semi-conservative theories	Value in fully-conservative theories
γ	How much space curvature is produced by unit rest-mass?	1	γ	γ
β	How much 'non-linearity' is there in the superposition law for gravity?	1	β	β
ξ	Are there preferred-location effects?	0	ξ	ξ
α_1		0	α_1	0
α_2	Are there preferred-frame effects?	0	α_2	0
α_3		0	0	0
ζ_1		0	0	0
ζ_2	Is there violation of conservation of total momentum?	0	0	0
ζ_3		0	0	0
ζ_4		0	0	0

[a] For a compendium of PPN parameter values in alternative theories together with derivations, see TEGP, chapter 5.

describe the 'classical' tests of general relativity; ξ is non-zero in any theory of gravity that predicts preferred-location effects such as a galaxy-induced anisotropy in the local gravitational constant G_L (also called 'Whitehead' effects); α_1, α_2, α_3 measure whether or not the theory predicts post-Newtonian preferred-frame effects (TEGP, section 4.3); α_3, ζ_1, ζ_2, ζ_3, ζ_4 measure whether or not the theory predicts violations of global conservation laws for total momentum (TEGP, section 4.4). In Table 5.8, we show the values these parameters take (i) in general relativity, (ii) in any theory of gravity that possesses conservation laws for total momentum, called 'semiconservative' (any theory that is based on an invariant action principle is semiconservative), and (iii) in any theory that in addition possesses six global conservation laws for angular momentum, called 'fully conservative' (such theories automatically predict no post-Newtonian preferred-frame effects). Semiconservative theories have five free PPN parameters ($\gamma, \beta, \xi, \alpha_1$, α_2) while fully conservative theories have three (γ, β, ξ). In Brans–Dicke theory, $\gamma = (1 + \omega)/(2 + \omega)$, where ω is the adjustable 'coupling constant' of the theory, while the other PPN parameters are the same as in general relativity.

References

Aitken, R. G. (1963). *The Binary Stars*. Dover: New York.

Bartlett, D. F. and van Buren, D. (1986). *Physical Rev. Lett.*, **57**, 21.

Bertotti, B., Brill, D. R. and Krotkov, R. (1962). In *Gravitation: An Introduction to Current Research*, ed. L. Witten, pp. 1–48. Wiley: New York.

Cajori, F. (1928). In *Sir Isaac Newton 1727–1927*, pp. 127–88. Waverley Press: Baltimore.

Cajori, F. (1934). *Sir Isaac Newton's Mathematical Principles of Natural Philosophy and his System of the World*. University of California: Berkeley.

Campbell, L., McDow, J. C., Moffat, J. W. and Vincent, D. (1983). *Nature*, **305**, 508.

Damour, T. and Deruelle, N. (1986). *Annales de l'Institut Henri Poincare*, **44**, 263.

De Rujula, A. (1986). *Nature*, **323**, 760.

Dicke, R. H., Kuhn, J. R. and Libbrecht, K. G. (1985). *Nature*, **316**, 687.

Duvall, T. L., Jr., Dziembowski, W. A., Goode, P. R., Gough, D. O., Harvey, J. W. and Liebacher, J. W. (1984). *Nature*, **310**, 22.

Fischbach, E., Sudarsky, D., Szafer, A., Talmadge, C. and Aronson, S. H. (1986a). *Phys. Rev. Lett.*, **56**, 3.

Fischbach, E., Sudarsky, D., Szafer, A., Talmadge, C. and Aronson, S. H. (1986b). *Phys. Rev. Lett.*, **56**, 2424, 2426.

Fomalont, E. B. and Sramek, R. A. (1975). *Astrophys. J.*, **199**, 749–55.

Fomalont, E. B. and Sramek, R. A. (1976). *Phys. Rev. Lett.*, **36**, 1475–8.

Fomalont, E. B. and Sramek, R. A. (1977). *Comments Astrophys.*, **7**, 19.

Futamase, T. and Schutz, B. F. (1985). *Phys. Rev.*, **D32**, 2557.

Gillies, G. T. (1982). *The Newtonian Gravitational Constant: An Index of Measurements.* Bureau International des Poids et Mesures: Sevres.

Gough, D. O. (1982). *Nature*, **298**, 334.

Halpern, L., ed. (1978). *On the Measurement of Cosmological Variations of the Gravitational Constant.* University Presses of Florida: Gainesville.

Haugan, M. P. (1985). *Astrophys. J.*, **296**, 1.

Haugan, M. P. (1986a). In *Fourth Marcel Grossman Meeting on General Relativity*, ed. R. Ruffini. North Holland: Amsterdam.

Haughan, M. P. (1986b), in preparation.

Hellings, R. W. (1984). In *General Relativity and Gravitation*, ed. B. Bertotti, F. de Felice and A. Pascolini, p. 365. Reidel: Dordrecht.

Hill, H. A., Bos, R. J. and Goode, P. R. (1982). *Phys. Rev. Lett.*, **49**, 1794.

Hill, H. A., Rabaey, G. R. and Rosenwald, R. D. (1986). In *Relativity in Celestial Mechanics and Astrometry*, ed. J. Kovalevsky and V. A. Brumberg, pp. 345–54. Reidel: Dordrecht.

Holding, S. C., Stacey, F. D. and Tuck, G. J. (1986). *Phys. Rev.*, **D33**, 3487.

Hoskins, J. K., Newman, R. D., Spero, R. and Schultz, J. (1985). *Phys. Rev.*, **D32**, 3084.

Hulse, R. A. and Taylor, J. H. (1975). *Astrophys. J. Lett.*, **195**, L51.

Jaki, S. L. (1978). *Found. Phys.*, **8**, 927.

Keyser, P. T., Niebauer, T. and Faller, J. E. (1986). *Phys. Rev. Lett.*, **56**, 2425.

Kreuzer, L. B. (1968). *Phys. Rev.*, **169**, 1007.

Lamoreaux, S. K., Jacobs, J. P., Heckel, B. R., Raab, F. J. and Fortson, E. N. (1986). *Phys. Rev. Lett.*, **57**, 3125.

Mackenzie, A. S. (1900). *The Laws of Gravitation.* American Book Co.: New York.

Moulton, F. R. (1914). *Celestial Mechanics*, 2nd edn. Macmillan: New York.

Neufield, D. A. (1986). *Phys. Rev. Lett.*, **56**, 2344.

Newman, R. D. (1983). In *Proceedings of the Third Marcel Grossman Meeting on General Relativity*, ed. H. Ning. North-Holland: Amsterdam.

Nordtvedt, K., Jr. (1968). *Phys. Rev.*, **169**, 1017.

Nussinov, S. (1986). *Phys. Rev. Lett.*, **56**, 2350.

Prestage, J. D., Bollinger, J. J., Itano, W. M. and Wineland, D. J. (1985). *Phys. Rev. Lett.*, **54**, 2387.

Reasenberg, R. D. (1983). *Phil. Trans. Roy. Soc.* (*London*), **310**, 227.

Reasenberg, R. D., Shapiro, I. I., MacNeil, P. E., Goldstein, R. B., Breidenthal, J. C., Brenkle, J. P., Cain, D. L., Kaufman, T. M., Komarek, T. A. and Zygielbaum, A. I. (1979). *Astrophys. J. Lett.*, **234**, L219.

Robertson, D. S. and Carter, W. E. (1984). *Nature*, **310**, 572.

Roseveare, N. T. (1982). *Mercury's Perihelion from Le Verrier to Einstein.* Oxford University Press: Oxford.

Schutz, B. F. and Will, C. M. (1987). In preparation.

Shapiro, I. I. (1964). *Phys. Rev. Lett.*, **13**, 789.

Shapiro, I. I., Counselman, C. C. III and King, R. W. (1976). *Phys. Rev. Lett.*, **36**, 555.

Stacey, F. D., Tuck, G. J., Moore, G. I., Holding, S .C., Goodwin, B. D. and Zhou, R. (1986). *Rev. Mod. Phys.*, in press.

Thieberger, P. (1986). *Phys. Rev. Lett.*, **56**, 2347.

Thodberg, H. H. (1986). *Phys. Rev. Lett.*, **56**, 2423.

Thorpe, E., ed. (1921). *The Scientific Papers of the Honourable Henry Cavendish, F.R.S.*, Vol. II, p. 437. Cambridge University Press: Cambridge.

Turneaure, J. P., Will, C. M., Farrell, B. F., Mattison, E. M. and Vessot, R. F. C. (1983). *Phys. Rev.*, **D27**, 1705.

Vessot, R. F. C., Levine, M. W., Mattison, E. M., Blomberg, E. L., Hoffman, T. E., Nystrom, G. U., Farrell, B. F., Decher, R., Eby, P. B., Baugher, C. R., Watts, J. W., Teuber, D. L. and Wills, F. O. (1980). *Phys. Rev. Lett.*, **45**, 2081.

Weisberg, J. M. and Taylor, J. H. (1984). *Phys. Rev. Lett.*, **52**, 1348.

Wesson, P. S. (1978). *Cosmology and Geophysics*. Adam Hilger: Bristol.

Will, C. M. (1981). *Theory and Experiment in Gravitational Physics*. Cambridge University Press: Cambridge.

Will, C. M. (1984). *Phys. Rep.*, **113**, 345.

Will, C. M. (1986). *Was Einstein Right?* Basic Books: New York.

Williams, J. G., Dicke, R. H., Bender, P. L., Alley, C. O., Carter, W. E., Currie, D. G., Eckhardt, D. H., Faller, J. E., Kaula, W. M., Mulholland, J. D., Plotkin, H. H., Poultney, S. K., Shelus, P. J., Silverberg, E. C., Sinclair, W. S., Slade, M. A. and Wilkinson, D. T. (1976). *Phys. Rev. Lett.*, **36**, 551.

Winicour, J. (1983). *J. Math. Phys.*, **25**, 1193.

6

The problem of motion in Newtonian and Einsteinian gravity

THIBAULT DAMOUR

6.1 Introduction

The problem of motion, i.e. the problem of describing the dynamics of N gravitationally interacting extended bodies, is the cardinal problem of any theory of gravity. From the publication of Newton's *Principia* to the beginning of the twentieth century this problem has been thoroughly investigated within the framework of Newton's dynamics and theory of gravity. This has led to the formulation of many concepts and theoretical tools which have been applied to other fields in physics. For instance, the early development of quantum mechanics has been greatly facilitated by the existence of several structures of classical mechanics: canonical formalism, the Hamilton–Jacobi equation, action-angle variables, the eigenvalue problem of secular perturbations of the solar system .... In fact, quantum mechanics was successfully modelled on the formal structures of classical mechanics.

In contrast, Einstein's theory of gravity has developed within a conceptual and mathematical framework which was completely alien to the Newtonian framework, as is indicated by the name given to it by its founder: the general theory of relativity ('Die allgemeine Relativitätstheorie', Einstein, 1916a). As a consequence, the relationship between Einstein's and Newton's theories of gravity has been, and still is, very peculiar. On the one hand, from the technical point of view, the existence of Newton's theory facilitated the early development of Einstein's theory by suggesting an approximation method (called 'post-Newtonian') which allowed the theorists to draw quickly some observational consequences of general relativity. Indeed, this post-Newtonian approximation method, developed by Einstein, Droste and De Sitter within one year of the discovery of general relativity, has led to the predictions of: the relativistic advance of the perihelion of planets, the gravitational redshift, the deflection of light, the relativistic precession of the Moon's orbit, ... (see the survey of De Sitter, 1916). On the other hand, as emphasised recently by Eisenstaedt (1986, section 6), the use of this post-Newtonian approximation method has had, from a conceptual point of view, the adverse side-effect of introducing implicitly a 'neo-Newtonian' interpretation of general relativity. Indeed, technically this approximation method follows the Newtonian way of tackling gravitational problems as closely as possible. But this technical reduction of Einstein's theory into the Procrustean bed of Newton's theory surreptitiously entails a corresponding conceptual reduction: the Einsteinian problem of motion being, in fact, conceived within the Newtonian framework of an absolute (coordinate) space and an absolute (coordinate) time. However, some recent developments oblige us to reconsider the problem of motion within Einstein's theory.

On the one hand, the discovery of the binary pulsar PSR 1913 + 16 by Hulse and Taylor in 1974, and its continuous observation by Taylor and coworkers (see references in Weisberg and Taylor, 1984), has led to an impressively accurate tracking of the absolute orbital motion of a neutron star in a binary system. Now, a neutron star is so condensed that it creates a very strong self-gravitational field, so the study of the motion of such a strongly self-gravitating body poses a new theoretical problem that cannot be treated by the usual post-Newtonian methods which assume that the gravitational field is everywhere weak. This has obliged the theorists to develop new methods for studying the motion of 'condensed bodies' which draw directly on the physical and mathematical richness of Einstein's theory.

On the other hand, the extreme accuracy of modern observations of the motion of the Earth, the Moon and the planets obliges the theorists to reconsider the traditional (post-Newtonian) way of tackling the relativistic problem of motion. Indeed, although in principle the post-Newtonian methods can be used in the solar system because the gravitational field is everywhere weak and the velocities always small, the way they have been developed is somewhat unsatisfactory, or at least incomplete.

All this means that it is worth looking back in detail at the foundations of the problem of motion. The most natural way to do so is to begin by discussing the Newtonian way of tackling the problem of motion which will be the subject of the next six sections (Sections 6.2–6.7). Our principal aim in doing so will be to clarify the basic *ideas* which allow one to put the Newtonian problem of motion into a manageable form. In doing so we shall go into the details of the way these ideas are implemented in the Newtonian framework, some of which will seem fairly trivial to the reader; but we think it is important to go through them for the following reason: precisely because these details are technically trivial (or too well known), one is liable to miss the ideas which underlie them, so that, when tackling the corresponding relativistic problem, one has the tendency to *transplant*, within Einstein's mathematical framework, exactly the *same technical developments*, instead of trying to *transmute*, within Einstein's conceptual framework, the *ideas* underlying them.

In Section 6.2 we pose the Newtonian problem of the motion of N extended bodies in a general form. In Section 6.3 we discuss the splitting of this problem into two sub-problems: the external problem (the motion of a body as a whole), and the internal problem (the intrinsic motion of each body). This splitting is useful because the two sub-problems are only weakly coupled. Very fundamental facts underlie the remarkable weakness of this coupling, and these are discussed in Sections 6.4 and 6.5. In Section 6.4 we study one aspect of this weak coupling, namely the remarkable smallness of the influence of the internal structure on the 'external' motion. Reviving the terminology introduced by M. Brillouin (1922) and T. Levi-Civita (1937a), we shall speak of an 'effacement' of the internal structure in respect of the external problem. In Section 6.5 we study the other aspect, namely the remarkable smallness of the influence of the external motion on the internal one. Extending the use of Brillouin's and Levi-Civita's terminology, we shall speak of an 'effacement' of the external structure in the internal problem. These 'effacement' properties are closely linked with the so-called 'principle of equivalence' (in a sense they can be regarded as precise formulations of some aspects of the principle of equivalence, the latter often being expressed either in a vague manner or in a precise but restricted one). While preparing this article I went back to read those sections of the *Principia* in which Newton discusses the equivalence of inertial and gravitational mass. Newton was fully aware of the remarkable character of this equivalence, and he gives as many arguments as possible in its favour. In particular I found, much to my surprise, that Newton gives a very interesting argument, about the relative motion of celestial bodies, which anticipates ideas recently put forward by Nordtvedt (1968) within the framework of relativistic theories of gravity (see Section 6.6). We conclude our discussion of the Newtonian problem of motion by outlining the way the two sub-problems can be effectively solved (Section 6.7).

In the second part of this article (Sections 6.8–6.15) we discuss the

Einsteinian problem of motion. In Section 6.8 we pose this problem in a general form. In Section 6.9 we discuss the dimensionless parameters of the relativistic problem which can be used to concoct various approximation methods. In particular, we discuss the post-Newtonian methods (Section 6.10), the post-Minkowskian ones (Section 6.11) and some classes of 'singular perturbation' methods which have been found useful in general relativity (Section 6.12). In Section 6.13 we survey the various approaches which have been taken to try to 'transplant' or to 'transmute', within Einstein's theory, the classical splitting into external and internal problems. In particular, we discuss the problem of the definition of a 'centre of mass' (Section 6.13.1). We then give references to the explicit results which have been obtained for the external problem (Section 6.13.2), and we conclude by discussing the relativistic internal problem which is still poorly understood (Section 6.13.3). In Section 6.14 we indicate what is known about the effacement of internal structure in the relativistic external problem. For pedagogical reasons we give a detailed treatment of another derivation of the Newtonian effacement property (already discussed in Section 6.4). Indeed, this new derivation allows one to grasp the physical and mathematical mechanisms at work in the Einstein–Infeld–Hoffmann approach, simplified by the use of complex analytic continuation, which has been recently used to prove the strongest existing result of 'effacement' within Einstein's theory. Finally, in Section 6.15, we present the recent progress which has been achieved in the relativistic two-body problem, and we show how the secular acceleration effect observed in the binary pulsar can be considered as a relativistic descendant of an effect conceived by Laplace within a modification of the Newtonian framework. A brief conclusion ends this article (Section 6.16).

As our presentation of the problem of motion within Einstein's theory will necessarily be incomplete, and will probably also be biased towards the author's point of view, the reader is urged to consult other more complete, or differently biased, reviews: e.g. Ehlers (1977, 1980), the proceedings of the 67th Enrico Fermi School (Ehlers, 1979) (especially the reviews of D'Eath, Dixon and Havas), Walker and Will (1980), Will (1981), Cooperstock and Hobill (1982), the proceedings of the 1982 Les Houches-Nato Advanced Study Institute on Gravitational Radiation (Deruelle and Piran, 1983) (especially the reviews of Thorne, and Damour, and the round table on the equations of motion reported by Ashtekar), the proceedings of the 10th international conference on General Relativity and Gravitation (Bertotti *et al.*, 1984) (especially the reviews of Damour, and Walker, and the workshop report of Ehlers and Walker), Schutz (1984, 1985), Will (1986), and the proceedings of the 114th IAU symposium (Kovalevsky and Brumberg, 1986) (in particular see Brumberg, and Grishchuk and Kopejkin).

6.2 The N-extended-body problem in Newtonian gravity

The problem we have in mind is to describe, as accurately as possible, the dynamics of a system of separate extended bodies, each one having finite dimensions and being endowed with some internal structure governed by some equations of state. We shall first assume that it is possible to deal with a *finite* system, say the solar system, a star cluster, or a galaxy, and to consider it as isolated from the rest of the universe. This assumption will have to be justified or qualified later, using, on a higher hierarchical level, the same tools as the ones we are going to review concerning the simplification of the description of a given finite system: for instance, we would justify the possibility of considering the solar system as isolated by arguing that the stars of our Galaxy will cause only negligible tidal forces in the solar system, thereby leaving its internal dynamics unaffected, although it could certainly affect the overall motion of the solar system. However, such hierarchical reasoning leaves the issue of the global justification of our assumption unanswered, but this should not worry us too much because such an assumption is made in nearly all other scientific investigations, and might even be indispensable for any 'falsifiability' of predictions (in Popper's sense).

Having thereby restricted our attention to the dynamics of a finite system of bodies, let us further assume that our bodies are made of some perfect fluid with a given (isentropic) equation of state linking the pressure, p, to the mass density ρ:

$$p = p(\rho). \tag{1}$$

However, this assumption is made only to simplify our presentation and should be replaced, in each case of interest, by the actual equations of state governing the internal structure of the constituent bodies (for instance the Hooke equations for an elastic solid). Then the equations describing the full Newtonian dynamics of the system are: the equation of continuity

$$\frac{\partial \rho}{\partial t} + \frac{\partial(\rho v^i)}{\partial x^i} = 0, \tag{2}$$

where $v^i(\hat{x}, t)$ is the velocity field (in cartesian coordinates x^i; $i, j, \ldots = 1, 2, 3$);

the Euler equations for the local fluid motion

$$\rho\left(\frac{\partial v^i}{\partial t} + v^j \frac{\partial v^i}{\partial x^j}\right) = -\frac{\partial p}{\partial x^i} + \rho \frac{\partial U}{\partial x^i},\tag{3}$$

where $U(x^i, t)$ is the (positive) gravitational potential; and the Poisson equation

$$\Delta U = -4\pi G\rho,\tag{4a}$$

where G denotes Newton's constant $(G=(6.672\pm0.004)\times 10^{-8}\ \mathrm{cm}^3\ \mathrm{g}^{-1}\ \mathrm{s}^{-2})$. The former assumption of 'isolation' of the finite system under consideration can be interpreted as meaning that the gravitational potential U should fall off outside the system,

$$\lim_{\substack{|\check{x}| \to \infty \\ t=\mathrm{const.}}} (U(\check{x}, t))=0\tag{4b}$$

(where $|\check{x}|$ is the Euclidean norm of $\check{x}$). Then the only acceptable solution of Poisson's equation is:

$$U(\check{x}, t)= G \int \frac{\rho(\check{x}', t)}{|\check{x}-\check{x}'|}\, \mathrm{d}^3x',\tag{5}$$

where $|\check{x}-\check{x}'|$ denotes the Euclidean distance between the field point $\check{x}$ and the source point $\check{x}'$, and d^3x' denotes the Euclidean volume element in cartesian coordinates, $\mathrm{d}x'^1\,\mathrm{d}x'^2\,\mathrm{d}x'^3$.

Now the dynamics of the system are described by the evolution in time of $\rho(\check{x}, t)$ and $\check{v}(\check{x}, t)$, the evolution of the latter variables being governed by eqs. (2) and (3) in which p and U must be replaced by their expressions in terms of ρ, as given by eqs. (1) and (5). Moreover, the mass density $\rho(\check{x}, t)$ is further constrained to have a compact support consisting of N non-overlapping connected components ('N-extended-body problem'). From the mathematical point of view the latter constraint greatly complicates the problem of the evolution of the system (2)–(3). In fact, as far as I know, there exist no theorems which assure the existence of solutions to this system. (Even in the absence of gravitational interaction the fact that ρ takes the value zero outside the bodies creates serious problems when trying to prove the existence of solutions.) From the physical point of view, what complicates the problem of controlling the evolution of the system (2)–(3) is that it is both non-linear and non-local. One therefore needs to resort to approximation methods.

6.3 The external and the internal problems of motion (Newtonian case)

Tisserand (1960) in his famous *Traité de Mécanique Céleste* (vol. I, pp. 51–2) clearly spelled out the traditional way of tackling the N-extended-body problem in Newtonian gravity. It consists of decomposing the problem into two sub-problems, namely (using the terminology of Fock (1959)):

(i) *the external problem:* to determine the motion of the centres of mass of the N bodies;

(ii) *the internal problem:* to determine the motion of each body around its centre of mass.

Let us label the bodies by the latin letters, $a, b, c = 1, 2, \ldots, N$. For the ath body, which occupies the volume V_a, we define its total mass

$$m_a := \int_{V_a} \rho(\check{x}, t)\, \mathrm{d}^3 x \tag{6}$$

(m_a being constant thanks to the equation of continuity (2)), and the position of its centre of mass

$$z_a^i := m_a^{-1} \int_{V_a} x^i \rho(\check{x}, t)\, \mathrm{d}^3 x. \tag{7}$$

Then the well-known consequence of the continuity equation,

$$\frac{\mathrm{d}}{\mathrm{d}t} \int_{V_a} F(\check{x}, t)\rho(\check{x}, t)\, \mathrm{d}^3 x = \int_{V_a} \frac{\mathrm{d}F(\check{x}, t)}{\mathrm{d}t} \rho(\check{x}, t)\, \mathrm{d}^3 x, \tag{8}$$

where

$$\frac{\mathrm{d}F(\check{x}, t)}{\mathrm{d}t} := \frac{\partial F(\check{x}, t)}{\partial t} + v^i \frac{\partial F(\check{x}, t)}{\partial x^i} \tag{9}$$

denotes the 'convective derivative' of F, implies, when applied twice to the definition (7),

$$m_a \frac{\mathrm{d}^2 z_a^i}{\mathrm{d}t^2} = \int_{V_a} \rho \frac{\mathrm{d}v^i}{\mathrm{d}t}\, \mathrm{d}^3 x. \tag{10}$$

By the local equations of motion $\rho\, \mathrm{d}v^i/\mathrm{d}t = \mathscr{F}^i$, where $\mathscr{F}^i$ denotes the local force density, so that

$$m_a \frac{\mathrm{d}^2 z_a^i}{\mathrm{d}t^2} = \int_{V_a} \mathscr{F}^i\, \mathrm{d}^3 x, \tag{11}$$

('centre-of-mass theorem'). In our perfect fluid model, eq. (3), one has

$$\mathscr{F}^i = -\frac{\partial p}{\partial x^i} + \rho \frac{\partial U}{\partial x^i}. \tag{12}$$

Within each body the force density (12) can be decomposed in an internal force, or self-force,

$$\mathcal{F}^i_{(s)a} := -\frac{\partial p}{\partial x^i} + \rho \frac{\partial U^{(s)a}}{\partial x^i}, \tag{13}$$

and an external force,

$$\mathcal{F}^i_{(e)a} := \rho \frac{\partial U^{(e)a}}{\partial x^i}, \tag{14}$$

where the self-part $U^{(s)a}$ of the gravitational potential is $(\check{x} \in V_a)$

$$U^{(s)a}(\check{x}, t) := G \int_{V_a} \frac{\rho(\check{x}', t)}{|\check{x} - \check{x}'|} d^3x'; \tag{15}$$

its external part, $U^{(e)a} = U - U^{(s)a}$, resulting from an integration over the other bodies, is

$$U^{(e)a}(\check{x}, t) := \sum_{b \neq a} G \int_{V_b} \frac{\rho(\check{x}', t)}{|\check{x} - \check{x}'|} d^3x'. \tag{16}$$

In studying the internal motion of the ath body it is also traditional (although often implicit) to introduce an accelerated 'centre-of-mass frame of reference' with axes parallel to the global cartesian axes and with origin the centre of mass of the ath body. The position of any point, P, with respect to this frame, $\check{y}_a$, will be linked to its position in the global cartesian frame, $\check{x}$, by

$$y^i_a = x^i - z^i_a(t). \tag{17}$$

Time being absolute in Newtonian mechanics, the relative velocity,

$$w^i_a := \frac{dy^i_a}{dt}, \tag{18}$$

and the relative acceleration of the (material) point P will be, respectively, $w^i_a = v^i - dz^i_a/dt$, and $dw^i_a/dt = dv^i/dt - d^2z^i_a/dt^2$.

Using this notation the basic equation of the *external problem* (eq. (11)) reads

$$m_a \frac{d^2 z^i_a}{dt^2} = \int_{V_a} \mathcal{F}^i_{(e)a} d^3x + \int_{V_a} \mathcal{F}^i_{(s)a} d^3y_a, \tag{19}$$

while the basic equation of the ath *internal problem* (motion in the centre-of-mass frame of the ath body) is (from eqs. (3) and (17)–(18))

$$\rho \left(\frac{\partial w^i_a}{\partial t} + w^j_a \frac{\partial w^i_a}{\partial y^j_a} \right) = \mathcal{F}^i_{(s)a} + \mathcal{F}^i_{(e)a} - \rho \frac{d^2 z^i_a}{dt^2} \tag{20}$$

or, equivalently, by using eq. (19):

$$\rho\left(\frac{\partial w_a^i}{\partial t}+w_a^j\frac{\partial w_a^i}{\partial y_a^j}\right)=\mathcal{F}_{(s)a}^i-\frac{\rho}{m_a}\int_{V_a}\mathcal{F}_{(s)a}^i\,\mathrm{d}^3y_a+\mathcal{F}_{(e)a}^i-\frac{\rho}{m_a}\int_{V_a}\mathcal{F}_{(e)a}^i\,\mathrm{d}^3x. \qquad (21)$$

From these equations it is apparent that the external and internal problems are, *a priori*, intricately coupled. On the one hand the external problem, eq. (19), contains, as second term in its right-hand side, a term of purely internal origin, the total self-force:

$$F_{(s)a}^i:=\int_{V_a}\mathcal{F}_{(s)a}^i\,\mathrm{d}^3y_a, \qquad (22)$$

and, moreover, the first term in its right-hand side is of mixed external–internal origin (as discussed later). On the other hand, the internal problem, eq. (21), contains, as third and fourth terms in its right-hand side, terms of external origin, the 'tidal' force–density:

$$\mathcal{F}_{(t)a}^i(\hat{y}_a):=\mathcal{F}_{(e)a}^i(\hat{z}_a+\hat{y}_a)-\frac{\rho_a(\hat{y}_a)}{m_a}\int_{V_a}\mathcal{F}_{(e)a}^i(\hat{z}_a+\hat{y}_a')\,\mathrm{d}^3y_a', \qquad (23)$$

where we have introduced the notation

$$\rho_a(\hat{y}_a):=\rho(\hat{z}_a+\hat{y}_a), \qquad (24)$$

which will be (implicitly or explicitly) extended to all internal quantities.

6.4 The effacement of internal structure in the external problem

A priori, one would think that the previous decomposition of the problem of motion into two sub-problems is of no value because the main equations governing the sub-problems are coupled via the terms (22) and (23). However, there are two circumstances which strongly diminish the coupling between the external and internal problems. The first, if one may speak thus of a basic physical fact, is that the total self-force (22), must vanish on account of Newton's third law of motion ('action and reaction principle'). In our explicit perfect-fluid model where the local self-force density is given by eq. (13) so that

$$F_{(s)a}^i=\int_{V_a}\left[-\frac{\partial p_a}{\partial y_a^i}+\rho_a\frac{\partial U^{(s)a}}{\partial y_a^i}\right]\mathrm{d}^3y_a; \qquad (25)$$

this is easily checked by using Gauss's theorem and the vanishing of the pressure outside the body, for the first term on the right-hand side of eq. (25), and by replacing the value (15) for the self-gravitational potential in the second term, leading to

$$F^i_{(s)a} = 0 + \int_{V_a} \rho_a \frac{\partial U^{(s)a}}{\partial y^i_a} \, d^3 y_a$$

$$= -G \int_{V_a} \int_{V_a} \frac{\rho_a(\check{y}_a)\rho_a(\check{y}'_a)}{|\check{y}_a - \check{y}'_a|^3} (y^i_a - y'^i_a) \, d^3 y_a \, d^3 y'_a = 0 \qquad (26)$$

because of the antisymmetry of the integrand in $\check{y}_a$ and $\check{y}'_a$.

The second circumstance which strongly reduces the influence of the internal structure on the external motion is that the various bodies making up the system are usually widely separated. It is convenient to measure the strength (in fact the weakness) of the 'geometric coupling' between the N extended bodies by means of a dimensionless parameter

$$\alpha := \frac{L}{R}, \qquad (27)$$

where L is the characteristic linear dimension of the bodies and R the characteristic separation between them. We are here following Fock (1959) in characterising the Newtonian coupling between the bodies by only one parameter. There are situations where the consideration of more parameters (such as the ratio between the masses) is called for. Spyrou (1978) has attempted such an improved description by recording, in order of magnitude, the influence of each body $b \neq a$ on the motion of the ath one. We will not try to do so here because later in the article we will be mainly interested in the interplay between the Newtonian parameter (α) and the Einsteinian ones ($\beta_e, \beta_i, \gamma_e, \gamma_i$).

Assuming from now on that

$$\alpha \ll 1, \qquad (28)$$

we can follow the practice of fluid mechanicians (see, e.g. Lagerstrom *et al.*, 1967) and consider that the actual system we are interested in (corresponding to a particular value, say α_0, of α) is embedded in a (fictitious) sequence of systems parametrised by α such that $0 < \alpha \leqslant \alpha_0$. One is then looking for solutions of the external problem in the form of asymptotic expansions when α tends to zero.

As a first step towards finding such solutions one in fact starts by expanding the basic equation of the external problem itself, i.e. eq. (19), which reads, taking into account the vanishing of the self-force (eq. (26)), and the definition (14) of the external force:

$$m_a \frac{d^2 z^i_a}{dt^2} = \int_{V_a} d^3 y_a \rho_a(\check{y}_a, t) U^{(e)a}_{,i}(\check{z}_a + \check{y}_a), \qquad (29)$$

where, for notational simplicity, we have denoted partial differentiation by a comma: $U_{,i} := \partial U/\partial x^i$. Now, the Taylor expansion of $U^{(e)a}_{,i}(\vec{z}_a + \vec{y}_a)$ about $\vec{y}_a = \vec{0}$,

$$U^{(e)a}_{,i}(\vec{z}_a + \vec{y}_a) = U^{(e)a}_{,i}(\vec{z}_a) + y_a^j U^{(e)a}_{,ij}(\vec{z}_a) + \tfrac{1}{2} y_a^j y_a^k U^{(e)a}_{,ijk}(\vec{z}_a) + O(|\vec{y}_a|^3), \tag{30}$$

yields the desired asymptotic expansion about $\alpha = 0$:

$$m_a \frac{d^2 z_a^i}{dt^2} = m_a U^{(e)a}_{,i}(\vec{z}_a) + \tfrac{1}{2} I_a^{jk} U^{(e)a}_{,ijk}(\vec{z}_a) + O^{\text{ext}}(\alpha^3), \tag{31}$$

where we have used the vanishing of the first-order relative mass moment

$$I_a^i(t) := \int_{V_a} d^3 y_a \rho_a(\vec{y}_a, t) y_a^i = 0 \tag{32}$$

(as deduced from definitions (7) and (17)), where

$$I_a^{ij}(t) := \int_{V_a} d^3 y_a \rho_a(\vec{y}_a, t) y_a^i y_a^j \tag{33}$$

denotes the 'inertia tensor' (second-order relative mass moment) and where $O^{\text{ext}}(\alpha^3)$ denotes a remainder term (mainly due to the third-order moment coupling) which is of the order of $mL^3 \, \partial^4 U \sim m(L/R)^3 \, \partial U$, i.e. α^3 smaller than the first term in the right-hand side of eq. (31). In other words, if one uses a system of units adapted to the external problem, i.e. such that a characteristic mass, m, a characteristic separation, R, and Newton's constant (or a characteristic orbital time), are all equal to one:

$$m = R = G = 1, \tag{34}$$

then the first term on the right-hand side of eq. (31) is numerically of order one, the second term is numerically $\lesssim \alpha^2$, while the remaining term is $\lesssim \alpha^3$. Note that by its definition (16) the Laplacian of $U^{(e)a}$ is zero within the ath body so that we can replace, in eq. (31), the inertia tensor, I_a^{ij}, by its (symmetric and) trace-free part, the 'quadrupole tensor':

$$Q_a^{ij}(t) := I_a^{ij} - \tfrac{1}{3} \delta^{ij} I_a^{kk}, \tag{35}$$

and thus the equation for the external motion reads:

$$m_a \frac{d^2 z_a^i}{dt^2} = m_a U^{(e)a}_{,i}(\vec{z}_a) + \tfrac{1}{2} Q_a^{jk} U^{(e)a}_{,ijk}(\vec{z}_a) + O^{\text{ext}}(\alpha^3). \tag{36}$$

If we introduce an 'ellipticity parameter', ε,

$$\varepsilon := \sup_a \left(\frac{|Q_a^{ij}|}{|I_a^{ij}|} \right) \tag{37}$$

(the vertical bars denoting the Euclidean norm of tensors), then $0 \leqslant \varepsilon \leqslant 1$ and the magnitude of the second term on the right-hand side of eq. (36) is in fact $\lesssim \varepsilon \alpha^2$, which can be much smaller than α^2 if, as is frequently the case (for reasons to be discussed later), $\varepsilon \ll 1$.

Finally, we must replace the external potential $U^{(e)a}$ (eq. (16)) in eq. (36), i.e.

$$U^{(e)a}(\check{x}, t) = \sum_{b \neq a} G \int_{V_b} d^3 y_b \frac{\rho_b(\check{y}_b, t)}{|\check{x} - \check{z}_b - \check{y}_b|}, \tag{38}$$

by its α-expansion. The following Taylor expansion in powers of $\check{y}_b$ (with $A_{,i} := \partial A / \partial x^i$),

$$\frac{1}{|\check{x} - \check{z}_b - \check{y}_b|} = \frac{1}{|\check{x} - \check{z}_b|} - \left(\frac{1}{|\check{x} - \check{z}_b|}\right)_{,i} y_b^i + \frac{1}{2}\left(\frac{1}{|\check{x} - \check{z}_b|}\right)_{,ij} y_b^i y_b^j + O(|\check{y}_b|^3), \tag{39}$$

yields, again using the vanishing of the first-order relative moment and the nullity of the Laplacian of $|\check{x} - \check{z}_b|^{-1}$,

$$U^{(e)a}(\check{x}, t) = \sum_{b \neq a} \left\{ \frac{Gm_b}{|\check{x} - \check{z}_b|} + \frac{1}{2} G Q_b^{ij} \left(\frac{1}{|\check{x} - \check{z}_b|}\right)_{,ij} \right\} + O^{\text{ext}}(\alpha^3). \tag{40}$$

Hence we get for the α-expanded equation of external motion:

$$m_a \frac{d^2 z_a^i}{dt^2} = \sum_{b \neq a} \left\{ G m_a m_b \frac{\partial}{\partial z_a^i} \left(\frac{1}{|\check{z}_a - \check{z}_b|}\right) \right.$$

$$\left. + \frac{1}{2} G (m_a Q_b^{jk} + m_b Q_a^{jk}) \frac{\partial^3}{\partial z_a^i \partial z_a^j \partial z_a^k} \left(\frac{1}{|\check{z}_a - \check{z}_b|}\right) \right\} + O^{\text{ext}}(\alpha^3). \tag{41}$$

Now the system of eqs. (41), with $a = 1, \ldots, N$, does not yield an autonomous system for the external motion because it contains, along with the variables $\check{z}_a(t)$ describing the external motion, and the constant parameters m_a, the time-varying quadrupole tensors $Q_a^{ij}(t)$ which depend on the internal motion of the bodies. However, in eq. (41) the contribution of the internal structure $Q^{ij}(t)$ to the net force is only $\sim G \varepsilon m^2 L^2 / R^4$, which is $\varepsilon \alpha^2$ smaller than the Newtonian force between two point-masses $\sim Gm^2/R^2$ (first term on the right-hand side of eq. (41)). On the other hand, an *a priori* order of magnitude estimate of the influence of the internal structure on the first form of the equation of the external problem, eq. (19), would have yielded an internal-structure-dependent contribution $\sim Gm^2/L^2$, possibly coming from the self-force. Therefore, in Newtonian theory there is a drastic reduction of the influence of the internal structure on the external motion by a factor $\sim \varepsilon \alpha^4$ from *a priori* estimates, and by a factor $\sim \varepsilon \alpha^2$ in the final result. Using a terminology introduced by Brillouin (1922) and Levi-Civita (1937a), and revived recently in a study of the motion of two compact bodies (Damour, 1983a), we shall speak of an effacement of the internal structure in the external problem. When one can neglect the relative order of magnitude $\varepsilon\alpha^2 + \alpha^3$ the 'effacement' allows one to reduce the external problem to the problem of the motion of N point-masses fully characterised by their positions, $\vec{z}_a(t)$, and their (constant) masses, m_a, and subject to the autonomous differential system:

$$m_a \frac{d^2 z_a^i}{dt^2} = \sum_{b \neq a} Gm_a m_b \frac{\partial}{\partial z_a^i} \left(\frac{1}{|\vec{z}_a - \vec{z}_b|} \right). \tag{42}$$

6.5 The effacement of external structure in the internal problem

The exact equation governing the internal motion is (from eq. (21), taking into account the vanishing of the self-force),

$$\rho_a \left(\frac{\partial w_a^i}{\partial t} + w_a^j \frac{\partial w_a^i}{\partial y_a^j} \right) = -p_{a,i} + \rho_a U_{,i}^{(s)a} + \mathscr{F}_{(t)a}^i, \tag{43}$$

where the 'tidal' force density, $\mathscr{F}_{(t)a}$ (eqs. (14) and (23)), is

$$\mathscr{F}_{(t)a}^i(\vec{y}_a) := \rho_a U_{,i}^{(e)a}(\vec{z}_a + \vec{y}_a) - \frac{\rho_a}{m_a} \int_{V_a} d^3 y'_a \rho_a(\vec{y}'_a) U_{,i}^{(e)a}(\vec{z}_a + \vec{y}'_a). \tag{44}$$

A priori, the relative order of magnitude of the 'external' force $\mathscr{F}_{(t)a}$, as compared with the gravitational self-force density, $\rho_a U_{,i}^{(s)a} \sim \rho_a Gm/L^2$, would be thought to be given by the order of magnitude of the terms explicitly appearing on the right-hand side of eq. (44), which are $\sim \rho_a Gm/R^2$, so that, a priori, the influence of the external motion $(R(t))$ on the internal evolution would be of relative order $(L/R)^2 = \alpha^2$. However, a 'peculiar circumstance' (which was raised to the status of a fundamental physical fact by Einstein) causes a reduction between the two terms of the right-hand side of (44), leading to a further 'effacement' of the influence of the external motion on the internal one. Indeed, using the same Taylor expansion, eq. (30), as in the external problem and again taking into account the vanishing of the first-order relative moment, eq. (32), we find:

$$\mathscr{F}_{(t)a}^i(\vec{y}_a) = \rho_a U_{,ij}^{(e)a}(\vec{z}_a) y_a^j + \frac{1}{2}\rho_a U_{,ijk}^{(e)a}(\vec{z}_a) \left[y_a^j y_a^k - \frac{I_a^{jk}}{m_a} \right]$$

$$+ O^{\text{int}}(\alpha^5) \tag{45}$$

(where again I_a^{jk} can be replaced by Q_a^{jk}).

The first term in the right-hand side of eq. (45) is now of order α^3 (without factors ε) as compared with the gravitational self-force, while the second and third terms are, respectively, of order α^4 and α^5. In other words, if one uses a system of units adapted to the internal problem, i.e. such that a characteristic mass, m, a characteristic linear dimension, L, and Newton's constant (or a characteristic internal dynamical time) are all equal to 1:

$$m = L = G = 1, \tag{46}$$

then the three terms in the right-hand side of eq. (45) become numerically $\sim \alpha^3$, α^4 and α^5 respectively. When one can neglect the relative order of magnitude α^3 one can completely 'efface' the external motion and reduce the internal problem to solving the following autonomous integro-differential system:

$$\rho_a \left(\frac{\partial w_a^i}{\partial t} + w_a^j \frac{\partial w_a^i}{\partial y_a^j} \right) = -\frac{\partial p_a}{\partial y_a^i} + \rho_a \frac{\partial}{\partial y_a^i} \left\{ \int_{V_a} d^3 y_a' \frac{G \rho_a(\vec{y}_a', t)}{|\vec{y}_a - \vec{y}_a'|} \right\}, \tag{47}$$

$$\frac{\partial \rho_a}{\partial t} + \frac{\partial}{\partial y_a^i} (\rho_a w_a^i) = 0, \tag{48}$$

with some given equation of state (1) and the condition that ρ_a and p_a vanish outside some compact domain V_a.

6.6 Newton and the strong principle of equivalence

The 'peculiar circumstance' which gave rise, as discussed in the preceding section, to a stronger-than-expected effacement of the influence of the external structure on the internal motion is that the coefficient preceding $U_{,i}^{(e)a}$, which measures the response of the matter to an external gravitational field, i.e. the 'passive gravitational mass density', happens to be equal to the coefficient in front of the acceleration in the local equation of motion (see the original form (20)), i.e. the 'inertial mass density'. This equality

$$\rho^{\text{pass grav}} = \rho^{\text{inert}}, \tag{49}$$

together with the equality, implied by the principle of action and reaction, between the 'passive gravitational mass density' and the 'active' one (which creates the gravitational field, right-hand side of Poisson's equation (4a)),

$$\rho^{\text{pass grav}} = \rho^{\text{act grav}}, \tag{50}$$

was also partly responsible for the effacement of the internal structure in the external motion with reduction to the simpler problem of the motion of N points endowed with only one parameter each, their masses:

$$m_a^{\text{inert}} = m_a^{\text{pass grav}} = m_a^{\text{act grav}}. \tag{51}$$

Newton was fully aware of the very remarkable, and unexpected, character of the equality (49) and he spends several pages of the *Principia* (Newton, 1947, pp. 411–13, book III, proposition VI, theorem VI) discussing the experimental evidence suggesting it (the equality (50) being then deduced from the action and reaction principle in the proposition VII, theorem VII). He gives essentially four different arguments in favour of (49):

(i) one concerning the behaviour of test bodies around the Earth (free fall and pendulum experiments);
(ii) one comparing the acceleration of free-fall at the surface of the Earth with the (centrifugal) acceleration of the Moon in its motion around the Earth;
(iii) one comparing among themselves the accelerations of a set of satellites orbiting a common massive object (using the system of the Jovian satellites, and the solar planetary system);
(iv) one concerning the internal dynamics of a system of satellites around a body which itself orbits around the Sun (using mainly the Jovian system but alluding also to the Saturnian one and to the Earth–Moon system).

The second and third arguments, based on the comparison of several independent absolute measurements, are of poor accuracy (even nowadays) and therefore provide only a weak confirmation of the equality (49). By contrast the first and the fourth arguments are of differential type and provide, according to Newton, two independent confirmations of the equality (49) with a relative precision of one-thousandth, i.e.,

$$|\delta_{ab}| < 10^{-3}, \tag{52}$$

where, for a pair of bodies:

$$\delta_{ab} := \left(\frac{m_a^{\text{pass grav}}}{m_a^{\text{inert}}} \right) - \left(\frac{m_b^{\text{pass grav}}}{m_b^{\text{inert}}} \right). \tag{53}$$

All textbooks on gravitational theory quote the pendulum experiments of Newton and their result (52) but, surprisingly, textbooks and workers in the field seem to have ignored Newton's fourth argument in spite of the fact that, according to Newton, this argument leads to the same (remarkably good) result (52). In order to show Newton's profound premonitory achievement let us go a little bit into his reasoning.

The basis of Newton's reasoning is contained in corollary VI of his third law of motion (Newton, 1947, p. 21, axioms, or laws of motion):

If bodies, moved in any manner among themselves, are urged in the direction of parallel lines by equal accelerative forces, they will all continue to move among themselves, after the same manner as if they had not been urged by those forces.

This corollary immediately follows the famous corollary V (*loc. cit.* p. 20):

The motions of bodies included in a given space are the same among

themselves, whether that space is at rest, or moves uniformly forwards
in a right line without any circular motion,

which is Newton's form of the principle of (special) relativity. It seems clear
to me, from the proximity of these two statements, and from the similarity of
their formulations, that Newton guessed corollary VI was a kind of
generalisation of corollary V, i.e. a kind of principle of 'general' relativity (in
Einstein's sense). And, indeed, in all the uses of corollary VI in book III he
applies it to the common accelerations impressed on a system of bodies by a
very distant external body when the equality (49) is assumed to hold. In
other words, if we allow ourselves (although this will certainly displease the
science historians) to reinterpret Newton's thought using modern
terminology, we can say that corollary VI predates the Einsteinian principle
of equivalence which tries to raise the equality (49) to the status of a
generalisation of the principle of (special) relativity. Now, the recent
literature on theories of gravity (see especially Will, 1979, 1981) has
distinguished several formulations of the principle of equivalence. The
strongest formulation ('Strong Equivalence Principle') states essentially:

(i) that the equality (49) between passive-gravitational and inertial masses
   holds also for self-gravitating bodies, and
(ii) the outcome of local experiments, including gravitational ones,
   involving interactions internal to a system embedded in a homogeneous
   external gravitational field are the same as in the absence of external
   gravitational fields.

Now, the reasoning used by Newton in his fourth argument can be said to
predate the 'Strong Equivalence Principle'. Indeed, this reasoning runs as
follows (*a* representing for instance Jupiter, and *b* one of its satellites): if δ_{ab},
defined by eq. (53), is zero then by corollary VI the (gravitational) internal
dynamics of the system $\{a, b\}$ would be the same as in the absence of the Sun;
but if δ_{ab} is other than zero then this will perturb the internal dynamics of the
system $\{a, b\}$. Newton estimates that the main consequence of a non-zero δ_{ab}
will be to make the orbit of *b* around *a* eccentric, its centre *c* being displaced
(towards the Sun if $\delta_{ab} > 0$) by the amount:

$$\delta|\vec{z}_c - \vec{z}_\odot| = -\tfrac{1}{2}\delta_{ab} \cdot |\vec{z}_c - \vec{z}_\odot|. \tag{54}$$

In modern phraseology one can say that Newton predicts a 'polarisation'
of the orbit of the satellite in the direction of the Sun. Nordtvedt (1968) (see
also Will, 1981, p. 186) has recently 'rediscovered' such an effect and
proposed to test its existence in the Earth–Moon system using the Lunar
Laser Ranging data. Nordtvedt's intention was, similarly to Newton, to test

the 'equivalence principle for massive bodies'. As we shall discuss later, the strong form of the effacement of internal structure which holds within general relativity guarantees that in particular $\delta_{ab} \equiv 0$ (even for strongly self-gravitating bodies), but Nordtvedt pointed out that in other relativistic theories of gravity δ_{ab} might become non-zero for self-gravitating bodies.

To be fair it should be added that Newton's estimate (54) (given without intermediate steps) is incorrect both in magnitude and in sign and that the first correct calculation of the 'polarisation' of the orbit of a satellite by a relative acceleration field towards the Sun ('gravitational Stark effect') was achieved by Nordtvedt (1968). However, even if Newton's conclusion (based on the overestimate (54)) that the observations of the Jovian system (in his time) allow one to conclude that $|\delta_{ab}| < 10^{-3}$ (a value consistent with Newton's best pendulum experiments) is too optimistic, we can still admire Newton's remarkable insight which led him not only to ask profound questions about the strong equivalence principle but also to suggest where to look for a possible astronomical violation of the equality (49).

6.7 Solving the Newtonian problems of motion

It is outside the scope of this article to review, even sketchily, the work which has been done, and which still continues, towards solving the equations of the external and/or internal problems. The reader is referred to the classic treatises of Tisserand (1960) and Poincaré (1957), and for more recent developments to Brouwer and Clemence (1961) and Hagihara (1972) which contain many references. Jardetzky (1958) is fully devoted to the internal problem. For general results on dynamical systems see e.g. Arnold (1978), Gallavotti (1983) and references therein.

Traditionally, it was thought sufficient to deal both with the reduced external problem (42) (with complete effacement of the internal structures) and with a simple model for the internal problem (of the type (47)–(48)) including some account of the coupling with the external motion ('precession of equinoxes', ...). However, the tremendous increase in precision brought about by recent techniques (radar ranging, laser ranging, VLBI, ...) calls for extensive refinement of our solutions of the Newtonian problems of motion. But basically one can use the traditional splitting into external and internal problems as a very good zeroth approximation. Indeed, using the 'external units' (34) for the external problem, and the 'internal units' (46) for the internal one, and denoting symbolically by X_e (resp. X_i) a matrix of variables fully describing the external (resp. internal) evolution, the structure of the exact problem of motion is:

$$0=\frac{\mathrm{d}X_\mathrm{e}}{\mathrm{d}t_\mathrm{e}}+E_0(X_\mathrm{e})+\alpha^2 E_2(X_\mathrm{e},X_\mathrm{i})+\alpha^3 E_3(X_\mathrm{e},X_\mathrm{i})+\cdots, \tag{55a}$$

$$0=\frac{\mathrm{d}X_\mathrm{i}}{\mathrm{d}t_\mathrm{i}}+I_0(X_\mathrm{i})+\alpha^3 I_3(X_\mathrm{i},X_\mathrm{e})+\alpha^4 I_4(X_\mathrm{i},X_\mathrm{e})+\cdots, \tag{55b}$$

where t_e (resp. t_i) is the dimensionless external (resp. internal) time (in eq. (55b) the matrix X_i describing, *a priori*, a set of fields, can be infinite). Therefore, in principle, one can solve the coupled system (55) by an iteration process which starts with knowledge of the solution of the uncoupled system obtained as the $\alpha=0$ limit of (55). Note also that under normal conditions the internal motion will be nearly stationary and, in fact, slowly 'driven' by external influences, so that one expects X_i to vary with the external time

$$t_\mathrm{e}=\alpha^{3/2}t_\mathrm{i}. \tag{56}$$

Using t_e instead of t_i in eq. (55b) leads to

$$0=I_0(X_\mathrm{i})+\alpha^{3/2}\frac{\mathrm{d}X_\mathrm{i}}{\mathrm{d}t_\mathrm{e}}+\alpha^3 I_3(X_\mathrm{i},X_\mathrm{e})+\cdots, \tag{57}$$

which, together with (55a), can be iteratively solved starting with a stationary internal trial solution.

Another way of building a zeroth approximation to the system (55) is to start with the assumption that the N bodies undergo rigid motions. This allows one to reduce the system (55) (either taken exactly or truncated at some finite approximation in α) to an autonomous system of $6N$ differential equations for the $6N$ parameters describing the translational and the rotational degrees of freedom of the N bodies (see e.g. Fock, 1959, and Dixon, 1979, for detailed accounts of this approach).

However, the extreme accuracy (sometimes even at the centimetre level) now needed in both the translational motion of the planets (especially the Moon and the Earth) and the internal motion of the Earth makes it necessary to go beyond the approximation of rigid bodies, while still taking into account the smaller α-dependent couplings in the system (55). For recent reviews of the state of the art on the problem of Newtonian motion the reader is referred to Kovalevsky and Brumberg (1986) and references therein.

6.8 The *N*-extended-body problem in Einsteinian gravity

By analogy with the Newtonian case one usually restricts attention to the problem of the gravitational dynamics of a *finite* system of extended bodies. As recalled in Section 6.2, the assumption can, more or less, be checked *a posteriori* in Newtonian mechanics by computing the tidal forces due to the higher hierarchical level of construction of the universe. In Einstein's theory the problem is more difficult for two reasons: (1) as discussed later, one does not yet have an explicit way of separating the gravitational field, within a given system, into an internally generated part and an externally generated one, so that one often relies on Newtonian estimates to justify neglect of the influence of external matter, and (2) one would like to demand more from Einstein's theory than from Newton's, namely not only a consistent hierarchical modelisation of the universe but even a *global* description of the universe, because one feels that Einstein's theory has the capacity to yield one. However, for the time being, a general relativistic global description of our vast clumpy universe poses a formidable problem (see e.g. Ellis's discussion, 1984) so that we shall follow the usual approach and limit ourselves, from the start, to considering an *isolated* system made of a *finite* number of (compactly supported) extended bodies.

As in the Newtonian case (Section 6.2), let us further assume, for the sake of definiteness, that our bodies are made of some perfect fluid with a given (isentropic) equation of state linking the proper pressure (measured in a local rest frame) p_0, to the proper energy density ε_0:

$$p_0 = p_0(\varepsilon_0). \tag{58}$$

Then the stress–energy–momentum tensor of the fluid reads (in contravariant form):

$$T^{\mu\nu} = (\varepsilon_0 + p_0)u^\mu u^\nu + p_0 g^{\mu\nu}, \tag{59}$$

where Greek indices take the values $0, 1, 2, 3$, u^ν denotes the 4-velocity of the fluid and $g_{\mu\nu}(x^\lambda)$ (inverse of $g^{\mu\nu}$) are the components of the metric tensor in some coordinate system x^μ. Our metric conventions are those of Misner, Thorne and Wheeler (1973) (in particular the signature is $-+++$).

Then the equations describing both the dynamics of the matter and the generation of the 'relativistic gravitational potentials', $g_{\mu\nu}$, by the matter are the Einstein equations ($c =$ velocity of light)

$$E_{\mu\nu} = \frac{8\pi G}{c^4} T_{\mu\nu}, \tag{60}$$

where $E_{\mu\nu}$ denotes the Einstein tensor

$$E_{\mu\nu} := R_{\mu\nu} - \tfrac{1}{2} R g_{\mu\nu}. \tag{61}$$

The Einstein equations (60) replace eqs. (2) (continuity), (3) (Euler), and (4a) (Poisson) of the corresponding Newtonian problem. As they are purely local they must be supplemented by boundary conditions expressing the assumption that the system is 'isolated'. In contradistinction to the Newton case where the simple and unique fall-off condition (4*b*) was sufficient to determine a physically reasonable solution of the Poisson equation (4*a*) uniquely, it is still not known for sure how to formulate supplementary boundary conditions which ensure the existence and uniqueness of a physically reasonable solution of the Einstein equations (60). What is known, however, at least if one stays within the 'isolated system' approach (see, however, Schutz (1980, 1985) and Futamase (1983) for a different approach), is that one needs something like a dual-purpose boundary condition able to express: (1) that the gravitational field falls off far from the system, and (2) that there were no 'incoming gravitational waves' converging towards the system in the remote past. But there is no general agreement as to how to formulate these conditions in a technically precise way (see, however, Ehlers (1977, 1980) for the presentation of a tentative but elegant formulation).

Postponing the enforcement of 'correct' boundary conditions, we are faced with the problem of somehow solving the Einstein equations (60) with a right-hand side having the structure (59) with the constraint (58) and the further constraint that ε_0 must have as support (in spacetime) N non-intersecting world tubes. This problem is of a higher order of difficulty than the corresponding Newtonian problem because: (1) eqs. (60) are highly non-linear in $g_{\mu\nu}$, and (2) the phenomena of the generation of the gravitational field by the matter and of the action of the gravitational field on the matter are intimately coupled in eqs. (60). Indeed, it is well known that the Bianchi identities imply that a necessary consequence of (60) is

$$T^{\mu\nu}{}_{;\nu}=0, \tag{62}$$

which, in the case of a perfect fluid (59), yields the relativistic Euler equations,

$$(\varepsilon_0+p_0)u^\nu u^\mu{}_{;\nu}=-(g^{\mu\nu}+u^\mu u^\nu)p_{0;\nu}, \tag{63}$$

and the relativistic equation of continuity

$$(r_0 u^\mu)_{;\mu}=0. \tag{64}$$

In eq. (64) we have introduced a proper rest-mass density, r_0, defined (modulo a constant factor) as

$$r_0:=\exp\int\frac{\mathrm{d}\varepsilon_0}{\varepsilon_0+p_0(\varepsilon_0)}. \tag{65}$$

One also defines a proper specific internal energy density Π_0 by

$$\varepsilon_0:=r_0c^2\left(1+\frac{\Pi_0}{c^2}\right). \tag{66}$$

At this point one possible way to investigate the relativistic problem of motion is to postpone the problem of solving the Einstein field equations (60) (thought to correspond to Poisson's equation (4a)) and to fully exploit the 'dynamical equations' (62) (which for a perfect fluid yield (63)–(64)) assuming the metric as given. This approach has been developed mainly by W. G. Dixon, and has led to several remarkably elegant results (see Dixon, 1979; Ehlers and Rudolph, 1977). Recently Bailey and Israel (1980) have succeeded in giving this approach a high degree of formal perfection by means of a Lagrangian formulation. Moreover, their work leads to the completion of Dixon's approach (based on the sole consideration of (62)) with a correspondingly elegant covariant formulation of the field equations (60). Unfortunately, it seems to me that this formulation (which makes use of multipolar 'point sources') has only a formal meaning, and that further hard work is needed to give it a sound mathematical meaning. Moreover, it seems to me that this whole approach to the dynamics of extended bodies is not suited to dealing with the motion of strongly self-gravitating bodies which, however, has become a problem of considerable astrophysical interest.

As we are forced to tackle the Einstein field equations we now ask the advice of the mathematicians to see how, at least in principle, one can construct solutions of the partial differential system (60). The most common approach, due to A. Lichnerowicz and Y. Choquet-Bruhat (see e.g. Lichnerowicz, 1955; Bruhat, 1962; Choquet-Bruhat and York, 1980; York, 1983), starts by considering the system of the field equations (60) together with the dynamical equations (62). This system is in involution, in the sense of E. Cartan, and its integration consists of two separate problems: that of initial conditions and that of time development (or evolution). We shall not enter here into the problem of initial conditions because it is, *a priori*, not well suited to consideration of the supplementary boundary conditions alluded to above (see, however, Schutz (1980, 1985), Futamase (1983), Friedrich (1981, 1985), Choquet-Bruhat and Christodoulou (1983)). On the other hand, the constructive method often used by mathematicians to prove the existence and uniqueness of solutions of the evolution problem can be used as a guide to tackling the problem of motion. This method consists essentially in 'relaxing' the field equations (60) (which, *a priori*, do not form a partial differential system of well-defined type) by using special coordinate conditions. The most common conditions (though not the only ones) are the harmonic coordinate conditions:

$$H^\mu := g^{\mu\nu}{}_{,\nu} = 0, \tag{67}$$

where we have introduced the Gothic contravariant metric variables

$$\mathfrak{g}^{\mu\nu} := \sqrt{g}\, g^{\mu\nu}, \tag{68a}$$

$$g := -\det(g_{\mu\nu}). \tag{68b}$$

Then the relaxed system governing the evolution of the coupled matter–gravitational field becomes the following partial differential system:

$$\mathfrak{g}^{\alpha\beta}\mathfrak{g}^{\mu\nu}{}_{,\alpha\beta} + Q^{\mu\nu}(\mathfrak{g}, \partial\mathfrak{g}) = \frac{16\pi G}{c^4}\, g\{(\varepsilon_0 + p_0)u^\mu u^\nu + p_0 g^{\mu\nu}\}, \tag{69a}$$

$$(\varepsilon_0 + p_0)u^\nu u^\mu{}_{;\nu} = -(g^{\mu\nu} + u^\mu u^\nu)p_{0;\nu}, \tag{69b}$$

$$(r_0 u^\mu)_{;\mu} = 0, \tag{69c}$$

where $Q^{\mu\nu}$ is a complicated quadratic form in the first derivatives of $\mathfrak{g}$.

The problem of the motion of N extended bodies in general relativity is then reduced to finding the solutions of the system (69) (with (58) and (65)) which satisfy the 'harmonicity condition' (67), the constraint $u^\mu u_\mu = -1$, some reasonable boundary conditions for $g_{\mu\nu}$, and such that the support of ε_0 (and p_0) is made of N spatially compact world tubes. This problem is very difficult, on the one hand because of its strong non-linearity (both in $g_{\mu\nu}$ and in u^μ) and, on the other hand, because the condition on the support of ε_0 makes eqs. (69b, c) non-hyperbolic, so that it is not known, mathematically, if they admit any solutions at all. One therefore needs, for even stronger reasons than in the Newtonian case, to resort to approximation methods.

6.9 Approximation methods

The first step towards setting up approximation methods consists of identifying some small (dimensionless) parameters. We shall assume, as in the Newtonian case (Section 6.3), that the N bodies making up the system are widely separated so that the 'geometric coupling' parameter,

$$\alpha := \frac{L}{R}, \tag{70}$$

measuring the ratio between a characteristic linear dimension, L, of the bodies and a characteristic separation, R, between them is much smaller than one.

In Einsteinian gravity there exist other parameters which may be small. Let us introduce β_e: the ratio between a characteristic 'external' velocity (orbital velocity) and the velocity of light.

$$\beta_e := \frac{v^{\text{ext}}}{c}. \tag{71}$$

Similarly, let β_i denote the ratio between a characteristic 'internal' velocity (velocity of spinning motion) and the velocity of light:

$$\beta_i := \frac{v^{\text{int}}}{c}. \tag{72}$$

If m denotes a characteristic mass let us also introduce two other relativistic dimensionless parameters, an 'external' one,

$$\gamma_e := \frac{Gm}{c^2 R}. \tag{73}$$

and an 'internal' one

$$\gamma_i := \frac{Gm}{c^2 L}. \tag{74}$$

Together with the previously introduced Newtonian 'ellipticity' parameter (eq. (37)),

$$\varepsilon \sim \frac{(\text{quadrupole moment})}{mL^2} \sim (\text{relative deviation from sphericity}), \tag{75}$$

this makes a total of six parameters $(\alpha, \beta_e, \beta_i, \gamma_e, \gamma_i, \varepsilon)$ that can be used to concoct some approximation methods.

First let us note that some of these parameters are linked by equalities or inequalities. Indeed, first,

$$\gamma_e = \alpha \cdot \gamma_i, \tag{76}$$

and as it is generally expected that $L \gtrsim Gm/c^2$ (the equality, in order of magnitude, being reached only for 'condensed bodies', i.e. neutron stars or black holes) we shall always have at least

$$\gamma_i \lesssim 1. \tag{77}$$

Hence, from our previous minimal assumption about the relative separation of the bodies we shall have, in general:

$$\gamma_e \lesssim \alpha \ll 1 \quad \text{(minimal assumption)}. \tag{78}$$

On the other hand, we know, a priori, only that $\beta_e \lesssim 1, \beta_i \lesssim 1, \gamma_i \lesssim 1$ and $\varepsilon \lesssim 1$. However, if we further assume that the system is gravitationally bound, so that by the virial theorem $(v^{\text{ext}})^2 \sim Gm/R$ (or nearly gravitationally bound as in wide-angle scattering) then we shall have,

$$\beta_e^2 \sim \gamma_e \ll 1$$

$$\text{(gravitationally bound or wide-angle scattering)}, \tag{79a}$$

but in the case of small-angle gravitational scattering we would have

$$\gamma_e \ll \beta_e^2 \lesssim 1 \quad \text{(small-angle scattering)}. \tag{79b}$$

In some special situations one can encounter other small parameters, such as the ratio between some masses (which plays an important role in the mechanics of the solar system), but we will discuss only the more general cases (see e.g. Brumberg, 1986, and references therein for the proposal of a relativistic approximation method based on such a small parameter m/M).

Before embarking on a quick survey of the main approximation methods which have been used in the problem of motion, let the reader be warned that there are many different points of view concerning these methods and that my viewpoint is probably biased. He is therefore invited to consult other reviews, e.g. Ehlers *et al.* (1976), Ehlers (1977, 1980), Walker (1984), Ehlers and Walker (1984), Schutz (1984, 1985), Will (1986).

6.10 The post-Newtonian approximation methods

The best-known and most investigated methods of approximation in general relativity are based on the assumptions that

$$\sup(\beta_e^2, \beta_i^2) \lesssim \sup(\gamma_e, \gamma_i) \ll 1, \tag{80}$$

i.e. of an everywhere-weak gravitational field, and of slow motions. From eq. (76) we see that $\gamma_e < \gamma_i$ so that γ_i can be taken as the main relativistic small parameter of these methods, the (independent) parameter α being introduced only at a later stage.

The philosophy of these methods is, so to say, to follow in Newton's footsteps, i.e. to try to set up an approximation method for Einstein's theory which follows as closely as possible, both technically and conceptually, the Newtonian way of tackling gravitational problems. In these methods the role of Einstein's theory is reduced to that of providing small numerical corrections to the predictions of Newton's theory. As we shall see later, the danger of these approaches, rightly called 'Post-Newtonian Approximation' (PNA) methods, lies not so much in technical errors or ambiguities (which arise only at a high approximation level, see the discussion of radiation damping below), but in their overall conceptual framework which misses the richness and suppleness of the full Einsteinian theory and which amounts nearly, as was pointed out recently in a somewhat different context (Eisenstaedt, 1986, section 6), to a 'neo-Newtonian interpretation' of Einstein's theory.

Technically speaking, the post-Newtonian approaches seek to model the solution of the relativistic system (69) on the solution of the Newtonian system (2), (3), (4a). For instance, eq. (69c) is rewritten as a Newtonian-type equation of continuity:

$$\frac{\partial r_*}{\partial t} + \frac{\partial (r_* v^i)}{\partial x^i} = 0, \tag{81}$$

where the 'time' $t := x^0/c$, the 'velocity' $v^i = dx^i/dt = cu^i/u^0$ and,

$$r_* := r_0 \sqrt{g}\, u^0, \tag{82}$$

is a 'coordinate rest-mass density'.

Note that I have intentionally refrained from using the letter ρ to denote any of the relativistic densities (of energy or rest-mass, proper or 'coordinate'). Indeed, the fact that many authors use the same letter, ρ, to denote different things is a source of confusion (and error) in the literature on equations of motion and I did not wish to compound the confusion. Moreover, in keeping with what I said in the preceding paragraph, I think that this confusion of notions pertaining to different theories is not innocent so that the use of different letters helps to contrast Einstein's theory with Newton's.

Now the modelling of the reduced Einstein equations (69a) on the Poisson equation (4a) is achieved at a price of several Ansätze. First one writes

$$\mathfrak{g}^{\mu\nu} = f^{\mu\nu} + h^{\mu\nu}, \tag{83}$$

where $f^{\mu\nu} = f_{\mu\nu} = \text{diag}(-1, +1, +1, +1)$ are the usual components of a 'flat' metric. Then $h^{\mu\nu}$ is assumed to be everywhere small ('weak gravitational field') and admits an asymptotic expansion when $\gamma_i \to 0$ of the type:

$$h^{\mu\nu}(\vec{x}, t) = \gamma_i h^{\mu\nu}_{(2)}(\vec{x}, t) + \gamma_i^{3/2} h^{\mu\nu}_{(3)}(\vec{x}, t) + \gamma_i^2 h^{\mu\nu}_{(4)}(\vec{x}, t) + \cdots + \gamma_i^{n/2} h^{\mu\nu}_{(n)}(\vec{x}, t) + \cdots \tag{84}$$

The necessity to introduce half-integer powers of γ_i comes from the basic assumptions (80) which imply that the spatial part of the 4-velocity $u^k \sim \beta_e \lesssim \gamma_i^{1/2}$ (they also imply that $\Pi_0/c^2 \sim p_0/r_0 c^2 \sim \beta_i^2 \lesssim \gamma_i$). Another basic ingredient of all PNA methods is to require that

$$\left| \frac{1}{c} \frac{\partial h^{\mu\nu}}{\partial t} \right| \sim \gamma_i^{1/2} \left| \frac{\partial h^{\mu\nu}}{\partial x^k} \right|. \tag{85}$$

It is assumed that the expansion coefficients $h^{\mu\nu}_{(n)}$ are always and everywhere of order unity (or less), and that their spatial derivatives are everywhere of order $1/L$ (or less). Note also that, in general, the asymptotic expansion (84) is of the generalised type where some dependence on γ_i is implicitly contained in each coefficient $h^{\mu\nu}_{(n)}$. In order to give to (84) the rigorous meaning of a Poincaré-type asymptotic expansion one should first specify which of the parameters determining the gravitational field are fixed in the limit $\gamma_i \to 0$. For one possible approach along these lines see Futamase and Schutz (1983).

Then one easily sees that if one uses a system of units adapted to the study of the considered gravitationally interacting system, i.e. such that

$$G = m = L = 1, \tag{86}$$

then the numerical value of the velocity of light in these units is such that

$$\frac{1}{c^2} = \gamma_i \ll 1, \tag{87}$$

and all the $h_{(n)}$s, their time or space derivatives, and all the internal or external velocities become of numerical order unity (or less). With this in mind, one can summarise the post-Newtonian assumptions by saying that one looks for solutions of the system (69) in the form of formal expansions in powers of $1/c$ (and it is convenient to use this language even if one does not explicitly use the system of units (86)).

Among the field equations (69a) the dominant one is the zero–zero component which reads, at lowest order:

$$\Delta(\gamma_i h_{(2)}^{00}) = \frac{16\pi G}{c^2} r_0 + O(\gamma_i^2/L^2). \tag{88}$$

Eq. (88) is very similar to the Poisson equation (4a), but like the latter it does *not* determine $h_{(2)}^{00}$ unless we supplement it by some boundary conditions. Many authors working with PNAs have implicitly assumed that it was sufficient to use the same boundary condition as in the Newtonian case (eq. (4b)) namely:

$$\lim_{\substack{|\check{x}| \to \infty \\ t = \text{const}}} (h_{(n)}^{\mu\nu}(\check{x}, t)) = 0. \tag{89}$$

At the lowest level of approximation, (88), this uniquely determines (if we decide to use r_0 as our basic 'mass density')

$$\gamma_i h_{(2)}^{00}(\check{x}, t) = -4\frac{G}{c^2} \int \frac{r_0(\check{x}', t)}{|\check{x} - \check{x}'|} d^3x', \tag{90}$$

which leads, consistent with the previous assumptions, to a $h_{(2)}^{00}$ whose maximum value is of order unity.

If we extend this program to higher orders in $\gamma_i^{1/2}$ one finds that it leads to a formal hierarchy of Poisson equations for the $h_{(n)}^{\mu\nu}$s of the type:

$$\Delta(\gamma_i^{n/2} h_{(n)}^{\mu\nu}) = T^{\mu\nu} \text{ terms} + (\text{terms known from preceding approximations}). \tag{91}$$

For the first few steps this hierarchy admits a unique solution fulfilling the boundary conditions (89), and satisfying everywhere $h_{(n)}^{\mu\nu} \lesssim 1$. However, the terms resulting from lower approximations on the right-hand side of eq. (91) quickly lead to badly behaving Poisson equations which do not admit any solutions fulfilling the boundary conditions (89) (in the present scheme using harmonic coordinates this problem arises first for $h_{(5)}^{0i}$ and $h_{(6)}^{00}$). The reason for this incompatibility between the hierarchy (91) and the boundary conditions (89) does not, however, mean that there is a fundamental breakdown of the post-Newtonian approach at the level of $h_{(5)}^{0i}$ or $h_{(6)}^{00}$ (called second-post-Newtonian, or 2PN, level). Neither does it mean that the harmonic coordinates become intrinsically 'bad' at the 2PN level. The blame for this incompatibility lies with the boundary conditions (89). Indeed, as was first realised clearly by Fock (1959), the post-Newtonian expansion (84)–(85) is basically a *near-zone expansion* of the exact $h^{\mu\nu}(\dot{x}, t)$, i.e. an expansion valid only up to a radius, r, around the system which is much smaller than a characteristic wavelength, λ, of the gravitational radiation emitted by the system: $r \ll \lambda$. As a consequence the asymptotic behaviour for $|\dot{x}| \to \infty$, t fixed, of each term of the right-hand side of eq. (84) has, *a priori*, nothing to do with the real asymptotic behaviour of the exact $h^{\mu\nu}(\dot{x}, t)$ (because, indeed, the expansion on the right-hand side of eq. (84) becomes invalid when r gets large enough to reach the wave-zone $r \gtrsim \lambda$). This is clearly understood for the trivial example of the near-zone (or $1/c$) expansion of the simplest radiative field:

$$\frac{S(t-r/c)}{r} = \frac{S(t)}{r} - \frac{1}{c}\dot{S}(t) + \frac{1}{2c^2}\ddot{S}(t)r - \frac{1}{6c^3}\dddot{S}(t)r^2 + \cdots \tag{92}$$

The program we described above (solving the post-Newtonian expanded field equations using, often implicitly, the fall-off conditions (89) at each step) has been developed by many authors starting with the initiator of the PNA methods: J. Droste. See e.g. Droste (1916), de Sitter (1916), Chazy (1928, 1930), Levi-Civita (1937a, 1950), Einstein, Infeld and Hoffmann (1938), Papapetrou (1951), Infeld and Plebanski (1960) (and references therein), Chandrasekhar (1965), Chandrasekhar and Nutku (1969) (and other papers published in the *Astrophysical Journal*), Ohta *et al.* (1973, 1974a, 1974b). Note that those of the preceding authors who worked at the 2PN level had to use a non-harmonic coordinate condition in order to satisfy (89).

Other authors have tried to by-pass the difficulty of incorporating a 'good' boundary condition by replacing the hierarchy of differential systems

(91) with a formal iteration of integrals. One such method consists in keeping on the left-hand side of the (expanded) field equations (91) a d'Alembertian, $\Box$, instead of a Laplacian, the right-hand side being known from preceding approximations, and then to solve

$$\Box h_{(n)}(\check{x}, t) = -4\pi \Sigma_{(n)}(\check{x}, t) \tag{93}$$

by means of the formal near-zone expansion of the *retarded* integral of $\Sigma_{(n)}$:

$$h_{(n)}(\check{x}, t) = \sum_{p=0}^{\infty} \frac{(-)^p}{c^p p!} \frac{\partial^p}{\partial t^p} I_{p-1}(\Sigma_{(n)}), \tag{94}$$

where for any function $f(\check{x}, t)$

$$I_p[f](\check{x}, t) := \int f(\check{x}', t) |\check{x} - \check{x}'|^p \, d^3 x'. \tag{95}$$

The idea behind this formal procedure was to retain both the technical simplicity (eq. (85) neglecting many terms) and the conceptual ease of the post-Newtonian approach ($h_{(n)}$ being expressed as a functional of the *instantaneous* state of the source, as in Newtonian theory) while still trying to incorporate the 'real physics' (which states that in Einsteinian theory gravity propagates with the velocity of light). This programme has been developed by many authors, notably Peres (1959, 1960), Carmeli (1964, 1965), Synge (1969, 1970), Hogan and McCrea (1974), Anderson and Decanio (1975). An improvement of the method (leaving the time derivatives inside the integral operators in eq. (94) and 'reducing' them by means of the equations of motion) has been proposed by Ehlers (1977, 1980) and developed by Kerlick (1980) and Caporali (1981a). However, the use of the near-zone expanded retarded potentials (94) causes the appearance of divergent integrals. This means that all these schemes break down (because they become undefined) at the 3PN level (Kerlick, 1980), or even before in the case of the non-improved schemes. This breakdown is a direct consequence of the limitation of validity of the post-Newtonian expansions to the near-zone.

Other authors, conscious of the necessity to incorporate the propagating character of gravity (both in the near- and the wave-zone), but aware of the simplifications brought by the PNA assumptions, attempted to devise some mixed post-Minkowskian (see next section)-post-Newtonian approaches. See e.g. Fock (1959), Persides (1971a, 1971b), Winicour (1983), Gürses and Walker (1984), Schäfer (1985). However, in order to really face the problem of the near zone limitation of the post-Newtonian approach it seems necessary to have recourse to new methods (see Sections 6.11 and 6.12).

On a more fundamental level there are also the problems of understanding in general terms the transition between Einstein's and Newton's theories (see e.g. Ehlers (1981, 1984) and references therein), and of clarifying the mathematical, and numerical, validity of the post-Newtonian expansion (84). Concerning this last point Futamase and Schutz (1983) attempted to prove that the first few terms of the PN expansion (84) constituted a genuine asymptotic expansion (in the mathematical sense), but the assumptions they used (about the existence and uniformity of the $\gamma_i \to 0$ limit, plus some other implicit assumptions) were very strong so that it is not clear, to me, what they have actually proven. Another interesting problem is to clarify at what order the post-Newtonian expansion fundamentally breaks down and why (see Blanchet and Damour, 1986b, and references therein).

6.11 The post-Minkowskian approximation methods

This type of method is based solely on the assumed weakness of the gravitational field without any further assumptions concerning the magnitude of the internal or external velocities. Using the notation of Section 6.9 this can be written in general as:

$$\sup(\gamma_e, \gamma_i) \ll 1. \tag{96}$$

As, from eq. (76), $\gamma_e < \gamma_i$ one can take γ_i as the main relativistic parameter of these methods.

I favour the terminology 'Post-Minkowskian Approximation' (PMA) methods instead of the 'Fast Motion Approximation', 'Weak Field Expansion' or 'Non-Linearity Expansion' which are, or have been, also used. The reason for this preference is that, just as the Post-Newtonian Approximation methods were based on the idea of staying as close as possible to the conceptual framework of Newtonian theory (separating (absolute) time and (absolute) space, endowed with an (auxiliary) Euclidean metric, and using instantaneous potentials), so the Post-Minkowskian Approximation methods try to stay as close as possible to the conceptual framework of the Minkowskian interpretation of special relativity (using an absolute spacetime endowed with a (auxiliary) Minkowskian metric and retarded potentials). As in the post-Newtonian case, the use of post-Minkowskian methods exposes one to the risk of forgetting the richness of the full Einstein theory, and surreptitiously introducing a 'neo-Minkowskian' interpretation of General Relativity.

Technically speaking, the post-Minkowskian approaches start by writing, in a given coordinate system, the decomposition (83) of the 'real' (Gothic) metric, $\mathfrak{g}^{\mu\nu}$, into an auxiliary 'flat' metric, $f^{\mu\nu}$, and a weak 'gravitational field' $h^{\mu\nu}$. Then $h^{\mu\nu}$ is assumed to admit a formal asymptotic expansion when $\gamma_i \to 0$ of the type:

$$h^{\mu\nu}(x^\lambda) = \gamma_i h_1^{\mu\nu}(x^\lambda) + \gamma_i^2 h_2^{\mu\nu}(x^\lambda) + \cdots + \gamma_i^n h_n^{\mu\nu}(x^\lambda) + \cdots, \tag{97}$$

where the $h_n^{\mu\nu}(x^\lambda)$ are assumed to be of order unity (or less) all over spacetime, their first derivatives $\partial h_n^{\mu\nu}/\partial x^\lambda$ being assumed to be of order $1/L$ (or less). Then if one uses a system of units such that

$$c = m = L = 1, \tag{98}$$

the numerical value of Newton's constant in these units is

$$G = \gamma_i \ll 1, \tag{99}$$

and all the $h_n^{\mu\nu}$s, and $\partial h_n^{\mu\nu}/\partial x^\lambda$s, as well as all the internal or external physical quantities (masses, radii, velocities), become of numerical order unity (or less). Therefore, analogously to the post-Newtonian case, one can summarise the post-Minkowskian assumptions by saying that one looks for solutions of the Einstein equations (60) in the form of formal expansions in powers of G (and it is convenient to use this language even if one does not really use the unit system (98)).

In the PMA method it is convenient to use a coordinate condition which is formally Poincaré invariant (so that the coordinates can still undergo a Poincaré transformation); this is the case with the harmonic condition (67) which is used most often (note, however, that one could use $(g^w g^{\mu\nu})_{,\nu} = 0$ with any given power w, or many other conditions). Using the harmonic condition (67) we are led to look for solutions of the system (69) of the form (97). We can re-write eq. (69c) as a Minkowskian-type continuity equation,

$$(\hat{r}_0 \hat{u}^\mu)_{,\mu} = 0, \tag{100}$$

with a Minkowski-4-velocity $\hat{u}^\mu = u^\mu/(-f_{\alpha\beta}u^\alpha u^\beta)^{1/2}$ (satisfying $f_{\mu\nu}\hat{u}^\mu\hat{u}^\nu = -1$) and a Minkowski-proper-rest-mass density

$$\hat{r}_0 := r_0 \sqrt{g}(-f_{\alpha\beta}u^\alpha u^\beta)^{1/2}. \tag{101}$$

Eq. (69b) will then look like a Minkowski–Euler equation containing an explicit 'gravitational force density' starting at the order γ_i. Finally, the field equation (69a) leads to a formal hierarchy of inhomogeneous wave equations for the $h_n^{\mu\nu}$s of the type ($\Box_f := f^{\mu\nu} \partial_{\mu\nu}^2$):

$$\Box_f (\gamma_i^n h_n^{\mu\nu}) = T^{\mu\nu} \text{ terms} + \text{(terms known from preceding PM approximations).} \tag{102}$$

One must now complete the partial differential equation (102) by a prescription for choosing the 'physically correct' solutions. This is often (implicitly) done by requiring the satisfaction of the 'Fock conditions'

$$\lim_{\substack{r \to \infty \\ t+r/c=\text{const}}} (h_n^{\mu\nu}) = 0, \tag{103a}$$

$$\lim_{\substack{r \to \infty \\ t+r/c=\text{const}}} \left(r \left(\frac{\partial h_n^{\mu\nu}}{\partial r} + \frac{1}{c} \frac{\partial h_n^{\mu\nu}}{\partial t} \right) \right) = 0, \tag{103b}$$

where $r =: (\delta_{ij}x^i x^j)^{1/2}$ and $t =: x^0/c$ (these conditions must be satisfied along all past Minkowski-light-cones). Fock (1959) has proven that the fall-off condition (103a), together with the 'no-incoming-radiation condition' (103b) necessarily implies that

$$\gamma_i^n h_n^{\mu\nu}(x) = \int d^4x' G_{\text{ret}}^{(f)}(x-x') S^{\mu\nu}(x'), \tag{104}$$

where $G_{\text{ret}}^{(f)}(x-x')$ is the retarded Green function of the (Minkowski) wave operator $\Box_f = f^{\mu\nu} \partial^2/\partial x^\mu \partial x^\nu$, and where $S^{\mu\nu}$ denotes the right-hand side of eq. (102) ('effective source of the gravitational field'). Therefore, an alternative prescription for solving the hierarchy of differential systems (102) would be to replace it by the hierarchy of iterated retarded integrals (104). Note, however, that it is not guaranteed that the 'retarded' h_ns defined by (104) will satisfy the Fock conditions (103) (see Leipold and Walker, 1977).

The first step in the hierarchy (104) was introduced and explicitly worked out by Einstein (1916b); this is the famous 'linearized approximation' to general relativity:

$$\gamma_i h_1^{\mu\nu}(x) = \frac{16\pi G}{c^4} \int d^4x' G_{\text{ret}}^{(f)}(x-x') \hat{T}^{\mu\nu}(x'), \tag{105}$$

where, following the notation of eqs. (100)–(101), $\hat{T}^{\mu\nu}$ denotes a Minkowski-like stress–energy tensor of the matter.

For a long time the post-Minkowskian approximation stayed dormant, while the post-Newtonian one was developed. It was revived in 1956 and then began to develop in turn: see Bertotti (1956), Havas (1957), Kerr (1959), Bertotti and Plebanski (1960) (first attempt to explicitly tackle the non-linear approximation), Havas and Goldberg (1962), Kühnel (1963, 1964), Stephani (1964), Goenner (1970), and Bennewitz and Westpfahl (1971). A new approach to the post-linear formalism, based on an attempt to use retarded Green's functions in *curved* space-time (however, the final formulas use only the *flat* Green function), was devised by Thorne and Kovács (1975) and Crowley and Thorne (1977). Recently several authors have succeeded in computing *explicit* expressions for the equations of motion and gravitational field in the second post-Minkowskian approximation: Westpfahl and Goller (1979), Westpfahl and Hoyler (1980), Bel, Damour, Deruelle, Ibanez and Martin (1981), Deruelle (1982), Westpfahl (1985); and even partly for the third post-Minkowskian approximation: Damour (1982, 1983a, 1983c). More recently a combination of the post-Minkowskian approach with multipole expansions has allowed one to control the post-Minkowskian expansion (97) at *all* orders of approximation, see Blanchet and Damour (1984a, 1986a) and references therein. This result strengthens the hope that the post-Minkowskian approach is a 'good' approximation method which, contrary to the post-Newtonian one, can be unambiguously iterated to all orders, and which yields a useful sequence of approximate metrics all over spacetime (the fact that this sequence might not be convergent but only asymptotic (Christodoulou and Schmidt, 1979) is not a problem for its use in physics, and the weakly unsatisfactory behaviour of PM expansions in the very far wave-zone is easy to correct, see next section).

It should also be noted that Schäfer (1986) has recently introduced a non-linear expansion which is based on the Arnowitt–Deser–Misner hamiltonian approach (Arnowitt *et al.*, 1962) and which seems to combine some of the advantages of the post-Newtonian and post-Minkowskian approaches without having the defects of the post-Newtonian one (near-zone-validity only).

Finally, one can remark that the post-Minkowskian approach can be thought of as a 'meta-approximation-theory' for the post-Newtonian one. Indeed, when the assumptions (80) are satisfied one can use units such that

$$\sup(v^{\mathrm{int}}, v^{\mathrm{ext}}) = m = L = 1, \tag{106}$$

for which the numerical value of the velocity of light is very large, the numerical value of Newton's constant is $\lesssim 1$, and

$$\gamma_i = \frac{G}{c^2} \ll 1. \tag{107}$$

We can then try to re-expand each term of the PM expansion,

$$h^{\mu\nu}(x^\lambda) = \sum_{n \geq 1} \frac{G^n}{c^{2n}} h_n^{\mu\nu}\left(\hat{x}, t, \frac{1}{c}\right), \tag{108}$$

in a second (post-Newtonian or near-zone) asymptotic expansion when $1/c \to 0$. It is precisely from this point of view (descending from PMA to PNA) that one 'understands why' the PN expansions have a restricted spatial range of validity $(r \ll \lambda)$. Using this point of view one can similarly investigate the domain of validity of the null-cone-improved post-Newtonian approaches of Persides (1971a, b) and Winicour (1983). These methods essentially consist of looking for expansions of the type

$$h^{\mu\nu}(x^\lambda) = \sum_{m \geqslant 2} \frac{1}{c^m} k_m^{\mu\nu}(\check{x}, u), \qquad (109)$$

where u is some retarded time $(\simeq t - r/c)$ which is kept fixed in the limit process $1/c \to 0$. Now the expansion (109), if it exists, must 'descend' from the PM expansion (108), re-written in terms of $(\check{x}, u)$, and re-expanded for $1/c \to 0$. Then explicit calculations at the 2PM level (in general relativity, and in some scalar field models with $\lambda\phi^n$ non-linearities) show that these improved approaches have a range of spatial validity which is still restricted to the near-zone $(r \ll \lambda)$ (only the linearised approximation sees its range of validity extended to the wave-zone). In particular, this can be used to show that the 'logarithmic violation' of peeling found by Isaacson, Welling and Winicour (1984) is only a near-zone effect (behaviour when $r \gg R$ but $r \ll \lambda$) and says nothing about the 'real' asymptotic behaviour of the gravitational field at future null infinity.

6.12 Singular perturbation methods

The 'breakdown' of the post-Newtonian expansion (84)–(85) (formal expansion in powers of the small parameter $1/c$, see eq. (87)) outside of the near-zone (see eq. (92)) is an example of what goes under the general name of 'singular perturbations': i.e. an asymptotic expansion, when a 'small' parameter ε tends to zero, of some field quantity,

$$f(x; \varepsilon) = \sum_n \delta_n(\varepsilon) f_n(x) \qquad (110)$$

(where $\delta_{n+1}(\varepsilon)/\delta_n(\varepsilon) \to 0$ when $\varepsilon \to 0$), such that some of the coefficients $f_n(x)$ are not bounded on the domain of variation of the configuration-space variables x. Problems of this type are quite common in physics, particularly in fluid mechanics, and several methods have been developed to cope with them (see e.g. Lagerstrom et al., 1967; van Dyke, 1975). There are essentially two classes of methods. In one class one looks for a generalised asymptotic expansion,

$$f(x; \varepsilon) = \sum_n \delta_n(\varepsilon) g_n(x; \varepsilon), \qquad (111)$$

admitting generalised coefficients $g_n(x; \varepsilon)$ which are bounded all over the configuration space (this class includes the use of ε-dependent changes of configuration space variables $g_n = g_n(y)$ with $y = y(x; \varepsilon)$). In another class one uses, simultaneously, several ordinary (Poincaré-type) asymptotic expansions corresponding to the use of several different configuration–space variables related via ε-dependent transformations, for instance

$$f = \sum_n \delta_n(\varepsilon) f_n(x), \tag{112a}$$

$$f = \sum_n \delta_n(\varepsilon) F_n(X), \tag{112b}$$

where $X = X(x, \varepsilon)$. The subtlety of the latter approach lies in the fact that when $\varepsilon \to 0$ the ε-dependent domains of the configuration-space, where the expansions (112a, b) are *a priori* valid, are not required to overlap (and in general they do not).

One can say that, as seen from the post-Newtonian point of view, the post-Minkowskian expansions constitute a first class 'cure' for the wave-zone breakdown of PN expansions (see eq. (108)). Now the usual post-Minkowskian expansion itself exhibits a kind of (weak) breakdown in the 'exponentially far-wave zone'

$$r \gtrsim \lambda \exp\left\{\frac{c^2 \lambda}{4\pi Gm}\right\}, \tag{113}$$

because of the appearance of $\log r/r$ terms, when $r \to \infty$, in $h_2^{\mu\nu}$. However, as argued already by Fock (1959), Isaacson and Winicour (1968), and more recently by Anderson (1980) and others, this can easily be cured by a first-class method based on a (G/c^2)-dependent change of coordinates (for a complete proof that this cure works at all orders in G, see e.g. Blanchet (1987)).

The singular-perturbation methods of the second class have been found useful in several problems of relativistic gravity. They have been used, in an intuitive way, by Fock (1959, section 87) and Manasse (1963); and were introduced in a more formal way in the late sixties by Burke (1971) and Thorne (1969) to tackle the wave-zone breakdown of PN expansions. Recently they have been used by several authors: e.g. Demianski and Grischchuk (1974), D'Eath (1975a, b), Kates (1980a, b), Anderson, Kates, Kegeles and Madonna (1982), Damour (1983a), Anderson and Madonna (1983); Blanchet and Damour (1984b), Thorne and Hartle (1985), Futamase (1986). A general discussion of the formal way in which these methods fit into the framework of General Relativity has been attempted (Kates, 1981;

Damour, 1987). However, in my opinion, there are still fundamental issues concerning this second class of methods which have not been clarified. Indeed, the basic paradox of these methods is that the domains of *a priori* validity of, for instance, the two asymptotic expansions (112a) and (112b) become completely disjoint when $\varepsilon \to 0$, but still they are supposed to describe fully the physical problem at hand so that they must somehow fit together. Now the literature which deals with precise rules for the 'matching of asymptotic expansions' is full of ambiguities and controversies. In the context of General Relativity some of these ambiguities have been pointed out recently (Blanchet and Damour, 1984b; Damour, 1987). In particular, a re-examination of the problem initially studied by Burke (1971) has shown the necessity to take non-linear effects into account with great care (Blanchet and Damour, 1984b, and references therein). This is why, when using this type of method in the problem of the motion of condensed bodies, care has been taken to control in sufficient detail the structure of the non-linear corrections so as to be able to extract minimal, but reliable, information from the matching of the two expansions (Damour, 1983a, sections 5 and 6).

6.13 The external and the internal problems of motion (Einsteinian case)

Returning to the problem of motion of N widely separated bodies in general relativity, it is clear from the study of the corresponding Newtonian case (Section 6.3) that we should, at some point, try to split the problem into two sub-problems, one concerning the motion of each body as a whole ('external problem') and the other concerning the intrinsic evolution of each body ('internal problem'). However, the fact that there are more potentially small parameters in the relativistic case ($\alpha, \gamma_e, \gamma_i, \beta_e, \beta_i$) than in the Newtonian case (α) means that there are several different ways of doing this, which correspond, roughly speaking, to considering the non-relativistic and relativistic parameters in a different order.

The most commonly used way, implicitly influenced by the 'neo-Newtonian' (or the 'neo-Minkowskian') reinterpretation of general relativity, first considers the relativistic parameters and uses them to 'force' Einstein's theory to fit the Procrustean beds of Newtonian theory or of special relativity. Having thus recast the problem in the framework of an auxiliary absolute spacetime ($E_3 \times \mathbb{R}$ or M_4) the further consideration of the separation parameter α leads one 'naturally' to first seek a 'good' generalisation of the Newtonian centre of mass (7), and then to use this generalised centre of mass to formulate the external problem (the internal problem being supposed to follow without requiring special attention).

6.13.1 The centre of mass problem

Several different approaches to the definition of a 'good' centre for each body have been pursued. Some authors start by imposing some 'symmetry' (spherical or spheroidal) on each of the bodies and then consider the 'centre of symmetry'. See e.g. De Sitter (1916), Levi-Civita (1937–50), Papapetrou (1951), Misner, Thorne and Wheeler (1973), Hogan and McCrea (1974), Caporali (1981*b*). The danger with this approach is that the 'symmetry' is defined with respect to the auxiliary absolute space. But this means, in 'reality', that it is a symmetry with respect to the arbitrary global coordinate system. Therefore, it is just a fictitious 'coordinate symmetry' which has, *a priori*, nothing to do with the real physical symmetry of each object. On the contrary it is clear, because of the anisotropic 'Lorentz contraction' of moving bodies, that to impose a coordinate symmetry on moving objects means that, 'in reality', they are non-spherical (even when real tidal or spin effects are negligible) so that this requirement gives them in fact an unwanted 'ellipticity' of order β_e^2, and thereby an intrinsic quadrupole moment of order $mL^2\beta_e^2$ which gives an (erroneous) contribution to the equations of motion of the bodies of the order:

$$\delta \text{ Force} \sim \frac{Gm^2}{R^2}\alpha^2\beta_e^2. \tag{114}$$

Note that this contribution is of order $1/c^2$ (1PN order) and can become important if α is not very small or if one works to a higher post-Newtonian order. Note also that some authors are free of the preceding criticism, either because they take into account the Lorentz contraction explicitly (Lorentz and Droste, 1917), or because they endow the bodies with internal stresses suitable to give them the required coordinate symmetry (Hogan and McCrea, 1974; McCrea and O'Brien, 1978).

Other approaches to the definition of a 'good' centre of mass seek to imitate the Newtonian definition (6)–(7). The problem is thus reduced to finding a 'good' relativistic analogue, say $\hat{\rho}$, of the Newtonian mass density such that the centre of mass reads:

$$\hat{z}_a^i := \hat{m}_a^{-1}\int_{V_a} x^i\hat{\rho}(\check{x}, t)\, \mathrm{d}^3x, \tag{115}$$

with

$$\hat{m}_a := \int_{V_a} \hat{\rho}(\check{x}, t)\, \mathrm{d}^3x. \tag{116}$$

One of the initiators of this approach is Fock (1959). He begins by using the coordinate rest-mass density, eq. (82) (denoted ρ by him):

$$\hat{\rho}_1 := r_*, \tag{117}$$

but after having obtained the external equations of motion he introduces an 'effective mass' and an effective centre of mass (Fock, 1959, eqs. (78.11), (78.12); see also Brumberg, 1972, chapter 7) which correspond to modifying eq. (117) by taking into account both internal energy and self-gravitational energy relativistic corrections. More recently, it has become customary to start directly with such improved effective definitions of the 'mass' and the 'centre of mass' with the help of the following effective relativistic mass-density for the ath body (Will, 1974; Contopoulos and Spyrou, 1976; Spyrou, 1977):

$$\hat{\rho}_2 := r_* \left[1 + \frac{1}{c^2} \left\{ \frac{1}{2} (w_a^i)^2 - \frac{1}{2} U^{(s)a} + \Pi_0 \right\} \right], \tag{118}$$

using, respectively, the definitions (17)–(18), (15) and (66) for the internal velocity, self-gravitational potential and specific internal energy density.

Other authors use a purely formal approach to the centre-of-mass problem in the sense that they introduce from the start a distributional energy–momentum tensor with support located on some world-line (see e.g. Infeld and Plebanski (1960), Ohta et al. (1973), Havas (1979) and references quoted in these works). However, the crucial issue of justifying the (a priori meaningless) use of such a distributional energy–momentum tensor on the right-hand side of the non-linear Einstein field equations has, strangely enough, rarely been seriously considered. Until recently the only 'justification' for these approaches was a posteriori, in the sense that they gave final results in agreement with other 'orthodox' methods, at least at the lowest relativistic orders (1PN or 1PM). And still these methods seem a priori attractive because they simplify the calculations considerably and allow one either to quickly recover results known from complicated orthodox approaches or make explicit computations at the higher relativistic orders (2PN (Ohta et al., 1973, 1974); 2PM (Bel et al., 1981; Westpfahl, 1985)). A preliminary study of the inner consistency of the use of 'delta functions' in General Relativity has been initiated by Bel et al. (1981). But, even if one stays at the purely formal level, there seem to be obstacles to the use of 'delta functions' if one works at the third post-Minkowskian level. However, because of the efficiency of these approaches an effort has recently been made to bridge the gap between orthodox and distributional approaches. For the moment there exists no result which is valid at all orders, but it has been possible to prove (Damour, 1980, 1983a) that the calculations within a sound orthodox approach (combining a minimal matching of asymptotic expansions and an Einstein–Infeld–Hoffmann–Kerr-type approach) could be very much simplified by a mathematical trick (complex analytic continuation) which makes the calculations *resemble* formal 'delta function' calculations (see Section 6.14). However, I wish to emphasise that this resemblance is used only to simplify the explicit computations within a theoretically sound approach and does not mean that one is using a powerful, but otherwise arbitrary 'regularisation procedure' (there seems to have been some misunderstanding of this point by some authors, e.g. Futamase (1983), Schutz (1985), Grishchuk and Kopejkin (1986)).

One should also mention the work of Dixon (1979), Ehlers and Rudolph (1977) and Schattner (1978) on a covariant approach to the definition of a centre-of-mass world-line, although it is not clear how to relate this approach to the other, less covariant but more explicit, approaches.

Finally, a very different way of attacking the centre-of-mass problem proceeds the other way around. Instead of, first, defining the centre of mass as a tool to separate the problem of motion into external and internal parts, there are approaches which initially separate the external problem of motion, and then pin-point, within their external scheme, a convenient 'central point'. The initiators of this type of approach were Einstein, Infeld and Hoffmann (1938) (see also Einstein and Infeld, 1940, 1949). Modern disciples of the approach use the 'method of matched asymptotic expansions' (see e.g. D'Eath, 1975b; Damour, 1983a; and for a somewhat different approach Thorne and Hartle, 1985). In these approaches the 'central point' appears more as a 'centre of (gravitational) field' than as a 'centre of mass'.

After choosing some definition of the centre of mass of each body one can then proceed to derive its equations of motion and to discuss the internal problem.

6.13.2 Results in the external problem

It is outside the scope of this article to give a detailed account of all the results which have been obtained in the Einsteinian external problem of motion. Some of the recent results in the two-body problem will be discussed below (Section 6.15). We shall content ourselves here with giving a bird's-eye view of the work done and some references.

First let us recall that the general structure of the external equations of motion will be:

$$\frac{d^2 z}{dt^2} = G A(\alpha, \gamma_e, \gamma_i, \beta_e, \beta_i). \tag{119a}$$

Now the parameter α must necessarily be small in all situations where one wants to talk about an external problem in contrast to an internal one. Moreover, as $\gamma_e \lesssim \alpha$ (eq. (78)) there are certainly at least two small parameters among the five appearing in eq. (119a). Therefore we can expect that, in the most general case, it should be possible to write the following expanded form of the external equations of motion:

$$\frac{d^2z}{dt^2} = G \sum_{m,n \geq 0} A_{mn}(\gamma_i, \beta_e, \beta_i)\alpha^m\gamma_e^n. \tag{119b}$$

In order to simplify our presentation we shall introduce only powers of the small parameters, but the presence of more complicated scale functions for the expansion (such as $\alpha^m(\log \alpha)^{m'}, \ldots$) can easily be coped with. This kind of expansion (119b) does not require that the internal gravity γ_i, or the velocities, β_e, β_i, be small. It is this kind of expansion which appears in some post-Minkowskian approximation schemes for the external gravitational field, completed by some matching to the possibly strong internal gravitational field (sometimes formally modelled by the use of distributional sources). A term of the type $GA_{mn}\alpha^m\gamma_e^n$, which contains $(n+1)$ powers of the coupling constant G, is said to belong to the $(n+1)$-PM approximation level. The usual post-Minkowskian schemes (requiring everywhere-weak gravitational fields without conditions on the velocities) would mean that one is further expanding eq. (119b) in powers of γ_i:

$$\frac{d^2z}{dt^2} = G \sum_{m,n,p \geq 0} B_{mnp}(\beta_e, \beta_i)\alpha^m\gamma_e^n\gamma_i^p. \tag{119c}$$

Such schemes have never been developed in detail. On the other hand, the usual post-Newtonian schemes (requiring everywhere-weak gravitational fields and small velocities) lead to a fully expanded form:

$$\frac{d^2z}{dt^2} = G \sum_{m,n,p,q,r \geq 0} C_{mnpqr}\alpha^m\gamma_e^n\gamma_i^p\beta_e^q\beta_i^r. \tag{119d}$$

A term of the type $GC_{mnpqr}\alpha^m\gamma_e^n\gamma_i^p\beta_e^q\beta_i^r$, which contains $(2n+2p+q+r)$ powers of $1/c$, is said to belong to the $(n+p+\frac{1}{2}(q+r))$-PN approximation level.

Explicit results of the minimally expanded form (119b) have been obtained at the 1PM level by several authors, starting with the pioneering works of Bertotti (1956) and Havas (1957), see references in Havas (1979), Damour (1983a) and Westpfahl (1985). For explicit results at the 2PM level see Westpfahl and Goller (1979), Bel et al. (1981) and Westpfahl (1985). The 3PM level has been partially tackled (Damour, 1982, 1983c) keeping only the terms necessary to derive the 2.5PN equations of motion (of the form (119d) without γ_i-expansion). The 1PM equations of motion of spinning bodies have been studied by Ibanez and Martin (1982).

Explicit results of the fully expanded form (119d) have been obtained at the 1PN level by many authors starting with the pioneering work of Lorentz and Droste (1917). Note that the latter work is the first one to give the correct 1PN equations of motion, the earlier results of Droste and De Sitter and the later results of Chazy and Levi-Civita are incorrect. For references to more recent works see e.g. Brumberg (1972), Ehlers (1977, 1984), Caporali (1981b), Damour (1983a), Grischchuk and Kopejkin (1986) and a long-promised (but not yet published) review article by P. Havas. Special mention should be given to the problem of the motion of spinning bodies which involves several subtle issues (centre of mass, ordinary versus acceleration-dependent Lagrangian). This problem has been reviewed by Barker and O'Connell (1979). Further references to the works of Michalska, Gabosz, Lanzano and Brumberg will be found in Brumberg (1972). Recent works are due to Damour (1978, 1982), Dallas (1977), Bel and Martin (1981), Ibanez and Martin (1982) and Thorne and Hartle (1985). Concerning the issue of the presence of accelerations in the Lagrangian for spinning bodies and of their avoidance by use of a suitable non-covariant spin-condition see e.g. Damour (1982).

Explicit results at the 2PN level have been obtained by Ohta *et al.* (1974b) for N point-masses, by Damour and Deruelle (1981a) and Damour (1982, 1983a) for two condensed objects, by Schäfer (1985) for N point-masses, and by Kopejkin (1985) and Grishchuk and Kopejkin (1986) for two weakly self-gravitating fluid balls. The 2PN equations of motion can be derived from a Hamiltonian as first shown by the work of Ohta, Okamura, Kimura and Hiida (1974b). However, there are several subtle issues connected with this Hamiltonian and with its Legendre-transform associated Lagrangian. On the one hand, one term in the potential part of this Lagrangian, for the two-particle case, has been wrongly evaluated in Ohta *et al.* (1974a), as was first remarked in Damour (1982). The correct evaluation has been given in Damour (1983a) and Damour and Schäfer (1985). More important is the issue of the influence of the choice of the coordinate system on the Lagrangian. Indeed, it has been shown by Damour and Deruelle (1981b) and Damour (1982) that the 2PN motion of two condensed bodies *in harmonic coordinates* could be deduced only from a *generalised* Lagrangian function involving not only positions and velocities but also accelerations. This issue has been recently clarified by Damour and Schäfer

(1985), following a remark of Schäfer (1984) that velocity-dependent coordinate transformations introduce an acceleration dependence in the Lagrangian. In particular it has been explicitly checked that the ordinary Lagrangian of Ohta *et al.* (1974*b*), corresponding to a non-harmonic coordinate system, is indeed equivalent (in the two-body case) to the generalised Lagrangian of Damour (1982), corresponding to a harmonic coordinate system. On the other hand, the ordinary Lagrangian of Ohta *et al.* (1974*a*) is incorrect; it should depend on accelerations.

Explicit results at the 2.5PN level have been obtained by Damour (1982, 1983*a*), Schäfer (1983, 1985), Kopejkin (1985) and Grishchuk and Kopejkin (1986). These will be discussed in Section 6.15.

An important issue which distinguishes the fully expanded post-Newtonian equations of motion (119*d*) (as well as the intermediate post-Minkowskian expansion (119*c*)) from the minimally expanded eqs. (119*b*) is the explicit appearance of the internal field-strength parameter γ_i. This means, therefore, that the expansions (119*c*) or (119*d*) can be applied *a priori* only to weakly self-gravitating bodies. This limitation is very serious because modern astrophysics often has to deal with the motion of strongly self-gravitating bodies (also called condensed bodies, like neutron stars or black holes). However, several authors (McCrea, Grishchuk and Kopejkin, ...) have argued that it was justified to use the post-Newtonian method even for strongly self-gravitating bodies because explicit calculations showed that the way the parameter γ_i entered the eqs. (119*d*) was such that it could be incorporated into some 'effective' mass $m = m_0 + \gamma_i m_1 + \cdots$ (this 'effacement' property will be discussed in more detail below (Section 6.14)). After this 'renormalisation' of the mass the resulting equations of motion contain the external relativistic parameter γ_e but no longer contain the internal one γ_i, so that they seem to apply as well to weakly as to strongly self-gravitating objects. However, I wish to point out that, although this argument is certainly correct (except for black holes) if one proves the 'effacement' of the γ_is to all orders of approximation, it is definitely insufficient if one proves the effacement of the γ_is only up to some finite order. More precisely, we are going to show the surprising result that the usual way of truncating and then renormalising the PN expansion (119*d*) is valid only at the *first* post-Newtonian approximation and is already inconsistent at the 2PN level!

Indeed, in all explicit post-Newtonian calculations of the external equations of motion, working say at the *n*PN order, one is discarding not only all the terms coming from the higher PN levels but also many terms depending on the separation parameter, α, so that the uncontrolled relative error term in an explicit nPN equation of motion will be

$$\varepsilon \gtrsim \alpha^5 + \alpha^2 \beta_e^2 + \gamma_i^{n+1}, \tag{120a}$$

where the first term comes from the tidally induced quadrupole force between deformable bodies (see Section 6.14), the second term is the effect discussed in eq. (114) which should be included in a consistent calculation using coordinate-spherical bodies (as often assumed), and the third term is expected to be the dominant term coming from the $(n+1)$PN approximation (because $\beta_e^2 \sim \gamma_e \lesssim \gamma_i$). We are here neglecting the fact that at higher PN approximations the error terms seem no longer to be simple powers of the PN parameters but to include some logarithms. Now the point is that α and γ_i are *not* independent because $\gamma_i = Gm/c^2 L = (Gm/c^2 R) \times (L/R)^{-1} = \gamma_e \alpha^{-1}$, so that for a given *external* relativistic parameter, $\gamma_e \sim \beta_e^2$, the internal gravitational field strength, γ_i, increases when the bodies become more separated (note that this means that the expansions (119c) and (119d) were not real power expansions with respect to several independent parameters but were defined, for instance, by the re-expansion of the original coefficients $A_{mn}(\gamma_i)$ as $\gamma_i \to 0$). For simplicity's sake let us consider only the first and the third term on the right-hand side of (120a) because they are there independently of the wrong use of coordinate symmetry (the taking into account of the second term strengthens the conclusions below without changing them qualitatively). Then the relative error, ε, will certainly exceed a minimum value reached when $\alpha^5 \sim \gamma_i^{n+1} = \gamma_e^{n+1} \alpha^{-n-1}$, i.e. for an optimal $\alpha \sim \gamma_e^{(n+1)/(n+6)}$. Hence

$$\varepsilon \gtrsim \gamma_e^p, \tag{120b}$$

with

$$p = 5 \frac{n+1}{n+6}. \tag{120c}$$

Now if $n = 1$ (1PN approximation), $p = \frac{10}{7} > 1$, so that it is consistent to retain the 1PN corrections in the external equations of motion $(\sim \beta_e^2 + \gamma_e)$ and to discard all the other uncontrolled terms hidden in ε. 'Consistency' means here that there exists a limit process, in which γ_e, γ_i, and α tend to zero, for which ε becomes asymptotically negligible with respect to the terms explicitly retained. On the other hand, if $n = 2$ (2PN level), $p = \frac{15}{8} < 2$, so that for *any* limit process the last terms retained in the external equations of motion $(\sim (\beta_e^2 + \gamma_e)^2)$ are necessarily negligible compared with the uncontrolled error terms contained in ε. And this situation gets worse as n increases. This conclusion does not mean that the final 2PN or 2.5PN results obtained by several authors (explicitly dealing with extended fluid bodies) are incorrect, but it means that they have not yet been justified within the PN-fluid schemes. More work is needed to control the effect of the higher-order self-field effects.

6.13.3 The internal problem

In Newtonian gravity the internal problem means: 'to determine the motion of each body around its centre of mass'. Technically this means (Section 6.3) introducing the relative coordinates, y_a^i, expressing the position of each material point of the ath body with respect to the chosen centre of mass, $z_a^i(t)$, so that (see eq. (17))

$$x^i = z_a^i(t) + y_a^i. \tag{121}$$

Many authors, working in the post-Newtonian framework, have, more or less implicitly, assumed that the mathematical formula (121) was sufficient to express the passing to a 'centre-of-mass frame' in Einsteinian gravity. In principle this view is admissible because the coordinate systems are arbitrary in General Relativity. This is the view which is often taken in relativistic celestial mechanics to express, for instance, the motion of the Moon, or of an artificial satellite, with respect to the Earth (see e.g. Brumberg, 1972; Lestrade *et al.*, 1982; Martin *et al.*, 1985). However, in order to be able to use this viewpoint consistently one must also *write* the equations for the internal problem of each body using the spacetime coordinates $(y_a^i, t \equiv x^0/c)$ and *solve* them with the adequate precision. For a long time this internal problem was neglected because the accuracy of the measurements allowed one to use the results of the Newtonian internal problem, the relativistic 'corrections' being taken into account only in the external problem (and considered only through their *secular* effects). However, the great increase in accuracy of the measurements now available (concerning both the dynamics of natural or artificial bodies in the solar system, and the dynamics of binary pulsars) obliges us to tackle seriously the problem of a fully consistent relativistic description of the internal problem. Then if one tries to really use the coordinate system (y_a^i, t) defined by eq. (121) (with $z_a^i = \hat{z}_a^i$ being one's preferred relativistic centre of mass) this will lead to an internal problem which will have a structure of the type:

$$0 = \frac{dX_i}{dt} + I_0(X_i) + \beta_e^2 I'(X_i, X_e) + \gamma_e I''(X_i, X_e) + \cdots + \alpha^3 I_3(X_i, X_e) + \cdots, \tag{122}$$

to be compared with the Newtonian equation (55b). In the Newtonian case the influence of the external structure on the internal problem was *effaced* up to the order α^3. But here we see that there already is a strong influence of the external motion on the internal one of order β_e^2 or γ_e. The presence of the β_e^2-coupling is readily understood if one remembers that with the transformation (121) one is using the external coordinate time, t, as internal time. Then this means that each body will suffer an apparent 'Lorentz-contraction' in the direction of external motion by a factor $\sqrt{\{1-(v^{\text{ext}}/c)^2\}} \simeq 1-\frac{1}{2}(v^{\text{ext}}/c)^2$. Similarly, the γ_e-coupling in eq. (122) can be related to what I will call the apparent 'Einstein-contraction', i.e. the factor linking the coordinate length to the proper length, as was first discussed by Einstein (1916a). In harmonic coordinates this apparent 'Einstein-contraction' of a body embedded in an external gravitational potential $U^{(e)a}$ is isotropic and is given by a factor $\simeq 1 - U^{(e)a}/c^2$. The Lorentz contraction effects mean that (when real tidal effects are negligible or subtracted) a point on the Earth's surface suffers, in the global post-Newtonian coordinate system, an anisotropic (coordinate) displacement of maximal amplitude (in the direction of velocity of the Earth $\hat{v}_\oplus$)

$$\frac{1}{2}\left(\frac{v_\oplus}{c}\right)^2 R_\oplus \sim 3 \text{ cm}.$$

Such a displacement is not negligible in some of the very precise modern measurements.

As for the 'Einstein-contraction' effects they are less dangerous because they are isotropic. Because of the small eccentricity of the Earth's motion they produce an apparent radial contraction of Earth of mean value ~ 7 cm with a yearly variable part of millimetric amplitude. It is, however, historically interesting that it is essentially neglect of this Einstein contraction which caused De Sitter (1916) to make a mistake in his derivation of the equations of motion of N bodies at the first post-Newtonian approximation. On the other hand, Lorentz and Droste (1917) carefully take into account both the 'Lorentz' and the 'Einstein' contraction, correctly pointing out that, when real tidal effects are negligible, intrinsically spherical bodies will appear, in the coordinate system, as ellipsoids of coordinate volume $V \simeq V_0(1 - \frac{1}{2}(v^{\text{ext}}/c)^2 - 3U^{(e)}/c^2)$. See Martin *et al.* (1985), and Hellings (1986) for heuristic discussions of these effects in the modern context of global post-Newtonian descriptions of laser tracking of satellites and VLBI.

All the preceding subtleties (see also (114)) indicate that the usual global post-Newtonian way of approaching the internal problem may be, in the long term, more trouble than help. At this point it is worth looking back carefully at eq. (121) which, beyond its simple mathematical structure, hides a non-trivial physical phenomenon, namely an accelerated frame of reference with respect to which the external gravitational effects are strongly effaced, leaving only small tidal effects (Section 6.5). Now, if we were considering the internal problem of a *negligibly self-gravitating* body, the relativistic analogues of such an 'external-gravitational-field-effacing' frame would be the well-known 'locally inertial frames' which can be explicitly constructed by means of Fermi coordinates based on the centre-of-mass world-line (see e.g. Synge, 1960; Misner, Thorne and Wheeler, 1973). If we consider, however, a self-gravitating body '*a*', then there exists, at present, no general definition of such an 'external-gravitational-field-effacing' frame, and in fact no proof that such a definition is possible. However, it has been remarked by Bertotti (1954) that, at the first post-Newtonian approximation, the external motion of a self-gravitating body, '*a*', could be interpreted as the external motion of a test body (non-self-gravitating) in some effective external gravitational field, say $g^{(e)a}_{\mu\nu}$ (this property is related to the effacement of internal structure in the external problem, see next section). This result suggests that a good definition of an 'external-gravitational-field-effacing' frame consists of constructing a 'locally inertial frame' in the effective external field $g^{(e)a}_{\mu\nu}$ based on the centre-of-mass world-line of the *a*th body. At the heuristic level such a construction has been more or less explicitly assumed by many authors (see e.g. Will, 1974; Misner *et al.*, 1973; Will and Eardley, 1977; Mashhoon, 1984, 1985; Thorne and Hartle, 1985). The study of the motion of strongly self-gravitating bodies (neutron stars or black holes) has also made use of such 'external-field-effacing frames' (Y^i, T) linked with the external coordinate system x^μ by transformation formulas of the type:

$$x^\mu(T, Y^i) = z^\mu(T) + e^\mu_i(T)Y^i + O((Y^i)^2) + \cdots, \tag{123}$$

instead of the Newtonian-type formula (121) (see e.g. D'Eath (1975*b*), eq. (2.10), Damour (1983*a*), section 6, eq. (7)). Recently Ashby and Bertotti (1984, 1986) have explicitly constructed, within the first post-Newtonian approximation, the Fermi coordinate transformation (of the type (123)) based on the effective external metric $g^{(e)a}_{\mu\nu}$ for the motion of a self-gravitating body, and have verified that it has the good 'external-field-effacing' properties. This explicit result supports one's expectation that a good internal frame exists in the general case, but much work remains to be done to prove this expectation. The 'internal frame problem' has also recently become important in the relativistic celestial mechanics of the solar system (see the contributions of Bertotti, of Boucher, of Fukushima, Fujimoto, Kinoshita and Aoki, and of others in Kovalevsky and Brumberg (1986)).

6.14 The effacement of internal structure in the external problem (Einsteinian case)

As recalled in Sections 6.4–6.6, several independent circumstances (Action and Reaction Principle, Equivalence Principle, properties of multipole expansions) imply that the influence of the internal structure on the external motion is strongly reduced by a factor α^4 from *a priori* order of magnitude expectations. It was remarked by M. Brillouin (1922) that the arguments by which one can 'efface' the internal structure in Newton's theory cannot be immediately translated into Einstein's theory (especially because of the strong-non-linearity of the latter). This problem has been clearly formulated by Levi-Civita (1937*a*, 1950). He proved by detailed arguments within Einstein's theory that this 'effacement' property was still valid at the first post-Newtonian approximation, in the sense that: all large direct self-action effects cancel or contribute terms in the equations of motion which can be renormalised away, so that the final equations of motion can be written in terms of only some 'centres of mass' and some 'effective masses'. Actually these results had been previously obtained, though only implicitly, in the first correct work on 1PN equations of motion: Lorentz and Droste (1917). On the other hand, many works on the equations of motion (De Sitter, 1916; Chazy, 1930; Eddington and Clark, 1938) had assumed the effacement property without proof. Later, Fock (1959) remarked that the effective mass of each body, at 1PN order, can be written as

$$m_a = \int_{V_a} \mathrm{d}^3x \, r_* \left[1 + \frac{1}{c^2} \left\{ \frac{1}{2}(w_a^i)^2 - \frac{1}{2} U^{(s)a} + \Pi_0 \right\} \right], \qquad (124)$$

i.e. as the sum of its total rest-mass energy, of its internal kinetic energy, of its self-gravitational energy and of its internal elastic energy.

Recently the extension of the property of effacement beyond the first-post-Newtonian (1PN) approximation has been studied further. On the one hand, some authors, staying within the post-Newtonian framework (i.e. for weakly self-gravitating, slowly moving bodies) have shown that it still holds at the $2\frac{1}{2}$th post-Newtonian approximation (Grishchuk and Kopejkin, 1986; and Kopejkin, 1985, improving on previous partial results of Rudolph and Börner; and Breuer and Rudolph, 1982). They have shown (in the absence of internal motion, and for 'spherically symmetric' bodies) that the effective mass is equal, up to order c^{-6}, to the 'Tolman mass' of each body.

On the other hand, other authors (D'Eath, 1975b; Damour, 1983a; Thorne and Hartle, 1985) have shown that the effacement property still holds for strongly self-gravitating bodies. In particular, it has been proven that the equations of motion of non-rotating 'elastic' condensed bodies (neutron stars or black holes), which would be spherically symmetric if isolated, can be written in terms of only some 'centres of field' and some 'Schwarzschild masses' (Damour, 1983a) up to the appearance of very small internal-structure dependent terms of relative order γ_e^5 in the equations of motion (i.e. *tenth* post-Newtonian order in the usual counting!). The intuitive reason for this extremely good effacement is that each 'elastic' body is supposed to be in a stable spherically symmetric state when isolated so that when it is not isolated it becomes distorted by 'tidal forces' GmL/R^3, thereby acquiring a small (structure dependent) ellipticity

$$\varepsilon \sim \frac{\text{tidal gravity}}{\text{self gravity}} = \frac{GmL/R^3}{Gm/L^2} = \left(\frac{L}{R}\right)^3 = \alpha^3.$$

Then the corresponding tidally induced quadrupole moments of each body, εmL^2, introduce some (structure dependent) interbody forces $\sim \varepsilon m^2 L^2/R^4$ which are therefore $\varepsilon L^2/R^2 \sim (L/R)^5 = \alpha^5$ smaller than the main Newtonian forces. Now, by definition, condensed bodies are such that $L \sim Gm/c^2$, so that

$$\alpha = \frac{L}{R} \sim \frac{Gm}{c^2 R} = \gamma_e,$$

hence the very small relative correction $\alpha^5 \sim \gamma_e^5 = (Gm/c^2 R)^5 = O(1/c^{10}) =$ 10PN order.

It has been noted that the property of effacement of the internal structure in the external motion of condensed bodies seemed to be a very peculiar property of Einstein's theory and that in other relativistic theories of gravity the equations of motion of condensed bodies already contain, at the 1PN level ($O(1/c^2)$ relative corrections), some internal-structure-dependent parameters (see Will, 1981, and references therein). It is therefore worth trying to understand in a non-complex way what are the elements of structure of general relativity which allow the effacement of internal structure. In the proofs quoted above these elements of structure are somewhat lost in the midst of long calculations or complex arguments. However, it is possible to present a proof of the effacement property within Newtonian theory which is equivalent but different from the usual straightforward one presented in Section 6.5 and which shows up some of the basic elements of structure which allow one to prove the effacement property within Einstein's theory. I shall therefore spell out some details of this proof which can serve as a nice model problem for understanding some of the 'inner wheels' of general relativity.

The basic trick is to rewrite the Euler equations (3) in the form of a conservation law (using the continuity equation (2) and Poisson's law (4a)):

$$\frac{\partial}{\partial t}(\rho v^i) + \frac{\partial}{\partial x^j}(\rho v^i v^j) + \frac{\partial}{\partial x^j} T^{ij} + \frac{\partial}{\partial x^j} t^{ij} = 0, \tag{125}$$

where

$$T^{ij} = +p\delta^{ij}, \tag{126}$$

is the material stress tensor, and where

$$t^{ij} = +\frac{1}{4\pi G}\left[\frac{\partial U}{\partial x^i}\frac{\partial U}{\partial x^j} - \frac{1}{2}\delta_{ij}\frac{\partial U}{\partial x^k}\frac{\partial U}{\partial x^k}\right] \tag{127}$$

is a quadratic form in the gravitational field, which is the gravitational analogue of Maxwell's electromagnetic stress tensor. Note that t^{ij} satisfies the identity

$$\frac{\partial t^{ij}}{\partial x^j} \equiv +\frac{1}{4\pi G}\Delta U\frac{\partial U}{\partial x^i}. \tag{128}$$

Now if we integrate eq. (125) over *any* volume, V, containing the ath body and no other body in the system, the second and third terms of the left-hand side will give no contributions because of Gauss's theorem and the vanishing of ρ and p outside the body. Using eqs. (7)–(8) we therefore obtain the following equations of external motion:

$$m_a\frac{d^2 z_a^i}{dt^2} = F^i_{(e)}, \tag{129a}$$

where the external force is given as an integral over *any* surface S enclosing the ath body (and no other body):

$$F^i_{(e)} = -\int_S dS_j t^{ij}. \tag{129b}$$

Eqs. (128)–(129) are the Newtonian analogues of the famous Einstein–Infeld–Hoffmann surface-integral approach to the problem of motion in general relativity. Note that because of Gauss's theorem and the identity (128) the value of the surface integral (129b) is independent of the choice of the surface, S, as long as it stays outside the matter (so that U satisfies Laplace's equation).

Now the expression (129b) for the external force depends only on the gravitational field *outside* the bodies, say U^{out}. From eq. (40) U^{out} can be written as:

$$U^{\text{out}} = U^m + U^Q + O^{\text{ext}}(\alpha^3),\qquad(130)$$

with

$$U^m := \sum_a \frac{Gm_a}{|\vec{x} - \vec{z}_a(t)|},\qquad(131)$$

and

$$U^Q := \sum_a \frac{1}{2} GQ_a^{ij}(t)\, \frac{\partial^2}{\partial x^i\, \partial x^j}\left(\frac{1}{|\vec{x} - \vec{z}_a(t)|}\right)\qquad(132)$$

(actually we are now cheating a little because we are using the vanishing of the dipole term in U^{out} which, however, has been proven only by an argument using some knowledge of the internal structure). The effacement of internal structure in the equation for the external motion (129) is now clearly apparent if we choose for the arbitrary surface S a surface of dimensions $\sim R$, the interbody separation, and if we use (for a while) the external units (eq. (34)). Then both U^m and $U^m{}_{,i}$ are of (numerical) order unity, while U^Q and $U^Q{}_{,i}$ are of order α^2. Therefore

$$F^i_{(e)} = F^i_m + O(\alpha^2),\qquad(133)$$

where

$$F^i_m := -\int_S dS_j t^{ij}_m\qquad(134)$$

is of numerical order unity, with

$$t^{ij}_m := +\frac{1}{4\pi G}\left[U^m{}_{,i}U^m{}_{,j} - \tfrac{1}{2}\delta_{ij}U^m{}_{,k}U^m{}_{,k}\right].\qquad(135)$$

Therefore the preceding surface-integral approach allows one, without any calculation, to conclude that, up to order α^2, the equations of motion involve only the centres of mass and the masses (because F^i_m cannot depend on anything else). However, we need an explicit calculation to work out F^i_m. There are several ways of doing this; let us spell out one way which has been found useful in the relativistic generalisation of this approach.

First we notice that, although in the first formulation (129b) it was necessary for S to stay outside the ath body, therefore a finite distance away from $\vec{z}_a$, in the final (approximate) formulation (134) S can be any surface containing the point $\vec{z}_a$ (and no $\vec{z}_b$, $b \neq a$) because of Gauss's theorem and the identity (128). In particular, if convenient, we can let S shrink towards $\vec{z}_a$ without changing the value of F^i_m. But then we see that t^{ij}_m will tend to infinity because of the singularity of U^m near $\vec{x} = \vec{z}_a$. Evidently, it is guaranteed *a priori* that F^i_m will be finite and that the pole-like singularity is not 'physically real'. But still, at the computational level, past experience in many fields of physics has shown that it is efficient to use some idealised mathematical tools (point particles, delta functions, plane waves, ...) to deal with real physical situations. However, one then needs to have a clear mathematical framework to use such tools (distribution theory, Fourier transform, ...). In our case distribution theory is not appropriate to deal with t_m^{ij} which is quadratic in U^m. We can, however, use the following trick (inspired by the work of M. Riesz, 1949). Let us consider, if we are studying the motion of the ath body,

$$U_A^m := Gm_a r_a^{A-1} + U_{(e)a}^m, \tag{136}$$

where $r_a := |\dot{x} - \dot{z}_a(t)|$, A is a complex number, and

$$U_{(e)a}^m := \sum_{b \neq a} \frac{Gm_b}{|\dot{x} - \dot{z}_b(t)|}. \tag{137}$$

Now let us define, for a given surface S, the following complex function of the complex variable A:

$$F_m^i(A) := -\int_S dS_j t_m^{ij}(A), \tag{138}$$

with

$$t_m^{ij}(A) := \frac{1}{4\pi G} \left[\frac{\partial U_A^m}{\partial x^i} \frac{\partial U_A^m}{\partial x^j} - \frac{1}{2} \delta_{ij} \frac{\partial U_A^m}{\partial x^k} \frac{\partial U_A^m}{\partial x^k} \right]. \tag{139}$$

It is easily verified that $F_m^i(A)$ is differentiable with respect to A (because r_a is always non-zero on S) and that

$$F_m^i(0) = F_m^i. \tag{140}$$

(Note that for $A \neq 0$, $F_m^i(A)$ depends also on the choice of S.)

If we consider now $t_m^{ij}(A)$ as a field defined within S (and not only on S), it is a smooth function of the field point, $\dot{x}$, *except perhaps at* $\dot{x} = \dot{z}_a$ near which, using

$$\frac{\partial r_a^{A-1}}{\partial x^i} = (A-1)(x^i - z_a^i)r_a^{A-3}, \tag{141}$$

it is seen to behave like

$$(x^i - z_a^i)(x^j - z_a^j)r_a^{2A-6} - \tfrac{1}{2}\delta_{ij}r_a^{2A-4}.$$

Hence if the real part of A is large enough (for instance, Real$(A) > 3$) $t_m^{ij}(A)$ will define a field which is differentiable at all points within S, including $\dot{x} = \dot{z}_a$. Then for such large A we can apply Gauss theorem to transform the surface-integral expression (138) into an integral over the volume V

contained within S:

$$F^i_m(A) = -\int_V d^3x \, \frac{\partial t^{ij}_m(A)}{\partial x^j}.$$

(142)

Now the identity (128) also holds for $t^{ij}_m(A)$; hence

$$\frac{\partial t^{ij}_m(A)}{\partial x^j} = \frac{1}{4\pi G} \, \Delta U^m_A \frac{\partial U^m_A}{\partial x^i}$$

$$= \frac{m_a}{4\pi} \, \Delta(r_a^{A-1}) \left[Gm_a \frac{\partial r_a^{A-1}}{\partial x^i} + \frac{\partial U^m_{(e)a}}{\partial x^i} \right],$$

(143)

where we have used $\Delta U^m_{(e)a} = 0$ within V.

Using also

$$\Delta(r_a^{A-1}) = A(A-1)r_a^{A-3},$$

(144)

we see that if we develop the right-hand side of eq. (143) the first term will contribute to $F^i_m(A)$ the quantity:

$$-\int_V d^3x \, \frac{Gm_a^2}{4\pi} A(A-1)^2 (x^i - z_a^i) r_a^{2A-6},$$

(145)

which is clearly zero by symmetry arguments if we choose for S a sphere of radius R_a centred on $\overset{\circ}{z}_a$ (the method works as well for arbitrary surfaces but we wish to simplify the presentation). We are left with

$$F^i_m(A) = -\frac{m_a}{4\pi} A(A-1) \int_V d^3x \, r_a^{A-3} \frac{\partial U^m_{(e)a}}{\partial x^i}.$$

(146)

Both the left-hand and the right-hand sides of eq. (146) are analytic functions of A in the domain Real$(A) > 0$ (such that the volume integral is convergent). As they coincide for Real$(A) > 3$, as has been just shown, they will also necessarily coincide for Real$(A) > 0$. Now when $A \to 0$, the factor in front of the right-hand side of eq. (146) shows that only the infinitesimal neighbourhood of $\overset{\circ}{x} = \overset{\circ}{z}_a$ contributes to the limit $F^i_m(0) = F^i_m$ (see (140)). Plugging into eq. (146) the Taylor expansion of $U^u_{(e)a,i}$ around $\overset{\circ}{x} = \overset{\circ}{z}_a$ it is easily seen that only the first term contributes to the limit:

$$F^i_m(0) = F^i_m = \lim_{A \to 0} \left\{ -\frac{m_a}{4\pi} A(A-1) \frac{\partial U^m_{(e)a}}{\partial x^i} (\overset{\circ}{z}_a) \int_0^{R_a} 4\pi r_a^2 \, dr_a r_a^{A-3} \right\}$$

$$= +m_a \frac{\partial U^m_{(e)a}}{\partial x^i} (\overset{\circ}{z}_a),$$

(147)

as shown by an easy calculation.

The final result, combining eqs. (128), (133), (137), (147), is

$$m_a \frac{\mathrm{d}^2 z_a^i}{\mathrm{d}t^2} = m_a \sum_{b \neq a} \frac{\partial}{\partial z_a^i} \left(\frac{Gm_b}{|\vec{z}_a - \vec{z}_b|} \right) + O(\alpha^2), \tag{148}$$

in agreement with eq. (41).

In summary, the pith of this 'complex analytic continuation' consists of replacing the initial computation of the 'real' total force acting on the ath body,

$$F_a^i = \int_{V_a} \mathrm{d}^3 x \rho U_{,i} = -\frac{1}{4\pi G} \int_{V_a} \mathrm{d}^3 x \, \Delta U U_{,i}, \tag{149}$$

where ρ is the 'real' density and U the real total gravitational potential (self + external), by the computation of the analytic continuation to $A=0$ of

$$F_m^i(A) = -\frac{1}{4\pi G} \int_V \mathrm{d}^3 x \, \Delta U_A^m U_{A,i}^m, \tag{150}$$

which contains an A-modified gravitational potential associated with the 'point-mass' $(\vec{z}_a, m_a)$. But, although, computationally speaking, this method can be seen as a 'point-mass + analytic regularisation' method, we recall that the real origin of this method lay in the surface-integral (129b) of the 'real' gravitational potential outside the body. A generalisation of this method to the Einsteinian case has recently been used in the computation of the equations of motion of two condensed bodies (Damour, 1983a). And similarly, although the method resembles a 'regularised point-mass' calculation, it is in fact based on a rigorous surface-integral formulation of the external equations of motion essentially due to Einstein, Infeld and Hoffmann (1938) and to Kerr (1959).

The lesson of the preceding discussion is that both in Newton's and in Einstein's theory we can replace the usual Newtonian view of the gravitational interaction as a set of forces acting on, and being produced by, each element of matter, by a view, say *à la* Faraday, where the gravitational field *between* the bodies contains a stress distribution which acts on the bodies via its integrated effects on surfaces enclosing them. The advantage of the second formulation is that it allows one to prove more simply (because one does not need to look at the detailed cancellation of self-forces), and in a more general case (allowing strongly self-gravitating bodies instead of the only weakly self-gravitating bodies which are within the reach of post-Newtonian methods), the property of effacement of internal structure in the general relativistic external problem of motion (for different implementations of this approach 'from outside', see e.g. D'Eath, 1975b;

Damour, 1983a; Thorne and Hartle, 1985, and references therein to the earlier works of Weyl, Einstein and others).

6.15 The problem of gravitational radiation damping and the relativistic Laplace effect

The discovery of the binary pulsar 1913 + 16 (Hulse and Taylor, 1975), a seemingly clean system of two condensed objects swiftly orbiting around each other, has raised the hope of observing effects due to gravitational radiation damping. Indeed, the measurement of the arrival times on Earth of the radio pulses emitted by one of the objects (the pulsar) allows one to track the orbital motion of the pulsar with remarkable accuracy (now achieving ± 6 km both on short time scales, 5 min; and on very long ones, ten years). Therefore, even very small damping forces in the system will become observable by their secular effects on the orbital motion. Such an effect has indeed been observed by Taylor and coworkers (Taylor, Fowler and McCulloch, 1979; Taylor and Weisberg, 1982) with sign and magnitude in excellent agreement (now better than 4 per cent) with the so-called 'quadrupole formula', a formula which had been previously derived (Peters and Mathews, 1963; Peters, 1964) by inferring the secular effects on the orbital motion of the loss of energy in the form of gravitational radiation. This suggests the conclusion that, for the first time, a clear proof of the existence of gravitational radiation, with the properties predicted by General Relativity, has been obtained.

Because of the importance of this conclusion one must critically examine both the hypothesis and the chain of deductions leading to it. We shall not examine here the hypotheses concerning the structure (two condensed bodies) and the 'cleanness' (no other effects perturbing the orbital motion) of the system (see Will, 1981; Taylor and Weisberg, 1982; Damour and Deruelle, 1986). Instead we shall concentrate on the theoretical arguments used in the derivation and application of the 'quadrupole formula'.

We do not wish to review in detail the controversy about the meanings and the derivations of several 'quadrupole formulae' and the related problem of gravitational radiation damping (see e.g. Ehlers et al., 1976; Damour, 1983a; Walker, 1984; Ehlers and Walker, 1984; Schutz, 1985; Will, 1986, and references therein). However, we wish to emphasise that most of the work which has been done concerning the quadrupole formulae has only an indirect bearing on the problem of explaining the secular acceleration of the orbital motion of PSR 1913 + 16 observed by Taylor and coworkers. Indeed, one class of quadrupole formulae gives information about a 'gravitational wave energy flux', say F, in the wave-zone. However, this outgoing flux has never been convincingly related to the time derivative of some 'local energy of the system' given as an explicit functional of the *instantaneous* state of the system with an accuracy comparable with the (very small) outgoing flux. Indeed, if we know nothing about the functional dependence of some 'energy of the system', E, the equation,

$$\frac{dE}{dt} = -F, \tag{151}$$

says nothing about the evolution of the state of the system (see e.g. sections 13 and 15 of Damour, 1983a, for a more precise discussion of this point). Moreover, even if we had in hand a well-defined (and well-proven) conservation law of the type (151), we would have still to face the problem that all derivations of eq. (151) deal with the *total* energy flux, F, and the *total* 'energy of the system', E. Now in the case of the binary pulsar there is, roughly speaking, as much energy flux in the form of electromagnetic waves as there is in the form of gravitational waves, and as much spin kinetic energy as orbital energy. Therefore, a total (non-detailed) energy balance equation (151) says, *a priori*, nothing about the *orbital* evolution of the system. A second class of quadrupole formulae gives information about the presence of 'time-odd' terms in the local equations of motion (terms that change sign under time reversal). However, the mere knowledge of the presence of these terms in the equations of external motion,

$$\frac{dX_e}{dt} = A^{\text{even}}(X_e) + A^{\text{odd}}(X_e), \tag{152}$$

does not allow one to conclude that the type of 'secular' effects which are observed must come only from A^{odd}. Indeed, A^{even} could induce time-symmetric effects of long period which, on the time scale of the observations (ten years) could look like real secular effects (there are many instances of this in the mechanics of the solar system).

Therefore, in order to be able to meaningfully compare the theory and the observations it is necessary to derive the full equations of external motion for a binary system up to the appearance of time-odd terms, and to control their solution by a method powerful enough to clearly exhibit all secular (or long-period) effects in the orbital motion. In other words, what is needed is a complete relativistic celestial mechanical description of a binary system. Few works have attempted such a complete description. Many works on radiation damping (e.g. by Anderson and Decanio, Kerlick, Papapetrou and Linet, Futamase, ...) have stayed at the level of a global hydrodynamical description without explicitly considering a system of N separated bodies. Other works (e.g. by Breuer and Rudolph, Linet, ...) have considered a system of N separated bodies but have only partially computed the equations of orbital motion of the bodies. References to these works will be found in Damour (1983a), Walker (1984), Ehlers and Walker (1984), Schutz (1985), Grishchuk and Kopejkin (1986). Here we shall focus our attention on three different lines of work which have sought a complete *kinematical* description of a two-body (or N-body) system up to the level where radiation damping first appears.

On the one hand, a new method has been set up to obtain the equations of external motion of two condensed bodies (Damour, 1980, 1983a). This method combines the best tools now available to treat the problem: (1) a post-Minkowskian approximation scheme to obtain the gravitational field outside the bodies incorporating a natural 'no-incoming-radiation condition' and giving rise to no 'near-zone only' problems, (2) a matched asymptotic expansions scheme used to prove the property of effacement (see Section 6.14) and to uniquely determine the gravitational field external to N condensed bodies, and (3) an Einstein–Infeld–Hoffmann–Kerr-type approach to compute the equations of orbital motion from the knowledge of the external field only (the explicit calculation of the surface integrals being simplified by the use of the analytic continuation trick explained in Section 6.14). Then this method has been implemented, in the case of two bodies, to compute the external gravitational field (Bel *et al.*, 1981; Damour, 1983c) and the equations of orbital motion, obtained first in retarded integro-differential form and then expanded in v^{ext}/c to get a conventional differential system. The final equations of motion (at the $2\frac{1}{2}$th post-Newtonian level) are expressed only in terms of the *instantaneous* positions, velocities and spins in a given harmonic coordinate system. 'Time', t, means here the zeroth harmonic coordinate of the post-Minkowskian external scheme. 'Position', $\dot{z}, \dot{z}'$, means the spatial harmonic coordinate of each 'centre of field', 'velocity' means $v^i := \mathrm{d}z^i/\mathrm{d}t$, $v'^i := \mathrm{d}z'^i/\mathrm{d}t$. 'Spin', S_{ij}, S'_{ij}, means the spatial components of the antisymmetric (Minkowskian) tensors, $S_{\mu\nu}, S'_{\mu\nu}$ describing in the post-Minkowskian-external-field approach the intrinsic angular momentum of the bodies. They are constrained to satisfy the following Minkowski-covariant spin condition:

$$S_{\mu\nu}v^\nu = 0 = S'_{\mu\nu}v'^\nu. \tag{153}$$

Let us introduce the notation, $RN^i := z^i(t) - z'^i(t)$ (with $N^iN^i = 1$), $V^i := v^i(t) - v'^i(t)$, $(Nv) := N^iv^i$ (etc. ...), $v^2 := v^iv^i$ (etc. ...) and $a^i := \mathrm{d}v^i/\mathrm{d}t$ for the accelerations. Then each body must satisfy the following equation of motion (Damour and Deruelle, 1981a; Damour, 1982):

$$a^i = A_0^i(\vec{z}-\vec{z}') + c^{-2}A_2^i(\vec{z}-\vec{z}',\vec{v},\vec{v}') + c^{-4}A_4^i(\vec{z}-\vec{z}',\vec{v},\vec{v}',\vec{S},\vec{S}')$$
$$+ c^{-5}A_5^i(\vec{z}-\vec{z}',\vec{v}-\vec{v}') + O(c^{-6}), \tag{154}$$

with

$$A_0^i = -Gm'R^{-2}N^i, \tag{155}$$

$$A_2^i = Gm'R^{-2}\{N^i[-v^2 - 2v'^2 + 4(vv') + \tfrac{3}{2}(Nv')^2 + 5(Gm/R) + 4(Gm'/R)]$$
$$+ (v^i - v'^i)[4(Nv) - 3(Nv')]\}, \tag{156}$$

$$A_4^i = B_4^i + C_4^i + D_4^i, \tag{157}$$

$$B_4^i = Gm'R^{-2}\{N^i[-2v'^4 + 4v'^2(vv') - 2(vv')^2 + \tfrac{3}{2}v^2(Nv')^2 + \tfrac{9}{2}v'^2(Nv')^2$$
$$- 6(vv')(Nv')^2 - \tfrac{15}{8}(Nv')^4 + (Gm/R)(-\tfrac{15}{4}v^2 + \tfrac{5}{4}v'^2 - \tfrac{5}{2}(vv') + \tfrac{39}{2}(Nv)^2$$
$$- 39(Nv)(Nv') + \tfrac{17}{2}(Nv')^2) + (Gm'/R)(4v'^2 - 8(vv') + 2(Nv)^2$$
$$- 4(Nv)(Nv') - 6(Nv')^2)]$$
$$+ (v^i - v'^i)[v^2(Nv') + 4v'^2(Nv) - 5v'^2(Nv') - 4(vv')(Nv)$$
$$+ 4(vv')(Nv') - 6(Nv)(Nv')^2 + \tfrac{9}{2}(Nv')^3$$
$$+ (Gm/R)(-\tfrac{63}{4}(Nv) + \tfrac{55}{4}(Nv')) + (Gm'/R)(-2(Nv) - 2(Nv'))]\}, \tag{158}$$

$$C_4^i = G^3m'R^{-4}N^i[-\tfrac{57}{4}m^2 - 9m'^2 - \tfrac{69}{2}mm'], \tag{159}$$

$$D_4^i = \left(\frac{S^{ik}}{m} + 2\frac{S'^{ik}}{m'}\right)(v^l - v'^l)\left(\frac{Gm'}{R}\right)_{,kl} + \left(2\frac{S^{kl}}{m} + 2\frac{S'^{kl}}{m'}\right)(v^l - v'^l)\left(\frac{Gm'}{R}\right)_{,ik}, \tag{160}$$

and

$$A_5^i = \tfrac{4}{5}G^2mm'R^{-3}\{V^i[-V^2 + 2(Gm/R) - 8(Gm'/R)]$$
$$+ N^i(NV)[3V^2 - 6(Gm/R) + \tfrac{52}{3}(Gm'/R)]\}. \tag{161}$$

The two parameters m and m' appearing in eqs. (154)–(161) are the 'Schwarzschild masses' of the condensed bodies. They are two constants which appear in the external gravitational field, in which are hidden many internal structure effects (see the discussion of the 'effacement of internal structure' in Section 6.14). On the other hand, the spin tensors undergo a slow evolution (on the post-Newtonian time scale, i.e. β_e^{-2} times the orbital period) which is also obtained in the Einstein–Infeld–Hoffmann–Kerr-type approach (Damour, 1982, and references therein). Introducing, *à la* Schiff, a suitable spin-vector, $\vec{S}$, associated with $S_{\mu\nu}$, the law of evolution ('spin precession') reads for the first body (see also references in Section 6.13.2)

$$\frac{d\vec{S}}{dt} = \left[\frac{Gm'}{c^2R^2}\vec{N} \times \left(\frac{3}{2}\vec{v} - 2\vec{v}'\right)\right] \times \vec{S} + O\left(\frac{1}{c^4}\right). \tag{162}$$

The following equations of orbital motion for a binary system, obtained from eq. (154) by leaving out the spin–orbit coupling term and the term of order c^{-5}:

$$\begin{cases} a^i = \mathcal{A}^i(z, z', v, v') := A_0^i + c^{-2}A_2^i + c^{-4}(B_4^i + C_4^i), & (163a) \\ a'^i = \mathcal{A}'^i(z, z', v, v') := A_0'^i + c^{-2}A_2'^i + c^{-4}(B_4'^i + C_4'^i), & (163b) \end{cases}$$

have been explicitly shown not to be deducible from a conventional Lagrangian (function of positions and velocities) but, instead, from a generalised Lagrangian (depending on accelerations) (Damour and Deruelle, 1981b; Damour, 1982):

$$L(z, v, a) = L_0(z - z', v, v') + c^{-2}L_2(z - z', v, v') + c^{-4}L_4(z - z', v, v', a, a'),$$
(164)

$$L_0 = \Sigma \left(\frac{1}{2}mv^2 + \frac{1}{2}\frac{Gmm'}{R} \right),$$
(165)

and

$$L_2 = \Sigma \left(\frac{1}{8}mv^4 + \frac{Gmm'}{R} \left[\frac{3}{2}v^2 - \frac{7}{4}(vv') - \frac{1}{4}(Nv)(Nv') - \frac{1}{2}\frac{Gm}{R} \right] \right).$$
(166)

$$L_4(z, v, a) = M_4 + N_4,$$
(167)

$$M_4(z, v, a) = \Sigma(\tfrac{1}{16}mv^6) + \Sigma Gmm'R^{-1}[\tfrac{7}{8}v^4 + \tfrac{15}{16}v^2v'^2 - 2v^2(vv') + \tfrac{1}{8}(vv')^2$$
$$- \tfrac{7}{8}(Nv)^2v'^2 + \tfrac{3}{4}(Nv)(Nv')(vv') + \tfrac{3}{16}(Nv)^2(Nv')^2]$$
$$+ \Sigma G^2m^2m'R^{-2}[\tfrac{1}{4}v^2 + \tfrac{7}{4}v'^2 - \tfrac{7}{4}(vv') + \tfrac{7}{2}(Nv)^2$$
$$+ \tfrac{1}{2}(Nv')^2 - \tfrac{7}{2}(Nv)(Nv')]$$
$$+ \Sigma Gmm'[(Na)(\tfrac{7}{8}v'^2 - \tfrac{1}{8}(Nv')^2) - \tfrac{7}{4}(v'a)(Nv')],$$
(168)

and

$$N_4(z) = \frac{G^3mm'}{R^3} \left[\frac{1}{2}m^2 + \frac{1}{2}m'^2 + \frac{19}{4}mm' \right].$$
(169)

The generalised Lagrangian (164) is (approximately) invariant under the Poincaré group. This allows one to construct ten Noetherian quantities which would be conserved during the motion (163) (Damour and Deruelle, 1981c; Damour, 1983a), E^{Noet}, $\vec{P}^{\text{Noet}}$, $\vec{J}^{\text{Noet}}$, $\vec{K}^{\text{Noet}}$, which are *single-valued* functions of the *instantaneous* state of the system (they can be reduced so as to depend only on simultaneous positions and velocities).

Now if we go back to the more exact equations of motion (154) for the binary system, they read:

$$\begin{cases} a^i = \mathcal{A}^i + c^{-4}D_4^i + c^{-5}A_5^i, & (170a) \\ a'^i = \mathcal{A}'^i + c^{-4}D_4'^i + c^{-5}A_5'^i, & (170b) \end{cases}$$

They can be solved by treating $c^{-4}D_4^i + c^{-5}A_5^i$ as a perturbative force superimposed on the Lagrangian system (163). The use of the method of variation of arbitrary constants allows one to control the effects of these perturbative forces in great detail. For the effects of $c^{-4}D_4^i$ see Barker and O'Connell (1975, 1979), and for those of $c^{-5}A_5^i$ see Damour (1983b, 1985). In particular it has been shown that the 'relativistic mean anomaly' l (an angle such that $l_N = 2\pi N$ for the Nth return of each body to its (proper) periastron) has the following (approximate) time dependence:

$$\frac{l}{2\pi} = \frac{c_0}{2\pi} + \frac{t}{P_0} - \frac{1}{2}\frac{\dot{P}_0}{P_0^2}t^2 + \text{periodic terms}, \qquad (171)$$

where the periodic terms have *only* a *short* period $\simeq P_0$ ('orbital period'). This proves that the *observable* secular effect is given only by the third term on the right-hand side of eq. (171), where, as shown by a direct variation-of-constant calculation:

$$\dot{P}_0 = -\frac{192\pi}{5c^5}\left(\frac{2\pi G}{P_0}\right)^{5/3}\frac{mm'}{(m+m')^{1/3}}\frac{1 + \frac{73}{24}e^2 + \frac{37}{96}e^4}{(1-e^2)^{7/2}}, \qquad (172)$$

e denoting a 'relativistic eccentricity' (for details of the meaning and proof of eqs. (171)–(172) see Damour, 1985, for earlier heuristic proofs see Esposito and Harrison (1975) and Wagoner (1975)).

Finally, these results concerning the coordinate motion of the binary system in a special harmonic coordinate system geared to the binary system have to be related to the quantities actually measured by the observers. This problem has been re-examined recently with a view to correcting, simplifying and completing previous works (see Damour and Deruelle, 1986, and references therein). This allows one to conclude that the 'theoretical parameters' introduced above (such as m, m', P_0, e, $\dot{P}_0$, ...) coincide with their corresponding 'observed parameters', denoted m_p, m_c, P_b, e, $\dot{P}_b$, ...

On the other hand, a recently developed program (Schäfer, 1983, 1985, 1986) based on a Hamiltonian approach to the interaction of N spinless point-particles with the gravitational wave-field has allowed one to reach elegantly (in a different gauge more adapted to the separation of conservative and damping effects) the main result of the previous method: i.e. a force, equivalent to $c^{-5}A_5^i$ (coming from the interaction with the dynamical degrees of freedom of the gravitational field) acting on the Hamiltonian sub-system of the instantaneously interacting N particles. However, the treatment of the problems associated with the use of point-particles has not been fully justified so that the physical relevance of this work is not obvious on its own.

Another line of work aimed at treating the problem of the motion of a binary system with the same completeness as the first method discussed above, has also been recently reported (Grishchuk and Kopejkin, 1983, 1986; Kopejkin, 1985). It is based on: (1) a post-Newtonian approximation scheme of the type of eq. (94), and (2) the assumption that the bodies are non-rotating, 'spherically symmetric' fluid balls. Spherical symmetry is taken in the coordinate sense, so that the centre of mass defined by eq. (115), with $\hat{\rho}$ given by eq. (117), coincides with the centre of symmetry. Then the equations for the motion of the centre of mass of each body are obtained by integrating the local post-Newtonian equations of motion (69b). They have been explicitly calculated, retaining all the higher derivatives that appear. If one 'reduces' these higher derivatives by using the lower-order equations of motion (as was done in the first method) the explicit result of this post-Newtonian calculation is (Kopejkin, 1985):

$$
\begin{aligned}
(\mu + c^{-2}\mu_2 + c^{-4}\mu_4)a^i = {} & [(\mu + c^{-2}\mu_2 + c^{-4}\mu_4)A_0^i(\mu' + c^{-2}\mu_2' + c^{-4}\mu_4')] \\
& + c^{-2}[(\mu + c^{-2}\mu_2)A_2^i(\mu + c^{-2}\mu_2, \mu' + c^{-2}\mu_2')] \\
& + c^{-4}\mu(B_4^i(\mu, \mu') + C_4^i(\mu, \mu')) \\
& + c^{-5}\mu A_5^i(\mu, \mu') + O(c^{-6}),
\end{aligned}
\tag{173}
$$

where the square brackets mean that one should expand in *explicit* powers of c^{-1} and discard all powers greater than or equal to the sixth. In eq. (173) the parameter μ denotes the 'rest mass' of the body:

$$
\mu := \int_V \mathrm{d}^3 x r_*,
\tag{174}
$$

while

$$
\mu_2 := \int_V \mathrm{d}^3 x r_* (\Pi_0 - \tfrac{1}{2}U^{(\mathrm{s})}),
\tag{175}
$$

$$
\mu_4 := \int_V \mathrm{d}^3 x r_* U^{(\mathrm{s})}(\tfrac{1}{2}U^{(\mathrm{s})} - \Pi_0 - 3p_0/r_*).
\tag{176}
$$

Now, formally speaking, if we set

$$
m := \mu + c^{-2}\mu_2 + c^{-4}\mu_4,
\tag{177}
$$

and if we forget about the truncation implicit in the square brackets, eq. (173) becomes, identical, modulo $O(c^{-6})$, with eq. (154) minus the spin–orbit term (160). This possibility of sweeping the various relativistic corrections, $c^{-2}\mu_2$ and $c^{-4}\mu_4$, into the 'effective' mass (177) is a direct post-Newtonian proof of the effacement of internal structure in the external motion, and was discussed in the preceding section. Finally, the kinematics of the binary system has been studied along the same lines as the variation-of-constant approach sketched above. In particular, Grishchuk and Kopejkin (1986) also introduce a generalised Lagrangian function of positions, velocities and accelerations. Their Lagrangian is a member of the general class of generalised Lagrangians described in Damour and Deruelle (1981b), i.e.,

$$L^{gen}(z, v, a) = L(z, v, a) + \frac{d}{dt} \{F(z, v)\} + c^{-4}Q(a - A_0), \tag{178}$$

where $L(z, v, a)$ is the Lagrangian (164), $dF/dt = \partial F/\partial z \cdot v + \partial F/\partial v \cdot a$ is a total time derivative which does not contribute to the equations of motion, and $Q(a - A_0)$ is a quadratic form of the differences $\vec{a} - \vec{A}_0(z, v)$, $\vec{a}' - \vec{A}'_0(z, v)$ which does *not* contribute to the order-reduced equations of motion at the order considered. They would contribute at the order c^{-6}, but then it is sufficient to replace $a - A_0$ by $a - A_0 - c^{-2}A_2$ to postpone the influence of such 'double-zero' terms (using the terminology of Barker and O'Connell (1980)). The statement of Grishchuk and Kopejkin that the Lagrangian (164) is 'only valid in a non-harmonic frame of reference' is incorrect and is based on a confusion between the effects of a 'double-zero' term such as $c^{-4}Q(a - A_0)$ (which does not contribute) and those of what one might call a 'simple-zero' term, i.e. a linear function, $c^{-4}\mathscr{L}(a - A_0)$, of the differences $a - A_0$. As remarked recently by Schäfer (1984), a simple-zero term *does* contribute to the equations of motion but in a way which is equivalent to a change of coordinate system at order c^{-4}. For a general investigation of the effect of infinitesimal coordinate transformations on generalised Lagrangians see Damour and Schäfer (1985) and references therein. In particular, it has been shown in the latter work that the acceleration-dependent two-body Lagrangian (164) is equivalent, modulo a suitable coordinate transformation, to the special $N = 2$ case of an *ordinary N*-body Lagrangian, $L(z, v)$, derived earlier by Ohta, Okamura, Kimura and Hiida (1974b), who were using a non-harmonic coordinate system.

The fact that two independent methods (post-Minkowskian + Einstein–Infeld–Hoffmann and post-Newtonian + perfect fluid) give formally identical equations of motion, in harmonic coordinates, is a strong confirmation of the validity of the numerical coefficients in eqs. (155)–(159) and (161). So much so that the coefficients in (165)–(169) are further confirmed by comparison with the works of Ohta *et al.* (1974a, b) and Schäfer (1985) based on a third independent method (post-Newtonian + delta functions) (after correction of a wrongly evaluated two-body integral in Ohta *et al.* (1974a); see Damour (1983a, section 13); Damour and Schäfer (1985)). However, only the first method has been tailored to deal with the astrophysical problem at hand, i.e. the non-linear-retarded gravitational interaction between two *condensed* objects. Indeed, the first method, as sketched above, essentially uses only *one* small parameter which can be taken as $\gamma_e = Gm/c^2R$. The parameter $\alpha = L/R$ is of the order of γ_e because $L \sim Gm/c^2$, by definition of condensed bodies, and the parameter β_e is used only in the last step of the method and is then related to γ_e by $\beta_e \sim \gamma_e^{1/2}$. By construction this method does not require that the internal gravitational field be weak; on the contrary, it uses $\gamma_i \sim 1$ to simplify things (however, it has also been assumed that $\beta_i^2 \lesssim \gamma_e$ in order to be able to efficiently use the effacement property).

On the other hand, the third method sketched above essentially uses the usual post-Newtonian + perfect fluid approach. This means: (1) that there are problems with incorporating boundary conditions (which show up as divergent integrals in the next approximation), and (2) that it can be applied only to weakly self-gravitating bodies. We have even shown in Section 6.13.2 that, at the level at which explicit calculations have been performed, retention of the 2 and 2.5PN terms in the external equations of motion was logically inconsistent because the uncontrolled error term formally denoted '$O(c^{-6})$' in eq. (173) was in fact at least $\sim \gamma_e^{1\,5/8}$, i.e. larger even than the 2PN corrections. This means that application of the PN equation (173) to the binary pulsar cannot, at present, be justified. The fact that the 'sweeping' of the self-field effects into the effective mass (177) works at the 2PN level is an *indication*, but certainly not a proof, that eq. (173) can be applied to strongly self-gravitating objects. In fact, the argument above says that, at present, there is no proof that eq. (173) can be meaningfully applied to any objects at all! Note also that the divergence problems appearing at the 3PN approximation (see e.g. Kerlick, 1980), which is an indication that the 3PN error is bigger than expected, can only strengthen our argument.

Finally, it should be emphasised that this issue of strong self-field effects is not only important from the theoretical point of view of the justification of the derivation of the equations of motion, but it is also essential from the purely numerical point of view. Indeed, in the case of the binary pulsar one expects self-fields $\gamma_i \sim 10$ per cent so that, for instance, the last term of eq. (173) as obtained explicitly by the post-Newtonian method, $c^{-5}\mu A_5^i(\mu, \mu')$, differs by ~ 10 per cent from the corresponding explicit result of the previous method, eq. (161), because of the difference in mass $(m - \mu)/\mu \sim \gamma_i$. Now the variation-of-constants-calculation sketched above shows that the observed effect (172) comes directly from the term of order c^{-5} in the equations of motion. Therefore, if one takes at face value the equations of motion (173) as they come out of the post-Newtonian calculation they predict a secular acceleration of the mean anomaly which differs by, say, more than 10 per cent from the prediction (172) of the other method (consistently taking into account that within all methods the *observed* masses should be identified with the ones appearing in the 1PN equations of motion, i.e. m in the first method, and $\mu + c^{-2}\mu_2$ in the third method). Now the observed value of $\dot{P}$ coincides with the predicted value to within less than 4 per cent, so that such a difference in prediction makes the post-Newtonian + perfect fluid approach, truncated at the 2PN level, of dubious use from the numerical point of view.

In concluding this section, we can say that, in General Relativity, the 2.5PN equations governing the orbital motion of two condensed bodies can be considered as a perturbed Lagrangian system. The Lagrangian system (163) (deducible from the generalised Lagrangian (164)) gives rise, in the bound case, to a multi-periodic motion (with a radial period $\simeq P_0$, and an angular period $\sim P_0\beta_e^{-2}$, 'periastron precession'). Then the most interesting perturbation is the last term of eq. (154):

$$\underset{\substack{\text{Einstein}\\\text{damping}}}{A^i} := c^{-5}A_5^i, \tag{179}$$

which can rightly be considered as a 'radiation damping' term because it leads to a secular decrease of the Noetherian energy of the 'unperturbed' Lagrangian (and because, in Schäfer's approach, it does appear as a direct coupling with the 'radiative degrees of freedom' of the gravitational field).

Indeed, using

$$\frac{\mathrm{d}E^{\text{Noet}}}{\mathrm{d}t} = c^{-5}(mv^i A_5^i + m'v'^i A_5'^i) + O(c^{-6}), \tag{180}$$

and,

$$A_5^i = \tfrac{3}{2}z^j Q_{ij}^{(5)} + 2v^j I_{ij}^{(4)} + \tfrac{10}{3}a^i I^{(3)} + \tfrac{1}{5}I_{iss}^{(5)} - J_{iss}^{(4)} + O(c^{-2}), \tag{181}$$

with

$$I_{ij} := \sum Gmz^i z^j, \quad I := I_{ss}, \quad Q_{ij} := I_{ij} - \tfrac{1}{3}I\delta_{ij},$$

and

$$I_{iss} := \sum Gmz^2 z^i, \quad J_{iss} := \sum Gmz^2 v^i, \quad Q^{(n)} := \mathrm{d}^n Q/\mathrm{d}t^n, \tag{182}$$

one finds

$$\frac{\mathrm{d}E^{\text{Noet}}}{\mathrm{d}t} = -c^{-5}\frac{\mathrm{d}}{\mathrm{d}t}E_5 - \frac{1}{5Gc^5}Q_{ij}^{(3)}Q_{ij}^{(3)} + O(c^{-6}), \tag{183}$$

where both E^{Noet} and E_5 are univalued functions of the instantaneous state of the binary system: $\vec{z}(t), \vec{z}'(t), \vec{v}(t), \vec{v}'(t)$. Therefore E^{Noet} experiences a secular decrease,

$$\left\langle \frac{\mathrm{d}E^{\mathrm{Noet}}}{\mathrm{d}t} \right\rangle = -\frac{1}{5Gc^5} \langle Q_{ij}^{(3)} Q_{ij}^{(3)} \rangle, \tag{184}$$

whose value coincides with the so-called 'quadrupole flux formula' giving the outgoing flux of gravitational radiation in the wave zone. Actually, this 'quadrupole flux formula' is, in my opinion, rather less well established than the 'secular acceleration (quadrupole) formula' (172) (because of the problem of justifying the 'good' quadrupole (182) for condensed bodies), but I quote it here only to justify the name 'gravitational radiation damping'. However, I wish to stress that what is actually observed is an orbital effect, so that A_{damping} should rather be thought of as the relativistic descendant of the Laplace effect.

Indeed, Laplace (1805), in his famous *Traité de Mécanique Céleste* which studied 'les altérations que le mouvement des planètes et des comètes peut éprouver [...] par la transmission successive de la pesanteur' (i.e. the orbital perturbations due to a finite velocity of propagation of gravity), introduced the idea that if gravity propagates with a finite velocity, say c_g, then there will be a small modification of Newton's law (due to aberration effects) amounting to the addition of a small 'damping' term of the type:

$$A^i_{\substack{\mathrm{Laplace} \\ \mathrm{damping}}} = -\frac{Gm'}{R^2} \frac{V^i}{c_g}. \tag{185}$$

He then showed that this damping term would cause a shrinkage and a circularisation of the orbit, together with a *secular acceleration* of the angular motion. He was aware of the fact that the last effect is the easiest to observe; and even concluded from the 'known' secular acceleration of the Moon that the velocity of propagation of gravity must be

$$c_g \geqslant 7 \times 10^6 \, c_{\mathrm{light}}! \tag{186}$$

The last conclusion of Laplace is not quite correct, as we now know, but we can still admire his profound insight into what would be the main orbital effects caused by a finite velocity of propagation of gravity. Indeed, it is clear from eq. (161) that the Einsteinian damping term (179) (in harmonic coordinates) contains many terms of the Laplace damping type (185). However, the particular structure of Einstein's theory has resulted in considerable reduction of the relative magnitude of this damping term from the (not so) 'naive' expectation, V/c, due to aberration effects, down to the present $(V/c)^5$. Still, the qualitative conclusions of Laplace are correct, the main effects of the damping term (179) being, indeed, a shrinkage and a circularisation of the orbit (Peters, 1964) and a secular acceleration of the mean anomaly given by the eqs. (171)–(172) (which can be called the 'secular acceleration quadrupole formula' because its demonstration shows that $\dot{P}_0$ is proportional to the secular decrease of E^{Noet} which, according to eq. (184) can be expressed in terms of the 'quadrupole moment', Q_{ij}, of eq. (182)).

Such an acceleration effect has indeed been observed in the binary pulsar PSR 1913 + 16 with a sign and a magnitude which is in remarkable agreement with the theoretical formula (172) (within 4 per cent, Weisberg and Taylor, 1984). This excellent agreement provides an impressive confirmation of the structure (170) of the equations of motion. As this structure was derived from eqs. (154)–(161), which were themselves derived by using both the fully non-linear structure (strong self-gravity, effacement property) and the radiative structure (retarded integrals) of Einstein's theory, we can say that this constitutes a profound confirmation of the non-linear hyperbolic structure of Einstein's theory and thereby an indirect proof of the existence of gravitational radiation. Moreover, some (semi-heuristic) investigations within other relativistic theories of gravity (see Will, 1981, and references therein) indicate that this is also a very sensitive test for a gravity theory which might allow us to discard many alternative theories which, like Einstein's, are compatible with the weak-field quasi-static relativistic effects in the solar system.

6.16 Conclusion

Henri Poincaré (1934) once remarked that real problems can never be classified as solved or unsolved ones, but that they are always *more or less solved* ('il y a seulement des problèmes plus ou moins résolus'). This remark applies particularly well to the problem of motion which has had a chequered history. Even the Newtonian problem of motion, which appeared to be well understood after the development of the powerful methods of classical celestial mechanics (codified around 1889 in the treatise of Tisserand), embarked on an entirely new career after the work of Poincaré (1892) which has led to many further developments (see e.g. Arnold, 1978; Gallavotti, 1983). In my opinion the Einsteinian problem of motion has not even reached a classical stage where the basic problems appear (provisionally) as 'well understood'. At first sight the best-developed approximation method in general relativity, the 'post-Newtonian' one, would seem to constitute such a classical stage, but the literature on the post-Newtonian problem of motion is full of repetitions, errors or ambiguities. Several problems (especially at the 1PN approximation) have been done over and over again with slightly modified approaches, and different notation (or worse, the same notation for different quantities), and still it seems to me that some of the basic issues have not really been tackled. Moreover, the development of other approximation methods ('post-Minkowskian', 'singular perturbation') has rather complicated the picture by giving rise to many hybrids (some of which have been discussed above).

My original intention was to conclude this survey by giving a list of the issues that need to be clarified. I renounced this project because, if one wishes to look at the work done with a critical eye, nearly all aspects of the problem of motion need to be thoroughly re-investigated for mathematical, physical or conceptual reasons; so that the list of open problems would, consistent with the remark of Poincaré, include all the issues discussed above. Another reason for renouncing this project is that Brumberg (1986) has very recently given such a list, to which I gladly refer the reader.

One thing is certain: the Einsteinian problem of motion is no longer a purely theoretical problem, thanks to the dramatic improvement in the precision of position measurements in the solar system, and to the discovery of the binary pulsar 1913 + 16 which is a marvellous relativistic laboratory; the Einsteinian problem of motion has become an important tool of modern astrophysics. It is therefore of some urgency, not only to complete and unify the work already done, but also to develop new approaches which will keep the best aspects of existing methods while freeing themselves from their conceptual and/or technical drawbacks. These new approaches should aim at both formal and conceptual clarification of the basic issues, and at securing more accurate explicit results. Let us hope that such methods will be developed before the (first) centenary of the publication of Einstein's classic paper on 'the foundation of the general theory of relativity' ('Die Grundlage der allgemeinen Relativitätstheorie', Einstein, 1916a) which is comparable with Newton's *Philosophiae Naturalis Principia Mathematica* both in its importance for physics and its structure which consists, after a conceptual introduction, of a formal mathematical part followed by a physical part where connection is made with the natural world.

References

Anderson, J. L. (1980). *Phys. Rev. Lett.*, **45**, 1745–8.
Anderson, J. L. and Decanio, T. C. (1975). *Gen. Rel. Grav.*, **6**, 197.

Anderson, J. L., Kates, R. E., Kegeles, L. S. and Madonna, R. G. (1982). *Phys. Rev.*, **D25**, 2038–48.

Anderson, J. L. and Madonna, R. G. (1983). *Gen. Rel. Grav.*, **15**, 1121–9.

Arnold, V. I. (1978). *Mathematical Methods of Classical Mechanics*. Springer: New York.

Arnowitt, R., Deser, S. and Misner, C. W. (1962). In *Gravitation: An Introduction to Current Research*, ed. L. Witten, pp. 227–65. Wiley: New York.

Ashby, N. and Bertotti, B. (1984). *Phys. Rev. Lett.*, **52**, 485–8.

Ashby, N. and Bertotti, B. (1986). *Phys. Rev. D.*, **34**, 2246–59.

Bailey, I. and Israel, W. (1980). *Ann. Phys. (N.Y.)*, **130**, 188–214.

Barker, B. M. and O'Connell, R. F. (1975). *Phys. Rev.*, **D12**, 329–35.

Barker, B. M. and O'Connell, R. F. (1979). *Gen. Rel. Grav.*, **11**, 149–75.

Barker, B. M. and O'Connell, R. F. (1980). *Can. J. Phys.*, **58**, 1659–66.

Bel, L., Damour, T., Deruelle, N., Ibanez, J. and Martin, J. (1981). *Gen. Rel. Grav.*, **13**, 963–1004.

Bel, L. and Martin, J. (1981). *Ann. Inst. H. Poincaré (Phys. Théor.)*, **34A**, 231 and references therein.

Bennewitz, F. and Westpfahl, K. (1971). *Commun. Math. Phys.*, **23**, 296–318.

Bertotti, B. (1954). *Nuov. Cim.*, **12**, 226–32.

Bertotti, B. (1956). *Nuov. Cim.*, **4**, 898–906.

Bertotti, B., de Felice, F. and Pascolini, A., eds. (1984). *General Relativity and Gravitation* (GR 10, Padua 1983). Reidel: Dordrecht.

Bertotti, B. and Plebanski, J. (1960). *Ann. Phys. (N.Y.)*, **11**, 169–200.

Blanchet, L. (1987). *Proc. Roy. Soc. (Lond.)*, **A409**, 383–99.

Blanchet, L. and Damour, T. (1984a). *C.R. Acad. Sci. Paris* (II), **298**, 431–4.

Blanchet, L. and Damour, T. (1984b). *Phys. Lett.*, **104A**, 82–5.

Blanchet, L. and Damour, T. (1986a). *Phil. Trans. R. Soc. London*, **A320**, 379–430.

Blanchet, L. and Damour, T. (1986b). To be submitted for publication.

Breuer, R. A. and Rudolph, E. (1982). *Gen. Rel. Grav.*, **14**, 181–211 (this work is an extension of the work of E. Rudolph and G. Börner (1978a, b), *Gen. Rel. Grav.*, **9**, 809–20 and 821–33).

Brillouin, M. (1922). *C.R. Acad. Sci. Paris*, **175**, 1009–12.

Brouwer, D. and Clemence, G. M. (1961). *Methods of Celestial Mechanics*. Academic Press: New York.

Bruhat, Y. (1962). In *Gravitation: An Introduction to Current Research*, ed. L. Witten, pp. 130–68. Wiley: New York.

Brumberg, V. A. (1972). *Relativistic Celestial Mechanics*. Nauka: Moscow (in Russian).

Brumberg, V. A. (1986). In *Relativity in Celestial Mechanics and Astrometry* (IAU Symposium 114), ed. J. Kovalevsky and V. A. Brumberg, pp. 5–17. Reidel: Dordrecht.

Burke, W. L. (1971). *J. Math. Phys.*, **12**, 401–18.

Caporali, A. (1981a). *Nuov. Cim.*, **61B**, 181–204.

Caporali, A. (1981b). *Nuov. Cim.*, **61B**, 205–12.

Carmeli, M. (1964). *Phys. Lett.*, **9**, 132–4.

Carmeli, M. (1965). *Nuov. Cim.*, **37**, 842–75.

Chandrasekhar, S. (1965). *Astrophys. J.*, **142**, 1488–1512.

Chandrasekhar, S. and Nutku, Y. (1969). *Astrophys. J.*, **158**, 55–79.

Chazy, J. (1928, 1930). *La Théorie de la Relativité et la Mécanique Céleste*, vols. 1 and 2. Gauthier-Villars: Paris.

Choquet-Bruhat, Y. and Christodoulou, D. (1983). In *Advances in Mathematics, Supplementary Volume 8*, ed. V. Guillemin, pp. 73–91. Academic Press: New York.

Choquet-Bruhat, Y. and York, J. W. Jr (1980). In *General Relativity and Gravitation*, ed. A. Held, vol. 1, pp. 99–172. Plenum Press: New York.

Christodoulou, D. and Schmidt, B. (1979). *Commun. Math. Phys.*, **68**, 275–89.

Contopoulos, G. and Spyrou, N. (1976). *Astrophys. J.*, **205**, 592–8.

Cooperstock, F. I. and Hobill, D. W. (1982). *Gen. Rel. Grav.*, **14**, 361–78.

Crowley, R. C. and Thorne, K. S. (1977). *Astrophys. J.*, **215**, 624–35.

Dallas, S. S. (1977). *Cel. Mech.*, **15**, 111–23.

Damour, T. (1978). In *Physics and Astrophysics of Neutron Stars and Black Holes*, ed. R. Giacconi and R. Ruffini (65th E. Fermi School, Varenna, 1975), pp. 547–9. North-Holland: Amsterdam.

Damour, T. (1980). *C.R. Acad. Sci. Paris*, **291**, série A, 227–9.

Damour, T. (1982). *C.R. Acad. Sci. Paris*, **294**, série II, 1355–7.

Damour, T. (1983*a*). In *Gravitational Radiation*, ed. N. Deruelle and T. Piran (Les Houches 1982), pp. 59–144. North-Holland: Amsterdam.

Damour, T. (1983*b*). *Phys. Rev. Lett.*, **51**, 1019–21.

Damour, T. (1983*c*). In *Proceedings of the Third Marcel Grossmann Meeting on General Relativity*, ed. Hu Ning (Shangai, 1982), part A, pp. 583–97. Science Press: Beijing; North-Holland: Amsterdam.

Damour, T. (1985). In *Proceedings of Journées Relativistes 1983*, ed. S. Benenti, M. Ferraris and M. Francaviglia (Torino, 1983), pp. 89–110. Pitagora Editrice: Bologna.

Damour, T. (1987). In *Géométrie et Physique*, ed. Y. Choquet, B. Coll, R. Kerner and A. Lichnerowicz (Journées Relativistes 1985, Luminy), pp. 72–83. Hermann: Paris.

Damour, T. and Deruelle, N. (1981*a*). *Phys. Lett.*, **87A**, 81–4.

Damour, T. and Deruelle, N. (1981*b*). *C.R. Acad. Sci. Paris*, **293** (II), 537–40.

Damour, T. and Deruelle, N. (1981*c*). *C.R. Acad. Sci. Paris*, **293** (II), 877–80.

Damour, T. and Deruelle, N. (1986). *Ann. Inst. H. Poincaré (Physique Théorique)*, **44**, 263–92 (this work is a continuation of T. Damour and N. Deruelle (1985), *Ann. Inst. H. Poincaré (Phys. Théor.)*, **43**, 107–32).

Damour, T. and Schäfer, G. (1985). *Gen. Rel. Grav.*, **17**, 879–905.

D'Eath, P. D. (1975*a*). *Phys. Rev.*, **D11**, 1387–1403.

D'Eath, P. D. (1975*b*). *Phys. Rev.*, **D11**, 2183–99.

Demianski, M. and Grishchuk, L. P. (1974). *Gen. Rel. Grav.*, **5**, 673.

Deruelle, N. (1982). Thèse de doctorat d'Etat, unpublished, and (1983) in *Proceedings of the Third Marcel Grossmann Meeting*, ed. Hu Ning, pp. 955–8. Science Press: Beijing; North-Holland: Amsterdam.

Deruelle, N. and Piran, T., eds. (1983). *Gravitational Radiation* (NATO ASI, les Houches, 1982). North-Holland:Amsterdam.

De Sitter, W. (1916). *Mon. Not. R. Astr. Soc.*, **76**, 699–728; **77**, 155–84.

Dixon, W. G. (1979). In *Isolated Gravitating Systems in General Relativity*, ed. J. Ehlers (Varenna, 1976, course 67), pp. 156–219. North-Holland: Amsterdam.

Droste, J. (1916). *Versl. K. Akad. Wet. Amsterdam*, **19**, 447–55.

Eddington, A. and Clark, G. L. (1938). *Proc. Roy. Soc. (Lond.)*, **A166**, 465–75.

Ehlers, J. (1977). In *Proceedings of the International School of General Relativistic Effects in Physics and Astrophysics*, ed. J. Ehlers (Erice, 1977, course 3), pp. 45–73. MPI: München (annotated in 1983).

Ehlers, J., ed. (1979). *Isolated Gravitating Systems in General Relativity* (67th Enrico Fermi School, Varenna, 1976). North-Holland: Amsterdam.

Ehlers, J. (1980). *Ann. N.Y. Acad. Sci.*, **336**, 279–94.

Ehlers, J. (1981). In *Grundlagen-probleme der modernen Physik*, ed. J. Nitsch *et al.*, pp. 65–84. Bibliogr. Inst.: Mannheim.

Ehlers, J. (1984). In *Proceedings of the 7th International Congress of Logic, Methodology and Philosophy of Science*. North-Holland: Amsterdam.

Ehlers, J., Rosenblum, A., Golberg, J. N. and Havas, P. (1976). *Astrophys. J.*, **208**, L77–81.

Ehlers, J. and Rudolph, E. (1977). *Gen. Rel. Grav.*, **8**, 197.

Ehlers, J. and Walker, M. (1984). In *General Relativity and Gravitation*, ed. B. Bertotti *et al.* (GR 10, Padua, 1983), pp. 125–37. Reidel: Dordrecht.

Einstein, A. (1916*a*). *Ann. der Physik*, **49**, 769 (English translation in H. A. Lorentz *et al.*, *The Principle of Relativity*, pp. 109–64. Dover: New York, 1952).

Einstein, A. (1916*b*). *Preuss. Akad. Wiss. Berlin, Sitzber.*, 688–96.

Einstein, A. and Infeld, L. (1940). *Ann. Math.*, **41**, 455.

Einstein, A. and Infeld, L. (1949). *Can. J. Math.*, **1**, 209.

Einstein, A., Infeld, L. and Hoffmann, B. (1938). *Ann. Math.*, **39**, 65–100.

Eisenstaedt, J. (1986). *Arch. for Hist. of Exact Sciences*, **35**, 115–85.

Ellis, G. F. R. (1984). In *General Relativity and Gravitation*, ed. B. Bertotti *et al.* (GR 10, Padua, 1983), pp. 215–88. Reidel: Dordrecht.

Esposito, L. W. and Harrison, E. R. (1975). *Astrophys. J.*, **196**, L1–2.

Fock, V. (1959). *The Theory of Space Time and Gravitation*. Pergamon Press: Oxford.

Friedrich, H. (1981). *Proc. R. Soc. (Lond.)*, **A378**, 401–21.

Friedrich, H. (1985). Preprint, Hamburg.

Futamase, T. (1983). *Phys. Rev.*, **D28**, 2373–81.

Futamase, T. (1986). 'The strong field point particle limit and the quadrupole formula in the binary pulsar system'. Preprint, Washington University.

Futamase, T. and Schutz, B. F. (1983). *Phys. Rev.*, **D28**, 2363–72.

Gallavotti, G. (1983). *The Elements of Mechanics*. Springer: New York.

Goenner, H. (1970). *J. Math. Phys.*, **11**, 1645.

Grishchuk, L. P. and Kopejkin, S. M. (1983). *Pis'ma Astron. Zh.*, **9**, 436–40 (*Sov. Astron. Lett.*, **9**, 230–2).

Grishchuk, L. P. and Kopejkin, S. M. (1986). In *Relativity in Celestial Mechanics and Astrometry* (IAU Symposium 114, Leningrad, 1985), ed. J. Kovalevsky and V. A. Brumberg, pp. 19–34. Reidel: Dordrecht.

Gürses, M. and Walker, M. (1984). *Phys. Lett.*, **101A**, 15–19.

Hagihara, Y. (1972). *Celestial Mechanics*. MIT Press: Cambridge, Mass.

Havas, P. (1957). *Phys. Rev.*, **108**, 1351.

Havas, P. (1979). In *Isolated Gravitating Systems in General Relativity*, ed. J. Ehlers, pp. 74–155. North-Holland: Amsterdam.

Havas, P. and Goldberg, J. N. (1962). *Phys. Rev.*, **128**, 398–414.

Hellings, R. W. (1986). *Astron. J.*, **91**, 650–9.

Hogan, P. A. and McCrea, J. D. (1974). *Gen. Rel. Grav.*, **5**, 79–113.

Hulse, R. A. and Taylor, J. M. (1975). *Astrophys. J. (Lett.)*, **195**, L51–53.

Ibanez, J. and Martin, J. (1982). *Gen. Rel. Grav.*, **14**, 439–51, and *Phys. Rev.*, **D26**, 384–9.

Infeld, L. and Plebanski, J. (1960). *Motion and Relativity*. Pergamon Press: Oxford.

Isaacson, R. A., Welling, J. S. and Winicour, J. (1984). *Phys. Rev. Lett.*, **53**, 1870–2.

Isaacson, R. A. and Winicour, J. (1968). *Phys. Rev.*, **168**, 1451–6.

Jardetzky, W. S. (1958). *Theories of Figures of Celestial Bodies*. Interscience: New York.

Kates, R. E. (1980*a*). *Phys. Rev.*, **D22**, 1853–70.

Kates, R. E. (1980*b*). *Phys. Rev.*, **D22**, 1871–8.

Kates, R. E. (1981). *Ann. Phys.* (*NY*), **132**, 1–17.

Kerlick, G. D. (1980). *Gen. Rel. Grav.*, **12**, 467, 521.

Kerr, R. P. (1959). *Nuov. Cim.*, **13**, 469–91, 492–502, 673–89.

Kopejkin, S. M. (1985). *Astron. Zh.*, **62**, 889–904 (in Russian).

Kovalevsky, J. and Brumberg, V. A., eds. (1986). *Relativity in Celestial Mechanics and Astrometry* (IAU Symposium 114, Leningrad, 1985). Reidel: Dordrecht.

Kühnel, A. (1963). *Acta Phys. Polon.*, **24**, 399.

Kühnel, A. (1964). *Ann. Phys.* (*NY*), **28**, 116–33.

Lagerstrom, P. A., Howard, L. N. and Liu, C. S. (1967). *Fluid Mechanics and Singular Perturbations: A Collection of Papers by Saul Kaplun.* Academic Press: New York.

Laplace, P. S. (1805). *Traité de Mécanique Céleste*, vol. 4, pt 2, Book X, chapt. VII. Courcier: Paris, an XIII (reprinted in *Oeuvres complètes de Laplace*, vol. 4, pp. 314–27. Gauthier-Villars: Paris, 1880).

Leipold, G. and Walker, M. (1977). *Ann. Inst. H. Poincaré A* (*Phys. Theor.*), **27**, 61–71.

Lestrade, J. F., Chapront, J. and Chapront-Touzé, M. (1982). In *High-Precision Earth Rotation and Earth–Moon Dynamics*, ed. O. Calame, pp. 217–24. Reidel: Dordrecht.

Levi-Civita, T. (1937*a*). *Am. J. Math.*, **59**, 9–22.

Levi-Civita, T. (1937*b*). *Am. J. Math.*, **59**, 225–34.

Levi-Civita, T. (1950). 'Le problème des n corps en relativité générale, *Mémorial des Sciences Mathématiques*, **116**. Gauthier-Villars: Paris.

Lichnerowicz, A. (1955). *Théories Relativistes de la Gravitation et de l'Electromagnétisme.* Masson: Paris.

Lorentz, H. A. and Droste, J. (1917). *Versl. K. Akad. Wet. Amsterdam*, **26**, 392 (part I) and 649 (part II). English translation in H. A. Lorentz, *Collected Papers*, ed. P. Zeeman and A. D. Fokker, vol. V, pp. 330–55. Martinus Nijhoff: The Hague, 1937.

McCrea, J. D. and O'Brien, G. (1978). *Gen. Rel. Grav.*, **9**, 1101–18.

Manasse, F. K. (1963). *J. Math. Phys.*, **4**, 746–61.

Martin, C. F., Torrence, M. H. and Misner, C. W. (1985). *J. of Geophys. Res.*, **90**, 9403–10.

Mashhoon, B. (1984). *Gen. Rel. Grav.*, **16**, 311.

Mashhoon, B. (1985). *Found. Phys.*, **15**, 497.

Misner, C. W., Thorne, K. S. and Wheeler, J. A. (1973). *Gravitation*. Freeman: San Francisco (the many-body system is treated in Exercise 39.15, pp. 1091–5).

Newton, I. (1947). *The Mathematical Principles of Natural Philosophy*, ed. F. Cajori. University of California Press: Berkeley.

Nordtvedt, K., Jr (1968). *Phys. Rev.*, **170**, 1186–7.

Ohta, T., Okamura, H., Kimura, T. and Hiida, K. (1973). *Prog. Theor. Phys.*, **50**, 492–514.

Ohta, T., Okamura, H., Kimura, T. and Hiida, K. (1974*a*). *Prog. Theor. Phys.*, **51**, 1220–38.

Ohta, T., Okamura, H., Kimura, T. and Hiida, K. (1974*b*). *Prog. Theor. Phys.*, **51**, 1598–1612.

Papapetrou, A. (1951). *Proc. Phys. Soc.*, **A64**, 57–75.

Peres, A. (1959). Nuov. Cim., **11**, 617, 644; **13**, 437.

Peres, A. (1960). *Nuov. Cim.*, **15**, 351–69.

Persides, S. (1971*a*). *J. Math. Phys.*, **12**, 2355–61.

Persides, S. (1971*b*). *Astrophys. J.*, **170**, 479–98.

Peters, P. C. (1964). *Phys. Rev.*, **136**, B1224–32.

Peters, P. C. and Mathews, J. (1963). *Phys. Rev.*, **131**, 435–40.

Poincaré, H. (1934). *Science et Méthode*, Book 1, chapt. II, p. 34. Flammarion: Paris.

Poincaré, H. (1957). *Les méthodes nouvelles de la mécanique céleste*. Dover: New York (reprint of the first edition by Gauthier-Villars, Paris, 1892–99).

Riesz, M. (1949). *Acta Mathematica*, **81**, 1–223.

Schäfer, G. (1983). *Mitt. Astron. Ges.*, **58**, 135–7.

Schäfer, G. (1984). *Phys. Lett.*, **100A**, 128–9.

Schäfer, G. (1985). *Ann. Phys. (NY)*, **161**, 81–100.

Schäfer, G. (1986). *Gen. Rel. Grav.*, **18**, 255–70.

Schattner, R. (1978). *Gen. Rel. Grav.*, **10**, 377–93, 395–9.

Schutz, B. F. (1980). *Phys. Rev.*, **D22**, 249–59.

Schutz, B. F. (1984). In *Relativistic Astrophysics and Cosmology*, ed. X. Fustero and E.Verdaguer, pp. 35–97. World Scientific: Singapore.

Schutz, B. F. (1985). In *Relativity, Supersymmetry and Cosmology*, ed. O. Bressan *et al.* (SILARG V, Bariloche, 1985), pp. 3–80. World Scientific: Singapore.

Spyrou, N. (1977). *Gen. Rel. Grav.* **8**, 463–89 (this work has been extended by N. Spyrou (1977–85) in a series of papers in the same journal).

Spyrou, N. (1978). *Cel. Mech.*, **18**, 351–70.

Stephani, H. (1964). *Acta Phys. Polon.*, **26**, 1045–60.

Synge, J. L. (1960). *Relativity: The General Theory*. North-Holland: Amsterdam.

Synge, J. L. (1969). *Proc. Roy. Ir. Acad.*, **A67**, 47.

Synge, J. L. (1970). *Proc. Roy. Ir. Acad.*, **A69**, 11.

Taylor, J. H., Fowler, L. A. and McCulloch, P. M. (1979). *Nature*, **277**, 437–40.

Taylor, J. H. and Weisberg, J. M. (1982). *Astrophys. J.*, **253**, 908–20.

Thorne, K. S. (1969). *Astrophys. J.*, **158**, 997–1019.

Thorne, K. S. and Kovács, S. J. (1975). *Astrophys. J.*, **200**, 245–62 (see also two subsequent papers by S. J. Kovács and K. S. Thorne (1977, 1978) in the same journal).

Thorne, K. S. and Hartle, J. B. (1985). *Phys. Rev.*, **D31**, 1815–37.

Tisserand, F. (1960). *Traité de Mécanique Céleste*, vol. I. Gauthier-Villars: Paris (reprint of the first edition of 1889).

Van Dyke, M. (1975). *Perturbation Methods in Fluid Mechanics*. Parabolic Press: Stanford, California (annotated edition).

Wagoner, R. V. (1975). *Astrophys. J.*, **196**, L63–5.

Walker, M. (1984). In *General Relativity and Gravitation*, ed. B. Bertotti *et al.* (GR 10, Padua, 1983), pp. 107–37. Reidel: Dordrecht.

Walker, M. and Will, C. M. (1980). *Astrophys. J. (Lett.)*, **242**, L129–33.

Weisberg, J. M. and Taylor, J. H. (1984). *Phys. Rev. Lett.*, **52**, 1348–50.

Westpfahl, K. (1985). *Fortschritte der Physik*, **33**, 417–93.

Westpfahl, K. and Goller, M. (1979). *Lett. Nuov. Cim.*, **26**, 573–6.

Westpfahl, K. and Goller, M. (1980). *Lett. Nuov. Cim.*, **27**, 161.

Westpfahl, K. and Hoyler, H. (1980). *Lett. Nuov. Cim.*, **27**, 581.

Will, C. M. (1974). In *Experimental Gravitation*, ed. B. Bertotti (56th E. Fermi School, Varenna, 1972), pp. 1–110. Academic Press: New York (see pp. 64–5).

Will, C. M. (1979). In *General Relativity: An Einstein Centenary Survey*, ed. S. W. Hawking and W. Israel, pp. 24–89. Cambridge University Press: Cambridge.

Will, C. M. (1981). *Theory and Experiment in Gravitational Physics*, Cambridge University Press: Cambridge.

Will, C. M. (1986). *Can. J. Phys.*, **64**, 140–5.

Will, C. M. and Eardley, D. M. (1977). *Astrophys. J.*, **212**, L91–4.
Winicour, J. (1983). *J. Math. Phys.*, **24**, 1193–8.
York, J. W. Jr (1983). In *Gravitational Radiation*, ed. N. Deruelle and T. Piran, pp. 175–201. North-Holland: Amsterdam.

7

Dark stars: the evolution of an idea

WERNER ISRAEL†

7.1 Introduction

Theory may often delay understanding of new phenomena observed with new technology unless theorists are quite open-minded as to what types of physical laws may need to be applied: conservatism is unsafe. ... In astrophysics, historically, theories have only seldom had predictive usefulness as guides to experimenters.

(Greenstein, 1984)

How difficult were these advances? If we now acquaint ourselves with some of the detours, errors and psychological blocks, it is not to console ourselves with the thought that even great physicists do not always move in a straight line, but rather to grasp in some measure how difficult it really was.

(Hund, 1971)

To pause halfway up a steep hill, look down at one's starting-point barely half a mile below and recall the arduous, winding path one has actually followed, is a salutary exercise in humility. The subject of this essay – the evolution of our ideas since the age of Laplace concerning the dark objects that populate the universe – is now at that middle distance where a retrospective look begins to discern broad contours without yet losing all of the detail that will in time become laundered and streamlined by Darwinian selection of citations into the handful of names and dates of the potted histories.

Tracing this evolution, with its meanderings and vicissitudes, as a continuous process, means abandoning attempts at a 'balanced' account which gives consideration to ideas in proportion to their evolutionary

† Work supported by Canadian Institute for Advanced Research and by Natural Sciences and Engineering Research Council of Canada.

fitness. Some dinosaurs have to be resurrected. The old and discarded is often scaffolding for the new: Faraday's concept of field grew out of the aether, Carnot's thermodynamics from the notion of a caloric fluid. Conversely, the attempt to trace the broad current of ideas and opinions without getting mired in technicalities has meant that many technical advances that have been central to the development are mentioned only in passing, or not at all, just because they were neither controversial nor wrong in an interesting way.

As basic science moves further from everyday reality, a number of questions increasingly solicit attention. How does the collective scientific mind evolve toward new concepts? What keeps the scientific wagon on the right track? With science increasingly under the sway of fads, how far dare one trust scientific consensus as an autostabilizing mechanism? The evolutionary picture suggested by the present account is not altogether reassuring on this score. The black hole idea is more than 200 years old: general relativity more than 70. Yet, if our present ideas about gravitational collapse and the existence of neutron stars and black holes happen to be near the truth, a scarce 25 years ago virtually the entire scientific community was complacently wandering in the darkness. It is not easy to detect in this story anything that often resembles the Kuhnian cycle of paradigm and revolution. Rather, one sees a meandering path that resolves into a Brownian motion at the microlevel of individual scientists. Perhaps it is not altogether flippant to suggest that the evolutionary picture that naturally presents itself is of a substantially classical motion (but with considerable quantum spread) that arises by constructive interference from a probing of all possible paths, very occasionally interrupted by barrier tunneling at places where the incident current becomes sufficiently large. Whether this motion will ultimately approach a fixed point or limit cycle, and whether there is more than one, are questions for future generations.

This narrative comes to an end around 1974, when the main pieces of the pattern, as we now construe it, were in place. As it moves nearer to the present, the perspective inevitably becomes more subjective. Others, located at diverse research centres around the globe, will have experienced the events recounted here in their own very different ways. While I have striven to keep technicalities to a minimum, this is basically a historical sketch and not an exposition of the subject itself. For the newcomer, Shklovsky (1978) on stellar physics, and Israel (1983), on black holes, are recent semipopular expositions with extensive references to collateral reading.

The many people who patiently endured my questions and who are not to blame for my misconstruction of their replies or the way it has all been pieced together, include J. D. Barrow, S. Chandrasekhar, D. Finkelstein, S. W. Hawking, D. P. Hube, M. D. Kruskal, W. H. McCrea, M. J. Rees, K. S. Thorne and G. M. Volkoff, and I should like to extend to them my deepest thanks. My thanks are also due to Professor Z. Maki and Professor K. Nishijima, the successive Directors of the Research Institute for Fundamental Physics, Kyoto University, and their colleagues, for their kind hospitality at Yukawa Hall, where this work was begun. I am indebted to the staff of the Scientific Periodicals Library, Cambridge University, for making their unique resources available, and uncomplainingly bringing out an endless stream of volumes from their stacks, and to Yoshiko Fujinaka, Mary Yiu and Cheryl Torbett for their superb technical assistance.

7.2 Early speculations (1784–1921)

Unseen bodies may, for aught we can tell, predominate in mass over the sum-total of those that shine: they supply possibly the chief part of the motive power of the universe.

(Clerke, 1903)

In regions where our ignorance is great, occasional guesses are permissible.

(Lodge, 1921)

The name of the Reverend John Michell (MA (Cantab.) 1752, BD 1761) fell so quickly into obscurity after his death that two of the English scientific classics of our century, Eddington's *Internal Constitution of the Stars* (1926) and Hawking and Ellis's *Large Scale Structure of Space–Time* (1973) give credit to a Frenchman, P. S. Laplace (1796), as having been first to suggest that 'the attractive force of a heavenly body could be so large that light could not flow out of it'.

Yet in his day Michell's reputation as a natural philosopher stood second in English circles only to that of his contemporary and friend, Henry Cavendish. The paper communicated to the Royal Society by Cavendish on November 27, 1783, in which Michell expounded this and many related ideas, caused such a stir in London circles that it overshadowed exciting news coming from Paris about Coulomb's electrical experiments,† as extant

† Coulomb's invention of the torsion balance had actually been anticipated by Michell, and the inverse-square law by Cavendish, using the now celebrated null method of measuring the force inside a hollow charged conductor. Both Cavendish and Michell were reclusive personalities who published very little of their wide-ranging work. Many of Cavendish's electrical researches were found after his death in a number of sealed packets of papers, eventually published, under the editorship of Maxwell, in 1879. None of Michell's unpublished work survives.

correspondence of Sir Joseph Banks (then President of the Royal Society) bears witness.

Michell's (1784) paper is far from airy speculation, but a concrete proposal, worked out in detail, for obtaining information about stellar distances, masses and sizes by observing the retardation of light emitted from their surfaces.† This was at a time when, despite efforts extending over a century, no semi-annual stellar parallax had yet been found, and some astronomers doubted whether they would ever be detectable.

Michell's work is of course based on two of Newton's major hypotheses, universal gravitation and the corpuscular theory of light. He says:

> Let us now suppose the particles of light to be attracted in the same manner as all other bodies with which we are acquainted; ... gravitation being, as far as we know, or have any reason to believe, a universal law of nature.

Even Michell's geometrical style of reasoning is quintessentially Newtonian. Russell McCormmach (1968), in his moving account of the friendship and collaboration of Michell and Cavendish, describes their deep attachment to the Newtonian scheme of things:

> Newton wrote in the *Principia* that all bodies are to be regarded as subject to the principle of gravitation. Every body, however great or small, is related to every other body in the universe by a mutual attraction. It was this postulated universality of the force of gravity which contributed so greatly to the order and unity of the Newtonian world. This unity was, for its followers, an untested article of faith for nearly a century after the *Principia*. During this time the evidence of gravitational attraction continued to be drawn from the motions of the earth, moon, planets, comets, and falling bodies – phenomena which span an intermediate range of masses, sizes and distances. In three domains of experience, involving the extreme upper and lower limits of masses and dimensions, the action of gravity had not yet been observed; the gravity of the 'fixed' stars; the mutual attraction of terrestrial bodies; and the gravitation of light. The task of deducing observable consequences from each of these supposed instances of universal gravitation fell to the Reverend John Michell (1724–1793), a teacher at Cambridge (1749–1763) and afterwards Rector of Thornhill, Yorkshire. His renowned London friend Henry Cavendish (1731–1810) encouraged him in these researches and became involved in the

† Michell calculated that a retardation of 5 per cent would change the refractive index sufficiently to make it easily measurable with a prism.

resulting observational and experimental questions. The immersion of Michell and Cavendish in gravitational studies was an essential feature of their commitment to a unified Newtonian world. Their commitment had yet a broader significance: Newton's theory of gravitation inspired their image of physical reality, and it served as their model of an exact science ....

The freshness and scope of Michell's scientific imagination help to explain Cavendish's unique interest in their correspondence. But there is more to it than this. Michell's isolation was not wholly physical; it was intellectual too, and Cavendish shared that isolation; their true bond was philosophical. They believed with Newton that the purpose of natural philosophy is to discover the mathematical laws of the attracting and repelling forces between the particles of bodies, and to deduce new phenomena from the forces. They stood out against the current of British natural philosophy, which by then was neither mathematical nor primarily concerned with laws of force. The best science of the day was one-sidedly experimental; and the speculative interest was generally focused on the ether, and on force-denying mechanisms, and on the physical connections between the post-Newtonian weightless fluids.

While observational questions are always paramount in Michell's paper, he also contemplated the possibility that there might exist stars much larger than the sun for which the velocity of escape exceeds the speed of light. Even here, he has a prescient suggestion of how such stars might be detectable:

> If there should really exist in nature any bodies whose density is not less than that of the sun, and whose diameters are more than 500 times the diameter of the sun, since their light could not arrive at us ... we could have no information from sight; yet, if any other luminous bodies should happen to revolve about them we might still perhaps from the motions of these revolving bodies infer the existence of the central ones with some degree of probability, as this might afford a clue to some of the apparent irregularities of the revolving bodies ...; but as the consequences of such a supposition are very obvious, and the consideration of them somewhat beside my present purpose, I shall not prosecute them any further.

A remarkable development of this line of thought occurs in the work of the German astronomer Johann Georg von Soldner (1801; see Jáki, 1978). Soldner calculated the Newtonian deflection of light corpuscles passing near a star. He also speculated on the possibility that the stars of the Milky Way might be orbiting a central, very massive dark object of the sort postulated by Laplace, but decided against it on the grounds that the resulting proper motions would be too large not to be noticed.

Speculation about invisible stars went quickly out of favour as the wave theory of light gained ascendancy especially after the discovery of interference by Thomas Young in 1801, and removed any reason to believe that light should be affected by gravity. References to this hypothesis were deleted from the 1808 and later editions of Laplace's *Exposition du Système du Monde*. But it was natural that it would resurface, albeit in a significantly different form, after the success of the 1919 eclipse expedition.

In February 1920, A. Anderson of University College, Galway, made an intriguing speculation in the *Philosophical Magazine*:

> We may remark, though perhaps the assumption is very violent, that if the mass of the sun were concentrated in a sphere of diameter 1.47 kilometres, the index of refraction near it would become infinitely great, and we should have a very powerful condensing lens, too powerful indeed, for the light emitted by the sun itself would have no velocity at its surface. Thus if, in accordance with the suggestion of Hemholtz, the body of the sun should go on contracting there will come a time when it is shrouded in darkness, not because it has no light to emit, but because its gravitational field will become impermeable to light.

This was an extraordinary anticipation of the gravitational collapse scenario which was to be so adamantly resisted in the 1930s and for a quarter of a century afterward.

The convenient fiction, due to Eddington, that the paths of light rays in a gravitational field could be formally described and visualized as propagation in an 'aether' of variable refractive index, was imaginatively taken up by Sir Oliver Lodge, whose firm belief in the aether had somehow withstood the assault of the 1905 relativistic revolution (see, e.g. his contributions to the discussion of the results of the 1919 Sobral expedition, Lodge, 1920). In a remarkable address to a Student's Science Club at the University of Birmingham in 1921, he presented on this basis an essentially correct physical picture of the full range of black holes that are of astrophysical interest today:

> If light is subject to gravity, if in any real sense light has weight, it is natural to trace the consequences of such a fact. One of these consequences would be that a sufficiently massive and concentrated body would be able to retain light and prevent its escaping. And the body need not be a single mass or sun, it might be a stellar system of

exceedingly porous character so that light could penetrate freely into the interior . . . .

. . . we find that a system able to control and retain its light must have a density and size comparable to

$$\rho R^2 = \cdots = 1.6 \times 10^{27} \text{ c.g.s.}$$

It is hardly feasible for any single mass to satisfy this condition; either the density or the size is too enormous . . .

For a body of density 10^{12}, – which must be the maximum possible density, as its particles would be then all jammed together, – the radius need only be 400 kilometres . . . .

If a mass like that of the sun (2.2×10^{33} grammes) could be concentrated into a globe of about 3 kilometres in radius, such a globe would have the properties above referred to; but a concentration to that extent is beyond the range of rational attention . . . .

But a stellar system – say a super spiral nebula – of aggregate mass equal to 10^{15} suns . . . might have a group radius of 300 parsecs . . . with a corresponding average density of 10^{-15} c.g.s. without much light being able to escape from it. This really does not seem an utterly impossible concentration of matter.

(Lodge, 1921)

At a time when astronomers still believed, in spite of their awareness that stellar material was highly ionized, that this material could not be compressed to a density much higher than that of water (e.g. Eddington, 1921), Lodge's casual acceptance of the possibility of attaining nuclear density is truly noteworthy. This imaginative contribution was yet another stillbirth, and it was to be not the last before the black hole idea finally came into its own more than 40 years later.

7.3 White dwarfs: the first compact massive objects (1910–26)

The message of the companion of Sirius when it was decoded ran: 'I am composed of material 3000 times denser than anything you have ever come across; a ton of my material would be a little nugget that you could put in a matchbox'. What reply can one make to such a message? The reply which most of us made in 1914 was – 'Shut up. Don't talk nonsense'.

(Eddington, 1927)

The first astronomical *bête noire* of our century, the white dwarfs, cast their shadow on a generation still grounded in a Victorian sense of the permanence of things and the Lucretian doctrine of atoms that could 'not be swamped by any force, for they are preserved indefinitely by their absolute solidity'. As the web of observation and theory slowly tightened, the scientific reaction – first disregard, then dismay, yielding only gradually to the beginnings of acceptance – set a pattern for the disclosures yet to come.

The companions of Procyon and Sirius had puzzled astronomers since before the turn of the century. Comparable in mass with the sun, they were more than a hundred times dimmer, 'a condition that we are powerless to explain' (Campbell, 1913). Superficially, there seemed to be two alternatives:

> Either they have a far less surface brilliancy than the sun or their density
> is much greater. There can be no doubt that the former is the case.
>
> (Newcomb, 1908)

This emphatic conclusion would have caused no misgivings if these objects had been unmistakably red and thus of low surface temperature: their dimness would have been explicable and not at all unusual. But, as far as it could be distinguished from the overwhelming brilliance of Sirius itself, the companion appeared nearly as white to the eye.† In December 1915, Walter S. Adams announced that he had finally succeeded in securing a spectrogram with the Cassegrain reflector at Mt Wilson as the companion passed to its furthest distance from the primary in its 49-year orbit. It showed a spectrum identical with that of Sirius. Puzzlement now gave way to some embarrassment, since spectral intensity measurements of 109 stars by Wilsing and Scheiner at Potsdam during the previous five years had demonstrated (as had earlier theoretical work by Schwarzschild) that to a fair approximation stars radiate like black bodies, with effective temperatures well correlated with colour index and spectral type.‡ If the observations were taken at face value, the dimness of Sirius B could not be attributed to a low surface temperature. Presumably the reaction of the anonymous reporter for *The Observatory* (1916) was not untypical: 'The results rather suggest that the spectrum obtained is due to light from Sirius

---

† In the case of Procyon, the spectral class of the companion remains somewhat uncertain to this day due to the proximity of the bright component (Liebert, 1980).

‡ According to the classification system perfected at the Harvard College Observatory in the early 1900s, stellar spectra may be divided into several reasonably well-defined types, labelled O, B, A, F, G, K, M in order of decreasing surface temperature. Sub-classes are denoted by the digits 0 to 9, so that A0 is the hottest star of type A. The colour of a star is a clue to its spectral type. Stars of types O and B (characterized by prominence of helium lines) are bluish-white, type A (hydrogen lines) white, types F and G (calcium lines) yellow, types K and M (metallic lines) orange to red.

reflected by the companion'. Adams had already anticipated the objection:

It is ... by no means necessary to have recourse to this explanation, since in the case of o_2 Eridani, where there can be no question of reflected light, we know of a similar case of a star of very low intrinsic brightness which has a spectrum of type A_0.

(Adams, 1915)

Some months earlier, the Danish astronomer Ejnar Hertzprung (1915) had taken note of this system as the single exception known to him to the general rule that 'stars between absolute magnitudes $+3$ and $+8$ were approximately all of the same colour', and commented, 'This exception is, in fact, very strange'.

The anomaly of o_2 Eridani had been noticed by others even before this. Years later, Henry Norris Russell of Princeton was to recall a day in 1910 when the first white dwarf made itself known to an inner circle of three:

I was visiting my friend and generous benefactor, Prof. Edward C. Pickering [director of the Harvard College Observatory (1876–1919)]. With characteristic kindness he had volunteered to have the spectra observed for all the stars – including comparison stars – which had been observed in the observations for stellar parallax which Hinks and I made at Cambridge, and I discussed. This piece of apparently routine work proved very fruitful – it led to the discovery that all the stars of very faint absolute magnitude were of spectral type M. In conversation on this subject (as I recall it), I asked Pickering about certain other faint stars, not on my list, mentioning in particular 40 Eridani B. Characteristically, he sent a note to the Observatory office and before long the answer came (I think from Mrs Fleming) that the spectrum of this star was A. I knew enough about it, even in those paleozoic days, to realize at once that there was an extreme inconsistency between what we would then have called 'possible' values of the surface brightness and density. I must have shown that I was not only puzzled but crestfallen, at this exception to what looked like a very pretty rule of stellar characteristics; but Pickering smiled upon me and said: 'It is just these exceptions that lead to an advance in our knowledge', and so the white dwarfs entered the realm of study.

(Russell, 1939, 1944)

It was, however, left to a 23-year-old Estonian astrophysicist, Ernest J. Öpik, then working in Moscow, to take upon himself the odium of confronting the paradox openly. In May 1916 he submitted a paper in which he calculated mean densities for 40 stars in 39 binary systems whose spectral class was known at the time. (With communication impeded by the war, Öpik as yet was unaware of Adams's work on the spectrum of Sirius B.) He wrote:

> Among the binaries with known orbital elements there is one which is not included in the preceding discussion, o_2 Eridani; ... the density according to equation (6) ... would be 25 000. This impossible result indicates that in this case our assumptions are wrong; the only possible explanation is that, however high the temperature, the surface brightness or the radiating power is very low; probably o_2 Eridani is a pair of very rarefied nebulae.
>
> (Öpik, 1916)

The key to the true explanation was forged just six months later and in a closely related context, but it took another eight years before anyone saw the connection. On 8 December 1916 Eddington presented to the Royal Astronomical Society the first of his pioneering papers on radiative transport in stellar interiors. In the ensuing discussion, Jeans pointed out that the value $\mu = 54$ which Eddington had adopted for the mean molecular weight of the stellar material was almost certainly an overestimate, since 'for these temperatures and energy we have very hard Röntgen radiation, and so the atoms in the gas will be smashed up' (*The Observatory*, 1917). Jeans (1928) later traced this idea back to Descartes (1644), who had conjectured that the sun and fixed stars were made of matter 'which possesses such violence of agitation that, impinging upon other bodies, it gets divided into indefinitely minute particles'. Under such conditions, Jeans was now pointing out, electrons normally bound to much heavier nuclei would be set free as independent particles, thus greatly reducing the mean molecular weight of the material.

In his subsequent work, Eddington (1918, 1921) revised his estimate for μ downwards to values in the range 2 to 4, but, like everyone else (e.g. Jeffreys, 1918), he still believed that, in stars like the sun, material would deviate drastically from a perfect gas. It was more than just 'common sense' that deterred the astronomers from questioning this axiom. Breakdown of the gas laws at ordinary densities was a pillar of the giant-and-dwarf evolutionary theory of Russell (1914), which held sway well into the 1920s. In Russell's version of the theory, stars were pictured as beginning their lives as gaseous red giants, contracting and heating up as they (supposedly) evolve towards the blue end of the main sequence, at which point their material becomes liquid and begins to cool. The stars then evolve along the main sequence to end their luminous phase as red dwarfs. The discovery of the mass–luminosity relationship by Eddington (1924*a*) cut the ground from under this theory, since it showed that different locations on the main sequence correspond to different masses. It had an even more far-reaching and remarkable result:

We must recall that the theory was developed for stars in the condition of a perfect gas .... But in the left half of the diagram we have the sun whose material is denser than water, Krueger 60 denser than iron .... What business have they on the curve reserved for a perfect gas?... The shock was even greater than I can well indicate to you, because the great drop in brightness when the star is too dense to behave as a perfect gas was a fundamental tenet in our conception of stellar evolution. On the strength of it the stars had been divided into two groups known as giants and dwarfs, the former being the gaseous stars and the latter the dense stars.

(Eddington, 1927)

Eddington communicated these results to a memorable meeting of the Royal Astronomical Society (RAS) on 14 March 1924. Even now, he did not yet feel free to take it quite for granted that ionization would reduce atomic sizes:

Does losing an electron mean that the boundary of the atom is gone? Is the boundary or merely the boundary-stone removed? I think that most physicists would consider that the removal of the outer electrons means the removal of the boundary.

(Eddington, 1924*a*)

Today such hesitancy makes curious reading until one recalls that quantum mechanics was still two years in the future, and Bohr's *ad hoc* quantization rules for the placement of electronic orbits offered no clue to a dynamical explanation for the size and rigidity of atoms. Eddington now proceeded to the climax of his address:

There is an interesting class of stars known as the 'white dwarfs', of which the companion to Sirius is a famous example. It presents a curious puzzle .... If ... it has the ordinary radiative power associated with F-type stars, we can work out its surface area and its diameter, which comes out not very much larger than the earth; its mass is about $\frac{4}{5}$ that of the sun, so that the density would be enormous – about 50 000 gm per c.c. That argument is well-known, but I think most of us have mentally added 'which is absurd'. According to the present conclusions, however, it is not absurd; so that, unless the spectroscopic classification has deceived us, the companion of Sirius may be an actual example of how disrupted atoms may pack together to a far higher density than ordinary matter.

(Eddington, 1924*a*)

Eddington felt sufficiently bold not to wait for his colleagues' reactions to this 'fascinating train of ideas' (as Jeans was to phrase it) before following up the consequences. About six weeks earlier, he had written to Walter Adams at Mt Wilson:

I have lately been wondering if you would find it possible with your great instrument to measure the radial velocity of the companion of Sirius – with a view to determining its density by means of the Einstein shift of the spectral lines (compared with Sirius).

I take it that it is about 10^m fainter than Sirius and as it is of the same spectral type it should have a surface 10 000 times smaller and a radius 100 times less .... Of course this involves a density of about 100 000 times the density of water, which seems incredible. But I have recently been entertaining the wild idea that it may be just possible ....

I should scarcely venture to suggest your following this wild idea if I did not regard a negative result as also of interest ... a very definite challenge to thermodynamicists ....

(Eddington, 1924*b*)

Adams replied on 12 February, promising to attempt measurements, and added:

I had realised the shift would be large but had not attempted to compute it .... I have looked at these plates many times with a view to measurement but have always laid them aside ....

On 2 March he wrote again:

We have obtained some results on the companion of Sirius which I believe will interest you .... The simple mean is + 20 km per sec; or if the best line is given double weight the mean is + 23 km .... I should very much appreciate your opinion on these results. The probable error is, of course, large. ...

Eddington's initial response on 22 March was cautious:

I feel pretty well convinced that large densities are possible, but whether we have been lucky enough to find an actual specimen is another question. It seems almost too good to be true. It looks all right, but owing to the difficulty of the measurement and the rather large probable error I regard your result with some caution – as I think you would wish, and shall not say anything about it publicly at present. I do not know whether you are proposing to take further spectrograms or measures before you publish it. It is evidently of exceptional theoretical importance besides being a great practical triumph over difficulties.

Fifteen months later Adams (1925) was ready to publish his conclusions and Eddington's satisfaction was unalloyed:

I am very delighted that your measures have turned out so well and am deeply interested in the details of the paper. But it is very staggering that it should turn out like this . . . .

(Eddington, 1925)†

Eddington's classic, *The Internal Constitution of the Stars*, was a survey of the subject as it stood just prior to the discovery of wave mechanics. Discussing the white dwarfs he poses, in his characteristic Carrollian fashion, a conundrum which has come to be known as Eddington's paradox:

I do not see how a star which has once got into this compressed condition is ever going to get out of it. So far as we know, the close packing of matter is only possible so long as the temperature is great enough to ionise the material. When the star cools down and regains the normal density associated with solids, it must expand and do work against gravity. *The star will need energy in order to cool* . . . . We can scarcely credit the star with sufficient foresight to retain more than 90 per cent [of its internal energy] in reserve for the difficulty awaiting it. It would seem that the star will be in an awkward predicament when its supply of sub-atomic energy ultimately fails. Imagine a body continually losing heat but with insufficient energy to grow cold!

(Eddington, 1926)

This dilemma, in one form or another, would dog the theorists for more than three decades. In 1926, however, it appeared that deliverance was at hand. R. H. Fowler pointed out that, on the basis of the new quantum statistics, even an absolutely cold assembly of electrons confined to a finite volume must retain a finite spread of momenta, and hence a pressure, since the Pauli exclusion principle would not allow more than two electrons (with opposite spins) to occupy the same quantum phase cell. Milne (1930) showed that this zero-point pressure can balance a cold star against gravity at a uniquely determined radius (a decreasing function of mass) that corresponds well with the actual sizes of the white dwarfs. Fowler gave a graphic description of the condition of matter in such a star:

† Since the 1950s it has been generally admitted that both the theoretical and observed redshifts for Sirius B are highly uncertain. However, there is reasonably good agreement in the case of o_2 Eridani.

The black dwarf material is best likened to a single gigantic molecule in its lowest quantum state. On the Fermi–Dirac statistics its high density can be achieved in one way only, in virtue of a corresponding great energy content. But this energy can no more be expended in radiation than the energy of a normal atom or molecule. The only difference between black dwarf matter and a normal molecule is that the molecule can exist in a free state while black dwarf matter can only so exist under very high external pressure.

(Fowler, 1929)

It thus seemed for a while, at the end of the 1920s, that one of the most bizarre chapters in the history of astrophysics had been brought to a rounded and satisfying conclusion. However, even before the decade was quite out, the demon exorcised by Fowler had covertly wormed its way back.

7.4 The Chandrasekhar limit (1929–35)

The ground will fall away from our feet, its particles dissolved amid the mingled wreckage of heaven and earth. The whole world will vanish into the abyss, and in the twinkling of an eye no remnant will be left but empty space and invisible atoms. At whatever point you allow matter to fall short, this will be the gateway to perdition.

(Lucretius, 55 BC)

Edmund C. Stoner of Leeds University is best known for his pioneering work on the application of Fermi–Dirac statistics to the theory of para- and ferromagnetic phenomena. His 1934 monograph *Magnetism and Matter*, and his long series of research and review articles, became standard references. His paper, 'The distribution of electrons among atomic levels', written in 1924 at the age of 25, strongly influenced Pauli's formulation of the exclusion principle a year later, and in part anticipated it. In 1928, stimulated by some speculative ideas of Jeans on 'liquid stars', Stoner's interest turned briefly to astrophysics.

Jeans (1926, 1928) had suggested that material deep inside a star might deviate from Boyle's law – a condition he deemed necessary for stability† –

---

† According to Jean's calculations, stars in radiative equilibrium would be vibrationally unstable if they are gaseous throughout (a conclusion strongly qualified by later work). He hoped thereby to characterize untenanted portions of the Hertzprung–Russell diagram as regions of instability where jamming is relieved by the stripping-off of successive layers of electrons. Not for some years was it recognized that stars whose energy sources are concentrated toward the centre will have convective cores, for which stability is not a problem (e.g. Jeffreys, 1930; Biermann, 1932; Cowling, 1934).

due to jamming of incompletely ionized atoms of the heavier elements. This idea at once ran into the difficulty that, for effective jamming, the size of the inner electronic shells needed to be very much larger than radii estimated on the basis of Bohr theory and, later, quantum mechanics. Stoner (1929) accordingly set out to examine whether significant deviations would result from Fermi–Dirac degeneracy, which is a jamming of electrons in phase space, and, specifically, what kind of upper bound this jamming imposes on the density of stellar material. Stoner derived an upper bound (attained in the limit of complete degeneracy) which varied as the square of the star's mass and was about an order of magnitude larger than the mean density typical of white dwarfs.

This work was quickly followed up by Wilhelm Anderson (1929) of Tartu University, Estonia, who pointed out that near Stoner's upper limit the electrons are approaching the speed of light, and the equation of state therefore needs correction for the increase of particle mass with velocity. Allowing for this, Anderson found that a completely degenerate star of given density has a mass that is significantly lower than Stoner's non-relativistic estimate. In fact, it is clear from his tabulation and formulae that his equations predict the existence of a critical mass (close to one solar mass for a star composed of helium) for which the density becomes infinite, and above which no completely degenerate configurations exist. However, Anderson refrains from any direct reference to this, and merely comments that his equations must break down at high densities because they fail to take account of effects such as the contribution of the gravitational potential energy of the star to its total mass and the 'compressibility of the electrons and protons' at very high densities.

Acknowledging Anderson's criticism, Stoner now attacked the problem afresh, and laid the groundwork for a systematic treatment, in a paper, submitted in December 1929, that contains what came to be known as the Stoner–Anderson equation of state for a completely degenerate relativistic gas.† Stoner (1930) explicitly confirms the existence of a limiting mass (his estimate of 1.7 solar masses is somewhat higher than Anderson's) 'above

---

† Nearly two years earlier the statistical thermodynamics of relativistic Bose and Fermi gases had received a general and comprehensive formulation at the hands of Ferencz Jüttner (1928) of Breslau. However, Jüttner did not explicitly consider the limit of complete degeneracy (which greatly simplifies the general formulae), and he was unaware of the astrophysical relevance of relativistic degeneracy. He considered his work to be of theoretical interest only. His paper appeared in a German physics journal and did not come to the attention of astronomers for some years.

which the gravitational kinetic equilibrium considered will not occur'. On the significance and implications of this he offers no comment beyond the remark that the known white dwarfs fall below the critical limit.

In the calculations of Stoner and Anderson the astrophysical aspects were handled very schematically: the star was simply idealized as a sphere of uniform density. In view of the crudeness of this approximation it must be considered fortuitous that their estimates for the critical mass came so close to the correct value of 1.44 solar masses (for a helium star). The definitive value was obtained by the 19-year-old Subramanyan Chandrasekhar in July 1930 (without knowing at the time of Stoner's work) during a sea voyage from India to Cambridge. He showed his result to Fowler, who was doubtful, and sent it to Milne for his opinion: the reaction was not encouraging. It nevertheless did eventually find its way into print in the USA as a brief paper in the *Astrophysical Journal* (Chandrasekhar, 1931a). As he recalled in 1977, 'I didn't understand at the time what this limit meant, and I didn't know how it would end. But it is very curious that Fowler did not think the result very important'.†

Although the concept of the atom as the indestructible building block of matter had broken down, the belief remained firm that there must be a 'maximum density of which matter is capable', now generally supposed to be 'given by known physical properties of material such as the radii of electrons and nuclei' (Milne, 1930; Fowler, 1926).‡ Under the influence of these ideas, Chandrasekhar (1931b, 1932) for a while attempted to show that white dwarfs can have arbitrarily large masses if they contain central cores in which matter has reached nuclear densities and become incompressible. This idea encounters numerous difficulties and it did not survive the discovery of the neutron a year later.

In January 1932, Lev Landau submitted a short paper, entitled 'On the theory of stars', in which he independently recovered Chandrasekhar's limiting mass. Landau's work has great sweep but there are some oddities. He proposes that

> ... to attack the problem of stellar structure by methods of theoretical physics ... we must at first investigate the statistical equilibrium of a given mass without generation of energy, . . . the quantum effects for the

† Chandrasekhar's reminiscences (Chandrasekhar, 1969, 1977; Wali, 1982) offer a fascinating behind-the-scenes glimpse into the unfolding of the dramatic developments of the 1930s.

‡ The evolution of scientific views on an 'ultimate' density of matter has been traced by W. Anderson (1936) and E. R. Harrison (1972).

non-relativistic Fermi-gas ... lead to the existence of a stable
equilibrium. The presence of sources would produce only an additional
expansion due to the radiation pressure.

From this and his subsequent remarks it would appear that Landau had
entirely overlooked the key role of thermal gas pressure in supporting the
weight of hot stars. Apparently he was at this time under the impression that
the mass limit which he proceeded to derive for cold stars would apply to all
stars, at least to a good approximation.

Thus we get an equilibrium state only for masses greater [*sic*] than a
critical mass $M_0 = \ldots$ about $1.5\odot$ (for $m = 2$ protonic masses). For
$M > M_0$ there exists in the whole quantum theory no cause preventing
the system from collapsing to a point .... As in reality such masses exist
quietly as stars and do not show any such ridiculous tendencies we must
conclude that all stars heavier than $1.5\odot$ certainly possess regions in
which the laws of quantum mechanics (and therefore of quantum
statistics) are violated.

(Landau, 1932)

During the next three years, as Chandrasekhar completed the first of his
monumental contributions to astrophysics – a comprehensive theory of the
internal structure of white dwarf stars and of stars with degenerate cores
based on the Stoner–Anderson formula – the issue of the critical mass
continued to cast a pall over the whole subject. In October 1934 he
summarized the state of knowledge as follows.

Finally, it is necessary to emphasize one major result of the whole
investigation, namely, that it must be taken as well established that the
life-history of a star of small mass must be essentially different from ... a
star of large mass. For a star of small mass the natural white-dwarf
stage is an initial step towards complete extinction. A star of large mass
($> \mathcal{M}$) cannot pass into the white-dwarf stage and one is left speculating
on other possibilities.

(Chandrasekhar, S., 1934*b*)

One possibility in particular was engrossing his thoughts at the time.
Coronal evaporation from Wolf–Rayet and similar stars, first noted by the
Canadian astronomer, Carlyle S. Beals (1929), at the Dominion
Astrophysical Observatory, had become an active focus of interest, and
Chandrasekhar (1934*a*) himself had worked on the theory. Earlier that year
he had presented a report to the regular journal seminar conducted by
Eddington on a paper by Kosirev (1934) dealing with physical conditions in
the extended photospheres of Wolf–Rayet stars. At the end of his lecture he

ventured – in violation of the rules: speakers were not supposed to discuss their own work! – to slip in the suggestion that mass shedding might be the safety valve that provides relief for overweight stars. Eddington, however, could not accept this as a panacea. In conversations with Chandrasekhar, he repeatedly expressed the conviction that, at least in some cases, evolution must continue to the point where pressure support failed and the star was obliged to contract.

Chandrasekhar outlined his tentative hypothesis in the concluding section of a paper submitted to the Monthly Notices on 1 January 1935:

> A possibility is that at some stage in the process of contraction the flux of radiation in the outer layers of the stars will become so large that a profuse ejection of matter will begin to take place. This ejection of matter will continue till the mass of the star becomes small enough for central degeneracy to be possible. ... Thus, whether the star is of large or small mass, the final stage in its evolution is always the white-dwarf stage ....

>   (Chandrasekhar, 1935*a*)

This scenario, at first advanced and treated with reserve (e.g. Russell, Dugan and Stewart, 1938), gradually took root over the next two decades (e.g. Gamow and Schönberg, 1941; Chandrasekhar, 1951; Hoyle, 1955; Payne-Gaposchkin, 1957; Schatzman, 1958; Struve, 1962; O'Dell, 1963). It finally developed into a widespread implicit faith (especially outside the Soviet Union) in continuous or catastrophic mass ejection as a universal regulating mechanism which brings every star in its final stages below the critical weight. From 1963, when it became attractive to contemplate other possibilities, the notion was viewed more critically:

> The conventional view is that in all such cases a process of mass ejection takes place, conveniently reducing the final mass below one or other of the mass limits just mentioned. We wish to emphasize that this view is no more than a superstition. ... Indeed, it would be a curious situation if astrophysical processes occurring long before the onset of the final implosion crisis were to operate always to prevent the crisis from arising. This would imply an unlikely 'foreknowledge' on the part of natural processes.

>   (Hoyle, Fowler, Burbidge and Burbidge, 1964)

But we are running ahead of our story.

7.5 Eddington's intervention (1935)

There is evidently need for a re-survey of this subject, which had become confused by the distractions of the Stoner–Anderson formula.

(Eddington, 1939)

Early in 1935, as Chandrasekhar was quietly putting the finishing touches to his work on white dwarfs, there was a bombshell. Eddington, whose time had been occupied with the development of his Fundamental Theory, quite unexpectedly entered the arena with a characteristically dramatic opening salvo.

Dr Chandrasekhar has been referring to degeneracy. There are two expressions used in this connection, 'ordinary' degeneracy and 'relativistic' degeneracy, and perhaps I had better begin by explaining the difference. They refer to formulae expressing the electron pressure P in terms of the electron density σ. For ordinary degeneracy $P = K\sigma^{5/3}$. But it is generally supposed that this is only the limiting form of a more complicated relativistic formula, which shows P varying as something between $\sigma^{5/3}$ and $\sigma^{4/3}$, approximating to $\sigma^{4/3}$ at the highest densities. I do not know whether I shall escape from this meeting alive, but the point of my paper is that there is no such thing as relativistic degeneracy!

I would remark first that the relativistic formula has defeated the original intention of Prof. R. H. Fowler. . . . In 1924 . . . I was troubled by a difficulty that there seemed to be no way in which a dense star could cool down. . . . Soon afterwards Fermi–Dirac statistics were discovered, and Prof. Fowler applied them to the problem and showed that they solved the difficulty; but now Dr Chandrasekhar has revived it again. Fowler used the ordinary formula; Chandrasekhar, using the relativistic formula which has been accepted for the last five years, shows that a star of mass greater than a certain limit $\mathcal{M}$ remains a perfect gas and can never cool down. The star has to go on radiating and radiating and contracting and contracting until, I suppose, it gets down to a few km radius, when gravity becomes strong enough to hold in the radiation, and the star can at last find peace.

Dr Chandrasekhar has got this result before, but he has rubbed it in in his last paper; and, when discussing it with him, I felt driven to the conclusion that this was almost a *reductio ad absurdum* of the relativistic degeneracy formula. Various accidents may intervene to save the star, but I want more protection than that. I think there should be a law of Nature to prevent a star from behaving in this absurd way!

. . . I feel satisfied myself that the current formula is based on a partial relativity theory, and that if the theory is made complete the relativity corrections are compensated, so that we come back to the 'ordinary' formula.

Suppose we are dealing with a cubic centimetre of material in the middle of a star. Ordinarily we analyse this into electrons, protons, etc., travelling about in all directions. . . . In the ordinary analysis of matter into electrons one is dealing with progressive waves; but in the analysis which leads to the Exclusion Principle (used in deriving the degeneracy formula) the electron is represented by a standing wave. . . . The electron represented by a progressive wave can be brought to rest by a Lorentz transformation, and it then becomes a standing wave. This transformation introduces a factor into the equation, which is not needed if the waves referred to are standing waves originally. . . .

(Eddington, 1935a)

It is not difficult to recapture the spell cast by Eddington's charisma on the audience in Burlington House on 11 January (Wali, 1982). But in cold print his logic must have appeared somewhat less pellucidly clear.

Eddington never wavered in his rejection of the Stoner–Anderson equation of state, though his ground for opposing it seemed to shift this way and that in the half-dozen or so publications he devoted to this subject over the next few years (e.g. Eddington, 1935b, 1939, 1940). I shall not attempt to recapitulate his various arguments here; Schatzman (1958) has summarized and critically analysed some of them.

Apart from some efforts at rebuttal (e.g. Møller and Chandrasekhar, 1935; Peierls, 1936; Dirac et al., 1942), Eddington's objections attracted little notice in the literature. The comprehensive review articles of Hund (1936) and Strömgren (1937), the standard astronomy text of the day (Russell, Dugan and Stewart, 1938) and Gamow's (1937, 1940) books refer to the Stoner–Anderson formula and the Chandrasekhar limit without mentioning Eddington's reservations. In the 1936 edition of his massive book, *Statistical Mechanics*, R. H. Fowler, whose non-relativistic formula Eddington was stoutly and gratuitously 'defending', presents the standard derivation with a courteous nod to Eddington in a footnote:

If his contentions are correct then the formulae derived from [the relativistic relation between energy and momentum] are meaningless and we must always use formulae derived from $E = p_*^2/2m$ (i.e. Schrödinger's equation) for electrons represented by standing waves in an enclosure.

The events of 1935 have special significance for the light they shed on the generally confused state of the subject at the time and because of

Eddington's key role and position of influence in the development of twentieth-century astrophysics – 'our incomparable pioneer' in the words of Milne's obituary. With his unrivalled command of the two key theories, stellar structure and general relativity, it would seem that Eddington was uniquely placed at this critical juncture to point the way to future developments – to predict the existence of a new class of celestial objects. But events took a different turn. The roots and repercussions of this curious and sad episode will inevitably remain a focus of speculation for years (Sullivan, 1979; Wali, 1982; Blandford, this volume, Chapter 8). The questions involved are complex and subtle and may never be fully resolved. Here I shall venture only some scattered comments that appear relevant to the issue.

The scenario which seemed a *reductio ad absurdum* to Eddington was not yet the ultimate catastrophe of collapse to zero volume that confronts physicists today. As we shall see, Eddington, like virtually every relativist of the time, considered the Schwarzschild radius to be both a singularity and an impassable barrier. The image that he conjures up of the star 'at last finding peace' is of a body frozen at the Schwarzschild radius; it was not yet appreciated that a co-moving observer would see something radically different.

No one was better placed than Eddington to recognize that asking for an equation of state that would *in principle* prevent the collapse of an arbitrary configuration is asking the impossible. It had been known since the time of Schwarzschild (1916b) that, once general relativity is taken into account, a sphere of radius smaller than $1\frac{1}{8}$ times its Schwarzschild radius cannot hold itself in equilibrium even if it is made of incompressible material. (Only a few months earlier Eddington had communicated to the RAS a paper (Sen, 1934) on this very subject which made use of Eddington's own definition of 'incompressibility' (see below).) However, if, as Eddington believed, Fowler's equation of state remains accurate at high densities, white dwarf radii would all be much larger than their Schwarzschild radii unless they happen to be more than 100 times heavier than the sun, and in the 1930s no stars as massive as this were definitely known to exist. Thus, Eddington believed that no actual star would ever get into a situation where general relativistic effects could become significant.

There is evidence suggesting that Eddington was set on a collision course with the Stoner–Anderson formula even if he had not been put off by its consequences. This actually goes back to an unorthodox definition of particle density which he proposed in his 1923 book, *The Mathematical*

Theory of Relativity (p. 121). The presently accepted formulation, first clearly set forth by van Dantzig (1939), considers n to be the magnitude (or temporal component in the rest-frame) of a conserved 4-vector N^μ, the particle flux vector. Eddington, however, believed that n should be extractable from the stress-energy tensor $T^{\mu\nu}$ (at least for a monatomic system) and proposed the identification $nm = T^\mu_\mu$, where m is the particle rest-mass. This idea, enshrined as it was in a major textbook, remained a source of confusion for years (e.g. Sen, 1934; Curtis, 1950). From a conventional standpoint it is plainly wrong, since it would require the internal energy density $T^0_0 - nm$ to be three times the pressure, which is too large by a factor of two in the case of a non-relativistic gas. In his later work Eddington (1936a) was well aware of this discrepancy, but believed it could be attributed to an extra potential or 'interchange' energy associated with an effective repulsive 'force' describing the action of the Pauli exclusion principle. (Weisskopf (1975) has recently described how the concept of such a phenomenological exclusion 'force' offers a powerful heuristic viewpoint for a semi-quantitative understanding of the bulk properties of matter.)

To bring out the connection with Eddington's stand on the Stoner–Anderson formula, let us evaluate $T^{\mu\nu}$ and N^ν for a gas. A particle with 4-velocity v^μ has the ordinary velocity $dx^\mu/dx^0 = \gamma^{-1} v^\mu$ with respect to an arbitrarily chosen Lorentz frame, where $\gamma = v^0 = (1 - v^2)^{-1/2}$ is the Lorentz dilation factor. (We have set $c = 1$.) According to the conventionally accepted formulation, $T^{\mu\nu}$ and N^μ are defined as the fluxes of 4-momentum mv^ν and particle number, i.e.

$$T^{\mu\nu} = \sum \gamma^{-1} v^\mu m v^\nu, \quad N^\mu = \sum \gamma^{-1} v^\mu.$$

The summation is over all particles in a cell of unit volume which is at rest in the chosen frame. (The factors γ^{-1}, which appear to spoil Lorentz covariance, actually guarantee it: $\gamma^{-1} d^3 p$ is the Lorentz-invariant measure of volume in momentum space.) Specializing to the rest-frame of the medium we correctly obtain $n = N^0 = \sum 1$.

Eddington, by contrast, always held that $T^{\mu\nu}$ is to be calculated by summing, *in the macroscopic rest-frame*, the stress-energy tensors of the individual particles. This prescription yields a *different* tensor, $T^{\mu\nu}_{\text{Edd}} = \sum m v^\mu v^\nu$, whose trace, $(T^\mu_\mu)_{\text{Edd}} = m \sum 1 = nm$, indeed determines the particle density, as Eddington claimed.

The effect of Eddington's proposal is to exaggerate the stability of massive stars: absence of the contraction factor γ^{-1} means that, for a given particle density, his stress tensor predicts a higher pressure than the conventional one. Specifically, consider an electron gas in its lowest quantum state: each of the lowest quantum cells (with phase volume h^3), up to a momentum ceiling p_* (say), will be filled by a pair of electrons having opposite spins. The summation $\sum$ is thus to be replaced by the integral $\int_0^{p_*} \ldots 2(4\pi p^2 \, dp)/h^3$. Eddington's expressions then give $P_{\text{Edd}} \sim p_*^5$, $n \sim p_*^3$, i.e. $P_{\text{Edd}} \sim n^{5/3}$, which is precisely Fowler's formula.

Eddington's own presentation of his arguments never took quite this simple form, and he never referred to their 1923 pre-quantum antecedents, presumably because he considered that enumeration of quantum states is a procedure different in principle from the counting of classical particles.†
Nevertheless, his classical and quantum procedures conform to the same mould. (The resemblance is plainest in the 1940 paper.)

I have gone into this point in painstaking detail because it seems to have gone generally unnoticed and because it reveals the underlying continuity and inner consistency of Eddington's thought over a time-span stretching well before the events of 1935. By his own account (Eddington, 1936b), the *motivation* which first caused him to question the Stoner–Anderson result was the 'stellar buffoonery' to which it led; but there can be little doubt that the *reason* for his sustained opposition was grounded in purely technical considerations whose seeds went back more than a decade. Indeed, it is interesting that after its debut as the launch-pad of his 1935 paper, the contraction scenario and its 'absurdity' make no further appearance in his published work.

To speculate about a hypothetical replay of the famous RAS talk, with all reference to a '*reductio ad absurdum*' magically removed, is idle but hard to resist (Chandrasekhar, 1969, 1972; Sullivan, 1979; Wali, 1982). Could it have changed astrophysical history? Given the limited optical window of pre-war astronomy and the many psychological hurdles, the possibility appears remote of any radical shift in pre-war opinion on this issue.

In the first instance, there were Eddington's own preconceptions. By 1936 his unconventional stress tensor had so permeated his thinking on 'molar relativity' and Fundamental Theory that a change of course would have meant dismantling a complex interlocking structure. Attempts to rebut his contentions must have been a source of deep frustration to both sides, since they were often speaking different languages. In the manner typical of creative minds, his way to assimilate a new body of ideas, such as relativity or quantum mechanics, was to re-create it in his own terms. It made

---

†See the remarks on p. 254 of Eddington (1936a).

communication not always easy. Once, defending the Eddington model for main-sequence stars against a rival theory of Milne's, he declared,

> Prof. Milne did not enter into detail as to why he arrives at results so widely different from my own; and my interest in the rest of the paper is dimmed because it would be absurd to pretend that I think there is the remotest chance of his being right .... I maintain that the interior determines the state of the photosphere; Prof. Milne somehow reaches the opposite view.
>
> (Eddington, 1929)

It stung Milne to the retort:

> I recognize that Sir Arthur Eddington has dug a most valuable trench into unknown territory. But he has encountered a rocky obstacle which he cannot get round. If he would make the mental effort to scramble up the sides of the trench he would find the surrounding country totally different from what he had imagined and the obstacle entirely an underground one.
>
> (Milne, 1930)

Suppose, nevertheless, that Eddington's technical concerns could have been sufficiently allayed to make the Stoner–Anderson formula and the Chandrasekhar limit logically acceptable to him. Would his belief in the contraction scenario as an inescapable consequence have held firm? This is the hardest and most hypothetical question of all: we do not know how deep his convictions on this issue really went. Others who accepted the Chandrasekhar limit did not feel impelled to this extremity. He himself speaks of 'various accidents intervening to save the star'. One has to judge whether it is likely that, lacking any observational support for collapse, he would have been prepared to abandon an astronomically conservative stance to take a position that even the arch-radical Landau at that time considered untenable. If the inner conviction was there, it is more than probable that he would have done so. The courage and integrity with which he defended his beliefs were legendary. Eddington's failure in 1935 was not a failure of nerve, but an aberration of a soaring imagination.

One is on much safer ground in predicting the probable reaction of his contemporaries. In 1935, the astronomical community was not yet ready to 'buy' the idea of gravitational collapse, not even if a master salesman like Eddington had been ready to exert all of his persuasion. This is plain from the apathy that greeted Oppenheimer's proposal of the idea four years later. Even today, after half a century, it would be rash indeed to claim that the black hole's battle for acceptance is finally over.

The year 1935 marks the end of a glorious period in British astronomical history. For an unbroken span of 20 years the meeting rooms of the Royal Astronomical Society in Burlington House had held the centre of the world's astrophysical stage. But now the spotlight was moving westward.

7.6 Neutron stars and gravitational collapse (1934–59)

Once man is in a rut he seems to have the urge to dig ever deeper.

(Zwicky, 1957)

It was natural enough that Chadwick's discovery of the neutron in February 1932 should invite speculation about stars consisting wholly or partly of neutrons. The question was indeed quickly taken up by a few adventurous spirits among the astronomers and the nuclear physicists, though with entirely different motives. Conjectures about the nature of such objects accordingly took two quite distinct forms during the 1930s.

The astronomically motivated view was that neutron stars are remnants of supernovae (and hence very rare objects). The originator of this view was the volatile Caltech astronomer Fritz Zwicky (born in Bulgaria and educated in Switzerland), who remained for 25 years its sole advocate in the face of a conventional wisdom which maintained that astronomy does not need anything more exotic than white dwarfs as endproducts of stellar evolution.

The second line of speculation was prompted by the obstacles faced by nuclear physicists in the mid-1930s in their attempt to account for stellar energy generation in terms of thermonuclear reactions. The idea, promoted mainly by Landau (1938) and Gamow (1937), was that many or possibly all stars contain degenerate neutron cores. It was supposed that a slow growth of the core fuels the star's radiation, as normal material becomes neutronized and drastically reduced in volume, thus releasing gravitational energy. (That there were reasons to expect that neutronization would actually be a runaway process, attended by a rapid collapse of the central regions of the star, was not appreciated until after the work of Gamow and Schönberg (1941), who argued that cooling due to neutrino losses by the 'urca-process' in inverse β-decay would result in a catastrophic failure of pressure support near the centre.) With Bethe's discovery of the carbon cycle in 1939, this line of thinking came to an abrupt end. But it did survive just long enough to spawn the classic contributions of Robert Oppenheimer and his students George Volkoff and Hartland Snyder at the Berkeley campus of the University of California.

It was, however, the astronomers who introduced both the concept and the term 'neutron star' in its presently understood sense. It came in a brief paragraph, appended as an 'additional remark', to a visionary pair of papers by Baade and Zwicky in which supernovae were named and distinguished as a class of objects fundamentally different from ordinary novae:

> In addition, the new problem of developing a more detailed picture of the happenings in a supernova now confronts us. With all reserve we advance the view that a supernova represents the transition of an ordinary star into a *neutron star*, consisting mainly of neutrons. Such a star may possess a very small radius and an extremely high density. As neutrons can be packed much more closely than ordinary nuclei and electrons, the 'gravitational packing' energy in a *cold* neutron star may become very large, and under certain circumstances, may far exceed the ordinary nuclear packing fractions. A neutron star would therefore represent the most stable configuration of matter as such. The consequences of this hypothesis will be developed in another place, where also will be mentioned some observations that tend to support the idea of stellar bodies made up mainly of neutrons.
>
> (Baade and Zwicky, 1934)

Zwicky (1938, 1939) was well aware that general relativity was required for a detailed understanding of neutron star structure. But his attempts to develop a general relativistic theory were overtaken by the definitive (and differently motivated) work of Oppenheimer and Volkoff (1939).

It should be mentioned here that the general idea that stellar explosions were associated with collapse to a superdense configuration was not original with Baade and Zwicky but had been in the air for several years (e.g. Menzel, 1926; Milne, 1931), and appeared to derive observational support from the resemblance of old novae and the nuclei of planetary nebulae to white dwarf stars (e.g. Gerasomovic, 1931). Perhaps the earliest reference to the idea is to be found in a paper by H. N. Russell (1925). This was a last-ditch attempt to reconcile the giant-and-dwarf evolutionary theory with Eddington's mass–luminosity relation, but it anticipates in a very generalized yet remarkably judicious way, considering the state of nuclear physics at the time, a number of modern ideas about the ignition of thermonuclear reactions in stars and about stellar evolution.

It would appear that Lev Landau began to think about neutron stars on the very day that news of Chadwick's discovery reached Copenhagen. Leon Rosenfeld (1973) has recalled how, on that evening, as he, Bohr and Landau sat discussing its various implications, Landau came out with the idea of 'weird stars' (*unheimliche Sterne*). But it was five years before Landau committed any of his thoughts to paper. In a brief article entitled 'The origin of stellar energy' he pointed out that

> ... in spite of the fact that the 'neutronic' state of matter is, in usual conditions, energetically less favourable, since the reaction of neutron formation is strongly endothermic, this state must nevertheless become stable when the mass of the body is large enough. In this case the gravitational energy gained in going over to the neutronic state with its greater density, compensates the losses of internal energy.†

Landau's estimate of $10^{-3}M_\odot$ for the minimum mass of a neutron core (using Fowler's non-relativistic equation of state) was later revised to a few tenths of a solar mass by Oppenheimer and Serber (1938). His motivation emerges in the concluding paragraphs:

> ... we can regard a star as a body which has a neutronic core whose steady growth liberates the energy which maintains the star at its high temperatures. ...

> ... the author has shown in a previous article [Landau, 1932] that the formation of a core must certainly take place in a body with a mass greater than $1.5\odot$. In stars with smaller mass the conditions which could make possible the formation of the initial core have yet to be made clear.

(Landau, 1938)

Similar sentiments were expressed by Gamow (1937) in the final chapter of his book, *Atomic Nuclei and Nuclear Transformations*, although at the time of writing the outlook for a thermonuclear origin of stellar energy was not as dim as it came to seem two years later. The situation as it appeared in the fall of 1938 was summarized by Oppenheimer and Robert Serber:

> It would seem that the formation of deuterons by proton collision, and at the least partially regenerative capture of protons by elements between carbon and oxygen could be made to account successfully for the main-sequence stars. ... Nevertheless it has been clear that these reactions could in no way account for the enormously greater radiation of such stars as Capella, and for these one would either have to invoke other and readier nuclear reactions with a correspondingly reduced time-scale, or one would be led, as in the earlier arguments of Milne, to expect serious deviations from the Eddington model.

---

† T. E. Sterne (1933) had shown earlier that, even without taking gravitational energy into account, transition to the neutronic phase is energetically favourable for cold matter compressed to densities exceeding $10^{10}\,g/cm^3$.

In this connection . . . the suggestion [Gamow, 1937; Landau, 1938] of a condensed neutron core, which would make essential deviations from the Eddington model possible even for stars so light that without a core a highly degenerate central zone could not be stable, still seems of some interest.

It was against this background that Oppenheimer and his students embarked upon their remarkable investigations. The first question of interest was 'to investigate whether there is an upper limit to the possible size of such a neutron core'. It was obvious that Newtonian gravitational theory would give a generous upper limit of four times the Chandrasekhar limit for white dwarfs, or about six solar masses, since the star's pressure would be carried by neutrons instead of mainly by the electrons stripped off from helium nuclei, thus reducing the mean mass per pressure – producing particle by a factor 2. However, because of the much smaller radius of a neutron star, general relativistic corrections would become appreciable, as would the weight associated with the kinetic energy of the particles, and both effects could be expected to bring down this upper limit substantially. In their paper, 'On massive neutron cores', Oppenheimer and Volkoff (1939) reported the results of their numerical integrations of the general-relativistic equations of hydrostatic equilibrium for various central pressures. For masses larger than 0.7 solar masses – the 'Oppenheimer–Volkoff limit' – there were no equilibrium configurations.

Since almost nothing was known at the time about nuclear forces, Oppenheimer and Volkoff had little choice but to use the Stoner–Anderson equation of state for a gas of free neutrons in their calculations. But they went on to show that their estimate for the limiting mass would not be appreciably affected by any modification of this equation at supranuclear densities that was physically reasonable in the sense that the trace of the stress-energy tensor did not become negative.†

Six months later, on 10 July 1939, Oppenheimer and Snyder followed this with the submission of a manuscript that has strong claims to be considered the most daring and uncannily prophetic paper ever published in the field. Its scope and breathtaking sweep may be gathered from the abstract:

When all the thermonuclear sources of energy are exhausted a sufficiently heavy star will collapse. Unless fission due to rotation, the radiation of mass, or the blowing off of mass by radiation, reduce the star's mass to the order of that of the sun, this contraction will continue

---

† This was widely considered to be a rigorous requirement, whose proof was attributed to von Neumann (cf. Chandrasekhar, 1935b). However, Zel'dovich (1961) has produced a counter-example showing that it is possible, at least in principle, to circumvent it.

indefinitely ... the radius of the star approaches asymptotically its gravitational radius; light from the star is progressively reddened, and can escape over a progressively narrower range of angles. ... The total time of collapse for an observer co-moving with the stellar matter is finite, and for ... typical stellar mass of the order of a day; an external observer sees the star asymptotically shrinking to its gravitational radius.

There is nothing in this paper which needs revision today, even its terminology is undated. Nevertheless, the questions it left unanswered were to retard acceptance of the results when interest in the subject began to revive 20 years later.

Oppenheimer and Snyder had integrated the interior spherically symmetric Einstein equations for a case in which they could obtain an analytic solution, in the (as later work was to confirm, correct) belief that 'the general features of the solution obtained in this way give a valid indication even for the case where the pressure is not zero, provided that the mass is great enough to cause collapse'. They showed how this interior solution, expressed in co-moving coordinates, could be joined to the exterior Schwarzschild solution, expressed in the standard Schwarzschild coordinates. Unfortunately, with this choice of coordinates, it was impossible to obtain a global picture of the dynamics beyond the point where the star reaches its gravitational radius, since here the exterior time becomes infinite. To a reader of the paper, it was thus left obscure how the interior co-moving picture, in which the collapse proceeds to zero radius in a finite proper time, is to be reconciled with the external view, which sees the contraction 'freezing' at the gravitational radius.† The authors do not even make clear their views on the status of the gravitational radius, which was still generally considered to be some kind of singularity. The nature of the Schwarzschild 'singularity' was being discussed by H. P. Robertson and Einstein at Princeton early in 1939. Robertson's unpublished investigations (to be described in the following section) were carried out in 1936–1937 while on a sabbatical leave at Caltech. It is thus conceivable, though I have no evidence for it, that Oppenheimer had learnt from Robertson‡ about the

---

† A clear global picture first emerged from the work at Princeton of Misner's student Beckedorff (1962), who fitted the interior and exterior geometries together into a single spacetime diagram.

‡ Opportunities for personal contact between the two men were certainly available. Robertson's host at Caltech was Richard C. Tolman, a friend and colleague of Oppenheimer. Tolman was then engaged in completing his book, *The Principles of Statistical Mechanics* (1938), dedicated to Oppenheimer, who was consulted on its every aspect.

fictitious character of the singularity, and may even have supposed that this was common knowledge in relativistic circles.

By the time the Oppenheimer–Snyder work appeared in print in September 1939, Great Britain and France were at war with Germany. The authors had exhausted the topic as far as it could be pursued in the prewar 'optical' era of astronomy. No tie-up with observation was apparent in the scenario of a star that is invisible to begin with, and at the end of its nuclear resources, undergoing a total collapse. Even the premises on which the research was based had been undermined by the appearance, six months earlier, of Bethe's (1939) work, 'On energy generation in stars', which had dealt the finishing blow to ideas that stars need neutron cores in order to shine.

In the months that remained before they dispersed, the three authors moved on to more fashionable problems in nuclear and cosmic ray physics and meson theory. Early in 1940 Volkoff left for Princeton to work with Wigner on the theory of nuclear forces, then returned to the University of British Columbia to take up a teaching position; Snyder accepted a position at Northwestern University (Illinois); by the end of 1941 Oppenheimer had been recruited into the Manhattan Project. They never returned to active research in this field, though Oppenheimer sometimes attended the 'Texas' symposia on relativistic astrophysics in later years.

By 1940 the subject had entered its dark age. Interest remained active in the possible ignition mechanisms of supernovae (e.g. Gamow and Schönberg, 1941; Schönberg and Chandrasekhar, 1942; Hoyle, 1946), but the dynamics of the explosion was beyond the computational facilities then available. It was generally supposed that the remnant would be a white dwarf (e.g. Hoyle, 1955). Minkowski's (1942) analysis of the data for the Crab nebula, which pointed to a radius for the central star about twice that of the earth, could be seen as lending support to this view. Only the maverick Zwicky (e.g. 1941, 1957) still professed an interest in neutron stars, and he continued to press his views from time to time, but nobody paid attention.†

Post-war strides in computer technology and the understanding of nuclear forces eventually made it timely to reconsider the Oppenheimer–Volkoff problem using a more realistic equation of state. In 1957 this was undertaken by the Princeton nuclear physicist John Archibald Wheeler and

---

† In his popular book of 1940, Gamow devoted considerable space to Zwicky's ideas, but became noncommittal on this issue in his subsequent technical papers (Gamow and Schönberg, 1941; Gamow, 1944).

his assistants, B. Kent Harrison and Masami Wakano. It marked the entry of a charismatic new personality into the then moribund field of general relativity. Wheeler's interest was not primarily in the possibility of deriving observable predictions. He was disturbed by a question of principle – 'What is the final equilibrium state of an A-nucleon system when A is large?' – and he found the Draconian answer offered by Oppenheimer and Snyder difficult to accept.

Assembling all the available theoretical information, Harrison and Wheeler constructed a semi-empirical equation of state for 'cold matter, catalysed to the endpoint of thermonuclear evolution' at all stages of compression from $10 \, \mathrm{g/cm^3}$ to supranuclear densities. Using this, Wakano performed 44 numerical integrations, for different choices of central density, of the general-relativistic stellar equilibrium equations on Princeton's MANIAC computer. The resulting plot of mass vs. central density was a curve with two humps, corresponding to the white dwarf and neutron star configurations. The Chandrasekhar and Oppenheimer–Volkoff results were confirmed† and for the first time brought together into a single overall picture.

Wheeler's report on this work to the Solvay congress in Brussels on ₊0 June 1958 was only part of a panoramic survey of the entire field that did not shrink from dragging basic issues into the open:

Of all the implications of general relativity for the structure and evolution of the universe, this question of the fate of great masses of matter is one of the most challenging. Moreover, the issue cannot be escaped by appealing to stellar explosion or rotational disruption, for the issue as it presents itself today is one of principle, not one of observational astrophysics.

Won't the star explode? Let it! . . . Simply catch the ejected matter and extract its kinetic energy. Let it fall back on the star. Then the original number of nucleons, A, is restored; but the mass-energy of the system drops. Ultimately the star gets tired. It can't eject matter. It can't radiate photons. It can't emit neutrinos. It comes into the absolutely lowest state possible for an A-nucleon system under the dual action of nuclear and gravitational forces . . . this is the state we are interested in as a matter of principle.

---

† Equations of state currently considered viable give mass limits three or four times higher than the original Oppenheimer–Volkoff value (Baym and Pethick, 1979).

It may not be necessary to found this discussion entirely on idealized experimentation and issues solely of principle. It is conceivable that astrophysical evolution may occasionally or even often lead to contracting stars substantially more massive than the sun.

Regarding the 'final equilibrium state' he remarked:

Perhaps there is no final equilibrium state: this is the proposal of Oppenheimer and Snyder . . . .

A new look at this proposal today suggests that it does not give an acceptable answer to the fate of a system of A-nucleons under gravitational forces . . . .

No escape is apparent except to assume that the nucleons at the centre of a highly compressed mass must necessarily dissolve away into radiation – electromagnetic, gravitational, or neutrinos, or some combination of the three – at such a rate or in such numbers as to keep the total number of nucleons from exceeding a certain critical number.

Oppenheimer, who was present at the meeting, demurred:

Would not the simplest assumption about the fate of a star more than the critical mass be this, that it undergoes continued gravitational contraction and cuts itself off from the rest of the universe?

But Wheeler then, and for several years afterward, had strong reservations. He pointed out that in the Oppenheimer–Snyder scenario for the spherical collapse of dust *in vacuo*

(1) No mechanism of release of the gravitational energy into the surroundings is taken into account . . . . Therefore this approach rules out *by definition* any approach to an equilibrium.

(2) The particles are envisaged as falling into a K. Schwarzschild singularity. However . . . the forces between nucleons come in in a most vital way . . . *any answer is incomplete that does not deal with the ultimate constitution of a nucleon.*

                                                      (Adams *et al.*, 1958)

Five or six years later, when it became commonplace to consider seriously the collapse of astrophysical objects 10^8 times as massive as the sun, it was easy to allay these concerns. For masses of this order, as Zel'dovich and Novikov (1965) emphasized, densities are only a few g/cm^3 at the time of crossing the gravitational radius:

Under these conditions, which are in no way remarkable, certainly nothing fantastic can take place. The only thing that is unusually large is the gravitational field, but according to the principle of equivalence the gravitational field itself does not produce local changes in the laws that govern physical processes.

By the early 1960s the Soviet school had established a commanding position in this rapidly moving field. This development owes much to the towering figure of Ya. B. Zel'dovich. But an important contributory factor was the headstart in psychological preparedness which the Soviet scientists held on their Western colleagues. In Western circles the work of Oppenheimer and Snyder was a forgotten skeleton in the cupboard by the 1950s, and the notion of gravitational collapse, if anyone had even thought of raising it, would have been dismissed as the wildest speculation; but in the Soviet Union it was an orthodox textbook result. In the 1951 edition of their widely read and admired treatise, *Statistical Physics* – the English translation did not become available until after 1958 – Landau and Lifshitz, citing Oppenheimer and Snyder, stated the position clearly and unequivocally:

> The question arises of the behaviour of a body with mass greater than M_{max}. It is clear from the start that such a body must tend to contract indefinitely. ... From the point of view of a distant observer ... the sphere contracts so that its radius tends asymptotically to ... its gravitational radius. From the point of view of a 'local' observer the substance 'collapses' with a velocity approaching that of light, and it reaches the centre in a finite proper time.

Within a year of the Solvay conference there were signs that Wheeler's crusade was beginning to find echoes. The Canadian nuclear astrophysicist A. G. W. Cameron announced that, in disagreement with the establishment view, he had concluded from 'an examination of the physics of supernova explosions and of the formation of the elements ... that neutron stars are probable products of the supernova process'. His paper concludes with the words:

> It may also be noted that we are still faced with the basic philosophical dilemma associated with the finite number of nucleons contained in the limiting models for which the general relativistic equations of hydrostatic equilibrium admit a solution. As Wheeler (1958) has noted for similar cases, despite the fact that there are no solutions for larger numbers of nucleons, what is to prevent such nucleons from being added to a neutron star?

> (Cameron, 1959)

7.7 The Schwarzschild 'singularity' (1916–66)

There is something queer about a gravitating particle, it seems worthy of further study.

(Synge, 1949)

With the lid prised open, numerous investigators, encouraged to a certain extent by the work of Oppenheimer and Snyder on gravitational collapse, rushed to explore the bewildering region $r < 2m$. At the present time, the nature of bodies which have collapsed into the interior region and their attendant phenomena appear familiar and substantive to these investigators.

(Cooperstock and Junevicus, 1973)

Karl Schwarzschild (1916a) obtained his famous metric, representing the spherisymmetric gravitational field of a point mass, in December 1915, just before Einstein's theory reached its definitive, generally covariant form. For lack of incentive, the nature of the complete vacuum geometry described by this metric was still not generally understood 40 years later, although there had been scattered attempts to probe the 'bewildering region' $r < 2m$. The final breakthrough in understanding, which came in the late 1950s, emerged virtually by osmosis and was sparked less by concerns about gravitational collapse than by Wheeler's (1955) boldly imaginative ideas on multiply connected topologies in general relativity, based on his conviction that 'general relativity, quantized, leaves no escape from topological complexities'. The scientists chiefly responsible were, like Wheeler, specialists in other areas who were newcomers to the field of general relativity, and thus unburdened with the traditional belief that $r = 2m$ was somehow 'unphysical', a belief encoded in the very name 'Schwarzschild singularity'† and supported by two generations of textbooks and the authority of Einstein himself. Indeed, many professional relativists and theoretical physicists at first reacted conservatively to the new developments (e.g. Synge, 1960; McCrea, 1964; Morrison, 1965, 1977), an attitude that lingers in some quarters today.

† The Schwarzschild metric in its standard form

$$\mathrm{d}s^2 = f^{-1}\,\mathrm{d}r^2 + r^2(\mathrm{d}\theta^2 + \sin^2\theta\,\mathrm{d}\phi^2) - f\,\mathrm{d}t^2, \quad f = 1 - 2m/r$$

exhibits singularities at $r = 0$ and at $r = 2m$ (called the 'gravitational radius' for a particle of mass $c^2 m/G$. It is generally understood nowadays that, unlike $r = 0$ where the curvature actually becomes infinite, the 'Schwarzschild singularity' at $r = 2m$ is merely an apparent effect (like the singularity of polar coordinates at the origin) due to a breakdown of Schwarzschild's time coordinate at the gravitational radius. One symptom of this breakdown is that, although light rays moving radially inwards or outwards cross the gravitational radius within a finite proper time in the experience of local free-falling observers, in an (r,t) map their paths, given by $f^{-1}\mathrm{d}r + \mathrm{d}t = 0$, appear to approach $r = 2m$ asymptotically as $t \to \infty$. Kruskal's (1960) prescription for removing the apparent singularity is to replace (r,t) by new coordinates (u,v) which effectively 'straighten' the paths of radial light rays to $u \pm v$ const. Kruskal's coordinates are formally defined by

$$4m\,\mathrm{d}\ln(u \pm v) = f^{-1}\,\mathrm{d}r \pm \mathrm{d}t$$

The early investigators took a mainly pragmatic view: the Schwarzschild 'singularity' is physically significant – i.e. actually exposed as part of the exterior vacuum to which is appertains – only for bodies smaller than their gravitational radius, and no such bodies were known. In his second paper, Schwarzschild (1916*b*) simply mentions without comment that this radius is 3 km for the sun and 10^{-28} cm for a 1 g mass. However, he also shows that a static sphere of fluid with uniform energy density cannot, as a matter of principle, be compressed all the way to its gravitational radius: the central pressure will become infinite when the radius reaches $\frac{9}{8}$ ($2m$).

Another reason of principle why $r = 2m$ could apparently be ignored was adduced by a 30-year-old Dutch scientist, Johannes Droste (1916), a student of H. A. Lorentz, in a paper communicated to the Amsterdam Academy at the end of May. In this remarkable work, Droste independently derived the Schwarzschild exterior solution and proceeded at once to a thorough and elegant discussion of the orbits of particles and light rays in this field. He observed that at $r = 3m$ light rays would follow circular orbits. He also noted that an infalling particle would take an infinite (coordinate) time to reach $r = 2m$ and concluded that 'we may, in studying its motion, disregard the space $r < 2m$'. The theme of inaccessibility of $r < 2m$ was often re-echoed:

> There is a magic circle which no measurement can bring us inside. It is not unnatural that we should picture something obstructing our closer approach and say that a particle of matter is filling up the interior.
>
> (Eddington, 1920)
>
> Every gravitating particle has a ring-fence around it, which no other body can penetrate.
>
> (Whittaker, 1949)

Droste was the first to state clearly the operational significance of the Schwarzschild radial coordinate: 'The circumference of a circle $r = $ const. is $2\pi r \ldots$; this shows how r can be measured'. Not everyone understood so clearly that the operational meaning of the coordinates is (apart from global identifications) entirely contained in the metric itself, and the role of general covariance was a source of widespread confusion. How could the theory lead to unique physical predictions if the coordinates could be changed at will? Einstein himself had unhappily contributed to this confusion. In the final months of 1959, while still groping his way toward the generally covariant form of the field equations, he had calculated the perihelion precession of Mercury using a special coordinate condition $(-g)^{1/2} = 1$, which at that stage he still considered an essential element of the theory. (This condition was also used by Schwarzschild, who did not have access to

Einstein's concluding papers.) In 1921 two noted scientists, the French mathematician Paul Painlevé (1921) and the Swedish ophthalmologist and Nobel laureate Allvar Gullstrand (1922),† independently set out to demonstrate that Einstein's successful prediction was really an artefact of this coordinate condition. They attempted to show that the answer for the perihelion advance came out differently when calculated for the metric

$$ds^2 = dr^2 + r^2(d\theta^2 + \sin^2\theta\, d\phi^2) \pm 2(2m/r)^{1/2}\, dr\, dt - (1 - 2m/r)\, dt^2,$$

which is obtainable from the Schwarzschild metric by a simple change of the time coordinate. Unwittingly, they had for the first time hit upon a form of the metric in which $2m$ is manifestly non-singular. (In the 1930s essentially the same form was rediscovered by Lemaître and by Robertson, who introduced exactly the same time coordinate.) Understandably, this passed unnoticed at the time. Three years later, the opportunity slipped by again when Eddington (1924c), in a paper comparing Einstein's and Whitehead's theories of relativity, found it convenient to introduce a time coordinate in terms of which outgoing radial light rays in the Schwarzschild field travel with unit speed. He thus obtained, without realizing it, another form non-singular at $r = 2m$ that was rediscovered years later by Finkelstein (1958).

   Attempts to visualize how the geometry was distorted by the gravitational field focused at first on the spatial slices of constant Schwarzschild time. By ignoring one of the angular coordinates, a slice could be pictured as a surface in a fictitious three-dimensional Euclidean space. It was shown by Flamm (1916) that this surface is the upper half of a paraboloid obtained by revolving a parabola whose axis is horizontal about a vertical axis, i.e. a funnel-shaped surface terminating in a circle of circumference $2\pi(2m)$ at its narrow end. Since the spatial geometry here is non-singular, there appeared to be no obstacle to extending it to include the lower half of the paraboloid.‡ This step looked even more plausible if one introduced an 'isotropic' radial coordinate ρ by the substitution $r = (1 + m/2\rho)^2\rho$. Weyl (1917) thus arrived at a picture in which the region $0 < \rho < m/2$ 'interior' to the gravitational

---

† After a two-year public debate with C. W. Oseen and E. Kretschmann (e.g. 1923), Gullstrand at length conceded. It was somewhat unfortunate that it was Gullstrand who was chosen in 1921 to prepare a confidential report on relativity for the use of the Nobel committee (Pais, 1982). There was no physics award that year. The prize went to Einstein in 1922 'for his services to theoretical physics and especially for his discovery of the law of the photoelectric effect'. A valuable historical account of the Schwarzschild solution up to 1923 has been published by Eisenstaedt (1982).

‡ It is, however, not clear from this argument that the 4-geometry is non-singular at the throat. That the extended slice is a Cauchy surface for an initially regular vacuum spacetime was not proved until 40 years later (Misner and Wheeler, 1957).

radius $r = 2m$ fans out to a second asymptotically flat space, linked to the exterior space by a 'throat' or 'Einstein–Rosen (1935) bridge', as it was later called. Since the extended three-dimensional slice is nested by 2-spheres whose circumferential radius is never smaller than $2m$, von Laue (1953) inferred that 'points with $r < 2m$ do not exist'. Weyl (1917) deduced from the expression

$$(-g_{tt})^{1/2} = (\rho - m/2)/(\rho + m/2)$$

that 'cosmic' (i.e. coordinate) time and proper time would run in opposite directions for a stationary observer in the 'interior' region. He concluded, 'Of course, only a segment of the solution that does not extend as far as the singular sphere can actually be realized in Nature'.

The first explicit statement that $r = 2m$ is not singular came from Lemaître (1933):

La singularité du champ de Schwarzschild est donc une singularité fictive, analogue à celle qui se présentait à l'horizon du centre dans la forme originale de l'univers de Sitter.

He demonstrated this by exhibiting a coordinate system (essentially that of Painlevé and Gullstrand) in terms of which the metric at $r = 2m$ was manifestly regular.

Lemaître's statement was buried as an incidental remark in the middle of a long and somewhat inaccessible paper on cosmology and it went largely unnoticed. However, it did attract the attention of another cosmologist, Howard Percy Robertson. Examining the orbits of radially moving test particles in the Schwarzschild field Robertson noticed (see Robertson and Noonan, 1968) a second point of resemblance between $r = 2m$ and the de Sitter horizon: although a particle dropped from any finite radius takes an infinite coordinate time to reach $r = 2m$, the proper time measured by a comoving observer is finite. The Schwarzschild 'singularity' is neither singular nor inaccessible.

In 1939, Robertson gave a lecture on the Schwarzschild 'singularity' in Toronto. (In the audience was the Professor of Applied Mathematics, J. L. Synge, who was to retain vivid memories of the talk.) He also discussed the problem with Einstein, who was sufficiently intrigued that in May he sent off a paper whose declared aim was to examine 'whether it is possible to build up a field containing such singularities with the help of actual gravitating masses, or whether such regions with vanishing g_{44} do not exist in cases which have physical reality'.

The essential result of this investigation is a clear understanding as to

why the 'Schwarzschild singularities' do not exist in physical reality. . . .
The 'Schwarzschild singularity' does not appear for the reason that
matter cannot be concentrated arbitrarily.

(Einstein, 1939)

To bypass the uncertainties in the equation of state at high densities,
Einstein considered an idealized model whose microscopic dynamics he
could treat explicitly: in effect, a stationary relativistic 'globular cluster'
whose 'stars' follow circular Keplerian orbits. Einstein found that, for a
cluster of given mass there was a minimum size, $1\frac{1}{2}$ times larger than the
gravitational radius, below which the outermost 'stars' would have to travel
faster than light in order to stay in orbit. (This is really just the result of
Droste mentioned earlier.)

There is nothing wrong with this model, but in drawing such far-reaching
conclusions from it, Einstein slipped back into an oversight which had
vitiated his cosmological paper of twenty years before and which he
considered the biggest blunder of his life: he had again overlooked the
possibility of non-stationary matter configurations. That collapsing masses
which never reach a stationary condition are not only possible but
sometimes inevitable was not spelt out until two months later by
Oppenheimer and Snyder.

Bergmann's standard text gives a fair impression of the general state of
understanding in the early 1940s:

The Schwarzschild field has a singular spherical surface at $r = 2\kappa m$. . . .
Robertson has shown that, if a Schwarzschild field could be realized, a
test body ... would take only a finite proper time to cross the
'Schwarzschild singularity' ... and he has concluded that at least part
of the singular character of the surface $r = 2\kappa m$ must be attributed to the
choice of coordinate system.

In nature, mass is never sufficiently concentrated to permit a
Schwarzschild singularity to occur in empty space. Einstein considered
a system of many mass points. . . . The investigation showed that even
before the critical concentration of particles is reached, some of the
particles (the ones on the outside) begin to move with the velocity of
light . . . .

(Bergmann, 1942)

In 1949, John Lighton Synge, who had recently returned to Ireland as a
research professor at the Dublin Institute for Advanced Studies, decided
that it was time for a fresh assault on the problem. The purely formal
coordinate transformations which Lemaître and Robertson had used to

'remove' the singularity and extend the spacetime left many unanswered questions and obscurities. The original Schwarzschild metric is time-reversible and one would have expected this to carry over to any complete extension. Yet the extensions of Lemaître and Robertson were clearly not reversible: they allowed only 'one-way traffic' through $r = 2m$. Moreover, the old Flamm–Weyl–Einstein–Rosen picture of two asymptotically flat spaces connected by a throat seemed quite incompatible with these extensions. Synge saw this problem as an ideal proving ground for a cause he had long championed: the geometric approach to general relativity.

Synge's attitude to the possible physical significance of such investigations is made clear in the opening paragraphs of his paper. Remarking that the Schwarzschild line element appears to be valid only for $r > 2m$, he continues:

> This limitation is not commonly regarded as serious, and certainly is not so if the general theory of relativity is thought of solely as a macroscopic theory to be applied to astronomical problems, for then the singularity $r = 2m$ is buried inside the body, i.e. outside the domain of the field equations $R_{mn} = 0$. But if we accord to these equations an importance comparable to that which we attach to Laplace's equation, we can hardly remain satisfied by an appeal to the known sizes of astronomical bodies. We have a right to ask whether the general theory of relativity actually denies the existence of a gravitating *particle*, or whether [the standard Schwarzschild metric] may not in fact lead to the field of a particle in spite of the apparent singularity at $r = 2m$.
>
> (Synge, 1950)

He proceeds to exhibit the complete analytical extension of the Schwarzschild manifold in terms of coordinates u, v with the property that $v \pm u$const are the paths of radial light rays, i.e. Kruskal-like coordinates. Unfortunately, he obtained something extra. For $r < 2m$, Synge's coordinates are inverse trigonometric functions of the standard Kruskal coordinates. He was thus led to a picture of an infinite lattice of regions with $0 \leqslant r \leqslant 2m$ in which test particles which have fallen through $r = 2m$ will oscillate forever. If truncated at the first of the geometrical singularities $r = 0$ – which would represent its maximal extension as a manifold – Synge's extension is geometrically the same as Kruskal's. But, because of the complexity of its analysis and the bizarre conclusions, its essential significance was not understood.

In the view of many relativists the whole procedure of analytic extension of spacetimes was a mathematical game without clearly defined rules (e.g.

Bel, 1969). Is the extended manifold necessarily unique (even aside from topological ambiguities)? (Misner (1967) showed by a counter-example that the answer is no.) Why insist on analytic continuation for hyperbolic equations whose characteristic surfaces can admit discontinuities? At what stage of the continuation does the mathematics take leave of physical reality? The beginnings of a clear answer to these questions did not emerge until the early 1970s.

The decisive step in the unravelling of the extended Schwarzschild manifold was taken by an amateur unaware of previous work in the field, the noted plasma physicist Martin Kruskal. In the mid-1950s Kruskal and a few colleagues at Princeton formed a small study group to teach themselves general relativity, using one of the current textbooks as a guide. Kruskal noticed that the Schwarzschild metric near $r = 2m$ closely resembles the flat Minkowski metric expressed in uniformly accelerated ('Rindler') coordinates. By simply following for the Schwarzschild case the path that leads back from Rindler to Lorentz coordinates, he arrived at the well-known coordinates associated with his name.

Kruskal shelved his calculation when Wheeler, to whom he showed it, reacted without special interest. A couple of years later (probably in 1958), well aware now of its importance, Wheeler reacquainted himself with Kruskal's transformation and publicized it at the Royaumont conference on general relativity in June 1959. By the end of 1959 the result had still not appeared in print, though there were several references to its existence (e.g. Finkelstein (1958) and Fronsdal (1959) mention it in notes added in proof). Finally, in desperation, Wheeler was driven to writing up the work himself in a brief paper for which author's credit was given to Kruskal (1960).

Meanwhile, others had independently tackled the problem without knowledge of Kruskal's or other previous work. In a paper entitled 'Past–future asymmetry of the gravitational field of a point particle' David Finkelstein (1958) at the Stevens Institute of Technology showed that Schwarzschild's metric can be extended in two ways, using either retarded or advanced time. At the time of submitting his manuscript, Finkelstein believed that these were inequivalent extensions (rather than complementary parts of a single complete extension) but was able to correct this impression in a postscript added after Kruskal (whom he met at a plasma physics conference) explained his work to him. (Finkelstein's extension remains of practical importance because it covers in a simple explicit form as much of the exterior vacuum manifold as is needed for the description of spherical collapse.) Results equivalent to Kruskal's were derived in mid-1959 by Christian Fronsdal, a particle physicist at CERN, who followed the more complicated route of imbedding the Schwarzschild manifold in a six-dimensional flat spacetime. At the same time the Adelaide mathematician George Szekeres (1960) independently wrote down Kruskal's transformation – he modestly credits his result 'essentially' to Synge (1950) – in an article which languished in obscurity in a Hungarian journal of mathematics until 1966.

The scope of Szekere's article is actually much more general. It contains the first careful definition (using Riemannian normal coordinates) and classification of spacetime singularities. It also suggests for the first time – and rejects – the possibility of an 'elliptic interpretation' of the extended Schwarzschild manifold. Noting that each point of the original spacetime is 'represented twice' in the extension, he remarks,

> Hence in order to obtain a manifold that represents physical reality, it seems to be necessary to identify all pairs of opposite points (u, v) and $(-u, -v)$; ... but ... the Schwarzschild identification introduces an artificial singularity at $u=0$, $v=0$ essentially of the same kind as the singularity at the vertex of a cone obtained by identifying the points (x, y) and $(-x, -y)$ of the Euclidean plane ... it seems difficult to find any physical justification for this identification process.

Since it is, indeed, 'hard to believe that every mass point should have the effect of splitting the universe in two, thus necessitating a second copy of ours' (Rindler, 1965), the elliptic interpretation was resurrected six years later, with varying degrees of reservation, independently by a number of authors (Souriau, 1965; Rindler, 1965; Belinfante, 1966; Anderson and Gautreau, 1966; Israel, 1966), but was soon abandoned. Quite apart from the conical singularity there were obvious problems with causality which proved to be insurmountable.†

7.8 Quasars and relativistic astrophysics (1951–72)

Last week Minkowski got the spectrum of the disputed object [Cygnus A] and promptly paid off the bet (one bottle of Scotch) which I had made with him: the emission spectrum is indeed the outstanding feature (redshift ~ 1500 km/sec) and the high excitation indicated by the

---

† In the last few years, an analogous idea has been revived in a quantum context by t'Hooft (e.g. 1985; cf. Gibbons, 1986; Sanchez and Whiting, 1986).

presence of the [Ne$\overline{V}$] lines!! There the matter stands at present; the collision case appears to be well established now.

(Baade, 1952)

We have this family of chaps, less than a second of arc across. We don't know what to make of them.

(Hanbury Brown, 1961)

Radar expertise (and surplus equipment) acquired in World War II was the stimulus that enabled British and Australian scientists to spearhead the advance of radio astronomy in the post-war decade. This ushered in the most eventful era in the history of astronomy since the time of Galileo. An adequate account of these developments would require a book, and this brief section cannot do more than sketchily trace one strand of this rich tapestry. Numerous vivid accounts (many first-hand) that fill out the picture are now available (e.g. J. L. Greenstein, 1963, 1984; Robinson *et al.*, 1965; Hey, 1973; Bell Burnell, 1977; Sullivan, 1979; Hoyle, 1981; Smith and Lovell, 1983; Ginzburg, 1984; G. Greenstein, 1984; Hazard, 1985; and Blandford, this volume, Chapter 8).

In 1950 just one galaxy (Andromeda) was known to be a radio source, but several dozen other discrete sources (generally much more powerful) had been detected. Only three of these had been associated with visible objects: one was coincident with the Crab nebula and two others (NGC4486 and 5128) with 'nebulae' in Virgo (Messier 87) and Centaurus, whose status as external galaxies was not to be established for another couple of years. The prevailing consensus – Thomas Gold (1951) was the only vocal dissenter – was that the discrete radio sources were dark objects, 'radio stars' (e.g. Ryle, 1950) within our galaxy. To place them outside the galaxy would have ascribed to them an intrinsic radio power far exceeding Andromeda; moreover, they were not correlated in position with prominent galaxies. It even appeared that a local background of such discrete sources was actually required to explain the long wavelength component of the continuum radio emission from the Milky Way. Kiepenheuer (1950) had already explained this component as synchrotron emission from cosmic ray electrons spiralling in the galactic magnetic field, and these ideas were actively pursued by Ginzburg and others in the Soviet Union over the next few years. But they did not sink in in Western astronomical circles until the mid-1950s, when measurements of the optical polarization of the Crab nebula (proposed by I. M. Gordon) spectacularly vindicated Shklovsky's (1953) idea that the synchrotron mechanism was at work in the Crab.

Astronomical opinion then rapidly evolved towards a consensus (Ryle, 1955) on the extragalactic character of the 'radio stars'.

A major breakthrough in this development had come in October 1951 with the accurate positioning of the powerful radio source Cygnus A by F. Graham Smith at Cambridge, using interferometer methods. By the following summer Walter Baade at the Palomar Observatory identified it with a faint dumbbell-shaped object which he interpreted as a pair of colliding galaxies. Historically this was a turning point, the first observational evidence that seemed to point clearly at the extragalactic nature of the discrete sources and even to suggest a physical mechanism. But no one claimed that all the sources could be explained in this way. Regarding Messier 87 (now identified as an elliptical galaxy with a peculiar jet), Baade and Minkowski (1954) remarked that 'no explanation can be suggested as to why and how the pecularity in NGC4486 leads to strong radio emission'.

Following a suggestion by Shklovsky (1955), it was established by Baade in 1956 that the jet in M87 is optically polarized. The true scale of the energies involved at last began to emerge. According to synchrotron theory (Burbidge, 1956), the energy stored in the jet in the form of relativistic particles and magnetic field is equivalent to that released in 10^7 supernova explosions.

Even as an explanation for specific sources, the collision hypothesis remained in vogue for only a few years. (It seemed to be difficult, for instance, to obtain a high efficiency of conversion from low-grade collisional energy into relativistic particles (Burbidge, 1958).) By the turn of the decade (see Struve, 1960) only the original proponents of the idea were still ready to defend it (Minkowski, 1961).† It was slowly ousted from favour by the hypothesis – stemming from Viktor A. Ambartsumian (1958) and promoted vigorously by the Burbidges – of violent outbursts in the nuclei of galaxies, due to causes unknown.

In a forceful and provocative address to the 1958 Solvay Congress, Ambartsumian had marshalled an array of empirical evidence, from the existence of polarized jets to the circumstance that clusters of galaxies

† Mounting evidence of a correlation between quasar-like activity and membership of close interacting pairs is currently reviving the fortunes of the collision hypothesis, in particular for Cygnus A (Thompson, 1984) and similar galaxies (Gaskell, 1985; Begelman *et al.*, 1980). There is an interesting blend of the collision and accretion hypotheses in some recent quasar theories, which suggest that accretion may be fuelled through entanglement of the material in the overlapping nuclei (each containing a black hole) belonging to a giant elliptical and a satellite dwarf galaxy.

appeared to be gravitationally unbound, all of which pointed, in his opinion, to the inference that galactic nuclei were the seal of explosive events that gave birth to new galaxies:

> Apparently we must reject the idea that the nucleus of a galaxy is composed of common stars alone. We must admit that highly massive bodies are members of the nucleus which are capable not only of splitting into parts that move away at a great velocity but also of ejecting condensations of matter containing a mass many times exceeding that of the sun ....
>
> ... the primary nucleus of a galaxy for reasons unknown to us is divided into separate parts which give birth to independent galaxies that become components of the system. In this case the process of division must take place in [a] small volume with a diameter measuring some parsecs or tens of parsecs ...
>
> Radio galaxies Perseus A and Cygnus A are systems in which division of nuclei has taken place but the separation of galaxies is not yet complete.

The hypothesis of large-scale ejection of matter from galactic nuclei had already been invoked 30 years before by Jeans in an attempt to explain spiral structure:

> The type of conjecture which presents itself, somewhat insistently, is that the centres of the nebulae are of the nature of 'singular points', at which matter is poured into our universe from some other, and entirely extraneous, spatial dimension, so that, to a denizen of our universe, they appear as points at which matter is being continually created.
>
> (Jeans, 1928)

These sentiments foreshadow the 'white hole' quasar theories of the mid-1960s (Novikov, 1964; Ne'eman, 1965).

With the theory in this unsettled state, the observers were on the threshold of remarkable new developments. Optical counterparts of many radio sources had been found by 1960 and identified as distant galaxies. Minkowski found that one object, the source 3C295, was a galaxy in Boötes which had the unprecedented redshift of 0.46. The interest of the cosmologists was aroused. Towards the end of 1960, Allan Sandage at Palomar and a group of collaborators – the radio astronomers Thomas A. Matthews and T. G. Bolton and the spectroscopists Jesse Greenstein and Guido Munch – initiated a careful study of three unidentified sources whose large radio flux and small angular size (less than a few seconds of arc according to radio interferometer measurements at Jodrell Bank) encouraged the hope that they might be galaxies of very large redshift.

Sandage presented a preliminary report on this study in an unscheduled talk at the New York meeting of the American Astronomical Society (AAS) at the end of December 1960.

The first object studied was 3C48 .... A direct plate was taken on September 26, 1960 with every expectation of finding a distant cluster of galaxies, but measurement of the plate gave the unexpected result that the only object lying within the error rectangle of the radio position was one which appeared to be stellar. The stellar object was associated with an exceedingly faint wisp of nebulosity ....

(Matthews and Sandage, 1963)

The contemporary report on the AAS meeting in *Sky and Telescope* (1961) gave this account:

Spectrograms taken by Jesse L. Greenstein and Guido Munch of Caltech, and by Dr Sandage, show a combination of strong emission and absorption lines unlike that of any other star known .... The spectrum shows no hydrogen, the main constituent of all normal stars.

Since the distance of 3C48 is unknown, there is a remote possibility that it may be a very distant galaxy of stars; but there is general agreement among the astronomers concerned that it is a relatively nearby star with most peculiar properties. It could be a supernova remnant.

The spectrum of 3C48 remained an enigma for the next two years, although the possibility of 'a big redshift on the spectrum' was raised informally more than once (Hoyle, 1981). Matthews and Sandage observed variations in optical brightness of up to 30 per cent in less than a year, seemingly irrefutable evidence that 3C48 was not a galaxy. As late as February 1963 Burbidge was speculating on the possibility that 3C48 and the other 'radio stars' were neutron stars and 'that the spectral features might arise from elements which are normally rare and that there may be considerable gravitational redshifts'.

However, the real breakthrough came from a different direction. A rare and fortunate series of lunar occultations of the powerful source 3C273 which occurred in April, August and October, 1962, were observed by Hazard, Mackey and Shimmins (1963) with the 210-foot steerable dish at Parkes, Australia. They revealed a double radio structure and gave the position with the unprecented accuracy of less than a second of arc. An optical search by the Caltech astronomer Maarten Schmidt (1963) (to whom Hazard had sent his result at the suggestion of John Bolton), now showed that

> The only objects seen on a 200-inch plate near ... the radio source 3C273 ... are a star of about thirteenth magnitude and a faint wisp or jet. ...
>
> Spectra of the star were taken with the prime-focus spectrograph of the 200-in. telescope .... A redshift $\Delta\lambda/\lambda_0$ of 0.158 allows identification of four emission bands as Balmer lines ....

Greenstein (1963) immediately re-examined his 1961 plates of 3C48. He and Matthews were now able to identify six lines by applying a redshift of 37 per cent.

All these revelations appeared back-to-back in the March 16 issue of *Nature* along with a theoretical paper by Hoyle and Fowler (1963), proposing a collapsing superstar hypothesis, which concluded:

> Our present opinion is that only through the contraction of a mass of 10^7–10^8 $M_\odot$ to the relativity limit can the energies of the strongest sources be obtained.

An earlier suggestion that gravitational energy released in contraction might account for the radio sources had come from Ginzburg (1961):

> ... the gravitational contraction of a galaxy (or of its central region), which is accompanied by the appearance of motions and inhomogeneities in the gas (instability) and, subsequently star formation, must also lead to the generation of cosmic rays. It is easy to see the energy required (up to 10^{61} erg in the case of Cygnus A) may have a gravitational origin.

In June 1963, Ivor Robinson (1965), Alfred Schild, Engelbert Schucking and Peter Bergmann sent out an invitation to more than 300 scientists which read in part:

> The intriguing new discoveries and the theory put forward by Hoyle and Fowler open up the discussion of a wealth of exciting questions. Among the problems raised are the following:
>
> (a) The astronomers observed some unusual objects connected with radio sources. Are these the debris of a gravitational implosion?
>
> (b) By what machinery is gravitational energy converted into radio waves?
>
> (c) Does gravitational collapse lead, on our present assumptions, to indefinite contraction and a singularity in space time?

(d) If so, how must we change our theoretical assumption to avoid this catastrophe?

... After checking with some of the protagonists, we have come to the conclusion that a conference late this year at the Southwest Center for Advanced Studies in Dallas, Texas, might be well timed.

The First Texas Symposium on Relativistic Astrophysics took place in a downtown Dallas hotel from 16 to 18 December, 1963 (*Sky and Telescope*, 1964). It was an event which no one who participated is likely to forget. The opening mood was sombre as the scientists were welcomed by Governor John Connally, his arm in a cast and sling from the tragedy which had happened a few blocks away on November 22. But as the epic discoveries unfolded the rising excitement could not be contained. Although my own graduate research had been in general relativity, this was the first time I heard the work of Oppenheimer and Snyder being discussed. No doubt many others were in the same situation. I recall an ebullient Roy Kerr (1963) treating me in my hotel room to a preview of his talk on the topological structure of his recently published metric: 'Pass through this magic ring and – presto! – you're in a completely different universe where radius and mass are negative!' This was heady stuff, but it seemed to bear no relation to physics. None the less, these were heartening words from Thomas Gold at his after-dinner speech:

This, of course, is a historic meeting. It will be remembered as the meeting where these great new astronomical discoveries were first discussed. It will also be remembered for the display there of strong men wrestling with even stronger facts.

It was, I believe, chiefly Hoyle's genius which produced the extremely attractive idea that here we have a case that allowed one to suggest that the relativists with their sophisticated work were not only magnificent cultural ornaments but might actually be useful to science! Everyone is pleased: the relativists who feel they are being appreciated, who are suddenly experts in a field they hardly knew existed; the astrophysicists for having enlarged their domain, their empire, by the annexation of another subject, general relativity. It is all very pleasing, so let us all hope that it is right. What a shame it would be if we had to go and dismiss all the relativists again.

At one of the opening sessions Greenstein addressed the possibility that 3C48 might be a comparatively nearby object with a redshift of gravitational origin. It is reconcilable, he concluded, with the spectral features (sharp emission lines, presence of forbidden lines) if the source is nearer than the moon, but astronomers do not seem to need such an object to account for planetary motions!

Harlan Smith reported on a study of old Harvard plates which revealed optical variations in 3C273 on timescales of years and months, with even some indications of flashes in which the brightness doubled in less than ten days. It appeared that the primary source of energy could not be more than a light-week in diameter.

Most of the Soviet scientists invited were unable to attend, but they sent preprints which circulated freely. An extraordinary one, by Zel'dovich and Novikov (1964), envisaged a supermassive object powered by accretion (Zel'dovich, 1964; Salpeter, 1964), and argued that its mass must be at least $10^8 \, M_\odot$ if the observed luminosity is not to exceed the Eddington limit. The strongest source could be adequately fuelled by accretion of just a few solar masses per year. Ginzburg (1964), Ozernoy and Kardashev (1964) stressed the role of magnetic fields, which would be strongly amplified by flux-freezing when a compact object forms in a collapse. These ideas paved the way for the theoretical understanding of pulsars a few years later.

The basic difficulty with gravitational collapse as a power source was emphasized by Freeman Dyson in one of the concluding summaries. The timescale for collapse is only about a day; the lifetime of the sources, inferred, for example, from the length of the jets, is at least 10^6–10^7 years.

Attempts over the next couple of years to meet this difficulty head-on by replacing the Oppenheimer–Snyder scenario by an oscillatory picture (e.g. Gertsenshteîn, 1967) came to nothing. Modifying the Einstein field equations with a negative-energy cosmological – or 'C-term' – (Hoyle and Narlikar, 1964; Faulkner et al., 1964) introduces repulsive gravitational forces which can lead to a bounce at small radius, and a pulsation as seen by a comoving observer. On the other hand, it seemed obvious (Zel'dovich and Novikov, 1965) that an external observer could never see the bouncing object re-emerge from the event horizon $r = 2m$ that it has not yet entered in his remote future! The elliptic reinterpretation of the extended Schwarzschild spacetime predicts oscillatory behaviour for infalling test particles at the price of relinquishing ordinary ideas of causality within a sphere slightly larger than the gravitational radius, which, it could be – and was – argued, might lie beyond the range of conscious intervention (Israel, 1966). Even so, it proved impossible to devise a consistent dynamical picture of a gravitating source which revisits the same Cauchy surface.

There were, of course, many other hypotheses about quasars and active galactic nuclei: chain reactions of supernova explosions in distant galaxies, matter–antimatter annihilation, white holes, quasi-local hypotheses in which the quasars were massive objects ejected at relativistic speeds from the nucleus of our own galaxy, and so on. In retrospect it seems clear that many of these ideas were over-reactions to phenomena whose power (as distinct from total energy) requirements did not call for such extraordinary physics. Only one or two seem to have stood the test of time: the model suggested by Lynden-Bell (1969) for the more quiescent sources in which a super-massive black hole is fed by an accretion disc,† and the 'magnetoid' or 'spinar' model of Ginzburg and Ozernoy (1977), Morrison and others, in which the energy source is a magnetized supermassive disc stabilized by rotation. As seen from a present-day vantage point the situation in the mid-1960s has been well summarized by Rees (1984):

> There has been progress towards a consensus in that some bizarre ideas that could be seriously discussed a decade ago have been discarded. But if we compare present ideas with the most insightful proposals advanced when quasars were first discovered 20 years ago (such proposals being selected, of course, with benefit of hindsight), progress indeed seems meager. It is especially instructive to read Zel'dovich and Novikov's (1964) paper entitled 'The mass of quasi-stellar objects'. In this paper, on the basis of early data on 3C273, they conjectured the following: (a) Radiation pressure perhaps balances gravity, so the central mass is $\sim 10^8 M_\odot$. (b) For a likely efficiency of 10% the accretion rate would be $3 M_\odot \text{yr}^{-1}$. (c) The radiation would come from an effective 'photosphere' at a radius of $\sim 2 \times 10^{15}$ cm (i.e. $\gg r_g$), outside of which line opacity would cause radiation to drive a wind. (d) The accretion may be self-regulatory, with a characteristic time scale of ~ 3 yr. These suggestions accord with the ideas that remain popular today, and we cannot yet make any firmly based statements that are more specific.

By good fortune the fourth 'Texas' symposium in Dallas in December 1968 was perfectly timed to catch the excitement of the unravelling of the pulsar mystery. It was at the meeting itself that most delegates learned of the Crab pulsar's deceleration measured at Arecibo. Rotational energy was being lost at a rate which exactly accounted for the energy radiated (mainly in X-rays) by the nebula! At one stroke Gold's (1968) model of pulsars was vindicated and the existence of neutron stars established beyond reasonable doubt. To borrow a well-worn phrase, seldom in the history of astronomy has so bizarre a concept been so firmly established in so short a time, although the 'central star' in the Crab, whose spectrum showed a blank, had

† Similar models involving neutron stars or solar-mass black holes had been proposed a little earlier by Prendergast and Burbidge (1968) for galactic binary X-ray sources.

for 25 years been suspected of being the seat of mysterious, highly energetic processes (Minkowski, 1942; Hoyle, 1955). A hint of the true source of the Crab's radiation had come from Wheeler (1966) in a review of 'superdense stars':

> ... several workers have looked to the residual neutron star itself as a device to power this radiation, either through its heat energy or through its energy of vibration .... (Energy of rotation appears not yet to have been investigated as a source of power. Presumably this mechanism can only be effective – if then – when the magnetic field of the residual neutron star is well coupled to the surrounding ion clouds.)

Spinning neutron stars with oblique magnetic axes had been studied by Pacini (1967) several months before Hewish's (1968) group announced their discovery. Yet no one managed to predict that such objects would pulse, and the exact mechanism remains a matter of controversy to this day.

X-ray astronomy owes its inception in the late 1940s to another legacy of World War II: a small stock of V2 rockets captured from the Germans. X-rays from outside the solar system were first detected in 1962. In 1963, taking advantage of a lunar occultation, Herbert Friedman and his colleagues at the Naval Research Laboratory identified the Crab Nebula as a diffuse X-ray source, and, in 1964, they identified the irregularly variable source Cygnus X-1. By 1971, after it had been located to within a few minutes of arc by Uhuru, Cygnus X-1 was linked with a variable radio source which was then more accurately pinpointed by radio astronomers in Green Bank and Holland. It was finally identified with the one-line optical binary HDE 226868 by L. Webster and P. Murdin at the Royal Greenwich Observatory and by C. T. Bolton (1972) at the David Dunlap Observatory (Toronto). The mass of $\sim 6\,M_{\odot}$ estimated for the unseen component made it the first durable observational candidate for a black hole, and it remained the only one for more than a decade.[†]

7.9 Non-spherical collapse: from frozen star to black hole[‡] (1964–71)

> ... it becomes dimmer millisecond by millisecond, and in less than a second is too dark to see. What was once the core of a star is no longer

[†] In December 1982, a team at the Dominion Astrophysical Observatory found evidence for a massive invisible component in the extragalactic X-ray binary LMC X-3 (Cowley *et al.*, 1983).

[‡] The reader is cautioned that this section is little more than a personal reminiscence. Review articles by Carter (1973) and Tipler, Clarke and Ellis (1980) contain objective historical accounts, particularly of developments in the 1960s.

Fig. 7.1. Is there a black hole in Cygnus X-1? A bet between Stephen Hawking and Kip Thorne made at Caltech in December 1974, on which neither side has yet collected.

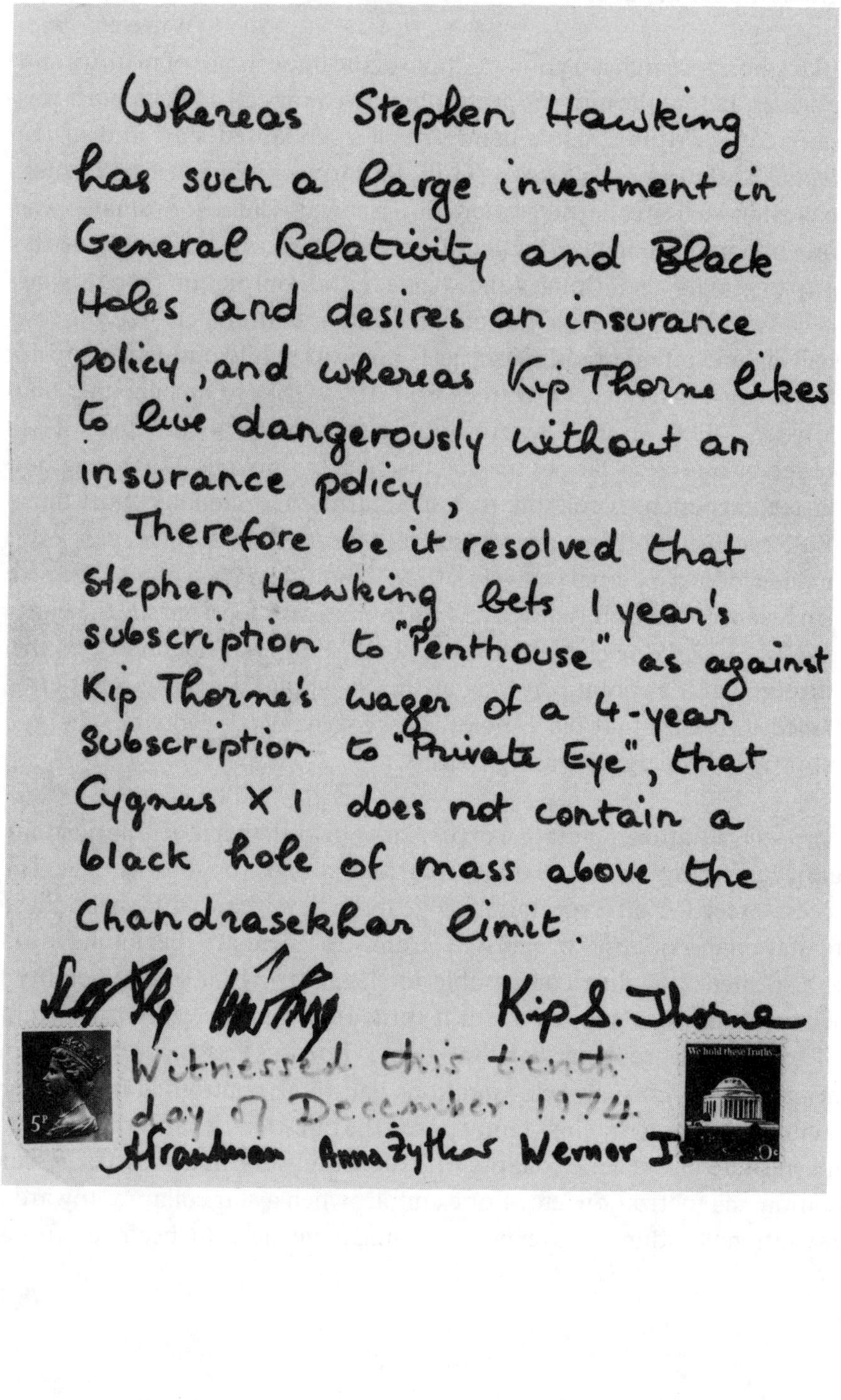

Whereas Stephen Hawking has such a large investment in General Relativity and Black Holes and desires an insurance policy, and whereas Kip Thorne likes to live dangerously without an insurance policy,

Therefore be it resolved that Stephen Hawking bets 1 year's subscription to "Penthouse" as against Kip Thorne's wager of a 4-year subscription to "Private Eye", that Cygnus X 1 does not contain a black hole of mass above the Chandrasekhar limit.

Stephen Hawking   Kip S. Thorne

Witnessed this tenth day of December 1974.

Braidman Anna Zytkow Werner Is...

visible. The core like the Cheshire cat fades from view. One leaves behind only its grin, the other, only its gravitational attraction .... Moreover light and particles incident from the outside emerge and go down the black hole only to add to its mass and increase its gravitational attraction.

(Wheeler, 1968)

Black hole research is a good example of the importance of pictures and phrases. Before the mid-1960s the object we now call a 'black hole' was referred to in the English literature as a 'collapsed star' and in the Russian literature as a 'frozen star'. The corresponding mental picture, based on stellar collapse as viewed in Schwarzschild coordinates was one of a collapsing star that contracts more and more rapidly as the grip of gravity gets stronger and stronger, the contraction then slowing because of a growing gravitational redshift and ultimately freezing to a halt at an 'infinite-redshift surface' (Schwarzschild radius), there to hover for all eternity. Of course, from the work of Oppenheimer and Snyder (1939) we were aware of an alternate viewpoint, that of an observer on the surface of the collapsing star who sees no freezing but instead experiences collapse to a singularity in a painfully short time. But because nothing inside the infinite-redshift surface can ever influence the external universe, that 'comoving viewpoint' seemed irrelevant for astrophysics. Thus astrophysical theorizing in the early 1960s (e.g. Zel'dovich and Novikov, 1964, 1965) was dominated by the 'frozen-star viewpoint'. As long as this viewpoint prevailed, physicists failed to realize that black holes can be dynamical, evolving, energy-storing and energy-releasing objects.

(Thorne et al., 1986)

In 1963, all at once putative experts in a field they had left untilled for a quarter of a century, the relativists scurried to make up for lost time. Their success exceeded all expectations. Within a few years, understanding of gravitational collapse progressed from its inchoate beginnings to a sophisticated discipline comparable in elegance, rigour and generality to thermodynamics, a subject which it turned out unexpectedly to resemble.

The array of problems was formidable. For the first time, the full non-linearity of the Einstein field equations had to be confronted for a generic situation. Astrophysics insistently posed the question: what is the evolution and endstate of a generic collapse with all complications included: pressure, rotation, magnetization, etc? For example, when a star collapses toward its gravitational radius, one expects its magnetic field to become strongly amplified by flux-freezing as its surface area contracts. Could the final collapsed object thus retain a permanent magnetosphere sufficiently powerful to accelerate particles in a surrounding plasma to the relativistic energies that one observes in the Crab nebula and quasar-like sources? (Ginzburg, 1964; Ginzburg and Ozernoy, 1965.)

To guide the way toward answers to such questions on the nature of a generic collapse only one analytical solution was available. This was the highly idealized and special Oppenheimer–Snyder scenario for the collapse of a spherical cloud of dust, in which the exterior field was represented by the Schwarzschild metric. The stationary axially symmetric vacuum metric which had recently been found by Kerr (1963) contained an event horizon (similar to $r = 2m$) and thus conceivably represented an exterior field for the endstate of some spinning collapse, though again (presumably) of a very special kind, since the solution involved only the two minimal parameters, mass and angular momentum. Both solutions had bizarre features which made it debatable whether they could be at all representative of a generic collapse.

In the first instance, both metrics are marked by singularities of infinite curvature. In the Schwarzschild case this singularity (at the centre, $r = 0$) is spacelike, thus obstructing attempts to follow the evolution past the moment of maximal compression. On the other hand, it seemed plausible to argue, and was generally believed in the early 1960s (e.g. Lindquist and Wheeler, 1957; Zel'dovich, 1965), that this central singularity was entirely artificial, an artefact of the special initial conditions (spherical symmetry) imposed on the solution, which had the effect of aiming all the particles at one point. Moreover, this view received support from a detailed analysis by Lifshitz and Khalatnikov (1961, 1963) which concluded that singularities are not a generic feature of solutions of the Einstein field equations:[†]

> Therefore the results that have been set forth exclude the possibility of the existence of a singularity in the future and mean that the contraction

---

[†] Specifically, the argument was that a generic solution involves eight arbitrary functions of the spatial coordinates (six components of the spatial metric tensor and their six time-derivatives less four degrees of freedom due to arbitrariness of the spacetime coordinates), whereas a singular solution appeared to allow only seven arbitrary functions. The conflict with the singularity theorems of Hawking and Penrose (1969) was later traced to a subtle point (Khalatnikov and Lifshitz, 1970; Misner, 1969): the generic solution does not have an analytic expansion near a singularity as Lifshitz and Khalatnikov had originally assumed, but exhibits a complicated stochastic ('mixmaster') behaviour.

of the world (if it should occur in general) should in final analysis alternate with its expansion.

(Lifshitz *et al.*, 1961)

The Kerr metric, analytically extended beyond its horizon, revealed even more alarming idiosyncrasies. The region within the inner horizon admits closed timelike curves, implying an irremediable breakdown of causality in this region. If the analytically completed solution was indeed relevant to collapse rather than just a mathematical curiosity, the outlook was grim:

All these things suggest that the breakdown in general relativity may be of a global rather than (or as well as) of a local nature, in which case it is very serious indeed. If this turns out to be the case then one will not be able to expect to cure the trouble by minor modifications significant only in regions of high curvature, so that the whole theory might have to be abandoned, or at least drastically reformulated.

(Carter, 1968)

The possibility that the available, very special models of collapse might be totally misleading as a basis for induction led to some drastic proposals for modifying the Oppenheimer–Snyder scenario or 'rationalizing' the region within the horizon, including the idea of non-analytic continuation (Komar, 1965) and others already mentioned in Section 7.7.

An essential stabilizing influence during this period was the appearance of a two-part review article of Zel'dovich and Novikov (1964–65), which surveyed and virtually defined the new field of 'relativistic astrophysics'. A mine of sound physical judgement and good sense whose incisive analyses drove straight to the heart of every problem, 'often doubtful but rarely wrong' (Wolfendale, 1986), it became the bible on gravitational collapse until the end of the decade, when it was succeeded by the 'New Testament' of Roger Penrose (1969). There could be no doubt, the authors stoutly insisted, that for an external observer the Oppenheimer–Snyder picture of collapse to a 'frozen' state must be generically correct, at least for small deviations from spherical symmetry and small angular velocities. As to what happens inside the event horizon (the '*T*-region'), they were again characteristically forthright:

We still have no complete answer to this question. We can only state the following. According to the derivations of Lifshitz, Sudakov and Khalatnikov the matter cannot contract to infinite density. As we have seen above, the star cannot expand again, even in an asymmetrical manner, so as to come out from under the Schwarzschild sphere into a region which can be accessible to an external observer. It is possible that the development of the asymmetry leads to a stronger variation of the geometry of space-time inside the T-region or even to change in the topology. At any rate, no matter what occurs inside the T-region, this will never be manifest in the spacetime region outside the Schwarzschild sphere and the external observer will never learn about it.

Soon after the appearance of this article, it was recognized (Novikov, 1966; de la Cruz and Israel, 1967) that a spherical star actually could rebound from inside the Schwarzschild sphere if it was endowed with an arbitrarily small amount of charge or by dint of such devices as the introduction of the repulsive, negative energy 'C-field' of Hoyle and Narlikar. The star would then re-emerge from a *past* horizon of the analytically extended manifold. To preserve causality, one was then forced to assume that this past horizon was part of a different universe! To many people this picture seemed at least as fantastic as the alternative of irreversible collapse to zero volume, although subscribers to white hole theories of quasars may have found it somewhat less difficult to swallow.

However, all bets were off concerning the possibility of a non-singular outcome to a collapse after the publication of the theorem by Roger Penrose (1965), which has claims to be considered the most influential development in general relativity in the 50 years since Einstein founded the theory. Penrose demonstrated that singularities were, after all, a generic feature of gravitational collapse and must appear soon after the formation of a trapped surface (a surface from which light cannot escape outwards). Although the theorem could offer no information about the nature of the singularity, there was a widespread belief that it would have an all-enveloping spacelike character, as in the Schwarzschild case, thus obstructing further development of the solution.

Penrose's paper was just as important for what it initiated as for what it accomplished. Powerful global techniques exploiting the causal structure of spacetime had been introduced into the theory for the first time. In the hands of Hawking and Penrose (1969), with essential input from Robert Geroch (e.g. 1970), these techniques were developed to yield far-reaching results on the occurrence of singularities in cosmology. Their application to gravitational collapse culminated in the elegant results of Hawking and Ellis (1973) on the global properties of event horizons and, in particular, Hawking's (1971) proof of 'the second law of black hole mechanics', which states that the area of the event horizon can never decrease.

An indispensable aid to clarity was the ingenious 'conformal diagram' invented by Penrose and extended to spacetimes with event horizons by Carter (1966). This is essentially a Kruskal diagram compactified so as to bring infinity and all horizons onto finite parts of the map. With its aid one could survey at a glance the global structure of the most complicated spacetimes and it quickly became a standard tool of the trade.

The rationale for analytically extending spacetime manifolds, and the physical status of the more bizarre extensions, like that of the Kerr metric, could now be clarified. Relativists began to understand that the natural guiding principle was not indiscriminate analytic continuation, but rather the maximal Cauchy development of *generic* regular initial data. In other words, one should be able to trust a solution of the field equations, evolving from regular initial data, up to the point where it, or any perturbation of it obtained by generically perturbing the initial data, becomes singular. This principle identified the natural boundaries of an extension as the compact sheets of Cauchy horizons, in particular the ingoing sheets of inner event horizons; these act as infinite blue-shift surfaces which magnify perturbations. Thus, as was first emphasized by Simpson and Penrose (1973), the oscillatory collapse of a charged sphere is an unstable idiosyncrasy of exact spherical symmetry; a generic collapse terminates at the inner horizon, which is generically singular.

However, we must return to the era preceding this age of enlightenment in order to follow another line of development. The jubilee year of general relativity saw the appearance of another major and richly suggestive paper, by Doroshkevich, Zel'dovich and Novikov (1965) (DZN), which launched a broad attack on the problem of non-spherical collapse.

The authors' first aim was to show that the Oppenheimer–Snyder scenario is stable, i.e. that collapse with small deviations from spherical symmetry proceeds qualitatively like spherical collapse and that an event horizon still forms. Their argument, in essence, was that, although asymmetries might grow without bound as the collapse proceeds to zero volume, the resulting blowup of the field, propagating causally along characteristics of the unperturbed spacetime, could never reach the horizon itself. Although this argument was plausible, it would have been impossible to make it rigorous with the mathematical techniques then (or now) available; these allow one to predict the behaviour of a solution from initial data for a finite time interval only, whereas the horizon must persist for an infinite time. This question, which remains today the outstanding unsolved mathematical problem of the field, was later codified as the 'cosmic censorship hypothesis' by Penrose (1969).

DZN next addressed the question of the nature of the final 'frozen' state, in which the external field becomes asymptotically stationary. Examination of special cases indicated that static fields with multipole moments become singular at the event horizon. They went on to argue as follows:

An analysis of the static solution outside the body shows that the deviation from the spherical solution, which is caused by a change in the source of the field, leads to the appearance of true singularities of spacetime on the Schwarzschild surface $g_{00} = 0$. On the other hand, in the comoving system of a contracting body with small initial deviations from sphericity in the density distribution, the Schwarzschild surface is in no way specially distinguished and is not accompanied by the appearance of true singularities either in the metric or in the density. A comparison of these results leads to the conclusion that the quadrupole and higher moments of the external gravitational field attenuate during the relativistic stages of collapse.[†]

This conclusion was highly intriguing but puzzling, since no dynamical explanation was evident or, indeed, offered for the decay of the multipole moments. One possibility, suggested later, was that it was an apparent effect, seen from the outside, due to 'Lorentz contraction of the bulges on the star's surface' (Thorne, 1967). DZN were willing to commit themselves only on what the cause was *not*:

The change in the multipole moments during the course of contraction of the body should be accompanied by radiation of gravitational waves, but the energy carried by this radiation is small. The radiation of waves is a consequence of the change of the multipole moments and should not be regarded as the cause of their total damping. We note that in Newtonian theory, the moments also vary during the course of the compression of the body, but for finite body dimensions they are finite. In Einstein's theory, a relativistic damping is superimposed on this change in the moments of the external field, due to the change in the dimensions of the contracting body.

The external observer 'sees' (for example, with the aid of neutrino and antineutrino radiation) in the ultimate cooled state the finite nonsphericity of the distribution of masses in the sources of the field. However, this nonsphericity is not at all manifest in the external field.

[†] Ginzburg and Ozernoy (1965) had noted a few months earlier that, for an external observer, the effective magnetic moment of a magnetized spherical star contracting quasistatically toward its gravitational radius appears to attenuate like t^{-1}.

The authors now proceeded to consider collapse with rotation. Here, unfortunately, they were diverted from an anticipation of the 'no hair conjecture' by the fact that the only form of Kerr's metric then known contained 'non-removable off-diagonal components $g_{\mu\nu}$ in addition to g_{03}'. They interpreted this to mean that any material source of the Kerr metric must have meridional circulation as well as spin. They concluded that there must be a great variety of endstates for spinning collapse, amongst which the Kerr fields were merely a special class. In fact, the unwelcome off-diagonal elements are actually removable, as Boyer and Lindquist (1967) were to show by an explicit coordinate transformation, and Papapetrou (1966) by a general theorem, a year or two later.

I recall being struck by this paper when I came across the English translation in the physics library at the Technion (where I was a visitor in the spring of 1966) and my bafflement at some of its passages. One part that I could immediately connect with was the discussion of static perturbations of the Schwarzschild metric. I had previously done some work on this (Israel and Khan, 1964), inspired by a preprint sent to me in 1963 by a former University of Alberta student, Lawrence Mysak, who had gone to Adelaide to work with George Szekeres (Mysak and Szekeres, 1966).† Our study had shown rather generally that static event horizons remain non-singular but become tidally deformed by the presence of external bodies. If, therefore, it was assumed that no external bodies were present, and if the DZN contention about the effect of multipole moments was indeed correct (as I suspected), then it should be possible to prove that every non-singular static event horizon must be spherical.

This was not followed up immediately because I was still caught up with the elliptic interpretation (Section 7.7) and possibilities for oscillatory collapse. In the fall of 1966, I began a sabbatical leave at the Dublin Institute for Advanced Studies, accompanied by my student Vicente de la Cruz, who, sensibly, did not share my brief infatuation with the elliptic interpretation. Toward the end of that year, confronted by the preposterous need to postulate a new universe every time a charged sphere bounced from under its gravitational radius (de la Cruz and Israel, 1967), and disgusted with the failure of the elliptic interpretation to provide a consistent way out of the mess, I turned to the possibility that the picture of an exactly spherical bounce might be generically unstable. This idea was good, but in my thinking I placed the locus of instability at the event horizon instead of at the

† Because of difficulties with referees, this paper had to wait three years for publication.

infinite-blueshift sheet of the inner horizon where it belonged. It then occurred to me that the property that only spherical static horizons were non-singular could be taken as indicating that the formation of horizons in gravitational collapse is a non-generic and unstable feature – in a realistic collapse no horizon should form at all.

At any rate it seemed worthwhile at this point to put the sphericity of static horizons on a firm footing. In a short time I was able to prove a theorem to this effect, using some identities of a generalized Green's type which the static field equations provided very obligingly – and rather surprisingly, since they are non-linear. The general method of the proof appeared to be naturally extendable to the stationary case, and the very similar algebraic structures of the Schwarzschild and Kerr metrics made it a good bet at this stage that one should be able (with considerably more labour!) to show that the Kerr horizons were the unique stationary horizons.

This theorem, and my attempt to interpret it as indicating instability of horizons, formed part of a seminar on 'gravitational bounce' which I presented at King's College, London, in February 1967. It was part of a double bill (J. P. Vigier was the other speaker) and there was a large turnout, with Bondi, Pirani, Carter and Misner (then on sabbatical leave in Cambridge) among the people in the front row.

My arguments for horizon instability aroused little interest. It was the theorem itself, which I had thought would be casually accepted as a straightforward codification of the DZN results, which caused surprise, and some skepticism. ('A remarkable result, *if* it is right' was one comment.) It appeared that the DZN paper was either not well known or had not found wide acceptance among British relativists.

Only Misner had no qualms about accepting the result.† His opening comment in the discussion period was, 'I suppose this could mean that collapses with angular momentum are also very restricted'. I agreed that it looked very plausible that the unique stationary fields with event horizons were the Kerr family. But since I was wedded to my own interpretation of this at the time, it is to Misner that the credit is due for the first apprehension of the 'no-hair conjecture' with its correct dynamical interpretation.

In the published version of the theorem I concluded (trying to hedge my

---

† Like me, he was predisposed to believing it by his previous experience with Schwarzschild-like particles in external fields (Manasse, 1963). Misner told me that he had actually attempted to prove such a theorem in collaboration with C. V. Vishveshwara, but they had, as he put it, 'given up too soon'.

bets):

> ... only two alternatives would be open – either the body has to divest itself of all quadrupole and higher moments by some mechanism (perhaps gravitational radiation), or else an event horizon ceases to exist.
>
> (Israel, 1967*a*)

In an article in *Nature*, I expressed my concerns more freely:

> Because of the slowing-down of processes for the external observer, it seems reasonable to suppose ... that it is permissible to treat the limiting external field as static. The foregoing theorem on singular event horizons then applies. Before reaching a singularity at the event horizon, the star will respond to the precursory large tidal forces in such a way as to subdue, if possible, the rise of the singularity itself. There are two possibilities.
>
> The first is that the star shakes off its quadrupole and higher moments, and is, so to speak, patted into spherical shape before crossing the event horizon. Some such idea has been suggested previously.† The chief difficulty here is to find a plausible mechanism ... which will accomplish this task in the finite proper time available.
>
> The second possibility, which has not been considered before, is that the event horizon becomes smudged out and obliterated.
>
> (Israel, 1967*b*)

This argument was an object lesson in how to arrive at nonsense by stretching the frozen star picture beyond its proper limits. What is important to remember is that it was not then clear just where these limits were, and no overall picture was available. Perhaps the biggest conceptual hurdle then facing the theorists was to reconcile the duality of the frozen star picture and the local dynamical view of free fall. It was like knowing about complementarity without having command over the mathematical apparatus of quantum mechanics. In some ways it was worse, because in problems like the decay of multipole moments both aspects of the duality were involved.

The naive version of the frozen star picture was essentially this: as the star freezes and the last gravitational waves depart,‡ the external field reduces to

---

† Here I was uncomfortably aware that DZN had not said exactly this. On the other hand, I did not know exactly what it was they had said, much less how to express it succinctly!

‡ With entrance to the hole apparently obstructed by the frozen star, it required some stretch of imagination to appreciate that gravitational radiation could freely propagate inwards as well as outwards.

its Coulomb-like part, which cannot fade away because it is anchored in the source still lingering just outside its gravitational radius. The guiding model is, of course, the electromagnetic field, where wavelike (transverse) components can be self-sustaining (i.e. source-free) but Coulomb-like (longitudinal) fields require a source. The defect of this naive version is that in a non-linear theory like gravitation a sufficiently strong field can be its own source and thus even Coulomb-like fields can be self-sustaining. It becomes possible to conceive that in a collapse the field and the material source completely part company, and their multipole moments become independent, as DZN (read with hindsight) seem perhaps to have implied. But a heuristic picture which needs such a sophisticated re-interpretation becomes practically almost useless. What the frozen star picture was in the last analysis telling us was that the frozen star is completely irrelevant to the final state!

There would be little point to this tedious rehash of personal errors and confusions if I did not think that it reflects, however distortedly, the significant shift that occurred during 1967–1969 in the general way of thinking about the end-state of a gravitational collapse. Although the idea of horizon instability found a few echoes in the literature (e.g. Janis *et al.*, 1968; Bel, 1969), generally the frozen star was giving way to a new image: an elemental, self-sustaining gravitational field which has severed all causal connection with the material source that created it, and settled, like a soap bubble,† into the simplest configuration consistent with the external constraints. It was, above all, Roger Penrose, at least in England, who inspired and guided the transition to this radically new viewpoint, so aptly encoded in Wheeler's (1968) coincidental and timely coinage of the term 'black hole'.‡

When we met in London near the end of 1967, Penrose explained this view to me essentially as follows:

Doubts have frequently been expressed concerning (the no-hair conjecture), since it is felt that a body would be unlikely to throw off all its excess multipole moments just as it crossed the Schwarzschild radius. But . . . I would certainly not expect the body itself to throw off its multipole moments. On the other hand, the gravitational field *itself* has a lot of settling to do after the body has fallen into the 'hole'. The

† This analogy is, I believe, essentially due to Misner (Thorne, 1970, 1972).
‡ The term was casually slipped into an address to the New York meeting of the American Association for the Advancement of Science on 29 December 1967. Wheeler had first used it a few months earlier in informal discussions at a conference at the Institute for Space Studies organized by V. Canuto.

asymptotic measurement of the multipole moments need have very little to do with the detailed structure of the body itself; the *field* can contribute very significantly. In the process of settling down, the field radiates gravitationally – and electromagnetically too, if electromagnetic field is present. Only the mass, angular momentum and charge need survive as ultimate independent parameters.

(Penrose, 1969)

I was fully converted to this view when, in the fall of 1969, our group at the University of Alberta carried out numerical integrations of idealized collapse models with small departures for spherical symmetry (de la Cruz *et al.*, 1970). Our results showed that the asymmetries are indeed radiated away, mostly *downwards* through the horizon. In effect, once the body that originally supported the asymmetric field energies against gravity has collapsed, the asymmetries have no alternative but to fall in after it (Israel, 1971). These calculations were greatly refined and extended in the work of Richard Price (1972), which revealed the details of how the multipole moments asymptotically decay with time.

In this way a coherent picture of the terminal stage of the evolution of a generic gravitational collapse emerged even before the no-hair conjecture on which it was founded was fully demonstrated. The proof involved technical complications that occupied several theorists in the next few years. My proof was restricted to static black holes, charged or uncharged (Israel, 1967, 1968). For stationary black holes the first key result was Carter's (1971) proof that the uncharged axisymmetric ones could depend only on two parameters (mass and angular momentum), which was compelling evidence that the Kerr black holes were probably the only ones. (The argument was clinched by David C. Robinson (1975).) Finally, Hawking (1972) demonstrated that all stationary black holes must be either static or axisymmetric, which meant that the preceding theorems had covered all possible cases.†

Those who began their groping exploration of the mysteries of gravitational collapse in the early 1960s would scarcely have credited that within a few years the essential features would be so elegantly and completely mapped out. Nevertheless, two key problems remain. One, the proper formulation and proof of a cosmic censorship principle, is by general consent probably of a purely technical character and will have to await the

---

† Actually it was not until 1983 that the charged axisymmetric case was settled, independently by P. O. Mazur and G. Bunting. Their methods also greatly simplified and clarified the structure of the proofs.

development of more sophisticated mathematical methods. The other, the question of what happens near the singularity predicted by the classical Einstein theory, is far deeper:

> When eventually we have a better theory of nature, then perhaps we can try our hands, again, at understanding the extraordinary physics which must take place at a space-time singularity.
>
> (Penrose, 1969)

7.10 Towards the quantum era: the thermodynamics of black holes (1970–74)

> In those days in 1973 when I was often told that I was headed the wrong way, I drew some comfort from Wheeler's opinion that 'black hole thermodynamics is crazy, perhaps crazy enough to work'.
>
> (Bekenstein, 1980)

The inaugural conference of the newly formed European Physical Society, held in Florence in the second week of April 1969, drew more than 800 participants who were treated to reviews by leading specialists to the current state of the art in virtually every active branch of physics. Penrose's (1969) contribution was a survey of gravitational collapse which for the first time consolidated what had been learnt since the mid-1960s from a coherent viewpoint and articulated the key problems. For specialists and non-specialists alike this lecture opened new vistas: the Penrose–Hawking singularity theorems and the no-hair conjecture were graphically described and their significance unfolded; the 'cosmic censor, who forbids the appearance of naked singularities, clothing each one in an absolute event horizon', made his debut. Of the many imaginative new insights in this lecture not the least was the notion that the energy content of a black hole is (in principle) not irrevocably lost if the hole has angular momentum. This idea launched two distinct lines of development – the thermodynamics and the quantum mechanics of black holes – which dramatically reconverged in 1974 with Stephen Hawking's discovery that black holes radiate thermally by a quantum tunneling process.

The details of Penrose's energy extraction process need not detain us here. Essentially, it exploits a layer just outside the event horizon (dubbed the 'ergosphere' by Wheeler and Ruffini) where (negative) gravitational potential energies are still sufficiently large that negative total energies are possible for particles whose motion lags behind the inertial dragging effect of the black hole's angular momentum. (Only spinning black holes possess an ergosphere.) The Penrose process imagines a particle which has been injected into this region splitting into two particles. One has negative energy and drops into the hole; the other escapes to the outside with greater energy than the particle that was injected. The nett effect is an extraction of some of the hole's rotational energy, with a corresponding reduction of its mass and angular momentum.

A 21-year-old Princeton graduate student, Demetrios Christodoulou (1970), undertook a detailed study of the efficiency of quasistatic processes of this type, in which the black hole's parameters are changed gradually by injecting a succession of test particles. He found that the efficiency is limited by an inequality which, in effect, states that a certain quantity can never decrease. This is a function of the mass and angular momentum of the black hole which Christodoulou called the 'irreducible mass'; its simple geometrical significance was not yet apparent.

At the Texas Symposium in Austin in December 1970, Stephen Hawking (1971) crowned the edifice of classical black hole theory with his elegant demonstration that Christodoulou's result was a special case of a very general law. Hawking's theorem states that the area of the event horizon can never decrease in any interaction of a black hole with its environment, provided the energy density of any accreted material is non-negative.

The obvious analogies with thermodynamics already foretokened in the title of Christodoulou's (1970) paper, 'Reversible and irreversible transformations in black-hole physics', were extensively developed by Bardeen, Carter and Hawking during the Les Houches Summer School on black holes in August 1972 (Carter, 1973). Black hole analogues were formulated for all four of the laws of thermodynamics, with the black hole's area playing the role of entropy and its surface gravity the role of temperature.

However, there was almost universal agreement that this analogy was purely formal:

> In fact the effective temperature of a black hole is absolute zero. One way of seeing this is to note that a black hole cannot be in equilibrium with black body radiation at any non-zero temperature, because no radiation could be emitted from the hole whereas some radiation would always cross the horizon into the black hole.
>
> (Bardeen, Carter and Hawking, 1973)

The real effective temperature of a black hole is well defined and unambiguously zero, as also are its chemical potentials. The ordinary particle conservation laws, and the ordinary second law of thermodynamics are unquestionably transcended by a black hole, in the sense that particles and entropy can be lost without trace from an external point of view. It is not possible to mitigate this transcendence by somehow relating the amount of entropy (or the number of particles) which have gone in to the subsequent increase in surface area.

(Carter, 1973)

These admonitions were partly intended for one person in particular, a young Les Houches student from Princeton, Jacob D. Bekenstein (1972, 1973), who was advocating the heretical idea that the area of a black hole actually was the same thing as its entropy, apart from a universal proportionality factor.† This proposal had no takers and was up against a seemingly impregnable counterexample given by Robert Geroch at a colloquium in Princeton in December 1971, which showed that the entropy of a black hole can be increased arbitrarily without changing its area: Lower a weightless box containing radiation to the horizon of a spherical black hole, extracting all of its mass-energy as work. Now open the box and allow the radiation to drain into the hole: the entropy of the hole grows, but its mass (and therefore surface area) remains unchanged.‡

With the benefit of hindsight it is easy to pinpoint the underlying difficulty with Bekenstein's proposal. As he conceived it, it was not quite 'crazy' enough. For all his radicalism concerning the interpretation of black hole area, he hesitated to relinquish the conventional view that the surface gravity had nothing to do with the real temperature of the black hole:

But we emphasize that one should not regard [the surface gravity] as *the* temperature of the black hole: such an identification can easily lead to all sorts of paradoxes, and is thus not useful.

(Bekenstein, 1973)

The struggle to keep this inherently contradictory scheme afloat led him into convoluted arguments which, at best, were just short of compelling and, in less fortunate instances, an easy target for critics (e.g. Israel, 1973).

Help was to come from an unexpected quarter, again traceable to Penrose's energy-extraction process. Zel'dovich (1970) and also Misner (1972)† pointed out that there is a wave analogue of this process, called

---

† Bekenstein's (1980) personal reminiscences offer a vivid account of this period.
‡ This paradox was not cleared up until ten years later, when W. G. Unruh and R. M. Wald (1982) showed that not all of the mass-energy can be extracted as work because of buoyancy effects due to the bath of Hawking radiation enveloping the hole.
† At this time Misner was examining various possibilities for an amplification mechanism involving a massive black hole at the galactic centre in an effort to explain the gravitational wave observations of his colleague at Maryland, Joseph Weber.

'superradiant scattering', in which an impinging wave mode is amplified as it scatters off a rotating black hole if its energy per unit angular momentum is smaller than the angular velocity of the hole. Regarding this as a form of induced emission, Zel'dovich argued that there should be spontaneous emission in the same modes by virtue of the well-known Einstein relations of the quantum theory of radiation. In other words, a spinning black hole should spontaneously emit radiation from its ergosphere and, in the process, slow down until it comes to a stop and the ergosphere disappears; then, according to Zel'dovich, the emission should stop. Quantum field-theoretical calculations by Zel'dovich and Starobinsky, using a somewhat *ad hoc* renormalization prescription, supported this prediction.

Here is Stephen Hawking's (1987) account of what happened next:

If black holes have entropy, then they also ought to have a temperature ... so black holes ought to emit radiation. But by their very definition, black holes are objects which do not emit anything. It therefore seemed that the identification of the area of the event horizon of a black hole with entropy could not be correct. In fact, in 1972, I wrote a paper with Brandon Carter and an American colleague, Jim Bardeen, in which we pointed out that, although there were many similarities between entropy and the area of the event horizon, there was this apparently fatal difficulty. I must admit that in writing this paper I was motivated partly by irritation with Bekenstein who, I felt, had misused my result about the increase of the area of the event horizon. However, in the end it turned out that he was basically correct.

In September 1973 I visited Moscow.† While I was there I discussed black holes with two of the leading Soviet experts. Yakov Zel'dovich and Alexander Starobinsky. They convinced me that the quantum mechanical Uncertainty Principle implied that rotating black holes should create and emit particles. I believed their arguments on physical grounds but I did not like the mathematical way in which they calculated the emission. I therefore set about devising a better treatment of the emission. I described the formulation of this treatment at an informal seminar that I gave at Oxford at the end of November 1973. At that time I had not done the calculations to find out how much would actually be emitted. I was expecting to discover just the radiation that Zel'dovich and Starobinsky had predicted from rotating black

† This was the first meeting of Hawking and Zel'dovich. When they were introduced, Hawking is reputed to have said, 'You are Zel'dovich? I thought you were a corporation'. This story is, unfortunately, apocryphal.

holes. However, when I did the calculation, I found, to my surprise and annoyance, that even non-rotating black holes should apparently create and emit particles at a steady rate. At first I thought that this emission was a pathological result which indicated that one of the approximations that I had used in my treatment was not valid. I was afraid that if Bekenstein found out about it, he would use it as a further argument to support his ideas about the entropy of black holes, which I still did not like. However, the more I thought about it, the more it seemed that the approximations ought to hold. What finally convinced me that the emission was a real effect was that the spectrum of the emitted particles was exactly that which would be emitted by a hot body, and that the black hole was emitting particles at just the rate required to prevent violations of the Second Law. Since then the calculations have been repeated in a number of different forms by other people. They all confirm that black holes ought to emit particles and radiation as if they were hot bodies with a temperature which is higher the smaller the mass of the black hole.

Hawking's preliminary report, 'Black hole explosions?', submitted to *Nature* on January 17, 1974, aroused strong opposition almost as soon as it was in print (Davies and Taylor, 1974); skepticism was prolonged and virtually unanimous. Over the next two years, as various alternative derivations appeared (e.g. Hartle and Hawking, 1976) which were less unconventional from the quantum field theorist's point of view, it slowly became clear that Hawking's result was genuine and inescapable.

The discovery that, at the quantum level, black holes are thermodynamical black bodies at one stroke crowned a classical theory which had reached an undreamed-of level of elegance and sophistication and opened the door to new horizons in the quantum domain. Many perceived in this mysterious result a clue to some possible future synthesis of general relativity, quantum theory and thermodynamics whose nature we cannot yet comprehend.

But science is not just a catalogue of ascertained facts about the universe; it is a mode of progress, sometimes tortuous, sometimes uncertain. And our interest in science is not merely a desire to hear the latest facts added to the collection; we like to discuss our hopes and fears, probabilities and expectations. I have told the detective story as far as it has unrolled itself. I do not know whether we have reached the last chapter.

(Eddington, 1927)

References

Adams, J. B., Harrison, B. K., Klauder, L. T. Jnr, Mjolsness, R., Wakano, M., Wheeler, J. A. and Willey, R. (1958). Some implications of general relativity for the structure and evolution of the universe. *La Structure et L'Evolution de L'Univers*, Onzième Conseil de Physique, Institut International de Physique Solvay. R. Stoops, Brussels, pp. 97–146.

Adams, W. S. (1915). The spectrum of the companion of Sirius. *Publ. Astron. Soc. Pacific*, **27**, 236–7.

Adams, W. S. (1925). The relativity displacement of the spectral lines in the comparison of Sirius. *Proc. Nat. Acad. Sci.*, **11**, 382–7.

Ambartsumian, V. A. (1958). On the evolution of galaxies. Contribution to 11th Solvay Congress, pp. 241–79 (see Adams *et al.*, 1958).

Anderson, A. (1920). On the advance of the perihelion of a planet, and the path of a ray of light in the gravitational field of the Sun. *Phil. Mag.*, **39**, 626–8.

Anderson, J. L. and Gautreau, R. (1966). Possible causal violations at radii greater than the Schwarzschild radius. *Phys. Lett.*, **20**, 24–5.

Anderson, W. (1929). Über die Grenzdichte der Materie und der Energie. *Z. Phys.*, **56**, 851–6.

Anderson, W. (1936). Existiert eine obere Grenze für die Dichte der Materie und der Energie? *Acta Comment. Univ. Tartuensis*, **A29**, 1–142.

Baade, W. (1952). Letter to Bernard Lovell (26 May). Reprinted in Smith and Lovell (1983).

Baade, W. and Minkowski, R. (1954). On the identification of radio sources. *Astrophys. J.*, **119**, 215–31.

Baade, W. and Zwicky, F. (1934). On supernovae; Cosmic rays from supernovae. *Proc. Nat. Acad. Sci.*, **20**, 254–63.

Bardeen, J. M., Carter, B. and Hawking, S. W. (1973). The four laws of black hole mechanics. *Commun. Math. Phys.*, **31**, 161–70.

Baym, G. and Pethick, C. (1979). Physics of neutron stars. *Ann. Rev. Astron. Astrophys.*, **17**, 415–43.

Beals, C. S. (1929). On the nature of Wolf–Rayet emission. *Mon. Not. R. Astron. Soc.*, **90**, 202–12.

Beckedorff, D. L. (1962). Terminal configurations of stellar evolution. A.B. Senior thesis, Princeton [reported in Harrison *et al.* (1965), p. 126].

Begelman, C., Blandford, R. D. and Rees, M. J. (1980). Massive black hole binaries in active galactic nuclei. *Nature*, **287**, 307–9.

Bekenstein, J. D. (1972). Black holes and the second law. *Lett. Nuovo Cim.*, **4**, 737–40.

Bekenstein, J. D. (1973). Black holes and entropy. *Phys. Rev.*, **D7**, 2333–46.

Bekenstein, J. D. (1980). Black hole thermodynamics. *Physics Today* (January), pp. 24–31.

Bel, L. (1969). Schwarzschild singularity. *J. Math. Phys.*, **10**, 1501–3.

Belinfante, F. J. (1966). Kruskal space with wormholes. *Phys. Lett.*, **20**, 25–7.

Bell Burnell, S. Jocelyn (1977). Petit Four (8th Texas Symposium on Relativistic Astrophysics). *Ann. NY Acad. Sci.*, **302**, 685–9.

Bergmann, P. G. (1942). *An Introduction to the Theory of Relativity*, pp. 203–4. Prentice-Hall: New York.

Bethe, H. A. (1939). On energy generation in stars. *Phys. Rev.*, **55**, 103(L), 434–56.

Biermann, L. (1932). Untersuchungen über den inneren Aufbau der Sterne. IV. Konvektionszonen im Innern der Sterne. *Z. Astrophys.*, **5**, 117–39.

Bolton, C. T. (1972). Identification of Cygnus X-1 with HDE 226868. *Nature*, **235**, 271–3.

Boyer, R. H. and Lindquist, R. W. (1967). Maximal analytic extension of the Kerr metric. *J. Math. Phys.*, **8**, 265–81.

Brown, R. Hanbury (1961). Cited in Hoyle (1981), p. 13.

Burbidge, G. R. (1956). On synchrotron radiation from Messier 87. *Astrophys. J.*, **124**, 416–29.

Burbidge, G. R. (1958). Possible sources of radio emission in clusters of galaxies. *Astrophys. J.*, **128**, 1–8.

Burbidge, G. R. (1963). The detection of stars with neutron cores. *Astrophys. J.*, **137**, 995–6.

Cameron, A. G. W. (1959). Neutron star models. *Astrophys. J.*, **130**, 884–94.

Campbell, W. W. (1913). *Stellar Motions*, p. 242. Yale University Press: New Haven.

Carter, B. (1966). Complete analytic extension of the symmetry axis of Kerr's solution of Einstein's equations. *Phys. Rev.*, **141**, 1242–7.

Carter, B. (1968). Global structure of the Kerr family of gravitational fields. *Phys. Rev.*, **174**, 1559–71.

Carter, B. (1971). An axisymmetric black hole has only two degrees of freedom. *Phys. Rev. Lett.*, **26**, 331–3.

Carter, B. (1973). Black hole equilibrium states. In DeWitt (1973), pp. 56–214.

Chandrasekhar, S. (1931*a*). The maximum mass of ideal white dwarfs. *Astrophys. J.*, **74**, 81–2.

Chandrasekhar, S. (1931*b*). The highly collapsed configurations of a stellar mass. *Mon. Not. R. Astron. Soc.*, **91**, 456–66.

Chandrasekhar, S. (1932). Some remarks on the state of matter in the interior of stars. *Z. Astrophys.*, **5**, 321–7.

Chandrasekhar, S. (1934*a*). On the hypothesis of the radial ejection of high-speed atoms from the Wolf–Rayet stars, and the novae. *Mon. Not. R. Astron. Soc.*, **94**, 522–38.

Chandrasekhar, S. (1934*b*). Stellar configurations with degenerate cores. *Observatory*, **57**, 373–77.

Chandrasekhar, S. (1935*a*). The highly collapsed configurations of a stellar mass (second paper). *Mon. Not. R. Astron. Soc.*, **95**, 207–25.

Chandrasekhar, S. (1935*b*). Stellar configurations with degenerate cores (second paper). *Mon. Not. R. Astron. Soc.*, **95**, 676–93.

Chandrasekhar, S. (1951). The structure, the composition, and the source of energy of the stars. In *Astrophysics*, ed. J. A. Hynek. McGraw-Hill: New York.

Chandrasekhar, S. (1969). Some historical notes. *Am. J. Phys.*, **37**, 577–84.

Chandrasekhar, S. (1972). The increasing role of general relativity in astronomy (Halley Lecture). *Observatory*, **92**, 160–74.

Chandrasekhar, S. (1977). Interview with Spencer Weart (Niels Bohr Library, American Institute of Physics), cited in Greenstein (1984), Chapter 12.

Chandrasekhar, S. (1980). The role of general relativity in astronomy: retrospect and prospect. In *Highlights of Astronomy*, vol. 5, ed. P. A. Wayman, pp. 45–61. Reidel: Dordrecht.

Chandrasekhar, S. (1983). *Eddington. The Most Distinguished Astrophysicist of his Time.* Cambridge University Press: Cambridge.

Christodoulou, D. (1970). Reversible and irreversible transformations in black-hole physics. *Phys. Rev. Lett.*, **25**, 1596–7.

Clerke, Agnes M. (1903). *Problems in Astrophysics*, p. 400. A. and C. Black: London.

Cooperstock, F. I. and Junevicus, G. J. G. (1973). Perspectives on the Schwarzschild singularity. *Nuovo Cimento*, **16B**, 387–97.

Cowley, A., Crampton, D., Hutchings, J. B., Remillard, R. and Penfold, J. (1983). Discovery of a massive unseen star in LMC X-3. *Astrophys. J.*, **272**, 118–22.

Cowling, T. G. (1934). The stability of gaseous stars. *Mon. Not. R. Astron. Soc.*, **94**, 768–82.

Cruz, V. de la, Chase, J. E. and Israel, W. (1970). Gravitational collapse with asymmetries. *Phys. Rev. Lett.*, **24**, 423–6.

Cruz, V. de la and Israel, W. (1967). Gravitational bounce. *Nuovo Cimento*, **51A**, 744–60.

Curtis, A. R. (1950). The velocity of sound in general relativity, with a discussion of the problem of the fluid sphere with constant velocity of sound. *Proc. R. Soc.*, **A200**, 248–61.

Dantzig, D. van (1939). Stress tensor and particle density in special relativity theory. *Nature*, **143**, 855–6.

Davies, P. C. W. and Taylor, J. G. (1974). Do black holes really explode? *Nature*, **250**, 37–8.

Descartes, R. (1644). *Principiorum Philosophae*, Part III, Chapter 52.

Detweiler, S. (1982). *Black Holes: Selected Reprints*. Am. Assoc. of Physics Teachers, Stony Brook, NY.

Dirac, P. A. M., Peierls, R. and Pryce, M. H. L. (1942). On Lorentz invariance in the quantum theory. *Proc. Camb. Phil. Soc.*, **38**, 193–200.

Doroshkevich, A. G., Zel'dovich, Ya. B. and Novikov, I. D. (1965). Gravitational collapse of nonsymmetric and rotating masses. *Sov. Phys. JETP*, **22**, 122–30.

Douglas, A. Vibert (1956). *The Life of Arthur Stanley Eddington*. Thomas Nelson: London.

Droste, J. (1916). The field of a single centre in Einstein's theory of gravitation, and the motion of a particle in that field. *Proc. Acad. Sci. (Amsterdam)*, **19**, 197–215.

Eddington, A. S. (1918). On the conditions in the interior of a star. *Astrophys. J.*, **48**, 205–13.

Eddington, A. S. (1920). *Space, Time and Gravitation*, p. 98. Cambridge University Press: Cambridge.

Eddington, A. S. (1921). Das Strahlungsgleichgewicht der Sterne. *Z. Phys.*, **7**, 351–97.

Eddington, A. S. (1924a). On the relation between the masses and luminosities of the stars. *Observatory*, **47**, 107–14.

Eddington, A. S. (1924b, 1925). Correspondence with W. S. Adams, cited in Douglas (1956), pp. 75–7.

Eddington, A. S. (1924c). A comparison of Whitehead's and Einstein's formulas. *Nature*, **113**, 192.

Eddington, A. S. (1926). *The Internal Constitution of the Stars*, p. 172. Cambridge University Press: Cambridge.

Eddington, A. S. (1927). *Stars and Atoms*, pp. 37, 50. Cambridge University Press: Cambridge.

Eddington, A. S. (1929). RAS Meeting, Nov. 8. *Observatory*, **52**, 345–54.

Eddington, A. S. (1935a). Relativistic degeneracy. *Observatory*, **58**, 37–9.

Eddington, A. S. (1935b). On 'relativistic degeneracy'. *Mon. Not. R. Astron. Soc.*, **95**, 194–206.

Eddington, A. S. (1936a). *Relativity Theory of Protons and Electrons*, §§13.3, 13.5, 13.8. Cambridge University Press: Cambridge.

Eddington, A. S. (1936*b*). Lecture, Harvard Tercentenary Conference, quoted in Chandrasekhar (1983), p. 47.

Eddington, A. S. (1939). The hydrogen content of white dwarf stars in relation to stellar evolution. *Mon. Not. R. Astron. Soc.*, **99**, 595–606.

Eddington, A. S. (1940). The physics of white dwarf matter. *Mon. Not. R. Astron. Soc.*, **100**, 582–94.

Einstein, A. (1939). On a stationary system with spherical symmetry consisting of many gravitating masses. *Ann. Math. (Princeton)*, **40**, 922–36.

Einstein, A. and Rosen, N. (1935). The particle problem in general theory of relativity. *Phys. Rev.*, **48**, 73–7.

Eisenstaedt, J. (1982). Histoire et singularités de la solution de Schwarzschild (1915–1923), *Arch. Hist. Exact Sci.*, **27**, 157–98.

Faulkner, J., Hoyle, F. and Narlikar, J. V. (1964). On the behaviour of radiation near massive bodies. *Astrophys. J.*, **140**, 1100–5.

Finkelstein, D. (1958). Past–future asymmetry of the gravitational field of a point-particle. *Phys. Rev.*, **110**, 965–7.

Flamm, L. (1916). Beiträge zur Einsteinschen Gravitations theorie. *Phys. Z.*, **17**, 448–54.

Fowler, R. H. (1926). On dense matter. *Mon. Not. R. Astron. Soc.*, **87**, 114–22.

Fowler, R. H. (1929). *Statistical Mechanics*, p. 522. Cambridge University Press: Cambridge.

Fronsdal, C. (1959). Completion and embedding of the Schwarzschild solution. *Phys. Rev.*, **116**, 778–81.

Gamow, G. (1937). *Structure of Atomic Nuclei and Nuclear Transformations*, pp. 234–8. Clarendon Press: Oxford.

Gamow, G. (1940). *The Birth and Death of the Sun*, pp. 170, 189–93. Viking Press: New York.

Gamow, G. and Schönberg, M. (1941). Neutrino theory of stellar collapse. *Phys. Rev.*, **59**, 539–47.

Gamow, G. (1944). The evolution of contracting stars. *Phys. Rev.*, **65**, 20–32.

Gaskell, C. M. (1985). Galactic mergers, starburst galaxies, quasar activity and massive binary black holes. *Nature*, **315**, 386.

Gerasomovic, B. P. (1931). The nuclei of planetary nebulae. *Observatory*, **54**, 108–10.

Geroch, R. P. (1970). The domain of dependence. *J. Math. Phys.*, **11**, 437–49.

Gertsenshteîn, M. E. (1967). The possibility of an oscillatory nature of gravitational collapse. *Sov. Phys. JETP*, **24**, 87–90; The nature of the central body in the Schwarzschild solution. *Ibid.*, **24**, 754–9.

Gibbons, G. W. (1986). The elliptic interpretation of black holes and quantum mechanics. *Nucl. Phys.*, **B271**, 497–508.

Ginzburg, V. L. (1961). The nature of the radio galaxies. *Sov. Astron.*, **5**, 282–3.

Ginzburg, V. L. (1964). On the magnetic fields of collapsing masses and on the nature of superstars. In Robinson *et al.* (1965), pp. 283–7; *Sov. Phys. Dokl.* **9**, 329–32.

Ginzburg, V. L. and Ozernoy, L. M. (1965). On gravitational collapse of magnetic stars. *Sov. Phys. JETP*, **20**, 689–96.

Ginzburg, V. L. and Ozernoy, L. M. (1977). On the nature of quasars and active galactic nuclei. *Astrophys. Space Sci.*, **48**, 401–20; **50**, 23–41.

Ginzburg, V. L. (1984). Remarks on my work in radio astronomy. In Sullivan (1984), pp. 289–302.

Gold, T. (1951). Meeting of Royal Astronomical Society, 12 Oct. 1951. *Observatory*, **72**, 210–14.

Gold, T. (1968). Rotating neutron stars as the origin of the pulsating radio sources. *Nature*, **218**, 731–2.

Greenstein, G. (1984). *Frozen Star*. Macdonald: London.

Greenstein, J. L. (1963). Quasi-stellar radio sources. *Scientific American*, **209** (December), 54–62.

Greenstein, J. L. (1984). Optical and radio astronomers in the early years. In Sullivan (1984), pp. 67–81.

Gullstrand, A. (1922). Allegemeine Lösung des statischen Einkörper-problems in der Einsteinschen Gravitations theorie. *Arkiv. Mat. Astron. Fys.*, **16**(8), 1–15.

Harrison, B. K., Thorne, K. S., Wakano, M. and Wheeler, J. A. (1965). *Gravitation Theory and Gravitational Collapse*. University of Chicago Press: Chicago.

Harrison, E. R. (1972). Particle barriers in cosmology. *Comment Astrophys. Space Phys.*, **4**, 187–92.

Hartle, J. B. and Hawking, S. W. (1976). Path-integral derivation of black-hole radiance. *Phys. Rev.*, **D13**, 2188–203.

Hawking, S. W. (1971). Gravitational radiation from colliding black holes. *Phys. Rev. Lett.*, **26**, 1344–6.

Hawking, S. W. (1972). Black holes in general relativity. *Commun. Math. Phys.*, **25**, 152–66.

Hawking, S. W. (1974). Black hole explosions? *Nature*, **248**, 30–1.

Hawking, S. W. (1987). *From the Big Bang to Black Holes: A Short History of Time*. Bantam Books: New York (in press).

Hawking, S. W. and Ellis, G. F. R. (1973). *The Large Scale Structure of Space-Time*. Cambridge University Press: Cambridge.

Hawking, S. W. and Penrose, R. (1969). The singularities of gravitational collapse and cosmology. *Proc. R. Soc. (London)*, **A314**, 529–48.

Hazard, C. (1985). The coming of age of QSOs. In *Active Galactic Nuclei*, ed. J. E. Dyson, pp. 1–19. Manchester University Press: Manchester.

Hazard, C., Mackey, M. B. and Shimmins, A. J. (1963). Investigation of the radio source 3C273 by the method of lunar occultations. *Nature*, **197**, 1037–9.

Hertzprung, E. (1905). Zur Strahlung der Sterne (giants and dwarfs). *Z. Wiss. Photog.*, **3**. Engl. transl. in *Source Book in Astronomy*, ed. H. Shapley. Harvard University Press: Cambridge, Mass.

Hertzprung, E. (1915). Effective wavelengths of absolutely faint stars. *Astrophys. J.*, **42**, 111–19.

Hewish, A. (1975). Pulsars and high-density physics. *Rev. Mod. Phys.*, **47**, 567–72.

Hewish, A., Bell, S. J., Pilkington, J. D. H., Scott, P. F. and Collins, R. A. (1968). Observation of a rapidly pulsating radio source. *Nature*, **217**, 709–13.

Hey, J. S. (1973). *The Evolution of Radio Astronomy*. Neale Watson: New York.

t'Hooft, G. (1985). On the quantum structure of a black hole. *Nucl. Phys.*, **B256**, 727–45.

Hoyle, F. (1946). The synthesis of the elements from hydrogen. *Mon. Not. R. Astron. Soc.*, **106**, 343–83.

Hoyle, F. (1955). *Frontiers of Astronomy*, p. 204. Heinemann: London.

Hoyle, F. (1981). *The Quasar Controversy Resolved*. University College Cardiff Press: Cardiff.

Hoyle, F. and Fowler, W. A. (1963). Nature of strong radio sources. *Nature*, **197**, 533–5.

Hoyle, F., Fowler, W. A., Burbidge, G. R. and Burbidge, E. M. (1964). On relativistic astrophysics. *Astrophys. J.*, **139**, 909–28.

Hoyle, F. and Narlikar, J. V. (1964). On the avoidance of singularities in C-field cosmology. *Proc. R. Soc.*, **A278**, 465–78.

Hund, F. (1936). Materie unter sehr hohen Drucken und Temperaturen. *Ergebnisse Exakten Naturwiss.*, **15**, 189–228.

Hund, F. (1971). Irrwege und Hemmungen beim Werden der Quantentheorie. In *Quanten und Felder*, ed. H. P. Dürr, pp. 1–10. F. Vieweg: Braunschweig.

Israel, W. (1966). Is gravitational collapse really irreversible? *Nature*, **211**, 466–7.

Israel, W. (1967a). Event horizons in static vacuum space-times. *Phys. Rev.*, **164**, 1776–9.

Israel, W. (1967b). Possible instability in the self-closure phenomenon in gravitational collapse. *Nature*, **216**, 148–9 and 312.

Israel, W. (1968). Event horizons in static electrovac space-times. *Commun. Math. Phys.*, **8**, 245–60.

Israel, W. (1971). Event horizons and gravitational collapse. *J. Gen. Rel. Grav.*, **2**, 53–61.

Israel, W. (1973). Entropy and black hole dynamics. *Lett. Nuovo Cimento*, **6**, 267–9.

Israel, W. (1983). Black holes. *Sci. Prog. (Oxford)*, **68**, 333–63.

Israel, W. and Khan, K. A. (1964). Collinear particles and Bondi dipoles in general relativity. *Nuovo Cimento*, **33**, 331–44.

Jaki, S. L. (1978). Johann Georg von Soldner and the gravitational binding of light, with an English translation of his essay on it published in 1801. *Found. Phys.*, **8**, 927–50.

Janis, A. I., Newman, E. T. and Winicour, J. (1968). Reality of the Schwarzschild singularity. *Phys. Rev. Lett.*, **20**, 878–80.

Jeans, J. H. (1926). On liquid stars and the liberation of stellar energy. *Mon. Not. R. Astron. Soc.*, **87**, 400–14.

Jeans, J. H. (1928). *Astronomy and Cosmogony*, pp. 72, 352 and Chap. 5. Cambridge University Press: Cambridge.

Jeffreys, H. (1918). The compressibility of dwarf stars and planets. *Mon. Not. R. Astron. Soc.*, **78**, 183–4.

Jeffreys, H. (1930). Convections in stars. *Mon. Not. R. Astron. Soc.*, **91**, 121–2.

Jüttner, F. (1928). Die relativistische Quantentheorie des idealen Gases. *Z. Phys.*, **47**, 542–66.

Kardashev, N. S. (1964). Magnetic collapse and the nature of intense sources of cosmic radio-frequency emission. *Sov. Astron. AJ*, **8**, 643–8.

Kerr, R. P. (1963). Gravitational collapse and rotation. In Robinson, Schild and Schucking (1965), pp. 99–102.

Khalatnikov, I. M. and Lifshitz, E. M. (1970). General cosmological solution of the gravitational equations with a singularity in time. *Phys. Rev. Lett.*, **24**, 76–9.

Kiepenheuer, K. O. (1950). Cosmic rays as the source of general galactic radio emission. *Phys. Rev.*, **79**, 738.

Komar, A. (1965). Bootstrap gravitational geons. *Phys. Rev.*, **137B**, 462–6.

Kosirev, N. A. (1934). Radiative equilibrium of the extended photosphere. *Mon. Not. R. Astron. Soc.*, **94**, 430–43.

Kretschman, E. (1923). Das statische Einkörperproblem in der Einsteinschen Theorie. Antwort an Hrn A. Gullstrand. *Arkiv Mat. Astron. Fys.*, **17**, 1–4.

Kruskal, M. D. (1960). Maximal extension of Schwarzschild metric. *Phys. Rev.*, **119**, 1743–5.

Landau, L. (1932). On the theory of stars. *Phys. Z. Sowjetunion*, **1**, 285–7.

Landau, L. (1938). The origin of stellar energy. *Nature*, **141**, 333–4.

Laplace, P. S. (1796). *Exposition du Système du Monde*, vol. 2, p. 305. J. B. M. Duprat: Paris. See also Hawking and Ellis (1973), pp. 365–8.

Laue, M. von (1953). *Die Relativitäts theorie* (3rd ed.), vol. 2, p. 135. F. Vieweg: Braunschweig.

Lemaître, G. (1933). L'univers en expansion. *Ann. Soc. Sci. (Bruxelles)*, **A53**, 51–85.

Liebert, J. (1980). White dwarf stars. *Ann. Rev. Astron. Astrophys.*, **18**, 363–98.

Lifshitz, E. M., Sudakov, V. V. and Khalatnikov, I. M. (1961). Singularities of cosmological solutions of gravitational equations III. *Sov. Phys. JETP*, **13**, 1298–303.

Lifshitz, E. M. and Khalatnikov, I. M. (1963). Problems of relativistic cosmology. *Sov. Phys. Uspekhi*, **6**, 495–522.

Lindquist, R. W. and Wheeler, J. A. (1957). Dynamics of a lattice universe by the Schwarzschild-cell method. *Rev. Mod. Phys.*, **29**, 432–43.

Lodge, Sir Oliver (1920). Discussion on theory of relativity. *Mon. Not. R. Astron. Soc.*, **80**, 96–118.

Lodge, Sir Oliver (1921). On the supposed weight and ultimate fate of radiation. *Phil. Mag.*, **41**, 549–57.

Lucretius (55 BC). *On the Nature of the Universe*, p. 59. Penguin Books: Harmondsworth.

Lynden-Bell, D. (1969). Galactic nuclei as collapsed old quasars. *Nature*, **223**, 690–4.

McCormmach, R. (1968). John Michell and Henry Cavendish: weighing the stars. *Brit. J. Hist. Sci.*, **4**, 126–55.

McCrea, W. H. (1964). Release of gravitational energy in general relativity. *Nature*, **201**, 589.

Manasse, F. K. (1963). Distortion in the metric of a small centre of gravitational attraction due to its proximity to a very large mass. *J. Math. Phys.*, **4**, 746–61.

Matthews, T. A. and Sandage, A. R. (1963). Optical identification of 3C48, 3C196 and 3C286 with stellar objects. *Astrophys. J.*, **138**, 30–56.

Menzel, D. H. (1926). The planetary nebulae. *Publ. Astron. Soc. Pacific*, **38**, 295–312.

Michell, John (1784). On the means of discovering the distance, magnitude, etc., of the fixed stars, in consequence of the diminution of their light, in case such a diminution should be found to take place in any of them, and such other data should be procured from observations, as would be further necessary for that purpose. *Phil. Trans. R. Soc. (London)*, **74**, 35–57. Reprinted in Detweiler (1982).

Milne, E. A. (1930). The analysis of stellar structure. *Mon. Not. R. Astron. Soc.*, **91**, 4–55; *Observatory*, **53**, 305–8.

Milne, E. A. (1931). On dense stars. *Observatory*, **54**, 140–5.

Minkowski, R. (1942). The Crab nebula. *Astrophys. J.*, **96**, 199–213.

Minkowski, R. (1961). Letter. *Sky and Telescope*, **21**, 23.

Misner, C. W. (1967). Taub-Nut space as a counterexample to almost anything. In *Relativity Theory and Astrophysics*, vol 1, ed. J. Ehlers, pp. 160–7. American Mathematical Society: Providence, Rhode Island.

Misner, C. W. and Wheeler, J. A. (1957). Classical physics as geometry: gravitation, electromagnetism, unquantized charge and mass as properties of empty space. *Ann. Phys.*, **2**, 525–603.

Misner, C. W. (1969). Mixmaster universe. *Phys. Rev. Lett.*, **22**, 1071–4.

Misner, C. W. (1972). Interpretation of gravitational wave observations. *Phys. Rev. Lett.*, **28**, 994–7.

Møller, Chr. and Chandrasekhar, S. (1935). Relativistic degeneracy. *Mon. Not. R. Astron. Soc.*, **95**, 673–6.

Morrison, P. (1965). Summary. In Robinson, Schild and Schucking (1965), pp. 437–8.

Morrison, P. (1977). Astronomy and the laws of physics. In *Highlights of Astronomy*, vol. 4, part 1, ed. E. A. Müller, p. 35. Reidel: Dordrecht.

Mysak, L. A. and Szekeres, G. (1966). Behaviour of the Schwarzschild singularity in superimposed gravitational fields. *Can. J. Phys.*, **44**, 617–27.

Ne'eman, Y. (1965). Expansion as an energy source in quasi-stellar radio sources. *Astrophys. J.*, **141**, 1303–5.

Newcomb, S. (1908). *The Stars: A Study of the Universe*, p. 205. John Murray: London.

Novikov, I. D. (1964). Delayed explosion of a part of the Friedman universe and quasars. *Sov. Astron. AJ*, **8**, 857–63.

Novikov, I. D. (1966). Change of relativistic collapse into anticollapse and kinematics of a charged sphere. *Sov. Phys. JETP Lett.*, **3**, 142–4.

Observatory, The (1916). Notes: The spectrum of the companion of Sirius. **39**, 142–3.

Observatory, The (1917). Proceedings at meeting of the Royal Astronomical Society, Friday, 8 Dec. 1916. **40**, 35–49.

O'Dell, C. R. (1963). The evolution of the central stars of planetary nebulae. *Astrophys. J.*, **138**, 67–78.

Öpik, E. J. (1916). On the densities of visual binary stars. *Astrophys. J.*, **44**, 292–302.

Oppenheimer, J. R. and Serber, R. (1938). On the stability of stellar neutron cores. *Phys. Rev.*, **54**, 540.

Oppenheimer, J. R. and Snyder, H. (1939). On continued gravitational contraction. *Phys. Rev.*, **56**, 455–9.

Oppenheimer, J. R. and Volkoff, G. (1939). On massive neutron cores. *Phys. Rev.*, **55**, 374–81.

Pacini, F. (1967). Energy emission from a neutron star. *Nature*, **216**, 567–8.

Painlevé, P. (1921). La Mécanique classique et la theorie de la relativité. *C.R. Acad. Sci. (Paris)*, **173**, 677–80.

Pais, A. (1982). *Subtle is the Lord ... The Science and the Life of Albert Einstein*, p. 509. Clarendon Press: Oxford.

Papapetrou, A. (1966). Champs gravitationnels stationnaires a symétrie axiale. *Ann. Inst. H. Poincaré*, **A IV**, 83–105.

Payne-Gaposchkin, C. (1957). *The Galactic Novae*, p. 312. North-Holland: Amsterdam.

Peierls, R. (1936). Note on the derivation of the equation of state for a degenerate relativistic gas. *Mon. Not. R. Astron. Soc.*, **96**, 780–4.

Penrose, R. (1965). Gravitational collapse and space-time singularities. *Phys. Rev. Lett.*, **14**, 57–9.

Penrose, R. (1969). Gravitational collapse: the role of general relativity. *Riv. Nuovo Cimento*, **1**, 252–76.

Prendergast, K. H. and Burbidge, G. R. (1968). On the nature of some galactic X-ray sources. *Astrophys. J.*, **151**, L83–8.

Price, R. H. (1972). Nonspherical perturbations of relativistic gravitational collapse. *Phys. Rev.*, **D5**, 2419–54.

Rees, M. J. (1984). Black hole models for active galactic nuclei. *Ann. Rev. Astron. Astrophys.*, **22**, 471–506.

Rindler, W. (1965). Elliptic Kruskal–Schwarzschild space. *Phys. Rev. Lett.*, **15**, 1001–2.

Robertson, H. P. and Noonan, T. W. (1968). *Relativity and Cosmology*, pp. 246–52. W. B. Saunders: Philadelphia.

Robinson, D. C. (1975). Uniqueness of the Kerr black hole. *Phys. Rev. Lett.*, **34**, 905–6.

Robinson, I., Schild, A. and Schucking, E. L. (eds.) (1965). *Quasistellar Sources and Gravitational Collapse*, Editors' introduction, pp. i–xvii. University of Chicago Press: Chicago.

Rosenfeld, L. (1973). Remarks in *Seizième Conseil de Physique Solvay*, p. 174. Stoops: Brussels.

Russell, H. N. (1912). Relation between the spectra and other characteristics of the stars. *Proc. Amer. Phil. Soc.*, **51**, 569.

Russell, H. N. (1914). Relations between the spectra and other characteristics of the stars. *Popular Astron.*, **22**, 275–94, 331–51.

Russell, H. N. (1925). The problem of stellar evolution. *Nature*, **116**, 209–12.

Russell, H. N. (1939). Introductory remarks to the Paris meeting: *Novae and White Dwarfs*. Hermann: Paris (1941). Reprinted in Schatzman (1958), p. 1.

Russell, H. N. (1944). Note on white dwarfs and small companions. *Astron. J.*, **51**, 13.

Russell, H. N., Dugan, R. S. and Stewart, J. Q. (1938). *Astronomy*, vol. 2, pp. 918, 958–9. Ginn: Boston.

Ryle, M. (1950). Radio astronomy. *Rep. Prog. Phys.*, **13**, 184–246.

Ryle, M. (1955). Meeting of Royal Astronomical Society, 13 May 1955. *Observatory*, **75**, 104–8.

Salpeter, E. E. (1964). Accretion of interstellar matter by massive objects. *Astrophys. J.*, **140**, 796–9.

Sanchez, N. and Whiting, B. (1986). Quantum field theory and the antipodal identification of black holes. Preprint.

Schatzman, E. (1958). *White Dwarfs*, pp. 68–73, 166–70. North-Holland: Amsterdam.

Schmidt, M. (1963). 3C273: a star-like object with a large redshift. *Nature*, **197**, 1040.

Schönberg, M. and Chandrasekhar, S. (1942). On the evolution of main-sequence stars. *Astrophys. J.*, **96**, 161–72.

Schwarzschild, K. (1916a). Über das Gravitationsfeld eines Masses nach der Einsteinschen Theorie. *Sitzungsberichte Königlich Preuss. Akad. Wiss., Physik-Math. Kl.*, 189–96. Translation in Detweiler (1982).

Schwarzschild, K. (1916b). Über das Gravitationsfeld einer Kugal aus inkompressibler Flüssigkeit nach der Einsteinschen Theorie. *Sitzungsberichte Königlich Preuss. Akad. Wiss., Physik-Math. Kl.*, 424–34.

Sen, N. R. (1934). On the equilibrium of an incompressible sphere. *Mon. Not. R. Astron. Soc.*, **94**, 550–64.

Shklovsky, J. S. (1953). On the nature of the Crab nebula's optical emission. *Dokl. Akad. Nauk. SSR*, **90**, 983.

Shklovsky, J. S. (1955). On the nature of the emission from the galaxy NGC4486. In *Radio Astronomy*, ed. H. C. van de Hulst, pp. 205–7. IAU Symposium No. 4, Jodrell Bank, August 1955. Cambridge University Press: Cambridge.

Shklovsky, J. S. (1978). *Stars: Their Birth, Life and Death*. Freeman: San Francisco.

Simpson, M. and Penrose, R. (1973). Internal instability in a Reissner–Nordström black hole. *Int. J. Theor. Phys.*, **7**, 183–97.

Sky and Telescope (1961). American astronomers report: First true radio star? *Sky and Telescope*, **21**, 148.

Sky and Telescope (1964). Dallas conference on super radio sources. *Sky and Telescope*, **27**, 80–4.

Smith, F. Graham and Lovell, B. (1983). On the discovery of extragalactic radio sources. *J. Hist. Astron.*, **14**, 155–65.

Souriau, J. M. (1965). Prolongements du champ de Schwarzschild. *Bull. Soc. Math. (France)*, **93**, 193–207.

Sterne, T. E. (1933). The equilibrium theory of the abundance of the elements: a statistical investigation of assemblies in equilibrium in which transmutations occur. *Mon. Not. R. Astron. Soc.*, **93**, 736–67 (see p. 750).

Stoner, E. C. (1929). The limiting density in white dwarf stars. *Phil. Mag.*, **7**, 63–70.

Stoner, E. C. (1930). The equilibrium of dense stars. *Phil. Mag.*, **9**, 944–63.

Strömgren, B. (1937). Die Theorie der Sterninnern und die Entwicklung der Sternen. *Ergebnisse Exakt. Naturwiss.*, **16**, 465–534.

Struve, O. (1960). The problem of Cygnus A. *Sky and Telescope*, **20**, 259–62.

Struve, O. (1962). *The Universe*, chap. 5. MIT Press: Cambridge, Mass.

Sullivan, W. (1979). *Black Holes*. Doubleday: New York.

Sullivan, W. T. III (1984). *The Early Years of Radio Astronomy*. Cambridge University Press: Cambridge.

Synge, J. L. (1949). Gravitational field of a particle. *Nature*, **164**, 148–9.

Synge, J. L. (1950). The gravitational field of a particle. *Proc. R. Irish Acad.*, **A53**, 83–114.

Synge, J. L. (1960). *Relativity: The General Theory*, p. 283. North-Holland: Amsterdam.

Szekeres, G. (1960). On the singularities of a Riemannian manifold. *Publ. Math. Debrecen*, **7**, 285–301.

Thompson, L. A. (1984). High-resolution imaging from Mauna Kea: Cygnus A. *Astrophys. J.*, **279**, L47–9.

Thorne, K. S. (1967). The general relativistic theory of stellar structure and dynamics. In *High-Energy Astrophysics*, vol. III, ed. C. DeWitt, E. Schatzman and P. Veron, p. 414. Gordon and Breach: New York.

Thorne, K. S. (1970). Nonspherical gravitational collapse: does it produce black holes? *Comment. Astrophys. Space Phys.*, **2**, 191–9.

Thorne, K. S. (1972). Nonspherical gravitational collapse: a short review. In *Magic without Magic*, ed. J. R. Klauder, pp. 231–58. W. H. Freeman: San Francisco.

Thorne, K. S., Price, R. H. and MacDonald, D. A. (eds) (1986). *Black Holes: The Membrane Paradigm*, chap. 1. Yale University Press: New Haven.

Tipler, F. J., Clarke, C. J. S. and Ellis, G. F. R. (1980). Singularities and horizons – a review article. In *General Relativity and Gravitation*, ed. A. Held, Vol. 2, pp. 97–206. Plenum Press: New York.

Unruh, W. G. and Wald, R. M. (1982). Acceleration radiation and the generalized second law of thermodynamics. *Phys. Rev.*, **D25**, 942–58.

Wali, K. C. (1982). Chandrasekhar vs. Eddington – an unanticipated confrontation. *Physics Today* (October), pp. 33–40.

Weisskopf, V. F. (1975). Of atoms, mountains and stars: a study in qualitative physics. *Science*, **187**, 605–12.

Weyl, H. (1917). Zur Gravitationstheorie. *Ann. Phys.*, **54**, 117–45.

Wheeler, J. A. (1955). Geons. *Phys. Rev.*, **97**, 511–36.

Wheeler, J. A. (1966). Superdense stars. *Ann. Rev. Astron. Astrophys.*, **4**, 393–432, see p. 418.

Wheeler, J. A. (1968). Our universe: the known and the unknown. *Amer. Sci.*, **56**, 1–20.

Whittaker, E. T. (1949). *From Euclid to Eddington*, p. 124. Cambridge University Press: Cambridge.

Wolfendale, A. W. (1986). Address on presentation of Gold Medal of the R.A.S. to Ya. B. Zel'dovich. *Q. J. R. Astron. Soc.*, **27**, 541–2.

Zel'dovich, Ya. B. (1961). The equation of state at ultrahigh densities. *Sov. Phys. JETP*, **14**, 1143–7 (1962).

Zel'dovich, Ya. B. (1964). The fate of a star and the evolution of gravitational energy upon accretion. *Sov. Phys. Dokl.*, **9**, 195–7.

Zel'dovich, Ya. B. (1965). Survey of modern cosmology. *Adv. in Astron. Astrophys.*, **3**, 241–379, see p. 335.

Zel'dovich, Ya. B. (1970). Generation of waves by a rotating body. *JETP Lett.*, **14**, 180–1.

Zel'dovich, Ya. B. and Novikov, I. D. (1964). Mass of quasi-stellar objects. *Sov. Phys. Dokl.*, **9**, 834–7.

Zel'dovich, Ya. B. and Novikov, I. D. (1965). Relativistic astrophysics. *Sov. Phys. Uspekhi*, **7**, 763–88; **8**, 522–77.

Zwicky, F. (1938). On collapsed neutron stars. *Astrophys. J.*, **88**, 522–5.

Zwicky, F. (1939). On the theory and observation of highly collapsed stars. *Phys. Rev.*, **55**, 726–43.

Zwicky, F. (1941). Neutron stars as gravitational lenses. *Phys. Rev.*, **59**, 221.

Zwicky, F. (1957). *Morphological Astronomy*, pp. 6, 255–6. Springer: Berlin.

Note added in proof

I am indebted to several colleagues, in particular J. Eisenstaedt, S. W. Hawking, W. H. McCrea, W. G. Unruh and J. A. Wheeler, for their constructive critical comments on the manuscript and suggestions, some of which it has been possible to incorporate in the proof stages. A book by Kip Thorne, which I understand is now actively in preparation, may be expected to shape into a definitive history of relativistic astrophysics.

Gordon Baym kindly informs me that Rosenfeld's dating and placing of his conversation on neutron stars with Bohr and Landau (see p. 224) was probably a slip of memory, very understandable after a lapse of forty years. At the time in question (February or March 1932) none of the three protagonists was in Copenhagen (Landau had moved to Kharkov). The conversation may have taken place during a meeting in the Soviet Union two years later.

8

Astrophysical black holes

R. D. BLANDFORD

8.1 Introduction

The idea that stars *might* exist with surface escape velocities in excess of the speed of light is at least two hundred years old (Michell, 1784; Laplace, 1795). However, it was not until the work of Chandrasekhar (1931) that there was a good reason that they *ought* to exist. As is well known, Chandrasekhar showed that white dwarf stars, supported by relativistic electron degeneracy pressure had a maximum mass of 1.4 $M_\odot$ (for a mean molecular weight per electron of 2; c.f. also Landau, 1932). Somewhat later, the discovery of the neutron led to the proposal that neutron stars also exist (Landau, cited in Shapiro and Teukolsky, 1983), and the prescient suggestion by Baade and Zwicky (1934) that 'supernovae represent the transitions from ordinary stars into *neutron stars*'. The maximum mass of a neutron star was calculated (Oppenheimer and Volkoff, 1939) to be 0.75 $M_\odot$. Although this is rather lower than the value calculated using a modern equation of state, it verified the general conclusion that cold stars were limited to a mass equivalent to $\sim (Gm_p^2/\hbar c)^{3/2}$ nucleons.

These discoveries motivated Oppenheimer and Snyder (1939) to study what would happen to aggregations of cold matter with mass in excess of this limit. They understood the kinematic essentials of gravitational collapse by analysing the radial infall of a pressure-free star. They showed that, although the collapse would formally take an infinite time when viewed from large distance, the time measured by an observer comoving with the star would be finite as would also be the time until a distant observer would find the star to be undetectably faint. (Interestingly, Einstein (1939) published a paper in the same year analysing quasi-static clusters of masses moving on circular orbits. He calculated that material particle orbits could not have radii less than one and a half Schwarzschild radii, in Schwarzschild coordinates, and on this basis asserted that 'Schwarzschild singularities' do not exist in physical reality. He believed that this would prove to be a general conclusion.)

After a long interval, these ideas were elaborated by Wheeler and his colleagues in the west (Harrison *et al.*, 1965; Thorne, 1966) and Zel'dovich and his colleagues in the Soviet Union (Zel'dovich and Novikov, 1965, 1966) during the 1960s. Wheeler (1966) in particular emphasised that the very possibility of a physical singularity (as opposed to the coordinate singularity at the Schwarzschild radius) posed a strong challenge to theoretical physics and declared that it was 'difficult to name any situation more enveloped in paradox than the phenomenon of gravitational collapse'. The singularity theorems by Penrose (1965) and Hawking (1965) implied that singularities could not be avoided by minor departures from spherical symmetry. Apparently, none of the 'accidents' desired by Eddington (1935), in his famous attack on the idea of gravitational collapse, would intervene to save a collapsing star. (It is interesting to speculate upon whether Eddington would have retained his prejudices against the formation of gravitational singularities had he understood the principle of 'cosmic censorship' (Penrose, 1969).)

These developments merged with parallel advances in studies of the solutions of the Einstein equation with the publication (Kerr, 1963) of the stationary axisymmetric metric that now bears his name. However, as described in Carter (1979), ten years were to elapse before the work of Israel, Carter, Robinson and Hawking showed that the Kerr metric almost certainly described the *only* uncharged, asymptotically flat vacuum spacetime that would remain after gravitational collapse, thus vindicating Wheeler's famous 'no hair' conjecture. It is perhaps not surprising that in the middle of these discoveries, 'collapsed stars', with their implied strong connection to their stellar parents should become the 'black holes' of Wheeler (1968); a name which embodies not just the absorbing properties of an event horizon, but the independence of the structure from the history of the collapse.

The theoretical framework for carrying out investigations of black hole astrophysics was therefore largely established in time for the discovery of galactic X-ray sources by Giacconi *et al.* (1962) and quasars by Schmidt (1963). However, less well developed were ideas about how the presence of black holes might be inferred by astronomers. The reason is not hard to uncover. In order to address this issue, the theoretical astrophysicist must consider ill-posed problems in radiative transfer, magnetohydrodynamics, plasma physics, stellar dynamics and stellar evolution, not to mention the vagaries of interpreting astronomical phenomenology. This is a qualitatively different type of research from formal investigations of the physical properties of an isolated black hole. In the absence of any strong observational stimulus, such research is inevitably highly uncertain and speculative. Nevertheless, there were some prophets and it is with the development of their ideas leading up to the present observational position on the existence of black holes that we are concerned in this article.

Black holes, by definition, are structures involving strong field general relativity and can only be properly described in these terms. Nevertheless, it is a feature of (and perhaps a sad commentary on) current progress in black hole astrophysics that virtually all the physical processes that are likely to be observable can be described with adequate precision using pre- general relativity (although not entirely Newtonian) models. Indeed in most of the relevant astronomical literature metrics, horizons, etc. are avoided and calculations are performed in a flat space terminated when necessary at the Schwarzschild radius. Much of the discussion that follows will be couched in this language, not as a reactionary attack on general relativity, but instead as a means of emphasising that astronomical observations of black holes are not very promising ways to verify that general relativity is the correct theory of gravitation. Solar system tests and the first binary pulsar PSR 1913 + 16 are far more important in this respect.

Much of the subject matter encompassed by this article has been more completely reviewed elsewhere. In particular, in an earlier volume celebrating the centenary of Einstein's birth, Blandford and Thorne (1979) reviewed this topic as it stood in 1978. In this volume, I will therefore just describe the key observational and theoretical discoveries, in keeping with the historical theme of the present volume and summarise comparatively recent developments.

In the following section, I describe some physical effects believed to be important for astronomically occurring black holes. In Section 8.3, I summarise the observational evidence for the existence of black holes in the two environments where we have good reason to believe that they can be found, i.e. active galactic nuclei and accreting binary X-ray sources, as well as list some of the more speculative possibilities. Prospects for the near future are briefly summarised in Section 8.4.

8.2 Black holes in astrophysics

8.2.1 General remarks

Although some physicists have regarded the quite genuine difficulties that the creation of a singularity poses for physics as somehow providing a teleological argument against the existence of black holes, and rather more observational astronomers have retained a perfectly proper scepticism about the evidence that has been advanced for their presence, most astrophysicists now see black hole formation as a natural and inevitable consequence of stellar and galactic evolution. No longer is it considered obligatory to insert adjectives like bizarre or exotic in popular accounts. The description of the usual foolhardy astronaut, although strange by everyday standards, now seems qualitatively similar to the gedanken experiments of special relativity. Besides, the principal properties of black holes are much easier to communicate than, for example, the behaviour of quarks and gluons. In short, astronomers and physicists have got used to the idea.

However, none of this familiarity removes the obligation to make good our hunches and demonstrate, beyond all reasonable doubt, that black holes really do exist. Here progress has been distressingly slow and no better individual candidate than Cygnus X-1, first discussed in 1972, has been forthcoming and to date only two more good cases can be claimed. This contrasts with the observational position on neutron stars, where, following the discovery of the first radio pulsar by Hewish *et al.* in 1967, we now have nearly 500 examples of neutron stars (Lyne, Manchester and Taylor, 1985). (Although, even here, the possibility that we might be observing quark stars cannot be completely dismissed (Alcock, 1986).)

The incentives for proving that black holes are really present in binary X-ray sources, active galactic nuclei and perhaps elsewhere are twofold. Firstly, measurement of their frequencies, masses and spins would certainly clarify our understanding of the endpoints of stellar evolution, and rather more remotely, the early phases of galactic evolution. Secondly, and of more interest to physicists, they offer the opportunity of observing some of the distinctive features of strong field gravity. Here, we should not be too optimistic. The 'no-hair' theorem condemns us to observing the gas orbiting a black hole and, unlike the case with a magnetised neutron star, it is extremely difficult to imagine making a clean, quantitative measurement of a relativistic effect that is not hopelessly confused by the erratic magnetohydrodynamic behaviour of the gas. Indeed, it is one of the peculiar features of this subject, that we find it fairly easy to include the twentieth century physics of general relativity and quantum mechanics in our calculations; it is the nineteenth century subjects of fluid mechanics and electromagnetic theory that cause the problems. Black holes may provide a useful quantitative test of general relativity. A corollary of this is that most serious modelling can, and is performed, using just 'Newtonian' physics. In the spirit of this volume, we now describe some relevant properties of relativistic black holes in Newtonian language.

8.2.2 *The gravitational field around a black hole*

As is well known, general relativity causes the gravitational force around a black hole to become infinite at a finite radius, the black hole horizon. In the case of a non-rotating hole, this is usually described by the Schwarzschild metric. However, this behaviour can be imitated by using a Newtonian potential $\phi = -m/(r-2m)$, setting $G=c=1$ (e.g. Paczyński and Wiita, 1980). The sum of the gravitational potential and the centrifugal potential define an effective potential for radial motion $\Phi = \phi + l^2/2r^2$, where l is the specific angular momentum. This effective potential has a minimum if $l \geqslant (27/2)^{1/2}m$. The most bound stable circular orbit has an angular momentum $l = (27/2)^{1/2}m$, a radius $r = 6m$ and a binding energy $1/24$, rather similar to the Schwarzschild values. Cold gas in orbit about a black hole will form a thin accretion disk extending down to $6m$, presumably evolving under internal viscous stresses. The thickness of the disk will be given by balancing the vertical gravitational acceleration $g_z = -mz/r(r-2m)^2$ with the internal pressure gradient divided by the gas density. If the gas is isothermal, with sound speed s, the disk will have a Gaussian density profile with half width $s(r-2m)(r/m)^{1/2}$. Again, this expression is similar to its relativistic counterpart (e.g. Novikov and Thorne, 1973).

Now, if the infalling gas is unable to cool fast enough, either by accreting so rapidly that the gas traps the escaping radiation, or so slowly that two-body processes cannot produce photons fast enough, then a thick accretion disk or torus may be produced. In this case, gas moving in the equatorial plane of the torus can orbit stably within $r = 6m$, held in place by a pressure gradient. If, for example, the gas has a constant specific angular momentum l, the torus will steadily inflate until its innermost equatorial orbit coincides with the saddle point in the effective potential where a cusp will be found. If the torus tries to inflate further then the gas will simply pour through the cusp onto the black hole. The gas at the cusp will be bound (i.e. have $\Phi < 0$) as long as $l > 4m$ and the radius of the associated marginally bound circular orbit is $4m$. The surface of marginal binding is given by $\Phi=0$ or the paraboloid

$$z=\frac{(\rho-4m)(\rho+4m)}{8m},\qquad(8.1)$$

where ρ is the cylindrical radius. This surface defines a funnel which may possibly collimate outflowing jets (Lynden-Bell, 1978; Fig. 8.1). If the gas is more tightly bound than this then it will form a torus in orbit about the black hole. Again very similar results are produced using a relativistic analysis (Abramowicz, Jaroszynski and Sikora, 1978).

This simple Newtonian model can only describe a non-rotating black

Fig. 8.1. Contours of constant effective potential Φ around a black hole described using a model Newtonian potential, $\phi=-m/(r-2m)$. (Distances are measured in units of m and only negative contours are shown.) The specific angular momentum is chosen to have a constant value $l=4m$ so that the isobars coincide with the equipotential surfaces and the contour passing through the cusp has zero binding energy. Gas can fill up the region occupied by the contours outside $r=4m$, and may pour through the cusp onto the black hole. The zero energy surface defines a pair of funnels which may be responsible for channelling some of the outflow and beaming some of the radiation in bright quasars and Seyfert galaxies.

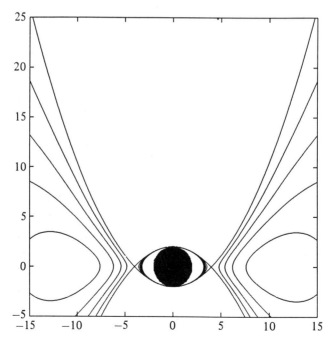

hole. However, we can describe the gravitational far field around a spinning black hole using an electromagnetic analogy and decomposing it into a gravitoelectric field **g** (the usual Newtonian gravity) and a gravitomagnetic contribution **H**. The Einstein equations for **g, H** for material of density μ and velocity $v \ll c$ approximates to a form strikingly similar to Maxwell's equation; the geodesic equation for the motion of a freely falling particle becomes equivalent to the Lorentz force law;

$$\nabla \cdot \mathbf{g} = -4\pi\mu,$$

$$\nabla \cdot \mathbf{H} = 0,$$

$$\nabla \times \mathbf{g} = 0,$$

$$\nabla \times \mathbf{H} = 4\left[-4\pi\mu v + \frac{\partial \mathbf{g}}{\partial t}\right],$$

$$\frac{d\mathbf{v}}{dt} = \mathbf{g} + \mathbf{v} \times \mathbf{H}, \tag{8.2}$$

(e.g. Braginsky *et al.*, 1977, where the equations are actually given to second order in v). **g** and **H** can be derived from scalar and vector potentials, respectively; $\mathbf{g} = -\nabla\phi, \mathbf{H} = \nabla \times \gamma$ where ϕ, γ are related to the metric tensor, $g_{\alpha\beta}$, by $\phi = (1 + g_{00})/2$, $\gamma_j = g_{0j}$. Note the minus sign in the 'Poisson' and 'Ampere' equations expressing the attractive character of gravity and the extra factor of 4, which relates to the spin-2 nature of the gravitational field.

Now, from our electromagnetic experience, we can infer immediately that at a distance $r \gg m$, where the space is approximately flat, a hole with spin angular momentum **S** will be surrounded by a gravitoelectric field $\mathbf{g} = -m\hat{\mathbf{r}}/r^2$ where the gravitoelectric charge is the hole's mass m. If we use the classical gyromagnetic ratio, we identify $S/2$ as the gravitomagnetic dipole moment. Now if we use the fact that the gravitomagnetic field is -4 times the usual dipolar field, we obtain an expression for the dipolar gravitomagnetic field surrounding a spinning black hole,

$$\mathbf{H} = 2[\mathbf{S} - 3(\mathbf{S} \cdot \hat{\mathbf{r}})\hat{\mathbf{r}}]/r^3, \tag{8.3}$$

g and **H** communicate information to a distant observer about the two parameters m, **S** which characterise an uncharged black hole.

We can now use these fields to calculate the rate of precession of a ring of gas in inclined orbit about a spinning black hole with specific angular momentum **l**. Pursuing our electromagnetic analogy, a torque per unit mass of $\mathbf{r} \times (\mathbf{v} \times \mathbf{H})$ will act upon elements of the ring. Using the formula for dipolar gravitomagnetic field and averaging over azimuth for a small angle of inclination gives a mean torque per unit mass of $2(\mathbf{S} \times \mathbf{l})/r^3$. The gravitomagnetic or Lense–Thirring (1918) precession frequency is therefore given by $2\,S/r^3$.

The importance of this precession for astrophysical black holes is that it may provide a mechanism for driving an orbital accretion disk into the equatorial plane of a spinning black hole and so communicating the gyroscopically stable spin direction of the hole to the direction of the outflowing jets (Bardeen and Petterson, 1975; Rees, 1978). Adjacent rings of a gas within an inclined accretion disk should precess at different rates and friction between them should drive the innermost rings into the equatorial plane. If we adopt the criterion that this mechanism is efficient when the precession period is shorter than the inflow time (but see Kumar and Pringle, 1985), we conclude that the disk should lie in the equatorial plane out to a radius $r_{BP} \sim (S/S_{max})^{2/3}(v_r/v)^{-2/3}m$, where $S_{max} = m^2$ is the maximum allowed spin angular momentum of the black hole and v_r is the inflow speed. (We discuss this further below.) A corollary is that the hole may change its spin substantially after it has increased its mass by a fractional amount $(S/S_{max})(m/r_{BP})^{1/2}$. Some extragalactic radio sources show jets that bend on both sides of the nucleus with a pronounced 'inversion' or 'S' symmetry (see Fig. 8.2). If jets are launched along the spin axis of the hole, then this type of distortion may be related to changes in the spin direction of the hole.

Fig. 8.2. 'Inversion symmetric' extragalactic radio source associated with the galaxy NGC326 (marked with a +). The apparent S-type symmetry of the radio contours might be caused by a systematic change in the orientation of the spin axis of a central black hole. (R. Ekers, personal communication.)

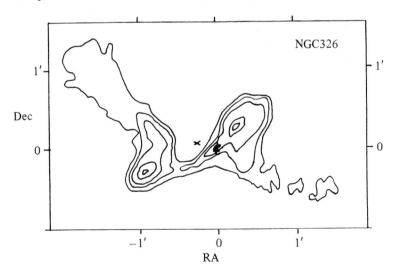

A related precessional effect is called geodetic precession. Consider a gyroscope, e.g. a smaller spinning black hole in orbit about a massive body (e.g. a large black hole). The motion of the gyroscope through the gravitoelectric field induces a gravitomagnetic field $\mathbf{H} = -\mathbf{v} \times \mathbf{g}$ in its frame. Again using our electromagnetic knowledge, we find that the gyroscope will precess with an angular frequency $\Omega_{SO} = -\mathbf{H}/2 = ml/2r^3$. This precession is the analog of spin–orbit coupling in a hydrogen atom. There is no Thomas term to be included because the gyroscope is freely falling rather than being accelerated relative to a local inertial frame by a non-gravitational force. (When the massive body is also spinning, there will be an additional contribution to the precession rate caused by its gravitomagnetic field

$$\Omega_{GM} = -\mathbf{H}/2 = [3(\mathbf{S} \cdot \hat{\mathbf{r}})\hat{\mathbf{r}} - \mathbf{S}/r^3]. \tag{8.4}$$

Usually, this will be small compared with the geodetic precession.)

However, there is a contribution to the precession rate attributable to the space curvature. Using an embedding diagram for the Schwarzschild metric (e.g. Misner, Thorne and Wheeler, 1973) it can be seen that the gyroscope will precess through an additional angle $\sim 2\pi m/r$ per orbital period. Adding this to the spin orbit precession gives a total geodetic precessional frequency of

$$\Omega_{geo} = 3\Omega_{SO} = 3ml/2r^3. \tag{8.5}$$

This formula has been derived assuming that the mass ratio is small. When the reduced mass of the two body system, μ is comparable with m then the total precession rate becomes

$$\Omega_{geo} = (3m + \mu)l/2r^3, \tag{8.6}$$

where $\mathbf{l}$ and $\mathbf{r}$ now refer to the relative orbit. (Barker and O'Connell, 1975). In the limit that the gyroscope becomes much heavier than the first body of mass m, the mean torque acting on it becomes equal to $2(\mathbf{l} \times \mathbf{S})/r^3$, equal and opposite to the Lens–Thirring torque computed above, as it should.

These arguments and the relationship to the proper relativistic calculations are discussed at length in Thorne, Price and MacDonald (1976).

Again there may be observational consequences of geodetic precession if, as proposed by Begelman, Blandford and Rees (1980), black hole binaries can form in the nuclei of galaxies. Typical precessional periods $\sim 10^4 (a/0.01 \text{ pc})^{5/2} (m/10^8 \text{ M}_\odot)^{-3/2}$ yr for typical orbital radii $a \sim 0.01$ pc and masses $m \sim 10^8 \text{ M}_\odot$ are deduced. Geodetic precession may also be responsible for those radio jets that exhibit pronounced inversion symmetry.

8.2.3 Electromagnetic effects

Another aspect of black hole astrophysics that is amenable to a classical or more specifically 'Maxwellian' interpretation is the interaction of a black hole with a magnetic field. Although charged, Kerr–Newman black holes are irrelevant to astronomy, uncharged Kerr holes interacting with magnetic fields supported by currents external to the horizon may be highly relevant. Some intuition about this interaction can be obtained using a sequence of four thought experiments.

Firstly, if a Schwarzschild black hole is placed in a uniform electric field, then, after the transients have decayed, a modified static field will remain in which the physical components of the electric field (as seen by an observer hovering over the horizon) becomes radial at the event horizon (e.g. Press, 1972). The observer might therefore think that the black hole is behaving like an electrical conductor and he can introduce a fictitious surface charge density on the horizon consistent with Gauss' theorem. This representation of an event horizon as a 'quasi-Newtonian' surface embodied with conductivity, as well as viscosity, surface pressure, momentum etc. (Damour, 1978; Znajek, 1978) is termed the membrane paradigm and has been extensively explored by Thorne and colleagues (Thorne, Price and MacDonald, 1976).

We ought to clarify what is meant by electric and magnetic fields at this stage. One way to do this is to define a set of fiducial observers. These observers will be at rest in Schwarzschild coordinates if the hole is non-rotating and coincide with the 'zero angular momentum' observers in the spacetime of a rotating hole (Bardeen, 1973). The electric and magnetic fields that we are talking about are the ones that they would measure. In the absence of free charges, these fields are solenoidal on the 3D hypersurfaces of constant Schwarzschild or Boyer–Lindquist time. We can therefore define electric and magnetic field lines (see Fig. 8.3).

In a second thought experiment, let a cloud of magnetised plasma fall into a black hole. As it approaches the horizon, the cloud will be tidally distorted and the electromagnetic field will vary rapidly. After the plasma has crossed the horizon, electromagnetic fields will linger in the neighbourhood of the horizon for roughly a light crossing time before decaying to leave behind a 'bald' black hole (consistent with the no-hair theorems). This tells us that if we regard the event horizon as possessing a conductivity, its value cannot be infinite. In flat space, a sphere of radius A, and electrical resistance R, will lose magnetic field in a time $\sim A/Rc^2$. Equating this to the light-crossing time $\sim A/c$, tells us that the resistance of the hole is $R_H \sim 1/c \equiv 30\ \Omega$.

More precisely, if we try to solve Maxwell's equations in the curved spacetime surrounding a Schwarzschild black hole, we find that the boundary conditions at the horizon dictate that the horizon must *appear* to have a surface resistivity of 377 Ω. Consider an observer freely falling from infinity with speed $v = (2m/r)^{1/2}$ relative to a stationary observer just hovering above the horizon. Let the Lorentz factor be $\gamma = (1 - 2m/r)^{-1/2}$ as usual. The electric and magnetic fields measured by the infalling observer, E', B', must remain finite near the horizon. However the stationary observer will see fields that look just like an electromagnetic wave propagating radially inwards. That is to say both the electric and the magnetic field will be equal, mutually perpendicular, and parallel to the horizon to a fractional accuracy $\sim \gamma^{-1}$. This is akin to the Weizsäcker–Williams method of virtual quanta. Let the components of electromagnetic field measured by a fiducial observer hovering just above the horizon be $E_\parallel, B_\parallel$. The magnitude of these field components, $(E_\parallel, B_\parallel) \sim \gamma \max(E', B')$ will diverge as the hovering observer approaches the horizon. The Poynting energy flux measured by this observer will diverge $\propto \gamma^2$ (one factor of γ for the blue shift of the photons, another to accommodate the change in photon rate). However, we can define renormalised horizon fields

$$\mathbf{E}_H, \mathbf{B}_H = \lim_{r \to 2m} \mathbf{E}_\parallel / \gamma, \, \mathbf{B}_\parallel / \gamma, \qquad (8.7)$$

Fig. 8.3. An isolated Schwarzschild black hole is immersed in a uniform electric field. (*a*) If we plot the electric field lines in Schwarzschild coordinates (in which a finite interval of proper radius close to the horizon corresponds to a very small interval of coordinate radius), the field lines appear to remain straight. (*b*) If we transform the radial coordinate to proper radius, then the field lines curve and cross the horizon normally. Observers hovering just above the horizon and falling through it radially will both measure a field perpendicular to the horizon. (Adapted from Thorne *et al.*, 1976.)

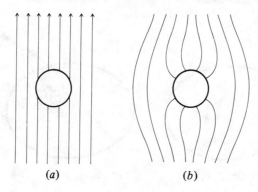

(*a*)   (*b*)

lying in the horizon, which remain finite. (The radial field components remain finite without any renormalisation.) Note that

$$\mathbf{E}_H = \mathbf{E}' + \hat{\mathbf{r}} \times \mathbf{B}',$$
$$\mathbf{B}_H = \mathbf{B}' - \hat{\mathbf{r}} \times \mathbf{E}', \tag{8.8}$$

verifying that $\mathbf{E}_H$, $\mathbf{B}_H$ are equal in magnitude and perpendicular (see Fig. 8.4).

Furthermore, the Poynting energy flux crossing the horizon, as measured by an observer at infinity, will be smaller than that measured by the

Fig. 8.4. (a) When an isolated black hole is penetrated by electric field lines, we can imagine that these terminate on surface charges located just above the horizon (i.e. on an imaginary surface called the stretched horizon). The associated surface charge density is given by $E_\perp/4\pi$. (b) A renormalised component of tangential magnetic field $\mathbf{B}_H$ just outside the horizon can likewise be thought of as terminating at a surface current density $\mathbf{J}_H = \hat{\mathbf{r}} \times \mathbf{B}_H/4\pi$. (c) The surface boundary condition discussed in the text implies that there is a renormalised electric field parallel to the horizon $\mathbf{E}_H = 4\pi\mathbf{J}_H$, equivalent to saying that the surface resistivity is 377 Ω. (d) When currents flow across the horizon, charge conservation can be thought of as being satisfied by surface current flowing in the horizon and surface charge density accumulating there. (Adapted from Thorne et al., 1976.)

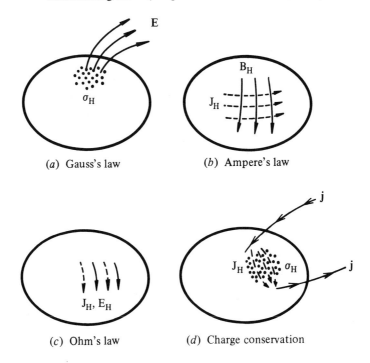

(a) Gauss's law   (b) Ampere's law

(c) Ohm's law   (d) Charge conservation

stationary observer by two redshift factors and given by $E_H B_H/4\pi = E_H^2/4\pi$ which will also be finite. Its integral over the area of the horizon will give the rate of increase of hole mass. Now if we imagine that there is no tangential magnetic field inside the horizon, we must terminate an exterior tangential magnetic field with a surface electric current in the horizon. $J_H = B_H/4\pi$ perpendicular to the tangential magnetic field just outside the horizon. Using our boundary condition $E_H = B_H$, we derive the constitutive relation $J_H = S_H E_H$ where S_H, the surface resistivity, equals $4\pi \equiv 377\,\Omega$. Not surprisingly, the resistance of the horizon approximately equals the impedance of free space. These results carry over to the Kerr metric. For more details of those arguments, see Carter (1979) and Thorne, Price and MacDonald (1976) and references therein.

In a third experiment (Damour, 1978), a potential difference V is applied between the equatorial and polar regions of a Schwarzschild black hole using a battery of internal resistance much smaller than R_H. Provided that both signs of charge are available to cross the horizon, current I will flow around the circuit and through the hole. Endowing the hole with resistance R_H implies that $I \sim V/R_H$ and that the hole will grow at a rate $\dot{m} \sim V^2/R_H$.

Now immerse the hole and circuit in a uniform field B_o parallel to the polar direction. There will be a Lorentz force acting on the hole and this will produce a torque $\sim m J_H B_o \sim V B_o/R_H$ which will start the hole spinning just like a simple electric motor. In this case there is a flux of electromagnetic angular momentum as well as energy crossing the horizon (see Fig. 8.5).

The fourth thought experiment is the one that concerns us most directly and is the converse of the last one (Blandford and Znajek, 1977). Suppose we have a spinning black hole immersed in a uniform field B_o, parallel to the spin axis. If the hole is surrounded by vacuum, then because the hole is a conductor, it will act as a unipolar inductor and a potential difference $V \sim \Omega_H m^2 B_o \sim \Omega_H \Phi$, where Φ is the flux threading the hole, will be induced between the poles and the equator and Ω_H is the angular velocity of the hole. (Ω_H must not exceed $\Omega_{max} = 1/2m$, the limiting angular velocity for a hole of total mass m.) As the hole is a conductor, the electric field measured by fiducial observers hovering just above the horizon and orbiting with an angular velocity very close of Ω_H will be radial and $E_H = 0$. Our boundary condition then implies that $B_H = 0$ as the full electromagnetic solution confirms (Wald, 1974). Just as we can introduce a fictitious surface current density to terminate a tangential magnetic field, so, in this example, we can introduce a quadrupolar surface charge density to terminate the radial electric field. It is interesting that the EMF can be thought of as arising from an interaction between the hole's gravitomagnetic potential and the external magnetic field (e.g. Thorne, Price and MacDonald, 1976; Fig. 8.6).

There will be no change in the mass or the angular momentum of the hole as long as it remains an open circuit and no current flows. However, if we now connect the pole to the equator via a load with large resistance $R_L \gg R_H$, then a small current will flow of magnitude $I \sim V/R_L$. The power that is dissipated in the resistive load, $V^2/R_L \sim \Omega_H^2/R_L$ is drawn from the spin of the hole. Now, as originally shown by Christodoulou (1970), the mass m of a spinning hole may be decomposed into an irreducible part, m_i, and a spin energy. This spin energy can, in principle, be extracted by Penrose-type processes such as the one we are analysing here. m is related to m_i by

$$m = m_i(1 - 4m_i^2\Omega_H^2)^{-1/2}, \tag{8.9}$$

a formula similar to that appropriate for rectilinear motion with the

Fig. 8.5. A potential difference V is applied across a Schwarzschild black hole so that a current $I \sim V/R_H$ will flow, inward at the pole and outward at the equator. The circuit can be regarded as being completed by a surface current $J_H \sim I/2\pi m$. Associated with this surface current is a horizon electric field $E_H \sim IR_H/2\pi m$ and a perpendicular magnetic field $B_H = E_H$. There is an inwardly directed Poynting flux $N \sim I^2 R_H/2\pi m^2$ and the hole gains mass at a rate $\dot{m} \sim 2\pi m^2 N \sim I^2 R_H$. Now if an external magnetic field B_0 is applied as shown, there will be a torque per unit area $\sim mJ_H B_0$ which will start the hole spinning. (Adapted from Thorne et al., 1976.)

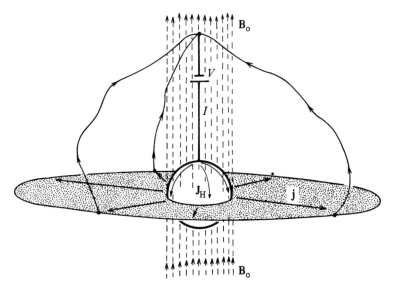

important difference that the 'velocity' $2m_i\Omega_H$ is limited to $2^{-1/2}$ rather than unity.

The way in which power is extracted is quite subtle. The boundary condition dictates that a fiducial observer just above the horizon will detect

Fig. 8.6. (a) When an isolated hole spins (with angular velocity Ω_H), parallel to an external magnetic field B_o, an electric field and an apparent quadrupolar surface charge density will be induced. The potential difference between the pole and the equator is $V \sim \Omega_H B_o m$. (b) When the pole and the equator are connected via a load of resistance R_L, a current will flow and power will be dissipated in the load. This power derives from the spin of the hole. (Adapted from Thorne *et al.*, 1976.)

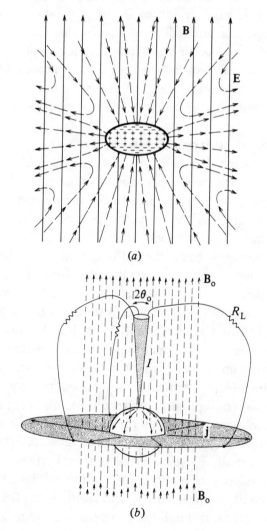

(a)

(b)

a small, inwardly-directed Poynting flux, diverging as he approaches the horizon as our boundary condition requires. However, he will also measure an outwardly directed electromagnetic angular momentum flux. Now, if we look at this from the point of view of a distant, non-rotating, observer (inserting the appropriate redshift factors), the energy and angular momentum fluxes combine so that the conserved energy flux at infinity will be outward. All but a fraction $\sim R_H/R_L$ of the reducible mass can be extracted in this manner.

As the value of the resistive load is reduced, the rate of power extraction will increase $\propto R_L^{-1}$. However, when $R_L \lesssim R_H$, we must take into account the internal resistance of the battery, and the power extracted will be $\sim V^2 R_L (R_L + R_H)^{-2}$. For a given emf, this is maximised when the load resistance matches the internal resistance of the black hole. (This is a circumstance which may occur in practice if the load is actually a relativistic electromagnetic wind in which the electric and magnetic fields are nearly equal.) In the limit when the hole is effectively short-circuited, the current will be $I \sim V/R_H$ and the hole will slow down at more or less constant total mass.

A black hole surrounded by a magnetised accretion disk may therefore act as a unipolar inductor. Up to 29 per cent (for $\Omega_H = \Omega_{max}$) of its mass is reducible and this may be dissipated in a sufficiently resistive load. The magnetic field, supported by currents in the surrounding disk, should be reasonably stationary and axisymmetric as the Bardeen–Peterson effect drives the disk into the equatorial plane of the hole. Instead of the wires invoked above, currents should flow into and out of the hole through the magnetosphere and connect at large distance from the hole where they presumably accelerate charged particles (the load). As particles have to flow in at the horizon and outward at large distance from the hole, they must have a source in the magnetosphere. If cross field diffusion is too slow to accomplish this, then, under conditions envisaged within galactic nuclei, electrons and positrons can be copiously created by γ-ray collisions.

The electromagnetic structure of a stationary, axisymmetric magnetosphere is conceptually quite simple. If the particles in the magnetosphere have negligible inertia, then the electromagnetic fields must be force-free, i.e. $\rho \mathbf{E} + \mathbf{j} \times \mathbf{B} = 0$. This implies that the axisymmetric magnetic surfaces must be equipotential and, by Ferraro's law have a well defined angular velocity ω specifying the frame in which the electric field vanishes. Stationarity of the magnetic field requires that the electric field be poloidal which in turn, through the force-free condition, implies that the current flows along the magnetic surfaces. These surfaces therefore replace the wires used in the circuit analysis.

Now decompose the magnetosphere into many nested elementary circuits in which the current flows along adjacent magnetic surfaces and connects across an annular ring of the horizon (resistance ΔR_H) and a part of the load (resistance ΔR_L). When the load supplies most of the resistance in the elementary circuit ($\Delta R_H \ll \Delta R_L$), the field lines will move with the hole ($\omega \sim \Omega_H$). When the load resistance is negligible (e.g. if it is a highly conducting accretion disk), the magnetic field lines will be frozen into it and will rotate with its angular velocity Ω_L (e.g. MacDonald and Thorne, 1982).

If we transform into a frame corotating with the magnetic field lying between two magnetic surfaces, with angular velocity ω, the potential difference across the annular ring of horizon is $\Delta V_H = (\Omega_H - \omega)\Delta\Phi/2\pi$, where $\Delta\Phi$ is the magnetic flux between the two surfaces. Similarly, the potential difference across the load is $\Delta V_L = (\omega - \Omega_L)\Delta\Phi/2\pi$. Now current conservation requires that $I = \Delta V_H/\Delta R_H = \Delta V_L/\Delta R_L$ or

$$\frac{\omega - \Omega_L}{\Omega_H - \omega} = \frac{\Delta R_L}{\Delta R_H}. \tag{8.10}$$

The magnetosphere acts as a sort of clutch that couples the hole to the load and rotates with a compromise angular velocity $\omega = (\Omega_H \Delta R_L + \Omega_L \Delta R_H)/(\Delta R_L + \Delta R_H)$. As it has no inertia itself, it transmits a torque $\Delta G = I\Delta\Phi/2\pi = I^2(\Delta R_H + \Delta R_L)/(\Omega_H - \Omega_L)$. This torque does work at a rate $-\Delta G\Omega_H$ on the hole, increasing its spin energy at this rate. It also does work at a rate $\Delta G\Omega_L$ on the load, increasing its spin energy. The sum of these two powers, which of course must be negative, represents dissipation at a rate $I^2\Delta R_H$ in the hole plus $I^2\Delta R_L$ in the load.

Next, let us idealise the problem further by imagining that the hole is rotating with a moderate angular velocity ($\Omega_H \sim 1/8m$) and that low latitude magnetic field lines connect to the inner parts of the disk which is an excellent conductor and rotates more rapidly than the hole with angular velocity $\Omega_D > \Omega_H$. Intermediate latitude field lines connect to the disk at large radii where the disk is rotating less rapidly than the hole. Finally, high latitude field lines are open and accelerate a relativistic wind.

Under these conditions, the black hole may evolve in several different ways. The disk-connected flux will exert a torque on the hole given by an integral over the horizon

$$G_D = \int \frac{(\Omega_D - \Omega_H)r_H^2 B_n}{S_H} \, d\Phi, \tag{8.11}$$

where $2\pi r_\mathrm{H}$ is the circumference of the horizon at that latitude and B_n is the normal component of magnetic field at the horizon. Clearly G_D can have either sign and in particular, it is possible for the disk to spin up the hole in an analogous manner to the spin up in pulsating X-ray binaries. The mass of the hole will increase at a rate that is the sum of the ohmic dissipation and the net work done on it by the disk torque. Conversely, a rapidly spinning hole may actually inhibit disk accretion if it is magnetically coupled to the inner edge of the accretion disk.

The open field lines can also change the mass and the spin of the hole. However, the boundary conditions at large distance are problematical. If we try to match the outflow onto a wind with slow speed $\sim v$, then the effective impedance will be $R_\mathrm{L} \sim v \ll R_\mathrm{H}$. Energy will only be extracted from the hole with low efficiency in this case. However, if the outflow is relativistic then we can have a good impedance matching and roughly half the spin energy can be liberated. Unfortunately, it is hard to be more quantitative than this, as we have to match the magnetospheric solutions onto an outflowing solution and thus determine the value of ω on the field lines.

In an important extension of these ideas, Phinney (1983) (cf. also Damour, 1975; Carter, 1979) has replaced the force-free equations with relativistic MHD theory. This allows particles, as well as electromagnetic fields to transport energy and angular momentum. In the limit that the fluid motion dominates the magnetic stresses, the Penrose process is recovered. Phinney argues that, although the particle stresses may be small, the requirement that the outflowing plasma pass through the necessary Alfvénic and magnetosonic critical points to give a supersonic wind may actually determine ω and therefore the efficiency of energy extraction, independent of the details of the dissipation in the load. Another aspect of the theory which must be treated self-consistently is the shape of the magnetic surfaces. These will be determined by imposing poloidal force balance and, if we are given a prescription for fixing ω, by the distribution of toroidal current in the accretion disk.

Electromagnetic extraction of energy from a spinning hole is a strong candidate for powering the jets of relativistic plasma that are observed in extragalactic radio sources. It may also be relevant to stellar mass black holes in mass transfer binary systems.

8.2.4 Accretion onto black holes

As this topic has been extensively reviewed elsewhere, I shall just consider a few recent developments.

Perhaps, the simplest type of accretion to treat is spherically symmetric accretion onto a Schwarzschild black hole, and not surprisingly, this still attracts the most theoretical interest. Like, the problem of the axisymmetric radio pulsar, it is deceptively simply posed, yet full of subtleties and ultimately of little relevance to interpreting the observations. Nevertheless, it has proved to be an important testbed for theoretical ideas with the bonus that numerical computations are relatively easy to carry out. Recent discussions include Krolik and London (1983), Schultz and Price (1985) and Colpi, Maraschi and Treves (1984).

Gas accreting onto a black hole in a binary system or an AGN will almost certainly have enough angular momentum to encircle the hole and presumably to form some sort of disk or torus around it. Close to a spinning hole, we expect the Lense–Thirring precession to drag the gas into the equatorial plane. Unfortunately, almost everything we want to know to interpret the observations depends upon the effective viscosity and we are still no nearer to having a good physical understanding of this (e.g. Pringle, 1981). Several different modes of disk accretion have been considered. When the rate of mass accretion is very low compared with the Eddington rate, the accreting gas may be unable to cool on the inflow timescale. The ions will therefore acquire a temperature comparable with the virial temperature and form a thick ring or torus. This can only happen if the electron temperature is much lower than the ion temperature. These conditions are probably optimal for the production of γ-rays and electromagnetic extraction of energy from the hole. In the case of AGN, the radio galaxies (which are able to produce more power in the form of relativistic plasma than apparently in the form of optical and UV photons) have been associated with this case. If this is correct then it might be interesting to search for some galactic counterparts.

At moderate rates of accretion, up to the Eddington rate, a fairly thin disk is expected to form. The structure of the outer disk at least is probably fairly independent of the nature of the central object and we can probably study them best in cataclysmic variable stars comprising a low mass 'red' star and a white dwarf. Indications are that the viscosity can be fairly large, compatible with the empirical α approximation in which the shear stress is proportional to the pressure with a constant of proportionality ~ 0.01–1 (Shakura and Sunyaev, 1973). As the inner parts of accretion disks can be radiation-pressure dominated, an important issue is whether to use the gas pressure or the radiation pressure in applying this prescription (Sakimoto and Coroniti, 1981). This can have a profound effect on the properties of disks around massive black holes in AGN. Limit cycle behaviour can be observed in accretion disks and may be explained by a variety of different mechanisms leading to episodic accretion (e.g. Lin *et al.*, 1985; Shields *et al.*, 1986). There is no guarantee that the accretion will be steady. In addition, as the disk is differentially rotating, it may be able to generate magnetic fields by some sort of dynamo process (e.g. Pudritz, 1983) as well as convect them inwards. These fields can energise a corona where non-thermal radiation mechanisms will predominate. This is particularly important in the context of quasars which seem to produce almost as much power through radio, infra-red, and X-ray 'non-thermal' channels as through 'thermal' emission in the UV with an effective temperature appropriate to an emitting area roughly 10 Schwarzschild radii across. Recent studies (e.g. Svennson, 1986) have explored the possibility that an electron positron plasma may be created by γ-ray collisions, and that this might be responsible for reprocessing the outflowing 'thermal' emission as well as producing the 'non-thermal' emission.

An important difference between accretion disks in binary X-ray sources and AGN is that in the former case the angular momentum of the inflowing gas which is generally assumed to be transported outwards through the disk by viscous stresses can be extracted by gravitational torques associated with the orbital motion. In the case of AGN there is no natural repository of angular momentum. One resolution of this difficulty is that the angular momentum be removed by hydromagnetic torques that operate in a similar manner to those just analysed in the case of the hole itself (e.g. Coroniti, 1985). This may be responsible for the collimation of double radio sources.

When the accretion rate is large, it has been proposed that an optically thick radiation-supported torus is produced in orbit around the black hole. Equilibrium models of these tori have been described and used to model quasars. In many respects they are similar to the original superstars of Hoyle and Fowler. In particular they appear to be prone to dynamical instability (Papaloizou and Pringle, 1984, 1985; Goldreich, Goodman and Narayan, 1986). The most unstable modes are non-axisymmetric and global in character and can be interpreted as edge waves coupled near to their co-rotation radii. Numerical simulations are essential to understand the non-linear development of these modes and in particular to determine if they can form a basis for a model of the viscosity (Hawley, 1986).

Accretion of stars can also be important. A spherical star cluster around a massive black hole will probably not be able to supply enough gas, via stellar collisions and tidal stripping to fuel a powerful quasar, although lower levels of activity may be supported in this way (e.g. Duncan and Shapiro, 1983; David, Durisen and Cohn, 1986). The rate of supply of gas may however be enhanced if the potential is triaxial (e.g. Norman and Silk, 1983) or the black hole is in a binary (e.g. Roos, 1981). An interesting mechanism proposed by Carter and Luminet (1982) is that the tidal shock a star receives on passage by a black hole may detonate a nuclear explosion and lead to a sudden release of energy. However, more detailed calculations have concluded that this does not occur in practice (Bicknell and Gingold, 1986). Massive black holes may actually be built up out of stars in a dense cluster. In an impressive series of relativistic stellar dynamical computations Shapiro and Teukolsky (1985) have shown how a collapse may begin. However, it remains to be seen if the collapse will continue and whether or not a large black hole can be made in this manner.

Extensive reviews of accretion onto black holes are to be found in the books by Shapiro and Teukolsky (1983) and Frank, King and Raine (1985) as well as the articles by Pringle (1981), Begelman, Blandford and Rees (1984), Begelman (1985) and especially Rees (1984).

8.3 Observational evidence for black holes

8.3.1 Active galactic nuclei

8.3.1.1 Early history

Although the discovery of quasars in 1963 is widely cited as initiating the modern era of astronomy, three parallel lines of investigation were all leading up to this discovery and already posed the problems with which astrophysicists were soon to be confronted. Together, and with the considerable benefit of hindsight, these observations also provided strong clues as to the solutions of these problems.

Firstly, Fath (1908) observed the spiral galaxy NGC 1068 (of course before it was known to be an external galaxy), and discovered unusual high excitation lines. Subsequent observations by Slipher (1917) demonstrated that it had a redshift of 1100 km s^{-1} and that the lines were broader than could be accounted for by rotation of the galaxy. Other galaxies like NGC 1068 showing bright condensed nuclei and broad, high excitation emission lines were studied by Seyfert (1943) and belong to a class which now bears his name. Interest in these galaxies waned somewhat until the late 1950s when Woltjer (1959) realised that Seyfert galaxies constituted a few per cent of all spirals and that consequently they must live for a few per cent of the age of the universe, at least 10^8 yr – much longer than the time it would take the emitting gas to escape from the nucleus. As the powers of nuclei of the brighter Seyfert galaxies were known to approach those of typical galaxies ($\sim 4 \times 10^{43}$ erg s^{-1}), it was already apparent that nuclear activity required considerable energy ($\geqslant 10^{59}$ erg per galaxy). Furthermore, the existence of smaller scale activity in our Galaxy and the Andromeda galaxy strongly hinted that most bright galaxies had active nuclei. Woltjer also deduced that a blue photo-ionising spectrum would be necessary to excite the emitting gas and suggested that it was produced by a cluster of hot stars, an idea that was contradicted by the spectroscopic measurements of NGC 1068 by Burbidge, Burbidge and Prendergast (1959), which required a smaller mass-to-ultra-violet light ratio than a *normal* distribution of stars.

Secondly, many radio 'stars', were identified with external galaxies following the demonstration by Baade in 1951 (using an accurate radio position determined by Smith) that the second most powerful radio source in the sky, Cygnus A (Reber, 1944) was identified with an unusual galaxy at redshift $z=0.05$ (Baade and Minkowski, 1954). The radio power was estimated to be 10^{44} erg s^{-1}, larger than the optical luminosity of a bright galaxy. What was equally remarkable was the discovery of high excitation emission lines in the galaxy spectrum. These were interpreted as evidence for a galaxy collision, an idea which had extraordinary longevity in view of the large power involved. In fact the energetics were strained even further when Shklovsky and Ginzburg proposed that the radio emission be produced by the synchrotron process and Burbidge (1959) showed that a total energy of 3×10^{59} erg was required to account for the intensity of the radio emission. Furthermore, if the ratio of relativistic protons to electrons in the source were to be similar to that found in galactic cosmic rays the minimum energy would have to increase by a factor ten to $\sim 3 \times 10^{60}$ erg, the rest mass equivalent of 2×10^6 M$_\odot$, enough to power the radio source for a billion years.

Strong evidence against the colliding galaxy model also came from the interferometric results of Jennison and Das Gupta (1953) who showed that Cygnus A in fact comprised two separate regions of emission that straddled the optical galaxy instead of being concentrated on it. This pattern was to be repeated in many more radio galaxies (Matthews, Maltby and Moffet, 1962). Powerful radio galaxies gave the first indications of the enormous energies involved in the most extreme examples of AGN.

The third line of enquiry, although limited to a single example, was to provide the link between the first two. In 1917, Curtis discovered that the elliptical galaxy M87 was unusual in that it had a narrow linear feature now called a jet protruding from its nucleus. In 1956, Hiltner discovered linear polarisation in the jet which enabled Burbidge (1956) to infer that the optical emission was synchrotron radiation by relativistic electrons and that at least 10^{56} erg of energy was required.

So, by the time of the announcement of the discovery of the quasar 3C273 (Schmidt, 1963; Hazard, MacKey and Shimmins, 1963), there was good evidence for ongoing nuclear activity, the copious production of relativistic electrons with aggregate energy far in excess of what might be expected from normal stars. This was even more strongly true of 3C273 which had an optical power of $\sim 10^{46}$ erg s^{-1} and even a longer jet than in M87 seen at both radio and optical wavelengths. The association of quasars with these milder phenomena was clearly understood (e.g. Burbidge, Burbidge and Sandage, 1964).

8.3.1.2 Early models of quasars

In addition, to these observational precursors of quasars, the theoretical ideas that were to prove useful were also being developed. In fact the first Texas conference on relativistic astrophysics (Robinson, Schild and Schucking, 1964) was convened prior to the discovery of the quasars that were to dominate its discussions, stimulated in part by the theoretical challenges posed by the double radio sources (Maran and Cameron, 1964). Hoyle and Fowler had previously proposed a model of AGN that involved a 'superstar' forming out of $\sim 10^6 \, M_\odot$ of radiation-dominated gas. This would radiate away its gravitational energy, at the Eddington limit, contracting in the process and was thought to be capable of powering the powerful radio galaxies.

Almost as soon as quasars were discovered, it was realised that unless the redshifts had a non-cosmological origin, their large powers required that matter be converted into energy with very large efficiency; probably larger than the maximum value ~ 0.007 associated with nuclear reactions. (In fact it is only comparatively recently that it has been possible to put this argument on a rigorous basis.) The release of gravitational energy was therefore implicated and several authors suggested that superstars might be responsible. If so, they would have to be very compact with surface escape speeds close to the speed of light. Unfortunately, as Feynman (unpublished work), Iben (1964) and others quickly showed, such a star would be driven unstable by post-Newtonian corrections to the equation of hydrostatic equilibrium unless the gravitational potential (divided by c^2) were less than the ratio of the gas pressure to the radiation pressure, already a small number. Superstars could therefore not form efficient machines for converting mass into energy; neither did they seem particularly promising for accelerating relativistic electrons. Nevertheless, this model did seem to be on the right track, because Smith and Hoffleit (1963) soon discovered that quasars were variable on timescales $\leqslant 1$ month implying that the energy was produced within a region $\leqslant 10^{17}$ cm across. (Curiously, the discovery of even more rapid variability in Seyfert galaxies, was not made until 1968 by Pacholczyk and Weymann (1968).)

Various efforts to stabilise superstars using rotation, turbulence, nuclear reactions and a 'C-field' proved unconvincing. Although it was shown that a low entropy (i.e. thin) disk could exist with a binding energy up to 0.38 times its mass (Bardeen and Wagoner, 1969) this structure was liable to serious instabilities (Salpeter and Wagoner, 1971; Salpeter, 1971) and would probably either fragment into stars or thicken to a high entropy disk radiating a significant fraction of the Eddington limit of unspecified structure and unknown stability properties.

Apart from the problem of not releasing much energy before collapsing there was the separate difficulty of accelerating relativistic electrons to produce non-thermal synchroton radiation. For this reason several authors proposed that the active agent within a quasar be a magnetised spinning superstar or disk, called a spinar or magnetoid (e.g. Sturrock, 1965; Ginzburg and Ozernoi, 1964). A simple estimate based on the virial theorem suggested that a mass M would be permeated by a magnetic flux $\sim 10^{30}(M/M_\odot)G$ cm^2. For a mass $\sim 10^8\,M_\odot$ and a size $\sim 10^{16}$ cm as indicated by the observations this suggested field strengths up to $10^6 G$ which would cause the disk to evolve rapidly dynamically (e.g. Woltjer, 1971). Again most of the energy would be released explosively near the end point of the evolution just prior to the formation of a black hole. Several suggestions were made that the double radio sources might be formed as a consequence of such an explosion or outburst from a magnetised spinning superstar.

Now these models essentially identified the formation of a black hole with the end of the activity. By contrast, Greenstein and Schmidt (1964) realised that it was '. . . important to know if continued energy and mass output from such a collapsed object are possible.' A clear answer to this question was provided by Zeldovich and Novikov (1964) and Salpeter (1964). These authors pointed out that gas accreting onto a black hole could release its energy reasonably efficiently. In the case of a Schwarzschild black hole, gas should orbit the hole in circular orbits and so the efficiency of energy release should be given roughly by the binding energy of the most bound stable circular orbit (i.e. ~ 0.06). They also realised that if gas were made freely available the accretion rate could build up to a limit setting in when the pressure of the escaping radiation equaled the attractive pull of gravity. Under either optically thick or optically thin conditions, this is given by

$$L_{\mathrm{Edd}} = \mathring{M}_{\mathrm{Edd}} c^2 = \frac{4\pi GMm_{\mathrm{p}}c}{\sigma_{\mathrm{T}}} \sim 10^{46} \left(\frac{M}{10^8 \, \mathrm{M}_\odot} \right) \mathrm{erg \, s^{-1}}, \qquad (8.12)$$

It is known as the Eddington limit. Associated with this accretion rate is a mass-independent timescale for increasing the mass of the hole,

$$t_{\mathrm{Edd}} = \frac{M}{\mathring{M}_{\mathrm{Edd}}} = \frac{\sigma_{\mathrm{T}}}{4\pi Gm_{\mathrm{p}}c} \sim 5 \times 10^8 \, \mathrm{yr}. \qquad (8.13)$$

It now seems rather surprising that black hole models of AGN were not developed more rapidly (although the subsequent discovery of the microwave background in 1965 and in 1967 must have distracted attention away from quasars). Part of the explanation was that it was hard to understand how the infall of gas could co-exist with outflow as apparently observed in the inner parts of several galaxies, including our own. Another reason was that it seems not to have been realised that radio sources, quasars and Seyfert galaxies required there to be a more or less steady release of energy over timescales $\geqslant 10^8$ yr. Most models seem to have envisaged impulsive energy release through a scaled up solar flare, black hole formation or binary coalescence (e.g. Hoyle and Fowler, 1963*a,b*; Burbidge, 1964). This contradiction was particularly acute for the double radio sources like Cygnus A which showed evidence for ongoing nuclear activity through the high excitation nuclear emission lines and yet were modelled by two anti-parallel 'jets' created in a single explosion (e.g. Le Blanc and Wilson, 1969). It is quite ironic that two of the staunchest defenders of the steady state cosmology Hoyle and Burbidge, should have believed in 'big bangs' in galactic nuclei.

The papers of Zeldovich, Novikov and Salpeter supplied the foundation for most current thinking about the centre powerhouse in AGN. However, they were not seriously developed until Lynden-Bell (1969) derived and applied a theory of thin accretion disks in orbit around massive Schwarzschild black holes. He argued that magnetic stresses within the disk would produce an effective viscosity which would cause angular momentum to be driven outwards at the same time as mass flowed radially inwards, releasing its gravitational binding energy in the process. Lynden-Bell initially followed Salpeter, in assuming that the hole was non-rotating.

However, as Bardeen (1970) was quick to point out, a black hole in an AGN would probably be rapidly rotating, having been spun up by the gas from the surrounding accretion disk. In this case the Kerr metric is appropriate and stable circular orbits have a binding energy per unit mass that can be as large as $0.42c^2$ when Ω_H approaches its maximal value of $1/2m$. (The efficiency of energy release is still 0.2 for a hole rotating with angular velocity equal to 0.75 of its maximal value.) Spinning black holes can therefore provide extremely efficient machines for converting the rest mass of accreted gas into radiative energy. Accretion disk models had a further advantage. As the magnetic field lines would be mostly frozen in to the differentially rotating disk and would therefore have to keep undergoing magnetic reconnection, strong, inductive electric fields would be created in the (relatively small) reconnection volumes and these would be able to accelerate relativistic electrons. However, in this and subsequent papers (Lynden-Bell, 1971; Lynden-Bell and Rees, 1971), accretion disks were still associated more with dead quasars, i.e. local, low power AGN. The active, high power objects were identified with massive, uncollapsed disks.

The idea that quasars in the prime of life might be powered by accretion disks around black holes gradually gained ground matching contemporary developments in the study of binary X-ray sources (e.g. Rees, 1977). At the same time, the principal competition, models involving dense clusters of stars, seemed to lose ground as the observational position developed.

8.3.1.3 Star clusters

A competitive class of quasar models to those involving single coherent objects developed in parallel to models involving massive black holes. This involved dense star clusters (e.g. Gold, Axford and Ray, 1964; Spitzer and Saslaw, 1966; Colgate, 1967; Sanders, 1970). In initial versions of this model it was proposed that quasars were the consequence of an exaggerated rate of supernovae within the cluster. $10^8 \, M_\odot$ of high mass stars in a volume $\sim 1 \, pc^3$ can qualitatively reproduce many of the observed features of a quasar, the ionising radiation coming mostly from the hot stars, the outbursts being attributed to the supernovae themselves and the non-thermal emission deriving from the overlapping supernova remnants. In a star cluster this dense, the relative velocity of the stars becomes comparable with their surface escape speeds. This implies that physical collisions become competitive with gravitational collisions. Unfortunately, the consequences of these collisions have never properly been understood. If the stars coalesce, then rapidly evolving high mass stars can be assembled. It was even suggested that supernovae might trigger one another in a sort of chain reaction. A cluster of this type would inevitably produce a lot of interstellar gas which may create additional stars and accelerate the dynamical evolution of the cluster (Bisnovatyi-Kogan and Sunyaev, 1972; Begelman and Rees, 1978).

The principal objection to star clusters was that, at their best, they were never very efficient. Typically, $\leqslant 10^{-3}$ of the mass of the stars can be converted into radiation over the life of the cluster. This difficulty may be ameliorated by replacing the stars with compact objects – black holes, neutron stars or possibly white dwarfs (e.g. Gunn and Ostriker, 1970; Arons, Kulsrud and Ostriker, 1975; Stoeger and Pacholczyk, 1983). A newly formed pulsar can store up to 1 per cent of its rest mass as spin energy (although observations of galactic pulsars argue against this actually occurring: Lyne, Manchester and Taylor, 1985). Stellar mass black holes accreting in a dense gaseous environment should be able to radiate with a similar efficiency per unit mass of accreted gas to that invoked for stellar mass black holes (i.e. ~ 10 per cent).

However, a second difficulty is harder to overcome. This is that large outbursts in powerful quasars radiate much more than 10^{52} erg, the maximum energy that can be derived from a single star. Attempts to explain this in terms of non-linear or collective interactions have been less than convincing.

A third and more direct problem with star cluster models arose when it became clear that most double radio sources had central components which could be resolved using VLBI. This realisation, in conjunction with the production of high resolution maps of the outer radio components which contained bright 'hot spots', encouraged Rees (1971), Longair, Ryle and Scheuer (1973), and Scheuer (1974) to develop models in which the radio components are fueled continuously through channels or jets. Blandford and Rees (1974) suggested a specific collimation mechanism for the jets – the formation of a pair of De Laval nozzles along the spin axis of a dense rotating cloud of gas. This model did not specify the nature or origin of the jet fluid, only requiring that it be relatively hot and light. Unfortunately, in a powerful source like Cygnus A, the nozzles must form on a scale ~ 100 pc, far larger than the size of the compact radio sources which are typically ~ 1 pc in size. This implies that the collimation of the jets occurs on scales $\leqslant 1$ pc. There was no straightforward way to accomplish this in the context of a massive star cluster, especially when it was discovered that the source axes on the VLBI scale were roughly parallel to the larger scale structure.

Attempts to concentrate a star cluster into a volume $\ll 1$ pc^3 only hasten its evolution into a massive black hole (Begelman and Rees, 1978; Shapiro and Teukolsky, 1985). A stable compact gyroscope, i.e. a single spinning object, rather than an assemblage of small, independent radiators was suggested.

8.3.1.4 Recent evidence for black holes in AGN

Further observational developments have strengthened the belief that AGN contain massive black holes. X-ray variability appears to be common in Seyfert galaxies, and in a minority of lower power examples, this has been seen to occur on timescales as short as minutes (e.g. Tennant et al., 1981; Marshall et al., 1983). In fact Barr and Mushotzky (1986) have claimed that the X-ray variability timescale is correlated with the X-ray luminosity in the form $(t/m) \sim 100(L_x/L_{Edd})$ which implies that the X-rays originate from within roughly a hundred Schwarzschild radii. (Even more rapid variability on timescales ~ 30 s has been reported from the BL Lac object H0323+02 (Feigelson et al., 1986), although here it is suspected that the variability timescale may be unrepresentatively short compared with the size of the emitting region on account of relativistic beaming.)

The stronger hypothesis, – that essentially *all* bright galaxies harbour one or more black holes in their nuclei – received a boost when it was found that low level emission line and radio activity could be detected in the majority of local galaxies (e.g. Keel, 1985). More recently, Filippenko and Sargent (1985) discovered that broad (i.e. high velocity) wings to the emission lines are commonplace and have therefore argued that low-level, quasar-like activity must generally be present in galaxies.

Superluminal expansion of radio components, first discovered in 1971 (Moffet et al., 1971; Cohen et al., 1971; Whitney et al., 1971) has been shown to be a fairly common property of compact radio sources (e.g. Cohen, 1986). This phenomenon, in which two or more distinct radio components separate with apparent speeds in excess of c, was actually predicted by Rees (1966). It is generally interpreted as an illusion caused when radio-emitting plasma moves with relativistic speed at a small angle to the observer's line of sight. Although the kinematic details remain an area of active controversy, the evidence that there is outflow at relativistic speed is strong and this, in turn, is circumstantial evidence for the presence of a relativistically deep gravitational potential well, i.e. a black hole.

Recent work on faint quasar counts (Koo, 1986; Boyle et al., 1987) has allowed an improved estimate of the overall efficiency of quasars (Soltan, 1982; Phinney, 1983). The number of quasars with optical flux in excess of some value F rises with decreasing flux $\propto F^{-2.3}$ down to a flux equivalent to 19^m. The number of fainter quasars rises far less rapidly so that the mean energy density of quasar light observed at earth converges at $\sim 22^m$ where there are $N_Q \sim 4 \times 10^6$ quasars on the sky. An average magnitude for these quasars of 20.5^m corresponds to a flux $F_Q \sim 2 \times 10^{-13}$ erg cm^{-2} s^{-1} if each quasar radiates like an A0 star. However, quasars must radiate significantly more UV radiation than an A0 star in order to account for the observed emission lines, and so we must multiply this flux by a bolometric correction which we estimate as $f_{UV} \sim 3$. The mean redshift of these quasars is $\langle z \rangle \sim 2$ and so each quasar photon will have its energy reduced by a factor $\sim (1 + \langle z \rangle)^{-1}$ as the universe expands. The mean energy density of quasar light observed at earth is then $U_Q \sim N_Q F_Q f_{UV}(1 + \langle z \rangle)/c \sim 1.5 \times 10^{-16}$ erg cm$^{-3} \sim 3000$ M$_\odot$ Mpc^{-3}. (This is a conservative estimate. Any increase in the number of quasars or their bolometric corrections would only increase the required efficiency.) Now, the local density of bright galaxies is roughly $\sim 10^{-2} h^{-3}$ Mpc^{-3}, where h is the Hubble constant in units of 100 km s^{-1} Mpc^{-1}. Therefore, at least 3×10^5 M$_\odot$ of energy equivalent must be released per bright galaxy. Now if we introduce an efficiency of conversion of rest mass into radiant energy $\varepsilon \sim 0.1$, the remnant nuclear masses per bright galaxy must be $\sim 3 \times 10^6 (\varepsilon/0.1)^{-1} h^{-3}$ M$_\odot$, if we assume that quasars are powered gravitationally.

As we discuss below, it appears that most bright galaxies have central masses $< 3 \times 10^6$ M$_\odot$. We can also probably already exclude the possibility that 10 per cent of nuclear remnants are heavier than 3×10^9 M$_\odot$. Hence, the required efficiency of energy production must exceed $\sim 10^{-3}$ ($h = 1$), or $\sim 10^{-2}$ ($h = 0.5$). This virtually rules out star cluster models, as well as black hole models that postulate super-critical accretion with low radiative efficiency (e.g. Abramowitz, Jaroszyński and Sikora, 1978).

As we discuss below, the strongest evidence for black holes in X-ray binaries is dynamical. Unfortunately, unlike in the stellar mass case, the alternatives to black holes (e.g. spinars, magnetoids etc.) are far less well specified than neutron stars, and therefore cannot be so easily ruled out on the basis of mass determinations. There have been several attempted dynamical determinations of the central masses of AGN.

The observational study of the nucleus of M87 by Young *et al.* (1978) and Sargent *et al.* (1978) showed that there was a central cusp of light over and above the central core of stars and that the measured velocity dispersion of the stars rises within the central ~ 100 pc (*cf. also* de Vaucouleurs and Nieto, 1979). They interpreted these observations in terms of a model involving a central black hole of mass $\sim 3 \times 10^9$–5×10^9 M$_\odot$. (This conclusion has been challenged by Dressler, 1980.) It is known that not all bright galaxies show evidence for central cusps (Young et al., 1978). If we take this evidence at face value, we find that we are only probing the stellar dynamics on a length scale equivalent to 3×10^5 Schwarzschild radii. There has therefore been some attention paid to dynamical explanations of these observations not involving a black hole. One possibility is that the central star cluster has an anisotropic velocity dispersion so that the preponderance of radial orbits near the centre is able to reproduce the observed cusp (Duncan and Wheeler, 1980). Newton and Binney (1984) and Richstone and Tremaine (1985) have shown that the observations can be consistently interpreted using a constant mass to light ratio ($M/L_B \sim 10$ and a $\sim 2:1$ radial velocity anisotropy. However, it is not clear that an equilibrium stellar distribution of this type is stable against velocity space instabilities (e.g. Barnes, Goodman and Hut, 1986). Related observational work has been carried out by Tonry (1984), Dressler (1984), Davies and Illingworth (1986) and especially Kormendy (1986) on the nuclei of the galaxies M31, M32, and NGC 1052. Rises in both the central velocity dispersions and the rotational velocity are reported, consistent with central black hole masses $\sim 10^6$–10^7 M$_\odot$. However, the worry about anisotropic velocity dispersions remains, and we can strictly only regard these masses as upper limits at this stage.

Further details on these and related topics are to be found in the proceedings of four recent conferences edited by Dyson (1985), Miller (1985), Giuricin et al. (1986), Swarup and Kapahi (1986).

8.3.1.5 Future observations

It is of interest to speculate upon what future observations might give firmer dynamical evidence for massive black holes in AGN. The most immediate hope lies with Space Telescope now, regrettably, scheduled for launch no earlier than 1988. Its linear resolution should improve on that of ground-based telescopes by at least a factor 10 and so it should be able to resolve the stellar cusps of galaxies quite well (if they are present). In particular, if the central velocity dispersions should rise by a further factor $\sim 10^{1/2} \sim 3$, this will probably rule out anisotropic velocity dispersion models.

More speculatively, stronger evidence for massive black holes would be forthcoming if a single example of a star or another black hole in orbit could be found. As discussed by Rees (1982) in the context of our Galactic centre, a single star orbiting outside its tidal radius close to an underluminous black hole might modulate the light output with a regular period $\leqslant 1$ h. The effects of Lense–Thirring precession might even be detectable. Searches for low power periodic signals in the continuum emission of selected objects might turn up an example of this.

Alternatively, when two galaxies (each possessing black holes in their nuclei) merge, the hole associated with the smaller galaxy will be dragged into the nucleus of the larger galaxy in a few orbital periods. Thereafter, the dynamical evolution time will increase to $\sim 10^9$ yr as the interaction with individual stars takes over (Begelman, Blandford and Rees, 1980; Roos, 1981). The holes will then spiral together until they become so close that gravitational radiation by the black hole binary takes over. Now, if we assume that the heavier black hole has a mass $m_1 = 10^6 m_{16}$ M$_\odot$, and the lighter one a mass $m_2 = 10^6 m_{26}$ M$_\odot$, then the lifetime to coalescence will be related to the orbital period P yr by $T \sim 4 \times 10^7 P^{8/3} m_{26}^{-2/3}$ yr. Searches for regular periodic behaviour in quasars (e.g. Ozernoi and Chertoprid, 1969) have not yet been fruitful and we would not expect them to contain binaries with measurable periods. However, it is just possible that a few lower power Seyferts may harbour binary black holes with detectable periods $\leqslant 1$ yr. A black hole binary would probably perturb the central stellar distribution so strongly that the black hole would be supplied with gas and the nucleus would be active at this time. For black hole masses $m_{16} \sim m_{26} \leqslant 1$, and an assumed merger rate of one per galaxy, roughly 1 in 300 galaxies could contain binaries and show periodic behaviour with $P \leqslant 1$ yr. (As we discuss below, Lacey and Ostriker (1975) propose that binaries of this type occur very frequently in galactic nuclei.) These binaries would probably not be seen as spectroscopic binaries because the sizes of the broad emission line regions are inferred to be significantly larger than the radii of these hypothetical binary orbits. However, they may be detectable as eclipsing binaries if both black holes are accompanied by extensive accretion disks. A regular monitoring program of nearby Seyfert and LINER nuclei might possibly uncover an example.

8.3.1.6 *Possible observations incompatible with the presence of massive black holes*

In the same spirit, it is worth discussing some conceivable future observational discoveries which might be capable of ruling out the presence of massive black holes in AGN. One of these is the detection of large amplitude, high Q pulsations with timescales of order a day or larger in a powerful quasar. This could probably not be attributable to modulation of the emission from an accretion disk by the orbit of a single star as discussed above and would instead probably represent the presence of a massive uncollapsed object. Another possibility is the detection of X-ray or γ-ray lines with a common redshift relative to the host galaxy of $z \sim 0.1$–0.2. This would certainly revive interest in neutron star cluster models of AGN. Alternatively, Space Telescope (operating with its advertised resolving power) may fail to detect any evidence for dormant central masses in the nuclei of local galaxies. As the arguments above imply, this would call into question far more than the black hole model (especially if $h \sim 0.5$).

However, perhaps the most telling argument against black hole models would come if it could be demonstrated that the compact optical and X-ray sources are generally displaced from the dynamical centres of their host galaxies as well as each other. An argument similar to this has already been used to limit the masses of black holes perhaps associated with X-ray sources in globular clusters. More relevantly, it has been applied to our Galactic centre and it is illustrative to discuss the current controversy as to whether or not our Galactic centre contains a massive black hole.

8.3.1.7 Our Galactic centre

The Galactic centre region, most recently reviewed by Brown and Liszt (1984) and Lo (1986) contains a compact radio source with unique spectral properties known as Sgr $A*$ (Balick and Brown, 1974). This source is more compact than $\sim 3 \times 10^{14}$ cm (Lo et al., 1985), and is surrounded by a ring of gas and dust of radius ~ 2 pc apparently connected to Sgr $A*$ by three curving radio features. These radio 'arms' can also be observed in $12.8\,\mu$ infra red Ne II lines and although the kinematical configuration is still unclear, a central mass of $\sim 3 \times 10^6$ M$_\odot$ is strongly implied (Crawford et al., 1985). An even higher velocity dispersion feature was discovered by Hall, Kleinmann and Scoville (1982) in He I lines originating from the vicinity of Sgr $A*$. Naive application of the virial theorem gives a similar central mass although this feature is also interpretable as an outflowing wind. In addition, Sgr $A*$ is located near the centre of a cusp of stars, traced out by their infra-red emission. It therefore might seem natural to assume that a $\sim 3 \times 10^6$ M$_\odot$ black hole is present in Sgr $A*$.

However, there is also an unusual infra-red source located nearby known as IRS16. This source which, is relatively blue, is argued to be the source of the photo-ionising flux required within the central pc and to comprise ~ 100 hot (B0) stars. However, Sanders and Allen (1986 and references therein) have shown that IRS16 is comprised of several subsources and is displaced from Sgr $A*$ by $\sim 10^{17}$ cm. So, we must choose between IRS16 and Sgr $A*$ as the centre of the galaxy. There are already some clues. We know on dynamical grounds that any million solar mass black hole must settle very close to the centre of the star cluster. If the black hole is identified with Sgr $A*$, then its tidal field would destroy a small star cluster like IRS16 in a timescale ~ 1000 yr, far shorter than the age of the stars. On these grounds, Allen and Sanders argue that the mass in Sgr $A*$ must be much less than 3×10^6 M$_\odot$, typically ~ 100 M$_\odot$. However, VLBI observations (Backer and Sramek, 1987 and private communication) have already shown that Sgr $A*$ is at rest (to less than ~ 50 km s^{-1}) with respect to the dynamical centre of the Galaxy. This measurement would already seem to rule out the possibility that Sgr $A*$ is in orbit about the Galactic centre (and also that it is an outflowing radio jet). This implies that Sgr $A*$ and not IRS16 is the Galactic centre. It seems that independent of the black hole hypothesis, we must consider that either there is an unlikely superposition of unusual objects or that IRS16 is a transient object. A final choice will probably be made on the basis of an improved limit on the proper motion and infra-red radial velocity mapping.

The Galactic centre is also a source of variable γ-rays, notably the 511 keV positron annihilation line. If we want to associate this emission with a relativistic electron-positron wind (with a discharge $N \sim 10^{43}$ s^{-1}) flowing out from the Galactic centre, then we would naively expect that the size of the source of the wind l would be determined by the condition that there is an optical depth of order unity to photon–photon pair production in the source, i.e. $l \sim N\sigma_T/4\pi c \sim 3 \times 10^7$ cm. Using this argument, Lingenfelter and Ramaty (1982) have suggested that the positron source is a ~ 100–1000 M$_\odot$ black hole for which l is comparable to the Schwarzschild radius and for which the total luminosity is similar to the Eddington limit.

Another difficulty with postulating the presence of a massive black hole in the Galactic centre has been emphasised by Gurzadyan and Ozernoi (1981 and references therein). This is that if we use the standard estimate of the rate of swallowing stars by a black hole (e.g. Lightman and Shapiro, 1977), then the galactic centre would be far more luminous than it is observed to be and the black hole would have grown to a far larger mass by now. However, Rees (1982) has argued that stars will only be captured occasionally, and that when they are, the stellar debris is mostly ejected very quickly by the energy liberated as a small fraction of the star's mass accretes on the hole. The mass of a black hole in a galaxy like ours could therefore stabilise at $\sim 10^6$ M$_\odot$ and the fraction of the time when it is emitting at a rate comparable with the Eddington limit will be correspondingly small.

8.3.2 Black holes in binary X-ray sources

8.3.2.1 Early history

As we described in Section 8.1, the idea that black holes are a natural end point of stellar evolution of massive stars is over fifty years old and, compared to the situation with AGN, there has been comparatively little resistance to claims that they exist in binary stars. However, the discovery of Sco X-1 in 1962 (Giacconni *et al.*, 1962) and its identification with a 13^m star were not immediately heralded as discoveries of neutron stars (Shklovsky, 1967) or black holes despite the fact that several physicists (e.g. Chiu, Salpeter, Wheeler) were getting interested in their theoretical properties. Neutron stars were (rightly) not regarded as promising prospects for optical observation because of their small ($\sim 10^{13}$ cm) surface areas. It was however, realised that newly formed neutron stars should be hot with temperatures ($\sim 10^7$ K) and should be observable for a cooling time ~ 1000 years as X-ray sources. Indeed it was explicitly suggested (e.g. Chiu, 1964) that the few discrete X-ray sources known at the time might be neutron stars made in recent supernovae.

Again, it was Zeldovich (1964*a*) who appears to have been the first to realise that a compact object in a mass transfer binary or just conceivably a dense interstellar cloud would be a more prodigious source of X-rays. The potential energy released when a gram of gas falls onto a neutron star or black hole is $\sim 10^{20}$ erg, far larger than the mean amount of energy per gram of neutron star that can be transported by degenerate electron conduction to the surface before neutrino emission cools the interior (i.e. $\sim 10^{12}$ erg). He also noted that this emission mode was available to black holes despite the absence of a hard surface, because gas would probably not approach the event horizon smoothly but, instead, separate gas streams could collide with one another, shock, heat and then radiate away their internal energy. Accretion with angular momentum is likely to be more radiatively efficient than spherical accretion (Zeldovich, 1964*b*). It was also realised that the best evidence for the presence of a black hole was likely to be dynamical and several binary star catalogues were searched for massive unseen comparisons (e.g. Zeldovich and Guseynov, 1965; Trimble and Thorne, 1969) of which several were found. (Unfortunately, none of these candidates has survived.)

8.3.2.2 Cyg X-1

The observational position improved dramatically with the launch of the Uhuru satellite in 1971. One of the first sources to be investigated was Cygnus X-1. It was a powerful X-ray source but, unlike several other powerful sources, it did not exhibit regular pulsation (Schreier *et al.*, 1971). Instead Cygnus X-1 had an unusually hard X-ray spectrum and displayed rapid flickering on the timescale of a few milliseconds. The optical identification was with a bright blue supergiant HD 226868 of spectral type 09.7Iab. Optical spectroscopy by Webster and Murdin (1972) and Bolton (1972) showed spectral lines whose Doppler shifts varied sinusoidally with a period $P_b = 5.6$ d and a velocity amplitude $K = 76$ km s^{-1} coming from the surface of the supergiant. (They also showed features that appear to originate from gas streaming toward the X-ray emitting companion.) Now if m_X is the mass of X-ray emitting star and m_o is the mass of the supergiant, we can deduce the value of mass function

$$f(m_o, m_X, i) = \frac{m_X^3 \sin^3 i}{(m_X + m_o)^2} = \frac{P_b K^3}{2\pi G} = 0.25 \, M_\odot,$$

where i is the angle of inclination between the orbital angular momentum vector and the line of sight.

Now, given the mass function we see that m_X increases monotomically from zero as m_o and $\cos i$ increase from zero. However, a normal blue supergiant of this spectral class would have a mass $\sim 30 \, M_\odot$. Furthermore, the X-ray emission also exhibits the 5.6 d period, confirming the identification. However, eclipses are not seen in the high energy X-rays. (Partial eclipse is seen at low energy which is attributed to photoelectric absorption in the outflowing wind from the supergiant.) The inclination angle is then probably less than 60°. Together these values imply that $m_X \geqslant 8 \, M_\odot$ comfortably larger than the Oppenheimer–Volkoff mass for a neutron star. Not surprisingly, Webster, Murdin and Bolton concluded that it was 'inevitable that we speculate' that the comparison to HD 226868 was a black hole.

Unfortunately, this argument is contestable as it stands. The main difficulty is that the supergiant may well have lost a lot of mass over the lifetime of the system (not necessarily to the X-ray source but possibly out of the binary altogether). It may therefore have a very much smaller mass than its spectral type would conventionally indicate and perhaps be a hot subdwarf (Trimble *et al.*, 1973). For this reason, Paczyński (1974) showed that it was possible to place a lower bound on m_X independent of the evolutionary history of its optical companion. To do this, we note that observations of the spectrum of the optical star allow one to deduce its effective temperature and its surface gravity. What we cannot measure is the distance d to the binary. Now the radius of the star will be $R_0 \propto d$ and the mass will scale $m_0 \propto R^{1/2} \propto d^{1/2}$. Now, as we know the binary period, we can rewrite the mass function as $m_X \sin i \propto a^2$ where a is the separation of the stars. However, the absence of eclipses implies $a > R_0 \sec i \propto d \sec i$. Hence, by minimising $\mathrm{cosec}\, i \sec^2 i$ with respect to the unknown inclination i we can put a distance-dependent lower bound on m_X, i.e.

$$m_X \geq 3.4\, \mathrm{M}_\odot (d/2\ \mathrm{kpc})^2. \tag{8.14}$$

Now detailed studies of the reddening of HD 226868 and nearby stars allow one to place a lower bound on d (Margon et al., 1973; Bregman et al., 1973), which indicates a distance in excess of 2 kpc which, in turn implies that $m_X > 3.4\, \mathrm{M}_\odot$. Of course local patchiness in the distribution may allow Cyg X-1 to be closer than 2 kpc, but already this is looking rather contrived. In fact the system can be further constrained by modelling the ellipsoidal light variations associated with the tidal deformation of the optical star and the rotational broadening of the spectral lines, and the results indicate that the star is indeed a high mass supergiant as indicated by its spectral type (Avni and Bahcall, 1975). The most recent spectroscopic study of this system, by Gies and Bolton (1986) concludes that the minimal (probable) masses are $m_X = 7\, \mathrm{M}_\odot$ (16 $\mathrm{M}_\odot$), $m_0 = 20\, \mathrm{M}_\odot$ (33 $\mathrm{M}_\odot$). More contrived models involving triple systems and X-ray generation by normal stars have been proposed, but the observations show none of the expected features (e.g. Bahcall, 1978).

In conclusion, the case that Cyg X-1 has a mass in excess of the Oppenheimer–Volkov limit is, by astronomical standards pretty strong and, importantly, has strengthened significantly since 1972. It seems that the case for calling it a black hole is as strong as the Oppenheimer–Volkoff limit which is computed to be ~ 1.6–2 $\mathrm{M}_\odot$ for reasonable equations of state (Arnett and Bowers, 1977) and $\sim 3\, \mathrm{M}_\odot$ for equations of state constrained solely by the requirement that general relativity be the correct theory of gravity (e.g. Rhoades and Ruffini, 1974).

8.3.2.3 LMC X-3

More recently, LMC X-3, an X-ray source in the Magellanic clouds has been associated with a young, hot star on the basis of an extremely good X-ray position. The optical star has a binary period $P_b = 1.7$ d and $K = 235$ km s^{-1}

(Cowley *et al.*, 1983). This source has two favourable features. Firstly, the mass function is computed to be 2.3 ± 0.3 M$_\odot$ and secondly the distance to the Magellanic clouds is known fairly accurately to be $d = 55$ kpc. Repeating the arguments listed above for Cyg X-1, we find that the B3V spectral type optical companion should have a mass ~ 7 M$_\odot$ if there has been no unusual evolution. Independent of the evolutionary history, the avoidance of X-ray eclipses requires that $m_X \geqslant 10 \pm 4$ M$_\odot$ (Paczynski, 1984).

Unfortunately, there is a possible complication in this system because the accretion disk around the X-ray source may contribute a significant portion of the optical emission from this binary allowing both the compact object and its companion to be of lower mass. In particular, the mass of the X-ray source may possibly be as low as ~ 2.5 M$_\odot$ (Mazeh *et al.*, 1986), uncomfortably close to the Oppenheimer–Volkoff limit. In this case the companion would have to be a $\leqslant 1$ M$_\odot$ helium star, a possibility that should eventually be ruled out using high dispersion spectroscopy.

8.3.2.4 A0620-00

Most recently, McClintock and Remillard (1986) have shown that the orbital period for an X-ray nova A0620-00 is a spectroscopic binary. (An X-ray nova is a transient X-ray source that becomes very bright for several months and then fades away.) Fortunately, A0620-00 also brightened at optical wavelengths and so there is no doubt about the identity of the companion. McClintock and Remillard found that $P_b = 7.75$ h and $K = 457$ km s^{-1} giving a mass function $f(m_o, m_X, i) = 3.2 \pm 0.2$ M$_\odot$. Again, there are no X-ray eclipses and from the spectral type of the optical companion (K5V) and modelling of the light curve, it is deduced that $m_X > 3.2$ M$_\odot$ (McClintock, 1986). This system is unusual in one other respect. During its 1975 outburst, it became the brightest X-ray source in the sky. However, during quiescence it is unobservably faint in X-rays and so appears to radiate far less in X-rays than would be expected on the basis of the optical emission from the accretion disk. Somehow or other the X-ray energy that should be released by gas as it spirals in towards the black hole is being absorbed or scattered out of the line of sight. (Alternatively, it is possible that the gas accumulates in a reservoir in the outer parts of the disk.) Until the reason for this behaviour is understood, there must be some small residual doubt about the inferred geometry of the binary.

8.3.2.5 LMC X-1

A fourth source, LMC X-1, has been put forward as a black hole candidate on dynamical grounds by Hutchings *et al.* (1983). The optical identification of this source, which is also found in the Large Magellanic Cloud is less certain. The companion is believed to be a bright 07 III-V star. Recent discovery of a soft X-ray photo-ionised nebula centered on this star (Pakull and Angebault, 1986) provides confirmatory evidence that the identification is correct because the high ionisation states observed in the nebula could not be produced by a normal star. Unfortunately no X-ray modulation at the optical period of $P_b = 3.9$ d has been seen. Using $K = 66$ km s^{-1}, a mass function $f(m_o, m_X, i) = 0.12$ M$_\odot$ is obtained, although there is still some uncertainty about this as other periods $P_b \leqslant 1$ d may fit the data as well. The absence of X-ray eclipses can then be used to argue that $i \leqslant 64°$. In addition, the requirement that the bolometric luminosity of the optical star be less than the Eddington limit leads to a lower bound on the companion mass $m_o \geqslant 8$ M$_\odot$ with a probable black hole $m_X \sim 4$ M$_\odot$. Future spectroscopic observations should either rule out or strengthen the black hole candidacy of this system.

8.3.2.6 SS433

An extensively studied galactic object which has also been argued to contain a black hole is SS433. SS433 originally appeared in a list of emission line stars. After some false trails (chronicled in Margon, 1980), Clark and Murdin (1978) realised that it was coincident with an X-ray source and a compact radio source at the centre of a large supernova remnant W 50 which they naturally presumed were related. They also noted that, in addition to the unusually strong and broad hydrogen and helium emission lines, several weaker unidentifiable lines were also present in the spectrum. In a more careful study of the spectrum, Margon et al. (1979) identified the weaker lines with redshifted and blueshifted hydrogen and helium transitions. Furthermore, the Doppler shifts changed from night to night in a systematic manner.

Among a host of ephemeral theoretical suggestions, was one by Fabian and Rees (1979) who suggested that the emission be produced by two oppositely directly jets, analogous to those found in the double radio sources, but in this case containing dense clouds of emitting gas moving with relativistic speed. Milgrom (1979) further speculated that the modulation of the Doppler shift may be due to precession of these jets about an axis fixed in space. Further observations by Abell and Margon (1979) verified this kinematic model. Using several periods of data, we now know that the jet speed is stable at $v = 0.26c$, the vertex angle of the cone on which the jet precesses is $\theta = 20°$, the inclination of the cone axis to the line of sight is $j = 80°$ and the precession period is $P_{\text{p}} = 163$ d (Margon, 1984).

SS433 also exhibits a 13.1 d, 0.7^m photometric double-peaked variation (as well as a slower 163 d variation). This is attributed to mutual eclipsing of an accretion disk surrounding a compact object and its binary companion orbiting with a 13.1 d period (Leibowicz, 1984). This same period is also detectable in small radial velocity variation of the brighter, 'stationary' emission lines. However, the hydrogen lines vary with a smaller velocity amplitude and different orbital phase to the higher excitation He II lines. It is important to determine which of these two radial velocity variations should be used to determine the mass function, because the former possibility gives $f(m_o, m_X, i) \sim 0.5 \, M_\odot$ and a mass $m_X \sim 1 \, M_\odot$ for the compact object (which could presumably be a neutron star) and the latter suggests $f(m_o, m_X, i) \sim 11 \, M_\odot$ implying a black hole companion. Now, Crampton and Hutchings (1981) have argued that the higher excitation helium lines should be produced closer to the photo-emission source, presumably the compact object, and that the hydrogen lines are associated with gas streams hitting the outer edge of the accretion disk in a hot spot. Furthermore, the phase of maximum He II radial velocity is roughly 90° different from the primary minimum in the photometric variation which is entirely consistent with the gas stream interpretation.

Further confirmatory evidence that the He II lines measure the velocity of the compact object and consequently that SS433 is a high mass system is provided by the discovery of orbital modulation of the 'moving' emission lines. The wavelength λ_o of a spectral line emitted by gas in either jet is related to the emitted wavelength λ_e by

$$\lambda_o = \frac{\lambda_e}{(c^2 - v^2)^{1/2}} [c \pm v(\sin j \sin \theta \cos \psi + \cos j \cos \theta)], \quad (8.15)$$

where the $\pm$ signs refer to the two jets and ψ is the precessional phase which can be written at $2\pi t / P_{\text{p}} + \psi_o$ in the absence of orbital modulation. Now, on quite general grounds, we expect the star in orbit about the compact object to nutate and to perturb the disk (together with the jets which are generally assumed to be launched perpendicular to the disk) at half the synodic period, $P_{\text{s}} = 1/2(P_{\text{b}}^{-1} \pm P_{\text{p}}^{-1})$. If, as we shall find, the precession is retrograde with respect to the orbital motion, we must choose the plus sign and $P_{\text{s}} = 6.06$ d. (Retrograde precession will result if the compact object exerts a torque on the star and the disk is slave to the star, or the star forces the disk to process directly. Lense–Thirring precession would be prograde and is therefore ruled out.) In general, we might expect j, σ and ψ to contain contributions $\propto \cos(2\pi t/P_s + \text{const.})$ which would create sidebands in a frequency analysis of the measured Doppler factors. Now in addition to the fundamental precessional period of 163 d, two additional periods are found at 6.28 d and 5.83 d $= (P_s^{-1} \pm P_p^{-1})^{-1}$. No signal is detected at $P_s = 6.06$ d. If we now inspect equation (8.15), we find that these sidebands can be reproduced if the precessional place ψ and the cone angle θ are modulated by the orbital motion. This confirms that the precession is retrograde (Katz et al., 1982; Newsom and Collins, 1982). Now, given this kinematical interpretation of the sidebands in the Doppler shift data, we can compare the phase of the modulation with the orbital phase and again we find that it is consistent with the phase of the He II lines (Margon, 1984).

We therefore have a fairly comprehensive *kinematical* description of the SS433 systems (Fig. 8.7). What is quite disturbing is the absence of convincing *dynamical* explanations for the jet collimation, unique speed, precession and orbital modulation. (For a critique of various possibilities, see Margon, 1984; Collins, 1985.) Nevertheless, if we take the kinematics at face value, and assume that the jet precesses about the orbital angular momentum (i.e. $i = j$) then the mass function implies that $m_o^3 = 11(m_o + m_X)^2$ M$_\odot$. The final stage in the argument is to put a lower bound on the mass of the companion star $m_o \geqslant 20$ M$_\odot$ from the requirement that it fill its Roche lobe and be able to radiate the observed luminosity of $\sim 8 \times 10^{38}$ erg s^{-1} (quoted by Band and Grindlay, 1985). The mass of the compact object is then $m_X \geqslant 7$ M$_\odot$, indicating a black hole and consistent with it radiating $\sim 10^{39}$ erg s^{-1} at the Eddington limit.

However, as was the case with Cyg X-1, the compact object could itself be a close binary with orbital period $P_{b2} \sim 1.5$ d and precessional period $P_p \sim 4 P_b^2 / 3 P_{b2} \cos \theta$. A recent study of this possibility by Fabian et al. (1986) shows that it is possible to devise evolutionary pathways to this configuration. This model allows the compact object to be a neutron star. However, it seems hard for this system to avoid producing a further modulation of the optical emission with a period $\sim P_{b2}$. So far, this period has not been detected.

In conclusion, the weight of evidence favours SS433 containing a black hole. However, the arguments for this are less direct than in Cyg X-1. The importance of SS433 is that it shares some properties with active galactic nuclei (relativistic precessing jets, winds, broad emission lines etc.) and it may be a miniature version of a quasar.

8.3.2.7 *Other black hole candidates*

We have described the evidence for black hole candidacy for five X-ray sources in decreasing order of reliability. It is surprising that more examples have not been found since 1972. We have seen that the identification, and orbit must both be secure and we must usually add an astrophysical argument to limit the range of inclination angle before we can assemble a strong case for having a black hole binary. In the past, various other criteria have been proposed for black holes but these have not turned out to be reliable. Nevertheless it is instructive to discuss why these candidates are no longer taken seriously.

Firstly, the X-ray quiet binary stars that have massive, apparently invisible companions such as ε Aur (e.g. Cameron, 1971), β Lyr (e.g.

Fig. 8.7. Kinematic model of SS433. Two antiparallel jets emanating from a compact object in a binary system precess on a cone with opening angle $\theta \sim 20°$ and period $P_p \sim 163$ d. Tidal torques produced by the orbiting companion impose a small nutation on the jets with a 6.06 d period. (Adapted from Collins, 1985.)

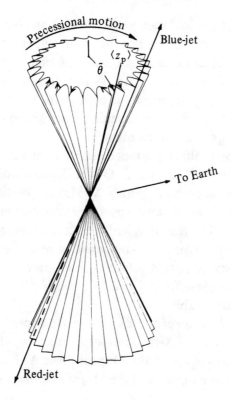

Devinney, 1971) and BM Ori (Wilson, 1972) can all be modelled using normal companion stars and so there is no obligation to postulate the presence of a black hole. However, the absence of X-rays, only forbids an *accreting* compact object and even here, the example of A0620-00 warns us that it is possible for X-rays from a disk to be beamed away from us or attenuated. It is still worth investigating other possibilities. Black hole binaries that are currently X-ray active, may constitute only a small fraction of the total.

Rapid (~ 1 ms), aperiodic X-ray variability (e.g. Sunyaev, 1973) was proposed as a signature of a black hole and for a long while Cir X-1 was held to be a black hole on these grounds. However, Cir X-1 undergoes X-ray bursting on timescales ~ 10 s and this is strongly believed to be a signature of a neutron star (Tennant, Fabian and Shafer, 1986). Conversely, the known neutron star binary, V0332 + 52 (as well as GX339-4 and Cir X-1) exhibits rapid flickering, just like Cyg X-1 (Stella *et al.*, 1985). It is rather surprising that rapid variability has been regarded for so long as a peculiar signature of a black hole as orbits near the surface of a ~ 1.4 M$_\odot$, 10 km neutron star have periods ~ 1 ms, similar to those for orbits around a more massive black hole.

Spectral criteria have also been proposed as an indication of the presence of a black hole. In particular, White and Marshall (1984) pointed out that the black hole candidates had unusually soft (i.e. steep) X-ray spectra. However, these sources are not always in this state and many of them (notably Cyg X-1) change frequently to emit a hard (i.e. flat) X-ray spectrum. Unfortunately Cir X-1 behaves in a similar manner and so the usefulness of this criterion is suspect at the moment.

As an illustration of the importance of obtaining a good X-ray position, consider the case of the X-ray source OAO1653-40 identified by Polidan *et al.* (1978) with the supergiant binary V861 Sco (Wolff and Beichmann, 1979), on the basis of two X-ray eclipses separated by four times the known binary period. If the identification had been correct, then the mass function of 0.5 M$_\odot$ and spectral type of the star would have implied a minimum mass for the secondary star of 12.5 M$_\odot$. However, Parmar *et al.* (1980) found that the X-ray emission came from a nearby pulsating X-ray source and that V861 Sco was undetectable.

A more direct indication of a black hole would appear to be the discovery of an X-ray source with an X-ray luminosity exceeding the Eddington limit for a star with mass equal to the Oppenheimer–Volkoff limit. SMC-X-1 (which is at the known distance of the Magellanic clouds) might have been just such a star with a 1–35 keV luminosity of 6×10^{38} erg s^{-1} (the Eddington limit for a $\sim 4\,M_\odot$ star) and yet it has regular pulsations with period of 0.7 s (Lucke *et al.*, 1976) indicative of a neutron star, so this argument is also unreliable.

More satellite X-ray observations with ($\sim$ arcsecond) positional accuracy will be necessary to find additional black hole binary candidates. It will be particularly advantageous if these are used in conjunction with an all-sky X-ray survey instrument that can be used to locate X-ray transients. Further discussion of X-ray binaries can be found in the review of McClintock (1986) and the conference proceedings edited by Sanford, Laskarides and Salton (1982).

8.3.3 *Other possibilities*

Active galactic nuclei and binary X-ray sources are not the only *possible* environments for astrophysical black holes; neither need black holes be restricted to the stellar or the million to billion $M_\odot$ mass range. However, there is as yet no observational evidence at all for *any* black holes outside AGN and binary stars. This has not deterred theorists from speculating further about the possibilities.

As reviewed by Blandford and Thorne (1979), Shapiro and Teukolsky (1983), there are expected to be at least as many black holes formed singly as are retained in binaries. (If the supernova explosion is spherically symmetric and more than half of the system mass is lost, then a newly formed black hole will escape from a binary.) If, for illustration, we assume that main sequence stars with initial masses in excess of $10\,M_\odot$ evolve to give black holes, whilst more numerous stars of intermediate mass evolve to give neutron stars, then we can use the stellar mass function (e.g. Bahcall and Soneira, 1980) to relate the number density of dead neutron stars to the number density of black holes. Estimates of the birth rate of radio pulsars indicate that there are between 10^8 and 4×10^8 defunct radio pulsars in the Galaxy (Lyne, Manchester and Taylor, 1985), and we might expect a third as many black holes to be formed. The nearest one would then be less than 5 pc away. Unfortunately, our understanding of the details of supernova explosions and mass loss in evolving stars, although it has progressed substantially, is still not at the point where we can confidently estimate what fraction of massive stars evolve through to give black holes (e.g. Bethe and Brown, 1985). (This estimate is in any case complicated by the additional uncertainties associated with mass transfer in binaries which may allow a neutron star to accrete enough mass to drive it over the Oppenheimer–

Volkoff limit.) Unfortunately, even if black holes are this common, the nearby ones will only accrete very slowly in the interstellar medium, and will be unobservably faint (e.g. Shapiro and Teukolsky, 1983).

From the dynamics of stars in the solar neighbourhood, we can estimate the density of matter in the galactic disk. This is known as the Oort limit and the stars that we can see account for slightly more than half of it. Black holes formed as stellar remnants in the first generation of Galactic stars may account for the difference ($\sim 0.1\,M_\odot\,pc^{-3}$, Bahcall, 1984). There is also a 'missing mass' problem associated with the halo of our Galaxy and again, there must be as much invisible mass as visible mass at the solar radius. These may be primordial ($\sim 10^5\,M_\odot$) black holes. However there are some constraints. Lacey and Ostriker (1985) have proposed that the halo mass comprise $\sim 10^6\,M_\odot$ black holes perhaps formed primordially from isothermal fluctuations in a hot, baryon-density universe (e.g. Carr and Rees, 1984). These holes would steadily heat the galactic disk and thereby explain the observed relationship between disk thickness and stellar age. Lacey and Ostriker demonstrate that their interactions with individual stars can change the local stellar velocity distribution function to a form somewhat similar to that measured. Dynamical friction would cause masses this heavy to sink into the galactic centre over the lifetime of a galaxy. In fact there is a problem in that far too much mass will be accreted in our Galaxy. These authors therefore propose that when three holes accumulate in a galactic nucleus a dynamical slingshot (e.g. Begelman, Blandford and Rees, 1980), will lead to the ejection of some or all of the black holes. Presumably, when the hole settling rate is large enough, a more massive black hole can be formed. A large cosmic density of $\sim 10^6\,M_\odot$ black holes, which this proposal would require, might produce some observable gravitational lens images. A difficulty with this model is that interstellar gas accreting onto these black holes might make them bright enough to have been detected. For instance, making somewhat optimistic assumptions about the accretion rate, McDowell (1985) has deduced, on the basis of their absence from IRAS source catalogues and proper motion surveys, that halo holes must be less massive than $\sim 1000\,M_\odot$. Another possible problem has been discussed by Bahcall, Hut and Tremaine (1984) who have argued that the existence of very wide binaries, that would otherwise be disrupted by dynamical interaction with heavy stars, constrains the missing mass *in the disk* to comprise objects less massive than $2\,M_\odot$, presumably excluding black holes. This may also constrain the masses of halo holes.

As reviewed in Blandford and Thorne (1979), globular clusters are roughly a hundred times more X-ray luminous per unit mass than the rest of the galaxy and when this was discovered, several authors proposed that $\sim 1000\,M_\odot$ black holes reside in their nuclei (e.g. Bahcall and Ostriker, 1975). However, when improved X-ray positions became available, it was clear that the X-ray sources were not located at the cluster centres as would be expected for a massive object. From the observed distribution of these sources, it can be concluded that the ratio of the mass of the X-ray-emitting object to the mass of the field stars (typically $\sim 0.5\,M_\odot$) lies in the range ~ 1.5–5 (Grindlay, 1985). They are therefore believed to be accreting neutron star binaries. Nevertheless, a problem may remain in that more globular clusters should show evidence for collapse of their cores. Larson (1984) has proposed that in fact central cusps are present in many globular clusters, but these comprise invisible black holes which sink to the bottom of the potential well.

Discovery of six pairs of quasars that are possibly gravitationally lensed (e.g. Turner, 1986), (three of which show no evidence for luminous lenses), has revived interest in cosmological $\geqslant 10^{12}\,M_\odot$ massive black holes (e.g. Paczynski, 1986). These have the additional advantage that they create only two images as generally observed to be the case, rather than an odd number of images as one would expect with a non-singular transparent lens (e.g. Narayan, 1986).

Planetary mass primordial black holes have been postulated by Freese, Price and Schramm (1983). These would have masses between the Hawking mass ($\sim 10^{15}$ g) and $1\,M_\odot$. They may stimulate galaxy formation.

8.4 Future prospects

Black holes, an inevitable consequence of general relativity, were predicted long before there were any observational indications that they exist. Although astronomers have treated them with some scepticism, their place as endpoints of stellar and galactic nuclear evolution now seems fairly secure. The arguments for the X-ray sources Cyg X-1, LMCX-3 and A0620-00 (and possibly also LMCX-1 and SS433), being black holes in mass transfer binaries are at least as strong as (to take one example) the reasons for believing that helium is made in the big bang, and, in both cases, rely essentially on Newtonian dynamics, the assumption that locally discovered physical laws operate elsewhere in the universe, and common sense. In active galactic nuclei, there is now an impressive body of circumstantial evidence that most of the power derives from a coherent spinning object of size not a lot larger than its Schwarzschild radius. The efficiency with which this object converts mass into radiation must exceed $\sim 10^{-3}$ and perhaps also 10^{-2}. A $\sim 10^6$–10^9 M$_\odot$ black hole fits this description completely. The reason why we do not reject competitors such as spinars with the conviction that we reject neutron stars in Cyg X-1 is that spinars are less well-defined physical entities, and we find it hard to describe their structure and examine their stability properties. Perhaps Nature will have the same difficulties. After all, we do not consider it reasonable that the X-ray-emitting object in Cyg X-1 be a rapidly spinning disk of gas, supported by turbulence and magnetic field for, say $\sim 10^{15}$ rotational periods. So, why should we take any more seriously the notion of a similar structure a million times larger in AGN surviving for $\sim 10^9$ periods?

Of course none of this *proves* that black holes exist. As I have argued, a rigorous proof is going to be extremely hard to come by. This is the normal state of affairs in most sciences. It is surely more productive at this stage to accept the evidence and proceed. To quote the third volume of the *Principia* (Newton, 1687), 'In experimental philosophy we are to look upon propositions inferred by general induction from phenomena as accurately or very nearly true, notwithstanding any contrary hypothesis that may be imagined, till such time as other phenomena occur, by which they may either be made more accurate, or liable to exceptions.'

Let me give some examples of ways of proceeding that are actively being pursued. Firstly, it is slightly surprising that so few stellar mass black holes have been found; far less than was expected in 1972. By contrast, almost 500 neutron stars have been discovered since 1967. Perhaps this is telling us something of importance about mass loss in the final stages of stellar evolution or the dynamics of supernova explosions and that the usual corpse of a massive star is a neutron star, black holes requiring somewhat unusual conditions for their formation. Secondly, black holes in binary systems display some common, though not exclusive, properties such as rapid variability, spectral changes etc. Perhaps this behaviour, although improperly understood, can be isolated in AGN and used to infer the masses of the black holes. Thirdly, SS433 shows features in common with some extragalactic double radio sources. Have we looked hard enough for a true extragalactic counterpart to SS433?

Excepting chance discoveries, the prospects for observational advance in this field lie mostly with space astronomy. X-ray telescopes are crucial to finding more stellar mass black holes and to probing the most rapid variability in AGN. Further advances must await the launches of ROSAT and AXAF. In addition, it will be a great disappointment if Space Telescope does not revolutionise our view of AGN by probing the central stellar cusps, finding optical jets and telling us what types of galaxy are associated with the different forms of activity. In particular, understanding the 'environmental impact' of an active nucleus on the surrounding galaxy should help us to determine the duty cycle for this activity. Finally, Gamma Ray Observatory may discover the high energy photons that should be emitted if electron–positron pair plasmas are a prominent feature of AGN.

On the theoretical front, the major obstacle to progress, has been our ignorance of the details of the viscosity in accretion disks and of the dynamical importance of large scale magnetic field. The best prospects for progress here probably lie with large scale numerical computation. This will be carried out in the form of experiments on the behaviour of simple systems which should give further insight into the general properties of relativistic magnetohydrodynamical flows. There is little prospect of performing detailed simulations of sources to which observational data can be fitted. Similar explorations of radiative transfer in high temperature plasmas are underway and should be interpreted in the same spirit.

Although progress on the astrophysical theory of black holes has been disappointingly slow, it is some comfort to recall that it took a hundred years for the work of Euler, Lagrange and Laplace to reap the full benefit of the scientific revolution brought about by the publication of the *Principia* and over 150 years for its most spectacular observational success, the discovery of Neptune. Perhaps we should be more patient in our attempts to derive comparable results from the general theory of reliativity.

Acknowledgements

I thank the Chief, Division of Radiophysics, CSIRO, Sydney and the Acting Director Anglo-Australian Observatory for hospitality during the writing of this review. I also thank David Allen for helpful discussions on the Galactic centre and Roger Romani for guidance on SS433. I am grateful to Ineke Stacey for assistance with the preparation of the manuscript and Robyn Shobbrook for bibliographic help. Support under the National Science Foundation grant AST84-15355 is gratefully acknowledged.

References

Abell, G. O. and Margon, B. (1979). *Nature*, **279**, 701.
Abramowicz, M. A., Jaroszyński, M. and Sikora, M. (1978). *Astron. Astrophys.*, **63**, 221.

Alcock, C. R. (1986). *The Origin and Evolution of Neutron Stars*, ed. D. Helfand. Reidel: Dordrecht. (in press).

Arnett, W. D. and Bowers, R. L. (1977). *Astrophys. J. Suppl.*, **33**, 415.

Arons, J., Kulsrud, R. M. and Ostriker, J. P. (1975). *Astrophys. J.*, **198**, 687.

Avni, Y. and Bahcall, J. N. (1975). *Astrophys. J.*, **197**, 675.

Baade, W. and Zwicky, F. (1934). *Phys. Rev.*, **45**, 138.

Baade, W. and Minkowski, R. (1954). *Astrophys. J.*, **206**, 14.

Backer, D. and Sramek, R. (1987). In preparation.

Bahcall, J. N., Hut, P. and Tremaine, S. D. (1984). *Astrophys. J.*, **290**, 15.

Bahcall, J. N. and Ostriker, J. P. (1975). *Nature*, **256**, 23.

Bahcall, J. N. (1978). *Ann. Rev. Astron. Astrophys.*, **16**, 241.

Bahcall, J. N. and Soneira, R. M. (1980). *Astrophys. J. Suppl.*, **44**, 73.

Bahcall, J. N. (1984). *Astrophys. J.*, **287**, 926.

Balick, B. and Brown, R. L. (1974). *Astrophys. J.*, **194**, 265.

Band, D. L. and Grindlay, J. E. (1985). *Astrophys. J.*, **285**, 702.

Bardeen, J. M. (1970). *Nature*, **226**, 64.

Bardeen, J. (1973). *Black Holes*, ed. B. DeWitt and B. S. DeWitt. Gordon and Breach: New York.

Bardeen, J. and Wagoner, R. V. (1969). *Astrophys. J. Lett.*, **158**, L65.

Bardeen, J. M. and Petterson, J. A. (1975). *Astrophys. J. Lett.*, **195**, 65.

Barker, B. M. and O'Connell, R. F. (1975). *Phys. Rev.*, **D12**, 329.

Barnes, J., Goodman, J. and Hut, P. (1986). *Astrophys. J.*, **300**, 112.

Barr, R. and Mushotzky, R. F. (1986). *Nature*, **322**, 421.

Begelman, M. C. (1985). *Astrophysics of Active Galaxies and Quasi Stellar Objects*, ed. J. Miller. University Science Books: California.

Begelman, M. C., Blandford, R. D. and Rees, M. J. (1980). *Nature*, **287**, 307.

Begelman, M. C. and Rees, M. J. (1978). *Mon. Not. R. Astron. Soc.*, **188**, 847.

Begelman, M. C., Blandford, R. D. and Rees, M. J. (1984). *Rev. Mod. Phys.*, **56**, 255.

Bethe, H. A. and Brown, G. (1985). *Sci. Amer.*, **252**, 5, 40.

Bicknell, G. V. and Gingold, R. A. (1986). Preprint.

Bisnovatyi-Kogan, G. S. and Sunyaev, R. A. (1972). *Sov. Astron. A.J.*, **16**, 206.

Blandford, R. D. and Thorne, K. S. (1979). *General Relativity*, ed. S. W. Hawking and W. Israel. Cambridge University Press: Cambridge.

Blandford, R. D. and Rees, M. J. (1974). *Mon. Not. R. Astron. Soc.*, **169**, 395.

Blandford, R. D. and Znajek, R. L. (1977). *Mon. Not. R. Astron. Soc.*, **179**, 433.

Bolton, C. T. (1972). *Nature*, **235**, 271.

Boyle, B. J., Fong, R., Shanks, T. and Peterson, B. A. (1987). *Mon. Not. R. Astron. Soc.*, In press.

Braginsky, V. B., Caves, C. M. and Thorne, K. S. (1977). *Phys. Rev.*, **D15**, 2047.

Bregman, J., Butter, D., Kemper, B., Koski, A., Kraft R. P. and Stone, R. P. S. (1973). *Astrophys. J. Lett.*, **186**, L117.

Brown, R. L. and Liszt, H. S. (1984). *Ann. Rev. Astron. Astrophys.*, **22**, 223.

Burbidge, G. R. (1956). *Astrophys. J.*, **129**, 849.

Burbidge, G. R. (1959). *Paris Symposium on Radio Astronomy*, ed. R. N. Bracewell. Stanford: California.

Burbidge, G. R. (1964). *Proceedings Solvay Conference*.

Burbidge, E. M., Burbidge, G. R. and Prendergast, K. H. (1959). *Astrophys. J.*, **130**, 26.

Burbidge, E. M., Burbidge, G. R. and Sandage, A. (1964). *Rev. Mod. Phys.*, **35**, 97.

Cameron, A. G. W. (1971). *Nature*, **229**, 178.

Carr, B. J. and Rees, M. J. (1984). *Mon. Not. R. Astron. Soc.*, **206**, 315.

Carter, B. (1979). *General Relativity. An Einstein Centenary Survey*, ed. S. W. Hawking and W. Israel. Cambridge University Press: Cambridge.

Carter, B. and Luminet, J. P. (1982). *Nature*, **296**, 211.

Chandrasekhar, S. (1931). *Mon. Not. R. Astron. Soc.*, **91**, 456.

Chiu, H-Y. (1964). *Ann. Phys.*, **26**, 364.

Christodoulou, D. (1970). *Phys. Rev. Lett.*, **25**, 1596.

Clark, D. H. and Murdin, P. (1978). *Nature*, **276**, 45.

Cohen, M. H., Cannon, W., Purcell, G. H., Shaffer, D. B., Broderick, J. J., Kellermann, K. I. and Jauncey, D. L. (1971). *Astrophys. J.*, **170**, 207.

Cohen, M. H. (1986). *Highlights of Modern Astrophysics: Concepts and Controversies*, ed. S. Shapiro and S. Teukolsky. Wiley: New York.

Colpi, M., Maraschi, L. and Treves, P. (1984). *Astrophys. J.*, **280**, 319.

Colgate, S. A. (1967). *Astrophys. J.*, **150**, 163.

Collins, G. W. II (1985). *Mon. Not. R. Astron. Soc.*, **213**, 279.

Coroniti, F. V. (1985). *Unstable Current Systems in Astrophysics*, ed. Kundu and Holman. Reidel: Dordrecht.

Cowley, A. P., Crampton, D., Hutchings, J. B., Remillard, R. and Penfold, J. E. (1983). *Astrophys. J.*, **272**, 118.

Crampton, D. and Hutchings, J. B. (1981). *Astrophys. J.*, **251**, 604.

Crawford, M. K., Genzel, R., Harris, A. I., Jaffe, D. T., Lacy, J. H., Layten, J. B., Serabyn, E. and Townes, C. H. (1985). *Nature*, **315**, 467.

Damour, T. (1975). *Ann. N.Y. Acad Sci.*, **263**, 113.

Damour, T. (1978). *Phys. Rev.*, **D18**, 3598.

David, L. P., Durisen, R. H. and Cohn, H. (1986). *Astrophys. J.*, In press.

Davies, R. L. and Illingworth, G. (1986). *Astrophys. J.*, **302**, 234.

de Vaucouleurs, G. and Nieto, J. L. (1979). *Astrophys. J.*, **230**, 697.

Devinney, E. J. (1971). *Nature*, **233**, 110.

Dressler, A. (1980). *Astrophys. J. Lett.*, **240**, 41.

Dressler, A. (1984). *Astrophys. J. Lett.*, **286**, L97.

Duncan, M. and Shapiro, S. L. (1983). *Astrophys. J.*, **268**, 565.

Duncan, M. J. and Wheeler, J. C. (1980). *Astrophys. J. Lett.*, **237**, L27.

Dyson, J. (ed.) (1985). *Active Galactic Nuclei*, Manchester University Press: Manchester.

Eddington, A. S. (1935). *The Observatory*, **58**, 38.

Einstein, A. (1939). *Ann. Math. Princeton*, **40**, 922.

Fabian, A. C. and Rees, M. J. (1979). *Mon. Not. R. Astron. Soc.*, **187**, 13p.

Fabian, A. C., Eggleton, P. P., Hut, P. and Pringle, J. E. (1986). *Astrophys. J.*, **305**, 333.

Fath, E. A. (1908). *Lick, Obs. Bull.*, **5**, 71.

Feigelson, E. D., Bradt, M., McClintock, J., Remillard, R., Urry, C. M., Tapia, S., Geldzahler, B., Johnston, K., Romanishin, W., Wehinger, P. A., Wycoff, S., Madejski, G., Schwartz, D. A., Thorstensen, J. and Schaefer, B. E. (1986). *Astrophys. J.*, **302**, 337.

Filippenko, A. L. and Sargent, W. L. W. (1985). *Astrophys. J. Suppl.*, **57**, 3.

Frank, J., King, A. R. and Raine, D. J. (1985). *Accretion Power in Astrophysics*. Cambridge University Press: Cambridge.

Freese, K., Price, R. and Schramm, D. (1983). *Astrophys. J.*, **275**, 405.

Giacconi, R., Gursky, H., Paolini, F. R. and Rossi, B. B. (1962). *Phys. Rev. Lett.*, **9**, 439.

Gies, D. R. and Bolton, C. T. (1986). *Astrophys. J.*, **304**, 371.

Ginzburg, V. N. and Ozernoi, L. M. (1964). *Sov. Phys. JETP*, **47**, 1030.

Gold, T., Axford, W. I. and Ray, E. C. (1964). *Quasi-stellar Sources and Gravitational*

Collapse, ed. Robinson, I., Schild, A., Schucking, E. L., Chicago University Press: Chicago.

Giuricin, G., Mardirossian, F., Mezzetti, M. and Ramella, M. (eds.) (1986). *Structure and Evolution of Active Galactic Nuclei*. Reidel: Dordrecht.

Goldreich, P., Goodman, J. and Narayan, R. (1986). *Mon. Not. R. Astron. Soc.* In press.

Greenstein, J. and Schmidt, M. (1964). *Astrophys. J.*, **140**, 161.

Grindlay, J. (1985). *Dynamics of Star Clusters*, ed. J. Goodman and P. Hut. Reidel: Dordrecht.

Gunn, J. E. and Ostriker, J. P. (1970). *Astrophys. J.*, **157**, 1395.

Gurzadyan, V. G. and Ozernoi, L. M. (1981). *Astron. Astrophys.*, **95**, 39.

Hall, D. N. B., Kleinmann, S. G. and Scoville, N. Z. (1982). *Astrophys. J. Lett.*, **260**, L53.

Harrison, B. K., Thorne, K. S., Wakano, M. and Wheeler, J. A. (1965). *Gravitational Theory and Gravitational Collapse*. Chicago University Press: Chicago.

Hawking, S. W. (1965). *Phys. Rev. Lett.*, **15**, 689.

Hawley, J. (1986). Preprint.

Hazard, C., MacKey, M. B. and Shimmins, A. J. (1963). *Nature*, **197**, 1037.

Hewish, A., Bell, S. J., Pilkington, J. D. H., Scott, P. F. and Collins, R. A. (1968). *Nature*, **217**, 709.

Hoyle, F. and Fowler, W. A. (1963*a*). *Mon. Not. R. Astron. Soc.*, **125**, 169.

Hoyle, F. and Fowler, W. A. (1963*b*). *Nature*, **197**, 533.

Hutchings, J. B., Crampton, D. and Cowley, A. P. (1983). *Astrophys. J. Lett.*, **275**, L43.

Iben, I. (1964). *Quasi-stellar Sources and Gravitational Collapse*, ed. Robinson, I., Schild, A. and Schucking, E. L. Chicago University Press: Chicago.

Jennison, R. C. and Das Gupta, M. K. (1953). *Nature*, **172**, 96.

Katz, J. I., Anderson, S. F., Margon, B. and Grandis, S. A. (1982). *Astrophys. J.*, **260**, 780.

Keel, W. (1985). *Astrophysics of Active Galaxies and Quasi-stellar Objects*, ed. J. Miller. University Science Books: Mill Valley, California.

Kerr, R. P. (1963). *Phys. Rev. Lett.*, **11**, 237.

Koo, D. C. (1986). *Structure and Evolution of Active Galactic Nuclei*, eds. G. Giuricin, F. Mardirossion, M. Mezzetti and M. Ramella. Reidel: Dordrecht.

Kormendy, J. (1986). (Preprint).

Krolik, J. H. and London, R. A. (1983). *Astrophys. J.*, **267**, 371.

Kumar, S. and Pringle, J. E. (1985). *Mon. Not. R. Astron. Soc.*, **213**, 435.

Lacey, C. G. and Ostriker, J. P. (1975). *Nature*, **256**, 23.

Landau, L. (1932). *Phys. Z. Sowjetunion.*, **1**, 285.

Laplace, P. S. (1795). *Le System du Monde*, Vol. II. Paris.

Larson, R. B. (1984). *Mon. Not. R. Astron. Soc.*, **210**, 763.

Le Blanc, J. M. and Wilson, J. R. (1969). *Astrophys. J.*, **161**, 541.

Leibowicz, E. (1984). *Mon. Not. R. Astron. Soc.*, **210**, 279.

Lense, J. and Thirring, H. (1918). *Phys. Z.*, **19**, 156.

Lightman, A. P. and Shapiro, S. L. (1977). *Astrophys. J.*, **211**, 244.

Lin, D. N. C., Papaloizou, J. C. B. and Faulkner, J. (1985). *Mon. Not. R. Astron. Soc.*, **212**, 105.

Lingenfelter, R. E. and Ramaty, R. (1982). *The Galactic Centre*, ed. G. Riegler and R. Blandford. American Institute of Physics, New York.

Lo, K-Y., Backer, D. C., Ekers, R. D., Kellermann, K. I., Reid, M. and Moran, J. M. (1985). *Nature*, **315**, 124.

Lo, K-Y. (1986). *Science*, **233**, 1394.

Longair, M. S., Ryle, M. and Scheuer, P. A. G. (1973). *Mon. Not. R. Astron. Soc.*, **164**, 243.

Lucke, R., Yentis, D., Friedman, H., Fritz, G. and Shulman, S. (1976). *Astrophys. J. Lett.*, **206**, L25.

Lynden-Bell, D. (1969). *Nature*, **223**, 690.

Lynden-Bell, D. (1971). *Nuclei of Galaxies*, ed. D. J. K. O'Connell. North Holland: Amsterdam.

Lynden-Bell, D. and Rees, M. J. (1971). *Mon. Not. R. Astron. Soc.*, **152**, 461.

Lynden-Bell, D. (1978). *Phys. Scripta*, **17**, 185.

Lyne, A. G., Manchester, R. N. and Taylor, J. H. (1985). *Mon. Not. R. Astron. Soc.*, **213**, 613.

MacDonald, D. A. and Thorne, K. S. (1982). *Mon. Not. R. Astron. Soc.*, **198**, 345.

McClintock, J. E. and Remillard, R. A. (1986). *Astrophys. J.*, 308.

McClintock, J. E. (1986). Preprint.

McDowell, J. (1985). *Mon. Not. R. Astron. Soc.*, **217**, 77.

Maran, S. P. and Cameron, A. G. W. (1964). *Physics of Non-thermal Radio Sources*. NASA: Washington.

Margon, B., Ford, H. C., Katz, J. I., Kwitter, K. B., Ulrich, R. K., Stone, R. P. S. and Klemola, A. (1979). *Astrophys. J. Lett.*, **230**, L41.

Margon, B. (1980). *Sci. Amer.*, **243**, 4, 44.

Margon, B. (1984). *Ann. Rev. Astron. Astrophys.*, **22**, 507.

Margon, B., Bowyer, S. and Stone, R. S. (1973). *Astrophys. J. Lett.*, **185**, L113.

Marshall, F. E., Holt, S., Mushotzky, R. F. and Becker, R. M. (1983). *Astrophys. J. Lett.*, **269**, L311.

Matthews, T. A., Maltby, P. and Moffet, A. T. (1962). *Astrophys. J.*, **137**, 153.

Mazeh, T., von Paradis, J., van der Heuvel, E. P. J. and Savonije, G. J. (1986). *Astron. Astrophys.*, **157**, 113.

Michell, J. (1784). *Philos. Trans.*, **74**, 35.

Milgrom, M. (1979). *Astron. Astrophys.*, **76**, L3.

Miller, J. (ed.) (1985). *Astrophysics of Active Galaxies and Quasi-Stellar Objects*, University Science Books: California.

Misner, C. W., Thorne, K. S. and Wheeler, J. A. (1973). *Gravitation*. Freeman: San Francisco.

Moffet, A. T., Gubbay, J., Robertson, D. S. and Legg, A. J. (1971). *External Galaxies and Quasi-stellar Objects*, ed. D. S. Evans. Reidel: Dordrecht.

Narayan, R. (1986). *Quasars*, ed. G. Swarup and V. Kapahi. Reidel: Dordrecht.

Newsom, G. H. and Collins, G. W. II (1982). *Astrophys. J.*, **262**, 714.

Newton, A. J. and Binney, J. (1984). *Mon. Not. R. Astron. Soc.*, **210**, 711.

Norman, C. A. and Silk, J. (1983). *Astrophys. J.*, **266**, 502.

Novikov, I. D. and Thorne, K. S. (1973). *Black Holes*, ed. B. DeWitt and B. S. DeWitt. Gordon and Breach: New York.

Oppenheimer, J. R. and Volkoff, G. M. (1939). *Phys. Rev.*, **55**, 374.

Oppenheimer, J. R. and Snyder, H. (1939). *Phys. Rev.*, **56**, 455.

Ozernoi, L. M. and Chertoprid, V. E. (1969). *Sov. Astr. A. J.*, **46**, 940.

Pacholczyk, A. G. and Weymann, R. E. (1968). *Astron. J.*, **73**, 850.

Paczyński, B. (1974). *Astron. Astrophys.*, **34**, 161.

Paczyński, B. (1984). *Astrophys. J. Lett.*, **273**, L81.

Paczyński, B. (1986). *Nature*, **321**, 419.

Paczyński, B. and Wiita, P. J. (1980). *Astron. Astrophys.*, **88**, 23.

Papaloizou, J. C. B. and Pringle, J. E. (1984). *Mon. Not. R. Astron. Soc.*, **208**, 721.

Papaloizou, J. C. B. and Pringle, J. E. (1985). *Mon. Not. R. Astron. Soc.*, **213**, 799.
Pakull, M. W. and Angebault, L. P. (1986). *Nature*, **322**, 511.
Parmar, A. N., Brandiardi-Raymont, G., Pollard, G. S. G., Sanford, P. W., Fabian, A. C., Stewart, G. C., Schreier, E. J., Poliden, R. S., Oegerle, W. E. and Locke, M. (1980). *Mon. Not. R. Astron. Soc.*, **193**, 49p.
Penrose, R. (1965). *Phys. Rev. Lett.*, **14**, 57.
Penrose, R. (1969). *Rev. Nuovo. Cimento*, **1**, 252.
Phinney, E. S. (1983). Unpublished thesis, University of Cambridge.
Polidan, R. S., Pollar, G. S. G., Sanford, P. W. and Locke, M. C. (1978). *Nature*, **275**, 296.
Press, W. H. (1972). *Astrophys. J.*, **138**, 211.
Pringle, J. E. (1981). *Ann. Rev. Astron. Astrophys.*, **19**, 137.
Pudritz, R. E. (1983). *Mon. Not. R. Astron. Soc.*, **195**, 881.
Reber, G. (1944). *Astrophys. J.*, **100**, 279.
Rees, M. J. (1966). *Nature*, **211**, 468.
Rees, M. J. (1971). *Nature*, **229**, 312.
Rees, M. J. (1977). *Ann. NY Acad. Sci.*, **302**, 613.
Rees, M. J. (1978). *Nature*, **275**, 516.
Rees, M. J. (1982). *The Galactic Centre*, ed. G. Riegler and R. D. Blandford. American Institute of Physics: New York.
Rees, M. J. (1984). *Ann. Rev. Astron. Astrophys.*, **22**, 471.
Rhoades, C. E. and Ruffini, R. (1974). *Phys. Rev. Lett.*, **33**, 324.
Richstone, D. and Tremaine, S. D. (1985). *Astrophys. J.*, **296**, 370.
Robinson, I., Schild, A. and Schucking, E. (1964). *Quasi-stellar Sources and Gravitational Collapse*. Chicago University Press: Chicago.
Roos, N. (1981). *Astron. Astrophys.*, **104**, 218.
Sakimoto, P. J. and Coroniti, F. V. (1981). *Astrophys. J.*, **247**, 19.
Salpeter, E. E. (1964). *Astrophys. J.*, **140**, 796.
Salpeter, E. E. (1971). *Nature Phys. Sci.*, **233**, 5.
Salpeter, E. E. and Wagoner, R. V. (1971). *Astrophys. J.*, **164**, 557.
Sanders, R. H. (1970). *Astrophys. J.*, **162**, 784.
Sanders, R. H. and Allen, D. A. (1986). *Nature*, **319**, 191.
Sanford, P. W., Laskarides, P. and Salton, J. eds. (1982). *Galactic X-ray Sources*. Wiley: London.
Sargent, W. L. W., Young, P. J., Boksenberg, A., Shortridge, K., Lynds, C. R. and Hartwick, F. D. A. (1978). *Astrophys. J.*, **221**, 731.
Scheuer, P. A. G. (1974). *Mon. Not. R. Astron. Soc.*, **166**, 513.
Schmidt, M. (1963). *Nature*, **197**, 1040.
Schultz, A. L. and Price, R. H. (1985). *Astrophys. J.*, **291**, 1.
Schreier, E., Gursky, H., Kellogg, E., Tananbaum, H., Giacconi, R. (1971). *Astrophys. J. Lett.*, **170**, L21.
Seyfert, C. K. (1943). *Astrophys. J.*, **97**, 28.
Shakura, N. and Sunyaev, R. A. (1973). *Astron. Astrophys.*, **24**, 337.
Shapiro, S. L. and Teukolsky, S. A. (1983). *Physics of Compact Objects: Black Holes, White Dwarfs and Neutron Stars*. Wiley: New York.
Shapiro, S. L. and Teukolsky. S. A. (1985). *Astrophys. J. Lett.*, **292**, L41.
Shields, G. A., McKee, C. F., Lin, D. N. C. and Begelman, M. C. (1986). *Astrophys. J.*, **306**, 90.
Shklovsky, I. S. (1967). *Astrophys. J. Lett.*, **148**, L1.

Slipher, V. (1917). *Lowell Observatory Bulletin*, **3**, 59.

Smith, H. J. and Hoffleit, D. (1963). *Nature*, **198**, 650.

Soltan, A. (1982). *Mon. Not. R. Astron. Soc.*, **200**, 115.

Spitzer, L. and Saslaw, W. C. (1966). *Astrophys. J.*, **143**, 400.

Stella, L., White, N. E., Davelaar, J., Parmar, A. N., Blissett, R. J. and van de Klis, M. (1985). *Astrophys. J. Lett.*, **288**, L45.

Stoeger, W. R. and Pacholczyk, A. G. (1983). *Proc. 3rd Marcel Grossmann Meeting*, ed. Hu, N.

Sturrock, P. A. (1965). *Nature*, **205**, 861.

Sunyaev, R. A. (1973). *Sov. Astron.*, **16**, 941.

Svennson, R. (1986). *Radiation Hydrodynamics and Compact Objects*, ed. D. Mihalas and K-H. A. Winkler. Springer: Berlin.

Swarup, G. and Kapahi, V. (eds.). (1986). *Quasars*. Reidel: Dordrecht.

Tennant, A. F., Mushotzky, R. F., Boldt, E. A. and Swenk, J. H. (1981). *Astrophys. J.*, **251**, 15.

Tennant, A. F., Fabian, A. C. and Shafer, R. A. M. (1986). *Mon. Not. R. Astron. Soc.*, **219**, 871.

Thorne, K. S. (1966). In *Proc. Intern. School Phys. 'Enrico Fermi' Course 35, Varenna, Italy*, ed. L. Gratton.

Thorne, K. S., Price, R. H. and MacDonald, A. (1976). *Black Holes: The Membrane Paradigm*. Yale University Press: New Haven.

Trimble, V. L. and Thorne, K. S. (1969). *Astrophys. J.*, **156**, 1013.

Trimble, V. L., Rose, W. K. and Weber, J. (1973). *Mon. Not. R. Astron. Soc.*, **162**, 1p.

Tonry, J. (1984). *Astrophys. J. Lett.*, **283**, L27.

Turner, E. L. (1986). *Dark Matter in the Universe*, ed. G. R. Burbidge. Reidel: Dordrecht.

Wald, R. M. (1974). *Phys. Rev.*, **D10**, 1680.

Webster, B. L. and Murdin, P. (1972). *Nature*, **235**, 37.

Wheeler, J. A. (1966). *Ann. Rev. Astron. Astrophys.*, **4**, 393.

Wheeler, J. A. (1968). *American Scientist*, **56**, 1.

White, N. E. and Marshall, F. E. (1984). *Astrophys. J.*, **281**, 349.

Whitney, A. R., Shapiro, I. I., Rogers, A. E. E., Robertson, D. S., Knight, D. A., Clark, T. A., Goldstein, R. M., Marandino, G. E. and Vandenberg, N. R. (1971). *Science*, **173**, 225.

Wilson, R. E. (1972). *Astrophys. Sp. Sci.*, **19**, 165.

Wolff, S. C. and Beichmann, C. A. (1979). *Astrophys. J.*, **230**, 519.

Woltjer, L. (1959). *Astrophys. J.*, **130**, 38.

Woltjer, L. (1971). *Nuclei of Galaxies*, ed. D. J. K. O'Connell. North Holland: Amsterdam.

Young, P. H., Westphal, J. A., Kristian, J., Wilson, C. P. and Landauer, F. P. (1978). *Astrophys. J.*, **221**, 721.

Zeldovich, Ya B., (1964a). *Sov. Phys. Dokl.*, **9**, 195.

Zeldovich, Ya B. and Guseynov, O. H. (1965). *Astrophys. J.*, **144**, 841.

Zeldovich, Ya B. and Novikov, I. D. (1964b). *Sov. Phys. Dokl.*, **158**, 811.

Zeldovich, Ya B. and Novikov, I. D. (1965). *Sov. Phys. Usp.*, **7**, 763.

Zeldovich, Ya B. and Novikov, I. D. (1966). *Sov. Phys. Usp.*, **8**, 522.

Znajek, R. L. (1978). *Mon. Not. R. Astron. Soc.*, **185**, 833.

9

Gravitational radiation

KIP S. THORNE†

9.1 Introduction

9.1.1 The motivations for gravitational-wave research

The discovery of cosmic radio waves in the 1930s and their detailed study in the 40s, 50s, and 60s created a revolution in our view of the universe (Kellerman and Sheets, 1983; Sullivan, 1982, 1984). Previously the universe, as viewed by light, was regarded as serene and quiescent – dominated by stars and planets that wheel smoothly in their orbits, shining steadily and evolving (with few exceptions) on timescales of millions or billions of years. By contrast, the universe as viewed by radio waves was violent: galaxies in collision, jets ejected from galactic nuclei, quasars with luminosities far greater than our galaxy varying on timescales of hours, pulsars with gigantic radio beams rotating at several or many rotations per second; these were the typical strong radio emitters.

The radio revolution was so spectacular because the information carried by radio waves is so different from that carried by light. A factor 10^7 difference in wavelength meant the difference between photons predominantly thermal in origin (light) and photons predominantly nonthermal (radio), the difference between the bremsstrahlung and atomic transitions of stellar and planetary atmospheres on one hand, and the synchrotron radiation of intergalactic magnetized plasmas on the other.

As different as cosmic radio and optical radiations may be, their differences pale by comparison with those of electromagnetic waves and gravitational waves: cosmic gravitational waves should be emitted by, and carry detailed information about, coherent bulk motions of matter (e.g., collapsing stellar cores) or coherent vibrations of spacetime curvature (e.g.,

† Supported in part by the National Science Foundation (AST-8514911).

black holes). By contrast, cosmic electromagnetic waves are usually incoherent superpositions of emission from individual atoms, molecules, and charged particles. Gravitational waves are emitted most strongly in regions of spacetime where gravity is relativistic and where the velocities of bulk motion are near the speed of light. But electromagnetic waves come almost entirely from weak-gravity, low-velocity regions, since strong-gravity regions tend to be obscured by surrounding matter. Gravitational waves pass through surrounding matter with impunity, by contrast with electromagnetic waves which are easily absorbed and scattered, and even by contrast with neutrinos which, although they easily penetrate normal matter, should scatter thousands of times while leaving the core of a supernova.

These differences make it likely that, if cosmic gravitational waves can be detected and studied, they will create a revolution in our view of the universe comparable to or greater than that which resulted from the discovery of radio waves.

It might be argued that we are now so sophisticated and complete in our electromagnetically based (radio, millimeter, infrared, optical, ultraviolet, X-ray, gamma-ray, cosmic ray) understanding of the universe, compared to the optically based astronomers of the 1930s and 1940s, that a gravitational-wave revolution will be far less spectacular than was the radio revolution. It seems unlikely to me that we are so sophisticated. I am painfully aware of our lack of sophistication when I contemplate the sorry state of present estimates of the gravity waves bathing the earth (Section 9.4 below): for each type of gravity-wave source that has been studied, with the exception of binary stars and their coalescences, either (i) the strength of the source's waves for a given distance from earth is uncertain by several orders of magnitude; or (ii) the rate of occurrence of that type of source, and thus the distance to the nearest one, is uncertain by several orders of magnitude; or (iii) the very existence of the source is uncertain.

Although these uncertainties make us unhappy when we try to plan for the design and construction of gravitational-wave detectors, we will be rewarded with great surprises when gravity waves are ultimately detected and studied: the waves will give us extensive information about the universe that we are unlikely ever to obtain in any other way.

Detailed studies of cosmic gravitational waves are also likely to yield experimental tests of fundamental laws of physics which cannot be tested in any other way.

The first discovery of gravitational waves would directly verify the predictions of general relativity, and other relativistic theories of gravity, that such waves should exist. (There has already been an indirect verification, in the form of the observed inspiral of the binary pulsar due to gravitational–radiation reaction; Weisberg and Taylor, 1984; Taylor, 1987.)

By comparing the arrival times of the first bursts of light and gravitational waves from a distant supernova, one could verify general relativity's prediction that electromagnetic and gravitational waves propagate with the same speed – i.e., that they couple to the static gravity (spacetime curvature) of our Galaxy and other galaxies in the same way. For a supernova in the Virgo cluster of galaxies (15 Mpc distant), first detected optically one day after the light curve starts to rise, the electromagnetic and gravitational speeds could be checked to be the same to within a fractional accuracy $(1 \text{ light day})/(15 \text{ Mpc}) = 5 \times 10^{-11}$.

By measuring the polarization properties of the gravitational waves, one could verify general relativity's prediction that the waves are transverse and traceless – and thus are the classical consequences of spin-two gravitons (Eardley, Lee and Lightman, 1973; Eardley *et al.*, 1973).

By comparing the detailed wave forms of observed gravitational wave bursts with those predicted for the coalescence of black-hole binaries (which will be computed by numerical relativity in the next few years, see Section 9.3.3(e) below), one could verify that certain bursts are indeed produced by black-hole coalescences – and, as a consequence, verify unequivocally the existence of black holes and general relativity's predictions of their behavior in highly dynamical circumstances. Such verifications would constitute by far the strongest test ever of Einstein's laws of gravity.

9.1.2 A brief history of gravitational-wave research

Einstein (1916) laid the foundations of gravitational-wave theory within months after his final formulation of general relativity – restricting himself to weak (linearized) waves emitted by bodies with negligible self-gravity and propagating through flat, empty spacetime. During the next few years Einstein (1918), Weyl (1922), and Eddington (1924) elaborated on Einstein's initial work so that by the mid-1920s the linearized theory of gravitational waves was fully understood. However, it was clear – at least to Eddington (1924) – that for sources with significant self-gravity (e.g. binary systems), the linearized analysis was invalid.

Landau and Lifshitz (1941) gave the first fairly satisfactory treatment of the emission of waves by self-gravitating systems; but a series of failed attempts to analyze radiation reaction in such systems in the late 1940s and 1950s (pp. 73 and 74 of Damour, 1983) shook physicists' faith in the ability of the waves to carry off energy, and even in the correctness of the Landau–Lifshitz formula for the emitted wave field. It required a clever thought experiment by Bondi (1957) to restore faith in the energy of the waves, and a series of beautiful and rigorous studies of the asymptotic properties of the waves at infinity by Bondi and collaborators (Bondi, 1960; Bondi, van der Burg, and Metzner, 1962; Sachs, 1962, 1963; Penrose, 1963*a,b*) and of the propagation of short-wavelength waves through a curved background spacetime by Isaacson (1968*a,b*) to restore faith that the fundamental theory of gravitational waves is soundly based.

The experimental search for cosmic gravitational waves was initiated by Joseph Weber (1960) at a time when almost nothing was known about possible cosmic sources and when nobody else had the vision to see that there were technological possibilities of ultimate success. After a decade of effort, Weber (1969) announced to the world tentative evidence that his resonant-bar gravity-wave detectors – one near Washington, DC, the other near Chicago – were being excited simultaneously by gravitational waves. There followed a six-year period of excitement and feverish effort as 15 other research groups around the world tried to construct and operate similar bar detectors (Tyson and Giffard, 1978; Amaldi and Pizella, 1979; de Sabbata and Weber, 1977; Weber, 1986, and references therein). Sadly, even with markedly improved sensitivities, these efforts gave no convincing evidence that gravity waves were actually being seen.

In parallel with this experimental effort, astrophysicists worldwide struggled through the early 1970s to milk, from electromagnetic observations of the universe and from fundamental theory, as much information as possible about the characteristics of the gravity waves that might be bathing the earth. By the mid-1970s a fuzzy but helpful picture had begun to emerge: While the sensitivities of the detectors to kilohertz-frequency bursts arriving, say, three times per year had improved during the early 70s from dimensionless amplitude $h_{3/yr} \sim 1 \times 10^{-15}$ to $h_{3/yr} \sim 3 \times 10^{-16}$ (a factor 10 improvement in energy flux), it seemed highly unlikely that such bursts bathing the earth would exceed $h_{3/yr} \sim 1 \times 10^{-16}$; a reasonable probability of success would require $h_{3/yr} \sim 10^{-20}$ or better; and a high probability would require $h_{3/yr} \sim 10^{-21}$ to $h_{3/yr} \sim 10^{-22}$ (Smarr, ed., 1979; Fig. 9.4 below). Although these estimates were discouraging, the theoretical efforts that produced them were making clear the enormous potential payoff that could follow the successful detection of gravity waves.

Fortunately, the experimental efforts of the early 1970s had pointed the way toward major possible detector improvements; and, consequently, although most of the first-generation experimental groups became discouraged and dropped out, a handful of highly talented groups continued onward into the 1980s with a second-generation effort involving major technological changes such as cooling the bars to liquid-helium temperatures, changing bar materials, switching from passive to active transducers, and even developing completely new types of detectors, most notably laser-interferometer gravity-wave detectors (called 'beam' detectors in this chapter). These second-generation efforts have reached fruition in the last few years: bars with kilohertz burst sensitivities $h_{3/yr} \sim 10^{-17}$ (30 times higher in amplitude than the first-generation and 1000 higher in energy) are now collecting data in coordinated searches (Section 9.5.2(d) below); and small-scale beam detectors with $h_{3/yr} \sim 5 \times 10^{-17}$ are now operating (Section 9.5.3(d) below) as prototypes for full-scale detectors with projected ultimate sensitivities in the 10^{-22} region (Section 9.5.3(g) below). The regime of possible success has been reached, and the regimes of reasonably probable success and highly probable success look reachable – though only with vigorous continuing efforts and the expenditure of non-trivial sums of money.

In parallel with these 1980s' second-generation efforts, theorists have redoubled their struggle to firm up our understanding of the waves bathing the earth, but with only modest results: the problem of knowing what kinds of sources actually occur, and how frequently, is hampered by the paucity of electromagnetic information; and, as a result, the apparent recent improvements in our knowledge (Section 9.4 below) might be little more than changes of fashion. On the other hand, given a specific scenario for how a postulated source behaves, theorists have become far more adept than before – thanks not least to supercomputers – at computing the details of the gravitational waves it should emit (Section 9.3.3 below). As a consequence, when waves are ultimately detected, the prospects have become reasonable for deciphering from them the details of their sources.

While the present, 1987, gravity-wave searches might bring success, it is not likely they will. Thus, we must anticipate a continued vigorous effort at technology development during the coming years, with the prospects of success improving significantly at each step along the way. In parallel, we must anticipate a continuing major effort by relativity theorists to refine their ability to decipher the source behaviors corresponding to postulated gravitational-wave forms, and a continuing effort by astrophysicists to give better guidance as to what kinds of sources actually exist and in what numbers. The efforts must be great; but jointly they are likely to give an extremely valuable payoff.

9.1.3 Overview of this chapter

This chapter reviews all aspects of gravitational-wave research – experimental, theoretical relativity, and theoretical astrophysics. This review is intended to be readable by physicists (including advanced students) who are not specialists in general relativity, in experimental gravity, or in astrophysics.

A word of warning: in preparing this review I have *not* done a thorough literature search (I lacked the necessary energy!); nor have I cited all major original references of which I am aware (that would have made the reference list even longer than it is!). However, I have attempted to present all significant ideas and issues with which I am familiar, citing wherever possible the earliest occurrence of the idea or issue and one or more recent, thorough discussions of it.

This review is divided into four major parts: the physical and mathematical description of gravitational waves (Section 9.2), the generation and propagation of gravitational waves (Section 9.3), astrophysical sources of gravitational waves (Section 9.4), and the detection of gravitational waves (Section 9.5).

The physical and mathematical description of gravitational waves (Section 9.2) is presented in a form that compactifies and updates the corresponding material in the textbook that I coauthored fourteen years ago (Misner, Thorne, and Wheeler, 1973; cited henceforth as MTW). Emphasis focuses on the 'shortwave approximation' as a tool for defining waves mathematically (Section 9.2.1), and on measurements in the proper reference frame of an observer as a tool for defining waves physically (Section 9.2.2). A special 'TT coordinate system' is then introduced (Section 9.2.3) for use in analyzing systems large compared to a wavelength of the waves; and the energy, momentum, and quantization of gravitational waves are discussed (Section 9.2.4).

The theory of the generation and propagation of gravitational waves (Section 9.3) is far more sophisticated today than when MTW was written: we now understand how, in realistic situations, to split the problem of wave generation off from that of wave propagation, and how to mesh generation and propagation together using the technique of 'matched asymptotic expansions' (Section 9.3.1). We also understand more deeply the most elementary of all ways to compute wave generation, the 'quadrupole formalism', and its relationship to radiation reaction in the emitting system (Section 9.3.2). Unfortunately, the most strongly emitting systems should violate the mathematical approximations that underlie the quadrupole formalism and thus can be analyzed only in rough order of magnitude using it; for more accurate analyses one must use a more sophisticated wave-generation formalism. A catalog of more sophisticated formalisms is given in Section 9.3.3 – which is an updated version of a review I wrote ten years ago (Thorne, 1977). The theory of the propagation of the waves from source to earth is sketched in Section 9.3.4, and various wave-propagation effects (absorption, scattering, dispersion, tails, gravitational focusing, diffraction, parametric amplification by background curvature, non-linear coupling of waves to themselves, and generation of background curvature by the waves' energy and momentum) are described in Section 9.3.5 – which with 9.3.4 is a shortened version of my recent (Thorne, 1983), long review of wave propagation. Section 9.3 concludes with very brief descriptions of elegant mathematical work on idealized wave-propagation situations: the asymptotic structure of waves propagating toward 'future null infinity' in an asymptotically flat spacetime (Section 9.3.6), and exact, analytic solutions to the Einstein equations for wave generation and propagation (Section 9.3.7).

It is nearly a decade since a detailed and thorough review has been written of astrophysical sources and detectors for gravitational waves (Smarr, ed., 1979; Douglass and Braginsky, 1979); and in the intervening time these topics have changed enormously. (For recent reviews of a number of subtopics see the chapters in Deruelle and Piran, 1983.) Sections 9.4 and 9.5 attempt a comprehensive review in a manner that closely ties the sources to the detection efforts. In these sections the sources and the detection strategies are split up into three categories: those for gravitational-wave 'bursts' (Section 9.4.1 and Fig. 9.4); those for periodic gravitational waves (Section 9.4.2 and Fig. 9.6); and those for a stochastic gravitational-wave background (Section 9.4.3 and Fig. 9.7). All previous reviews have been cavalier about factors of 2 in the definitions of wave strengths and detector sensitivities. This review tries to standardize the definitions, including factors of 2. The standardization is based on signal-to-noise-ratio analyses that are given at the beginnings of Sections 9.4.1, 9.4.2, and 9.4.3. For each source that looks favorable for wave detection, Section 9.4 gives a description of our current state of knowledge of the source and gives current estimates of the wave strengths and other wave characteristics.

The burst sources treated in Section 9.4.1 include supernovae (collapse of a normal stellar core to form a neutron star, subsection c), the collapse of a star or star cluster to form a black hole (subsection d), the inspiral and coalescence of compact binaries (neutron stars and black holes, subsection e), and the fall of stars and small holes into supermassive holes (subsection f). Because our knowledge of sources is so poor, it is useful to estimate how strong the strongest wave bursts bathing the earth could be without violating our cherished beliefs about the laws of physics and the nature of the universe; this is done in subsection g. The periodic sources treated in Section 9.4.2 include rotating neutron stars (rigidly rotating pulsars, and neutron stars spun up by accretion until they encounter a radiation-reaction-driven instability, subsection b), and binary stars (including unevolved binaries, WUMa stars, white-dwarf binaries, and neutron-star binaries, subsection c). The stochastic sources in Section 9.4.3 include large numbers of binary stars whose waves superpose stochastically (subsection b), pre-galactic, Population III stars (subsection c), the big-bang singularity in which the universe began – with subsequent parametric amplification of its waves by background curvature in inflationary and other scenarios (subsection d), phase transitions in the subsequent but still early universe (subsection e), and cosmic strings produced by phase transitions (subsection f). Present estimates of the strengths of the waves from all these sources are shown in Figs. 9.4 (burst), 9.6 (periodic), and 9.7 (stochastic) along with the sensitivities of present and proposed detectors.

The detectors in Section 9.5 are divided into those that operate in the high-frequency regime, $f \gtrsim 10$ Hz (Sections 9.5.2, 9.5.3, and 9.5.4), those for low frequencies, 10 Hz $\gtrsim f \gtrsim 10^{-5}$ Hz (Section 9.5.5), and those for very low frequencies, $f \lesssim 10^{-5}$ Hz (Section 9.5.6). The high-frequency detectors are all earth-based; but because of seismic and gravity-gradient noise, the low- and very-low-frequency detectors must be space-based. Sections 9.5.2 and 9.5.3 describe in great detail the earth-based, high-frequency bar and beam detectors which have been under development for many years and show great promise for the future. Section 9.5.4 describes briefly other types of earth-based, high-frequency detectors. Section 9.5.5 describes low-frequency detectors including doppler tracking of spacecraft (subsection a), beam detectors in space which hold great promise for the turn of the century (subsection b), the normal modes of the earth and sun (subsections c and d), the vibrations of blocks of the earth's crust (subsection e), and the earth-orbiting skyhook (subsection f). Section 9.5.6 describes very-low-frequency detectors including the timing of pulsars (neutron-star rotations), which recently has placed interesting observational limits on a stochastic background (subsection a), the timing of orbital motions of binary and planetary systems (subsection b), and astronomical observations of anisotropies in the temperature of the cosmic microwave radiation (subsection c), deviations from the Hubble flow (subsection d), and products of primordial nucleosynthesis (subsection d).

9.1.4 Notation and conventions

Throughout this chapter, unless otherwise stated, we shall assume that general relativity correctly describes classical gravitational waves. Our notation will be that of Misner, Thorne, and Wheeler (1973) – cited as MTW throughout – including, e.g., the use of Greek indices for spacetime (running from 0 to 3) and Latin for space (running from 1 to 3), the use of commas for partial derivatives and semicolons for covariant derivatives, the use of the Einstein summation convention, and the use of geometrized units in which Newton's gravitation constant G and the speed of light c are set equal to unity. Sometimes, particularly when discussing gravitational-wave detectors, we shall restore the Gs and cs to the equations and use cgs units.

9.2 The physical and mathematical description of a gravitational wave

9.2.1 Shortwave approximation

General relativistic gravitational waves are ripples in the curvature of spacetime that propagate with the speed of light. Because gravity is non-linear, it is not possible in a fully precise manner to separate the contributions of gravitational waves to the curvature from the contributions of the earth, the sun, the galaxy, or anything else; and since such a separation underlies the very concept of a gravitational wave, this means that gravitational waves are not precisely defined entities.

On the other hand, in realistic astrophysical situations the lengthscale λ on which the waves vary (their reduced wavelength, $\lambda = \lambda/2\pi$) is very short compared to the lengthscales $\mathscr{L}$ on which all other important curvatures vary; and this difference in lengthscale makes possible a high-accuracy, but approximate, split of the Riemann curvature tensor $R_{\alpha\beta\gamma\delta}$ into a 'background curvature' $R^{B}_{\alpha\beta\gamma\delta}$ plus a contribution $R^{GW}_{\alpha\beta\gamma\delta}$ due to gravitational waves: the background $R^{B}_{\alpha\beta\gamma\delta}$ is the average of $R_{\alpha\beta\gamma\delta}$ over several wavelengths

$$R^{B}_{\alpha\beta\gamma\delta} \equiv \langle R_{\alpha\beta\gamma\delta} \rangle; \tag{1a}$$

and the waves' curvature $R^{GW}_{\alpha\beta\gamma\delta}$ is the rapidly varying difference

$$R^{GW}_{\alpha\beta\gamma\delta} \equiv R_{\alpha\beta\gamma\delta} - R^{B}_{\alpha\beta\gamma\delta}. \tag{1b}$$

Heuristically, $R^B_{\alpha\beta\gamma\delta}$ is like the large-scale (10 cm) curvature of an orange, while $R^{GW}_{\alpha\beta\gamma\delta}$ is like the fine-scale granulation of the skin of the orange (lengthscale a few millimeters).

This method of defining a gravitational wave, introduced into general relativity by Wheeler (1955) and Power and Wheeler (1957), is a special case of a standard technique in mathematical physics variously called 'shortwave approximation' or 'two-timing' or 'two-lengthscale expansion' or 'two-variable expansion' (see e.g. Chapter 3 of Cole, 1968); and it is intimately connected to the 'WKB approximation'. There is an elegant shortwave-approximation formalism for gravitational-wave theory due largely to Brill and Hartle (1964) and to Isaacson (1968a,b). Not surprisingly the formalism reveals that general relativistic gravitational waves propagate through vacuum in essentially the same manner as light – with the same speed, with the same changes of amplitude due to curvature of the wavefronts, with the same diffraction effects when focussed by a gravitational lens, etc. Because that formalism has been reviewed extensively elsewhere (e.g. Thorne, 1983), I shall not delve into it here, except for a brief description in Section 9.3.4 and an enumeration and brief discussion of some of its predictions in Section 9.3.5 below.

9.2.2 Measurements in the proper reference frame of an observer

In general relativity the Riemann curvature tensor is defined, operationally, by the relative accelerations it produces between adjacent particles (e.g. the 'equation of geodesic deviation', Sections 8.7 and 11.3 of MTW). Correspondingly, a gravitational wave can be defined operationally in the following way:

Consider an observer (freely falling or accelerated, it doesn't matter so long as the acceleration is slowly varying). Let the observer carry with herself a small Cartesian latticework of measuring rods and synchronized clocks (a 'proper reference frame' in the sense of Section 13.6 of MTW, with spatial coordinates x^j that measure proper distance along orthogonal axes). She is to measure the 'force of gravity' F_j that acts on a particle of mass m, momentarily at rest at location x^j. For example, she might let the particle fall freely, measure its acceleration g_j in her proper reference frame, and multiply it by the particle's mass m to get the force $F_j = mg_j$. Alternatively, she might measure the force required to hold the particle fixed in the coordinate grid of her proper reference frame and equate F_j to minus that force. Of course, this is not different in any way from what she would do were she a Newtonian physicist rather than a relativistic physicist. The difference lies in how she analyzes and thinks about the force of gravity F_j. As a relativist, she recognizes (cf. Box 37.1 of MTW) that F_j is made up of a nearly steady, nearly position-independent component caused by her own failure to fall freely, plus a component proportional to the particle's Cartesian coordinate position x^j (relative acceleration of particle and origin of coordinates) which is caused by spacetime curvature $R_{\alpha\beta\gamma\delta}$. This latter contribution,

$$F_j = -mR_{j0k0}x^k \tag{2}$$

(where the index 0 denotes a component along her time basis vector), she splits up into a piece that changes slowly in time (background curvature contribution) plus a piece that is rapidly varying. She makes certain that there are no rapidly moving or rapidly changing nearby sources of gravity to account for the rapid variations; if there are none, then she can attribute the rapidly varying component of the force to gravitational waves

$$F_j^{\mathrm{GW}} = -mR_{j0k0}^{\mathrm{GW}}x^k. \tag{3}$$

It is conventional to use, as the primary entity for describing a gravitational wave, not the Riemann curvature tensor $R_{\alpha\beta\gamma\delta}^{\mathrm{GW}}$ which has dimensions $1/\mathrm{time}^2$ (or $1/\mathrm{length}^2$), but rather a dimensionless 'gravitational-wave field' h_{jk}^{TT}. In terms of the force-producing components of the Riemann tensor R_{j0k0}^{GW} and proper time t as measured by our observer, h_{jk}^{TT} is defined by (cf. Section 2.3 of Thorne 1983)

$$\frac{\partial^2 h_{jk}^{\mathrm{TT}}}{\partial t^2} \equiv -2R_{j0k0}^{\mathrm{GW}}. \tag{4}$$

The convenience of this gravitational-wave field lies in its simple relationship to displacements produced by the waves: if, for simplicity, the observer is freely falling and keeps the axes of her coordinate grid tied to gyroscopes and is in a region of spacetime where gravitational waves are the only source of spacetime curvature, then the waves will produce tiny oscillatory changes δx^j in the position of a test particle relative to the origin of her coordinate grid; and these changes will satisfy the equation of motion ('equation of geodesic deviation')

$$m\frac{\mathrm{d}^2\,\delta x^j}{\mathrm{d}t^2} = F_j^{\mathrm{GW}} = -mR_{j0k0}^{\mathrm{GW}}x^k = \frac{1}{2}\,m\,\frac{\partial^2 h_{jk}^{\mathrm{TT}}}{\partial t^2}\,x^k. \tag{5}$$

Because any realistic wave is so weak that the oscillatory changes δx^j are miniscule compared to the distance of the particle from the origin, x^k can be regarded as essentially constant on the right-hand side, and equation (5) can then be integrated easily to give

$$\delta x^j = \frac{1}{2} h_{jk}^{TT} x^k.$$ (6)

Thus, aside from a factor $\frac{1}{2}$, h_{jk}^{TT} plays the role of the 'dimensionless strain of space', or the 'time-integrated shear of space': it is the ratio of the wave-induced displacement of a free particle relative to the origin, to its orginal displacement from the origin. (In the above equations and below it does not matter whether a spatial index is up or down, since the spatial coordinates are Cartesian.)

The superscript TT on the gravitational-wave field is to remind us that, according to general relativity, the field is 'transverse and traceless'. More specifically: the Einstein field equations guarantee that $R_{\alpha\beta\gamma\delta}^{GW}$ and hence also h_{jk}^{TT} propagate with the speed of light. Since the sources of cosmic gravity waves are very far away, the waves look very nearly planar as they pass through the observer's proper reference frame. If we orient the x, y, z spatial axes so the waves propagate in the z direction, then the 'transversality' of the waves means that the only non-zero components of the wave field are h_{xx}^{TT}, $h_{xy}^{TT} = h_{yx}^{TT}$, and h_{yy}^{TT}; and the 'trace-free' property means that $h_{xx}^{TT} = -h_{yy}^{TT}$. Thus, the gravitational waves, like electromagnetic waves, have only two independent components – two polarization states.

Because of their TT nature, gravitational waves produce a quadrupolar, divergence-free force field (equation (5) and Fig. 9.1). This force field has two components corresponding to the two polarization states of the waves: the

Fig. 9.1. Lines of force for gravitational waves (equations (5) and (7)): (*a*) with '+' polarization, $m\,\delta\ddot{x} = \frac{1}{2}\ddot{h}_+ x$ and $m\,\delta\ddot{y} = -\frac{1}{2}\ddot{h}_+ y$; and (*b*) with '×' polarization, $m\,\delta\ddot{x} = \frac{1}{2}\ddot{h}_\times y$ and $m\,\delta\ddot{y} = \frac{1}{2}\ddot{h}_\times x$ – where dots denote time derivatives.

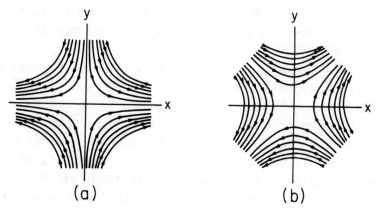

(a) (b)

quantity

$$h_+ \equiv h_{xx}^{TT} = -h_{yy}^{TT} \tag{7a}$$

produces a force field with the orientation of a '+' sign, while

$$h_\times \equiv h_{xy}^{TT} = h_{yx}^{TT} \tag{7b}$$

produces one with the orientation of a '×' sign. Thus, h_+ and $h_\times$ are called the 'plus' and 'cross' (or + and ×) gravity-wave amplitudes. From these amplitudes and the polarization tensors $e_{xx}^+ = -e_{yy}^+ = 1$, $e_{xy}^\times = e_{yx}^\times = 1$ (all other components zero), one can reconstruct the full wave field

$$h_{jk}^{TT} = h_+ e_{jk}^+ + h_\times e_{jk}^\times. \tag{7c}$$

It is straightforward to show that, if one rotates the x and y axes in the transverse plane through an angle $\Delta\psi$, the gravity-wave amplitudes are changed to

$$h_+^{new} = h_+^{old} \cos 2\Delta\psi + h_\times^{old} \sin 2\Delta\psi,$$
$$h_\times^{new} = -h_+^{old} \sin 2\Delta\psi + h_\times^{old} \cos 2\Delta\psi. \tag{7d}$$

The quadrupolar symmetry of the force field, together with the tenets of canonical field theory, tells us that general relativistic gravitational waves must be associated with quanta of spin two ('gravitons'). The spin is always the ratio of 360 degrees to the angle, about the propagation direction, through which one must rotate an instantaneous field to make it return to its original state – the 'return angle'. For electromagnetic waves the return angle is 360 degrees and the spin is one; for general relativistic gravitational waves the return angle is 180 degrees and the spin is two. In other relativistic theories of gravity the wave field has other symmetries and therefore other spins for its quanta – and in most relativistic theories, by contrast with general relativity, the symmetries are not even frame-invariant (local Lorentz-invariant); and, as a result, the waves of those theories cannot be incorporated into canonical field theory and cannot be quantized by canonical techniques. For details see Eardley, Lee and Lightman (1973), and Eardley *et al.* (1973).

The above definition of the gravitational-wave field h_{jk}^{TT} relies on a specific choice of reference frame. It turns out that, if one pursues this definition in two different reference frames related by a boost in some arbitrary direction, and if one chooses the spatial axes of the two frames so they are unrotated relative to each other, then the instantaneous wave fields h_{jk}^{TT} (and correspondingly h_+ and $h_\times$) will be the same. Stated more precisely – but in a language I shall not explain – the general-relativistic gravitational-wave field h_{jk}^{TT} has 'boost-weight zero' (it is a scalar under boosts); but, as discussed above, it has 'spin-weight two' (it behaves like a spin-two field under rotations). For further details see Section 2.3.2 of Thorne (1983), and for the mathematics that goes along with the concepts of 'spin-weight' and 'boost-weight', see Geroch, Held and Penrose (1973).

9.2.3 TT coordinate system

The above description of the force law and force fields produced by a gravitational wave is just the leading order in a power series expansion in distance $r = (\delta_{jk} x^j x^k)^{\frac{1}{2}}$ from the origin of the observer's proper reference frame. The higher-order fractional corrections to the forces are of order $(r/\lambda)^2$ and, correspondingly, the proper reference frame's coordinates x^j fail to measure proper distance by fractional amounts of order $h_{jk}^{TT}(r/\lambda)^2$ (see, e.g. Zhang, 1986, for discussion in the case of a freely falling observer). Thus, the above description of gravity-wave forces is accurate only if the region of interest is spatially small compared to a reduced wavelength. When the region is large, an alternative description is needed.

The nicest alternative description makes use of a 'TT coordinate system' (Section 35.4 of MTW), i.e. a coordinate system which is nearly Minkowski throughout the spacetime region of interest, and in which the contribution of the waves to the deviations from the Minkowski metric is embodied in the same h_{jk}^{TT} as we introduced above.

Because the TT coordinates must be nearly Minkowski, they cannot cover a spacetime region that is too large: their extent in both time and space must be far smaller than the background radius of curvature $\mathscr{R}_B \sim |R_{\alpha\beta\gamma\delta}^B|^{-\frac{1}{2}}$. In typical situations, this limitation is very mild compared to the limitation on the size of an observer's proper reference frame: one can stretch TT coordinates over any region small compared to the Hubble distance (cutting out holes in the vicinities of black holes and neutron stars), but if the waves have a frequency of a kilohertz, then any proper reference frame used to study their forces must be small compared to $\lambda \sim 50$ km.

In a TT coordinate system the spacetime metric coefficients take the form

$$g_{\alpha\beta} = \eta_{\alpha\beta} + h_{\alpha\beta}^B + h_{\alpha\beta}^{TT}, \tag{8}$$

where $\eta_{\alpha\beta}$ are the Minkowski metric coefficients (diagonal -1, $+1$, $+1$, $+1$), $h_{\alpha\beta}^B$ is the background metric perturbation which varies on a long lengthscale $\mathscr{L}$, and $h_{\alpha\beta}^{TT}$ is the gravity-wave metric perturbation which varies on the short lengthscale λ. The time-time and space-time components of the gravity-wave perturbation h_{00}^{TT} and $h_{0j}^{TT} = h_{j0}^{TT}$ vanish, and the space-space components h_{jk}^{TT} are the same as the gravity-wave field that would be computed from the Riemann tensor (equation (4)) in the proper reference frame of any observer who is nearly at rest in the TT coordinate system. For proofs and discussions, see e.g. Sections 35.4, 37.1, and 37.2 of MTW.

Whereas physics in a proper reference frame can be formulated in the Newtonian language of three-dimensional forces, including gravitational forces, physics in a TT coordinate system must be formulated in the relativistic language of geodesic motion and vanishing divergence of the stress-energy tensor; cf. Section 9.5.1 below.

9.2.4 Energy, momentum, and quantization of gravitational waves

In the 1940s and early 1950s a controversy raged over whether or not gravitational waves can carry energy. The controversy was ultimately resolved by Herman Bondi (1957) using a simple thought experiment: place several beads on a rough stick and let a gravitational wave pass. The above description of the forces produced by the waves (which was first understood fully by Bondi and by Felix Pirani 1956 and their colleagues) guarantees that – if the stick is not *too* rough – the wave will push the beads back and forth on the stick, heating it. Surely, if the wave can heat a stick, it must carry energy.

It was not until the late 1960s that a fully satisfactory mathematical description of the energy in a gravity wave was devised: Richard Isaacson (1968*a*,*b*), using the shortwave approximation which he had developed in detail on the basis of cruder earlier work of Wheeler (1955), Power and Wheeler (1957) and Brill and Hartle (1964), introduced a stress-energy tensor $T_{\alpha\beta}^{GW}$ for gravitational waves. This stress-energy tensor, like the background curvature $R_{\alpha\beta\gamma\delta}^{B}$, is smooth on the lengthscale λ; it is obtained, in fact, by averaging the squared gradient of the wave field over several wavelengths:

$$T_{\alpha\beta}^{GW} = \frac{1}{32\pi} \sum_{i,j} \langle h_{ij,\alpha}^{TT} h_{ij,\beta}^{TT} \rangle, \tag{9}$$

where $\langle \cdots \rangle$ means 'average over several wavelengths'. (For a pedagogical derivation and discussion see Sections 35.7–35.15 of MTW; for a beautiful rederivation by the method of averaged Lagrangians see MacCallum and Taub, 1973.) If the waves are propagating in the z direction, this stress-energy tensor takes the standard form for a bundle of zero-rest-mass particles (gravitons) moving at the speed of light in the z direction:

$$T_{00}^{GW} = -T_{0z}^{GW} = -T_{z0}^{GW} = T_{zz}^{GW} = \frac{1}{16\pi} \langle (\partial h_+/\partial t)^2 + (\partial h_\times/\partial t)^2 \rangle. \tag{10}$$

In order of magnitude, restoring the factors of G and c, the energy flux in the waves if they have frequency $f = c/2\pi\lambda$ is

$$-T_{0z}^{\text{GW}} \simeq \frac{\pi}{4}\frac{c^3}{G} f^2 \langle h_+^2 + h_\times^2 \rangle = 320 \frac{\text{erg}}{\text{cm}^2\,\text{s}} \left(\frac{f}{1\,\text{kHz}}\right)^2 \left\langle \frac{h_+^2 + h_\times^2}{(10^{-21})^2} \right\rangle. \tag{11}$$

The numbers in this equation correspond to a strongly emitting supernova in the Virgo cluster of galaxies, where there are several supernovae per year. Contrast this huge gravity-wave energy flux with the peak electromagnetic flux at the height of the supernova, $\sim 10^{-9}\,\text{erg cm}^{-2}\,\text{s}^{-1}$; but note that the gravity waves should last for only a few milliseconds, while the strong electromagnetic output lasts for days.

Corresponding to the huge energy flux (11) in an astrophysically interesting gravitational wave is a huge occupation number for the quantum states of the gravitational-wave field: it is not hard to show that for the above supernova burst only a handful of quantum states are occupied; and they each contain $n \sim 10^{75}$ gravitons (equations (6)–(8) of Thorne *et al.*, 1979). This means that the waves behave exceedingly classically; quantum-mechanical corrections to the classical theory have fractional magnitude $1/\sqrt{n} \sim 10^{-37}$. (Although the full quantization of the gravitational field is exceedingly difficult and not yet fully under control, the quantization of weak gravitational waves propagating through a smooth background spacetime – equivalent to weak waves in flat spacetime – has been well understood for decades; see, e.g., the most elementary aspects of Feynman, 1963; Dewitt, 1967a,b.)

Isaacson's stress-energy tensor (9) for gravitational waves has the same properties and plays the same role as the stress-energy tensor for any other field or form of matter in the background spacetime. For example, $T_{\alpha\beta}^{\text{GW}}$ generates background curvature through the Einstein field equations (averaged over several wavelengths of the waves); also $T_{\alpha\beta}^{\text{GW}}$ has vanishing divergence (conservation of gravity-wave energy and momentum) in spacetime regions where the waves are not being generated, absorbed, or scattered. For full details see Isaacson (1968b) or Section 35.15 of MTW.

9.3 The generation and propagation of gravitational waves

9.3.1 Wave propagation split off from wave generation

Turn, now, to the generation of gravitational waves and their propagation from their source to the earth. Mathematically, the wave-generation problem and the wave-propagation problem are each difficult – though for very different reasons, so that to handle the difficulties requires two very different sets of mathematical tools and approximations. For ease of analysis, then, it is important to split the propagation problem off from the generation problem and treat them separately. This is accomplished by dividing the space around the source into three regions (Section III of Thorne, 1980*b*): a 'wave-generation region' at distances from the source $r \lesssim r_1 =$ ('Inner radius'); a 'local wave zone' at distances $r_1 \lesssim r \lesssim r_0 =$ ('Outer radius'); and a 'distant wave zone' at distances $r \gtrsim r_0$. The theory of wave generation is developed with one set of mathematical tools in the wave-generation region and the local wave zone, i.e. at distances $r \lesssim r_0$; the theory of propagation to earth is developed with the other set of tools in the local wave zone and the distant wave zone, i.e. at distances $r \gtrsim r_1$; and the two theories are matched together in their domain of overlap, the local wave zone $r_1 \lesssim r \lesssim r_0$.

The inner radius r_1 is far enough out to be in the wave zone, $r_1 \gg \lambda$; far enough to be in a region where the source's gravity is weak, $r_1 \gg 2M \equiv$ (Schwarzschild radius of source) $= 2 \times$ (mass of source); and far enough to be outside the source, $r_1 \gg L \equiv$ (size of source). The outer radius r_0 is far enough beyond the inner radius to leave many wavelengths in the local wave zone, $r_0 - r_1 \gg \lambda$; but not so far that the gravitational redshift can produce a significant net phase shift during propagation through the local wave zone, $\delta\phi = (M/\lambda)\ln(r_0/r_1) \ll 1$; and not so far that the background curvature of the external universe can significantly affect the propagation, $r_0 - r_1 \ll \mathcal{R}_B \equiv |R^B_{\alpha\beta\gamma\delta}|^{-\frac{1}{2}} \equiv$ (radius of curvature of background spacetime). These choices of r_1 and r_0 permit one to ignore, in the local wave zone $r_1 \lesssim r \lesssim r_0$, the background curvature both of the source and of the external universe; i.e. they permit one to regard the waves in the local wave zone as propagating through flat spacetime. This greatly simplifies calculations.

For a given, astrophysically interesting source the wave-generation task consists of computing with reasonable accuracy the dynamical behavior of the source's gravitational field in the wave-generation region $r \lesssim r_1$, and further computing how that dynamical curvature develops into outward propagating waves in the local wave zone, $r_1 \lesssim r \lesssim r_0$. Once this is done, the theorist can switch tasks and mathematical formalisms from wave generation to wave propagation. The wave-propagation task takes as input the propagating waves in the local wave zone and carries them outward (typically using the shortwave approximation and formalism) through the universe from source to earth.

As an addendum to the wave-generation task, one often computes the back-reaction effects of the wave emission on the source, i.e. the 'radiation reaction'.

9.3.2 The quadrupole formalism for wave generation and radiation reaction

Of all the techniques for computing wave generation, one, the 'quadrupole formalism', is especially important because it is highly accurate for many sources and is accurate in order of magnitude for most. The quadrupole formalism was derived originally by Einstein (1916, 1918) for sources with negligible self-gravity and slow internal motions. Later, in a series of steps by Landau and Lifshitz (1941), Fock (1959), Ipser (1971) and Thorne (1980b, Sections VII and XII), it became clear that the quadrupole formalism requires for high accuracy no constraints whatsoever on the strength of the source's internal gravity; all that is required is slow motion – more specifically, that the source's size L be small compared to the reduced wavelength λ of the waves it emits. In other words, the quadrupole formalism is to gravity-wave generation what the 'poor-antenna approximation' (dipole formalism) is to radio-wave generation. Moreover, as in the radio-wave problem, the quadrupole formalism typically is accurate to within factors of order 2 even for sources with sizes of order a reduced wavelength λ (cf. equation (4.8) of Thorne, 1980b); and since very few astrophysical systems are larger than a reduced wavelength of the waves they emit, this justifies the use of the quadrupole formalism for order-of-magnitude astrophysical estimates in almost all situations.

The quadrupole formalism writes the gravitational-wave field in the source's local wave zone (where the background curvature can be ignored) in the following simple form:

$$h_{jk}^{TT} = \frac{2}{r} \frac{\partial^2}{\partial t^2} \left[\mathscr{I}_{jk}(t-r) \right]^{TT}. \tag{12}$$

Here r is the distance to the source's center, t is proper time as measured by an observer at rest with respect to the source, $t-r$ is retarded time, the superscript TT means 'algebraically project out and keep only the part that is transverse to the (radial) direction of propagation and is traceless' (Box 35.1 of MTW), and $\mathscr{I}_{jk}(t-r)$ is the source's mass quadrupole moment evaluated at the retarded time $t-r$.

The meaning of 'mass quadrupole moment' is well known when the source has weak internal gravity and small internal stresses, so Newtonian gravity is a good approximation to general relativity inside and near the source. Then $\mathscr{I}_{jk}$ is the symmetric, trace-free (STF) part of the second moment of the source's mass density ρ, as computed in a Cartesian coordinate system centered on the source:

$$\mathscr{I}_{jk}(t)=\left[\int\rho(t)x^jx^k\,d^3x\right]^{STF}=\int\rho(t)\left[x^jx^k-\frac{1}{3}r^2\delta_{jk}\right]d^3x; \qquad (13a)$$

equivalently, it is the coefficient of the $1/r^3$ part of the source's Newtonian gravitational potential

$$\Phi=-\frac{M}{r}-\frac{3}{2}\frac{\mathscr{I}_{jk}(t)x^jx^k}{r^5}-\frac{5}{2}\frac{\mathscr{I}_{ijk}x^ix^jx^k}{r^7}+\cdots. \qquad (13b)$$

If the source has strong internal gravity, one can no longer express its mass quadrupole moment as the simple integral (13a). However, so long as the source has slow internal motions ($L \ll \lambda$), there will be a region of space far enough from the source to be in vacuum ($r > L$) and far enough for gravity to be weak ($r \gg 2M$), yet near enough for retardation and wave behavior to be unimportant ($r \ll \lambda$). In this 'weak-field, vacuum, near-zone' gravity can be described with high accuracy as Newtonian; and the Newtonian potential is $\Phi \cong -\frac{1}{2}(g_{00}+1)$, where g_{00} is the time-time part of the metric in a coordinate system that is as Minkowski as possible throughout the weak-field, vacuum near zone. The mass quadrupole moment can then be read off this Newtonian potential using the standard formula (13b). For further discussion see the review in Section 3 of Thorne (1983) or the original treatment in Part Two of Thorne (1980b).

For order-of-magnitude calculations of h_{jk}^{TT} (equation (12)), one can approximate the TT part of the second time derivative of the mass quadrupole moment by that portion of the source's internal kinetic energy which is associated with non-spherical motions, E_{kin}^{ns}; and, unless the source is at very large cosmological redshifts $z \gg 1$, one can propagate the waves to earth as though the intervening spacetime were flat – with the result that the local wave-zone formula for the waves, equation (12), is valid also at earth. The result is the simple order-of-magnitude formula

$$h \sim \frac{E_{kin}^{ns}}{r} \qquad (14)$$

for the magnitude h of the gravity-wave field h_{jk}^{TT} at earth.

From the 'exact' quadrupole formula (12) for the wave field in the local wave zone and Isaacson's formula (9) for the stress-energy tensor of the waves, one can compute the fluxes of energy and angular momentum carried by the waves. By integrating those fluxes over a sphere surrounding the source in the local wave zone, one obtains for the rates of emission of energy (Einstein, 1916, 1918) and angular momentum (Peters, 1964)

$$\frac{\mathrm{d}E^{\mathrm{GW}}}{\mathrm{d}t} = \frac{1}{5} \sum_{j,k} \left\langle \left(\frac{\mathrm{d}^3 \mathcal{J}_{jk}}{\mathrm{d}t^3} \right)^2 \right\rangle, \tag{15}$$

$$\frac{\mathrm{d}J_i^{\mathrm{GW}}}{\mathrm{d}t} = \frac{2}{5} \sum_{j,k,l} \varepsilon_{ijk} \left\langle \frac{\partial^2 \mathcal{J}_{jl}}{\partial t^2} \frac{\partial^3 \mathcal{J}_{lk}}{\partial t^3} \right\rangle. \tag{16}$$

The linear momentum carried off by the waves vanishes when one computes it by the quadrupole formalism; but when one includes higher-order corrections to the field emitted by slow-motion sources, one finds for the rate of emission of linear momentum (first derived by Papapetrou, 1962, 1971; for a more modern derivation in the notation of this chapter see Section IV.C of Thorne, 1980*b*)

$$\frac{\mathrm{d}P_i^{\mathrm{GW}}}{\mathrm{d}t} = \frac{2}{63} \sum_{j,k} \left\langle \frac{\partial^3 \mathcal{J}_{jk}}{\partial t^3} \frac{\partial^4 \mathcal{J}_{jki}}{\partial t^4} \right\rangle + \frac{16}{45} \sum_{j,k,a} \varepsilon_{ijk} \left\langle \frac{\partial^3 \mathcal{J}_{ja}}{\partial t^3} \frac{\partial^3 \mathcal{S}_{ka}}{\partial t^3} \right\rangle. \tag{17}$$

Here $\mathcal{J}_{ijk}$ is the source's 'mass octupole moment' and $\mathcal{S}_{ij}$ is its 'current quadrupole moment' (gravitational analog of magnetic quadrupole moment). For sources with weak internal gravity and stresses (nearly Newtonian sources), these moments are computable from the simple volume integrals

$$\mathcal{J}_{ijk} = (\int \rho x^i x^j x^k \, \mathrm{d}^3 x)^{\mathrm{STF}}, \tag{18a}$$

$$\mathcal{S}_{ij} = (\int \rho \varepsilon_{ipq} x^p v^q x^j \, \mathrm{d}^3 x)^{\mathrm{STF}}, \tag{18b}$$

where, as in equation (15a), STF means 'make it symmetric and trace-free', i.e. 'symmetrize on all free indices and remove the traces on all pairs of free indices'. Independently of the strengths of the internal gravity and stresses, the moments can be read off the Newtonian potential $\Phi \cong -\frac{1}{2}(g_{00}+1)$ (equation (13b)) and off the 'gravitomagnetic potential' $\beta_i \equiv g_{0i}$ in the source's weak-field near zone:

$$\beta_i = -2\varepsilon_{ipq} \frac{J^p x^q}{r^3} - 4\varepsilon_{ipq} \frac{\mathcal{S}_{pa} x^q x^a}{r^5} - \cdots. \tag{19}$$

In equation (19) J^p, the moment in the leading, dipolar term, is the source's angular momentum. For further details, discussions, and derivations see Thorne (1983) or Thorne (1980*b*). For a discussion of the gravitomagnetic potential see, e.g., Chapter 3 of Thorne, Price and Macdonald (1986).

The laws of conservation of energy, angular momentum, and linear momentum imply that radiation reaction should deplete the source's energy, angular momentum, and linear momentum at just the right rates to compensate for the losses (15), (16), and (17); and a detailed analysis of the radiation reaction forces reveals that this is so (Peres, 1960). Particularly convenient in analyzing the radiation reaction in a source with weak self-gravity is a Newtonian-type radiation-reaction potential (Burke, 1969; Thorne, 1969; Chandrasekhar and Esposito, 1970; Section 36.8 of MTW).

Different physicists feel comfortable with different levels of rigor. In recent years these differences have shown up strongly and publicly in a controversy over derivations of the quadrupole wave-generation formula (12) and the formula for the energy sapped from a source by radiation reaction (negative of equation (15)). Many physicists – myself among them – were quite satisfied with derivations at the level of rigor, e.g., of Landau and Lifshitz (1941) and Peres (1960). Others (e.g. Ehlers *et al.*, 1976) felt that those early derivations were inadequately rigorous and, correspondingly, that the quadrupole formulae were suspect for sources with non-negligible self-gravity. The controversy was heightened by the fact that there were mathematical errors in some (but not all) of the early derivations (see Walker and Will, 1980 and Section 3.4.2 of Thorne, 1983 for discussions).

The controversy was still raging in the early 1980s; see, e.g. Ashtekar (1983). However, during the mid 1980s it has largely subsided. There are now many new derivations of the quadrupole formulae, with much improved rigor, and they all produce the same, standard results; see, e.g., Anderson *et al.* (1982), Blanchet and Damour (1984); Christodoulou and Schmidt (1979); Isaacson, Welling and Winicour (1984); and for reviews see Will (1986), Schutz (1986a) and Damour (1987).

Of particular interest is radiation reaction in the binary pulsar PSR 1913 + 16, which should cause the binary system's two neutron stars to spiral together slowly with a consequent gradual decrease in their orbital period. Because the duration of observations of the pulsar is short (12 years), the cumulative effects of radiation reaction on the orbit during those observations are 100 times smaller than post-Newtonian effects; and, consequently, the detailed effects of the radiation reaction could not be fully understood until the orbital equations were fully under control up through post-post-Newtonian order. Damour and Deruelle (1986) have now brought the orbital equations fully under control; and there is now a beautiful agreement between those equations – including the quadrupolar radiation reaction – and the observational data. For a detailed discussion see Chapter 6 of this book.

9.3.3 *A catalog of formalisms for computing wave generation*

The order-of-magnitude formula $h \sim E_{kin}^{ns}/r$ shows that of all sources at fixed distance r, the strongest emitters will be those with the largest non-spherical kinetic energies, i.e. those with the largest internal masses and with internal velocities approaching the speed of light. Thus, the strongest emitters are likely to violate the slow-motion assumption which underlies the quadrupole formalism, and will require for accurate analysis either higher-order corrections to the quadrupole formalism, or wave-generation formalisms that do not entail any slow-motion assumption.

There are a number of other wave-generation formalisms which can be applied to such sources. This section is a catalog of them, with references to detailed presentations and applications. For an out-of-date but unified presentation of most of these formalisms see Thorne (1977).

To be tractable with a minimum of numerical computation, a wave-generation formalism must break the extreme non-linearity of the Einstein field equations by imposing a power-series expansion in some small quantity and keeping only the lowest, linear order or the lowest few orders. Wave-generation formalisms can be classified according to their choice of the small expansion parameter. *Slow-motion formalisms* (of which the quadrupole formalism is an example) expand in $L/\lambda =$ (size of source)/(reduced wavelength of waves); subsection (a) below. *Post-Minkowski formalisms* expand in the strength of the gravitational field inside the source, i.e. in the magnitude of the deviations of the metric coefficients from their Minkowski values; subsection (b). *Post-Newtonian formalisms* expand simultaneously in L/λ and the strength of the internal gravitational field; subsection (c). *Perturbation formalisms* expand in the deviations of the metric from its form for some non-radiative, astrophysical system – e.g. from the Kerr metric for a rotating black hole, or from the metric for an equilibrium, rotating, relativistic stellar model, or from the Friedman–Robertson–Walker metric for a homogeneous, isotropic 'big-bang'; subsection (d).

Of all astrophysical sources, the very strongest emitters will entail gravitationally induced large-amplitude, high-velocity, non-spherical internal motions – e.g. the inspiral and coalescence of a binary black hole or binary neutron-star system. For such sources there is no small parameter in which one can expand. The only way to compute the full details of the wave field emitted by such sources is by the techniques of *numerical relativity*: the numerical solution of the full Einstein field equations on a supercomputer; subsection (e).

(a) *Slow-motion formalisms*

As for electromagnetic waves, so also for gravitational waves, slow-motion expansions give rise automatically to multipolar expansions: the electromagnetic vector potential is a sum of an electric dipolar term (typical magnitude $(Q/r)(L/\lambda)$ where Q is the charge of the source), plus an electric quadrupole and a magnetic dipole [magnitude $(Q/r)(L/\lambda)^2$], plus an electric octupole and a magnetic quadrupole [magnitude $(Q/r)(L/\lambda)^3$], etc. Similarly, the gravitational-wave field h_{jk}^{TT} is the sum of a mass-quadrupole term $[\sim(\partial^2/\partial t^2)\mathscr{I}_{jk}/r \sim (M/r)(L/\lambda)^2]$, plus a mass octupole $[\sim(\partial^3/\partial t^3)\mathscr{I}_{ijk}/r \sim (M/r)(L/\lambda)^3]$ and a current quadrupole $[\sim(\partial^2/\partial t^2)\mathscr{S}_{ij}/r \sim (M/r)(L/\lambda)^3]$, etc. In each case, electromagnetic and gravitational, there are two families of moments involved; and in each case as one goes to higher orders in L/λ one is driven to include higher-order moments of the source.

Early foundations for these expansions in the gravitational case will be found in Bonnor (1959) and Pirani (1964); full mathematical details will be found in Thorne (1980b); and major improvements and elucidations will be found in Blanchet and Damour (1986), Blanchet (1987a) and Damour (1987).

A good example which shows how, as the source's internal motions are speeded up, the higher moments gradually become more and more important, is the gravitational bremsstrahlung radiation emitted when two stars fly past each other with some high initial velocity; see Kovacs and Thorne (1978) and Turner and Will (1978) for full details.

In some slow-motion systems the mass quadrupole contribution may be suppressed, leaving the current quadrupole or the mass octupole to dominate. A good example is the torsional oscillation of a neutron star, in which the motions are slow because the shear modulus is weak, and the mass moments vanish because of parity considerations leaving the current quadrupole to dominate the radiation; for details see Schumaker and Thorne (1983). Another example is the Chandrasekhar (1970)–Friedman–Schutz (1978) (CFS) instability in neutron stars, which preferentially excites mass octupole ($l=3$) or hexadecapole ($l=4$) or $l=5$ modes of pulsation causing them to radiate more strongly than quadrupole; see Section 4.3(b) below.

(b) *Post-Minkowski formalisms*

Post-Minkowski wave-generation formalisms are sometimes called 'post-linear' because they entail expanding in the strength of the gravitational field beyond linear order; and they are sometimes called 'fast-motion' to contrast them with slow-motion formalisms.

There is a systematic way to take a post-Minkowski wave-generation formalism that is accurate to a given order in the strength of the source's internal gravity, and iterate it to obtain a formalism of higher accuracy (Thorne and Kovacs, 1975; Thorne, 1977).

The wave-generation formalism that is accurate to first post-Minkowski order (first order in the strength of internal gravity) is 'Linearized theory', i.e. the linear approximation to general relativity. Linearized theory is discussed in most textbooks, e.g. Chapter 18 and Sections 35.1–35.6 of MTW. Halpern and Desbrandes (1969) and, independently, Press (1977) have derived a particularly useful Linearized wave-generation formula for systems with sizes L large compared to a reduced wavelength λ. Examples of gravity-wave generation that have been analyzed by Linearized theory are: (i) the coherent (but painfully slow) transformation of electromagnetic waves into gravitational waves (first considered by Gertsenshtein, 1962, subsequent work reviewed in Section 4.1 of Grishchuk and Polnarev, 1980); and (ii) the waves emitted by the explosion of a non-spherical nuclear bomb (Wheeler, 1962; Wood *et al.*, 1970).

Linearized theory is completely ignorant of the source's internal gravity; it can correctly predict the emitted waves only if the source's motions are governed by non-gravitational forces – typically by electric or magnetic forces. For systems with significant but weak internal gravity (e.g. stellar pulsations, binary systems, and high-speed stellar encounters), one must use a wave-generation formalism accurate to the next, 'post-post-Minkowski' or 'post-Linear' order. For the details of such a formalism, see, e.g., Thorne and Kovacs (1975), Crowley and Thorne (1977); and for its application to high-speed stellar encounters (gravitational bremsstrahlung radiation) see Kovacs and Thorne (1977, 1978). For a recent review see Westpfahl (1985).

Thus far nobody has developed a post^2-Linear wave-generation formalism in detail – i.e. a formalism accurate to post^3-Minkowski order. There has been no great need for such a formalism, and the post-Linear formalism is sufficiently hard to work with in practice (cf. Kovacs and Thorne, 1977) that it is not clear whether post^2-Linear would be significantly easier than full-blown numerical relativity.

(c) *Post-Newtonian formalisms*

Post-Newtonian approximations to general relativity make the assumption – in accord with the virial theorem for gravitationally bound systems – that inside the source the deviations of the metric from Minkowski (i.e. the dimensionless strength of the source's gravity) have a magnitude ε of order $(L/\lambda)^2$; and accordingly they expand the Einstein field equations simultaneously in ε (post-Minkowski expansion) and L/λ (slow-motion expansion). See, e.g., Burke (1979) for a review of the method.

The lowest-order wave generation formalism that results from this expansion is called 'Newtonian' because it computes the evolution of the source using Newton's laws of gravity and mechanics, then evaluates the source's time-evolving quadrupole moment using the standard Newtonian volume integral (13a), then inserts that quadrupole moment into the standard quadrupole wave-generation formula (12). It is this Newtonian version of the quadrupole formalism that has been especially controversial (see the end of Section 3.2 above) but is now almost universally agreed to be highly reliable.

There have been a large number of important wave-generation calculations with this formalism. Some examples are: (i) the waves emitted by binary systems in Newtonian, elliptical orbits (Peters and Mathews, 1963); (ii) the waves emitted by a variety of models of stars that collapse to form neutron stars (Saenz and Shapiro, 1978, 1981); and (iii) the waves emitted in the head-on collision of two compact stars (Gilden and Shapiro, 1984).

The Newtonian wave-generation formalism starts losing accuracy when the source's internal gravity becomes too strong ($\varepsilon \sim 0.05$) and its internal velocities too high ($v \sim 0.2$) (Turner and Will, 1978) – e.g. in the late stages of the spiraling together of a neutron-star binary system. In such a situation it is useful to include the next higher-order corrections (one order higher in the strength of gravity ε, two higher in the speed L/λ). The result is the post-Newtonian formalism (Epstein and Wagoner, 1975; Wagoner, 1977; Tsvetkov, 1984). Examples of calculations that have been performed with the post-Newtonian formalism are the radiation from a system of bodies whose sizes are all small compared to their separations (Wagoner and Will, 1976), gravitational bremsstrahlung at moderate velocities (Turner and Will, 1978), and the radiation emitted by a slowly rotating star that collapses to a neutron star (Turner and Wagoner, 1979).

The foundations for a post²-Newtonian wave-generation formalism have also been worked out (Section V.E. of Thorne, 1980b); but it has never been developed in full detail or applied to any sources. Such a formalism – by contrast with the post²-Linear – would likely be far more tractable than full-blown numerical relativity; so it may one day prove useful in tying down the gravitational waves from large-amplitude processes involving neutron stars.

(d) *Perturbation formalisms*

Much can be learned about gravitational waves from neutron stars and black holes by studying weak, non-radial perturbations around their equilibrium structures. The theory of non-radial perturbations of non-rotating relativistic stars was developed by Campolattaro, Detweiler, Ipser, Price and Thorne (see pp. 195–201 of Thorne 1978 for a review and references); and extensions of the theory have been given by Schumaker and Thorne (1983), Detweiler and Lindblom (1985) and Finn (1986). Among its recent applications are computations by Lindblom and Detweiler (1983) of the normal modes of neutron stars with a variety of equations of state. For rotating stars the corresponding theory is due largely to Chandrasekhar, Friedman and Schutz (reviewed through 1977 on pp. 201–8 of Thorne 1978 and reviewed more recently by Schutz 1987). The most important recent applications are studies of the waves emitted by the 'Chandrasekhar–Friedman–Schutz instability' in rapidly rotating neutron stars; see Section 9.4.2(b) below.

For non-rotating (Schwarzschild) black holes the perturbation theory is due to Regge and Wheeler (1957) with major subsequent contributions by Bardeen, Chandrasekhar, Detweiler, Edelstein, Moncrief, Press, Vishveshwara and Zerilli; see pp. 180–8 of Thorne (1978) for a review. Recent applications include a definitive numerical evaluation of the eigenfrequencies and damping times of a Schwarzschild hole's quasinormal modes (Leaver, 1985, 1986a), and of the Green's function for arbitrary perturbations of a Schwarzschild hole (Leaver, 1986b). Finally, for rotating (Kerr) black holes the theory is due to Teukolsky (1972, 1973), Teukolsky and Press (1974), Wald (1973), Chrzanowski (1975), Cohen and Kegeles (1975), Chandrasekhar and Detweiler (1976), Detweiler (1977), Chandrasekhar (1983) and Sasaki and Nakamura (1982); and recent applications include the evaluation of the eigenfrequencies and damping times of a Kerr hole's quasinormal modes (Detweiler, 1980; Leaver, 1985, 1986a), and evaluations of the gravitational waves emitted when a compact body orbits a Kerr hole (Detweiler, 1978), scatters gravitationally off a Kerr hole (Kojima and Nakamura, 1984b), or plunges into a Kerr hole (Detweiler and Szedenits, 1979; Kojima and Nakamura, 1984a).

Primordial gravitational waves (waves created in or near the big-bang) cannot be analyzed by splitting space into a wave-generation region plus wave zones (Section 9.3.1), because the transition from wave generation to wave propagation is a temporal rather than a spatial one: the transition occurs when the size of the cosmological horizon expands to become much larger than a wavelength, thereby unfreezing a set of frozen-in initial perturbations. The only way, today, to analyze primordial waves is by a perturbation formalism somewhat akin to that used for stars and black holes: the unperturbed configuration is typically a non-radiative, Friedman–Robertson–Walker cosmological model, and the perturbations are studied by linearizing the Einstein equations – or a quantized variant of them – around that model. The resulting theory is due to Lifshitz (1946); see also Section 7.3 of Zel'dovich and Novikov (1983), and references cited in Section 9.4.3(d) below.

(e) *Numerical relativity*

Numerical solution of the full Einstein equations is the only way, today, to study wave generation in the strongest and most interesting of all gravity-wave sources: those with high internal velocities, strong internal gravity, and large deviations from a non-radiating spacetime. During the past decade several dozen researchers have worked vigorously to develop the field of numerical relativity. The results of this effort are: an analytic formulation of the initial value problem and the dynamical evolution problem for the Einstein field equations, in a form that facilitates numerical solutions (York, 1983, and references cited therein); for axisymmetric systems, viable ways to slice spacetime and choose the spatial coordinates so as to avoid pathologies and to compute the emitted gravitational waves efficiently (Smarr and York, 1978; Piran, 1983; Bardeen and Piran, 1983; Sasaki, 1984; Stewart and Friedrich, 1983; Isaacson, Welling and Winicour, 1983; Gomez *et al.*, 1986; Anderson and Hobill, 1986; Evans and Abrahams, 1987); and several good computer codes for evolving axisymmetric systems and computing their waves (Nakamura, 1983; Stark and Piran, 1986; Piran and Stark, 1986; Evans, 1986).

Among the problems that have been studied successfully with the codes thus far are the gravitational radiation produced by a head-on collision of two Schwarzschild black holes (Smarr, 1977a), by the collapse of a rotating star to form a Kerr black hole (Stark and Piran, 1986), and by the vibrations of a neutron star (Evans, 1986).

The experts in numerical relativity are now beginning to move on from axisymmetric systems, which have two non-trivial spatial dimensions and one non-trivial time, to asymmetric, generic systems with three non-trivial spatial dimensions and one time (e.g. Nakamura, 1987). This full '3 + 1' effort will require using the world's largest supercomputers, and will require new techniques for slicing spacetime into space plus time, for choosing the spatial coordinates, and for differencing the Einstein equations. The effort may absorb almost as many person-years as the development of gravitational-wave detectors; but it will be well worthwhile: the payoffs will include the ability to compute in detail the waveforms from the strongest gravity-wave sources in the universe, such as the spiraling together and coalescence of two black holes – waveforms that will be crucial to the interpretation of gravity-wave observations and to their use for strong-field, highly dynamical tests of general relativity.

One should not be misled into believing that numerical relativity will be the totally dominant tool for realistic gravitational-wave calculations in the coming years. On the contrary, we can expect a healthy interaction between numerical and analytical techniques; for discussion see Schutz (1986c).

9.3.4 Wave propagation in the real universe

Since I have recently written a detailed review of the theory of gravitational wave propagation (Section 2 of Thorne, 1983), I shall only sketch the main points briefly.

No matter how strong a source of gravity waves may be, once its waves are fully formed (once they have reached a location, e.g. in the local wave zone, where the inhomogeneity lengthscale $\mathscr{L}$ of the background curvature is large compared to their wavelength $\lambdabar$), they will have dimensionless amplitudes h small compared to unity. This can be seen as follows:

The stress-energy tensor of the gravitational waves, $T_{\alpha\beta}^{GW}$ (equation (9)), acts as a source for the background curvature through the Einstein equations $G_{\alpha\beta}^{B} = 8\pi(T_{\alpha\beta}^{GW} + T_{\alpha\beta}^{other})$. Since the background Einstein tensor $G_{\alpha\beta}^{B}$ has magnitude less than or of order $1/\mathscr{R}_{B}^{2}$ (where $\mathscr{R}_{B}$ is the background radius of curvature as defined from the Riemann tensor), and since the magnitude of the gravity-wave stress-energy tensor (9) is $h^{2}/\lambdabar^{2}$, the Einstein equations imply that

$$h \lesssim \lambdabar/\mathscr{R}_{B}. \tag{20}$$

Since the inhomogeneity lengthscale $\mathscr{L}$ of the background curvature is always less than or of order the radius of curvature $\mathscr{R}_{B}$, equation (20) implies the claimed result:

$$h \lesssim \lambdabar/\mathscr{R}_{B} \lesssim \lambdabar/\mathscr{L} \ll 1. \tag{21}$$

This permits one to study the subsequent propagation of the waves, once they are fully formed, using a linearized approximation to the Einstein field equations (Sections 35.13 and 35.14 of MTW): (i) one introduces a field $\bar{h}_{\alpha\beta}$ (which is actually the trace-reversed contribution of the waves to the spacetime metric in a suitable gauge). This field is defined to be equal to $h_{\alpha\beta}^{\mathrm{TT}}$ in the region from which the waves are propagating (the local wave zone in the case of isolated sources; the very early universe in the case of primordial waves). One then evolves the field $\bar{h}_{\alpha\beta}$ out into the surrounding universe and to earth using the curved-spacetime wave equation (equation (35.64) of MTW)

$$\bar{h}_{\alpha\beta|\mu}{}^{|\mu} + g_{\alpha\beta}^{\mathrm{B}}\bar{h}^{\mu\nu}{}_{|\nu\mu} - 2\bar{h}_{\mu(\alpha}{}^{|\mu}{}_{|\beta)} + 2R_{\mu\alpha\nu\beta}^{\mathrm{B}}\bar{h}^{\mu\nu} - 2R_{\mu(\alpha}^{\mathrm{B}}\bar{h}_{\beta)}{}^{\mu} = -16\pi\delta T_{\alpha\beta}.$$

(22)

Here $g_{\alpha\beta}^{\mathrm{B}}$ is the background metric, the subscript and superscript $|$ denote covariant derivatives with respect to the background metric, $R_{\alpha\beta}^{\mathrm{B}}$ and $R_{\alpha\beta\gamma\delta}^{\mathrm{B}}$ are the Ricci and Riemann curvature tensors of the background; and $\delta T_{\alpha\beta}$ is the perturbation in the non-gravitational stress-energy tensor produced by the trace-reversed metric perturbation $\bar{h}_{\alpha\beta}$ itself.

Although the field $\bar{h}_{\alpha\beta}$ initially is wavelike and thus has $\lambdabar \ll \mathscr{L} \lesssim \mathscr{R}_{\mathrm{B}}$, it might propagate into regions where the background has very short lengthscales, $\mathscr{L} \lesssim \lambdabar$. (For example, the waves produced by the Crab pulsar, with $\lambdabar \sim 1000$ km may propagate through a massive white dwarf with $\mathscr{L} \sim 1000$ km and $\mathscr{R}_{\mathrm{B}} \sim 30\,000$ km, or even through a neutron star with $\mathscr{L} \sim 10$ km and $\mathscr{R}_{\mathrm{B}} \sim 30$ km). If this happens, one need not worry. The wave equation (22), because it depends for its validity only on the weakness of the field $\bar{h}^{\alpha\beta}$ and not on the shortwave assumption, remains valid and carries the field through the region of short background lengthscale (where, strictly speaking, it is no longer a gravitational wave), and thence onward into regions of long lengthscale (where it once again is a gravitational wave).

In those regions where $\bar{h}_{\alpha\beta}$ is a wave, i.e. has $\lambdabar \ll \mathscr{L}$, one can compute from it the gravitational-wave field h_{jk}^{TT} by a very simple prescription: introduce the proper reference frame of a specific observer; and in that frame discard the time-time and time-space parts of $\bar{h}_{\alpha\beta}$, and algebraically project out from the space-space parts those pieces that are transverse to the propagation direction and are trace-free. The result will be h_{jk}^{TT}. For further discussion and justifications see Box 35.1 of MTW and Section 2.4.2 of Thorne (1983).

As I shall discuss below, the effects of the wave-stimulated stress-energy perturbations $\delta T_{\alpha\beta}$ are never large enough to be astrophysically important, so one can ignore them in propagation calculations. Moreover, in regions (nearly everywhere) where $\lambda \ll \mathscr{R}_{\mathrm{B}}$, the contributions of the background curvature tensors $R^{\mathrm{B}}_{\mu\alpha\nu\beta}$ and $R^{\mathrm{B}}_{\mu\alpha}$ to the wave equation can be ignored, and one can specialize the gauge so as to make $\bar{h}_{\alpha\beta}$ divergence free. The wave equation (22) then assumes the simplest form imaginable:

$$\Box \bar{h}_{\alpha\beta} \equiv \bar{h}_{\alpha\beta|\mu}{}^{\mu} = 0. \tag{23}$$

When, in addition, the radius of curvature of the wave fronts is large compared to λ (as is true everywhere except near the very rare focal points of gravitational lenses), the wave equation (23) can be solved easily by the techniques of geometric optics (Isaacson, 1968*a*; Exercise 35.15 of MTW; Section 2.5 of Thorne, 1983): the field $\bar{h}^{\alpha\beta}$ propagates along null rays; its polarization is parallel transported along the rays; and its amplitude, like the amplitude of light, varies along each ray as 1/(the radius of curvature of the wave front).

Because good gravitational lenses are so rare in the real universe, and because regions (black holes and neutron stars) with very strong curvature $(\mathscr{R}_{\mathrm{B}} \lesssim \lambda)$ are so rare and so small, the waves from almost every source will propagate to earth via pure geometric optics. Moreover, because the universe is almost globally Lorentz (flat) in its background geometry on lengthscales small compared to the Hubble distance, for sources at distances much less than Hubble (at cosmological redshifts $z \ll 1$), this geometric optics propagation will produce a simple $1/r$ falloff of amplitude and will preserve the wave form (the dependence on retarded time) and the polarization. More specifically, it will produce a gravitational-wave field in TT coordinates with the simple form

$$h^{\mathrm{TT}}_{jk} = \frac{A^{\mathrm{TT}}_{jk}(t-r, \theta, \phi)}{r}, \tag{24}$$

where r is distance to the source and θ, ϕ are spherical polar angles centered on the source. For a slow-motion source the function A^{TT}_{jk} is just twice the TT part of the second time derivative of the source's quadrupole moment, as one sees trivially by matching (24) onto (12) in the source's local wave zone.

For a source at a large cosmological redshift $z \gtrsim 1$, if one approximates the background spacetime geometry by that of a Friedmann–Robertson–Walker cosmological model, the geometric-optics propagation produces the same effects for gravity waves as for light: (i) the magnitude of h^{TT}_{jk} falls off with the same $1/R$ behavior as for light, where R is $(1/2\pi) \times$ (circumference of a sphere passing through the earth and centered on the source, at the time the waves reach earth) (equations (29.28)–(29.33) of MTW); (ii) the polarization, like that of light in vacuum, is parallel transported radially from source to earth; and (iii) the time dependence of the wave form is unchanged by propagation, except for a frequency-independent redshift $f_{received}/f_{emitted} = 1/(1 + z)$. For further details and derivations see Section 2.5.4 of Thorne (1983) or Section 7.2 of Thorne (1977).

9.3.5 A catalog of wave-propagation effects

In principle gravitational waves can experience almost all the familiar peculiarities of propagation that electromagnetic waves experience. Here I shall enumerate those that have been studied, mention briefly their importance or unimportance, and give references for further detail.

(a) *Absorption, scattering and dispersion by matter and electromagnetic fields*

Gravitational waves are so weakly absorbed by matter (Section 2.4.3 of Thorne, 1983) that absorption is astrophysically important only near the Planck era of the big-bang (Section 7.2 of Zel'dovich and Novikov, 1983). For experimenters, however, the tiny absorption that should occur in a gravity wave detector is very important; see Section 9.5.2(b) below.

The scattering of gravitational waves by matter is also so weak that it is never astrophysically important, except near the Planck era. In the analysis of resonant gravity wave detectors, however, scattering is conceptually important: the method of detailed balance (an idealized calculation in which a resonant detector is driven by monochromatic waves into vibrations of such unrealistically high amplitude that it reradiates at the same rate as it absorbs – i.e. it scatters the waves strongly) is a powerful way of computing the cross-section of a resonant detector; see Section 37.7 of MTW.

Coherent scattering by a medium produces dispersion – i.e. frequency-dependent propagation speeds. Although dispersion is very important for electromagnetic waves, it is never of any importance in the real universe for gravitational waves (Section 2.4.3 of Thorne, 1983). It is instructive, however, to imagine and study theoretically an unrealistic form of matter ('respondium') with such strong dispersion that it actually reflects gravitational waves (Press, 1979).

For detailed analyses of absorption, scattering, and/or dispersion by specific kinds of matter see the following references and the references cited therein: black holes, Matzner *et al.* (1985); De Logi and Kovacs (1977); neutron stars and other stars, Linet (1984); elementary particles, Section 7.2 of Zel'dovich and Novikov (1983); a magnetized plasma, Macedo and Nelson (1983); a uniform medium, Esposito (1971a,b), Papadopoulos and Esposito (1985), Szekeres (1971), Section 4.2 of Grishchuk and Polnarev

(1980), Section 2.4.3 of Thorne (1980); and a medium with boundaries, Dyson (1969), Carter and Quintana (1977).

Because gravitational and electromagnetic waves should propagate with the same speed, they can interact in a coherent way (Gertsenshtein, 1962). The interaction is so weak, however, that a substantial transformation of one into the other requires propagation over a distance of order the radius of curvature of the background spacetime which their own energy density produces. Thus, such coherent interaction is not likely ever to be important in the real universe – except possibly in gravity-wave detectors; see Section 9.5.4(a) below. For a review of the extensive literature on this subject see Grishchuk and Polnarev (1980). For a brief pedagogical discussion see Section 17.9 of Zel'dovich and Novikov (1983).

(b) *Scattering by background curvature, and tails of waves*

In regions where the background radius of curvature is comparable to or shorter than the reduced wavelength, $\mathcal{R}_B \lesssim \lambda$, the background strongly scatters the waves. This is very important in some sources of waves, as the waves are trying to form; for example, it is responsible for the normal-mode vibrations of black holes (Press, 1971), and it leads to the formation of 'tails' of the waves in a source's near zone (Price, 1972a,b; Thorne, 1972; Cunningham, Price and Moncrief, 1979) and to radiative tails in the wave zone (Leaver, 1986b; Blanchet, 1987) – which, however, are not likely ever to be observationally important.

(c) *Gravitational focusing*

Lumps of background curvature associated with black holes, stars, star clusters, and galaxies will focus gravitational waves in precisely the same manner as they focus electromagnetic waves; and just as this focusing is observationally important for the light and radio waves from a few very distant quasars, so it might also be important for very distant discrete sources of gravitational waves. Focusing by the sun, in the case of waves with sufficiently short wavelength can be significant, but not at earth; the focal point lies farther out in the solar system, near the orbit of Jupiter (Cyranski and Lubkin, 1974).

(d) *Diffraction*

Near the focal point of a gravitational lens the waves cease to propagate along null rays and begin to diffract, thereby lessening the strength of the focusing. The analysis of this is no different for gravitational waves than for electromagnetic or scalar waves, since polarization plays no important role. Diffraction causes focusing to be significant only when the lens has a gravitational radius $2M$ that is larger than or of order the waves' reduced wavelength λ. For an order of magnitude discussion see, e.g. Section 2.6.1 of Thorne (1983); for full details see Bontz and Haugan (1981).

(e) Parametric amplification by background curvature

In regions of a dynamical spacetime (e.g. the expanding universe) in which the characteristic wavelength λ of gravitational waves is larger than or comparable to the background radius of curvature $\mathcal{R}_\mathrm{B}$, $\lambda \gtrsim \mathcal{R}_\mathrm{B}$, the waves can be parametrically amplified by interaction with the dynamical background (Grishchuk, 1974, 1975a,b, 1977; Grishchuk and Polnarev, 1980). Viewed quantum mechanically, the interaction causes stimulated emission of new gravitons. This effect may well have enabled the expansion of the universe to amplify vacuum fluctuations from the big-bang singularity (from the Planck time) into a strong, stochastic background of gravitational waves today; see Section 9.4.3(d) below.

(f) Non-linear coupling of the waves to themselves (frequency doubling, etc.)

Because general relativistic gravity is non-linear, there is a non-linear coupling of gravitational waves to themselves; and in principle this leads to such non-linear conversion processes as frequency doubling. However, in practice these effects are not important in regions where the waves *are* waves (where $\lambda \ll \mathcal{L}$). This is because, in such regions, the dimensionless amplitude of the waves is very small compared to unity (equation (21) above). For a more detailed discussion see Section 2 of Thorne (1985).

(g) Generation of background curvature by the waves

The generation of background curvature by the stress-energy of the waves (Isaacson, 1968b, MTW Section 35.15) is important in cosmological models in any epoch when the waves are sufficiently strong that their energy density is comparable to that of matter; see, e.g. Hu (1978) and Chapter 17 of Zel'dovich and Novikov (1983). It is also important in a 'geon' – i.e. a bundle of gravitational waves that is held together by its own gravitational pull on itself (Wheeler, 1962; pp. 409–38 of Wheeler, 1964; Brill and Hartle, 1964). But geons surely do not exist in the real universe; they are only theoretical entities, useful for exploring issues in fundamental physics.

Wave-produced background curvature is also important in the idealized situation where one plane-fronted wave gets focused by passing through another ('plane-wave collision'): the focusing itself is produced by the wave-generated background curvature. Moreover, if the waves are precisely planar, a spacetime singularity forms at the focal plane (Khan and Penrose, 1971; Szekeres, 1972; Nutku and Halil, 1977; Tipler, 1980; Matzner and Tipler, 1984; Chandrasekhar and Xanthopoulos, 1986; Yurtsever, 1987a), and the generation of background curvature plays a key role in the singularity. In the more realistic case (which, however, almost certainly does not occur in the real universe except conceivably near the big-bang), in which the waves are almost planar but die out slowly at large transverse distances, if the transverse size is sufficiently large compared to the initial wave amplitude, then the focusing probably still drives the amplitude up far enough – before diffraction can act – to make background curvature generation become strong and force a singularity to form (Yurtsever, 1987b).

9.3.6 *Wave propagation in an idealized, asymptotically flat universe*

The split of wave propagation and wave generation into two separate calculations is worrisome to those physicists who seek the highest levels of rigor. Unfortunately, in analyses of waves in the real universe, where the background spacetime is complex, the split is essential. Progress cannot be made without it. However, the split can be avoided to some extent in an idealized 'universe', where the source of interest resides alone in an otherwise empty and asymptotically flat spacetime.

In such an idealized universe, and only there, it has been possible to treat wave generation and wave propagation simultaneously, with a single, elegant formalism that spans the weak-gravity portions of the wave-generation region, and all of the local wave zone and distant wave zone. In my opinion, the nicest version of this unified formalism is a variant of the post-Minkowski formalism due to Blanchet and Damour (1986) and Blanchet (1987a). See Damour (1986) for a review of this and of earlier work by others. This formalism is especially nice because, although it entails an expansion in the strength of the gravitational field, the expansion has been carried out to all orders, and a number of beautiful theorems have been proved about it.

Also of great interest, elegance, and beauty, in an idealized asymptotically flat universe, are expansions of the spacetime curvature along the outgoing light cone in inverse powers of the distance to the source (Bondi, 1960; Bondi, van der Burg and Metzner, 1962; Sachs, 1962, 1963; Penrose, 1963a,b; Newman and Penrose, 1965). Such expansions, carefully formulated and combined with conformal transformations that bring 'infinity' in to finite locations, reveal asymptotic structures of asymptotically flat spacetime that illuminate the properties of gravitational radiation. For recent reviews and references see Newman and Tod (1980), Walker (1983), Schmidt (1979, 1986), Ashtekar (1984).

9.3.7 Exact, analytic solutions for wave generation and propagation

Those who seek high rigor (and even less rigorous people like me) have also found pleasure in the few existing exact, analytic solutions of the Einstein field equations that describe sources which radiate into asymptotically flat spacetime (e.g. Bicak, 1968; Bicak, Hoenselaers and Schmidt, 1983; Bicak, 1985; and references therein); and in idealized exact solutions for cylindrical waves (Einstein and Rosen, 1936; Weber and Wheeler, 1957) and for planar waves (e.g. Rosen, 1937; Bondi, Pirani and Robinson, 1959; Ehlers and Kundt, 1962; Sections 35.9–35.12 of MTW). Also pleasing for its rigor is the extreme limit of geometric optics, where the wavelength becomes so short that the radiation is compacted into a *gravitational shock wave* (e.g. Pirani, 1957; Papapetrou, 1977 and references therein).

9.4 Astrophysical sources of gravitational waves

In the real universe it is useful to divide the anticipated gravitational waves into three classes, according to their temporal behaviors: *bursts*, which last for only a few cycles, or at most for times short compared to a typical observing run; *periodic waves*, which are superpositions of sinusoids with frequencies f_i that are more or less constant over times long compared to an observing run; and *stochastic waves*, which fluctuate stochastically and last for a time long compared to an observing run.

In this section I shall describe the present knowledge and speculations about various sources of gravitational waves. Attention will be restricted to those sources which are most promising for detection by present or planned detectors, with Section 9.4.1 focusing on burst sources, 9.4.2 on periodic sources, and 9.4.3 on stochastic sources.

As was emphasized in Section 9.1.1, for each source, with the single exception of binary stars and their coalescences (Section 9.4.1(e) below), either (i) the strength of the source's waves for a given distance from earth is uncertain by several orders of magnitude; or (ii) the rate of occurrence of that type of source, and thus the distance to the nearest one, is uncertain by several orders of magnitude; or (iii) the very existence of the source is uncertain. Despite these uncertainties, it is important in the wave-detection effort to estimate as best one can, for each source, the strengths of the waves bathing the earth. Such estimates will be stated below, with references; and they are collected together below in Figs. 9.6 (burst sources), 9.7 (periodic sources) and 9.6 (stochastic sources).

9.4.1 *Burst sources*

(a) *Bursts with and without memory*

As Braginsky and Grishchuk (1985) have recently emphasized, gravitational-wave bursts can be subdivided into two classes: *normal bursts*, in which h_{jk}^{TT} begins zero before the burst and returns to zero afterward, and *bursts with memory*, in which h_{jk}^{TT} (by convention) begins zero before the burst and then settles down into a non-zero, constant value Δh_{jk}^{TT} (the burst's 'memory') after the burst is over.

Physically, a burst with memory arises whenever the source, before the burst or afterward or both, consists of several free bodies that are moving with uniform velocities relative to each other. For example, the explosion of a star into several pieces will produce a burst with memory, as will the collision of two freely moving (not binary) stars or black holes, or the gravitational scattering of a star by a black hole; but the birth of a black hole in non-spherical stellar collapse will produce a burst without memory. In all cases, the 'memory' is the change in the total '1/r' coulomb-type gravitational field of the source (Braginsky and Thorne, 1987). For example, in the quadrupole approximation, when one takes account of the fact that one can add a time-independent constant to h_{jk}^{TT} ('gauge change' that moves stuff between the background field and the wave field) so as to make h_{jk}^{TT} zero initially, equations (12) and (13a) give

$$\Delta h_{jk}^{TT} = \Delta \sum_{A} \left(\frac{4m_A v_A^j v_A^k}{r} \right)^{STF}. \tag{25}$$

Here the summation is over the free bodies in the system, m_A and v_A^j are the mass and velocity of body A, and Δ denotes the change from before the burst is emitted to afterward.

As is discussed by Braginsky and Thorne (1987), the memory part of a burst, Δh_{jk}^{TT}, can be studied by any detector (with adequate sensitivity) that operates at a frequency lower than the burst's characteristic frequency, $f \lesssim f_c$. Put equivalently, one can think of a burst with memory as having a signal that extends down to all frequencies below f_c (cf. the 'zero-frequency limit' discussed by Smarr, 1977*b* and by Bontz and Price, 1979).

Current prejudice suggests that the strongest of burst sources (and thus the most interesting) may produce normal bursts rather than bursts with memory; but this prejudice could perfectly well be wrong. In accord with this prejudice, the remainder of this section will focus on normal bursts.

(b) *Characterization of the waves from a normal burst source and the noise in a detector searching for them*

Burst sources are best characterized by their full wave forms $h_{jk}^{TT}(t)$. However, when comparing with detector sensitivities, it is helpful to have a more compact characterization. Past discussions have used a loosely defined 'characteristic amplitude' h_c and 'characteristic frequency' f_c. However, the factors of order 3 that are glossed over by the loose definitions are beginning to be important in the planning of gravity-wave searches, especially in the case of the inspiral and coalescence of binary neutron stars (subsection (e) below); and I therefore shall be quite careful in my definitions of h_c and f_c and subsequently in my corresponding discussions of detector sensitivities.

As an aid to defining h_c and f_c with care, I show in Fig. 9.2 a diagram of the emission, propagation, and absorption of the wave. The source has

Fig. 9.2. The angles $\iota, \beta, \psi, \theta, \phi$ which characterize the emission, propagation and detection of a gravitational wave.

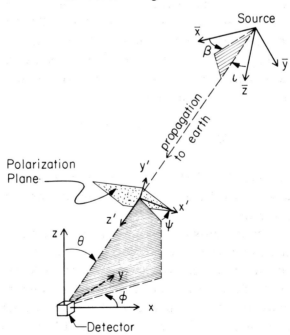

preferred local Cartesian axes $(\bar{x}, \bar{y}, \bar{z})$ with respect to which its internal structure is especially simple. We shall denote by (ι, β) the direction toward earth (spherical polar angles) relative to those axes. The detector, similarly, has preferred local Cartesian axes (x, y, z) with respect to which its internal structure is especially simple; and we shall denote by (θ, ϕ) the direction toward the source (spherical polar angles) relative to those detector axes. The waves themselves, as they pass through the detector, are most simply described in a third set of Cartesian axes (x', y', z') with origin at the detector's center of mass, z' axis along the waves' propagation direction, and x' and y' axes so oriented in the polarization plane as to make the wave forms $h_+ = h_{x'x'}^{TT} = -h_{y'y'}^{TT}$, and $h_\times = h_{x'y'}^{TT} = h_{y'x'}^{TT}$, especially simple. We shall denote by ψ the angle between the x' polarization axis and the $\phi = 0$ plane.

From a model for the source one can compute the wave forms $h_+(t - z'; \iota, \beta)$ and $h_\times(t - z'; \iota, \beta)$ arriving at the detector. If the detector is small compared to a reduced wavelength, as we shall assume throughout Section 9.4 and in Sections 9.5.1–9.5.3 but not 9.5.4–9.5.6, then it will feel a simple linear combination of h_+ and $h_\times$; i.e. it will detect the wave form

$$h(t) = F_+(\theta, \phi, \psi) h_+(t; \iota, \beta) + F_\times(\theta, \phi, \psi) h_\times(t; \iota, \beta). \tag{26}$$

Here F_+ and $F_\times$ are detector beam-pattern functions (to be discussed in Sections 9.5.2 and 9.5.3 below), which depend on the direction of the source (θ, ϕ) and the orientation ψ of the polarization axes relative to the detector's orientation (Fig. 9.2) and which have values in the range $0 \leqslant |F_A| \leqslant 1$. There will be some special choice of θ, ϕ, ψ for which $F_+ = 1$ and $F_\times = 0$; we shall call this the 'optimum source direction and polarization' for the $+$ mode of the waves.

In Section 9.5 we shall characterize the noise in a detector by a frequency-dependent spectral density $S_h(f)$ (with dimensions Hz^{-1}) defined as follows: if a precisely sinusoidal gravitational wave with known phase α, known frequency $f > 0$ and unknown rms amplitude h_o,

$$h = (2)^{\frac{1}{2}} h_o \cos(2\pi f t + \alpha), \qquad \alpha = \mathrm{const.}, \tag{27a}$$

impinges on the detector, and if the experimenters seek to detect the wave by Fourier analyzing the detector output with a bandwidth Δf (integration time $\hat{\tau} = 1/\Delta f$), then the amplitude signal-to-noise ratio will be

$$\frac{S}{N} = \frac{h_o}{[S_h(f)\Delta f]^{\frac{1}{2}}}. \tag{27b}$$

In much of the gravitational-wave literature this $S_h(f)$ is denoted $[h(f)]^2$; i.e. $h(f) \equiv [S_h(f)]^{\frac{1}{2}}$ (with dimensions $\mathrm{Hz}^{-\frac{1}{2}}$).

In this section we are interested not in periodic gravitational waves, but rather in bursts – i.e. in waves with complicated time dependences $h(t)$ and with durations short compared to the experimenters' observation time. We shall assume that the wave form $h(t)$ is known, and that our only question is whether the wave is really present or not – with some postulated starting time t_0. (Later we shall discuss the statistical consequences of an unknown starting time.) There is a well-known optimal strategy for searching for such a wave in the presence of noise with known spectral density:

(i) One constructs a *Wiener optimal filter* $K(t)$, that function of time whose Fourier transform $\tilde{K}(f)$ is the same as the transform $\tilde{h}(f)$ of the signal $h(t)$, weighted by $1/S_h(f)$ (so that noisy frequencies are suppressed):

$$\tilde{K}(f) \equiv \frac{\tilde{h}(f)}{S_h(f)}, \qquad \tilde{h}(f) \equiv \int_{-\infty}^{+\infty} h(t)\, e^{i2\pi ft}\, dt, \qquad (28a)$$

$$K(t) = \int_{-\infty}^{+\infty} \tilde{K}(f)\, e^{-i2\pi ft}\, df. \qquad (28b)$$

(ii) One then takes the output of the detector, which includes noise and possibly signal; from it one computes the gravitational wave form $h_{\text{output}}(t)$ that would have been required to produce the observed output if there were no noise present; and one then computes a quantity

$$W = \int_{-\infty}^{+\infty} K(t - t_0) h_{\text{output}}(t)\, dt. \qquad (28c)$$

(iii) This quantity will have root-mean-square contribution N from noise; and if the signal was actually present with starting time t_0, it will have a signal contribution S (i.e. $W = N + S$), with squared signal-to-noise ratio

$$\frac{S^2}{N^2} = \int_0^\infty \frac{2|\tilde{h}(f)|^2}{S_h(f)}\, df. \qquad (29)$$

See Wiener (1949), Sections 25–27 of Wainstein and Zubakov (1962), Kafka (1977), and Michelson and Taber (1984) for proofs and discussion. The integral in (29) is from 0 to ∞ rather than $-\infty$ to $+\infty$ and there is a factor 2 present because $S_h(f)$ is a 'one-sided spectral density' (defined only for $f \geq 0$; negative frequencies are folded into positive). The squared signal-to-noise ratio (29) will be the basis for our definitions of h_c and f_c.

Expression (29) is the squared signal-to-noise ratio for a specific ('fiducial') source at some specific distance r_0 from earth. Suppose that space is uniformly filled with sources, all identical to this fiducial source but with random directions (θ, ϕ), orientations (ι, β), and polarization angles ψ. Suppose, further, that inside the source's distance r_0 there is, on average, one burst each D_0 days. Then what, on average, will be the squared signal-to-noise ratio $(S^2/N^2)_{\text{strongest}}$ of the strongest burst that occurs each D_0 days? One might expect the answer to be the fiducial S^2/N^2 of (29), averaged over all the angles θ, ϕ, ψ, ι, β. Not so. There is a statistical preference for directions and polarizations that give larger values of S^2/N^2, because they can be seen out to greater distances where the event rate is greater. This effect gives, assuming the event rate goes up as r^3 and h_{jk}^{TT} goes down as $1/r$, $(S^2/N^2)_{\text{strongest}} = \langle S^3/N^3 \rangle^{\frac{2}{3}}$ where $\langle \cdots \rangle$ denotes an average over randomly distributed angles. Now, the $\frac{2}{3}$ power is a pain to deal with in subsequent calculations, so we shall switch to a straightforward angular average $\langle S^2/N^2 \rangle$ and to compensate we shall insert a multiplicative factor $\frac{3}{2}$, which is (approximately) the ratio of $\langle S^3/N^3 \rangle^{\frac{2}{3}}$ to $\langle S^2/N^2 \rangle$ if both source and detector have quadrupole beam patterns ('slow-motion, poor-antenna regime'). This, together with $\langle F_+^2 \rangle = \langle F_\times^2 \rangle$ and $\langle F_+ F_\times \rangle = 0$ (true for any quadrupole-beam-pattern gravity-wave detector), permits us to write $(S^2/N^2)_{\text{strongest}}$ in the following approximate form, where we omit the subscript 'strongest' for ease of notation and where $\langle |\tilde{h}_+|^2 + |\tilde{h}_\times|^2 \rangle$ is averaged over source angles (ι, β):

$$\frac{S^2}{N^2} \cong 3 \langle F_+^2(\theta, \phi, \psi) \rangle \int_0^\infty \frac{\langle |\tilde{h}_+|^2 + |\tilde{h}_\times|^2 \rangle}{S_h(f)} \, df. \tag{30}$$

We shall now specialize, for the remainder of this subsection, to gravity-wave searches with broad-band detectors, i.e. detectors for which the noise $S_h(f)$ is small over a band of frequencies $f_{\min} \lesssim f \lesssim f_{\max}$, with $f_{\max} \gtrsim 2 f_{\min}$. Narrow-band detectors will be treated separately in Section 9.5.2(b) below. For broad-band detectors equation (30) motivates us to define the characteristic frequency and amplitude of our fiducial source, at distance r_0, by

$$f_c \equiv \left[\int_0^\infty \frac{\langle |\tilde{h}_+|^2 + |\tilde{h}_\times|^2 \rangle}{S_h(f)} \, df \right]^{-1} \left[\int_0^\infty \frac{\langle |\tilde{h}_+|^2 + |\tilde{h}_\times|^2 \rangle}{S_h(f)} f \, df \right], \tag{31a}$$

$$h_c \equiv \left[3 \int_0^\infty \frac{S_h(f_c)}{S_h(f)} \langle |\tilde{h}_+|^2 + |\tilde{h}_\times|^2 \rangle f \, df \right]^{\frac{1}{2}}, \tag{31b}$$

where the average is over randomly distributed source orientation angles ι, β. Similarly, we shall define a characteristic detector noise amplitude at frequency f by

$$h_n(f) \equiv \frac{[f S_h(f)]^{\frac{1}{2}}}{\langle F_+^2(\theta, \phi, \psi) \rangle^{\frac{1}{2}}}. \tag{32}$$

In terms of these quantities equation (30) reads

$$\frac{S}{N} = \frac{h_c}{h_n(f_c)}. \tag{33}$$

This is the S/N of the strongest burst that arrives at earth, on average, at the same rate as bursts occur inside the fiducial source's distance r_o.

In Fig. 9.4 below we characterize detector burst sensitivities not by $h_n(f)$, but rather by a quantity $h_{3/yr}$ that answers the following question: *What is the characteristic strength $h_c = h_{3/yr}$ of a source with sufficiently large S/N (equation (33)) that, if it is seen three times per year (once each 10^7 seconds) by two identical detectors operating in coincidence, we can be 90% confident the detectors are not just seeing their own noise.* The use of two detectors permits (we shall presume) the elimination of non-Gaussian noise. Then the Gaussian distribution of the amplitude noise, together with the use of two detectors and the fact that in $\hat{\tau} = 10^7$ s the experimenters must try roughly $2\pi f_c \hat{\tau} \simeq f_c/10^{-8}$ Hz starting times t_o for their Wiener optimal filter, implies that we require $S/N \simeq [\ln(f_c/10^{-8} \text{ Hz})]^{\frac{1}{2}}$; and, correspondingly,

$$h_{3/yr} \simeq \left[\ln\left(\frac{f_c}{10^{-8} \text{ Hz}}\right)\right]^{\frac{1}{2}} \frac{[f_c S_h(f_c)]^{\frac{1}{2}}}{\langle F_+^2(\theta, \phi, \psi)\rangle^{\frac{1}{2}}} \simeq (3\text{--}5) h_n(f_c). \tag{34}$$

Here the range 3–5 corresponds to the range of frequencies of interest for most burst searches, 10^{-4} Hz–10^{+4} Hz. Note that because the events being sought are so far out on the tail of the Gaussian probability distribution, changing by a factor 10 or 100 the number of trial starting times t_o, or asking for 99% or 99.9% confidence rather than 90%, would have a negligible effect on this $h_{3/yr}$.

It is often useful to rewrite the f_c and h_c of equations (31) in terms of the energy flux per unit frequency $dE_{GW}/dA\, df$ carried past the detector by the waves. From equation (10) for the waves' stress-energy tensor together with Parseval's theorem we infer that

$$\frac{dE_{GW}}{dA\, df} = \frac{\pi}{2} f^2(|\tilde{h}_+(f)|^2 + |\tilde{h}_\times(f)|^2), \tag{35}$$

where the extra factor 2 comes from folding negative frequencies into positive (so $f > 0$). When this quantity is averaged over all directions ι, β, it gives $(4\pi r_o^2)^{-1} dE_{GW}/df$, where r_o is the distance to the source; and consequently equations (31) become

$$f_c = \left[\int_{-\infty}^{\infty} \frac{dE_{GW}/df}{f S_h(f)}\, d\ln f\right]^{-1} \left[\int_{-\infty}^{\infty} \frac{dE_{GW}/df}{S_h(f)}\, d\ln f\right], \tag{36a}$$

$$h_c = \left[\int_{-\infty}^{\infty} \frac{S_h(f_c)}{S_h(f)} \frac{3}{2\pi^2 r_o^2} \frac{dE_{GW}}{df} \, d\ln f \right]^{\frac{1}{2}}. \tag{36b}$$

For most burst sources (e.g. supernovae) the wave form is so uncertain that a careful calculation of h_c is unjustified. In such cases it is useful to reexpress h_c (equation (36b)), approximately, in terms of the total energy ΔE_{GW} radiated:

$$h_c \simeq \left(\frac{3}{2\pi^2} \frac{\Delta E_{GW}/f_c}{r_o^2} \right)^{\frac{1}{2}} = 2.7 \times 10^{-20} \left(\frac{\Delta E_{GW}}{M_\odot c^2} \right)^{\frac{1}{2}} \left(\frac{1 \text{ kHz}}{f_c} \right)^{\frac{1}{2}} \left(\frac{10 \text{ Mpc}}{r_o} \right), \tag{37}$$

where $M_\odot$ is the mass of the sun and 10 Mpc is the distance to the center of the Virgo cluster of galaxies (assuming a Hubble constant of 100 km s^{-1} Mpc^{-1}).

We turn now to a discussion of specific burst sources, and their characteristic frequencies f_c and wave strengths h_c.

(c) Supernovae (collapse to neutron star)

Supernovae of 'type II' are believed, with a high level of confidence, to be created by the gravitational collapse, to a neutron-star state, of the cores of massive, highly evolved stars. Supernovae of 'type I', by contrast, are thought to result from nuclear explosions of white dwarfs that are accreting mass from close companions – explosions in which the stellar core probably does not, but might, collapse to a neutron-star state (Woosley and Weaver, 1986; Evans, Iben and Smarr, 1987, and references therein). In addition to these two types of optically observed supernovae, there may well be stellar collapses to a neutron star that produce little optical display ('optically silent supernovae').

The rate of occurrence of supernovae of types I and II is fairly well determined observationally: in our Galaxy roughly one type I each 40 years and one type II each 40 years; out to the distance of the center of the Virgo Cluster of Galaxies (10 Mpc) several type I and several type II per year; rate increasing roughly as (distance)^3 near and beyond Virgo; see, e.g. Tammann (1981). Thus, to have an interesting event rate one must have adequate sensitivity to reach the center of Virgo, about 10 Mpc. Optically silent supernovae could be more frequent, since statistics on normal stars massive enough to form neutron stars when they die, permit a neutron-star birth rate that could be as high as one every four years in our galaxy if mass loss in the late stages of stellar evolution is smaller than normally thought. Alternatively, it is conceivable that optically silent supernovae never occur.

The strengths of the waves from a supernova depend crucially on the degree of non-sphericity in the stellar collapse that triggers it, and somewhat on the speed of collapse – i.e. on whether the collapse is nearly free-fall ('cold collapse') or is more gentle due to the resistance of thermal pressure ('hot collapse'). Perfectly spherical collapse will produce no waves; highly non-spherical collapse will produce strong waves. Little is known about the degree of non-sphericity in type II (which are surely due to stellar collapse), but current prejudice suggests that the typical type II might be quite spherical and thus poorly radiating. If type I are due to explosion of an accreting white dwarf and, contrary to current thought, the explosion is accompanied by collapse of the stellar core to a neutron star, then the white dwarf might be rapidly rotating due to the accretion, and centrifugal forces might then cause it to collapse very non-spherically and radiate strongly.

In the mid-1970s there was a swing of fashion from believing that the collapse is cold and fast to believing it is hot and slow (e.g. Wilson, 1974; Schramm and Arnett, 1975). Some astrophysicists were aghast at the consequence of this swing: for example, in the highly non-spherical but axisymmetric collapse models of Saenz and Shapiro (1978, their tables 1–4) the total energy radiated as gravitational waves was reduced from $\Delta E_{\mathrm{GW}}/M_\odot c^2 \sim 6 \times 10^{-3}$ for cold and fast to $\sim 1 \times 10^{-5}$ for hot and slow. This boded ill for attempts to detect gravitational waves, the astrophysicists thought. However, closer scrutiny revealed relatively little change in the prospects for detection: the total energy radiated ΔE_{GW} is a rather poor indicator of detectability. Much more relevant is the amplitude signal-to-noise ratio $S/N = h_{\mathrm{c}}/h_{\mathrm{n}}(f_{\mathrm{c}})$ (equation (33)). The planned LIGO beam detectors, when optimized for frequency f_{c}, have $h_{\mathrm{n}}(f_{\mathrm{c}}) \propto f_{\mathrm{c}}$ (Fig. 9.4 and equation (125a)); and this together with equation (37) gives

$$S/N \propto (\Delta E_{\mathrm{GW}}/f_{\mathrm{c}}^3)^{\frac{1}{2}}. \tag{38}$$

Although the new fashion (hot and slow) corresponded to a reduction in ΔE_{GW} by a factor 600, the characteristic frequency of the waves also went down (from $\simeq 3000$ Hz to $\simeq 500$ Hz; see Saenz and Shapiro, 1978); and, correspondingly, S/N was reduced by less than a factor 2.

This illustrates the importance of thinking in terms of h_{c}, f_{c}, and $S/N = h_{\mathrm{c}}/h_{\mathrm{n}}(f_{\mathrm{c}})$ rather than in terms of energy radiated, when evaluating the strengths of gravitational waves.

Corresponding to our poor knowledge of the strengths of the waves from supernovae is a similar poor knowledge of the wave forms. It seems unlikely that theorists will be able to firm up their predictions of the waveforms before observers detect and study them. Thus, it is best to think of the computed wave forms as giving us a tool for translating future observations into an understanding of what is happening in the stellar core. As an example consider Fig. 9.3, which shows the wave forms computed by Saenz and Shapiro (1978, 1981), using the quadrupole formalism, for two different stellar collapses. In each case – and in general, when one thinks the quadrupole formalism may be accurate – one can invert equation (12) to see how the source's quadrupole moment is behaving. The left curve (Saenz and Shapiro, 1978) shows several epochs labeled FF in which $h_+(t)$ varies approximately as $|t - t_o|^{-\frac{2}{3}}$, corresponding to free-fall motion; and these free-fall epochs are separated by three brief periods with sharply reversed peaks (labeled 'P' in the diagram) corresponding to a sharp acceleration in the direction opposite to the free fall. Considering the timescales of hundredths of a second for the free-fall epoch (characteristic of late stages of collapse to a neutron star) and ~0.5 milliseconds for the peak epochs (characteristic of neutron-star pulsation), the natural and correct interpretation is that these waves are from collapse to a neutron star in which the stellar core bounced sharply three times. The fact that the three sharp peaks are all in the same direction (up, not down) indicates that the sharp bounces were all along the same axis. Surely the other axis that projects on our sky should have bounced as well, or at least stopped its collapse; so there should be at least one sharp peak in the down direction. Indeed there is; it is superposed on the central up peak (region labeled E in the diagram). The natural and correct interpretation is that the star was centrifugally flattened by rotation; its pole collapsed fast and bounced three times (up peaks P) while its equator collapsed more slowly and bounced once (down peak E).

Fig. 9.3. Wave forms produced by two very different scenarios for the collapse of a normal star to form a neutron star. Wave form (*a*) is from Saenz and Shapiro (1978); (*b*) is from Saenz and Shapiro (1981).

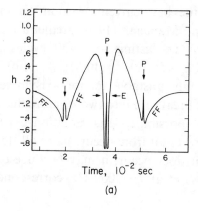

(a)

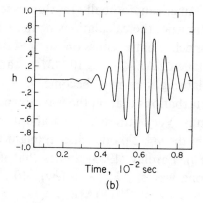

(b)

The very different right-hand curve in Fig. 9.3 (Saenz and Shapiro, 1981) implies a quadrupole moment that somehow is driven into sinusoidal oscillations which initially increase in amplitude and then die out. Again, the fact that the period ($\sim$0.6 ms) is that of a neutron-star pulsation suggests that something is triggering, then damping, such a pulsation. If this wave form was seen roughly one day before an optical supernova was found in the Virgo cluster, one would infer that the waves were from the supernova and one could deduce that the quadrupole-moment oscillations are so large in amplitude that they must have absorbed, say, 10^{-4} of the collapse energy. The natural explanation might be parametric amplification of quadrupole neutron star pulsations by a bouncing stellar collapse, followed by hydrodynamic damping – the process that gave rise to this computed wave form.

A large amount of effort has gone into model calculations of gravitational collapse to a neutron star and the waves it emits; but the effort has not produced a consensus by any means! For a detailed review of the literature up to 1982 see Eardley (1983); for an update on that review see Müller (1984). In the case of rapidly rotating collapse, where the emission should be strongest but the event rate is totally unknown (most collapses *could* be slowly rotating), there are three radically different scenarios and corresponding wave characteristics: (i) the star may remain axisymmetric throughout the collapse. In this case the best current 'wisdom' (but by no means a consensus) comes from calculations by Müller (1982) and is pessimistic. Those calculations predict the strongest emission to come in two different spectral regions: $f_c \sim 1000$ Hz where $\Delta E_{GW} \sim 1 \times 10^{-7} M_\odot c^2$ and $h_c \sim 1 \times 10^{-23}$ (10 Mpc/r_o) due to the initial collapse and bounce; and $f_c \sim 10^4$ Hz where $\Delta E_{GW} \sim 10^{-6} M_\odot c^2$ and $h_c \sim 1 \times 10^{-23}$ (10 Mpc/r_o) due to pulsations of the newly formed neutron star. (ii) The star may become unstable to an '$m=2$ bar-mode' deformation so it rotates end-over-end like an American football. In this case the best current 'wisdom' is more optimistic: the calculations of Ipser and Managan (1984) predict a highly monochromatic emission at $f \sim 1000$ Hz, lasting for ~ 30 cycles and producing $\Delta E_{GW} \sim 3 \times 10^{-4} M_\odot c^2$ and $h_c \sim 5 \times 10^{-22}$ (10 Mpc/r_o). (iii) The collapsing star may become so strongly unstable to non-axisymmetric perturbations that on the way down it breaks up into two or more discrete lumps. Very little is known about this possibility, and radiation reaction from the $m=2$ mode (case ii) might prevent it from occurring at all (Ipser, 1986). Eardley (1983) argues that if it does occur, it may produce quite strong waves: $\Delta E_{GW} \sim$ (a few) $\times 10^{-2} M_\odot c^2$ at $f_c \sim 1000$ Hz, corresponding to $h_c \sim 4 \times 10^{-21}$ (10 Mpc/r_o).

Because this best wisdom is so insecure, Fig. 9.4 shows wave strengths based not on these specific models, but rather on the general equation (37) for several possible values of ΔE_{GW} and r_o, and for the entire range of characteristic frequencies that have shown up in model calculations, $200\,\text{Hz} \lesssim f_c \lesssim 10\,000\,\text{Hz}$. Note that detectability depends strongly on whether the waves come off at low frequencies or high: a factor 8 reduction in ΔE_{GW} can be compensated by a factor 2 reduction in f_c.

Fig. 9.4. The characteristic amplitudes h_c (equation (31b)) and frequencies f_c (equation (31a)) of gravitational waves from several postulated *burst sources* (thin curves), and the sensitivities $h_{3/yr}$ of several existing and planned detectors (thick curves and circles) ($h_{3/yr}$ is the amplitude h_c of the weakest source that can be detected three times per year with 90% confidence by two identical detectors operating in coincidence). The abbreviations BH, NS and SN are used for black hole, neutron star and supernova. The sources are discussed in detail in the indicated subsections of Section 9.4.1, and the detectors in the indicated subsections of Section 9.5.

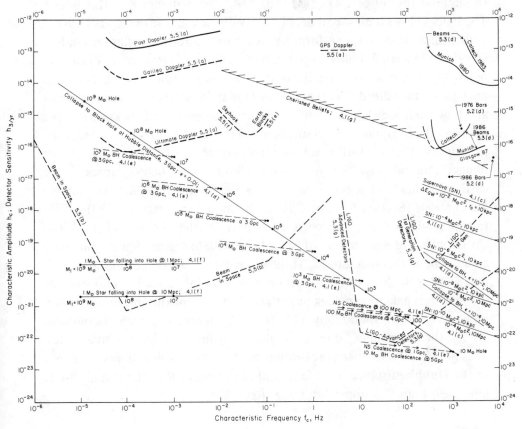

We note in passing that neutron-star pulsations might be excited not only as part of the star's birth throes, but also as a consequence of a sudden strain release or phase transition in an old neutron star (Thorne, 1978; Ramaty *et al.*, 1980; Haensel, Zdunik and Schaeffer, 1986). Since there are 10^8–10^9 old neutron stars in our galaxy, it is conceivable – though not highly likely – that our galaxy could produce an interesting event rate. To be detectable by the planned LIGOs, the waves would need to have $h_c \gtrsim 10^{-21}$ at $f_c \sim 3000$ Hz corresponding to

$$\Delta E_{GW} \gtrsim 7 \times 10^{45} \text{ erg} \times (10 \text{ kpc}/r_o)^2. \tag{39}$$

(d) *Collapse of a star or star cluster to form a black hole*

As with collapse to form a neutron star, so also for collapse to a black hole, the strengths of the waves produced are highly sensitive to the degree of non-sphericity, and the typical degree of non-sphericity is unknown. Equally unknown in the black-hole case, by contrast with the neutron star, is the frequency of occurrence of such collapses:

It is very likely that black holes exist in our universe with masses throughout the range $2 \text{ M}_\odot \lesssim M \lesssim 10^{10} \text{ M}_\odot$ (see Chapter 8 of this book). The holes of lowest mass can only form by direct collapse of a star. Those of higher mass, however, can form by many routes (direct collapse; gradual growth from a small hole by accretion; collision and coalescence of smaller holes; . . .). For discussions of the routes that might occur in a dense galactic nucleus see Blandford (1979) and Rees (1983). Which routes actually occur and how often are almost totally unknown. However, roughly known upper limits on the birth rates of the smallest and the largest holes give – under the assumption that all births are by direct collapse – corresponding upper limits on the rate at which gravity-wave bursts from such collapses hit the earth. For holes of a few solar masses the birth rate under reasonable assumptions (Section 8.3.3 of Chapter 8) should not exceed $\sim \frac{1}{3}$ the birth rate of neutron stars; and correspondingly, at the distance of the Virgo cluster it should not exceed $\sim 1/\text{year}$. Bethe (1986) argues that the rate may actually be of this magnitude. At the other end of the spectrum, holes with masses $M \gtrsim 10^6 \text{ M}_\odot$ probably occur only in galactic nuclei; and over its lifetime each galactic nucleus might give birth, at maximum, to only a few such holes. Correspondingly (Thorne and Braginsky, 1976; Blandford, 1979), the maximum rate of collapse-births of supermassive holes, $M \gtrsim 10^6 \text{ M}_\odot$, is a few per year throughout the observable universe – i.e. out to the Hubble distance. It is fashionable to believe that the actual rate is much less than this upper limit (e.g. Rees, 1983).

In one respect collapse to a black hole is better understood than collapse to a neutron star: the final object is far simpler, and correspondingly the waves from its vibrations, if they are triggered by the collapse, are far better understood. Detailed calculations suggest, in fact, that black-hole vibrations are rather easy to trigger (e.g. Detweiler, 1977) and that when they are triggered, the most slowly damped one or two quadrupole modes will dominate. Thus, while the details of the initial burst of waves may depend on unknown details of the collapse, the late-time behavior will have a well-established damped oscillatory form from which one can read off the mass of the hole with excellent accuracy and its angular momentum with modest accuracy (Detweiler, 1980; Leaver, 1985, 1986a; Stark and Piran, 1986; Piran and Stark, 1986).

As a specific example, Fig. 9.5 shows the waves produced by a specific model of a collapsing, rotating star as computed using numerical relativity techniques by Stark and Piran (1986). The solid curve is the computed waveform, and the dashed curve is a fit to it using a mixture of the two most weakly dampled quadrupole modes of a non-rotating hole. The fact that the fit is so good shows (i) that the final ringdown is, indeed, due to the black-hole vibrations; and (ii) that the hole was not rotating extremely rapidly, i.e. it had $(1-a/M) \gtrsim 0.3$, where a is the specific angular momentum and M the mass. If the hole had had $1-a/M < 0.3$, the 'Q' of the hole's oscillations would have been noticeably larger, and the ringdown of the waves would have been noticeably slower. (For details of the normal modes of black holes and the frequencies and damping times of the waves they should produce see Detweiler, 1980, and Leaver, 1985, 1986a.)

If collapse to a black hole radiates with an efficiency $\Delta E/Mc^2 \equiv \varepsilon$ and the hole is at a distance r_0 and has a mass M, then the characteristic frequency and amplitude of its waves will be (Stark and Piran, 1986; Piran and Stark, 1986; and equation (37) above)

$$f_c \cong \frac{1}{5\pi M} = (1.3 \times 10^4 \text{ Hz})\left(\frac{M_\odot}{M}\right), \tag{40a}$$

$$h_c \cong \left(\frac{15}{2\pi}\varepsilon\right)^{\frac{1}{2}}\frac{M}{r_0} = 7 \times 10^{-22}\left(\frac{\varepsilon}{0.01}\right)^{\frac{1}{2}}\left(\frac{M}{M_\odot}\right)\left(\frac{10 \text{ Mpc}}{r_0}\right)$$

$$= 1.0 \times 10^{-20}\left(\frac{\varepsilon}{0.01}\right)^{\frac{1}{2}}\left(\frac{10^3 \text{ Hz}}{f_c}\right)\left(\frac{10 \text{ Mpc}}{r_0}\right). \tag{40b}$$

If the collapse is axisymmetric, then the efficiency ε probably does not exceed 7×10^{-4} (Stark and Piran, 1986). However, in the non-axisymmetric case (e.g. formation of an elongated configuration due to rapid rotation, or bifurcation into one or more lumps during collapse (the 'collapse, pursuit, and plunge' scenario of Ruffini and Wheeler, 1971)), the efficiency might be in the range 0.01–0.1 (see, e.g., Eardley, 1983, and Rees, 1983). The source characteristics (40) are shown in Fig. 9.4 for black-hole births at the Hubble distance and at the distance of Virgo, with efficiencies of $\varepsilon = 10^{-2}$ and 10^{-4}.

(e) *Coalescence of compact binaries (neutron stars and black holes)*

Since a large fraction of all stars are in close binary systems, the dead remnants of stellar evolution may contain a significant number of binary

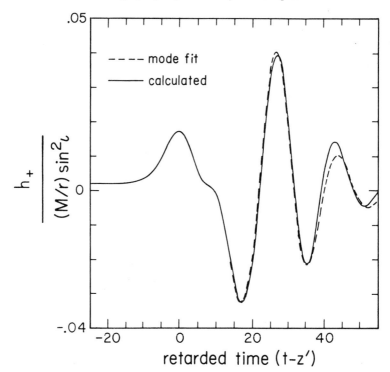

Fig. 9.5. The gravitational wave form produced by the gravitational collapse of an axisymmetric rotating star to produce a Kerr black hole, as computed by Piran and Stark (1986) using numerical relativity techniques. The star and the black hole it forms both have $J/M^2 \equiv a/M = 0.63$ (where J is angular momentum and M is mass). The dashed curve is a fit, to the wave form, of a superposition of the waves from the two most slowly damped quadrupolar normal modes of a non-rotating hole with mass M, $h_+ \sim \mathrm{Real}\{A_1 e^{-i\omega_1 t} + A_2 e^{-i\omega_2 t}\}$ with $\omega_1 = (0.374 - 0.089i)/M$, $\omega_2 = (0.348 - 0.274i)/M$. The fitting amplitudes are $A_1 = -0.9 - 1.1i$, $A_2 = 0.9 + 1.4i$. Thus, these two modes are roughly equally excited by the collapse.

systems whose components are neutron stars or black holes, and are close enough together to be driven into coalescence by gravitational radiation reaction in a time less than the age of the universe. The binary pulsar PSR 1913 + 16 is an example of such a system; it will coalesce 3.5×10^8 years from now.

As the two bodies in a compact binary spiral together, they emit periodic gravitational waves with a frequency that sweeps upward toward a maximum,

$$f_{max} \simeq 1 \text{ kHz for neutron stars,} \tag{41a}$$

$$f_{max} \simeq \frac{10 \text{ kHz}}{M_1/M_\odot} \text{ for holes with the larger having mass } M_1. \tag{41b}$$

The wave form during the nearly Newtonian part of the frequency sweep, $f \ll f_{max}$, is easily computed from the quadrupole formalism (Section 9.3.2). The post-Newtonian corrections to this waveform will become more and more important as f rises toward f_{max}; they are given in Wagoner and Will (1976); see also Gal'tsov, Matiukhin and Petukhov (1980). Ultimately, near f_{max}, higher-order corrections or full non-linear relativity are needed to get the wave form reasonably accurately. The final, coalescence stage will be especially interesting and complex in the case of a neutron-star binary, and may be quite sensitive to the masses of the two stars; and as with supernovae, we might not understand reliably what to expect until gravitational-wave observations show us. For a first, preliminary theoretical effort at understanding, see Clark and Eardley (1977). For black holes, by contrast, numerical relativity is likely to give us, within the next five years or so, a detailed and highly reliable picture of the final coalescence and the wave forms it produces, including the dependence on the hole's masses and angular momenta. Comparison of the predicted wave forms with observed ones will constitute the strongest test ever of general relativity. (The wave forms for the astrophysically unlikely cases of head-on collisions of two identical non-rotating holes or neutron stars have already been evaluated by numerical relativity; see Smarr, 1977a for black holes, and Gilden and Shapiro, 1984 for neutron stars.)

Because the binary system spends far more time in the early, low-frequency part of the sweep than in the later, high-frequency part or in the final coalescence, and because planned gravity wave detectors have less amplitude noise at low frequencies, $f \sim 100$ Hz, than at high, $f \gg 100$ Hz (cf. Fig. 9.4), it will be easier for detectors to see the Newtonian regime of the sweep than the post-Newtonian regime or the final coalescence – except in the case of black-hole binaries with $M_1 \sim 100{-}1000 \, M_\odot$ or $M_1 \gtrsim 10^6 \, M_\odot$ (Fig. 9.4). In the Newtonian regime, if we orient the polarization axes $\vec{e}_{x'}$ and $\vec{e}_{y'}$ along the major and minor axes of the projection of the orbital plane on the sky, then the wave form will be

$$h_+ = 2(1 + \cos^2 \iota)(\mu/r)(\pi M f)^{\frac{2}{3}} \cos(2\pi f t), \tag{42a}$$

$$h_\times = \pm 4 \cos \iota (\mu/r)(\pi M f)^{\frac{2}{3}} \sin(2\pi f t). \tag{42b}$$

Here it is assumed that the orbit is circular because radiation reaction long ago will have forced circularization (Peters and Mathews, 1963); ι is the angle of inclination of the orbit to the line of sight; M and μ are the total and reduced masses

$$M = M_1 + M_2, \qquad \mu = M_1 M_2 / M; \tag{42c}$$

and f, the frequency of the waves (equal to twice the orbital frequency), is given as a function of time by (MTW equation (36.17))

$$f = \frac{1}{\pi} \left[\frac{5}{256} \frac{1}{\mu M^{\frac{2}{3}}} \frac{1}{(t_0 - t)} \right]^{\frac{3}{8}}. \tag{42d}$$

The most promising detectors for coalescing neutron-star binaries and low-mass black-hole binaries are beam detectors in the planned multi-kilometer LIGOs. As we shall see in Section 9.5.3(e), a beam detector can be operated in several different optical configurations. The optimum configuration for searching for coalescing binaries is likely to be one with *light recycling*, for which the spectral density of shot noise (the dominant noise above some 'seismic cutoff' frequency f_s) will have the form

$$S_h(f) = \text{const} \times f_k[1 + (f/f_k)^2] \quad \text{at } f > f_s. \tag{43a}$$

Here f_k is a 'knee frequency' which the experimenters can adjust by changing the reflectivities of certain mirrors in their detectors; see equation (117c) and Fig. 9.13 below, and associated discussion. The constant in equation (43a) is independent of the choice of f_k. At frequencies below the 'seismic cutoff' f_s seismic noise is likely to come on very strong; accordingly, we shall make the approximation

$$S_h(f) = \infty \quad \text{for } f < f_s. \tag{43b}$$

By Fourier-transforming the wave forms (42), squaring, and averaging over the source orientation angle ι, we obtain

$$\langle |\tilde{h}_+|^2 + |\tilde{h}_\times|^2 \rangle = \frac{\pi}{12} \left(\frac{\mu}{r} \right)^2 \frac{M^3}{\mu} \frac{1}{(\pi M f)^{\frac{7}{3}}}. \tag{44}$$

By inserting this source strength (44) and the detector noise (43) into equation (30) and maximizing the resulting signal-to-noise ratio with respect to the knee frequency f_k, we find that the experimenter will do best to choose

$$f_k = 1.44 f_s. \tag{45}$$

For smaller choices of f_k there is not a wide enough frequency band between the seismic cutoff and the knee to take optimal advantage of the broad-band nature of the signal. For larger values of f_k the experimenter loses because the height $S_h(f)$ of the 'noise floor' at $f_s < f \lesssim f_k$ (equation (43a)) is proportional to f_k. With this choice of knee, equations (31a) and (43)–(45) give for the characteristic frequency $f_c = 0.909 f_k$. Below (equation (125a)) we shall characterize the sensitivites of beam detectors in full scale LIGOs, when searching for bursts, by the noise amplitude h_n at the knee; and, correspondingly, we here shall set f_c (which after all is somewhat arbitrary) to f_k rather than $0.909 f_k$:

$$f_c = f_k = 1.44 f_s. \tag{46a}$$

Equations (31b) and (43)–(45) then give for the characteristic amplitude of the waves from inspiraling binaries

$$h_c = 0.237 \frac{\mu^{\frac{1}{2}} M^{\frac{1}{3}}}{r_o f_c^{\frac{1}{6}}} = 4.1 \times 10^{-22} \left(\frac{\mu}{M_\odot}\right)^{\frac{1}{2}} \left(\frac{M}{M_\odot}\right)^{\frac{1}{3}} \left(\frac{100\,\text{Mpc}}{r}\right) \left(\frac{100\,\text{Hz}}{f_c}\right)^{\frac{1}{6}}.$$

$$\tag{46b}$$

The characteristic amplitude (46b) is enhanced over the actual rms wave strength $\langle h_+^2(t_c) + h_\times^2(t_c) \rangle^{\frac{1}{2}}$ the waves have at the time $t = t_c$ when they sweep through frequency f_c – enhanced by very nearly the square root of the number of periods, $n = (f^2/\dot{f})_{f=f_c} = (5/96\pi)(M/\mu)(\pi M f_c)^{-\frac{5}{3}}$, that the binary spends in the vicinity of the frequency f_c. This $\sqrt{n} \simeq 28(\mu/M_\odot)^{-\frac{1}{2}}(M/M_\odot)^{-\frac{1}{3}}(f_c/100\,\text{Hz})^{-\frac{5}{6}}$ enhancement corresponds to the enhancement in effective signal that the experimenters will achieve by optimal signal processing in their search for these frequency-sweeping bursts.

From a study of the waveform (42) using broad-band detectors at several widely spaced locations on the earth, one can deduce the following information: (i) the direction to the source (which comes from phase differences in the signal at different detectors in different locations); (ii) the inclination of the orbit to the line of sight (which comes from the amplitudes in the two different polarization modes); (iii) the direction the stars move in their orbit (which comes from the $+$ or $-$ sign in equation (42b)); (iv) the combination $(\mu^3 M^2)^{\frac{1}{5}}$ of the reduced and total masses; and (v) the distance r to the source. If the mass combination $(\mu^3 M^2)^{\frac{1}{5}}$ is $\lesssim 1.5 M_\odot$, one can be fairly sure the binary was made of neutron stars; if it is much larger, one can be fairly sure that at least one of the bodies was a massive black hole.

Especially intriguing is the possibility (Schutz, 1986b) that, in the case of neutron stars the coalescence will produce electromagnetic emission (e.g. due to an explosion of the less massive star, Blinnikov et al., 1984) that is strong enough to be detected at earth and thereby to pin down the source's location with far higher precision than can be obtained from the gravitational waves. In this case, a redshift will probably be obtainable from optical observations; and that redshift together with the gravitational-wave-determined distance r will give a value for the Hubble constant. Schutz (1986b), from a detailed study of the expected noise in future beam detectors, concludes that the prospects are good thereby to obtain a significantly better value for the Hubble constant than we now have. Even in the absence of electromagnetic signals from the coalescence, it may prove possible by statistical means to determine the Hubble constant using combined data for a number of coalescences; see Schutz (1986b) for details.

Clark, van den Heuvel and Sutantyo (1979) have estimated, from neutron-star observations in our own galaxy, that to see three coalescences of neutron-star binaries per year one must look out to a distance of 100^{+100}_{-40} Mpc, where the quoted uncertainties are at the 90% confidence level. Correspondingly, in Fig. 9.4 are shown the characteristic amplitude and frequency, for a range of values of the seismic cutoff f_s and corresponding values of $f_c = 1.44 f_s$, produced by the coalescence of two 1.4 solar mass neutron stars at 100 Mpc (estimated event rate about three per year) and at $\frac{1}{3}$ the Hubble distance (event rate about ten per day). Because much has been learned observationally about the statistics of neutron-star binaries since these estimates of event rates were made, a careful restudy of the estimates is much needed.

As Fig. 9.4 shows, future earth based beam detectors may be able to see black-hole coalescences throughout the universe, so long as the more massive of the two holes does not exceed $1000\,M_\odot/(1+z)$, where z is the hole's cosmological redshift. The coalescence rate for black-hole binaries of a few solar masses could be of order that for neutron-star binaries (a few per year at 100 Mpc), or might well be far lower. Particularly intriguing is a scenario, suggested as very plausible by Shapiro and Teukolsky (1985) and Quinlan and Shapiro (1987), in which a large fraction of galactic nuclei create, at some phase of their evolution, a dense cluster of neutron stars and small-mass black holes which – on a timescale of only a few years – form tight binaries that coalesce, with the coalesced holes then forming new tight binaries that coalesce, ... until the cluster goes unstable and collapses to form a single large hole. This scenario suggests that in typical years the earth might be hit by a number of spiraling wave bursts from coalescing binaries of masses $3 M_\odot$–$1000 M_\odot$ at the Hubble distance, $z \sim 1$. Also intriguing but much less likely is a scenario discussed by Bond and Carr (1984) in which a sizable fraction of the mass of the universe is in black-hole binaries with masses ~ 100–$1000 M_\odot$ for which the coalescence rate could be several per year in the local group of galaxies (distance ~ 1 Mpc).

One or more black hole binaries of any mass up to $\sim 10^8 M_\odot$ *might* have formed in the nuclei of a reasonable fraction of all galaxies during the past life of the universe, leading to event rates $\gtrsim 1$/year out to the Hubble distance (cf. Fig. 1 of Rees, 1983) – or they might never form. There is actually observational evidence for supermassive black-hole binaries formed by the coalescence of galactic nuclei (Begelman, Blandford and Rees, 1980; Rees, 1983); but the rate of such events probably does not exceed $\frac{1}{100}$ years out to the Hubble distance (Rees, 1983).

(f) *The fall of stars and small holes into supermassive holes*

The supermassive ($M_1 \gtrsim 10^5 M_\odot$) black holes thought to inhabit the nuclei of galaxies might typically grow by accretion on timescales as short as 10^8 years; see, e.g. Section 8.6 of Blandford and Thorne (1979). When such a hole grows larger than $10^9 M_\odot$, normal stars can pass near or plunge through its horizon without being torn apart tidally, and the number of stars that so scatter or plunge could well be of order one per year or more (e.g. Dymnikova, Popov and Zentsova, 1982). For smaller supermassive holes, scattering or plunging normal stars will be tidally disrupted, reducing the strength of their waves; but the reduction will not be great, at least in the case of radial infall, unless the hole is below $10^6 M_\odot$ (Nakamura and Sasaki, 1981; Haugan, Shapiro and Wasserman, 1982). For any hole, neutron stars and satellite holes can scatter or plunge through without enough disruption to strongly suppress their radiation; but the event rate (per supermassive hole) will typically be well below one per year.

The wave forms emitted when a star or small hole is scattered by or plunges into a supermassive hole have been evaluated with high precision using perturbation formalisms; see, e.g. Detweiler and Szedenits (1979), Kojima and Nakamura (1984a). The characteristic frequency and amplitude for typical (non-head-on) impact parameters are

$$f_c \cong \frac{1}{20 M_1} = 10^{-4} \, \text{Hz} \left(\frac{10^8 M_\odot}{M_1} \right), \tag{47a}$$

$$h_c \cong \frac{M_2}{2r_o} = 2 \times 10^{-21} \left(\frac{M_2}{M_\odot}\right)\left(\frac{10 \text{ Mpc}}{r_o}\right), \tag{47b}$$

where M_1 is the mass of the large hole, M_2 is that of the infalling body, and $r_o \sim 10$ Mpc might give a reasonable event rate since there are ~ 100 galaxies as massive as or more massive than our own inside this distance, including M87 for which observational data suggest a central black hole of mass $M_1 \sim 4 \times 10^9 M_\odot$ (Section 8.3.1.4 of Chapter 8). These h_c and f_c are plotted in Fig. 9.4 for several interesting sets of parameters. It is conceivable that such plunge bursts will be seen by beam detectors in space, if and when they are flown.

(g) Cherished beliefs

The above discussion makes clear how uncertain is our electromagnetically based knowledge of gravitational-wave sources. Correspondingly, it seems very likely that when gravitational waves are finally seen, they will come predominantly from sources we have not thought of or we have underestimated; and it seems quite possible that the waves will be stronger than the above estimates suggest.

In light of this, it is interesting to ask the following question: How strong could be the strongest bursts that strike the earth on average three times per year without violating our 'cherished beliefs' about the laws of physics and the universe? Zimmermann and Thorne (1980) have enumerated a set of cherished beliefs, including (i) that general relativity is correct, (ii) that we do not live in a special time or place in the universe, (iii) that there are no enormous primordial bursts, (iv) that no single, coherently radiating object in our galaxy has a mass exceeding $10^8 M_\odot$, and (v) that the strongest bursts were not beamed by their sources into solid angles $\ll \pi$. From these cherished beliefs they have derived the upper limit on the 3/year burst strength which is shown in Fig. 9.4. That limit could actually be achieved at frequencies $f \gtrsim 30$ Hz by an unlikely but not implausible scenario in which a large fraction of the mass of the universe was cycled, long ago, through a pre-galactic population of massive stars ('Population III' stars), leaving much of the universe's mass in compact binary systems that inhabit the halos of galaxies. If the mean lifetime of these binaries against spiraling together and coalescing is of order the age of the universe, then such coalescences in the halo of our own Milky Way galaxy would give bursts in the frequency domain $f \gtrsim 30$ Hz at the cherished belief level. For further discussion see Zimmermann and Thorne (1980), and Bond and Carr (1984).

While this scenario is unlikely, it is not totally implausible, and it serves to remind us that our best estimates of the waves bathing the earth could be grossly pessimistic.

9.4.2 Periodic sources

(a) *Characterization of the waves from periodic sources and the noise in a Detector searching for them*

The gravitational waves from a periodic source will be characterized by a discrete set of frequencies, and the waves at a given frequency will typically be right-hand or left-hand elliptically polarized; i.e. for some suitable choice of polarization axes $\hat{e}_{x'}$, $\hat{e}_{y'}$, they will have the form (similar to that of a decaying binary, equation (42), but with constant frequency)

$$h_+(t) = h_{o+} \cos 2\pi f t, \qquad h_\times(t) = \pm h_{o\times} \sin 2\pi f t. \tag{48}$$

If one wishes to quantify in a precise and standard manner all the properties of these waves, including the orientations of the 'preferred' x' and y' polarization axes, one might best do so using 'Stokes Parameters' analogous to those used in electromagnetic theory (Section 15 of Chandrasekhar, 1950). However, in this chapter we shall be concerned only with the frequencies f of the waves, and at a given emitted frequency, with a suitably defined characteristic amplitude h_c and a corresponding noise amplitude h_n in a detector searching for the waves.

As an aid in defining h_c and h_n, consider the following situation (analog of that for burst sources in Section 9.4.1(a)): a theorist tells us the frequency f, the phase, and the amplitudes $h_{o+}(\iota, \beta, r)$ and $h_{o\times}(\iota, \beta, r)$ to be expected from a specific model for a source, with orientation angles, ι, β and distance r. Suppose, further, that this type of source is distributed randomly throughout the universe and that the mean number of sources inside the distance r_o is n_o. What, then, will be (on average) the signal-to-noise ratio for the n_oth brightest source that a detector, broad-band or narrow, will see in a search (at known frequency and phase) lasting a time $\hat{\tau}$? By virtue of equations (27) and by analogy with equation (33) for burst sources, the answer turns out to be

$$\frac{S}{N} \cong \frac{h_c}{h_n(f)}. \tag{49}$$

Here

$$h_c \equiv (2/3)^{\frac{1}{2}} \langle |h_{o+}(\iota, \beta, r_o)|^2 + |h_{o\times}(\iota, \beta, r_o)|^2 \rangle^{\frac{1}{2}} \tag{50}$$

(with $\langle \cdots \rangle$ denoting an average over ι and β) is the characteristic amplitude of the periodic source (analog of (31b) for a burst source), and

$$h_n(f) \equiv \frac{[S_h(f)/\hat{\tau}]^{\frac{1}{2}}}{\langle|F_+(\theta, \phi, \psi)|^2\rangle^{\frac{1}{2}}} \tag{51}$$

is the noise amplitude. (The h_c of equation (50) is $(\frac{4}{3})^{\frac{1}{2}}$ larger than one naively would expect from equation (27). This factor $(\frac{4}{3})^{\frac{1}{2}}$ is an approximate correction for the fact that the angle averages in S/N should not be over squares but, rather, over squares associated with the rotation of the earth during data collection, then over (squares)$^{\frac{3}{2}}$ covering the rest of the sky and the orientation of the source, followed by a $\frac{1}{3}$ power after averaging; cf. the discussion preceding equation (30).) Correspondingly, if experimenters wish to be 90% confident of having seen that n_0th brightest source after $\frac{1}{3}$ year of search, then S/N must exceed $1.655 \simeq 1.7$ (Gaussian probability distribution), and h_c must exceed

$$h_{3/\mathrm{yr}} = 1.7h_n = \frac{1.7}{\langle|F_+|^2\rangle^{\frac{1}{2}}}[S_h(f) \times 10^{-7}\ \mathrm{Hz}]^{\frac{1}{2}} \text{ if } f \text{ and phase are known.} \tag{52a}$$

In the case that theory and electromagnetic observation have failed to tell us in advance the phase and frequency of the source, except to within Δf, the experimenters must try $\sim f/\Delta f$ values of the frequency, and correspondingly the Gaussian statistics of the noise will produce

$$h_{3/\mathrm{yr}} \simeq [2\ln(f/\Delta f)]^{\frac{1}{2}}h_n(f) = \frac{[2\ln(f/\Delta f)]^{\frac{1}{2}}}{\langle F_+^2\rangle^{\frac{1}{2}}}[S_h(f) \times 10^{-7}\ \mathrm{Hz}]^{\frac{1}{2}}$$

$$\text{if } f \text{ is known only to within } \Delta f \sim f. \tag{52b}$$

Fig. 9.6 shows h_c, for several postulated types of source (to be discussed in the following subsections), and correspondingly $h_{3/\mathrm{yr}}$ with f and the phase known, for several types of detectors (to be discussed in Section 9.5 below).

(b) Rotating neutron stars

A rotating neutron star (e.g. a pulsar) will emit gravitational waves at several frequencies as a result of deviations from symmetry around its rotation axis (deviations from 'axisymmetry'). The larger are those deviations and the more rapidly they rotate, the stronger will be the radiation.

Deviations from axisymmetry could arise in several ways: (i) The star's solid crust (well-established) or a solid core (not so well established) could support deformations that are residual remnants of the star's past history – a history that might be quite complex, including star quakes in which the crust or core cracks and deforms in much the manner of the solid earth in an earthquake; for detailed discussions see, e.g. Pandharipande, Pines and

Fig. 9.6. The characteristic amplitudes h_c (equation (50)) and frequencies f of waves from several postulated *periodic sources* (thin curves), and the sensitivities $h_{3/yr}$ of several existing and planned detectors (thick curves and circles) ($h_{3/yr}$ is the amplitude h_c of the weakest source detectable with 90% confidence in a $\frac{1}{3}$ yr $= 10^7$ s integration if the frequency and phase of the source are known in advance; equation (52a)). The sources shown in the high-frequency region, $f \gtrsim 10$ Hz, are all special cases of rotating, nonaxisymmetric neutron stars (Section 9.4.2(b)). The steeply sloping dotted lines labeled NS Rotation refer to rigidly rotating neutron stars with moment of inertial $I_{\bar{z}\bar{z}} = 10^{45}$ g cm^{-2}, and with various ellipticities ε and distances r labeled on the lines (equation (55)). The sources in the low-frequency region, $f \lesssim 0.1$ Hz, are all binary star systems in our galaxy (Section 9.4.2(c)): several specific, known binaries, which are indicated by name (μ Sco, V Pup, ...); the strongest six spectral lines from the famous binary pulsar PSR 1913 + 16; and the estimated strengths of the strongest white-dwarf ('WD') and neutron-star ('NS') binaries in our galaxy. The detectors are discussed in detail in the indicated subsections of Section 9.5.

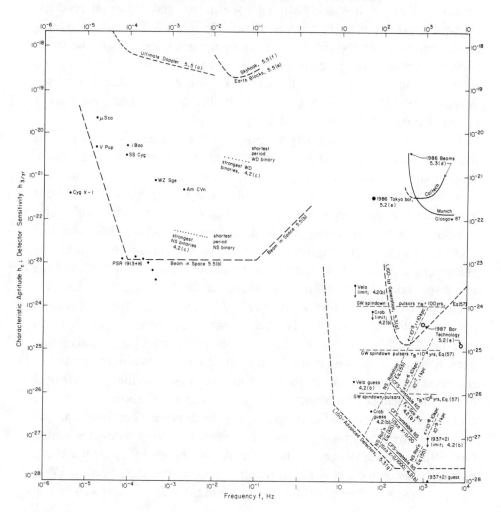

Smith (1976) and references therein. In old pulsars that have been spun up by accretion to near-millisecond rotation rates, theory and observational data suggest that the crust and core are quite well annealed into a nearly axisymmetric shape (Alpar and Pines, 1985); but in neutron stars that are only tens or hundreds or thousands of years old, it might well be otherwise (e.g. Zimmermann, 1978). (ii) The star's internal magnetic field, if sufficiently strong, could produce sufficient magnetic pressure to distort the star significantly (Zimmermann, 1978; Gal'tsov, Tsvetkov and Tsirulev, 1984). However, 'sufficiently strong' means, in the case, e.g., of the Crab Pulsar, ten times stronger than the star's measured surface field. (iii) If the star is rotating more rapidly than a critical rotation period, $P_{\text{crit}} \cong 0.7\text{–}1.7$ ms (which depends on the star's structure and its temperature-dependent viscosity), then an instability driven by gravitational radiation reaction ('Chandrasekhar (1970)–Friedman–Schutz' (1978), or 'CFS' instability) will create and maintain significantly strong hydrodynamic waves in the star's surface layers and mantle, propagating in the opposite direction to the star's rotation; and these will radiate strongly. For detailed discussions see Wagoner (1984), Lindblom (1986, 1987), Friedman, Ipser and Parker (1986), Schutz (1987), Cutler and Lindblom (1987).

At present we are extremely ignorant of the degree of asymmetry in rotating neutron stars, and accordingly we are ignorant of the strengths of the periodic waves to be expected from them. Pessimists will note that there is no observational evidence in any observed pulsar for sufficient non-axisymmetry to produce interestingly strong waves. Pessimists will point, especially, to the extremely small slowdown rate of the 1.6 ms pulsar PSR 1937 + 21, which implies such weak radiation reaction that the characteristic amplitude at earth cannot exceed 1×10^{-27} (Fig. 9.6), and the star's non-axisymmetric ellipticity cannot exceed 3×10^{-9}.

Optimists will also point to PSR 1937 + 21 and some other millisecond pulsars, and note a reasonably likely scenario for their origin (van den Heuvel, 1984; Wagoner, 1984): that they were spun up long ago by accretion from a binary companion until they hit the CFS instability, that they remained just beyond the instability point for awhile, with the spinup torque of accretion being counterbalanced by gravitational radiation reaction, and that the accretion stopped long ago leaving the stars plenty of time to anneal and settle down into their presently observed, highly axisymmetric states. Given the extreme observational difficulty of finding by electromagnetic means evidence for rapid neutron-star rotation (see, e.g., Section IV of Reynolds and Stinebring (1984) for searches in the radio), it may well be that there are a number of accreting neutron stars in our galaxy now in the CFS regime, radiating strong gravitational waves. For such a star the energy being radiated in gravitational waves and that being radiated as accretion-induced X-rays will both be proportional to the accretion rate; and consequently the characteristic amplitude of the gravitational waves at earth will be proportional to the square root of the X-ray flux arriving at earth, F_X:

$$h_c \simeq 2 \times 10^{-27} \left(\frac{300 \text{ Hz}}{f}\right)^{\frac{1}{2}} \left(\frac{F_X}{10^{-8} \text{ erg cm}^{-2} \text{ s}^{-1}}\right)^{\frac{1}{2}} \tag{53}$$

(Wagoner, 1984). The frequency f of the waves will be $f = l v_p/(2\pi R)$ where R is the star's radius, $l = 3$ or 4 or 5 is the spherical-harmonic order of the hydrodynamic wave, and v_p is the pattern speed of the wave as seen in the inertial frame of distant observers. The X-ray flux $F_X \simeq 10^{-8} \text{ erg cm}^{-2} \text{ s}^{-1}$ is $\frac{1}{20}$ that of Sco X-1, the brightest quasi-steady source in the sky and itself a candidate for a CFS-unstable object. As Fig. 9.6 shows, stars with X-ray fluxes as low as $\frac{1}{1000}$ Sco X-1 could be interestingly strong sources of gravitational waves. In such a star the density waves in the surface layers should modulate the emitted X-rays, but the sensitivities of past X-ray telescopes have been too poor to detect such rapid and weak modulations. There is an interesting proposal (Wood *et al.*, 1986) for a new, more sensitive X-ray telescope designed to search for such modulations in Sco X-1 and other, weaker X-ray sources. Such a telescope, operated in coordination with gravitational-wave detectors, might one day give a wealth of new information about neutron stars.

Even the most rapidly rotating of neutron stars will be smaller than a reduced wavelength of its emitted gravitational waves and thus can be described with reasonable accuracy by a slow-motion formalism (Ipser, 1971). If the emitter is CFS density waves, the multipoles involved are $l = 3$, 4, or 5. If the emitter is solidly supported or magnetic-field supported deformations, the waves should be largely quadrupolar, $l = 2$. Let us focus now on the latter case, i.e. on a star with rotation period large enough that the CFS instability does not act. Although gravity inside the star is significantly non-Newtonian, one can still define for the star a moment-of-inertia tensor I_{jk} equal to the ratio of its angular momentum to its angular velocity (both being vectors defined in the weak-gravity region well outside the star; see Thorne and Gürsel, 1983). If the star rotates about a principal axis of this moment-of-inertia tensor, i.e. if its angular momentum and angular velocity are parallel, then it will not precess, and (assuming a rotation period sufficiently long that it is CFS stable), its gravitational waves will be emitted at twice the rotation frequency. From the quadrupole variant of the slow-motion formalism we can then compute that the waves will have the standard periodic form of equation (48) with amplitudes

$$h_{o+} = 2(1 + \cos^2 \iota) \frac{(\mathcal{I}_{\bar{x}\bar{x}} - \mathcal{I}_{\bar{y}\bar{y}})(\pi f)^2}{r},$$

$$h_{o\times} = 4 \cos \iota \frac{(\mathcal{I}_{\bar{x}\bar{x}} - \mathcal{I}_{\bar{y}\bar{y}})(\pi f)^2}{r}. \tag{54}$$

Here ι is the angle between the neutron star's rotation axis and the line of sight from the earth, and $\mathcal{I}_{\bar{x}\bar{x}}$ and $\mathcal{I}_{\bar{y}\bar{y}}$ are the components of the star's quadrupole moment along the principal axes in its equatorial plane. The characteristic amplitude of these waves (equation (50)) is

$$h_c = 8\pi^2 \left(\frac{2}{15}\right)^{\frac{1}{2}} \frac{\varepsilon I_{\bar{z}\bar{z}} f^2}{r} = 7.7 \times 10^{-20} \varepsilon \left(\frac{I_{\bar{z}\bar{z}}}{10^{45} \text{ g cm}^2}\right) \left(\frac{f}{1 \text{ kHz}}\right)^2 \left(\frac{10 \text{ kpc}}{r}\right),$$

$$\tag{55}$$

where $I_{\bar{z}\bar{z}}$ is the moment of inertia of the star about its rotation axis (so $-\frac{1}{2}I_{\bar{z}\bar{z}}(\pi f)^2$ is its rotational energy), and

$$\varepsilon \equiv \frac{\mathcal{I}_{\bar{x}\bar{x}} - \mathcal{I}_{\bar{y}\bar{y}}}{I_{\bar{z}\bar{z}}} \tag{56}$$

is its 'gravitational ellipticity' in the equatorial plane. All neutron stars for which masses have been measured have M near $1.4 M_\odot$; and depending on the equation of state these masses correspond to $3 \times 10^{44} \text{ g cm}^2 \lesssim I_{\bar{z}\bar{z}} \lesssim 3 \times 10^{45} \text{ g cm}^2$. The likely values of the ellipticity ε are far less clear.

The observed slow-down rates of the Crab ($f = 60$ Hz, $r = 2$ kpc), Vela ($f = 22$ Hz, $r = 500$ pc), and PSR 1937 + 21 ($f = 1.25$ kHz, $r = 5$ kpc) pulsars, if due to gravitational radiation reaction (possible but not likely), correspond to $\varepsilon \simeq 6 \times 10^{-4}, 4 \times 10^{-3}$, and 3×10^{-9} respectively; and to $h_c \simeq 8 \times 10^{-25}, 3 \times 10^{-24}$, and 1×10^{-27}. Zimmermann (1978) argues that reasonable values for the Crab and Vela are $\varepsilon \sim 3 \times 10^{-6}$ and 3×10^{-5} corresponding to $h_c \sim 4 \times 10^{-27}$ and 2×10^{-26}. Alpar and Pines (1985) suggest reasonable values for PSR 1937 + 21 (which is old and well annealed) in the range $\varepsilon \sim 4 \times 10^{-10}$–$1 \times 10^{-11}$ corresponding to $h_c \sim 1 \times 10^{-28}$.

Blandford (1984) points out that, if there is a population of young pulsars (not yet discovered) that are spinning down by gravitational radiation reaction on a spin-down timescale τ_{GW}, then (i) the nearest will be at a distance $r \simeq R_G (\tau_B / \tau_{GW})^{\frac{1}{2}}$ (assumed $\lesssim \tau_{GW}$) is the mean time between births of these pulsars in our galaxy and R_G is the radius of the galaxy's disk;

(ii) the flux of gravitational-wave energy at earth from the nearest such pulsar $(3/64\pi)(2\pi f)^2 h_c^2$, will be equal to $[I_{\bar{z}\bar{z}}(\pi f)^2/\tau_{GW}](4\pi r^2)^{-1}$; and (iii) as a consequence, the characteristic amplitude of the waves from the nearest one will be

$$h_c \simeq \left[\frac{4}{3}\frac{I_{\bar{z}\bar{z}}}{r^2\tau_{GW}}\right]^{\frac{1}{2}} \simeq \left[\frac{4}{3}\frac{I_{\bar{z}\bar{z}}}{R_G^2\tau_B}\right]^{\frac{1}{2}} \sim 1.1 \times 10^{-25}\left(\frac{10^4 \text{ years}}{\tau_B}\right)^{\frac{1}{2}}, \tag{57}$$

independently of its frequency and ellipticity.

Fig. 9.6 shows values of h_c and f corresponding to some of the above possibilities.

If the star does not rotate about a principal axis of its moment of inertia, then it will precess. When one idealizes the star's interior as rigid, then although the interior gravity is significantly non-Newtonian, the precession is still described by the classic equations of Euler (see Thorne and Gürsel, 1983, for a proof); and the resulting waves will have a form that typically will entail significant spectral components at three frequencies: twice the rotation frequency, and the rotation frequency plus and minus the precession frequency (Zimmermann, 1980; Zimmermann and Szedenits, 1979). In reality, pliability of the neutron-star material will cause significant deviations from rigid rotation; but these three frequencies may still be dominant. If the star is idealized as a fluid body deformed by the pressure of an off-axis internal magnetic field, then the star does not precess and the radiation is emitted at the rotation frequency and twice the rotation frequency (Gal'tsov, Tsvetkov and Tsirulev, 1984).

If and when gravitational waves from rotating neutron stars are detected, they may carry a wealth of information about the star's structure and dynamics in the amplitudes and relative phasings of their various spectral components. Especially interesting may be the evolution of the various spectral components after a star quake; together with electromagnetic timing of the post-quake rotation, these may give us new insights into the coupling of the solid crust or core to the fluid mantle.

(c) *Binary stars*

Ordinary binary star systems are the most reliably understood of all sources of gravitational waves. From the measured mass and orbital parameters of a binary and its estimated distance, one can compute with confidence the details of its waves. Unfortunately, ordinary binaries have orbital periods no shorter than about an hour and, correspondingly, gravitational-wave frequencies $f \lesssim 10^{-3}$ Hz. The shortest known binary of all is a white-dwarf/

neutron-star system with orbital period 11 minutes and gravitational-wave frequency $f \simeq 3 \times 10^{-3}$ Hz (Priedhorsky, Stella and White, 1986). Because of seismic noise, detectors in earth laboratories cannot hope to see waves of such low frequency. However, beam detectors in space, tentatively planned for the turn of the century, should see them with relative ease (cf. Section 9.5.5(b) below).

Detailed formulas for the waves from a binary star, including the effects of the eccentricity and inclination of its orbit, have been derived from the quadrupole formalism by Peters and Mathews (1963) and Wahlquist (1987). See also Wagoner and Will (1976), and Gal'tsov, Matiukhin and Petukhov (1980) for post-Newtonian corrections. By virtue of the eccentricity of the orbit, waves will be emitted in equally spaced 'spectral lines' at twice the orbital frequency and harmonics thereof. For eccentricity $\varepsilon \lesssim 0.2$ the line at $f = 2f_{\text{orb}}$ is dominant; for $\varepsilon \simeq 0.5$ the lines at $f/f_{\text{orb}} \simeq 2$ through 8 are all strong; for $\varepsilon \simeq 0.7$ the lines at $f/f_{\text{orb}} \simeq 4$ through 20 are all strong. In the low-eccentricity case $\varepsilon \lesssim 0.2$ the waves have the form (42) with $f = 2f_{\text{orb}}$, which corresponds to a characteristic amplitude (equation (50))

$$h_{\text{c}} = 8 \left(\frac{2}{15} \right)^{\frac{1}{2}} \frac{\mu}{r} (\pi M f)^{\frac{2}{3}}$$

$$= 8.7 \times 10^{-21} \left(\frac{\mu}{M_{\odot}} \right) \left(\frac{M}{M_{\odot}} \right)^{\frac{2}{3}} \left(\frac{100 \text{ pc}}{r} \right) \left(\frac{f}{10^{-3} \text{ Hz}} \right)^{\frac{2}{3}}, \tag{58}$$

where μ is the reduced mass and M the total mass of the system. This amplitude is plotted in Fig. 9.6 for a few of the most strongly radiating known binaries. For lists of the most strongly radiating binaries and their characteristics see Braginsky (1965) and Douglass and Braginsky (1979).

White-dwarf and neutron-star binaries should also be important emitters – and they should extend to higher frequencies than ordinary binaries; but there is a paucity of observational data on them and the example of shortest known period has f only 3×10^{-3} Hz (see above). From the data that do exist (e.g. Iben and Tutukov, 1984), Lipunov and Postnov (1986), Lipunov, Postnov and Prokhorov (1987), and Hils et al. (1987) have estimated the characteristic amplitudes of the strongest white-dwarf and neutron-star binaries; see Fig. 9.6. For a very detailed treatment of the white-dwarf case see Evans, Iben and Smarr (1987). The highest frequency to be expected for any white-dwarf binary in our galaxy is 0.06 Hz since mass transfer from the less massive star to the more massive begins at or before this frequency; the highest for any neutron-star binary is 0.007 Hz since any binary of higher frequency than this would coalesce in a time less than the mean interval between coalescences, $\sim 10^4$ years.

Gravitational radiation reaction plays an important role in driving the evolution of close binary systems; see Paczynski and Sienkiewicz (1981) for details.

9.4.3 Stochastic sources

(a) *Characterization of stochastic gravitational waves*

It is useful to think about stochastic gravitational waves in terms of traveling-wave normal modes of the gravitational field. As with the electromagnetic field, there are two modes (because of two polarization states) for each volume $(2\pi\hbar)^3$ in phase space; and correspondingly, one can easily show, the energy density per unit logarithmic interval of frequency divided by the critical energy density $\rho_{crit} \sim 10^{-8}$ erg cm^{-3} to close the universe is

$$\Omega_{GW}(f) \equiv \frac{dE^{GW}/d^3x\,d\ln f}{\rho_{crit}} \sim \frac{\bar{n}}{10^{37}} \left(\frac{f}{1\text{ kHz}}\right)^3. \tag{59}$$

Here $\bar{n}$ is the average number of quanta in all modes with frequencies of order f. Below we shall see that $\Omega_{GW}(f)$ is likely to be $\gtrsim 10^{-14}$ at all frequencies of interest, and correspondingly the mean number of quanta in each mode is likely to be $\gtrsim 10^{20}$ – so large that a classical treatment is in order.

In TT coordinates the metric perturbation due to stochastic gravitational waves, evaluated at any chosen location x^i, will be a sum over contributions of all the modes of the field

$$h_{jk}^{TT}(t, x^i) = \sum_K h_{Kjk}^{TT}(t, x^i). \tag{60}$$

Here the index K labels modes of the field. For stochastic gravitational waves, the wave field h_{Kjk}^{TT} associated with mode K can be regarded as a 'random process' (i.e. a stochastically fluctuating function of time t); and the total field h_{jk}^{TT} at location x^i is the sum over all of the modes' random processes.

The field of a chosen mode K can be expressed as

$$h_{Kjk}^{TT} = h_K(t, x^i)e_{jk}^K, \tag{61}$$

where $h_K(t)$ is its scalar wave function and e_{jk}^K is its constant polarization vector, so normalized that $e_{jk}^K e_{jk}^K = 2$ in Cartesian coordinates; cf. equation (7c). Then h_K is a scalar random process in time (at fixed x^i) and its statistical properties can be characterized by a spectral density $S_{h_K}(f)$.

I shall assume that the modes are defined in such a way that there is no significant correlation between their wave fields h_{Kjk}^{TT}. As a result, when one averages the stress-energy tensor (9) of the waves at x^i over a sufficiently long time, all cross-terms between different modes get washed out and one obtains for the time-averaged specific intensity I_f at location x^i

$$I_f(t, x^i)\,\Delta\Omega \equiv \frac{dE}{dA\,dt\,df\,d\Omega}\,\Delta\Omega = \sum_{K \text{ in } \Delta\Omega} \frac{\pi f^2}{4} S_{h_K}. \tag{62}$$

Here E denotes energy, A denotes area, Ω denotes solid angle, and the sum is over all modes K with propagation directions in the infinitesimal solid angle $\Delta\Omega$. The total energy per unit logarithmic interval (used in defining $\Omega_{GW}(f)$ above) can be expressed in terms of the specific intensity in the standard way

$$\Omega_{GW}(f)\rho_{\text{crit}} = \frac{dE}{d^3x\,d\ln f} = \int f I_f\,d\Omega = \sum_K \frac{\pi f^3}{4} S_{h_K}, \tag{63}$$

where the integral is over the entire sphere and the sum is over all modes.

As for burst and periodic waves, so also for stochastic, we shall introduce a single characteristic amplitude h_c that is tied to a specific experimental situation: the experimenters use two identical gravity-wave receivers (broad-band or narrow), separated by a distance $\ll \lambda = c/2\pi f$, to search for *isotropic*, stochastic waves in the neighborhood of frequency f. The search is performed by a standard technique (Bendat, 1958; Section 9 of Drever, 1983): the outputs, $h_1(t)$ and $h_2(t)$ of detectors 1 and 2 are passed through identical filters which admit only Fourier components in a bandwidth $\Delta f \lesssim f$ centered on frequencies $\pm f$. The filtered outputs $w_1(t)$ and $w_2(t)$ are then multiplied together and integrated for a time $\hat{\tau}$ to get a single number $W = \int_{t_0}^{t_0+\hat{\tau}} w_1(t) w_2(t)$. This number will consist of a signal due to the identical stochastic backgrounds contained in $w_1(t)$ and $w_2(t)$, and a Gaussian noise due to the independent noises in the two detectors (each with the same spectral density $S_h(f)$). It turns out (e.g. Chapter 7 of Bendat 1958) that the ratio of the signal to the root-mean-square noise is

$$\frac{S}{N} = \frac{h_c(f)}{h_n(f)}, \tag{64}$$

where

$$h_c(f) \equiv \left[\sum_K f S_{h_K}(f) \right]^{\frac{1}{2}} = \left[\frac{4}{\pi f} \int I_f\,d\Omega \right]^{\frac{1}{2}} = \left[\frac{4}{\pi f^2} \Omega_{GW}(f)\rho_{\text{crit}} \right]^{\frac{1}{2}}$$

$$= 1.3 \times 10^{-18} \left(\frac{\rho_{\text{crit}}}{1.7 \times 10^{-8}\,\text{erg cm}^{-3}} \right)^{\frac{1}{2}} \left(\frac{1\,\text{Hz}}{f} \right) [\Omega_{GW}(f)]^{\frac{1}{2}} \tag{65}$$

is the characteristic amplitude of the isotropic, stochastic waves, and

$$h_n(f) = \frac{1}{(\frac{1}{2}\hat{\tau}\Delta f)^{\frac{1}{4}}} \frac{[f S_h(f)]^{\frac{1}{2}}}{\langle F_+^2 \rangle^{\frac{1}{2}}} \tag{66}$$

is the characteristic noise amplitude of the detectors. (Note: $\rho_{\text{crit}}/1.7 \times$

10^{-8} erg cm^{-3} = $(H_o/100$ km s^{-1} Mpc$^{-1})^2$ where H_o is the Hubble constant.) Correspondingly, if the experimenters wish to be 90% confident of having seen the stochastic background during a search of duration $\hat{\tau} = \frac{1}{3}$ year, the Gaussian probability distribution for S/N requires $S/N = 1.7$, which in turn means that h_c must exceed the noise level

$$h_{3/yr}(f) = 1.7h_n(f) = 2.0 \left(\frac{\Delta f}{10^{-7}\text{ Hz}}\right)^{-\frac{1}{4}} \frac{[fS_h(f)]^{\frac{1}{2}}}{\langle F_+^2 \rangle^{\frac{1}{2}}}. \tag{67}$$

The characteristic amplitudes h_c of various possible stochastic sources, and the noise levels $h_{3/yr}$ of various detectors are shown in Fig. 9.7.

(b) *Binary stars*

So many binary stars in our galaxy and in other galaxies radiate in the frequency region $f \lesssim 0.03$ Hz that they should superpose to produce a strong stochastic background. Lipunov and Postnov (1986), Lipunov, Postnov and Prokhorov (1987) and Hils *et al.* (1987) have made careful calculations of the characteristic amplitude of this stochastic background as a function of frequency; the results of Hils *et al.* are shown in Fig. 9.7 for the contribution of our own galaxy (which should be concentrated in the galactic plane). The contributions of all other galaxies (which should be isotropic) should be down from those of our own galaxy by $(h_c)_{\text{other}}/(h_c)_{\text{us}} \sim 0.15$.

The binary stochastic background in Fig. 9.7 is broken up into contributions from various types of binaries. Those shown as solid curves (unevolved binaries, WUMa stars (first discussed by Mironovskii, 1966), and cataclysmic variables (white-dwarf/normal-star systems)) are rather firmly based on optical studies of the statistics of these types of stars, and thus are rather reliable. Those shown dashed (close white-dwarf binaries (see Evans, Iben and Smarr, 1987, and the above references), and neutron-star binaries) are based on so little observational data and so much theory that they are highly uncertain.

The binary background presents a serious potential obstacle to searches for other kinds of waves in the frequency band 0.03 Hz $\lesssim f \lesssim 10^{-5}$ Hz where space-based beam detectors will operate. A broad-band burst can be seen above this background only if it has $h_{c\,\text{burst}} > h_{c\,\text{background}}$ – which means, e.g., that a $1M_\odot$ star falling into a supermassive black hole in the Virgo cluster will not be discernible unless the hole's mass is $M < 3 \times 10^5 M_\odot$ (cf. Figs. 9.4 and 9.7). A periodic source can be seen, after an integration time $\hat{\tau}$, only if it has $h_{c\,\text{periodic}} > (f\hat{\tau})^{-\frac{1}{2}} h_{c\,\text{background}}$ – which means, e.g., that if the close white-dwarf binaries are as numerous as estimated, the binaries ι Boo and SS Cyg will be discernible only after integration times of $\hat{\tau} > 10^7$ s. For further discussion see Evans, Iben and Smarr (1987) and Hils *et al.* (1987).

Fig. 9.7. The characteristic amplitudes h_c (equation (65)) and frequencies f of waves from several postulated *stochastic sources* (thin curves), and the sensitivities $h_{3/yr}$ of several existing and planned detectors (thick curves and circles) ($h_{3/yr}$ is the amplitude of the weakest source that can be detected with 90% confidence in a $\frac{1}{3}$ yr $= 10^7$ s integration). The very thin diagonal lines indicate values of h_c corresponding to constant $\Omega_{GW}(f) =$ (gravity wave energy in bandwidth $\Delta f = f$)/(energy to close the universe if $H_0 = 100$ km s^{-1} Mpc^{-1}). The sources shown as solid curves (background from various types of binary stars in our galaxy; Section 9.4.3(c)) are rather firm predictions. The sources shown as dashed and dotted curves are much less firm; see Sections 9.4.3(c)–(f) for discussion of specific sources. Not shown are primordial waves from the big-bang (Section 4.3(d)), which could have Ω_{GW} as large as 1 or as small as 10^{-14} or less according to various plausible scenarios – but which are limited by observations as discussed in Section 9.5.6. The detectors are discussed in detail in the indicated subsections of Section 9.5.

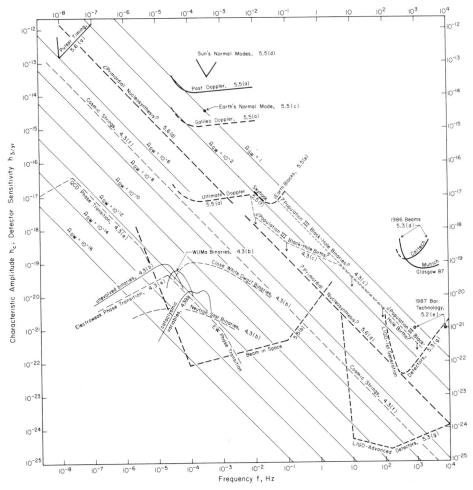

(c) *Population III stars*

If there was a pre-galactic population of massive stars ('Population III stars'; Carr, 1986), the violent events that terminated their lives (supernovae and collapse to black holes) might have produced gravitational waves that we today would see as isotropic and stochastic. Whereas existing binary stars are a firm source of stochastic background, these Population III stars are a highly speculative one. Carr (1980) has derived an upper limit on the characteristic amplitude that could have been produced by the deaths of such stars under any reasonable scenario; it is shown in Fig. 9.7 (upper short-dashed curve) along with the maximum characteristic amplitude that could come from the remnants of these stars if they became black-hole binaries that decay by radiation reaction on a timescale $\tau_{GW} \simeq 10^{10}$ years (Bond and Carr, 1984).

(d) *Primordial gravitational waves*

Photons coming from the big bang last scattered off matter at a cosmological redshift $z \sim 1000$ when the universe was roughly one million years old; and neutrinos last scattered at $z \sim 10^{10}$ when it was about 0.1 s old. An order-of-magnitude calculation shows that gravitons, by contrast, last scattered at roughly the Planck time, i.e. during the first 10^{-43} seconds when spacetime was quantized and the laws of physics were exceedingly different from today (Section 7.2 of Zel'dovich and Novikov, 1983). (An unlikely exception occurs if, at the epoch when the waves' reduced wavelength was $\lambda \sim$ (horizon size), much of the universe's energy density was in relativistic particles with mean free paths of order λ; then non-negligible absorption may occur; see Vishniac (1982).) Thus, in studying primordial gravitational waves (waves created in the big-bang), one usually can ignore their subsequent interactions with matter.

Not so for their subsequent interactions with the background spacetime curvature of the universe. Grishchuk (1974, 1975*a*,*b*, 1977) has shown that, as the primordial perturbations that give rise to present-day waves 'come inside the cosmological horizon' – and also before they enter the horizon – they can be parametrically amplified by their interaction with the dynamical background spacetime curvature; in other words, they can trigger further graviton creation. In this way exceedingly small initial fluctuations can be amplified into an interestingly strong stochastic background today.

Just how much stochastic background is produced depends crucially on ill-understood aspects of the initial singularity and on the equation-of-state-dependent and vacuum-dependent expansion rate in the very early universe. Some otherwise plausible models produce so much, $\Omega_{GW}(f) \gg 1$, as to be in violent conflict with the observed current state of the universe (e.g. p. 621 of Zel'dovich and Novikov, 1983). Other, equally plausible models can produce so little, $\Omega_{GW}(f) \ll 10^{-14}$, that there is no hope of detecting the waves in the foreseeable future.

In currently fashionable inflationary models of the universe vacuum fluctuations which initially are smaller than the horizon ($\lambda \ll \mathcal{R}_B =$ (background radius of curvature)) are driven outside the horizon ($\lambda \gg \mathcal{R}_B$) by the inflationary expansion. While outside the horizon, they are 'frozen' with constant amplitude h. Much later, after inflation ends, non-inflationary expansion brings them back inside the horizon. The number of quanta in each mode before entering and after leaving the horizon is

$$ n \sim \left[\frac{1}{16\pi} \left(\frac{h}{\lambda} \right)^2 \right] (2\pi\lambda)^3 \left(\frac{1}{\hbar/\lambda} \right) \sim \frac{\pi}{4} \frac{(h\lambda)^2}{\hbar}, $$

where the first factor is the waves' energy density, the second is the volume occupied by each mode, and the third is 1/(energy of one graviton). Before entering the horizon $n \simeq \frac{1}{2}$; so the above relation says that upon leaving it

$$ n_{out} \simeq n_{enter} \frac{\lambda_{leave}}{\lambda_{enter}} = \frac{1}{2} \frac{a_{leave}}{a_{enter}}, $$

where a is the expansion factor of the universe. Thus, the epoch of amplitude freezing is actually an epoch of parametric amplification (stimulated creation of new gravitons); and the total number of gravitons created depends on the total amount of expansion that occurs while the waves are outside the horizon. (There will be additional parametric amplification as the waves emerge from the horizon, $\lambda \sim \mathcal{R}_B$, but in inflationary models that is generally small compared to the amplification during freezing, $\lambda \gg \mathcal{R}_B$.) The total amount of inflationary expansion differs from one inflationary model to another; and correspondingly, the models can give Ω_{GW} as large as unity or Ω_{GW} too small ($\ll 10^{-14}$) for there to be hope of detecting the waves.

For discussions of the influence of the equation of state in the early universe on the spectrum of the amplified waves, see Grishchuk (1977) and Fig. 4 of Grishchuk and Polnarev (1980). For calculations of the waves produced by specific inflationary scenarios see Starobinsky (1979), Rubakov, Sazhin and Veryaskin (1982), Abbott and Wise (1984), Halliwell and Hawking (1985), Mijic, Morris and Suen (1986) and references therein. Because the range of possible strengths of primordial waves is so great, we do not bother to show it in Fig. 9.7 – aside from indicating the values of h_c corresponding to various values of $\Omega_{GW}(f)$.

(e) *Phase transitions*

During the early expansion of the universe, there may have been first-order phase transitions associated with QCD interactions and with Electroweak interactions. In each of these phase transitions the original phase would be supercooled, by the cosmological expansion, below the equilibrium temperature of the new phase. Bubbles of the new phase would then nucleate at isolated locations and expand at near-light velocity until they have compressed the original phase enough for the two phases to coexist in equilibrium. As Witten (1984) has pointed out, and Hogan (1986) has analyzed in detail, this 'cavitation' should have produced gravitational waves in two ways: (i) directly from expanding bubbles and the subsequent sound waves they generate, and (ii) subsequently from the inhomogeneities associated with the two co-existing phases (large-scale density inhomogeneities and corresponding inhomogeneities in the Hubble expansion rate). The resulting gravitational waves should possess a spectrum that peaks at wavelengths which were of order the horizon size when the cavitation occurred. Those wavelengths correspond to frequencies today $f_{max} \sim (2 \times 10^{-7} \, \text{Hz})(kT/1 \, \text{GeV})$, where T is the temperature of the phase transition. Hogan's (1986) predicted spectra, shown in Fig. 9.7, thus peak at $f_{max} \sim 2 \times 10^{-8} \, \text{Hz}$ (QCD, $T \sim 100 \, \text{MeV}$) and $f_{max} \sim 2 \times 10^{-5} \, \text{Hz}$ (Electroweak, $T \sim 100 \, \text{Gev}$). The $f^{+\frac{1}{2}}$ shape of the spectra at frequencies $f > f_{max}$ is a firm prediction, but the shape f^{-1} at $f < f_{max}$ is not (it could well be f^{-p} with $p > 1$). The amplitude shown is a reasonable upper limit, unless the phase transition is unusually catastrophic with very strong supercooling.

(f) *Cosmic strings*

Long before the QCD and Electroweak phase transitions – i.e. nearer the initial singularity – there may have been a phase transition associated with the grand-unified interactions, and that transition may have created cosmic strings – one-dimensional 'defects' in the vacuum with mass per unit length estimated to be $\mu \sim 10^{-6}$ and with tension equal to mass per unit length (Zel'dovich, 1980; Vilenkin, 1981a). As the universe's horizon expands to uncover the stochastic inhomogeneities in a string's shape, those inhomogeneities should begin to vibrate with speeds up to the speed of light. By self-intersection of the string, closed loops should form; and those loops could have acted as seeds for the condensation of galaxies and galaxy clusters (Zel'dovich, 1980; Vilenkin, 1981a; Turok and Brandenberger, 1986; Sato, 1987).

An unavoidable byproduct of this model for galaxy formation is huge amounts of stochastic gravitational waves produced by the vibrations of the closed loops (Vilenkin, 1981b). Detailed calculations by Vachaspati and Vilenkin (1985) (confirming earlier, less accurate calculations by many others) predict that, if the strings are not superconducting (for the superconducting case see Ostriker *et al.*, 1986),

$$\Omega_{GW}(f) \sim 10^{-7} \left(\frac{\mu}{10^{-6}} \right)^{\frac{1}{2}} \quad \text{for all } f \gtrsim 10^{-8} \text{ Hz} \left(\frac{10^{-6}}{\mu} \right). \tag{68}$$

(For the spectrum at lower frequencies see Hogan and Rees, 1984.) If μ is far less than 10^{-6} (i.e. if gravity-wave observations constrain $\Omega_{GW}(f)$ to be $\ll 10^{-7}$), then the non-superconducting cosmic-string theory of galaxy formation will face severe difficulties. Fig. 9.7 shows the predicted waves (68). From that diagram and the corresponding discussions in Section 9.5 it is clear that several different observational techniques have the prospect of placing cosmic string theory in jeopardy – or, hopefully, of discovering string-produced waves. (The apparent disproof of $\Omega_{GW} \sim 10^{-7}$ coming from 5°-scale anisotropy of the cosmic microwave radiation (Fig. 9.7 and Section 9.5.6(c)) does not in fact constrain cosmic strings, since this observational limit is sensitive only to waves that were present and had (reduced wavelength) $\sim$ (horizon size) at the epoch of recombination – before the strings that produce this wavelength began to vibrate and radiate.)

9.5 Detection of gravitational waves

9.5.1 *Methods of analyzing gravitational wave detectors*

When analyzing the performance of a gravitational wave detector, it is important to pay attention to the size L of the detector compared with a reduced wavelength λ of the waves it seeks.

If $L \ll \lambda$ then the detector can be contained entirely in the proper reference frame of its center, and the analysis can be performed using non-relativistic concepts augmented by the quadrupolar gravity-wave force field (3), (5). If one prefers, of course, one instead can analyze the detector in TT coordinates using general relativistic concepts and the spacetime metric (8). The two analyses are guaranteed to give the same predictions for the detector's performance, unless errors are made. However, errors are much more likely in the TT analysis than in the proper-reference-frame analysis, because our physical intuition about how experimental apparatus behaves is proper-reference-frame based rather than TT-coordinate based. As an example, we intuitively assume that if a microwave cavity is rigid, its walls will reside at fixed coordinate locations x^j. This remains true in the detector's proper reference frame (aside from fractional changes of order $(L^2/\lambdabar^2)h$, which are truly negligible if the detector is small and which the proper-reference-frame analysis ignores). But it is not true in TT coordinates; there the coordinate locations of a rigid wall are disturbed by fractional amounts of order h, which are crucial to analyses of microwave-cavity-based gravity wave detectors.

Thus, for small detectors, $L \ll \lambdabar$, the proper-reference-frame analysis is much to be preferred.

For large detectors, $L \gtrsim \lambdabar$, one cannot introduce a proper reference frame that covers the entire detector. Such detectors can only be analyzed using general relativistic concepts in TT coordinates (usually the best) or in some other suitable coordinate system (rarely as good).

9.5.2 *Resonant bar detectors*

More effort has been put into resonant bars than into any other type of gravity-wave detector. Weber's original detectors were of the resonant-bar type; all but one of the first-generation (pre-1977) earth-based detectors were of this type; and eight of the world's twelve research groups now building and operating earth-based detectors are working with bars. Of the current bar efforts three are in the United States (the University of Maryland (Weber, 1986), Stanford University (Boughn *et al.*, 1982; Michelson, 1983) and Louisiana State University (Hamilton *et al.*, 1986)); two are in Europe (the University of Rome with its detector sited at CERN (Amaldi *et al.*, 1984) and Moscow University (Braginsky, 1983)); and three are in the Far East (The University of Western Australia in Perth (Blair, 1983), Tokyo University (Owa *et al.*, 1986) and Guangzhou, China (Hu *et al.*, 1986)). The improvements in resonant-bar sensitivities since Weber's first detector have been a factor of roughly 200 in amplitude, corresponding to 40 000 in energy; and significant further improvements are yet to come.

(a) *How a resonant-bar detector works*

Schematically (Fig. 9.8), a resonant-bar detector consists of a large, heavy, solid bar whose mechanical oscillations are driven by gravitational waves, a transducer that converts information about the bar's oscillations into an electrical signal, an amplifier for the electrical signal, and a recording system. The transducer and amplifier together are sometimes called the sensor.

The transducer typically is mounted on one end of the bar (though other mountings are sometimes used), and it produces an output voltage or current proportional to the displacement $x(t)$ of the bar's end from equilibrium. Although $x(t)$ is a sum of contributions from all the $\sim 10^{29}$ normal modes of the bar, the transducer's output is filtered by the amplifier so that only the contribution of the bar's fundamental normal mode is passed on through. This is accomplished by a band-pass filter centered on the frequency f_0 of the fundamental mode, with bandwidth Δf somewhat smaller than the difference $f_1 - f_0$ between the bar's fundamental and its first harmonic. Thus, in effect, it is the fundamental mode of the bar that acts as the gravity-wave detector; and all the other normal modes are almost irrelevant.

Since the fundamental mode involves the relative in and out motion of the bar's left and right ends with just one node (at the bar's center), it corresponds to a standing sound wave with wavelength twice the length of the bar. Correspondingly, the bar's length must be

$$L \simeq \tfrac{1}{2} v_s / f_0, \tag{69}$$

where v_s is the speed of sound in the bar. Typical solid materials have longitudinal sound speeds of order 5 km s^{-1}; astrophysics suggests (Section 9.4) that 1 kHz is a reasonable frequency to search for gravitational waves; and correspondingly the lengths of typical resonant bar detectors are about 2 m and their masses are several tonnes. Notice that equation (69) gives for the ratio of the length of the bar to the reduced wavelength of the

Fig. 9.8. Schematic diagram of a *resonant-bar detector* for gravitational waves. The angles (θ, ϕ, ψ) characterizing the propagation and polarization directions of the waves relative to the detector are a specialization of the angles (θ, ϕ, ψ) shown in Fig. 9.2.

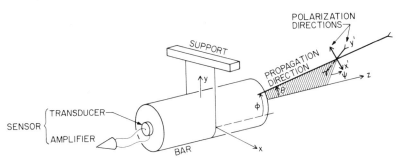

gravitational waves, $\lambda = c/2\pi f_0$,

$$L/\lambda \simeq \pi v_{\rm s}/c \simeq 5 \times 10^{-5}. \tag{70}$$

This justifies, with high accuracy, the use of a Newtonian-language, proper-reference-frame viewpoint in analyses of bar detectors. I shall adopt that viewpoint in this chapter.

The contribution of the fundamental mode to the displacement $x(t)$ of the bar's end can be expressed in the standard harmonic-oscillator form

$$x(t) = {\rm Real}[X(t)\,{\rm e}^{-{\rm i}2\pi f_0 t}], \tag{71}$$

where

$$X(t) = X_1 + {\rm i} X_2 \tag{72}$$

is the mode's complex amplitude and f_0 is its eigenfrequency. It is actually the quantity $X(t)$ that the sensor monitors. To the extent that one can ignore forces from the fundamental mode's environment (e.g. very weak coupling to the other modes), only gravitational waves will produce time changes of $X(t)$. When waves hit, they drive $X(t)$ to evolve in a complicated, time-dependent way, and that evolution in principle can be deconvolved to reveal some details of the waveform $h_{jk}^{\rm TT}(t)$.

Unfortunately, noise in the sensor is typically so severe that to control it the experimenter must use a bandwidth Δf that is far smaller than the frequency f_0 (typically $\Delta f/f_0 \sim 0.01$ in present detectors). Correspondingly, the sensor averages $X(t)$ for a time $\hat{\tau} \simeq 1/\Delta f$ that is long compared to the period $P_0 = 1/f_0$ of the gravitational waves being sought, before passing $X(t)$ on to the recording system. This means that for typical gravitational-wave bursts (e.g. from supernovae), which have durations of only a few times P_0, all that can be monitored is the total change ΔX in the complex amplitude from before the wave arrives until after it has passed. Only uncommonly long bursts, those lasting for more than $f_0/\Delta f \sim 100$ cycles, can be monitored in greater detail. In the future, however, there is hope of bringing the sensor noise under better control and thereby opening up the bandwidth to $\Delta f \simeq 0.2 f_0$ to permit detailed monitoring of much shorter bursts (Michelson and Taber, 1984). (For a description of several first-generation broad-band bars see Figure 2 of Drever, 1977 and associated text, and references therein.)

(b) *The sensitivity of bar detectors to short bursts*

A bar detector couples to the field (equation (26))

$$h(t) = F_+(\theta, \phi, \psi)h_+(t; \iota, \beta) + F_\times(\theta, \phi, \psi)h_\times(t; \iota, \beta), \tag{73}$$

where, if the bar is axially symmetric and the direction (θ, ϕ) and polarization angle ψ are defined as in Fig. 9.8, the beam-pattern factors are

$$F_+ = \sin^2 \theta \cos 2\psi, \qquad F_\times = \sin^2 \theta \sin 2\psi. \tag{74}$$

(See Chapter 37 of MTW for the key elements of a derivation.) We shall presume, throughout this subsection, that $h(t)$ is a burst of such short duration $\Delta t \lesssim \hat{\tau} = 1/\Delta f$ that the optimal way to search for it is to measure the mean square change $|\Delta X|^2$ it produces in the fundamental mode's complex amplitude. In this case the general formula (29) for the ratio S^2/N^2 of the burst's squared signal to the mean-square Gaussian noise in the detector can be reduced to the simple form (see, e.g., Giffard, 1976; Pallotino and Pizella, 1981; Michelson and Taber, 1984)

$$\frac{S^2}{N^2} = \frac{\frac{1}{2}M_{\text{eff}}(2\pi f_0)|\Delta X|^2}{kT_{\text{n}}} \tag{75}$$

Here k is Boltzmann's constant, T_{n} is a 'noise temperature' which characterizes the overall noise in the detector, and M_{eff} is an 'effective mass' associated with the fundamental mode, so defined that $\frac{1}{2}M_{\text{eff}}|X|^2(2\pi f_0)^2$ is the total energy in the mode when it is vibrating with complex amplitude X. (Since X is actually the amplitude of motion of the end of the bar, for a bar that has uniform cross-section and is long compared to its diameter, the effective mass is $M_{\text{eff}} = \frac{1}{2}(1 + v^2)M$ where v is the Poisson ratio of the bar's material and M is the bar's mass.)

Because the net wave-induced change ΔX in complex amplitude is independent of the mode's initial complex amplitude, the numerator of equation (75) is the energy that the wave would have deposited in the mode if the mode had been initially unexcited. This deposited energy can conveniently be expressed in terms of the cross-section $\sigma_0(f)$ that the mode would present to the wave if the wave had hit it from an optimal direction (broadside, $\theta = \pi/2$) and with an optimal polarization ($+$ mode with $\psi = 0$):

$$\frac{1}{2}M_{\text{eff}}(2\pi f_0)^2|\Delta X|^2 = \int_0^\infty \frac{\pi}{2} f^2|\tilde{h}(f)|^2 \sigma_0(f)\, df. \tag{76}$$

Here $\tilde{h}(f)$ is the Fourier transform of $h(t)$ (equation (73)), and for an optimal direction and polarization $(\pi/2)f^2|\tilde{h}(f)|^2$ would be the energy per unit area per unit frequency ($f \geq 0$) carried by the waves. Because $\sigma_0(f)$ is extremely sharply peaked around the resonant frequency f_0 (Section 37.5 of MTW), we can rewrite (76) and thence (75) in the form

$$\frac{S^2}{N^2} = \frac{\pi}{2} f_0^2|\tilde{h}(f_0)|^2 \frac{\int \sigma_0\, df}{kT_{\text{n}}}. \tag{77}$$

This is a special narrow-band-detector version of the general equation (29) for arbitrary detectors. Correspondingly, equation (30) for the strongest burst seen, on average, at the same rate as bursts occur inside the distance to our source, becomes

$$\frac{S^2}{N^2} \cong \frac{3}{2}\frac{\pi}{2} f_0^2 \langle |\tilde{h}_+(f_0)|^2 + |\tilde{h}_\times(f_0)|^2 \rangle \langle F_+^2 \rangle \frac{\int \sigma_o \, df}{k T_n}, \tag{78}$$

where, for the F_+ and $F_\times$ of (74),

$$\langle F_+^2 \rangle = \langle F_\times^2 \rangle = \frac{4}{15} = 0.267. \tag{79}$$

Equation (78) motivates us to define

$$h_c \equiv (3)^{\frac{1}{2}} f_0 \langle |\tilde{h}_+(f_0)|^2 + |\tilde{h}_\times(f_0)|^2 \rangle^{\frac{1}{2}} \tag{80}$$

as the *characteristic amplitude of the burst* (analog of (31b) for broad-band detectors) and

$$h_n \equiv \left[\frac{15}{\pi}\frac{k T_n}{\int \sigma_o \, df} \right]^{\frac{1}{2}} \cong 2.2 \left[\frac{G}{c^3}\frac{k T_n}{\int \sigma_o \, df} \right]^{\frac{1}{2}} \tag{81}$$

as the *detector's characteristic noise amplitude* (analog of (32) for a broad-band detector). Correspondingly, the characteristic amplitude $h_c = h_{3/yr}$ that a source must have to be detectable with 90 % confidence in a search lasting $\frac{1}{3}$ year is

$$h_{3/yr} \cong 5 h_n \cong 11 \left[\frac{G}{c^3}\frac{k T_n}{\int \sigma_o \, df} \right]^{\frac{1}{2}}$$

$$= 2.0 \times 10^{-16} \left[\frac{T_n/1K}{\int \sigma_o \, df/10^{-21} \, cm^2 \, Hz} \right]^{\frac{1}{2}}. \tag{82}$$

For a broad-band burst that is peaked near the detector's resonant frequency f_0 (so the f_c of equation (31a) is approximately f_0) and that lasts for a time not much longer than $\Delta t = 1/f_0$, the narrow-band characteristic amplitude (80) will be roughly equal to the broad-band characteristic amplitude (31b). For such bursts, and only for such bursts, it makes sense to plot a bar detector's narrow-band $h_{3/yr}$ on the same graph (Fig. 9.4) as a broad-band detector's $h_{3/yr}$. Inspiraling binaries do not belong to this class of bursts, so their detection by bars must be discussed separately from Fig. 9.4.

(c) *How to optimize the sensitivities of bar detectors*

Equation (82) shows that to optimize the sensitivity of a bar detector to short, broad-band bursts, one must achieve the largest possible frequency-integrated cross section $\int \sigma_0 \, df$ and the smallest possible noise temperature T_n.

The integrated cross section $\int \sigma_0 \, df$ can be computed by analyzing the response of the bar's fundamental mode to the force field (5) produced by a sinusoidal wave with optimal direction and polarization, and by then integrating over the wave's frequency. See MTW Box 37.4 for details. For a cylindrical bar with length L somewhat greater than radius R (the usual situation), the result depends only on the bar's mass M and internal sound velocity $v_s = (E/\rho)^{\frac{1}{2}}$ with $E = $ (Young's modulus) and $\rho = $ (density)(Paik and Wagoner, 1976):

$$\int_0 \sigma_0 \, df = \frac{8}{\pi} \frac{GMv_s^2}{c^3} [1 + \tfrac{1}{2}v(1-2v)(\pi R/L)^2 + \cdots]$$

$$= 1.6 \times 10^{-21} \, \text{cm}^2 \, \text{Hz} \left(\frac{M}{10^3 \, \text{kg}}\right) \left(\frac{v_s}{5 \, \text{km s}^{-1}}\right)^2. \tag{83}$$

(Here v is the Poisson ratio of the bar's material, and only the leading shape-dependent corrections are given.) Thus, it is desirable to build detectors that are as massive and have as high sound speed as possible.

The noise temperature T_n is determined by a combination of noise in the sensor and thermal noise in the bar.

The *thermal noise in the bar* is caused by weak coupling of the bar's fundamental mode to its environment – its $\sim 10^{29}$ other modes, the wire or cable or prongs that suspend the bar, and the residual gas in the vacuum chamber (Braginsky, Mitrofanov and Panov, 1985). If, as is normal, this environment is thermalized at some physical temperature T_b (subscript 'b' for bar or for thermal bath), then these couplings cause the mode's amplitude to execute a random walk (Brownian motion) in the domain $|X| \lesssim X_{th}$ corresponding to an energy kT_b:

$$X_{th} = \left[\frac{2kT_b}{M_{eff}(2\pi f_0)^2}\right]^{\frac{1}{2}}. \tag{84}$$

The fluctuation-dissipation theorem states that the timescale on which this random walk produces changes of order X_{th} is the same as the timescale $\tau^* = Q/\pi f_0$ for large-amplitude vibrations to be damped frictionally. (Here Q is the quality factor of the mode's large-amplitude oscillations: the number of radians of oscillation required for the energy to damp by a factor e.) Consequently, the mean square Brownian change in the mode's amplitude during the sensor's averaging time $\hat{\tau} = 1/\Delta f$ is

$$|\Delta X_{th}|^2 = X_{th}^2 \frac{\hat{\tau}}{\tau^*} = X_{th}^2 \frac{\pi}{Q} \frac{f_0}{\Delta f} \tag{85}$$

corresponding to

$$\tfrac{1}{2} M_{eff} (2\pi f_0)^2 |\Delta X_{th}|^2 = \pi \frac{kT_b}{Q} \frac{f_0}{\Delta f}. \tag{86}$$

The details of the *sensor noise* depend on a variety of factors, including: the detailed structure of the transducer (for a review of many transducer structures see Section 4.1.5 of Amaldi and Pizella, 1979), the strength of coupling of the transducer to the bar (characterized by a dimensionless coupling constant β that is roughly equal to the number of cycles of oscillation required for all the fundamental mode's energy to be fed into the transducer); the impedance mismatch between the transducer and the amplifier; the noise temperature T_a of the amplifier (converted to an equivalent noise temperature $(f_0/f_a)T_a$ at the bar's frequency f_0 if the amplifier operates at a different frequency f_a than the bar); and the sensor bandwidth Δf. Although the details vary from one sensor to another (see, e.g., Weiss, 1978; Pallotino and Pizella, 1980; Blair, 1983; Michelson and Taber, 1984), the spirit of the details is captured by the following approximate expression for the mean square change $|\Delta X_{sensor}|^2$ that the experimenter would infer from the sensor's output if (i) only the sensor noise were present, (ii) back-action forces of the sensor on the bar were negligible (see subsection (f) below), and (iii) impedances were properly matched:

$$\tfrac{1}{2} M_{eff} (2\pi f_0)^2 |\Delta X_{sensor}|^2 \simeq \frac{kT_a(f_0/f_a)}{\beta} \frac{\Delta f}{f_0}. \tag{87}$$

The sum of $\tfrac{1}{2} M_{eff} (2\pi f_0)^2 |\Delta X_{sensor}|^2$ (equation 87)) and $\tfrac{1}{2} M_{eff} (2\pi f_0)^2 |\Delta X_{th}|^2$ (equation (86)) is the mean square noise N^2 that appears in the denominator of equation (75); i.e., by definition, the detector's noise energy kT_n is this sum. If the experimenters choose too large a bandwidth Δf, then the sensor noise (87) will become inordinately large, producing a large T_n and masking the gravity-wave signal. If they choose too small a bandwidth, then the bar's thermal noise (86) will become inordinately large. There is an optimal bandwidth, typically $\Delta f \sim 10$ Hz in present detectors (corresponding to the averaging time $\hat{\tau} \sim 0.1$ s) at which roughly half the noise is from the sensor and half is from thermal motion of the bar. When the bandwidth is chosen optimally these two noise sources together produce a detector noise temperature

$$T_n \simeq \left[\frac{\pi}{\beta} \frac{T_b}{Q} \left(T_a \frac{f_0}{f_a} \right) \right]^{\frac{1}{2}}. \tag{88}$$

In practice, the experimenters choose a bar early in their experiments, thereby fixing the integrated cross section $\int \sigma_o \, df$; and they then struggle for many years to develop a sensor and its coupling to the bar, and a thermally cold environment, that will minimize the noise temperature T_n. Maximizing $\int \sigma_o \, df$ is achieved by maximizing the bar's mass and its velocity of sound (and, to a small extent, optimizing its shape subject to other experimental constraints such as available cryostats.) Minimizing T_n is achieved according to equation (88) by (i) maximizing the fundamental mode's quality factor Q (i.e. minimizing its coupling to the rest of the world), (ii) cooling the bar to as low a physical temperature T_b as possible, (iii) maximizing the strength β of coupling of the transducer to the bar, (iv) using an amplifier with as low a 'noise number' $kT_a/(2\pi\hbar f_a)$ as possible (the Heisenberg uncertainty principle limits the noise number to be $\gtrsim 1$; Weber, 1959; Heffner, 1962; Caves, 1982), and (v) struggling to get good impedance matching of the transducer and amplifier (a requirement for equation (88) to be valid).

(d) *Parameters of first- and second-generation bar detectors*

The first-generation bar detectors (pre-1977) were all made of aluminum, weighed roughly 1.5 tonnes, and had $f_0 \simeq 1.6$ kHz and $Q \sim 10^5$; they all operated at room temperature, $T_b \cong 300$ K; and most used piezo-electric transducers – i.e. crystals glued to the bar which produce small voltages when squeezed. For most the coupling of the transducer to the bar was weak, $\beta \lesssim 10^{-4}$; but those in Britain achieved strong coupling, $\beta \simeq 0.2$ and hence wide bandwidth $\Delta f/f_0 \sim 1$ at the price of reducing the bar's Q from $Q \sim 10^5$ to $Q \sim 2000$. The best amplifiers that could be impedance matched to the piezo-electric transducers had rather large noise numbers. For these first-generation bars the integrated cross-sections were $\int \sigma_o \, df \simeq 2 \times 10^{-21}$ cm^2 Hz, and the lowest detector noise temperatures were $T_n \simeq 4$ K corresponding to a minimum detectable burst amplitude with $\frac{1}{3}$ year of observation $h_{3/yr} \cong 3 \times 10^{-16}$. Despite great effort in the early 70s by excellent experimenters, there was great room for improvement. (For a thorough review of the first-generation experiments see Amaldi and Pizella, 1979; for other reviews, see Drever, 1977, Tyson and Giffard, 1978 and Weber, 1986.)

In moving into the second generation, almost all the groups cooled their bars to liquid helium temperatures ($T_b = 1.5$–4 K rather than 300 K).

Several of the groups (Maryland, Stanford, LSU) constructed massive bars ($M \cong 2$–5 tonnes) from a new alloy of aluminum that the Tokyo group discovered has a Q 10 times higher than the old one (5×10^7 vs 5×10^6 at 10^4 K temperature; Suzuki, Tsubono and Hirakawa, 1978). Perth and Moscow, by contrast, chose to use exotic bar materials. Perth used a 1.5 tonne Niobium bar with $Q = 2 \times 10^8$, while Moscow initially used a 10 kg sapphire bar with $Q = 4 \times 10^9$, and later when difficulty with cracking of the sapphire occurred, Moscow switched to a 10 kg silicon crystal bar with $Q = 2 \times 10^9$ – silicon and sapphire because of their very large Qs, an advantage bought at the price of low mass.

In the second generation each of the groups deemphasized piezo-electric transducers and developed some variant of one or another of two new transducer concepts: Moscow, LSU, Perth and Tokyo developed *parametric transducers* in which the bar's vibrations with frequency $f_0 \sim 10^3$ Hz modulate the capacitance in a microwave cavity (or rf circuit), thereby moving microwave photons from the frequency $f_D \sim 10^9$ Hz, at which the cavity is driven, into side bands at $f = f_D \pm f_0$, at which the amplifier measures the signal. The number of photons moved is proportional to the amplitude of vibration of the bar. Stanford and Maryland developed *resonant transducers* in which a mechanical diaphragm, with a mechanical resonant frequency very nearly that of the bar's fundamental mode, is attached to the bar's end. The vibration energy is quickly transferred back and forth between the bar and the diaphragm, with the diaphragm's displacement amplitude amplified over that of the bar by $|X_{\text{diaphragm}}|/|X_{\text{bar}}| \sim [(\text{bar mass})/(\text{diaphragm mass})]^{\frac{1}{2}}$. The vibrating diaphragm modulates the inductance of a superconducting circuit, and a SQUID (superconducting quantum interference device) amplifier is used to monitor the current in the circuit. Rome developed a similar resonant transducer with the diaphragm replaced by a toad stool which was one plate of a capacitor and with a FET transducer used to read out oscillations of the toad stool's voltage.

At present (late 1986) Stanford, Rome and LSU are all on the air with systems that have resonant frequencies $f_0 \cong 900$ Hz, cross-sections $\int \sigma_0 \, df \cong$ (4 to 8) $\times 10^{-21}$ cm^2 Hz, and noise temperatures $T_n \cong (0.01$–$0.04)$ K. Correspondingly the weakest burst they can detect with $\frac{1}{3}$ year of observation is $h_{3/\text{yr}} \cong 1.0 \times 10^{-17}$. Maryland and Perth are both likely to be on the air with similar characteristics before this book is published; and all five groups are hoping to push their noise levels down to $h_{3/\text{yr}} \sim 2 \times 10^{-18}$ within several years by further straightforward refinements. The Guangzhou group, having only entered the field very recently, is still at room temperature, but with a sensitivity $h_{3/yr} \simeq 1.6 \times 10^{-16}$, two times better than that of any of the first-generation room-temperature bars (Hu *et al.* 1986). The Moscow group with its small silicon and sapphire bars operates at a much higher frequency than any of the other groups, $f_0 \cong$ 8 kHz; by the time this book is published they will likely be on the air with $h_{3/yr} \sim 4 \times 10^{-17}$. The Tokyo group, on the other hand, has chosen a much lower frequency than the others, $f_0 \cong 60$ Hz and is now operating a narrow-band search for periodic waves from the Crab pulsar (see below). These sensitivities are shown in Fig. 9.4, along with those of other kinds of detectors. For details of the present detector configurations and near-term plans, see the gravitational-wave articles in the proceedings of recent conferences (Ruffini, ed., 1986; MacCallum, ed., 1987).

(e) *Sensitivies of bar detectors to periodic and stochastic waves*

Although I have discussed the detectors' performances entirely in the context of searching for short bursts of gravitational waves, bar detectors can be used also to search for periodic gravitational waves and a stochastic background at frequencies within the bandwidth Δf of their sensors.

In a search for periodic gravitational waves, the experimenters will typically use a bar with eigenfrequency f_0 slightly different from the expected frequency f of the waves. The waves then will drive the fundamental mode's complex amplitude X into oscillations with the beat frequency $f - f_0$. In searching for these oscillations the experimenters integrate for a long time; and correspondingly they can turn down the coupling strength β of the transducer to the bar, and/or narrow the bandwidth Δf, until the sensor's noise becomes negligible compared to the thermal noise of the bar (see, e.g., Michelson and Taber, 1981, 1984). (The maximum resulting bandwidth with the present Stanford detector would be $\Delta f \simeq 0.5$ Hz). When this is done, the general formulas of Section 9.4.2(a) are valid, with $S_h(f)$ the spectral density of the bar's thermal noise, converted into an equivalent gravity-wave spectral density

$$S_h(f) = \frac{4kT_b}{\int \sigma_o \, df} \frac{1}{f_0 Q}, \tag{89}$$

and with the beam-factor averages having the values (79). In particular, the detector's characteristic noise amplitude (equation (51)) is

$$h_n = \left[15 \frac{G}{c^3} \frac{kT_b}{\int \sigma_o \, df} \frac{1}{Q} \frac{1}{f_0 \hat{\tau}} \right]^{\frac{1}{2}}, \tag{90}$$

and the brightest source that can be seen with 90 % confidence in $\hat{\tau} = \frac{1}{3}$ year of integration is

$$h_{3/yr} = 1.7h_n = 3.9 \times 10^{-25} \left(\frac{T_b}{1\,K}\right)^{\frac{1}{2}} \left(\frac{10^{-21}\,\text{cm}^2\,\text{Hz}}{\int \sigma_o\,df}\right)^{\frac{1}{2}} \left(\frac{1000\,\text{Hz}}{f_0}\right)^{\frac{1}{2}} \left(\frac{10^7}{Q}\right)^{\frac{1}{2}}.$$

(91)

The Tokyo group is currently carrying out a search for gravitational waves from the Crab pulsar using the above technique (Owa *et al.*, 1986). Their 74 kg, cryogenically cooled antenna has $Q = 2.1 \times 10^7$, $T_b = 4\,K$, $f_0 = 60\,\text{Hz}$, and $\int \sigma_o\,df \simeq 2.2 \times 10^{-27}\,\text{cm}^2\,\text{Hz}$ (so low because for noncylindrical antennas with frequency f_0 lowered by special shaping, $\int \sigma_o\,df \propto f_0^2$; Hirakawa *et al.*, 1976). Correspondingly it has $h_{3/yr} \sim 3 \times 10^{-22}$. No other present bars are optimized for periodic waves, since there are no known sources in their frequency bands ($\sim 900\,\text{Hz}$ and $\sim 8\,\text{kHz}$). However, with the technology of present burst-optimized bars it should be possible to achieve the thermal-noise-limited sensitivity (91) with $T_b = 4\,K$, $\int \sigma_o\,df \simeq 8 \times 10^{-21}\,\text{cm}^2\,\text{Hz}$, $f_0 \simeq 900\,\text{Hz}$, and $Q \simeq 5 \times 10^6$ corresponding to $h_{3/yr} \simeq 4 \times 10^{-25}$ (Stanford, LSU, Rome, Maryland); and $T_b = 4\,K$, $\int \sigma_o df \simeq 2 \times 10^{-23}\,\text{cm}^2\,\text{Hz}$, $f_0 \simeq 8\,\text{kHz}$, and $Q \simeq 2 \times 10^9$ corresponding to $h_{3/yr} \simeq 1.4 \times 10^{-25}$ (Moscow). See Fig. 9.6.

When searching for stochastic background it is desirable to open up the bandwidth Δf until the sensor noise becomes almost as large as the bar's thermal noise ($\Delta f \simeq 0.5\,\text{Hz}$ for the present Stanford bar). With $S_h(f)$ then given (approximately) by the thermal-noise spectral density (equation (89)), the noise amplitude h_n and the 90%-confidence $\frac{1}{3}$-year sensitivity $h_{3/yr}$ for stochastic background become (equations (66) and (67))

$$h_n(f) \simeq \left[15 \frac{G}{c^3} \frac{kT_b}{\int \sigma_o\,df} \frac{1}{Q} \frac{1}{\sqrt{(\frac{1}{2}\hat{\tau}\Delta f)}}\right]^{\frac{1}{2}},$$

(92)

$$h_{3/yr} = 1.7h_n \cong 8 \times 10^{-22} \left(\frac{T_b}{1\,K}\right)^{\frac{1}{2}} \left(\frac{10^{-21}\,\text{cm}^2\,\text{Hz}}{\int \sigma_o\,df}\right)^{\frac{1}{2}} \left(\frac{10^7}{Q}\right)^{\frac{1}{2}} \left(\frac{1\,\text{Hz}}{\Delta f}\right)^{\frac{1}{4}}.$$

(93)

This $h_{3/yr}$ is shown in Fig. 9.7 for the parameters of 1987 bar technology (those given in the last paragraph, plus $\Delta f \simeq 1\,\text{Hz}$). For details of searches for stochastic background that were carried out using first-generation bar detectors, see Hough *et al.* (1975) and Hirakawa and Narihara (1975). For discussions of sensitivity that are more detailed and sophisticated than the above sketch, see Hirakawa, Owa and Iso (1985), Weiss (1979) and references therein.

(f) *Quantum limit and quantum non-demolition*

Quantum mechanics constrains the sensitivity that can be achieved by bar detectors using the present kinds of sensors: the fundamental mode of a bar, being highly decoupled from the rest of the world, can be regarded as a simple harmonic oscillator with mass M_{eff} and angular frequency $\omega_0 = 2\pi f_0$. As such, it is subject to the laws of quantum mechanics for oscillators: its generalized position x and momentum p must be regarded as hermitian operators that fail to commute, $[x, p] = i\hbar$; and, correspondingly, the real and imaginary parts of its complex amplitude,

$$X_1 = x \cos \omega_0 t - \left(\frac{p}{M_{eff}\omega_0}\right) \sin \omega_0 t,$$

$$X_2 = x \sin \omega_0 t + \left(\frac{p}{M_{eff}\omega_0}\right) \cos \omega_0 t, \tag{94}$$

are non-commuting hermitian operators with commutator

$$[X_1, X_2] = i \frac{\hbar}{M_{eff}\omega_0}. \tag{95}$$

From this commutation relation we infer, via the Heisenberg uncertainty principle, an absolute limit on the variances of X_1 and X_2 in any quantum mechanical state (Thorne *et al.*, 1978):

$$\Delta X_1 \Delta X_2 \geqslant \frac{\hbar}{2M_{eff}\omega_0}. \tag{96}$$

The principles of quantum mechanics guarantee that one can never measure X_1 and X_2 simultaneously with a precision that violates this uncertainty principle. In fact, it turns out (Caves, 1982; Yamamoto and Haus, 1986) that even the most ideal of measuring systems will introduce additional noise equal to (96), thereby producing a lower bound

$$\Delta X_1 \Delta X_2 \geqslant \frac{\hbar}{M_{eff}\omega_0} \tag{97}$$

on the products of the rms noise in the measured values of X_1 and X_2.

The sensors used in the present generation of bar detectors measure X_1 and X_2 simultaneously with equal accuracies; and, correspondingly, in searches for short bursts they are subject to the 'standard quantum limit' (Braginsky and Vorontsov, 1974; Giffard, 1976)

$$T_n = \tfrac{1}{2}M_{eff}\omega_0^2[(\Delta X_1)^2 + (\Delta X_2)^2]/k \geqslant \hbar\omega_0/k = 4.8 \times 10^{-8} \text{ K}\left(\frac{f_0}{1000 \text{ Hz}}\right). \tag{98}$$

Notice that this standard quantum limit places the severe constraint

$$h_{3/\text{yr}} \gtrsim 4.4 \times 10^{-20} \left(\frac{f_0}{1000 \text{ Hz}} \right)^{\frac{1}{2}} \left(\frac{10^{-21} \text{ cm}^2 \text{ Hz}}{\int \sigma_o \, df} \right)^{\frac{1}{2}} \tag{99}$$

on the detector's burst sensitivity (82). It is fairly likely, though far from certain, that the strongest kilohertz-frequency bursts striking the earth three times per year have characteristic amplitudes $h_c < 10^{-20}$ (see Section 9.4.1); and, correspondingly, it may turn out to be crucial for bar detectors of the future to circumvent the standard quantum limit (98), (99).

The uncertainty principle (96) suggests a promising method for circumventing the standard quantum limit (Thorne *et al.*, 1978; Braginsky, Vorontsov and Khalili, 1978): one should devise a new kind of sensor that measures X_1 with high accuracy, while giving up accuracy on X_2. Such sensors, called 'back-action-evading sensors' (a special case of 'quantum non-demolition sensors'), are now under development in a number of laboratories (Braginsky, 1983; Bocko and Johnson, 1984; Oelfke, 1983; Blair, 1982); and they may make possible bar sensitivities in the 1990s that will beat the standard quantum limit by modest factors. For a detailed review of quantum non-demolition measurements – i.e. measurements that do not change the quantum state of the system being measured – see Caves (1983).

Although a back-action-evading sensor gives up accuracy on one of the wave's two quadrature components, that accuracy can be regained by looking at the same wave with two different detectors: on one detector, with complex amplitude $X = X_1 + iX_2$, measure X_1 with high accuracy and X_2 with poor; on the other, with complex amplitude $Y = Y_1 + iY_2$, measure Y_2 with high accuracy and Y_1 with poor. From X_1 infer the detailed evolution of one of the wave's two quadrature components; from Y_2 infer the evolution of the other. In this way, in principle, the quantum mechanical properties of the detector can be completely circumvented and the only constraints of principle on the accuracy of measurement are associated with quantization of the waves themselves. For further discussion and details see the reviews by Caves *et al.* (1980); Caves (1983); Braginsky, Vorontsov and Thorne (1980).

It is worth noting that a back-action-evasion measurement, ideally performed, should drive the bar's fundamental mode into a 'squeezed state' (Hollenhorst, 1979). Squeezed states have been studied extensively in recent years in the context of quantum optics (see, e.g. Schumaker, 1986; and Walls, 1983); and we shall return to them in Section 9.5.3(f) below when discussing beam detectors.

(g) *Looking toward the future*

It may be that bar detectors in the distant future will find their greatest applications at much lower frequencies than are now common: at lower frequencies the bar can be much longer and more massive, and correspondingly more sensitive; and good bandwidth, $\Delta f \sim f_0$ will correspond to a longer averaging time and thus will be easier to achieve. For discussion, see Michelson (1986).

Although individual bar detectors in the foreseeable future will be limited to moderately small bandwidths $\Delta f/f \lesssim 0.2$, and therefore will be able to acquire only very limited information about the wave form $h_{jk}^{TT}(t)$ of a gravity-wave burst, once waves are being detected in profusion it may be possible cheaply to replicate the detectors with a variety of sizes and hence a variety of fundamental-mode frequencies, thereby creating a 'xylophone' of networked detectors with a large overall bandwidth (Michelson and Taber, 1984).

9.5.3 *Beam detectors*

(a) *A brief history of beam-detector research*

The germ of the idea of a laser-interferometer gravitational-wave detector ('beam detector') can be found in Pirani (1956); but – so far as I am aware – the first explicit suggestion of such a detector was made by Gertsenshtein and Pustovoit (1962). In the mid-1960s Joseph Weber, unaware of the Gertsenshtein–Pustovoit work, reinvented the idea but left it lying in his laboratory notebook unpublished and unpursued. In 1970 Rainer Weiss at MIT, unaware of Gertsenshtein–Pustovoit or Weber, reinvented the idea and carried out a detailed design and feasibility study (Weiss, 1972) in which many of the techniques now being used were conceived. Unfortunately, Weiss was unable to obtain funding to push forward with a significant experimental effort.

Robert Forward at Hughes Research Laboratories in Malibu, California, having learned the concept of the beam detector from Weber (his former thesis advisor), was motivated indirectly by Weiss in 1971 to construct a prototype detector with funding from Hughes. By 1972 Forward and his colleagues at Hughes were operating the world's first prototype beam detector – an instrument that demonstrated the idea could really work, and that was remarkably sensitive considering the modest effort put into it: $[S_h(f)]^{\frac{1}{2}} \simeq 2 \times 10^{-16}$ Hz$^{-\frac{1}{2}}$ between 2500 Hz and 25 000 Hz, corresponding to $h_{3/yr} \simeq 1 \times 10^{-13}$ for 2500 Hz bursts (Moss, Miller and Forward, 1971;

Forward and Moss, 1972; Forward, 1978). Regrettably, Forward could not obtain funds to move from this first prototype to a more sophisticated instrument; so his project was shut down.

With the completion of the first generation of bar detectors in 1975, each experimental group that decided to stay in the field looked carefully at a variety of possibilities for sensitivity improvement. While most groups decided to stick with bars, two switched to beam detectors: Munich, led by H. Billing, and Glasgow, led by Ronald Drever with Jim Hough second in command. The Munich group was strongly influenced by a proposal to develop beam detectors that Weiss had submitted to NSF, and that NSF had refused to fund; and so Munich pushed forward (Winkler, 1977) along the lines that Weiss had hoped to follow, using a Michelson interferometer design (see below). The Glasgow group first built a small Michelson interferometer (Drever *et al.*, 1977), then switched in 1977 to a new Fabry–Perot design invented by Drever (Drever *et al.*, 1980).

In 1979 Caltech managed to attract Drever away from Glasgow (part-time at first, full-time later), leaving Hough as the Glasgow leader. At Caltech Drever started up a beam-detector project; and NSF, finally recognizing that beam detectors were worth funding, agreed to support both Weiss at MIT to develop his original idea of a Michelson system and Drever at Caltech to develop his Fabry–Perot system. More recently, in 1983, Alain Brillet initiated a beam-detector effort in Orsay, France (Brillet and Tourrenc, 1983; Brillet, 1985).

Munich, Glasgow, Caltech and MIT all now have working beam detectors with amplitude sensitivities ~ 2000 times better than that of Forward's first prototype but ~ 5 times worse than the best bars. These detectors are small-scale (1–40 m) prototypes for the full-scale (several kilometer) beam detectors that will be required for real success. Design and costing studies are now underway for the full-scale systems (called 'Laser Interferometer Gravity Wave Observatories' or LIGOs); for details of these studies, see Linsay *et al.* (1983), Drever *et al.* (1985), Maischberger *et al.* (1985), Winkler *et al.* (1986), Hough *et al.* (1986). There is hope that full-scale LIGOs will be constructed in the late 1980s and early 1990s and will be operating in the mid- to late-1990s with sensitivities in the region where gravity waves are expected.

(b) *How a beam detector works*

The current and planned beam detectors are designed to operate at frequencies below 10 kHz because astrophysical arguments suggest that the waves will be weak above this frequency (see page 158 of Thorne, 1978); and they have their best sensitivities at frequencies below 1 kHz. Correspondingly, the waves they seek all have reduced wavelengths $\lambda > 5$ km and most have $\lambda > 50$ km. Since the planned detectors all have sizes $L \leqslant 4$ km, the condition $L \ll \lambda$ for use of a 'proper-reference-frame analysis' is satisfied, though only marginally in extreme cases. I shall use such an analysis in the discussion below. For an outline of the alternative, TT analysis, see, e.g., Exercise 37.6 of MTW.

A beam *detector* consists of one or more *receivers* that are operated simultaneously, with cross-correlated outputs – the cross-correlation, as usual, being the key to removing spurious, non-Gaussian noise. A simple version of a Michelson-type receiver is shown in Fig. 9.9, three-dimensionally in part (*a*) and as seen from above in part (*b*). (Ignore for the moment the propagation and polarization pieces of part (*a*).) The receiver consists of three masses which hang on wires from overhead supports and swing like pendula. The masses are arranged at the ends and corner of a right-angle L. When a gravitational wave propagates vertically through the receiver with polarization axes along the L ('+' polarization), its

Fig. 9.9. Schematic diagram of a *Michelson-type beam receiver* for gravitational waves (part (*b*)), and of the waves' propagation and polarization angles (θ, ϕ, ψ) relative to the receiver (part (*a*); cf. Fig. 9.2).

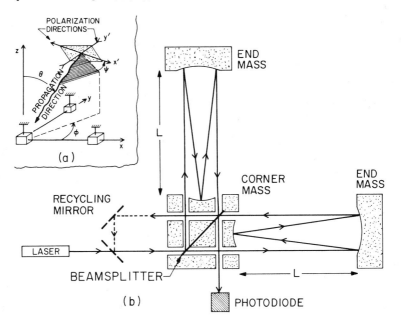

quadrupolar force field (5) pushes together the masses on one arm of the L while pushing apart the masses on the other arm. In the next half-cycle of the wave, the directions of the pushes are reversed. Since the waves being sought have frequencies f far above the 1 Hz swinging frequency of the pendula, the pendular restoring forces have no opportunity to make themselves felt: the masses respond to the gravity-wave pushes as though they were free. With the origin of the proper reference frame placed on the central mass, the central mass is left unaffected while the end masses oscillate longitudinally with displacements

$$\delta x(t) = \tfrac{1}{2}Lh_+(t) \qquad \text{for mass on } x \text{ axis,} \tag{100a}$$

$$\delta y(t) = -\tfrac{1}{2}Lh_+(t) \quad \text{for mass on } y \text{ axis} \tag{100b}$$

(equation (6)). Here L is the (approximately equal) length of each arm. Correspondingly, there is an oscillation in the difference $l(t)$ of the arm lengths, $\delta l(t) = \delta x(t) - \delta y(t)$, given by

$$\delta l(t) = h_+(t)L. \tag{101}$$

It is straightforward to show that in the more general case of a wave which impinges from a direction (θ, ϕ) on the sky with polarization axes rotated at an angle ψ relative to the constant-ϕ plane (Fig. 9.9(a)), the difference in arm lengths l oscillates as

$$\delta l(t) = h(t)L, \tag{102}$$

where $h(t)$ has the standard form (26)

$$h(t) = F_+(\theta, \phi, \psi)h_+(t; \iota, \beta) + F_\times(\theta, \phi, \psi)h_\times(t; \iota, \beta), \tag{103}$$

with beam-pattern factors (cf. Forward, 1978; Rudenko and Sazhin, 1980; Estabrook, 1985; Schutz and Tinto, 1987)

$$F_+(\theta, \phi, \psi) = \tfrac{1}{2}(1 + \cos^2 \theta) \cos 2\phi \cos 2\psi - \cos \theta \sin 2\phi \sin 2\psi,$$
$$\tag{104a}$$

$$F_\times(\theta, \phi, \psi) = \tfrac{1}{2}(1 + \cos^2 \theta) \cos 2\phi \sin 2\psi + \cos \theta \sin 2\phi \cos 2\psi.$$
$$\tag{104b}$$

The difference of arm lengths $l(t)$ is monitored by Michelson interferometry: a beam splitter and two mirrors are attached to the corner mass as shown in Fig. 9.9(b), and one mirror is attached to each end mass. A laser beam shines through a hole in the corner mass and onto the beam splitter, which directs half the beam toward each end mass. The mirrors on the end masses reflect the beams back toward the corner-mass mirrors, which in turn reflect the beams back to the end masses, which reflect the beams back through holes in the corner-mass mirrors and onto the beam

splitter where they are recombined. Part of the recombined beam goes out
one side of the beam splitter toward the laser (ignore for now the 'recycling
mirror' in (b), it is absent in the simple version of the receiver being discussed
here); the other part of the recombined beam goes out the other side toward
a photodetector. Oscillations in the arm-length difference $\delta l(t)$ produce
oscillations in the relative phases of the recombining light, and thence
oscillations in the fraction of the light which goes to the photodetector
versus that which goes back toward the laser. The photodetector, by
monitoring the oscillations in received intensity, in effect is monitoring the
oscillations $\delta l(t)$ of arm-length difference and thence the gravity-wave
oscillations $h(t)$.

In practice the laser beams are made to bounce back and forth in the arms
not just twice as shown in Fig. 9.9(b), but rather a large number of round-
trip times B, making B distinct spots on each end mirror. In the simple case
that the gravity-wave-induced arm-length difference $\delta l = Lh$ does not change
much during these many round trips (see subsection (e) below for the case of
large change), the bouncing light beam will build up during its B trips a total
phase delay

$$\Delta\Phi = \frac{2B\,\delta l}{\lambdabar_e} = \frac{2BL}{\lambdabar_e}h \qquad (105)$$

where $\lambdabar_e = \lambda_e/2\pi$ is the reduced wavelength of the light ($\lambdabar_e = 0.0818$ microns
for the light from the argon ion lasers currently being used). This phase delay
can be monitored, by the photodetector, with a precision $\Delta\Phi = 1/(N_\gamma\eta)^{\frac{1}{2}}$,
where N_γ is the total number of photons that the laser puts out during the
time $\hat{\tau}$ over which the photodetector intensity is averaged, and η is the
photon counting efficiency of the photodetector ($\eta \sim 0.4$–0.9). When
searching for a gravity-wave burst with characteristic frequency f, it is
optimal to average the photodetector intensity for half a gravity-wave
period, $\hat{\tau} \cong 1/2f$; and correspondingly the phase delay can be inferred with a
photon-counting-noise ('shot-noise') precision

$$\Delta\Phi_{shot} = \frac{1}{(N_\gamma\eta)^{\frac{1}{2}}} \simeq \left(\frac{\hbar c/\lambdabar_e}{I_o\eta(1/2f)}\right)^{\frac{1}{2}}, \qquad (106)$$

where I_o is the laser output power. By comparing equations (105) and (106)
we obtain a rough estimate of the amplitude of a gravity-wave burst that
produces a signal of the same strength as the rms shot noise

$$h_{\text{shot}} \cong \left[\frac{2\hbar c \lambda_e}{I_o \eta} \frac{f}{(2BL)^2} \right]^{\frac{1}{2}}$$

$$\cong 7.2 \times 10^{-21} \frac{50}{B} \frac{1 \text{ km}}{L} \left(\frac{\lambda_e}{0.082 \ \mu\text{m}} \right)^{\frac{1}{2}} \left(\frac{10 \text{ Watts}}{I_o \eta} \right)^{\frac{1}{2}} \left(\frac{f}{1000 \text{ Hz}} \right)^{\frac{1}{2}}. \qquad (107)$$

This formula and these numbers give some indication of the potential sensitivities of beam receivers.

Fabry–Perot beam receivers have essentially the same potential sensitivities as Michelson receivers. Fig. 9.10 shows a Fabry–Perot receiver viewed from above. Here, by contrast with Fig. 9.9, the corner mass has been broken into three separate pieces, one carrying each mirror and one carrying the beam splitter. This breaking up of the corner mass, initiated in Munich, reduces spurious forces on the mirrors; since 1985 it has been standard in all beam receivers. In the Fabry–Perot system of Fig. 9.10 each arm is operated as a resonant Fabry–Perot cavity: light from the laser is split at the beam splitter and enters the two cavities through the backs of the corner masses' partially transmitting mirrors. The two arms are arranged to have equilibrium lengths very nearly equal to a half-integral multiple of the wavelength $2\pi\lambda_e$ of the laser light. Consequently, the entering light finds itself in resonance with a mode of each cavity; and it gets resonantly

Fig. 9.10. Schematic diagram of a *Fabry–Perot-type beam receiver* for gravitational waves.

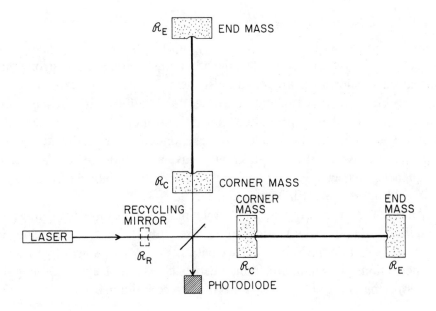

trapped in the cavity, building up to high intensity before exiting back toward the beam splitter.

Slight changes in the length of each cavity drive the cavity slightly off resonance and thereby produce sharp changes in the phase of the exiting light. Consequently, when the exiting light beams from the two cavities recombine at the beam splitter, their relative phase is highly sensitive to slight modulations δl of the two cavities' length difference l; and correspondingly the intensity of light onto the photodetector is highly sensitive to δl. If the corner mirrors have a probability for reflecting photons $\mathscr{R}_C$ and a transmission probability $1 - \mathscr{R}_C$ (and no scattering), and if the end mirrors reflect much more efficiently, then this Fabry–Perot sensitivity is described by the same formulas (105)–(107) as for a Michelson receiver, with the number of round-trips B in each Michelson arm replaced by $4/(1 - \mathscr{R}_C)$:

$$B \to 4/(1 - \mathscr{R}_C). \tag{108}$$

It is also possible (and, in fact, is current practice) to operate a Fabry–Perot receiver in an alternative mode where, instead of recombining and interfering the beams, the laser's frequency is locked to the eigenfrequency of one arm and the difference between the laser's frequency and that of the other arm is the gravity-wave signal; see, e.g. Hough $et\ al.$ (1983) or Spero (1986a) for details. This mode of operation is technically easier than beam recombination, but the ease is bought at the price of some debilitation in the ultimately achievable shot noise.

(c) Noise in beam detectors

Photon shot noise is but one of many noise sources that plague beam detectors. Almost always, thus far, the other noise sources are so strong that the effects of shot noise are lost amidst them. Typically the experimenters struggle for a long time to reduce the other noises sufficiently that shot noise shows up; then they improve the shot noise somewhat by increasing the laser power; then they begin a long struggle once again with the other noise sources. In this way the overall gravity-wave amplitude noise at kilohertz frequencies has been reduced during the period 1980–86 by a factor ~ 1000 (10^6-fold improvement in energy noise).

Among the most serious of the other noise sources are the following: amplitude and phase fluctuations in the laser output; imperfect matching of the laser beam into the Fabry–Perot cavities or the Michelson mirror system (imperfections in beam direction, beam shape, and beam wavelength); imperfect matching of the wave fronts of the recombining beams at the beam splitter; fluctuations in the index of refraction of the gas inside the arms (even though the arms are inside vacuum pipes, residual gas can cause problems); scattering of light from one part of the optical system into another; thermal noise (thermal vibrations) in the end and corner masses and in the wires that suspend them; imperfect alignment of the mirrors; and seismic and acoustic noise from the outside world, which cause vibrations of the wires from which the masses hang. Almost all of these effects produce a displacement noise δl that is independent of the arm length L; and, correspondingly, their effects on gravity-wave amplitude sensitivities scale as $h \propto 1/L$. It is this scaling that motivates the move from short prototypes to long LIGOs.

A beam receiver is intrinsically broad band: its output, the intensity of the light into the photodetector $I_{\mathrm{pd}}(t)$, is averaged (by filtering) over the shortest timescale, $\hat{\tau} \sim 10^{-4}$ s, that gravity waves are likely to contain, and then is recorded for future analysis. The future analysis can include searching in the data for signals of all frequencies $f \lesssim 1/\hat{\tau} \sim 10^4$ Hz, and for signals with complicated time dependences that embody a broad range of frequencies. Correspondingly, a beam receiver's net noise depends on the frequency f at which one studies the recorded output.

More specifically, the noise (excluding non-Gaussian, spurious events which are removed by coincident operation of two independent receivers) is characterized by the spectral density $S_l(f)$ of the output-inferred armlength difference l; or, equally well, by the spectral density $S_h(f)$ of the gravity wave $h(t)$ (equation (103)). Because the effect of the gravity wave on the arm-length difference is $\delta l(t) = L h(t)$, $S_h(f)$ is also called the spectral density of strain, and these two spectral densities are related by

$$S_h(f) = (1/L^2) S_l(f). \tag{109}$$

It is $S_h(f)$ or its square root, or $S_l(f)$ or its square root, that experimenters generally quote when discussing the overall performance of their beam detectors.

In Sections 9.4.1(b), 9.4.2(a) and 9.4.3(a) we derived expressions in terms of $S_h(f)$ for the minimum-amplitude signal $h_{3/\mathrm{yr}}$ that can be detected with 90% confidence in a $\frac{1}{3}$ year search using a detector composed of two identical cross-correlated receivers. Those expressions involve averages over the receiver's beam-pattern factors. For the L-shaped beam receivers of Figs. 9.9 and 9.10, with beam-pattern factors (104), the appropriate averages are

$$\langle F_+^2 \rangle = \langle F_\times^2 \rangle = \tfrac{1}{5}, \tag{110}$$

and, correspondingly,

$h_{3/yr}(f_c) = 11[f_c S_h(f_c)]^{\frac{1}{2}}$ for bursts (equation (34)),   (111)

$h_{3/yr}(f) = 3.8[S_h(f) \times 10^{-7} \text{ Hz}]^{\frac{1}{2}}$

for periodic waves (equation (52a)),   (112)

$$h_{3/yr}(f) = 4.5\left(\frac{\Delta f}{10^{-7} \text{ Hz}}\right)^{-\frac{1}{4}} [fS_h(f)]^{\frac{1}{2}}$$

for stochastic waves (equation (67)).   (113)

Here in the burst equation the numerical factor in equation (34) has been taken to be 5 corresponding to a characteristic frequency $f_c \sim 10$ Hz to 10^4 Hz – the largest band that earth-based receivers are likely ever to operate in; and in the periodic equation it is assumed that the frequency and phase are known in advance (equation (52a) rather than (52b)). Expressions (111)–(113) are the bases of the beam-detector noise amplitudes shown in Figs. 9.4, 9.6 and 9.7.

When the two receivers that make up a detector do not lie in the same plane (because of curvature of the earth between them) or do not have their arms parallel, their joint sensitivity is somewhat debilitated. Detailed analyses of this have been carried out by Whitcomb and Saulson (1984) and by Schutz and Tinto (1986).

(d) The noise in the present prototypes

At present the MIT group is developing Michelson receivers using a prototype with $L = 1.5$ m; the Munich group is developing Michelsons using $L = 30$ m (Shoemaker et al., 1986); the Glasgow group is developing Fabry–Perot receivers using $L = 10$ m; and the Caltech group is developing Fabry–Perots using $L = 40$ m. Although the arm lengths and optics of these four prototypes are very different, their displacement sensitivities are roughly the same – and have been so throughout the past six years, during which all have improved roughly in step by about three orders of magnitude. Throughout these improvements Munich has maintained a slight (factor 2 or 3 in amplitude) lead over the other groups. (Note added in proof. Since this was written Glasgow has forged ahead of Munich by a factor 3 in displacement sensitivity, making them equal in h sensitivity.) Fig. 9.11 shows the spectral density of strain noise for the Munich prototype as of February 1986 (Schilling, 1986). The noise is due, predominantly, to seismic noise (inadequate isolation) below 250 Hz, probably seismic noise between 250 and 1000 Hz, photon shot noise between 1000 and 6000 Hz, and thermal noise in the mirrors and pockels cells above 6000 Hz. In Figs. 9.4, 9.6 and 9.7 (upper right) are shown the sensitivities $h_{3/\text{yr}}$ (equations (111)–(113)) for the Munich and Caltech prototypes in 1986 (Schilling, 1986; Spero, 1986b) and Glasgow in early 1987. As an illustration of the sensitivity progress, Fig. 9.4 also shows $h_{3/\text{yr}}$ for bursts in the Munich and Caltech prototypes a few months after each was first turned on (1980 and 1983).

(e) *Spectral density of shot noise for simple, recycling and resonating receivers*

In the present prototypes it is advantageous to store the light in the arms as long as possible, thereby building up the largest possible phase shift and gravity-wave sensitivity; cf. the B-dependence in equations (105) and (107). However, in a kilometer-scale LIGO one easily can store the light longer than half a gravity-wave period, i.e. for more than $B = 75(1 \text{ km}/L) \times (1000 \text{ Hz}/f)$ round-trip traverses of an arm. Such long storage is self-defeating: the phase shift built up so laboriously during the first half-period of the wave gets removed during the second half-period because $h(t)$ reverses sign.

This shows up clearly in the spectral densities of shot noise $S_h(f)$ for the idealized Michelson and Fabry–Perot receivers of Figs. 9.9 and 9.10 (still without the 'recycling mirrors' in place). Assuming as above that the Michelson mirrors have negligible losses during B round trips in each arm, and the Fabry–Perot end mirrors have negligible transmission compared to the corner mirrors

$$1 - \mathcal{R}_E \ll 1 - \mathcal{R}_C \equiv 4/B \tag{114}$$

(cf. equation (108)), the spectral densities of shot noise are (cf. Gürsel *et al.*,

Fig. 9.11. Square root of spectral density of noise $[S_h(f)]^{\frac{1}{2}}$ plotted against frequency f for the Munich Michelson-type beam detector with 30 m arms, as of February 1986.

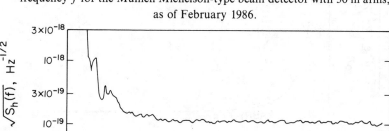

1983; Meers, 1983; Brillet and Meers, 1987)

$$S_h(f) = \frac{2\hbar c \lambda_e}{I_o \eta} \left(\frac{1}{2BL}\right)^2 \times \begin{cases} 1 + (2\pi BLf/c)^2 & \text{for Fabry–Perot} & (115a) \\[2ex] \left[\dfrac{2\pi BLf/c}{\sin(2\pi BLf/c)}\right]^2 & \text{for Michelson.} & (115b) \end{cases}$$

These spectral densities are shown in Fig. 9.12 as a function of $2BLf/c =$ (light storage time)/(gravity-wave period), with (gravity-wave period) $= 1/f$ held fixed. When (storage time)/(period) $\ll 1$ the two receivers have the same shot noise, and that shot noise improves with increasing storage time as $S_h(f) \propto$ (storage time)$^{-2}$. When (storage time)/(period) $\gtrsim 1$ the Fabry–Perot shot noise stops improving, while the Michelson shot noise undergoes a series of oscillations with minima equal to the Fabry–Perot shot noise. Note that the minima of the Michelson oscillations occur when $2BLf/c =$ (light-storage time)/(gravity-wave period) $= 0.5, 1.5, 2.5, \ldots$, while their

Fig. 9.12. The improvement of photon shot noise with increasing number of bounces of light in a simple beam receiver. The receiver is of the Fabry–Perot (Fig. 9.10) or Michelson (Fig. 9.9) type with recycling mirrors absent, with $1 - \mathcal{R}_E \ll 1 - \mathcal{R}_C = 4/B$ in the Fabry–Perot case and with B round trips in each arm in the Michelson case. Plotted vertically is spectral density of shot noise in units proportional to f^2 but independent of the light-storage time $2BL/c$. Plotted horizontally is (light-storage time)/(gravity-wave period) $= (2BL/c)f$. The Fabry–Perot shot noise is given by the solid curve (equation (115a)); the Michelson by the dashed curve (equation (115b)).

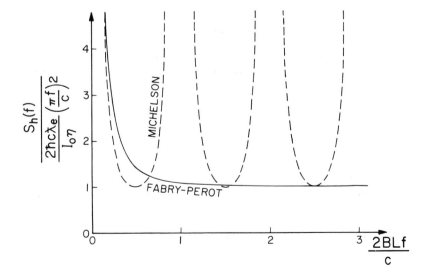

maxima ($S_h = \infty$) occur when the light is stored for an integral number of gravity-wave periods. This is just what we would expect from the fact that $h(t)$ reverses sign every half-period, thereby removing during a second half-period the signal put onto the light during a first half-period. In the Fabry–Perot receiver there are no oscillations of $S_h(f)$ because different photons experience different storage times – i.e. because of the probabilistic nature of the reflectivity $\mathscr{R}_C$.

Drever (1983) has devised a method for improving the sensitivity of either a Michelson or a Fabry–Perot when mirror reflectivities permit storing light for much longer than a half-period. The basic idea is to extract the light after a half-period, when further storage is self-defeating, and then reinsert it into the cavity along with and in phase with new laser light. More specifically, for the Michelson of Fig. 9.9(b), one adjusts the relative arm lengths so that very little of the recombined light emerges from the beam splitter toward the photodiode where the gravity-wave signal is read out (a mode of operation that optimizes the sensitivity, it turns out). Then most of the recombined light emerges toward the laser; and 'recycling mirrors' are inserted to direct that emergent light back into the beam splitter along with fresh laser light. For the Fabry–Perot of Fig. 9.10 the same effect is achieved with a single recycling mirror.

As an aid in quantifying the recycling-induced improvement in shot noise, consider the (realistic) situation in which the technology of mirror coatings limits the mirror reflectivities to some maximum value $\mathscr{R}_{max}$ (0.9999 at present; perhaps 0.999 99 a few years from now). It obviously is optimal to place at the end of each arm a mirror with this maximum, so $\mathscr{R}_E = \mathscr{R}_{max}$. Consider, for concreteness, the Fabry–Perot receiver of Fig. 9.10. If the corner mirrors are chosen also to have the maximum reflectivity, $\mathscr{R}_C = \mathscr{R}_E$, then essentially all the unused light leaks out the end mirrors and nothing is gained by recycling. In this case of a simple, non-recycling Fabry–Perot with $\mathscr{R}_C = \mathscr{R}_E = \mathscr{R}_{max}$ the spectral density of shot noise is (cf. Gürsel *et al.*, 1983; Meers, 1983; Brillet and Meers, 1987)

$$S_h(f) = S_o[1 + (f/f_o)^2],\tag{116a}$$

where

$$
\begin{aligned}
S_o &= \frac{2\hbar c \lambda_e}{I_o \eta}\left(\frac{1 - \mathscr{R}_E}{2L}\right)^2 \\
&= \left[\frac{3.6 \times 10^{-25}}{(\text{Hz})^{\frac{1}{2}}}\left(\frac{\lambda_e}{0.0818\ \mu\text{m}}\right)^{\frac{1}{2}}\left(\frac{100\ \text{W}}{I_o \eta}\right)^{\frac{1}{2}}\left(\frac{1 - \mathscr{R}_E}{10^{-4}}\right)\left(\frac{1\ \text{km}}{L}\right)\right]^2,
\end{aligned}
\tag{116b}
$$

$$f_o = \frac{(1-\mathcal{R}_E)c}{4\pi L} = 2.4 \text{ Hz}\left(\frac{1-\mathcal{R}_E}{10^{-4}}\right)\left(\frac{1 \text{ km}}{L}\right). \qquad (116c)$$

(Note that, because so much light is lost out the end mirrors, this noise is worse than the $\mathcal{R}_C \gtrsim \mathcal{R}_E$ limit of equation (115a) – worse by a factor 4 at frequencies $f \gg f_o$ and by a factor 16 at $f \ll f_o$.) This non-recycled ('simple') Fabry–Perot noise is shown as a solid curve in Fig. 9.13.

An experimenter who wishes to improve on this noise level by recycling must choose a frequency $f_k \gg f_o$ near which the noise level is to be minimized. The minimal noise level near $f = f_k$ will then be achieved by setting

$$1 - \mathcal{R}_C = 8\pi L f_k/c, \qquad (117a)$$

so the effective storage time in each arm is $2BL/c = 8Lc^{-1}/(1-\mathcal{R}_C) = (\pi f_k)^{-1} = (1/\pi) \times$ (period of a gravity wave with the optimal frequency f_k). Minimal noise also requires a special choice for the reflectivity $\mathcal{R}_R$ of the recycling mirror (Fig. 9.10)

$$1 - \mathcal{R}_R = \frac{4(1-\mathcal{R}_E)}{(1-\mathcal{R}_C)}. \qquad (117b)$$

With these choices of reflectivity, the light-recycling Fabry–Perot receiver of Fig. 9.10 has shot noise (cf. Gürsel et al., 1983; Meers, 1983; Brillet and Meers, 1987)

$$S_h(f) = \frac{f_k}{2f_o} S_o\left[1+\left(\frac{f}{f_k}\right)^2\right]. \qquad (117c)$$

This noise is depicted in Fig. 9.13 for two choices of f_k (dashed curves). Note that this noise has a 'knee' at the frequency f_k; for this reason f_k is called the *knee frequency*. Note further that recycling produces an overall improvement in $S_h(f)$, at frequencies $f \gtrsim f_k$, by a factor $f_o/2f_k$ relative to a non-recycled Fabry–Perot with equal reflectivities for corner and end mirrors (equations (116)) and an improvement by $2f_o/f_k$ relative to the very best non-recycled Fabry–Perot – one with $1-\mathcal{R}_{max} = 1-\mathcal{R}_E \ll 1-\mathcal{R}_C \ll 8\pi L f_k/c$ (equation (115a)). This improvement at $f \gtrsim f_k$ is bought at the price of worsened noise at $f \lesssim (f_k f_o/2)^{\frac{1}{2}}$. Note further that the improvement factor, over an optimal non-recycled Fabry–Perot at $f \gtrsim f_k$, is

$$\frac{S_h^{\text{recycled}}}{S_h^{\text{non-recycled}}} = \frac{2f_o}{f_k} = \frac{1-\mathcal{R}_E}{1-\mathcal{R}_C}, \qquad (118)$$

which is 1/(the mean number of times that the light can be recycled before it is lost by leakage through the high-reflectivity end mirrors). This is just what physical intuition should suggest.

For the recycled Michelson receiver of Fig. 9.9, recycling produces essentially the same $S_h(f)$ as for the Fabry–Perot of Fig. 9.10 at frequencies $f \lesssim f_k$. However, above the knee frequency the recycled Michelson exhibits a

Fig. 9.13. Spectral density of shot noise for Fabry–Perot beam receivers operated in various optical configurations. The vertical and horizontal scales are both logarithmic. The solid curve is for a simple Fabry–Perot with all mirrors – corner and end – having the maximum achievable reflectivity, $\mathscr{R}_C = \mathscr{R}_E = \mathscr{R}_{max}$ (equations (116)). The dashed curves are for a Fabry–Perot with light recycling (equation (117c)), in which the end mirrors have the maximum achievable reflectivity $\mathscr{R}_E = \mathscr{R}_{max}$, the corner mirrors are adjusted to produce the desired knee frequency (f_{k1} and f_{k2} for the two curves shown; equation (117a)), and the recycling mirror is adjusted to minimize the noise at the knee frequency (equation (117b)). The dotted curves are for a Fabry–Perot with light resonating (equation (119c); Fig. 9.14) in which the end mirrors have the maximum achievable reflectivity $\mathscr{R}_E = \mathscr{R}_{max}$; the corner mirrors are adjusted to produce the desired resonant frequency for gravity waves (f_{R1} and f_{R2} for the two curves shown; equation (119a)); and the recycling mirror is adjusted to minimize the noise at the resonant frequency (equation (119b)).

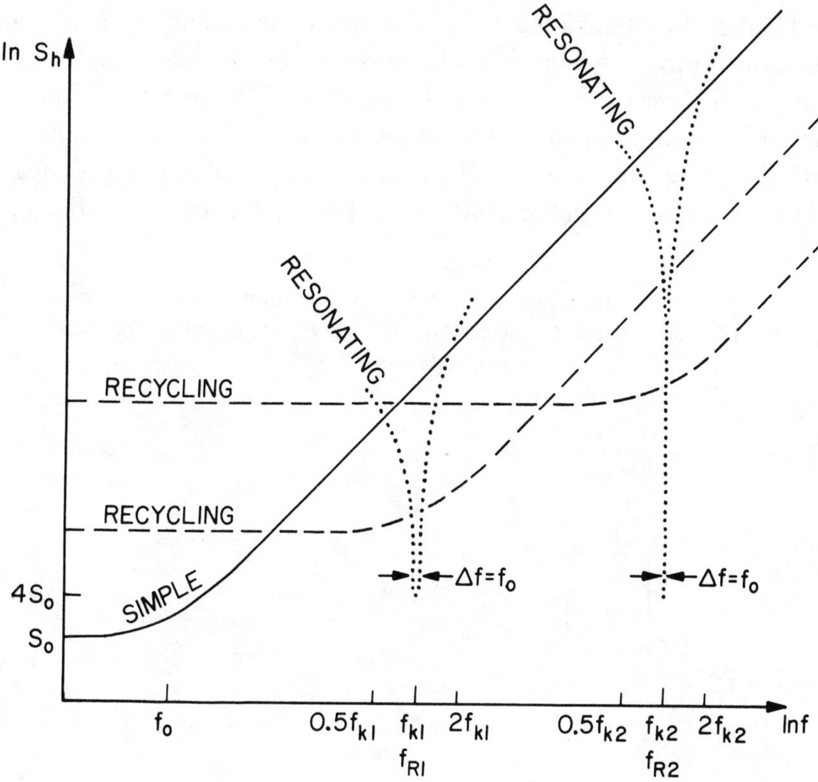

sequence of noise peaks and troughs analogous to those for a non-recycled Michelson (equation (115b) and Fig. 9.12).

Of all optical configurations yet invented, recycling is the best when one is searching for broad-band gravitational waves (waves with $\Delta f \gtrsim f$). However, when one is searching for narrow-band waves (waves with $\Delta f \ll f$, e.g. the periodic waves from a pulsar), recycling gives worse noise than a configuration called *light resonating* (invented by Drever, 1983), which is shown in Fig. 9.14. As usual, the method is conceptually simplest in the Michelson case but is described by the simplest formulas in the Fabry–Perot case.

The basic idea, as shown for a Michelson in Fig. 9.14(a), is to store the light in an arm for one half-cycle of the gravity-wave frequency of interest f_R (i.e. for $B = c/4f_R L$ round trips); then, instead of extracting it through the beam splitter and reinjecting it, simply move it into the other arm using a high-reflectivity, Fabry–Perot-type mirror – i.e. exchange the light between the arms. Then during the next half-cycle, with arms interchanged and the sign of the gravity wave reversed, each piece of light will experience a phase shift in the same direction as during the first half-cycle. At the end of the second half-cycle, interchange the light in the two arms, and repeat as many times as mirror losses will permit. The arrows in Fig. 9.14(a) show the light, as a result of these exchanges, circulating around the interferometer in a clockwise direction, but there is an equal amount of light circulating in a counterclockwise direction. The light circulating one way experiences a continual increase in phase shift throughout its circulation; the light

Fig. 9.14. Schematic diagrams of optical configurations that could be used for *resonating light* in Michelson-type (a) and Fabry–Perot-type (b) beam receivers (Dreaver, 1983).

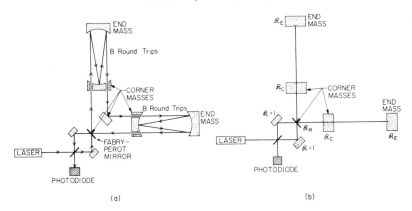

(a)   (b)

circulating oppositely experiences a continual decrease in phase shift; and these oppositely circulating beams, upon re-emerging through the back of the Fabry–Perot-type mirror and recombining at the beam splitter, produce a much enhanced interference signal in the photodetector.

For the Fabry–Perot receiver of Fig. 9.14(*b*) a single 'resonating mirror', denoted $\mathscr{R}_R$, produces the resonant interchange of the light between the two arms: the lengths of the cavities are so adjusted as to produce in each cavity a resonance at a frequency f_e very near the laser frequency; the reflectivities $\mathscr{R}_C$ of the corner mirrors, and the path lengths between them and the resonating mirror, are then so adjusted as to produce two resonant modes of the coupled cavities at light frequencies $f_e \pm f_R/2$ – which requires

$$1 - \mathscr{R}_C = 4\pi L f_R/c; \tag{119a}$$

the laser frequency is then adjusted so that it drives the mode at $f_e - f_R/2$; and the gravity waves with frequency f_R, by wiggling the end mirrors, then upconvert photons into the mode at $f_e + f_R/2$. In order to maximize the resulting gravity-wave-induced signal at the photodetector, the reflectivity $\mathscr{R}_R$ of the resonating mirror must be adjusted to

$$1 - \mathscr{R}_R = 2\frac{1 - \mathscr{R}_E}{1 - \mathscr{R}_C} = 2\frac{f_o}{f_R}. \tag{119b}$$

With these optimizations the spectral density of shot noise near f_R is (cf. Gürsel *et al.*, 1983; Meers, 1983; Vinet, 1986; Brillet and Meers, 1987)

$$S_h(f) = 4S_o\left[1 + \left(\frac{f - f_R}{f_o}\right)^2\right] \quad \text{for } |f - f_R| \ll f_o. \tag{119c}$$

As is seen in Fig. 9.13 (dotted curves), this represents an enormous improvement in noise over either a simple Fabry–Perot or a recycling Fabry–Perot. However, this improvement is bought at the price of an enormous narrow-banding of the receiver response: at frequencies $f \sim f_R/2$ and $f \sim 2f_R$ the noise is comparable to that in a simple Fabry–Perot; and at $f \ll f_R/2$ and $f \gg 2f_R$ it is far worse.

For an optimally configured, resonating Michelson the sensitivy near resonance, $|f - f_R| \ll f_o$, is similar to that of a Fabry–Perot (equation (119c)).

Neither recycling nor resonating is potentially useful in present prototypes because present mirror reflectivities limit the prototypes to short storage times. However, in the planned full-scale LIGOs where 100-fold longer arms permit 100-fold greater storage times, recycling and resonating both give promise of substantial improvements in beam-detector sensitivities. The first attempts to implement recycling on a prototype were carried out by the Munich group in 1986, with moderate success. Resonating has not yet been attempted.

(f) *Quantum limit, quantum non-demolition, and squeezed vacuum*

A beam receiver is similar to a 'Heisenberg microscope', in which a photon is used to measure the position of a particle. The analog of the Heisenberg photon is the laser beam and the analogs of the Heisenberg particle are the end and corner masses. Just as the Heisenberg photon kicks the particle it is trying to measure thereby enforcing the Heisenberg uncertainty principle, $\Delta x \, \Delta p \geqslant \hbar/2$, so fluctuations in the beam's light pressure kick the end and corner masses thereby enforcing a quantum limit on the sensitivity of the beam receiver. As I shall discuss below, these quantum fluctuations have their ultimate source in the vacuum fluctuations of quantum electrodynamics (Caves, 1980, 1981).

Caves (1987b) has used Feynman path-integral techniques (Caves, 1986, 1987a) to derive a pseudospectral density of displacement noise that characterizes the quantum limit of any harmonic oscillator (such as a beam receiver's swinging masses) when it is studied at frequencies high above the resonant frequency f_0, using techniques that produce the minimum possible noise:

$$S_x(f) = \frac{2\hbar}{m(2\pi f)^2},\qquad(120)$$

where m is the oscillator's mass and x is its displacement from equilibrium. For a beam receiver, whose swinging masses have resonant frequencies $f_0 \sim 1 \, \text{Hz}$ far below the region of interest, this pseudospectral density for their displacements can be translated into a pseudospectral density for the gravity-wave field $h(t)$ that one reads off the receiver's output:

$$S_h(f) = \frac{8\hbar}{m(2\pi f)^2 L^2}.\qquad(121)$$

Here each of the masses on which a mirror is attached is assumed to have the same mass m. This pseudospectral density can be used, in the same manner as an ordinary classical spectral density (subsection (c) above), to compute the standard quantum limit on the performance of a beam detector in various situations.

As for a bar detector, so also for a beam detector, there should be methods of circumventing this standard quantum limit. Unfortunately nobody has yet found a remotely practical way of doing so, except in one special case: Unruh (1982) has invented a method, which Caves' path-integral formalism says should work (Caves, 1987*b*), for beating the standard quantum limit when performing a narrow-band measurement of periodic gravitational waves. The key idea (which was invented independently and earlier in the classical domain by Gordienko, Gusev and Rudenko, 1977) is to place a spring between each mirror and the companion mass on which it rides, thereby turning the mirror and companion into a two-mode system. By setting the ratio of the spring frequency over the gravity-wave frequency equal to the square root of the mass of the big companion over the mass of the little mirror, one can force the laser beam's fluctuating light pressure to act on the motion of the big mass and not the mirror, while the interferometer reads out the mirror's motion. The result is an improved immunity of the receiver to Heisenberg's quantum noise over a narrow frequency band. While this scheme is clever and conceptually satisfactory, whether it can be implemented in a practical manner is not yet clear.

Although, for broad-band measurements with beam receivers, there as yet is no known way of beating the standard quantum limit, Caves (1981) has invented a clever scheme for getting closer to the quantum limit when inadequate laser power causes the shot noise to exceed it. The key to Caves' idea is his discovery that the ultimate source of the photon shot noise and the fluctuating radiation-pressure noise is *not*, as people previously thought, fluctuations in the laser output. Rather, it is fluctuations in the quantum-electrodynamic vacuum state ('vacuum fluctuations') that enter the beam splitter at right-angles to the incoming laser light and superpose on the laser light as it heads toward the two arms. As Caves has shown, by 'squeezing the vacuum' (i.e. reducing the vacuum fluctuations in the $\cos[(ct-x)/\lambda_e]$ part of the light while increasing them in the $\sin[(ct-x)/\lambda_e]$ part; achievable in principle by sending the vacuum through a pumped, non-linear medium), one can reduce the beam receiver's shot noise at the expense of increasing the fluctuations in its light-pressure noise. The net result is the same as if one were using a laser of higher power: in the typical case, where the actual power is too low to permit achieving the quantum limit, one improves the sensitivity and moves toward the quantum limit.

Recently Caves (1987*c*) has elucidated the ultimate sensitivities achievable when one combines his squeezed-vacuum technique with light recycling and resonating. He finds that because, in a resonating system, noise associated with losses in the end mirrors is as important on resonance as vacuum fluctuations entering the beam splitter, squeezing of the vacuum is not useful. By contrast, when combined with light recycling, squeezed-vacuum techniques can reduce $S_h(f)$, over a broad band of frequencies $\Delta f \gtrsim f$, to S_o

or to the quantum limit, whichever is larger:

$$S_h(f) = \max \begin{cases} S_o = \dfrac{2\hbar c \lambda_e}{I_o \eta} \left(\dfrac{1 - \mathscr{R}_E}{2L} \right)^2 \\[2ex] \dfrac{8\hbar}{m(2\pi f)^2 L^2} \end{cases} \tag{122}$$

Squeezed states of light have been produced by experimenters recently (Slusher *et al.*, 1985; Wu *et al.*, 1986), triggering great excitement in the field of non-linear quantum optics. This achievement has triggered an effort by the Munich group to implement Caves' squeezed-vacuum technique – an effort that is likely to succeed within the next few years.

(g) *Anticipated sensitivities of full-scale LIGO receivers*

The present Munich burst sensitivity $h_{3/yr} \cong 4 \times 10^{-17}$ at $f \sim 1000$ Hz, when scaled from the prototype arm length of 30 m to the planned Munich LIGO arm length of 3 km, would become $h_{3/yr} \cong 4 \times 10^{-19}$. Although scaling up is not straightforward (mirror diameters must be increased a factor ten, for example), the experimenters are quite confident of doing somewhat better than this scaled-up number by the early 1990s when the first LIGOs start operating.

The upper thick, dashed curves in Figs. 9.4, 9.6 and 9.7 (earlier in this chapter) show possible burst, periodic and stochastic sensitivities (based on a $\frac{1}{3}$-year search time and 90% confidence levels), for a first generation of detectors in the full-scale LIGOs. These curves correspond, at frequencies above ~ 500 Hz, to receivers that are shot-noise-limited, without recycling or resonating, with Argon-ion laser light $\lambda_e = 0.0818$ μm, with (laser power) $\times$ (photodetector efficiency) $= I_o \eta = 10$ W, and with light storage times of a half-gravity-wave period or longer, so [cf. equation (115) and Fig. 9.12]

$$[S_h(f)]^{\frac{1}{2}} = \left[\frac{2\hbar c \lambda_e}{I_o \eta} \left(\frac{\pi f}{c} \right)^2 \right]^{\frac{1}{2}} = (2.4 \times 10^{-22} \text{ Hz}^{-\frac{1}{2}}) \left(\frac{f}{1000 \text{ Hz}} \right); \tag{123a}$$

and, correspondingly (equations (111)–(113))

$$h_{3/yr} = 11(f_c S_h)^{\frac{1}{2}} = 8 \times 10^{-20} \left(\frac{f_c}{1000 \text{ Hz}} \right)^{\frac{3}{2}} \quad \text{for bursts,} \tag{123b}$$

$$h_{3/yr} = 3.8(S_h \times 10^{-7} \text{ Hz})^{\frac{1}{2}} = 2.9 \times 10^{-25} \left(\frac{f}{1000 \text{ Hz}} \right)$$

$$\text{for periodic sources,} \tag{123c}$$

$$h_{3/yr} = 4.5 \left(\frac{f}{10^{-7} \, \text{Hz}} \right)^{-\frac{1}{4}} (fS_h)^{\frac{1}{2}} \tag{123c}$$

$$= 1.1 \times 10^{-22} \left(\frac{f}{1000 \, \text{Hz}} \right)^{\frac{1}{4}} \quad \text{for stochastic sources,} \tag{123c}$$

where the search for stochastic waves is presumed to use a bandwidth $\Delta f = f$. Below $\sim 500 \, \text{Hz}$ it is presumed that seismic noise debilitates the performance of these first-generation LIGO receivers.

During the years following the first gravity-wave searches in the LIGOs, the experimenters plan to run a sequence of detectors with ever improving sensitivities, pushing the sensitivities ever downward, and improving the seismic isolation so the detectors can operate at ever decreasing minimum frequencies. A reasonable goal by the end of the 1990s is to approach the lower-most thick-dashed curves in Figs. 9.4, 9.6 and 9.7. These curves correspond to *advanced detectors* with the following characteristics:

$L = (\text{arm length}) = 4 \, \text{km}$,
$\lambda_e = 0.0818 \, \mu\text{m}$ (Argon-ion laser light),
$I_o \eta = (\text{laser power}) \times (\text{photodetector sensitivity}) = 100 \, \text{W}$,
$\mathscr{R}_E = \mathscr{R}_{max} = (\text{maximum mirror reflectivity}) = 0.9999$,
$m = (\text{mirror mass}) = 1000 \, \text{kg}$,

  photon shot noise dominant at $100 \, \text{Hz} \lesssim f \lesssim 10^4 \, \text{Hz}$,
  quantum-limit noise dominant at $10 \, \text{Hz} \lesssim f \lesssim 100 \, \text{Hz}$,
  seismic noise dominant at $f \lesssim 10 \, \text{Hz}$. (124)

It is presumed that light recycling is used for burst searches, with the knee frequency adjusted to equal the frequency of interest, so the photon shot noise (equations (111), (117c) and (116b,c) with $f_k = f_c$) is

$$h_{3/yr} = 11(f_c S_h)^{\frac{1}{2}} = 11(S_o f_c^2 / f_o)^{\frac{1}{2}} = 11 \left[\frac{2\pi \hbar \lambda_e}{I_o \eta} \frac{(1 - \mathscr{R}_E)}{L} f_c^2 \right]^{\frac{1}{2}}$$

$$= 1.3 \times 10^{-21} \left(\frac{f_c}{1 \, \text{kHz}} \right) \quad \text{for bursts;} \tag{125a}$$

and it is presumed that light resonating is used for periodic and stochastic searches, with the resonant frequency adjusted to equal the frequency of interest, so the photon shot noise (equations (112), (113), (119c) and (116b,c) with $f_R = f$) is

$$h_{3/yr} = 3.8(4S_o \times 10^{-7} \, \text{Hz})^{\frac{1}{2}} = 3.8 \left[\frac{2\hbar c \lambda_e}{I_o \eta} \times 10^{-7} \, \text{Hz} \right]^{\frac{1}{2}} \left(\frac{1 - \mathscr{R}_E}{L} \right)$$

$$= 2.1 \times 10^{-28} \quad \text{for periodic waves,} \tag{125b}$$

$$h_{3/\mathrm{yr}} = 4.5 \left(\frac{f_\mathrm{o}}{10^{-7}\,\mathrm{Hz}}\right)^{-\frac{1}{4}} (4S_\mathrm{o}f)^{\frac{1}{2}} = 4.5 \left[\frac{2\hbar c\lambda_\mathrm{e}}{I_\mathrm{o}\eta}\,f\right]^{\frac{1}{2}} \left[\frac{4\pi}{c}\left(\frac{1-\mathscr{R}_\mathrm{E}}{L}\right)^3 \times 10^{-7}\,\mathrm{Hz}\right]^{\frac{1}{4}}$$

$$= 5.2 \times 10^{-25} \left(\frac{f}{1\,\mathrm{kHz}}\right)^{\frac{1}{2}} \quad \text{for stochastic waves.} \tag{125c}$$

Here it is assumed that the stochastic search is restricted to the narrow bandwidth $\Delta f = f_\mathrm{o}$ over which the resonating receiver has good performance. The quantum-limit noise, which exceeds these shot noise levels at $f \lesssim 100$ Hz, is given by (equations (121) and (111)–(113))

$$h_{3/\mathrm{yr}} = 11\left[\frac{2}{\pi^2}\frac{\hbar}{mL^2}\frac{1}{f}\right]^{\frac{1}{2}} = 1.3 \times 10^{-23}\left(\frac{1000\,\mathrm{Hz}}{f}\right)^{\frac{1}{2}} \quad \text{for bursts,} \tag{126a}$$

$$h_{3/\mathrm{yr}} = 3.8\left[\frac{2}{\pi^2}\frac{\hbar}{mL^2}\frac{10^{-7}\,\mathrm{Hz}}{f^2}\right]^{\frac{1}{2}}$$

$$= 4.4 \times 10^{-29}\left(\frac{1000\,\mathrm{Hz}}{f}\right) \quad \text{for periodic waves,} \tag{126b}$$

$$h_{3/\mathrm{yr}} = 4.5\left[\frac{f_\mathrm{o}(S_\mathrm{QL}/S_\mathrm{o})^{\frac{1}{2}}}{10^{-7}\,\mathrm{Hz}}\right]^{-\frac{1}{4}}\left[\frac{4}{3}fS_\mathrm{QL}\right]^{\frac{1}{2}}$$

$$= 4.5\left(\frac{4}{3\pi}\right)^{\frac{1}{2}}\left(\frac{4\hbar}{mL^2}\right)^{\frac{3}{8}}\left[\frac{\hbar\lambda_\mathrm{e}}{I_\mathrm{o}\eta c}\left(\frac{10^{-7}\,\mathrm{Hz}}{f}\right)^2\right]^{\frac{1}{8}}$$

$$= 1.5 \times 10^{-25}\left(\frac{1000\,\mathrm{Hz}}{f}\right)^{\frac{1}{4}} \quad \text{for stochastic waves.} \tag{126c}$$

Here for stochastic waves S_QL is the quantum-limit spectral density (equation (121)), and the chosen bandwidth $\Delta f = f_\mathrm{o}(S_\mathrm{QL}/S_\mathrm{o})^{\frac{1}{2}}$ is that which makes shot noise (equation (119c) integrated over Δf, with $f_\mathrm{o} < \Delta f < f_\mathrm{R}$) equal to $\frac{1}{3}$ the quantum noise and minimizes their sum. At frequencies $f \lesssim 10$ Hz, seismic noise is presumed to strongly debilitate the receiver performances (cf. Saulson, 1984).

Figs. 9.4, 9.6 and 9.7 show, together with the above receiver sensitivities, the characteristic strengths of the gravity waves from a variety of sources that were discussed in Section 9.4 above, and the present and projected sensitivities of other kinds of detectors. As these diagrams suggest, the prospects for successful detection of gravity waves with the planned LIGOs are high: at the sensitivities of the 'advanced detectors' one could see the coalescence of neutron-star binaries at $\frac{1}{4}$ the Hubble distance where the event rate is estimated to be several per day, the coalescence of $10M_\odot$ black-hole binaries throughout the universe, supernovae in our galaxy if they put out $10^{-9}M_\odot$ of energy near 1000 Hz frequency and in the Virgo cluster if they put out $10^{-3}M_\odot$, millisecond pulsars whose frequencies are known in advance from electromagnetic measurements if they have ellipticities $10^{-8.5}$ or larger, CFS-unstable neutron stars with X-ray luminosities as small as $\frac{1}{1000}$ that of Sco X-1, and a stochastic background at $f \sim 30$ Hz with Ω_{GW} as small as 10^{-10}. It seems likely to me that such sensitivities are more than adequate for success; waves are likely to be detected at more modest sensitivities than these – somewhere between those of the 'first-generation' detectors and those of the 'advanced detectors' (Figs. 9.4, 9.6 and 9.7).

(h) *Ideas for other types of beam detectors*

The Michelson and Fabry–Perot configurations, on which almost all experimental work to date has focussed, are not the only possible types of beam receivers. A number of others have been conceived of, but have not been pursued vigorously. Since the LIGOs are intended to house several different receivers simultaneously and to have lifetimes of $\gtrsim 20$ years, during which a number of generations of receivers will be built and operated, some of these other configurations might one day operate in a LIGO. These other configurations include: (i) A *frequency-tagged interferometer* or *Michelson with overlapping beams* (Drever and Weiss, 1983), in which the light beam in each arm is made to shift in frequency with each round-trip pass, so that the beams of successive passes can overlap each other but not interfere with each other; such an interferometer is basically a Michelson but with small, Fabry–Perot-size mirrors. (ii) *Active interferometers* (Bagaev *et al.*, 1981; Brillet and Tourrenc, 1983), in which an active medium (atoms or molecules with transitions near the frequency of the laser light) resides in the arms (or arm – there might be only one) of the interferometer. (iii) *Spectroscopic detectors* (Nesterikhin, Rautian and Smirnov, 1978; Brillet and Tourrenc, 1983; Borde *et al.*, 1983), in which laser spectroscopy is used to monitor the frequency shifts produced by the gravitational waves.

Although in principle these alternative types of beam detectors can achieve sensitivities comparable to those of the standard (Michelson and Fabry–Perot) types, practical issues make the active and spectroscopic detectors look substantially less promising (Brillet and Tourrenc, 1983); and the frequency-tagged detector has not yet been pursued in sufficient depth to venture a tentative verdict.

9.5.4 *Other types of earth-laboratory detectors*

In addition to bar detectors and beam detectors, a number of other types of gravity-wave detectors that could operate in an earth-bound laboratory have been conceived of. Although none of these has looked sufficiently promising to justify substantial experimental effort, some are question marks (because they have not been pursued far enough for a clear verdict) rather than discards. Because I have not looked at most of them in enough detail, I shall not venture to sort out the question marks from the total rejects.

(a) Electromagnetically coupled detectors

One type of transducer used on bar detectors (e.g. by the Moscow and Perth groups) is a re-entrant microwave-cavity resonator whose capacitance is modulated by the bar's vibrations (Section 10 of Braginsky, Mitrofanov and Panov, 1985; Blair, 1983). There is an obvious similarity of this to a Fabry–Perot beam receiver in which each arm is an optical cavity with length modulated by the motions of the swinging masses. In fact, one can imagine a continuous sequence of detector configurations that leads from one to the other, by splitting the bar into pieces and gradually moving them apart, with the microwave cavity gradually being distorted and expanded until it becomes the optical cavity of the Fabry–Perot (Caves, 1978).

This argument leads to the recognition that bar detectors and beam detectors are but two examples of a large class of possible 'electromagnetically coupled detectors', in which gravitational waves drive the motions of masses, and electromagnetic fields measure those motions; or, when the detector gets larger than a reduced wavelength, gravity waves drive vibrations of both the electromagnetic fields and the masses, which are coupled together.

Other earth-laboratory-scale electromagnetically coupled configurations, besides resonant bars and optical beams, that have been considered theoretically and show some promise but have not been pursued in a serious experimental way, include: (i) large microwave cavities in which wall motion, driven by gravitational waves with $\lambda \gg L$, upconverts microwave quanta from one mode to another mode of slightly higher frequency (Pegoraro, Picasso and Radicati, 1978; Caves, 1979); (ii) optical or microwave cavities with $L \lesssim \lambda$, intended for detection of high-frequency waves $f \gg 10^4$ Hz, in which the gravitational waves interact directly with the resonating electromagnetic field to move quanta from one mode into another, or interact directly with a DC electric or magnetic field to create quanta at the gravity-wave frequency (Braginsky et al., 1973); (iii) as a specific, much studied example of such a cavity: an optical or microwave ring resonator in which circularly polarized gravitational waves propagating orthogonal to the resonator plane resonate with the circulating electromagnetic field, producing a linearly growing phase shift in the field. This scheme, proposed by Braginsky and Menskii (1971), was incorrectly analyzed by them and by me (Box 37.6 of MTW) – a source of some embarrassment. My error was in thinking I could cover the detector with a single proper reference frame, when its diameter L is larger than the reduced wavelength λ of the gravitational waves it sees. My incorrect conclusion, and that of Braginsky and Menskii (1971) was a quadratically growing phase shift, $\Delta\Phi \propto h(ct/\lambda)(ct/\lambda_e)$. The correct conclusion (Linet and Tourrenc, 1976) is a linearly growing phase shift, $\Delta\Phi \propto h(ct/\lambda_e)$, which makes this scheme no better in principle than a standard beam detector. WARNING: The literature is full of similarly incorrect analyses of the various detectors described in this section, 9.5.4; (iv) an optical cavity filled with an isotropic medium, or an optical fiber, in which gravity-wave-produced strains induce optical effects such as birefringence (Iacopini *et al.*, 1979; Vinet, 1985); (v) detectors using the Mossbauer effect (Kauffman, 1970).

For a general review of electromagnetically coupled detectors, see Grishchuk (1983), and for general analyses of them, see Tourrenc and Grossiord (1974), Tourrenc (1978) and Teissier du Cros (1985).

Analyses of such detectors sometimes produce wildly overoptimistic conclusions because they overlook the mundane issue of thermal noise in the mechanical parts of the detector (e.g. the walls of an electromagnetic cavity). For an example of an analysis that *does* take thermal noise into account properly, see Caves (1979).

(b) *Superfluid interferometers and superconducting circuits*

Anandan (1981), Chiao (1982) and Anandan and Chiao (1982) have suggested that *if* it is possible to construct a superfluid weak link, analogous to the superconducting weak link (Josephson junction) on which SQUIDS are based, then such a link could form the basis for a superfluid ring interferometer that would be sensitive to gravitational waves. Unfortunately, such weak links do not yet exist. Schrader (1984) and independently Anandan (1985, 1986) have suggested a gravity-wave detector based on the direct interaction between the gravitational wave and the magnetic field in superconducting solenoids, with the resulting current change monitored by a SQUID.

9.5.5 *Low-frequency detectors* $(10$–$10^{-5}\,Hz)$

As one goes to lower and lower frequencies it becomes harder and harder to isolate a gravity-wave detector in an earth-bound laboratory from seismic and acoustic vibrations and from fluctuating gravity gradients due to people, animals, trucks, etc. Ultimately, somewhere around 10 Hz, isolation will become impossibly difficult (see, e.g., Saulson, 1984). Thus, to operate in the 'low-frequency region' $10\,\mathrm{Hz} \gg f \gtrsim 10^{-5}\,\mathrm{Hz}$ will require putting detectors in space, or using normal modes of the earth or the sun as the detectors.

A number of space-borne or earth- or sun-normal-mode detectors have been conceived of for operation at low frequencies; and several have been constructed and have produced interesting limits on cosmic gravitational waves. In this section I shall describe these briefly.

(a) Doppler tracking of spacecraft

At periods between a few minutes and a few hours the best present gravity wave detector is the doppler tracking of spacecraft.

In doppler tracking a highly stable clock on earth ('master oscillator') is used to control the frequency of a monochromatic radio wave, which is transmitted from earth to the spacecraft. This 'uplink' radio wave is received by the spacecraft and 'transponded' back to the earth; i.e. the uplink wave is used by the spacecraft to control, in a phase-coherent way, the frequency of the 'downlink' radio wave that it transmits back to earth. When the down link radio wave is received at earth, its frequency is compared with that of the master oscillator; and from that comparison a doppler shift is read out.

When gravitational radiation sweeps through the solar system – most interestingly with a reduced wavelength of order the earth-spacecraft distance so it must be analyzed by TT methods – it perturbs the earth, the spacecraft, and the propagating radio wave. The net result is to produce fluctuations in the measured doppler shift with magnitude $\delta v/v \sim h_{jk}^{\mathrm{TT}}$. Each feature in the gravitational waveform $h_{jk}^{\mathrm{TT}}(t)$ shows up three times in the doppler shift, in a manner that can be regarded as due to interaction of the radio wave with the gravity wave at the events of emission from earth, transponding by spacecraft, and reception at earth (Estabrook and Wahlquist, 1975). This triplet structure can be used to help distinguish the effects of the gravitational wave from noise in the doppler data.

The use of doppler data for gravity-wave searches was first proposed by Braginsky and Gertsenshtein (1967), and was first pursued with preexisting data by Anderson (1971). The response of doppler tracking to a gravity wave was worked out by Davies (1974) for special cases and by Estabrook and Wahlquist (1975) for the general case. Experimental feasibility and noise thresholds were first discussed by Wahlquist *et al.* (1977). Experimental results have been obtained with the Viking spacecraft (Armstrong *et al.*, 1979), the Voyager spacecraft (Hellings *et al.*, 1981), Pioneer 10 (Anderson *et al.*, 1984), and Pioneer 11 (Armstrong *et al.*, 1987). A future experiment will entail coordinated observations with the Galileo and Ulysses spacecraft (Estabrook, 1987).

In all past experiments the most serious noise source was fluctuations in the interplanetary plasma (solar wind), which cause fluctuations in the dispersion of the radio waves and thence fluctuations in the doppler shift. These fluctuations have limited the sensitivity under optimal circumstances to $[fS_h(f)]^{\frac{1}{2}} \sim 3 \times 10^{-14}$ in the band $10^{-4}\,\mathrm{Hz} \lesssim f \lesssim 10^{-2}\,\mathrm{Hz}$. Fortunately, dispersion in an ionized plasma becomes less serious when one moves to higher radio-wave frequencies – a move that is also desired to produce higher bit rates for information transfer between the spacecraft and earth. Some future spacecraft, including the Galileo mission, will use X-band radio signals (10 GHz frequency) on both the uplink and the downlink, rather than S-band (2 GHz) up and X-band down as in the past; and correspondingly their gravity-wave sensitivities should improve to $[fS_h(f)]^{\frac{1}{2}} \sim 3 \times 10^{-15}$. In fact, on such spacecraft the ionized-plasma dispersion will be sufficiently low that fluctuations in dispersion due to water vapor in the earth's atmosphere may become the most serious noise source (Armstrong and Sramek, 1982). Plans for monitoring the water vapor as a means of reducing this noise are being developed (Resch *et al.*, 1984). In the more distant future, doppler tracking may be improved by using dual-frequency X-band/K-band (30 GHz) tracking to reduce and monitor the ionized plasma dispersion, by installing a highly stable clock on board the spacecraft (Smarr *et al.*, 1983) and/or by moving the earth-based antenna into earth orbit. These improvements might ultimately produce $[fS_h(f)]^{\frac{1}{2}}$ as low as 10^{-17} (Estabrook, 1987).

Figs. 9.4, 9.6 and 9.7 show the sensitivities $h_{3/\mathrm{yr}}$ corresponding to these noise levels (equations (34), (52a) and (67) with $\Delta f = f$).

At frequencies $f \sim 3\,\mathrm{Hz}$, far above the regime 10^{-2}–$10^{-4}\,\mathrm{Hz}$ of normal doppler tracking, but below the regime of earth-based detectors, one might hope to use doppler tracking of spacecraft in earth orbit (distances $\sim c/4f \sim 25\,000\,\mathrm{km}$). The best such spacecraft are those of the Global Positioning System (GPS). Unfortunately, even with the best projected improvements of the GPS we cannot anticipate $[fS_h(f)]^{\frac{1}{2}}$ better than $\sim 10^{-14}$ corresponding to a burst sensitivity $h_{3/\mathrm{yr}} \sim 10^{-13}$; see Fig. 9.4 and see Hansen, Chiu and Chao (1986) for a detailed study.

(b) Beam detectors in space

Several conceptual designs for optical beam detectors in space were suggested by Weiss et al. (1976) and are discussed in Weiss (1979). More recently Faller and Bender (1981), Bender et al. (1984) and Faller et al. (1985) have been carrying out a detailed design and feasibility study for such detectors; and it is possible they will fly sometime around the turn of the century. As now conceived (Faller et al., 1985), such a detector might consist of three 'drag-free' satellites, one at the corner and two at the ends of an L. The satellites might be in the same solar orbit as the earth–moon system, but far from the earth and moon; and their arm length might be $L \sim 1$ million km. Each satellite might carry its own 100 mW laser, with the end lasers phase locked to the light that arrives from the corner laser. With 50 cm diameter telescopes on each satellite to focus the laser beams, this could result in a one-pass $(B=1)$ Michelson interferometer system with shot-noise-limited sensitivity $[S_h(f)]^{\frac{1}{2}} = 10^{-20} \, \mathrm{Hz}^{-\frac{1}{2}}$ over the range $10^{-1} \, \mathrm{Hz} \gtrsim f \gtrsim 10^{-4} \, \mathrm{Hz}$. Above $10^{-1} \, \mathrm{Hz}$ the sensitivity would be degraded because of too-long an arm length $(L > \lambda)$. Below $10^{-4} \, \mathrm{Hz}$ stochastic non-gravitational forces might degrade the sensitivity. This projected sensitivity translates into the amplitude levels $h_{3/\mathrm{yr}}$ (equations (34), (52a) and (67) with $\Delta f = f$) shown in Figs. 9.4, 9.6 and 9.7.

(c) Earth's normal modes

Forward et al. (1961) and Weber (1967) pioneered the use of the earth's quadrupolar normal modes as 'resonant-bar' gravity-wave detectors. More recently Boughn and Kuhn (1984) have given a careful formulation of the method for inferring, from earth-normal-mode observations, limits on any stochastic gravitational waves that might be exciting the earth. Using their method on data from the IDA seismic network, Boughn and Kuhn (1987) have derived a slightly stronger limit on stochastic background than the early Weber (1967) limit: they find $\Omega_{\mathrm{GW}}(f) \lesssim 1$ at $f = 0.31 \, \mathrm{mHz}$ (Fig. 9.7). They expect, further, that a larger amount of data may permit this limit to be lowered to $\Omega_{\mathrm{GW}} \lesssim 0.1$.

(d) Sun's normal modes

Walgate (1983) stimulated people to think about the sun's quadrupolar normal modes as gravity-wave detectors, when he speculated (falsely; see, e.g., Kuhn and Bough, 1984, and Anderson et al., 1984) that an observed mode was being excited by monochromatic gravitational waves from the binary star Geminga.

Boughn and Kuhn (1984) have carried out a detailed analysis of the gravity-wave implications of observed solar pulsations. From the data of Isaac (1981) on low-order p- and g-modes they find $\Omega_{GW}(f) \lesssim 100$ at $f \simeq 3 \times 10^{-4}$ Hz; and they argue that better observational data may produce orders of magnitude improvement, bringing $\Omega_{GW}(f)$ well below unity.

(e) *Vibrations of blocks of the earth's crust*

Braginsky *et al.* (1985) have recently pointed out that blocks of the earth's crust with sizes 50 km $\leqslant L \leqslant$ 70 km could function as resonant-bar detectors for waves of frequency $f \simeq 0.03$ Hz. They propose monitoring such a block with an array of seismic stations and cross-correlating the data to pull out the seismic motions associated with the block's quadrupolar modes. Their calculations suggest a possible sensitivity $[fS_h(f)]^{\frac{1}{2}} \sim 2 \times 10^{-17}$ corresponding to $h_{3/yr} \sim 2 \times 10^{-16}$ for bursts.

(f) *Skyhook*

Braginsky and Thorne (1985) have suggested an earth-orbiting 'skyhook' gravity-wave detector that would operate in the 0.1–0.01 Hz region with sensitivity $[fS_h(f)]^{\frac{1}{2}} \sim 3 \times 10^{-17}$, which is much better than present or near-future doppler tracking and roughly comparable to that hoped for from blocks of the earth's crust (see Figs. 9.4, 9.6 and 9.7). The skyhook would consist of two masses, one on each end of a long thin cable with a spring at its center. As it orbits the earth, the cable would be stretched radially by the earth's tidal gravitational field. Gravitational waves would pull the masses apart and push them together in an oscillatory fashion; their motion would be transmitted to the spring by the cable; and a sensor would monitor the spring's resulting motion.

If it ever flies, the skyhook's role will be to provide, with a simple and inexpensive device, a moderate-sensitivity coverage of the 0.1–0.01 Hz region during the epoch before far more sensitive beam detectors are built and installed in space.

9.5.6 *Very-low-frequency detectors (frequencies below 10^{-5} Hz)*

At frequencies below about 10^{-5} Hz the only sources of gravitational waves are probably stochastic background from the early universe (Sections 9.4.3(d,e,f)); and the best detectors involve use of distant astronomical bodies.

(a) *Pulsar timing*

Remarkably good limits on very-low-frequency gravitational waves have come from the timing of pulsars (Taylor, 1987). The basic idea for pulsar timing as a gravity-wave detector (Sazhin, 1978; Detweiler, 1979) is this: the rotation of the pulsar's underlying neutron star is a highly stable clock, and pulsar timing compares that clock's ticking rate with the ticking of the very best atomic clocks in earth-bound laboratories. When gravity waves sweep over the pulsar, they affect the ticking rate of the pulsar's clock relative to clocks elsewhere in the universe, including earth; and those effects show up as fluctuations in the pulsar timing data. Similarly, when gravity waves sweep over the earth, they affect the ticking rates of all our clocks relative to those elsewhere in the universe; and those effects show up in pulsar timing. To the extent, then, that the pulsar timing data are free of fluctuations, one can infer that gravity waves of a given strength are not sweeping over either the pulsar or the earth. (See Blandford, Narayan and Romani, 1984, for a discussion of how to extract the gravity-wave information from the timing data and the pitfalls one encounters in doing so.) If fluctuations are seen in the timing data and are actually due to gravity waves sweeping over the pulsar, we probably will never be able to confirm that the source is gravity waves rather than fluctuations in the neutron star's rotation. However, if fluctuations are seen and are due to gravity waves sweeping over the earth, we might be able to learn their cause and study the waves' direction of propagation, polarization and wave form by simultaneously timing several pulsars on different parts of the sky.

In response to the Sazhin–Detweiler idea, Hellings and Downs (1983) and Romani and Taylor (1983) used timing data on four especially quiet pulsars (which Downs and Reichley (1983) had tracked for over a decade with the Goldstone antenna of NASA's deep space tracking network) to place a limit $\Omega_{GW}(f) \lesssim 1 \times 10^{-3}(f/10^{-8}\text{ Hz})^4$ (90% confidence) on any stochastic background in the frequency range $4 \times 10^{-9} \lesssim f \lesssim 10^{-7}$ Hz; Bertotti, Carr and Rees (1983) used data from Helfand *et al.* (1980) to obtain a similar limit.

Intrinsic noise in all the pulsars used in these analyses would have made it difficult to improve on these limits at fixed frequency (though, by further observations the region covered could have been extended to lower frequencies; the lowest being of order 1/(total time since the measurements began)). Fortunately, while these analyses were in process, they were put out of business by the discovery (Backer *et al.*, 1982) of a pulsar far quieter than any previously known: the millisecond pulsar PSR 1937+21. From three years of 1937 + 21 timing data taken with the Arecibo radio telescope, Davis *et al.* (1985) and Taylor (1987) have now placed the limit

$$\Omega_{GW}(f) \leqslant 1 \times 10^{-6} \left(\frac{f}{10^{-8} \text{ Hz}} \right)^4 \quad \text{for } f \gtrsim f_{min} = 10^{-8} \text{ Hz} \qquad (127)$$

on any isotropic stochastic background: see Fig. 9.7.

This level of sensitivity is so good that further progress at fixed frequency is limited by the long-term frequency stability of the world's best atomic clocks. Thus, unless clocks improve, we can expect the coefficient in (127) to improve at best as t^{-1}, while the lower frequency limit f_{min} decreases as t^{-1} producing $\Omega_{GW}(f_{min}) \propto t^{-5}$ (with $t = 0$ in 1982). Ultimately, when clocks have improved by one or two orders of magnitude, noise due to interstellar scintillation may become a problem (Armstrong, 1984).

Recently two other quiet, fast pulsars PSR 1855 + 09 and PSR 1953 + 29 have been discovered (Segelstein *et al.*, 1986). Together with PSR 1937 + 21 and others that we can hope for, they may one day form a network for gravitational-wave searches and observations. Such a network would alleviate problems with interstellar scintillations and atomic clock fluctuations.

(b) *The timing of orbital motions*

A gravitational wave with period long compared to observation times will produce a gradual secular change in the relative ticking rates of clocks in and out of the wave: $\dot{\omega}/\omega \sim (f/2\pi)h$, where f is the gravity-wave frequency, h is its amplitude, and ω is the ticking rate of the clock in the wave as monitored by any clock that is outside of the wave. For stochastic waves this gives $\dot{\omega}/\omega \sim [\Omega_{GW}(f)]^{\frac{1}{2}}(10^{10} \text{ yr})^{-1}$. Because the physical torques on pulsars are so large $(\dot{\omega}/\omega \gg (10^{10} \text{ yr})^{-1})$, pulsars cannot be useful in searches for waves with periods longer than the observation time. However (Mashhoon, Carr and Hu, 1981; Bertotti, Carr and Rees, 1983), the orbit of the binary pulsar is such a good clock, with such a well-understood slow down, that it can be beat against earth-based clocks to give interesting limits on waves with frequencies 10^{-8} Hz $\lesssim f \lesssim 10^{-13}$ Hz. (The low-frequency limit, 10^{-13} Hz, corresponds to waves with λ of order the distance between the earth and the pulsar.) The most recent observational data from Weisberg and Taylor (1984) give the limit

$$\int_{10^{-8} \text{ Hz}}^{10^{-13} \text{ Hz}} \Omega_{GW}(f) f^{-1} \, df \leqslant 0.5 H_{100}^{-2}, \qquad (128)$$

where H_{100} is the Hubble constant in units of 100 km/s Mpc^{-1}.

Orbital motions are also useful as detectors of waves with frequencies f of order the orbital frequency. The orbiting bodies respond to such waves in the same manner as does a resonant-bar detector. (After all, the orbit is, in a sense, nothing but a two-dimensional oscillator.) The idea of using a lunar or planetary or binary-star orbit in this way as a gravity-wave detector has been proposed or discussed by Braginsky and Gertsenshtein (1967), Anderson (1971), Bertotti (1973), Rudenko (1975) and many others. Mashhoon, Carr and Hu (1981) argue that Viking doppler data on the relative orbits of Mars and Earth place a limit $\Omega_{GW}(f) \lesssim 0.1$ at $f \simeq 3 \times 10^{-8}$ Hz (period of one year), and that laser ranging data on the moon's orbit – which now give $\Omega_{GW}(f) \lesssim 10$ – have the potential in some years to give $\Omega_{GW}(f) \lesssim 0.1$ at $f \simeq 3 \times 10^{-7}$ Hz (period of one month).

(c) Anisotropies in the temperature of the cosmic microwave radiation

Large-scale (quadrupolar) anisotropies in the cosmic microwave radiation would be produced by gravitational waves in the vicinity of the earth today (Sachs and Wolfe, 1967; Dautcourt, 1969). Present observational limits on such anisotropies imply $\Omega_{GW}(f) \lesssim 10^{-6}(f/3 \times 10^{-18} \text{ Hz})^2$ for any f, even $f \lesssim 1/(\text{Hubble time})$ (Grishchuk and Zel'dovich, 1978). Small-scale anisotropies (angular scales of order 5 degrees) would have been produced by gravitational waves during the epoch of plasma recombination, when the microwave radiation we now see last interacted with matter. The limit from these is remarkably good: $\Omega_{GW}(f) \lesssim 10^{-13}$ in a narrow frequency window at $f \simeq 10^{-16}$ Hz (Carr, 1980; Zel'dovich and Novikov, 1983; Starobinsky, 1985); but this limit is much less firmly based than others. It presumes we understand thoroughly the propagation of the cosmic microwave radiation during recombination and from the epoch of recombination to the present. Note: this limit does not constrain waves from cosmic strings or any other source that emitted most of its waves later than the epoch of recombination; see Traschen, Turok and Brandenberger (1986) for the effects of strings on the microwave anisotropy.

(d) Other astronomical observations

In addition to the timing of pulsars and of orbital motions, and microwave anisotropies, several other astronomical observations can be used as probes of low-frequency gravitational waves: (i) *the observed velocities of galaxies and clusters of galaxies*, i.e. deviations from the Hubble flow (Rees, 1971; Burke, 1975; Dautcourt, 1977). These give $\Omega_{GW}(f) \lesssim (f/3 \times 10^{-17})^2$ for $10^{-17} \text{ Hz} \lesssim f \lesssim 10^{-15}$ Hz (Carr, 1980). (ii) *Peculiarities in primordial nucleosynthesis*, produced by the influence of gravitational wave energy on the expansion rate of the universe during nucleosynthesis (the 'first three minutes') (Schvartzman, 1969). These produce a limit $\Omega_{GW}(f) \lesssim 10^{-4}$ at $f \gtrsim 10^{-10}$ Hz, which is not fully firmly based because it presumes we thoroughly understand conditions in the first few minutes of the universe. Note that this limit is irrelevant for waves emitted since the epoch of nucleosynthesis.

For detailed reviews and references on these and other 'astronomical detectors' of gravity waves, see Dautcourt (1974), Chapter 17 of Zel'dovich and Novikov (1983), and Section 4 of Carr (1980).

9.6 Conclusion

As I look back over this review of gravitational waves, I am struck by the enormous changes in our theoretical understanding – or at least in theoretical fashion – that have occurred over the past 5, 10 and 15 years; and I am impressed even more by the progress that experimenters have made in the quest to invent, design and build detectors of ever greater sensitivity. That the quest ultimately will succeed seems almost assured. The only question is when, and with how much further effort. Five years ago Jerry Ostriker and I made a bet, to wit:

Whereas both Jeremiah P. Ostriker and Kip S. Thorne believe that Einstein's equations are valid

And both are convinced that these equations predict the existence of gravitational waves

And both are confident that Nature will provide what physical law predicts

And both have faith that scientists can ultimately observe whatever Nature does supply

Nevertheless, they differ on the likely strengths of natural sources and on the probability of a near-future and verifiable detection.

Therefore they agree to wager one case of good red wine (JPO to supply French wine, KST to supply California) on the detection of extraterrestrial gravitational waves before the next Millennium (January 1, 2000). KST wins the wager if at least two experimental groups observe phenomena which they agree are gravitational waves. If not, JPO wins.

Signed and officially sealed
this sixth day of May 1981
Jeremiah P. Ostriker
Kip S. Thorne

I expect to win – but I won't guarantee it.

Acknowledgments

For helpful comments on the manuscript of this chapter I thank Thibault Damour, Ron Drever, John Armstrong, Peter Bender, Jiři Bičak, David Blair, Roger Blandford, Herman Bondi, Carlton Caves, Frank Estabrook, Charles Evans, Sam Finn, Craig Hogan, Jim Hough, Ed Leaver, Brian Meers, Roger Romani, David Schoemaker, Dan Stinebring, and Kimio Tsubono.

References

Abbott, L. and Wise, M. (1984). *Nuclear Physics B*, **244**, 541.

Alpar, M. A. and Pines, D. (1985). *Nature*, **314**, 334.

Amaldi, E. and Pizella, G. (1979). In *Relativity, Quanta, and Cosmology in the Development of the Scientific Thought of Einstein*, Volume 1, 9.

Amaldi, E., Coccia, E., Cosmelli, C., Ogawa, Y., Pizzella, G., Rapagnani, P., Ricci, F., Bonifazi, P., Castellano, M. G., Vannaroni, G., Bronzoni, F., Carelli, P., Foglietti, V., Cavallari, G., Habel, R., Modena, I. and Pallottino, G. V. (1984). *Il Nuovo Cimento*, **7C**, 338.

Anandan, J. (1981). *Physical Review Letters*, **47**, 463.

Anandan, J. (1985). *Physics Letters*, **105A**, 280.

Anandan, J. (1986). In Ruffini, ed. (1986).

Anandan, J. and Chiao, R. Y. (1982). *General Relativity and Gravitation*, **14**, 515.

Anderson, A. J. (1971). *Nature*, **229**, 547.

Anderson, J. D., Armstrong, J. W., Estabrook, F. B., Hellings, R. W., Law, E. K. and Wahlquist, H. D. (1984). *Nature*, **308**, 158.

Anderson, J. L. and Hobill, D. W. (1986). In *Dynamical Spacetimes and Numerical Relativity*, ed. J. M. Centrella, p. 389. Cambridge University Press.

Anderson, J. L., Kates, R. E., Kegeles, L. S. and Madonna, R. G. (1982). *Physical Review D*, **25**, 2038.

Armstrong, J. W. (1984). *Nature*, **307**, 527.

Armstrong, J. W., Estabrook, F. B. and Wahlquist, H. D. (1987). *Astrophysical Journal* (in press).

Armstrong, J. W. and Sramek, R. A. (1982). *Radio Science*, **17**, 1579.

Armstrong, J. W., Woo, R. and Estabrook, F. B. (1979). *Astrophysical Journal*, **230**, 570.

Ashtekar, A. (1983). In Deruelle and Piran, eds. (1983), p. 421.

Ashtekar, A. (1984). In *General Relativity and Gravitation*, ed. B. Bertotti *et al.*, p. 37. Reidel: Dordrecht.

Backer, D. C., Kulkarni, S. R., Heilis, C., Davis, M. M. and Gross, W. M. (1982). *Nature*, **300**, 615.

Bagaev, S. N., Chebotaev, V. P., Dychkov, A. S. and Goldort, V. G. (1981). *Applied Physics*, **25**, 161.

Bardeen, J. M. and Press, W. H. (1973). *Journal of Mathematical Physics*, **14**, 7.

Bardeen, J. M. and Piran, T. (1983). *Physics Reports*, **96**, 205.

Begelman, M. C., Blandford, R. D. and Rees, M. J. (1980). *Nature*, **287**, 307.

Bender, P. L., Faller, J. E., Hall, J. L., Hils, D. and Vincent, M. A. (1984). Abstract for *Fifth International Laser Ranging Instrumentation Workshop*; Herstmonceaux, 10–14 September, 1984.

Bendat, J. S. (1958). *Principles and Applications of Random Noise Theory*, Wiley: New York.

Bertotti, B. (1973). *Astrophysical Letters*, **14**, 51.

Bertotti, B., Carr, B. J. and Rees, M. J. (1983). *Monthly Notices of the Royal Astronomical Society*, **203**, 945.

Bethe, H. (1986). In *Highlights in Modern Astrophysics*, eds. S. L. Shapiro and S. A. Teukolsky, p. 45. Wiley: New York.

Bicak, J. (1968). *Proceedings of the Royal Society of London A*, **302**, 201.

Bicak, J. (1985). In *Galaxies, Axisymmetric Systems, and Relativity*, ed. M. A. H. MacCallum, Cambridge University Press.

Bicak, J., Hoenselaers, C. and Schmidt, B. G. (1983). *Proceedings of the Royal Society of London A*, **390**, 411.

Blair, D. (1982). *Physics Letters A*, **91**, 197.

Blair, D. G. (1983). In Deruelle and Piran (1983), p. 339.

Blanchet, L. (1987a). *Proceedings of the Royal Society of London A*, **409**, 383.

Blanchet, L. (1987b). Paper in preparation.

Blanchet, L. and Damour, T. (1984). *Physics Letters*, **104A**, 82.

Blanchet, L. and Damour, T. (1986). *Philosophical Transactions of the Royal Society* (in press).

Blandford, R. D. (1979). In Smarr, ed. (1979), p. 191.

Blandford, R. D. (1984). Private communication.

Blandford, R. D., Narayan, R. and Romani, R. W. (1984). *Journal of Astrophysics and Astronomy*, **5**, 369.

Blandford, R. D. and Thorne, K. S. (1979). In *General Relativity: An Einstein Centenary Survey*, eds. S. W. Hawking and W. Israel, Cambridge University Press.

Blinnikov, S. I., Novikov, I. D., Perevodchikova, T. V. and Polnarev, A. G. (1984). *Soviet Astronomy Letters*, **10**, 177.

Bocko, M. F. and Johnson, W. W. (1984). *Physical Review A*, **30**, 2135.

Bond, J. R. and Carr, B. J. (1984). *Monthly Notices of the Royal Astronomical Society*, **207**, 585.

Bondi, H. (1957). *Nature*, **179**, 1072.

Bondi, H. (1960). *Nature*, **186**, 535.

Bondi, H., Pirani, F. A. E. and Robinson, I. (1959). *Proceedings of the Royal Society of London A*, **251**, 519.

Bondi, H., van der Burg, M. G. J. and Metzner, A. W. K. (1962). *Proceedings of the Royal Society of London A*, **269**, 21.

Bonnor, W. B. (1959). *Philosophical Transactions of the Royal Society of London A*, **251**, 233.

Bontz, R. J. and Haugan, M. P. (1981). *Astrophysics and Space Space Science*, **78**, 204.

Bontz, R. J. and Price, R. H. (1979). *Astrophysical Journal*, **228**, 560.

Bordé, C. J., Sharma, J., Tourrenc, P. and Damour, T. (1983). *Journal de Physique-Lettres*, **44**, L983.

Boughn, S. P. and Kuhn, J. R. (1984). *Astrophysical Journal*, **286**, 387.

Boughn, S. P. and Kuhn, J. R. (1987). Paper in preparation.

Boughn, S. P., Fairbank, W. M., Giffard, R. P., Hollenhorst, J. N., Mapoles, E. R., McAshan, M. S., Michelson, P. F., Paik, H. J. and Taber, R. C. (1982). *Astrophysical Journal*, **261**, L19.

Braginsky, V. B. (1965). *Uspekhi Fizicheskikh Nauk*, **86**, 433.

Braginsky, V. B. (1983). In Deruelle and Piran (1983), p. 387.

Braginsky, V. B. and Gertsenshtein, M. E. (1967). *Soviet Physics – JETP Letters*, **5**, 287.

Braginsky, V. B. and Grishchuk, L. P. (1985). *Zhurnal Eksperimentalnoi i Teoreticheskoi Fiziki*, **89**, 744. English translation: *Soviet Physics – JETP*, **62**, 427.

Braginsky, V. B. and Menskii, M. B. (1971). *Soviet Physics – JETP Letters*, **13**, 417.

Braginsky, V. B. and Nazarenko, V. S. (1971). In *Proceedings of the Conference on Experimental Tests of Gravitational Theories, November 11–13, 1970, California Institute of Technology*, ed. R. W. Davies. Jet Propulsion Laboratory Technical Memorandum, 33–499: Pasadena, California.

Braginsky, V. B., Grishchuk, L. P., Doroshkevich, A. G., Zel'dovich, Ya. B., Novikov, I. D. and Sazhin, M. V. (1973). *Soviet Physics – JETP*, **38**, 865.

Braginsky, V. B., Gusev, A. V., Mitrofanov, V. P., Rudenko, V. N. and Yakimov, V. N. (1985). *Uspekhi Fizicheskikh Nauk*, **147**, 422.

Braginsky, V. B., Mitrofanov, V. P. and Panov, V. I. (1985). *Systems with Small Dissipation*. University of Chicago Press: Chicago.

Braginsky, V. B. and Thorne, K. S. (1985). *Nature*, **316**, 610.

Braginsky, V. B. and Thorne, K. S. (1987). *Nature* (in press).

Braginsky, V. B. and Vorontsov, Yu. I. (1974). *Soviet Physics—Uspekhi*, **17**, 644.

Braginsky, V. B., Vorontsov, Yu. I. and Khalili, F. Ya. (1978). *Soviet Physics – JETP Letters*, **27**, 276.

Braginsky, V. B., Vorontsov, Yu. I. and Thorne, K. S. (1980). *Science*, **209**, 547.

Brill, D. R. and Hartle, J. B. (1964). *Phys. Rev. B*, **135**, 271.

Brillet, A. (1985). *Annales de Physique*, **10**, 219.

Brillet, A. and Meers, B. (1987). Paper in preparation.

Brillet, A. and Tourrenc, P. (1983). In Deruelle and Piran (1983).

Burke, W. L. (1969). Unpublished Ph.D. thesis, Caltech.

Burke, W. L. (1975). *Astrophysical Journal*, **196**, 329.

Burke, W. L. (1979). In *Isolated Gravitating Systems in General Relativity*, ed. J. Ehlers, p. 220. North Holland: Amsterdam.

Carr, B. J. (1980). *Astronomy and Astrophysics*, **89**, 6.

Carr, B. J. (1986). In *Inner Space/Outer Space*, ed. E. W. Kolb, University of Chicago Press: Chicago.

Carter, B. and Quintana, H. (1977). *Physical Review D*, **16**, 2928.

Caves, C. M. (1978). Private communication.

Caves, C. M. (1979). *Physics Letters*, **80B**, 323.

Caves, C. M. (1980). *Physical Review Letters*, **45**, 75.

Caves, C. M. (1981). *Physical Review D*, **23**, 1693.

Caves, C. M. (1982). *Physical Review D*, **26**, 1817.

Caves, C. M. (1983). *Foundations of Quantum Mechanics*, ed. S. Kamefuchi *et al.*, p. 195. Physical Society of Japan: Tokyo.

Caves, C. M. (1986). *Physical Review D*, **33**, 1643.

Caves, C. M. (1987*a*). *Physical Review D*, **35**, 1815.

Caves, C. N. (1987*b*). In *Quantum Measurement and Chaos*, ed. E. R. Pike. Plenum: New York.

Caves, C. M. (1987*c*). Paper in preparation.

Caves, C. M., Thorne, K. S., Drever, R. W. P., Sandberg, V. D. and Zimmermann, M. (1980). *Reviews of Modern Physics*, **52**, 341.

Chandrasekhar, S. (1950). *Radiative Transfer*. Oxford University Press: London.

Chandrasekhar, S. (1970). *Physical Review Letters*, **24**, 611.

Chandrasekhar, S. (1975). *Proceedings of the Royal Society of London A*, **343**, 289.

Chandrasekhar, S. (1983). *The Mathematical Theory of Black Holes*. Oxford University Press: Oxford.

Chandrasekhar, S. and Detweiler, S. L. (1975). *Proceedings of the Royal Society of London*, **344**, 441.

Chandrasekhar, S. and Detweiler, S. L. (1976). *Proceedings of the Royal Society of London A*, **350**, 165.

Chandrasekhar, S. and Esposito, F. P. (1970). *Astrophysical Journal*, **160**, 153.

Chandrasekhar, S. and Friedman, J. L. (1972). *Astrophysical Journal*, **176**, 745.

Chandrasekhar, S. and Friedman, J. L. (1973). *Astrophysical Journal*, **181**, 481.

Chandrasekhar, S. and Xanthopoulos, B. C. (1986). *Proceedings of the Royal Society of London*, **408**, 175.

Chiao, R. Y. (1982). *Physical Review B*, **25**, 1655.

Christodoulou, D. and Schmidt, B. G. (1979). *Commun. Math. Phys.*, **68**, 275.

Chrzanowski, P. L. (1975). *Physical Review D*, **11**, 2042.

Clark, J. P. A., van den Heuvel, E. P. J. and Sutantyo, W. (1979). *Astronomy and Astrophysics*, **72**, 120.

Clark, J. P. A. and Eardley, D. M. (1977). *Astrophysical Journal*, **215**, 315.

Cohen, J. M. and Kegeles, L. S. (1975). *Physics Letters*, **54A**, 5.

Cole, J. D. (1968). *Perturbation Methods in Applied Mathematics*. Blaisdell: Waltham, Mass.

Crowley, R. J. and Thorne, K. S. (1977). *Astrophysical Journal*, **215**, 624.

Cunningham, C. T., Price, R. H. and Moncrief, V. (1979). *Astrophysical Journal*, **230**, 870.

Cutler, C. and Lindblom, L. (1987). *Astrophysical Journal* (in press).

Cyranski, J. F. and Lubkin, E. (1974). *Annals of Physics*, **87**, 205.

Damour, T. (1983). In Deruelle and Piran (1983). p. 59.

Damour, T. (1986). In *Proceedings of the Fourth Marcel Grossman Meeting on the Recent Developments of General Relativity*, ed. R. Ruffini. North Holland: Amsterdam

Damour, T. (1987). In *Gravitation in Astrophysics*, eds. B. Carter and J. B. Hartle. Plenum: New York.

Damour, T. and Deruelle, N. (1986). *Annales Institut Henri Poincaré (Physique Théorique)*, **44**, 263.

Dautcourt, G. (1969). *Monthly Notices of the Royal Astronomical Society*, **144**, 255.

Dautcourt, G. (1974). In *Confrontation of Cosmological Theories with Observational Data*, Proceedings of IAU Symposium 63, ed. M. S. Longair, p. 299. Reidel: Dordrecht.

Dautcourt, G. (1977). *Astronomische Nachrichten*, **298**, 81.

Davies, R. W. (1974). In *Colloque Internationaux CNRS No. 220, 'Ondes et Radiations Gravitationelles'*. Institut Henri Poincaré: Paris, p. 33.

Davis, M. M., Taylor, J. H., Weisberg, J. M. and Backer, D. C. (1985). *Nature*, **315**, 547.

De Logi, W. K. and Kovacs, S. J. (1977). *Physical Review D*, **16**, 2331.

Deruelle, N. and Piran, T. (1983). *Gravitational Radiation*. North Holland: Amsterdam.

de Sabbata, V. and Weber, J. eds. (1977). *Topics in Theoretical and Experimental Gravitation Physics*. Plenum: London.

Detweiler, S. L. (1975). *Astrophysical Journal*, **197**, 203.

Detweiler, S. L. (1977). *Proceedings of the Royal Society of London A*, **352**, 381.

Detweiler, S. L. (1978). *Astrophysical Journal*, **225**, 687.

Detweiler, S. L. (1979). *Astrophysical Journal*, **234**, 1100.

Detweiler, S. L. (1980). *Astrophysical Journal*, **239**, 292.

Detweiler, S. L. and Ipser, J. R. (1973). *Astrophysical Journal*, **185**, 685.

Detweiler, S. L. and Lindblom, L. (1985). *Astrophysical Journal*, **292**, 12.

Detweiler, S. L. and Szedenits, E. (1979). *Astrophysical Journal*, **231**, 211.

DeWitt, B. S. (1967a). *Phys. Rev.*, **160**, 1113.

DeWitt, B. S. (1967b). *Phys. Rev.*, **162**, 1239.

Douglass, D. H. and Braginsky, V. B. (1979). In *General Relativity, an Einstein Centenary Survey*, eds. S. W. Hawking and W. Israel, Cambridge University Press.

Downs, G. S. and Reichley, P. E. (1983). *Astrophysical Journal Supplement Series*, **53**, 169.

Drever, R. W. P. (1977). *Quarterly Journal of the Royal Astronomical Society*, **18**, 9.

Drever, R. W. P. (1983). In Deruelle and Piran (1983). p. 321.

Drever, R. W. P., Hough, J., Edelstein, W., Pugh, J. R. and Martin, W. (1977). In *Gravitazione Sperimentale*, ed. B. Bertotti, p. 365. Accademia Nazionale dei Lincei: Rome.

Drever, R. W. P., Ford, G. M., Hough, J., Kerr, I., Munley, A. J., Pugh, J. R., Robertson, N. A. and Ward, H. (1980). *Proceedings of the Ninth International Conference on General Relativity and Gravitation*, ed. E. Schmutzer, p. 265. VEB Deutscher Verlag der Wissenschaften: Berlin.

Drever, R. W. P. and Weiss, R. (1983). Unpublished work reported briefly in Appendix B.2(c) of Drever *et al.* (1985).

Drever, R. W. P., Weiss, R., Linsay, P. S., Saulson, P. R., Spero, R. and Schutz, B. F. (1985). A Detailed Engineering Design Study and Development and Testing of Components for a Laser Interferometer Gravitational Wave Observatory. Caltech: Pasadena, California, and MIT: Cambridge, Massachusetts (unpublished).

Dymnikova, I. G., Popov, A. K. and Zentsova, A. S. (1982). *Astrophysics and Space Science*, **85**, 231.

Dyson, F. J. (1969). *Astrophysical Journal*, **156**, 529.

Eardley, D. M. (1983). In Deruelle and Piran (1983). p. 257.

Eardley, D. M., Lee, D. L. and Lightman, A. P. (1973). *Physical Review D*, **8**, 3308.

Eardley, D. M., Lee, D. L., Lightman, A. P., Wagoner, R. V. and Will, C. M. (1973). *Physical Review Letters*, **30**, 884.

Eddington, A. S. (1924). *The Mathematical Theory of Relativity*, 2nd edn, Cambridge University Press; see especially supplementary note 8.

Edelstein, L. A. and Vishveshwara, C. V. (1970). *Physical Review D*, **1**, 3514.

Ehlers, J. and Kundt, W. (1962). In *Gravitation: An Introduction to Current Research*, ed. L. Witten. Wiley: New York.

Ehlers, J., Rosenblum, A., Goldberg, J. N. and Havas, P. (1976). *Astrophysical Journal Letters*, **208**, L77.

Einstein, A. (1916). *Preuss. Akad. Wiss. Berlin, Sitzungsberichte der physikalisch-mathematischen Klasse*, p. 688.

Einstein, A. (1918). *Preuss. Akad. Wiss. Berlin, Sitzungsberichte der Physikalisch-mathematischen Klasse*, p. 154.

Einstein, A. and Rosen, N. (1936). *Journal of the Franklin Institute*, **223**, 43.

Epstein, R. and Wagoner, R. V. (1975). *Astrophysical Journal*, **197**, 717.

Esposito, F. P. (1971a). *Astrophysical Journal*, **165**, 165.

Esposito, F. P. (1971b). *Astrophysical Journal*, **168**, 495.

Estabrook, F. B. (1985). *General Relativity and Gravitation*, **17**, 719.

Estabrook, F. B. (1987). *Acta Astronautica* (in press).

Estabrook, F. B. and Wahlquist, H. D. (1975). *General Relativity and Gravitation*, **6**, 439.

Evans, C. R. (1984). PhD thesis, University of Texas at Austin, unpublished.

Evans, C. R. (1986). In *Dynamical Spacetimes and Numerical Relativity*, ed. J. M. Centrella, p. 3. Cambridge University Press.

Evans, C. R. and Abrahams, A. M. (1987). *Physical Review D* (in preparation).

Evans, C. R., Iben, I. and Smarr, L. (1987). *Astrophysical Journal*, submitted.

Faller, J. E. and Bender, P. L. (1981). Abstract for the *International Conference on Precision Measurements and Fundamental Constants*, 8–12 June 1981.

Faller, J. E., Bender, P. L., Hall, J. L., Hils, D. and Vincent, M. A. (1985). In *Proceedings of the Colloquium 'Kilometric Optical Arrays in Space'*, Cargese (Corsica) 23–5 October 1984. ESA SP-226.

Feynman, R. P. (1963). *Acta Physica Polonica*, **24**, 697: also published in *Proceedings of the 1962 Warsaw Conference on the Theory of Gravitation*. PWN-Editions Scientifiques de Pologne: Warsaw (1964).

Finn, L. S. (1986). *Monthly Notices of the Royal Astronomical Society*, **222**, 393.

Fock, V. A. (1959). *Theory of Space, Time, and Gravitation*, Section 87. Pergamon: London.

Forward, R. L. (1978). *Physical Review D*, **17**, 379.

Forward, R. L. and Moss, G. E. (1972). *Bulletin of the American Physical Society*, **17**, 1183(A).

Forward, R. L., Zipoy, D., Weber, J., Smith, S. and Benioff, H. (1961). *Nature*, **189**, 473.

Friedman, J. L., Ipser, J. R. and Parker, L. (1986). *Astrophysical Journal*, **304**, 115.

Friedman, J. L. and Schutz, B. F. (1975a). *Astrophysical Journal (Letters)*, **199**, L157.

Friedman, J. L. and Schutz, B. F. (1975b). *Astrophysical Journal*, **200**, 204.

Friedman, J. L. and Schutz, B. F. (1978). *Astrophysical Journal*, **222**, 281.

Futamase, T. and Schutz, B. F. (1983). *Physical Review D*, **28**, 2363.

Gal'tsov, D. V., Matiukhin, A. A. and Petukhov, V. I. (1980). *Physics Letters*, **77A**, 387.

Gal'tsov, D. V., Tsvetkov, V. P. and Tsirulev, A. N. (1984). *Soviet Physics – JETP*, **59**, 472.

Geroch, R., Held, A. and Penrose, R. (1973). *Journal of Mathematical Physics*, **14**, 874.

Gertsenshtein, M. E. (1962). *Soviet Physics – JETP*, **14**, 84.

Gertsenshtein, M. E. and Pustovoit, V. I. (1962). *Soviet Physics – JETP*, **16**, 433.

Gilden, D. L. and Shapiro, S. L. (1984). *Astrophysical Journal*, **287**, 728.

Giffard, R. P. (1976). *Physical Review D*, **14**, 2478.

Gomez, R., Isaacson, R. A., Welling, J. S. and Winicour, J. (1986). In *Dynamical Spacetimes and Numerical Relativity*, ed. J. M. Centrella, p. 236. Cambridge University Press.

Gordienko, N. V., Gusev, A. V. and Rudenko, V. N. (1977). *Vestnik Moskovskovo Universiteta*, **18**, 48.

Grishchuk, L. P. (1974). *Soviet Physics – JETP*, **40**, 409.

Grishchuk, L. P. (1975a). *Lettere al Nuovo Cimento*, **12**, 60.

Grishchuk, L. P. (1975b). *Soviet Physics – JETP*, **40**, 409.

Grishchuk, L. P. (1977). *Annals of the New York Academy of Sciences*, **302**, 439.

Grishchuk, L. P. (1983). In *Proceedings of the Ninth International Conference on General Relativity and Gravitation*, ed. E. Schmutzer. Cambridge University Press.

Grishchuk, L. P. and Polnarev, A. G. (1980). In *General Relativity and Gravitation*, vol. 2, ed. A. Held, p. 393. Plenum: New York.

Grishchuk, L. P. and Zel'dovich, Ya. B. (1978). *Soviet Astronomy – AJ*, **22**, 125.

Gürsel, Y., Linsay, P., Saulson, P., Spero, R., Weiss, R. and Whitcomb, S. (1983). Unpublished Caltech/MIT manuscript.

Gusev, A. V. and Rudenko, V. N. (1976). *Radiotekhnika i Electronika*, **3**, 1865.

Haensel, P., Zdunik, J. L. and Schaeffer, R. (1986). *Astronomy and Astrophysics*, **160**, 251.

Halliwell, J. J. and Hawking, S. W. (1985). *Physical Review D*, **31**, 1777.

Halpern, L. E. and Desbrandes, R. (1969). *Annales Institut Henri Poincaré*, **9**, 309.

Hamilton, W. O., Xu, B.-X., Solomonson, N., Mann, A. G. and Sibley, A. (1986). Performance of the LSU Tuned Bar Gravitational Wave Detector. Unpublished technical memorandum, Louisiana State University.

Hansen, P. M., Chiu, Y. T. and Chao, C.-C. (1986). Aerospace Report No. ATR-86(8421)-2. The Aerospace Corporation: El Segundo, CA.

Hawking, S. W. (1985). *Physics Letters*, **150B**, 339.

Haugan, M. P., Shapiro, S. L. and Wasserman, I. (1982). *Astrophysical Journal*, **257**, 283.

Heffner, H. (1962). *Proceedings of the IRE*, **50**, 1604.

Helfand, D. H., Taylor, J. H., Backus, R. R. and Cordes, J. M. (1980). *Astrophysical Journal*, **237**, 206.

Hellings, R. W. and Downs, G. S. (1983). *Astrophysical Journal (Letters)*, **265**, L39.

Hellings, R. W., Callahan, P. S., Anderson, J. D. and Moffett, A. T. (1981). *Physical Review D*, **23**, 844.

Hils, D. L., Bender, P., Faller, J. E. and Webbink, R. F. (1987). Paper in preparation.

Hirakawa, H. and Narihara, K. (1975). *Physical Review Letters*, **35**, 330.

Hirakawa, H., Narihara, K. and Fujimoto, M.-K. (1976). *Journal of the Physical Society of Japan*, **41**, 1093.

Hirakawa, H., Owa, S. and Iso, K. (1985). *Journal of the Physical Society of Japan*, **54**, 1270.

Hogan, C. J. (1986). *Monthly Notices of the Royal Astronomical Society*, **218**, 629.

Hogan, C. J. and Rees, M. J. (1984). *Nature*, **311**, 109.

Hollenhorst, J. N. (1979). *Physical Review D*, **19**, 1669.

Hough, J., Meers, B. J., Newton, G. P., Robertson, N. A., Ward, H., Schutz, B. F. and Drever, R. W. P. (1986). A British Long Baseline Gravitational Wave Observatory, unpublished report submitted by Glasgow University to the SERC.

Hough, J., Pugh, J. R., Bland, R. and Drever, R. W. P. (1975). *Nature*, **254**, 498.

Hough, J., Drever, R. W. P., Munley, A. J., Lee, S.-A., Spero, R., Whitcomb, S. E., Pugh, J., Newton, G., Meers, B., Brooks, E. and Gursel, Y. (1983). In *Quantum Optics, Experimental Gravity, and Measurement Theory*, eds. P. Meystre and M. O. Scully, p. 515. Plenum: New York.

Hu, B. L. (1978). *Physical Review D*, **18**, 969.

Hu Enke, Guan Tongren, Yu Bo, Tang Mengxi, Chen Shusen, Zheng Qingzhang, Michelson, P. F., Moskowitz, B. E., McAshan, M. S., Fairbank, W. M. and Bassan, M. (1986). *Chinese Physics Letters* No. 11–12, 1.

Iacopini, E., Picasso, E., Pegoraro, F. and Radicati, L. A. (1979). *Physics Letters*, **73A**, 140.

Iben, I. and Tutukov, A. V. (1984). *Astrophysical Journal Supplements*, **54**, 335.

Ipser, J. R. (1971). *Astrophysical Journal*, **166**, 175.

Ipser, J. R. (1986). Private communication.

Ipser, J. R. and Managan, R. A. (1984). *Astrophysical Journal*, **282**, 287.

Ipser, J. R. and Thorne, K. S. (1973). *Astrophysical Journal*, **181**, 181.

Isaac, G. R. (1981). *Solar Physics*, **74**, 43.

Isaacson, R. A. (1968a). *Physical Review*, **166**, 1263.

Isaacson, R. A. (1968b). *Physical Review*, **166**, 1272.

Isaacson, R. A., Welling, J. S. and Winicour, J. (1983). *Journal of Mathematical Physics*, **24**, 1824.

Isaacson, R. A., Welling, J. S. and Winicour, J. (1984). *Physical Review Letters*, **53**, 1870.

Kafka, P. (1977). In de Sabbata and Weber (1977), p. 161.

Kahn, K. and Penrose, R. (1971). *Nature*, **229**, 185.

Kaufmann, W. J. (1970). *Nature*, **227**, 157.

Kellerman, K. I. and Sheets, B. (1983). *Serendipitous Discoveries in Radio Astronomy.* National Radio Astronomy Observatory: Green Bank, West Virginia.

Kojima, Y. and Nakamura, T. (1984*a*). *Progress of Theoretical Physics*, **71**, 79.

Kojima, Y. and Nakamura, T. (1984*b*). *Progress of Theoretical Physics*, **72**, 494.

Kovacs, S. J. and Thorne, K. S. (1977). *Astrophysical Journal*, **217**, 252.

Kovacs, S. J. and Thorne, K. S. (1978). *Astrophysical Journal*, **224**, 62.

Kuhn, J. R. and Boughn, S. P. (1984). *Nature*, **308**, 164.

Landau, L. D. and Lifshitz, E. M. (1941). *Teoriya Polya*. Nauka: Moscow. (English translation of a later edition, *The Classical Theory of Fields*. Addison-Wesley: Cambridge, Mass. (1951).)

Leaver, E. W. (1985). Solutions to a Generalized Spheroidal Wave Equation in Molecular Physics and General Relativity, and an Analysis of the Quasi-Normal Modes of Kerr Black Holes, unpublished PhD thesis, University of Utah: Salt Lake City.

Leaver, E. W. (1986*a*). *Proceedings of the Royal Society of London A*, **402**, 285.

Leaver, E. W. (1986*b*). *Physical Review D*, **34**, 384.

Lifshitz, E. M. (1946). *Zhurnal Eksperimentalnoi i Teoreticheskoi Fiziki*, **16**, 587.

Lindblom, L. (1986). *Astrophysical Journal*, **303**, 146.

Lindblom, L. (1987). *Astrophysical Journal* (in press).

Lindblom, L. and Detweiler, S. L. (1983). *Astrophysical Journal*, **53**, 73.

Linet, B. (1984). *General Relativity and Gravitation*, **16**, 89.

Linet, B. and Tourrenc, P. (1976). *Canadian Journal of Physics*, **54**, 1129.

Linsay, P., Saulson, P., Weiss, R. and Whitcomb, S. (1983). A Study of a Long Baseline Gravitational Wave Antenna System; report prepared for The National Science Foundation. MIT: Cambridge, Massachusetts (unpublished).

Lipunov, V. M. and Postnov, K. A. (1986). *Soviet Astronomy Letters*. In press.

Lipunov, V. M., Postnov, K. A. and Prokhorov, X. (1987). Paper in preparation.

Lyamov, V. E. and Rudenko, V. N. (1975). *Soviet Physics – JETP*, **40**, 787.

MacCallum, M. A. H., ed. (1987). *Proceedings of the Tenth International Conference on General Relativity and Gravitation*. Cambridge University Press.

MacCallum, M. A. H. and Taub, A. H. (1973). *Communications in Mathematical Physics*, **30**, 153.

Macedo, P. G. and Nelson, A. H. (1983). *Physical Review D*, **28**, 2382.

Maischberger, K., Rudiger, A., Schilling, R., Schnupp, L., Shoemaker, D. and Winkler, W. (1985). Vorschlag zum Bau eines grossen Laser-Interferometers zur Messung von Gravitationswellen. Max-Planck Institut fur Quantenoptik: Munich (unpublished).

Mashhoon, B., Carr, B. J. and Hu, B. L. (1981). *Astrophysical Journal*, **246**, 569.

Matzner, R. A., DeWitt-Morette, C., Nelson, B. and Zhang, T.-R. (1985). *Physical Review D*, **31**, 1869.

Matzner, R. A. and Tipler, F. J. (1984). *Physical Review D*, **29**, 1575.

Meers, B. (1983). Unpublished PhD thesis, University of Glasgow.

Michelson, P. F. (1983). In Deruelle and Piran (1983), p. 465.

Michelson, P. F. (1986). *Physical Review D*, **34**, 2966.

Michelson, P. F. and Taber, R. C. (1981). *Journal of Applied Physics*, **52**, 4313.

Michelson, P. F. and Taber, R. C. (1984). *Physical Review D*, **29**, 2149.

Mijic, M., Morris, M. and Suen, W.-M. (1986). *Physical Review D*, **34**, 2934.

Mironovskii, V. N. (1966). *Soviet Astronomy – AJ*, **9**, 752.

Misner, C. W., Thorne, K. S. and Wheeler, J. A. (1973). *Gravitation*. W. H. Freeman & Co.: San Francisco. Cited in text as MTW.

Moncrief, V. (1974). *Annals of Physics*, **88**, 323.

Moss, G. E., Miller, L. R. and Forward, R. L. (1971). *Applied Optics*, **10**, 2495.

Müller, E. (1982). *Astronomy and Astrophysics*, **114**, 53.

Müller, E. (1984). In *Problems of Collapse and Numerical Relativity*, eds. D. Bancel and M. Signore, p. 271. Reidel: Dordrecht.

Nakamura, T. (1983). In Deruelle and Piran (1983).

Nakamura, T. (1986). In *Gravitational Collapse and Relativity*, eds. H. Sato and T. Nakamura, p. 295. World Scientific: Singapore.

Nakamura, T. and Sasaki, M. (1981). *Physics Letters*, **106B**, 69.

Nesterikhin, Y. E., Rautian, S. G. and Smirnov, G. I. (1978). *Soviet Physics – JETP*, **48**, 1.

Newman, E. T. and Penrose, R. (1965). *Physical Review Letters*, **15**, 231.

Newman, E. T. and Tod, K. P. (1980). In *General Relativity and Gravitation, Volume 2*, ed. A. Held, p. 1. Plenum: New York.

Nutku, Y. and Halil, M. (1977). *Physical Review Letters*, **39**, 1379.

Oelfke, W. C. (1983). In *Quantum Optics, Experimental Gravitation and Measurement Theory*, eds. P. Meystre and M. O. Scully, p. 387. Plenum: New York.

Ostriker, J. P., Thompson, C. and Witten, E. (1986). *Physics Letters*, **B180**, 231.

Owa, S., Fujimoto, M.-K., Hirakawa, H., Morimoto, K., Suzuki, T. and Tsubono, K. (1986). In *Proceedings of Fourth Marcel Grossman Meeting on Recent Developments of General Relativity*, ed. R. Ruffini. North Holland: Amsterdam.

Paczynski, B. and Sienkiewicz, R. (1981). *Astrophysical Journal (Letters)*, **248**, L27.

Paik, H. J. and Wagoner, R. V. (1976). *Physical Review D*, **13**, 2694.

Pandharipande, V. R., Pines, D. and Smith, R. A. (1976). *Astrophysical Journal*, **208**, 550.

Pallotino, G. V. and Pizella, G. (1981). *Nuovo Cimento*, **4C**, 237.

Papadopoulos, D. and Esposito, F. P. (1985). *Astrophysical Journal*, **282**, 330.

Papapetrou, A. (1962). *Comptes Rendus Acad. Sci. Paris*, **255**, 1578.

Papapetrou, A. (1971). *Annales Institut Henri Poincaré*, **14**, 79.

Papapetrou, A. (1977). In deSabbata and Weber (1977), p. 83.

Pegoraro, F., Picasso, E. and Radicati, J. (1978). *Journal of Physics A*, **11**, 1949.

Penrose, R. (1963a). *Physical Review Letters*, **10**, 66.

Penrose, R. (1963b). *Relativity, Groups and Topology*, eds. C. DeWitt and B. DeWitt, p. 565. Gordon and Breach: New York.

Peres, A. (1960). *Nuovo Cimento*, **15**, 351.

Peters, P. C. (1964). *Physical Review*, **136**, B1224.

Peters, P. C. and Mathews, J. (1963). *Physical Review*, **131**, 435.

Piran, T. (1983). In Deruelle and Piran (1983), p. 203.

Piran, T. and Stark, R. F. (1986). In *Dynamical Spacetimes and Numerical Relativity*, ed. J. M. Centrella, p. 40. Cambridge University Press.

Pirani, F. A. E. (1956). *Acta Physica Polonica*, **15**, 389.

Pirani, F. A. E. (1957). *Physical Review*, **105**, 1089.

Pirani, F. A. E. (1964). In *Lectures on General Relativity*, eds. A. Trautman, F. A. E. Pirani and H. Bondi. Prentice-Hall: Englewood Cliffs, NJ.

Power, E. A. and Wheeler, J. A. (1957). *Reviews of Modern Physics*, **29**, 480.

Press, W. H. (1971). *Astrophysical Journal (Letters)*, **170**, L105.

Press, W. H. (1977). *Physical Review D*, **15**, 965.

Press, W. H. (1979). *General Relativity and Gravitation*, **11**, 105.

Price, R. H. (1972a). *Physical Review D*, **5**, 2419.

Price, R. H. (1972b). *Physical Review D*, **5**, 2439.

Price, R. H. and Thorne, K. S. (1969). *Astrophysical Journal*, **155**, 163.

Priedhorsky, W., Stella, L. and White, N. E., *International Astronomical Union Circular*, No. 4247, 28 August, 1986.

Quinlan, G. and Shapiro, S. L. (1987). *Astrophysical Journal*, in press.

Ramaty, R., Bonazzola, S., Cline, T. L., Kazanas, D. and Meszaros, P. (1980). *Nature*, **287**, 122.

Rees, M. J. (1971). *Monthly Notices of the Royal Astronomical Society*, **154**, 187.

Rees, M. J. (1983). In Deruelle and Piran (1983), p. 297.

Regge, T. and Wheeler, J. A. (1957). *Physical Review*, **108**, 1063.

Resch, G. M., Hogg, D. E. and Napier, P. G. (1984). *Radio Science*, **19**, 411.

Reynolds, S. P. and Stinebring, D. R. (1984). *Millisecond Pulsars*, National Radio Astronomy Observatory: Green Bank, West Virginia.

Romani, R. W. and Taylor, J. H. (1983). *Astrophysical Journal (Letters)*, **265**, L35.

Rosen, N. (1937). *Physikalische Zeitschift Sowjetunion*, **12**, 366.

Rubakov, V., Sazhin, M. C. and Veryaskin, A. (1982). *Physics Letters*, **115B**, 189.

Rudenko, V. N. (1975). *Soviet Astronomy – AJ*, **19**, 270.

Rudenko, V. N. and Sazhin, M. V. (1980). *Kvantovaya Elektronika*, **7**, 2344.

Ruffini, R. ed. (1986). *Proceedings of the Fourth Marcel Grossman Meeting on Recent Developments of General Relativity*. North Holland: Amsterdam.

Ruffini, R. and Wheeler, J. A. (1971). In *Proceedings of the Conference on Space Physics*. European Space Research Organization: Paris, p. 45.

Sachs, R. K. (1962). *Proceedings of the Royal Society of London A*, **270**, 103.

Sachs, R. K. (1963). In *Relativity, Groups and Topology*, eds. C. DeWitt and B. DeWitt, p. 521. Gordon and Breach: New York.

Sachs, R. K. and Wolfe, A. M. (1967). *Astrophysical Journal*, **147**, 73.

Saenz, R. A. and Shapiro, S. L. (1978). *Astrophysical Journal*, **221**, 286.

Saenz, R. A. and Shapiro, S. L. (1979). *Astrophysical Journal*, **229**, 1107.

Saenz, R. A. and Shapiro, S. L. (1981). *Astrophysical Journal*, **244**, 1033.

Sasaki, M. (1984). In *Problems of Collapse and Numerical Relativity*, eds. D. Bancel and M. Signore, p. 203. Reidel: Dordrecht.

Sasaki, M. and Nakamura, T. (1982). *Progress of Theoretical Physics*, **67**, 1788.

Sato, H. (1987). Paper in press.

Saulson, P. (1984). *Physical Review D*, **30**, 732.

Saulson, P. (1984). *Physical Review D*, **30**, 732.

Sazhin, M. V. (1978). *Soviet Astronomy – AJ*, **22**, 36.

Schmidt, B. G. (1979). In *Isolated Gravitating Systems*, ed. J. Ehlers, p. 11. North Holland: Amsterdam.

Schmidt, B. G. (1986). Gravitational Radiation Near Spatial and Null Infinity, preprint.

Schilling, R. (1986). Private communication.

Schoemaker, D., Winkler, W., Maischberger, K., Rudiger, A., Schilling, R. and Schnupp, L. (1986). In Ruffini, ed. (1986).

Schrader, R. (1984). *Physics Letters B*, **143B**, 421.

Schramm, D. N. and Arnett, W. D. (1975). *Astrophysical Journal*, **198**, 629.

Schumaker, B. L. (1986). *Physics Reports*, **135**, 318.

Schumaker, B. L. and Thorne, K. S. (1983). *Monthly Notices of the Royal Astronomical Society*, **203**, 457.

Schutz, B. F. (1972). *Astrophysical Journal Supplement Series*, **24**, 343.

Schutz, B. F. (1986a). In *Relativity, Supersymmetry, and Cosmology*, ed. O. Bressan, M. Castagnino and V. Hamity, p. 3. World Scientific: Singapore.

Schutz, B. F. (1986b). *Nature*, **323**, 310.

Schutz, B. F. (1986c). In *Dynamical Spacetimes and Numerical Relativity*, ed. J. Centrella, p. 446. Cambridge University Press.

Schutz, B. F. (1987). In *Gravitation in Astrophysics*, eds. B. Carter and J. Hartle. Plenum: New York.

Schutz, B. F. and Tinto, M. (1987). *Monthly Notices of the Royal Astronomical Society*, **224**, 131.

Schvartzman, V. F. (1969). *Soviet Physics – JETP Letters*, **9**, 184.

Segelstein, D. J., Rawley, L. A., Stinebring, D. R., Fruchter, A. S. and Taylor, J. H. (1986). *Nature*, **322**, 714.

Shapiro, S. L. and Teukolsky, S. A. (1985). *Astrophysical Journal (Letters)*, **292**, L41.

Slusher, R. E., Hollberg, L. W., Yurke, B., Mertz, J. C. and Valley, J. F. (1985). *Physical Review Letters*, **55**, 2409.

Smarr, L. (1977a). *Annals of the New York Academy of Sciences*, **302**, 569.

Smarr, L. (1977b). *Physical Review D*, **15**, 2069.

Smarr, L., ed. (1979). *Sources of Gravitational Radiation*. Cambridge University Press.

Smarr, L. and York, J. W. (1978). *Physical Review D*, **17**, 2529.

Smarr, L., Vessot, R. F. C., Lundquist, C. A., Decher, R. and Piran, T. (1983). *General Relativity and Gravitation*, **15**, 129.

Spero, R. (1986a). In Ruffini, ed. (1986).

Spero, R. (1986b). Private communication.

Stark, R. F. and Piran, T. (1985). *Physical Review Letters*, **55**, 891 and **56**, 97.

Stark, R. F. and Piran, T. (1986). In *Proceedings of the Fourth Marcel Grossman Meeting on Recent Developments of General Relativity*, ed. R. Ruffini (in press).

Starobinsky, A. A. (1979). *Soviet Physics – JETP Letters*, **30**, 682.

Starobinsky, A. A. (1985). *Soviet Astronomy Letters*, **11**, 133.

Stewart, J. M. and Friedrich, H. (1983). *Proceedings of the Royal Society of London A*, **384**, 427.

Suzuki, T., Tsubono, K. and Hirakawa, H. (1978). *Physics Letters A*, **67**, 10.

Sullivan, W. T. (1982). *Classics in Radio Astronomy*. Reidel: Dordrecht.

Sullivan, W. T. (1984). *The Early Years of Radio Astronomy*. Cambridge University Press.

Szekeres, P. (1971). *Annals of Physics*, **64**, 599.

Tammann, G. A. (1981). In *Supernovae: A Survey of Current Research*, ed. M. J. Rees and R. J. Stoneham, p. 371. Reidel: Dordrecht.

Taylor, J. H. (1987). In *General Relativity and Gravitation*, ed. M. A. H. MacCallum. Cambridge University Press (in press).

Teissier du Cros, F. (1985). *Annales de Physique*, **10**, 263.

Taylor, J. H. (1987). In *Proceedings of Eleventh International Conference on General Relativity and Gravitation*. Cambridge University Press (in press).

Teukolsky, S. A. (1972). *Physical Review Letters*, **29**, 1114.

Teukolsky, S. A. (1973). *Astrophysical Journal*, **185**, 635.

Teukolsky, S. A. and Press, W. H. (1974). *Astrophysical Journal*, **193**, 443.

Thorne, K. S. (1969a). *Astrophysical Journal*, **158**, 1.

Thorne, K. S. (1969b). *Astrophysical Journal*, **158**, 997.

Thorne, K. S. (1972). In *Magic Without Magic: John Archibald Wheeler*. Appendix A, p. 243. W. H. Freeman: San Francisco.

Thorne, K. S. (1977). In de Sabbata and Weber (1977), p. 1.

Thorne, K. S. (1978). In *Theoretical Principles in Astrophysics and Relativity*, eds. N. R. Lebovitz, W. H. Reid and P. O. Vandervoort, p. 149. University of Chicago Press: Chicago.

Thorne, K. S. (1980a). *Reviews of Modern Physics*, **52**, 285.

Thorne, K. S. (1980b). *Reviews of Modern Physics*, **52**, 299.

Thorne, K. S. (1983). Chapter 1 of Deruelle and Piran (1983).

Thorne, K. S. (1985). In *Nonlinear Phenomena in Physics*, ed. F. Claro, p. 280. Springer-Verlag: Berlin.

Thorne, K. S. and Braginsky, V. B. (1976). *Astrophysical Journal (Letters)*, **204**, L1.

Thorne, K. S. and Campolattaro, A. (1967). *Astrophysical Journal*, **149**, 591; and **152**, 673.

Thorne, K. S., Caves, C. M., Sandberg, V. D., Zimmermann, M. and Drever, R. W. P. (1979). In Smarr (1979). p. 49.

Thorne, K. S., Drever, R. W. P., Caves, C. M., Zimmermann, M. and Sandberg, V. D. (1978). *Physical Review Letters*, **40**, 667.

Thorne, K. S. and Gürsel, T. (1983). *Monthly Notices of the Royal Astronomical Society*, **205**, 809.

Thorne, K. S. and Kovacs, S. J. (1975). *Astrophysical Journal*, **200**, 245.

Thorne, K. S., Price, R. H. and Macdonald, D. M., eds. (1986). *Black Holes: The Membrane Paradigm*. Yale University Press: New Haven, Connecticut.

Tipler, F. J. (1980). *Physical Review D*, **22**, 2929.

Tourrenc, P. (1978). *General Relativity and Gravitation*, **9**, 123 and 141.

Tourrenc, P. and Crossiord, J.-L. (1974). *Nuovo Cimento*, **19B**, 105.

Traschen, J., Turok, N. and Brandenberger, R. (1986). *Physical Review D*, **34**, 919.

Tsvetkov, V. P. (1984). *Soviet Astronomy – AJ*, **28**, 394.

Turner, M. and Wagoner, R. V. (1979). In Smarr (1979), p. 383.

Turner, M. and Will, C. M. (1978). *Astrophysical Journal*, **220**, 1107.

Turok, N. and Brandenberger, R. H. (1986). *Physical Review D*, **33**, 2175.

Tyson, J. A. and Giffard, R. P. (1978). *Annual Reviews of Astronomy and Astrophysics*, **16**, 521.

Unruh, W. G. (1982). Unpublished work reported at NATO Advanced Study Institute on Gravitational Radiation, Les Houches, France, June 1982.

Vachaspati, T. and Vilenkin, A. (1985). *Physical Review D*, **31**, 3052.

van den Heuvel, E. P. J. (1984). In Reynolds and Stinebring, ed. (1984).

Vilenkin, A. (1981a). *Physical Review D*, **24**, 2082.

Vilenkin, A. (1981b). *Physics Letters*, **107B**, 47.

Vinet, J. Y. (1985). *Annales de Physique*, **10**, 253.

Vinet, J. Y. (1986). *Journal de Physique*, **47**, 639.

Vishniac, E. T. (1982). *Astrophysical Journal*, **257**, 456.

Wagoner, R. V. (1977). In *Gravitazione Sperimentale*, ed. B. Bertotti. Accademia Nazionale dei Lincei: Rome.

Wagoner, R. V. (1984). *Astrophysical Journal*, **278**, 345.

Wagoner, R. V. and Will, C. M. (1976). *Astrophysical Journal*, **210**, 764.

Wahlquist, H. D. (1987). *General Relativity and Gravitation* (in press).

Wahlquist, H. D., Anderson, J. D., Estabrook, F. B. and Thorne, K. S. (1977). In *Gravitazione Sperimentale*, ed. B. Bertotti, p. 335. Accademia Nazionale dei Lincei: Rome.

Wainstein, L. A. and Zubakov, V. D. (1962). *Extraction of Signals from Noise*, Prentice-Hall: London.

Wald, R. M. (1973). *Journal of Mathematical Physics*, **14**, 1453.

Walgate, R. (1983). *Nature*, **305**, 665.

Walker, M. (1983). In *Relativistic Astrophysics and Cosmology*, eds. X. Fustero and E. Verdaguer. World Scientific: Singapore.

Walker, M. and Will, C. M. (1980). *Astrophysical Journal (Letters)*, **242**, L129.

Walls, D. F. (1983). *Nature*, **306**, 141.

Weber, J. (1959). *Reviews of Modern Physics*, **31**, 681 (see especially Section V).

Weber, J. (1960). *Physical Review*, **117**, 306.

Weber, J. (1967). *Physical Review Letters*, **18**, 498.

Weber, J. (1969). *Physical Review Letters*, **22**, 1302.

Weber, J. (1986). In *Proceedings of the Sir Arthur Eddington Centenary Symposium. Volume 3 – Gravitational Radiation and Relativity*, eds. J. Weber and T. M. Karade, p. 1. World Scientific: Singapore.

Weber, J. and Wheeler, J. A. (1957). *Reviews of Modern Physics*, **29**, 509.

Weyl, H. (1922). *Space–Time–Matter*. Methuen: London.

Weisberg, J. M. and Taylor, J. H. (1984). *Physical Review Letters*, **52**, 1348.

Weiss, R. (1972). *Quarterly Progress Report of the Research Laboratory of Electronics of the Massachusetts Institute of Technology*, **105**, 54.

Weiss, R. (1978). In *Sources of Gravitational Radiation*, ed. L. Smarr, p. 7. Cambridge University Press.

Weiss, R. (1979). In Smarr (1979), p. 7.

Weiss, R., Bender, P. L., Misner, C. W. and Pound, R. V. (1976). *Report of the Sub-Panel on Relativity and Gravitation, Management and Operations Working Group for Shuttle Astronomy*. NASA: Washington, DC.

Westpfahl, K. (1985). *Fortschritte der Physik*, **33**, 417.

Wheeler, J. A. (1955). *Physical Review*, **97**, 511.

Wheeler, J. A. (1962). *Geometrodynamics*. Academic Press: New York.

Wheeler, J. A. (1964). In *Relativity, Groups, and Topology*, eds. C. DeWitt and B. DeWitt. Gordon and Breach: New York.

Whitcomb, S. and Saulson, P. (1984). Unpublished research at Caltech and MIT; a few of the results are given in Fig. E.3 of Drever, Weiss *et al.* (1985).

Wiener, N. (1949). *The Extrapolation, Interpolation and Smoothing of Stationary Time Series with Engineering Applications*. Wiley: New York.

Will, C. M. (1986). *Canadian Journal of Physics*, **64**, 140.

Wilson, J. R. (1974). *Physical Review Letters*, **32**, 849.

Winkler, W. (1977). In *Gravitazione Sperimentale*, ed. B. Bertotti. Academia Nazionale dei Lincei: Rome.

Winkler, W., Maischberger, K., Rüdiger, A., Schilling, R., Schnupp, L. and Shoemaker, D. (1986). In Ruffini, ed. (1986).

Witten, E. (1984). *Physical Review D*, **30**, 272.

Wood, K. S., Michelson, P. F., Boynton, P., Yearian, M. R., Gursky, H., Friedman, H. and Dieter, J. (1986). *A Proposal to NASA for an X-Ray Large Array (XLA) for the NASA Space Station*. Stanford: Palo Alto, California.

Wood, L., Zimmermann, G., Nukolls, J. and Chapline, G. (1971). *Bulletin of the American Physical Society*, **16**, 609.

Woosley, S. E. and Weaver, T. A. (1986). *Annual Reviews of Astronomy and Astrophysics*, **24**, 205.

Wu, L.-A., Kimble, H. J., Hall, J. L. and Wu, H. (1986). *Physical Review Letters*, **57**, 2520.

Yamamoto, Y. and Haus, H. A. (1986). *Reviews of Modern Physics*, **58**, 1001.

York, J. W. (1983). In Deruelle and Piran (1983), p. 175.

Yurtsever, U. (1986). *Abstracts of Contributed Papers, 11th International Conference on General Relativity and Gravitation*, Stockholm, Sweden, 6–12 July, 1986, p. 287. University of Stockholm: Stockholm.

Yurtsever, U. (1987a). *Physical Review D*, submitted.

Yurtsever, U. (1987b). Paper in preparation.

Zel'dovich, Ya. B. (1980). *Monthly Notices of the Royal Astronomical Society*, **192**, 663.

Zel'dovich, Ya. B. and Novikov, I. D. (1983). *Relativistic Astrophysics Vol. 2. The Structure and Evolution of the Universe*. University of Chicago Press: Chicago.

Zhang, X.-H. (1986). *Physical Review D*, **34**, 991.

Zerilli, F. J. (1970). *Physical Review D*, **2**, 2141.

Zimmermann, M. (1978). *Nature*, **271**, 524.

Zimmermann, M. (1980). *Physical Review D*, **21**, 891.

Zimmermann, M. and Szedenits, E. (1979). *Physical Review D*, **20**, 351.

Zimmerman, M. and Thorne, K. S. (1980). In *Essays in General Relativity*, ed. F. J. Tipler, p. 139. Academic Press: New York.

10

The emergence of structure in the universe: galaxy formation and 'dark matter'

MARTIN J. REES

10.1 Introduction

The first relativistic solutions for a homogeneous expanding universe were found by Friedmann (1922) before Hubble (1929) discovered the recession of the nebulae. Hubble's work, which showed that the universe did not resemble Einstein's (1917) static model, stimulated further studies of relativistic cosmology by Lemaître, Tolman and others. But there was then – and remained for several decades – a severe mismatch between the relative sophistication of the theory and the sparseness of the relevant data.

Hubble's work suggested that galaxies would have been crowded together in the past, and emerged from some kind of 'beginning'. But he had no direct evidence for cosmic evolution. Indeed the steady state theory, proposed in 1948 as a tenable alternative to the 'big bang', envisaged continuous creation of new matter and new galaxies, so that despite the expansion the overall cosmic scene never changed.

We would not expect to discern any cosmic evolutionary trend unless we can probe out to substantial redshifts. This entails studying objects billions of light years away with redshifts $z \approx 1$; although a programme to measure the cosmic deceleration was pursued for many years with the 200-inch Palomar telescope, the results were inconclusive, partly because normal galaxies are not luminous enough to be detectable at sufficiently large redshifts. It was Ryle and his colleagues, in the late 1950s, who found the first evidence that our entire universe was evolving. His radio telescope could pick up emission from some active galaxies (the ones thought to harbour massive black holes) even when these were too far away to be observed optically. One cannot determine a redshift or distance of these sources from radio measurements alone, but Ryle assumed that, statistically at least, the ones appearing faint were more distant than those appearing intense. He counted the numbers with various apparent intensities, and found that there were too many apparently faint ones (in other words, those at large distances) compared with brighter and closer ones (Ryle, 1958). This was discomforting to the 'steady statesmen', but compatible with an evolving universe if galaxies were more prone to violent outbursts in the remote past, when they were young. The subsequent discovery by *optical* astronomers of many hundreds of active galaxies at very large redshifts (quasars) has borne out this trend; but these objects, and their evolution, are still too poorly understood to be used for determining the geometry and deceleration of the universe.

The clinching evidence for a 'big bang' came when Penzias and Wilson (1965) detected the cosmic microwave background radiation. This radiation (whose thermal spectrum was quickly established) implied that intergalactic space was not completely cold, but at a temperature of 2.7 K. The corresponding photon density is $\sim 4 \times 10^8 \, \text{m}^{-3}$, implying that there are $\sim 10^9$ photons for every baryon. This discovery quickly led to a general acceptance of the so-called 'hot big bang' cosmology – a shift in the consensus among cosmologists as sudden and drastic as the shift of geophysical opinion in favour of continental drift that took place contemporaneously. There seemed no plausible way of accounting for the microwave background radiation except on the hypothesis that it was a relic of an epoch when the entire universe was hot, dense and opaque. Moreover, the high intrinsic isotropy of this radiation – better than one part in 10^4 – meant that the Robertson–Walker metric was a better approximation to the real universe than the theorists of the 1930s would have dared to hope.

In the late 1960s some theorists were emboldened to carry out a series of now-classic investigations of the early stages of a Friedmann universe composed of matter and radiation. Insofar as the present 'mix' of matter and thermal radiation was known, cosmologists could infer the appropriate equation of state at earlier times, and deduce the universe's 'thermal history'. Powerful corroborative support for the hot big bang came when the composition of material emerging from the 'fireball' was calculated, and found to be 25 per cent helium and 75 per cent hydrogen (Hoyle and Tayler, 1964; Peebles, 1966; Wagoner, Fowler and Hoyle, 1967). This was specially gratifying because the theory of stellar nucleosynthesis, which worked so well for carbon, iron, etc., was hard-pressed to explain why there was so much helium, and why its abundance was so uniform. Attributing most of the observed helium to the big bang therefore solved a long-standing problem in nucleogenesis, and bolstered cosmologists' confidence in extrapolating right back to the first few seconds of the universe's history, and assuming that the laws of microphysics were the same then as now. The applicability of the Friedmann equations at early times could not of course be taken for granted; the universe is certainly not completely homogeneous now, and could in principle have been more irregular, and more anisotropic, in the past. But the measured isotropy, together with the singularity theorems, implied that there must be some singularities in the universe's past (Hawking and Ellis, 1968), and the observed helium abundance (sensitive to the expansion rate when the temperature was $\sim 10^{10}$ K) constrains any possible early anisotropies.

More detailed calculations, combined with better observations of background radiation and of element abundances, have strengthened the consensus that the hot big bang model is basically valid. Several discrepant results *could* have emerged during the last 20 years, but did not: for instance, the standard Friedmann hot big bang would need drastic modification if any object were found to have a zero helium/hydrogen ratio, if any species of neutrino were found experimentally to have a mass in the range 1 keV– 1 MeV, or if there were a glaring discrepancy between the ages of the oldest stars and the largest plausible timescale in such models. The hot big bang is not yet firm dogma. Conceivably, our satisfaction will prove as transitory as that of a Ptolemaic astronomer who successfully fits a new epicycle. But the hot big bang model certainly seems more plausible than any equally specific alternative – most cosmologists would make the stronger claim that it has a more than 50 per cent chance of being essentially correct.

The main stages in the evolution of a standard hot big bang universe are depicted schematically in Fig. 10.1. Uncertainties about the relevant physics impede our confidence in discussing the extensive span of logarithmic time 10^{-43}–10^{-4} s when thermal energies exceed 100 MeV. When $t \gtrsim 10^{-4}$ s we can consistently utilise physics which is 'well known'; and, so long as the universe remains almost homogeneous, the evolution is straightforwardly calculable. However, at some stage small initial perturbations must have evolved into gravitationally bound systems (protogalaxies?); even though the controlling physics may then be Newtonian gravity and gas dynamics, the onset of non-linearity induces challengingly complex behaviour.

The progress of the last 20 years in delineating the big bang model has therefore brought two sets of questions into sharper focus.

(i) How did the universe behave during the very earliest phases when the physics is uncertain? Such fundamental properties as its scale, overall

homogeneity and matter content could have been imprinted during
these initial stages.

(ii) How did the dominant present-day structures in our universe – galaxies
and clusters – evolve from amorphous beginnings?

The first set of questions is addressed elsewhere in this volume. I shall
focus on the second topic, and outline some current ideas on the inter-
related issues of galaxy formation, dark matter, the density of the universe,
and the nature of the initial fluctuations.

Observational cosmology reveals that there may be just three (or perhaps
four) important constants which characterise the universe's overall
properties.

(i) *The Robertson–Walker curvature radius.* The fact that the universe is
still expanding, after $10^{60}t_{\text{Planck}}$, with a density within an order of
magnitude of the 'critical' density, implies that the initial curvature

Fig. 10.1. A schematic depiction of the history of a standard Friedmann hot
big bang model, from the Planck time to the present.

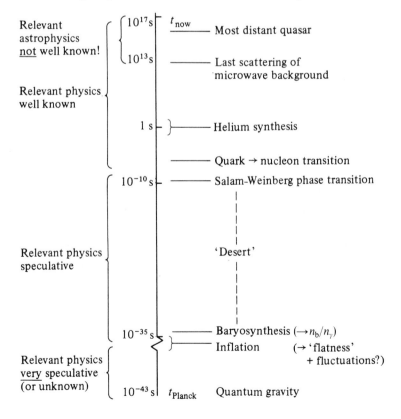

radius at t_{Planck} (or at the end of an 'inflationary' phase) is at least $\sim 10^{30}$ times larger than the horizon scale at that epoch. The precise value of this curvature radius depends on the density parameter Ω.

(ii) *The photon-to-baryon ratio.* This ratio is a measure of the entropy per baryon. Grand unified theories (GUTs) suggest that it can be explained in terms of baryon non-conservation processes.

(iii) *The fluctuation amplitude.* The prime mystery is perhaps why the large scale universe is so homogeneous. However, some fluctuations are essential in order to trigger galaxy formation. If these metric fluctuations have a scale-independent random-phase character, then the amplitude is pinned down to be 10^{-4}–10^{-5}, a number which theorists may hope to calculate; however, if the fluctuations have a more general character, few constraints can be set.

(iv) A possible fourth constant appears if the universe contains a dynamically-significant amount of non-baryonic matter. The *fraction of mass-energy in baryons* rather than more exotic species is then determined by physical processes in the high-density phases of the big bang.

Insofar as the aim of fundamental physics is to erode the number of independent underivable constants, it is gratifying that there is a serious chance of calculating the above numbers in terms of microphysical parameters. And then, given that the universe is sufficiently close to 'flatness', but nevertheless contains fluctuations (i.e. there are small-amplitude 'ripples' on the initial hypersurface), gravitationally bound structures are fated eventually to emerge. Such structures – the galaxies – then offer an environment where stars, some with planetary systems, form and evolve.

10.2 The constituents of the universe: 'dark' matter and 'luminous' matter

To the cosmologist, entire galaxies are just 'markers' or test particles scattered through space which indicate how the material content of the universe is distributed and how it is moving. But it is these 'small-scale' deviations from homogeneity, groups of galaxies and their constituent members, which constitute the main subject matter of astronomy. And it is a prime goal of our subject to understand how the universe has evolved from an initial dense fireball $\sim 10^{10}$ years ago to its present state, where galaxies dominate the large-scale cosmic scene.

An impediment to understanding galaxies and their spatial distribution is that the stars and gas we observe may be little more than a tracer for the material that is dynamically dominant. The evidence for 'dark matter' dates back more than fifty years, but has firmed up since the classic papers of Einasto, Kaasik and Saar (1974) and Ostriker, Peebles and Yahil (1974) and is now quite compelling (for a recent review, see the proceedings of IAU Symposium 117 (Knapp and Kormendy, 1986)); what the dark matter consists of is, however, still a mystery.

The masses inferred from *relative motions* of galaxies in apparently-bound groups and clusters exceed by a factor ~ 10 those inferred from the *internal dynamics* of the luminous parts of galaxies. This apparent discrepancy could be resolved if galaxies were embedded in extensive dark 'halos'. This hypothesis can be checked in some edge-on disc galaxies, where emission from gas can be observed at radii far exceeding the extent of the conspicuous stellar disc. The mass of this gas is itself negligible, but rotation velocities derived from its spectral lines do not fall off as $r^{-1/2}$, as would be expected if the gas were orbiting a mass distribution concentrated at much smaller radii. Instead, the velocity remains almost constant, implying $M(<r) \propto r$, out to ~ 80 kpc in some cases. Direct lower limits on the mass to light ratio M/L can be obtained for the matter in the outlying parts of galaxies with measured rotation curves, and for the haloes of edge-on galaxies. Values of M/L exceeding 300 solar units are sometimes found (see Rubin (1986) for a recent review).

Estimates of the masses of galaxy clusters come from the virial theorem – a technique first applied to the Coma cluster by Zwicky (1933). This method is now complemented by X-ray studies of thermal emission from hot gas in clusters, which probe the depth of the gravitational potential well. In well-studied clusters which appear to have reached virial equilibrium, M/L is typically 200 solar units. This matter must mainly be in some unknown form – neither ordinary stars nor the gas that emits X-rays. The data are summarised in Figure 10.2.

All things considered, the existence of dark matter is quite unsurprising – there are all too many forms it could take, and the aim of observers and theorists must be to narrow down the range of options. These topics have been discussed and reviewed at greater length elsewhere (e.g. Dekel, Einasto and Rees, 1987, and references cited therein).

The present Hubble timescale t_H is still uncertain, so in quoting numerical values I will follow a widespread convention and introduce a quantity $h = (3 \times 10^{17} \, s/t_H)$. The experts advocate values of h in the range 0.5–1; for a detailed assessment, see Hodge (1981) or Rowan-Robinson (1985). The

Fig. 10.2. (*a*) The apparent increase with scale of the mass-to-light ratio. This increase is due to the two distinct trends: (i) ordinary stars and gas are a decreasing proportion of the mass of the larger systems; and (ii) in the larger systems, even the 'ordinary' (star+gas) components have a larger M/L, because they consist primarily of elliptical galaxies with few young stars, and contain much hot gas revealed only by its X-ray emission. (*b*) The same data re-plotted, with effect (ii) subtracted out. One finds that the physically more fundamental ratio of 'ordinary' matter to 'dark' matter is *independent* of scale in all virialized systems larger than galaxies, and has a value consistent with $\Omega = 0.1$–0.2. The situation is less clear on still larger scales (superclusters) because the dynamics are uncertain and virial equilibrium does not prevail. This figure is adapted from Faber (1984) and Blumenthal *et al.* (1984); the latter paper gives fuller details of the data on which it is based. (The masses of dwarf spheroidal galaxies are uncertain and controversial; the diagrams show these systems plotted twice, depending on whether or not they contain dark matter.)

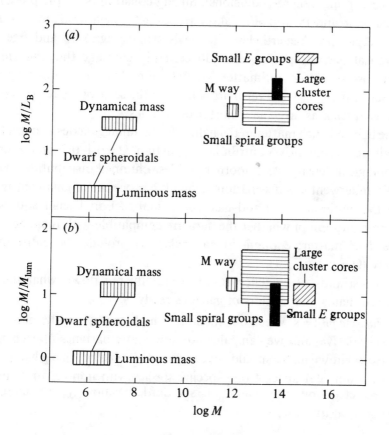

mean baryon density is then $n_b = 11\Omega_b h^2\,\mathrm{m}^{-3}$, where Ω_b is defined as the fraction of the critical density $\rho_{\mathrm{crit}} = (\frac{8}{3}\pi G t_H^2)^{-1}$ that is in baryonic form. An important number which is perhaps of fundamental significance is the ratio of the baryon density to the density n_γ of photons in the microwave background, apparently a black body with $T \approx 2.7\,\mathrm{K}$. This is then

$$\mathscr{S}^{-1} = n_b/n_\gamma \sim 3 \times 10^{-8} \left(\frac{T}{2.7\,\mathrm{K}}\right)^{-3} (\Omega_b h^2), \qquad (2.1)$$

$\mathscr{S}$ being a measure of the entropy per baryon; it is a number which GUT models must attempt to explain.

A direct lower limit on Ω_b of order 10^{-2} can be set from the observed mean density of baryons in conspicuously 'luminous' form (visible galaxies, and intergalactic gas revealed by its X-ray emission), implying a baryon-to-photon ratio of not less than about $3 \times 10^{-10}h^2$.

There is no firm evidence for any antimatter in the universe (apart from a small fraction of antiparticles in cosmic rays, which could have been produced in high-energy collisions). Strong constraints on the presence of antimatter in and around our Galaxy are set by the measured limits to the X-ray background. Nevertheless, if one is strictly agnostic and free from theoretical preconceptions, one can certainly envisage that the universe might possess matter–antimatter symmetry (i.e. that the overall net baryon number, and $\langle n_b/n_\gamma \rangle$, are zero) provided that the scale of the regions of each 'sign' is at least as large as a cluster of galaxies.

The inferred dark matter in the halos of individual galaxies and in clusters of galaxies apparently contributes a fraction $\Omega = 0.1$–0.2 of the critical cosmological density. Its smoother and less clumped distribution suggests that it underwent less dissipation during the processes of galaxy formation than the luminous stars and gas. (In a final section I shall address the separate question of whether the data are compatible with there being still more dark matter; enough, in particular, to provide the entire critical density ($\Omega = 1$).)

Three strands of evidence could eventually pin down what the dark matter in halos and clusters of galaxies really is.

(i) *Particle physics.* When our theories of high-energy physics become less speculative, and we can calculate how many particles of each species (with known mass) should have survived as relics of the big bang, it may turn out that at least one specific species with non-zero rest mass is predicted, on the basis of standard cosmology, to contribute significantly to Ω.

(ii) *Cosmogonic models.* Evidence on the clustering scales of galaxies, the density profiles in halos, etc., can be compared with the outcome of simulations of gravitational aggregation. The character of the clustering depends, in particular, on whether the dark matter is 'cold' (in the sense that its thermal velocities are unimportant on relevant scales) or 'hot'.

(iii) *Direct detection.* The individual entities that our galactic halo is composed of may reveal themselves by astronomical observations, or by direct Earth-based experiments.

10.2.1 *Baryonic dark matter*

The baryons observed in galaxies and gas, contributing $\Omega_b \approx 0.01$, could in principle be a dynamically *un*important fraction of all baryons in the universe. Most of the initial baryons might have condensed into a population of stars that were either pregalactic, or else formed during the initial collapse phase of protogalaxies: could these stars, or their remnants, contribute to the unseen mass? Ideally one would like to be able to calculate what happened when the first gravitationally bound clouds condensed from primordial material – to decide whether they form one (or a few) very massive objects (VMOs), or whether fragmentation proceeds efficiently down to low-mass stars. Our poor understanding of what determines the initial mass function (IMF) of stars forming now (in, for instance, the Orion nebula) gives us little confidence that we can calculate the nature of stars born in an environment very different from our present-day galaxy. Constraints imposed by background light, limits to the heavy element abundance and absence of dynamical effects in the halo (see, for example, Carr, Bond and Arnett, 1984; Rees, 1986) imply that the individual masses of halo objects are either less than $0.1 \, M_\odot$ or else in the range 10^2–$10^6 \, M_\odot$. These constraints are summarised in Fig. 10.3 and its caption.

Two mass-scales that emerge naturally in cosmogony are $10^6 \, M_\odot$ and $10^{-2} \, M_\odot$. The Jeans mass in the baryonic component of the universe after recombination is of order $10^6 \, M_\odot$; a similar mass appears in a different, but equally relevant, context, as the Jeans mass in $10^4 \, K$ gas that is in pressure balance with hotter gas in a collapsing protogalaxy (Fall and Rees, 1985). If a $10^6 \, M_\odot$ cloud collapses isothermally the Jeans mass falls, however, to $10^{-2} \, M_\odot$ before the onset of opacity effects; this is the case even in a primordial cloud containing no heavy elements. The IMF therefore depends crucially on the complexities of the fragmentation process. Does fragmentation of the first clouds proceed right down to the 'opacity limited' Jeans mass? Or, is fragmentation impeded by collisions (and coalescence) or protostars, or by tidal effects? I do not believe we can yet answer these questions with any confidence; it is therefore worth considering both 'Jupiters' and 'VMOs' seriously, in the hope that observations can offer some firmer clues than theory.

One way of discriminating between the 'Jupiter' and 'VMO' options is by searching for evidence of gravitational lensing. The probability of seeing lensing due to an object in our own halo is only of order 10^{-6} (Refsdal, 1964; Paczynski, 1986*b*); but the cross-section for effective lensing is proportional

Fig. 10.3. This diagram, from Carr, Bond and Arnett (1984) shows various constraints on the fraction Ω_* of the critical cosmological density that can be contributed by first-generation stars (or their remnants) in different mass ranges. The objects are assumed to form at a redshift z_f. There are dynamical constraints on the number of $\gtrsim 10^6 \, M_\odot$ black holes in galactic halos, because such massive bodies would have sedimented inwards via dynamical drag on the ordinary stars. The requirement not to overproduce heavy elements constrains the number of remnants of ordinary heavy stars which end their lives by exploding as supernovae. Stars in the mass range $0.1–1 \, M_\odot$ would still be shining after $\sim 10^{10}$ yr, producing too much background light. The possible options for dark matter are 'Jupiters' of below $0.1 \, M_\odot$, or the remnants of very massive objects (VMOs) in the mass range from a few hundred to $10^6 \, M_\odot$, which could have formed at a large z_f. (These latter objects do not necessarily eject any material processed beyond helium, and leave black hole remnants.) One way of deciding between these two possibilities is by seeking evidence for gravitational lensing, as discussed in the text.

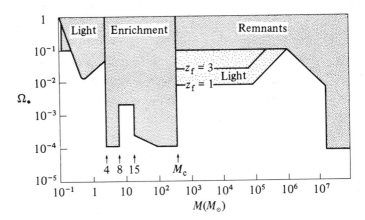

to distance, so there is, perhaps surprisingly, much *more* chance of detecting objects in the halos of galaxies (or in clusters) half way out to the Hubble radius. The probability that a compact source at redshift $z \gtrsim 1$ is significantly lensed by objects along the line-of-sight is of order Ω_l, independent of the individual lens mass involved (Refsdal, 1970; Press and Gunn, 1973); for a source at redshift $z_s < 1$ the probability is $\sim z_s^2 \Omega_l$. The angular separation, θ_l, of the lens images is, however, a diagnostic of the masses:

$$\theta_l \approx 2 \times 10^{-6} (M_l/M_\odot)^{1/2} \max[z_s^{-1/2}, 1]. \tag{2.2}$$

For $M_l \gtrsim 10^5 \, M_\odot$, very long baseline radio interferometers provide adequate resolution. Characteristic image shapes were discussed by Press and Gunn (1973), and Blandford and Jaroszynski (1981). We can probably exclude $\Omega = 1$ in $10^6 \, M_\odot$ objects.

For $M \lesssim 0.1 \, M_\odot$ ('Jupiters') the angular scale is less than $(10^{-6})''$. This cannot be directly resolved by any technique, until optical interferometers are deployed in space. There is nevertheless a genuine prospect of detecting lensing of this kind because of the variability that would ensue if the lens were to move transversely (Gott, 1981; Young, 1981): an object at the Hubble distance moving at $\gtrsim 10^2 \, \mathrm{km \, s^{-1}}$ takes only a few years to traverse an angle $(10^{-6})''$. The image structure and time variation are more complicated if the line-of-sight passes through, for example, a galactic halo, thereby encountering an above-average column density of dark matter. Several objects may then contribute to the imaging (Chang and Refsdal, 1979; Young, 1981; Paczynski, 1986a) yielding a 'frosted glass' effect. 'Minilensing' requires a source whose intrinsic angular size is smaller than the value of θ_l given by (2.2). Possibilities include distant supernovae (Efstathiou, Schneider and Wagoner, 1986) or the region that emits the optical continuum of quasars (but not the spectral lines, because the latter come from a more extended region). If there were a firm observational limit to the scatter in the equivalent widths of the lines from quasar to quasar (i.e. in the line–continuum ratio) this would constrain the value of Ω contributed by small compact objects. Canizares (1982) uses this argument to rule out $\Omega = 1$ in any objects with mass greater than $10^{-2} \, M_\odot$; $\Omega \approx 0.1$ cannot yet be excluded, but the method holds prospects of improvement.

10.2.2 *Constraints on Ω_b from primordial nucleosynthesis*

If we accept the concept of light element synthesis in the big bang (when kT is in the range 0.1–1 MeV) then an important constraint on the total baryonic density Ω_b comes from deuterium; this is an intermediate product in helium synthesis, the amount that survives being a steeply decreasing function of Ω_b. According to Yang *et al.* (1984) only if $\Omega_b h^2 \lesssim 0.04$ can enough deuterium be produced in a standard hot big bang to provide the observed abundance; and $\Omega_b h^2 \gtrsim 0.01$ is required in order not to *over*produce deuterium or ^{3}He. Also, weak constraints consistent with the other requirements are provided by ^{7}Li.

Because the relevant parameter in primordial nucleosynthesis is $n_b/n_\gamma \propto \Omega_b h^2$, more precise comparisons of models with observation must await a firmer value of the Hubble constant. If $h = 1$ (corresponding to a Hubble time of 10^{10} years) then the simplest inference would be that most of the dynamically detected dark matter – both in the halos of individual galaxies and in clusters and groups – was non-baryonic; but if $h = \frac{1}{2}$ (corresponding to a Hubble time of 2×10^{10} years) the lower limit to Ω_b set by the requirement not to overproduce $D + {}^3$He implies that some dark matter – maybe that in halos, if not in intergalactic space – *must* be baryonic, though only enough to contribute $\Omega_b \approx 0.1$. The strength of this limit (which of course precludes $\Omega_b = 1$ for any reasonable Hubble constant) depends on how compelling the case for 'standard' big bang nucleosynthesis actually is. Even in the context of this theory, Applegate, Hogan and Scherrer (1986) have recently pointed out that small scale inhomogeneities in the baryon–photon ratio, such as might arise at the quark–hadron transition, could modify the resultant abundances: if the scale of the inhomogeneities is smaller than the neutron diffusion scale at $t \approx 1$ s, then the neutron–proton ratio can vary from place to place in a manner that would be impossible in any homogeneous model, with the result that the ^{4}He abundance can be lower than in the standard model, and the observed deuterium could be primordial even if $\Omega_b = 1$.

10.2.3 Non-baryonic dark matter

If neutrinos have negligible rest mass, the present density expected for relic neutrinos from the big bang is $n_\nu = 1.1 \times 10^8 (T/2.7 \text{ K})^3 \text{ m}^{-3}$ for each two-component species. This conclusion holds for non-zero masses, provided that $m_\nu c^2$ is far below the thermal energy (5 MeV) at which neutrinos decoupled from other species and that the neutrinos are stable for the Hubble time. Comparison with the baryon density shows that neutrinos outnumber baryons by such a big factor ($\sim \mathscr{S}$) that they can be dynamically dominant over baryons even if their masses are only a few electron volts. In fact, a single species of neutrino would yield a contribution to Ω of $\Omega_\nu =$

$0.01\,h^{-3}(m_\nu)_{\text{eV}}$; so if $h=0.5$, only 25 eV is sufficient to provide the critical density.

The entire range $100h^2$ eV–3 GeV is incompatible with the hot big bang model (Gunn *et al.*, 1978). (For $m_\nu \gtrsim 3$ GeV, the rest mass term in the Boltzmann factor would kill off most of the neutrinos before they decoupled; the number surviving would be less than n_b). If any species of neutrino were discovered to have a mass in this excluded range, it would show that one cannot extrapolate the hot big bang back to $kT \gtrsim 5$ MeV, and that most of the photons must have been generated at later times.

More than 15 years ago, Cowsik and McLelland (1973) and Marx and Szalay (1972) conjectured that neutrinos could provide the 'unseen' mass in galactic halos and clusters. At that time the suggestion was not followed up very extensively; but in the 1980s physicists became more open-minded about the possibility of non-zero neutrino masses. A change of theoretical attitude, coupled with experimental claims that $m_\nu \approx 36$ eV (Lyubimov *et al.*, 1980) stimulated astrophysicists to explore scenarios for galaxy formation in which neutrino clustering and diffusion play a key role (see Section 10.4). More recently, other kinds of non-baryonic matter, such as supersymmetric particles and axions, have also been considered.

Provided that we know the mass and annihilation cross-section for any species of elementary particle, we can in principle calculate how many survive from the big bang, and the resultant contribution each species makes to Ω. Progress in experimental particle physics may therefore reveal a particle which must contribute significantly to Ω, unless we abandon the hot big bang theory entirely. Indeed, once we admit the possibility that non-baryonic matter may be dynamically important, the ratio $\Omega_{\text{non-baryonic}}/\Omega_b$ in the universe becomes another fundamental number as important for cosmology as n_b/n_γ.

Non-baryonic matter, if dynamically dominant, has an important influence on the cosmogonic process (see Section 10.4). Insofar as we come to understand galaxy formation, we may therefore acquire circumstantial evidence about whether such matter exists and what form it might take. However, it turns out that one may be able to detect such particles (even in the laboratory) if they contribute the mass of our own galactic halo.

Weakly interacting massive ($\gtrsim 1$ GeV) particles would have cross-sections of order 10^{-36} cm^{-2} for interactions with nucleons. Such particles, scattering elastically off nucleons in the Sun, would lose energy via the recoil, and could thereby become trapped (Steigman *et al.*, 1978; Spergel and Press, 1985). Despite being vastly outnumbered by ordinary nuclei, they could then contribute to energy transport in the solar core, because their mean free path is so long; the central temperature would then be slightly lower, with resultant observational consequences such as an alteration in the frequency of some modes of solar oscillation, and a reduction in the ^{8}B neutrino flux. Over the lifetime of the Sun, an isothermal core of 'inos' could build up a mass of 10^{-12} M$_\odot$ if annihilations did not occur. Annihilations would restrict this buildup, unless the cross-section for annihilation is far below that for scattering, or unless the big bang produces an excess of inos over anti-inos (as it does for baryons). However, even if annihilations prevent a dense enough core building up to affect the Sun's structure, energetic neutrinos from these annihilations may reveal their presence in the underwater detectors developed to search for proton decay. Already, scalar or Dirac neutrinos with mass exceeding 6 GeV can be excluded, and analogous limits come from considering annihilations in the Earth rather than the Sun (Silk *et al.*, 1985).

If our galactic halo were comprised of massive weakly interacting particles, then their density near the Earth would be $\sim 10^5 m_{\text{GeV}^{-1}}$ m^{-3}, and their typical velocities ~ 300 km s^{-1}. There is a genuine prospect of detecting this background in the laboratory. Whenever one of these particles collided with a nucleus, there would be a recoil, with associated energy in the keV range. This energy deposition could be detected if it occurred in a very cold solid (with low heat capacity $\propto T^3$). The detection could be achieved by maintaining an array of superconducting grains just below the transition temperature: the heat would then raise the temperature of one grain above the critical value, thereby allowing magnetic flux to penetrate (Goodman and Witten, 1985; Drukier, Freese and Spergel, 1986). Alternatively, the phonons could be detected before thermalisation, or the thermal pulse could be detected directly (Smith, 1986).

If the dark matter were in so-called 'invisible axions', there would be a chance of detecting these also via conversion into photons on interaction with matter or strong magnetic fields. The prospects here are perhaps rather less promising, because the photons would be in a narrow energy band, depending on the poorly known axion mass, so a broad range of photon energies (spanning the microwave, millimetre and infrared bands) would need to be searched (Sikivie, 1985).

Ingenious schemes for detecting a halo population of exotic particles (reviewed by Smith, 1986) seem to me among the most worthwhile and exciting high-risk experiments in physics or astrophysics – potentially as important as those that led to the discovery of the cosmic microwave background in the 1960s. A null result would surprise nobody; on the other hand, such experiments could reveal new supersymmetric particles (or axions, as the case may be), as well as determining what 90 per cent of our universe consists of. (Because the detection is sensitive to velocity, they would even reveal the halo's velocity dispersion and rotation. The mean velocity of halo particles relative to the detector would have an annual variation, because of the Earth's motion around the Sun. Such an annual modulation, with an amplitude of a few per cent and a peak in June, would be an unambiguous signature discriminating against spurious background.)

10.2.4 *Macroscopic 'cold dark-matter' candidates: nuggets of strange matter*

Witten (1984) conjectured that grains or nuggets of 'strange matter', containing up, down, and strange quarks, may survive stably from the quark–hadron transition at $t = 10^{-4}$ (see Fig. 10.1). Such objects, in some sense intermediate between elementary particles and lumps of astrophysical size, would count as non-baryonic matter in the context of nucleosynthesis. Recent work (Applegate and Hogan, 1985; Alcock and Fahri, 1985) suggests that neutrino heating when $kT \approx 1$ MeV would destroy nuggets unless they had a mass of planetary order; and it is unclear that any larger ones could even form, because this would involve coordination over a scale larger than the particle horizon at the relevant epoch. If they indeed had survived, these nuggets might be detectable (De Rejula and Glashow, 1984); interesting constraints on the number in various mass ranges could be set from the results of monopole searches and proton decay experiments, from the number of meteor showers, and from limits on the frequency of small-scale seismic events.

10.2.5 *The baryon–photon ratio*

Ideas on baryon synthesis – if the GUTs on which they are based are borne out by future developments – might allow us to test whether the Friedmann models apply back at temperatures of around 10^{15} GeV, corresponding to times $\sim 10^{-36}$ s. The observed baryon-to-photon ratio (eq. (2.1)), a measure of the fractional excess of baryons over their antiparticles at early times, is $\sim 10^{-9}$. (Were it much smaller than this, the universe would not be baryon-dominated when its age was of order a characteristic stellar lifetime.) The value of the net baryon excess arising from out-of-equilibrium decay of X and Y particles can be computed, given a specific GUT (Kolb and Wolfram, 1980); it involves a small parameter related to the CP-violation parameter in weak interactions. This work is not yet on the same footing as the calculations of primordial helium and deuterium; it is perhaps at the same level as nucleosynthesis was in the pioneering days of Gamow and Lemaitre. But if it could be firmed up it would represent an extraordinary triumph. The mixture of radiation and matter characterizing our universe would not be *ad hoc* but would be a consequence of the simplest initial conditions. Also, as well as vindicating a GUT, it would reassure us about extrapolating in one bound, based on a Friedmann model, right back to the threshold of classical cosmology, almost back to the Planck time. On a logarithmic scale, this is a bigger extrapolation from the nucleosynthesis era than is involved in going to that era from the present time. It would also place constraints on dissipative processes arising from viscosity, phase transitions, black hole evaporation, etc., which might occur as the universe cooled through the 'desert' between 10^{15} and $100\,\text{GeV}$. Although these ideas are still speculative, the 'prediction' of the photon–baryon ratio may turn out to offer one of the few empirical tests of GUTs. (If baryon number were strictly conserved and the universe actually possessed a conserved quantity of order 10^{80}, then the concept of inflation would lose its appeal. The advent of GUTs is therefore a prerequisite for the viability of such theories, irrespective of the mechanism that drives the phase of exponential growth.)

If the baryon–photon ratio could be calculated, this would determine Ω_b. If $\Omega_b < 1$, then a strictly flat universe would require some non-baryonic contribution. Of course, one may eventually have theoretical knowledge of the rest masses of all other relevant particles; such information, in conjunction with knowledge of n_b/n_γ, would determine their contribution to Ω_b also. Looked at from this point of view, it perhaps seems coincidental that non-baryonic matter should dominate, but only by an order of magnitude rather than a vastly larger factor.

10.3 Large-scale structure and isotropy

For galaxies to have formed, the early universe must have contained some inhomogeneities (despite its overall Friedmannian character); otherwise its baryon content would, after 10^{10} years, still just comprise uniform neutral H and He. Gas dynamical and dissipative effects must have been important in the formation of individual galaxies (so that the characteristics of galaxies are related only indirectly to the form of the primordial fluctuation spectrum). But inhomogeneities on much larger scales are more likely to have been induced primarily by gravitation. The clustering properties of galaxies therefore offer more direct evidence about the initial fluctuation amplitudes. Note, however, that we can use these to infer the total density fluctuations only insofar as galaxies are a good tracer for the overall mass distribution. Large-scale inhomogeneities would also induce 'peculiar velocities' – deviations from the Hubble flow – in galaxies and even entire clusters. The existing evidence on superclustering – though agreed by all to be of primary importance – is still tentative and ambiguous.

A quite separate line of attack on large-scale structure is offered by the background radiation. The microwave background isotropy (established via relative rather than absolute measurements) is already known to be amazingly precise: effects at the level of one part in 10^4 would be detectable in a total cosmic background which is $\sim 10^2$ weaker than the contribution from the Earth itself. Recent reviews of the data have been given by Partridge (1986), Wilkinson (1986) and Kaiser and Silk (1986). There are no confirmed anisotropies apart from a 'dipole' anisotropy, of $\Delta T/T \approx 1.2 \times 10^{-3}$, indicating a motion of our Local Group of galaxies towards a direction 45° from the Virgo cluster. The upper limits are of order 10^{-4} on all angles from a few arc minutes up to 90° (quadrupole).

The microwave photons were last scattered at a 'cosmic photosphere' whose redshift z_* and thickness depend on the thermal history of the matter. If the primordial plasma (re)combined as it cooled adiabatically below a temperature of a few thousand degrees, and there was no significant later reheating, then $z_* \approx 10^3$; the effective thickness of the photosphere, determined by the width Δz_* of the function $(\mathrm{d}/\mathrm{d}z)\mathrm{e}^{-\tau(z)}$, is then only $\sim 0.1 z_*$. This is because the reduction in the free electron density (which results in the 'fog lifting' and the microwave photons becoming free to travel uninterruptedly) is rather sudden. If reheating caused the intergalactic medium to 'fog up' again at a later epoch, with $\tau > 1$ at redshifts below 1000, then z_* would be smaller and $\Delta z_* \approx z$. Three cases are illustrated in Fig. 10.4. Note that reheating at $z < 10$ could never generate $\tau > 1$, even if the universe contained enough ionized intergalactic gas to give $\Omega = 1$.

A Friedmann model with $\Omega_0 < 1$ evolves very like a 'flat' model at redshifts $> \Omega_0^{-1}$. Provided that $z_* > \Omega_0^{-1}$, which is so for almost all relevant models, the angle subtended by a given comoving region on the last scattering shell is essentially independent of z_*. A comoving length l corresponds to an angle

$$\theta_l \approx \tfrac{1}{2}\Omega_0(l/ct_{\mathrm{H}}). \tag{3.1}$$

A region of present length 1 Mpc subtends an angle $\sim \tfrac{1}{2}\Omega_0 h$ arc minutes. A region that came within the horizon at z_* has present length $l_* \approx 2ct_{\mathrm{H}} \times (\Omega_0 z_*)^{-1/2}$ and subtends an angle $\theta_* \approx (\Omega_0/z_*)^{1/2}$.

Another important scale, if $\Omega_0 < 1$, is the length corresponding to the intrinsic curvature of the spacelike hypersurfaces in the Robertson–Walker metric; this subtends an angle $\theta_c \approx \Omega_0(1-\Omega_0)^{-1/2}$.

The fluctuation amplitude on a given scale is conveniently characterised by

$$\varepsilon(l) = \begin{cases} \text{value of } \langle(\delta\rho/\rho)^2\rangle^{1/2} \text{ when scale } l \text{ is first} \\ \text{encompassed within particle horizon.} \end{cases} \tag{3.2}$$

If the growth were unaffected by pressure, this spectrum leads to $\langle(\delta\rho/\rho)^2\rangle^{1/2} \propto \varepsilon(l) \times l^{-2}$ for all scales l within the horizon at a given time – density contrasts, in the linear regime would grow $\propto R(t)$, and would eventually condense out as virialised systems with gravitational binding energy $\sim \varepsilon c^2$ per unit mass. The quantity ε is the 'curvature fluctuation', which one would like to be able to calculate from first principles.

Any fluctuation straddling the cosmic photosphere whose comoving length-scale exceeds l_*, the horizon scale at z_*, but is less than ct_H would generate a temperature anisotropy $(\Delta T/T)$ which is simply of order ε, as was first calculated by Sachs and Wolfe (1967). Similar fluctuations along the line-of-sight with redshifts less than z_* have smaller effects (Rees and Sciama, 1968; Dyer, 1976). If $z_* \approx 1000$, then all fluctuations subtending angular scales $> (2\Omega_0^{1/2})$ degrees exceed the horizon scale; even if $z_* \approx 10$ (an implausibly extreme lower limit, attained only if most of the critical density

Fig. 10.4. Optical depth $\tau(z)$ (dot-dash lines) of the universe back to a redshift z, and 'visibility factors' $\tau e^{-\tau}$ (dotted lines) for two values of x_e, the fraction of the critical density in the form of ionized plasma (assumed independent of z for illustrative purposes). Even a small amount of reheating is sufficient to produce a last scattering surface that is more 'smeared out' in redshift than the standard model of recombination (labelled 'rec'). In some models for galaxy formation where energy (and maybe heavy elements) are generated at large redshifts, other kinds of opacity may be competitive with Thomson scattering (from Hogan, Kaiser and Rees, 1982).

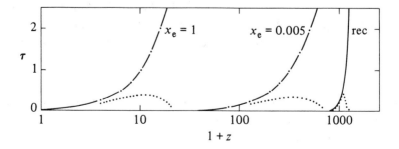

were in gas which had been reionized at $z > 10$), the horizon scale subtends $\lesssim 15°$. Consequently the high isotropy of the background ($\Delta T/T < 10^{-4}$) implies that, if $\Omega_0 \approx 1$

$$\varepsilon \lesssim 10^{-4} \quad \text{for} \quad \begin{cases} 1 > l/ct_H > \frac{1}{3} & (z_* \approx 10) \\ 1 > l/ct_H > \frac{1}{30} & (z_* \approx 10^3) \end{cases}. \tag{3.3}$$

There seem to be no ambiguities in this argument; there are only minor changes if other values of Ω_0 are taken. Fluctuations on scales $\sim ct_H$ would cause large angle (quadrupole and octopole) effects; the smaller angular scale observations probe smaller values of l (cf. (3.1)).

[If $\Omega_0 \lesssim 1$ then our particle horizon will eventually grow to encompass an infinite amount of matter that we cannot yet observe. Inhomogeneities on scales $\gg ct_H$ can induce gradients and shear across the region within our horizon, and may cause anisotropy in the microwave background. The observed limits to dipole and quadrupole anisotropy can thereby be used to set some limits on these still larger scales (Grishchuk and Zel'dovich, 1978; Kaiser, 1982). These limits, which are themselves somewhat model-dependent, become progressively less stringent (in terms of ε) on larger scales. We certainly cannot rule out a gross deviation from homogeneity (bubble wall? Minkowski space?) on scales $\gg 10^2 ct_H$. There can only be 'philosophical' reasons for extrapolating the observed inhomogeneity of the part of the universe we can now study to a domain which (if the universe is open) is infinitely larger.]

The scales to which (3.3) applies are larger than those on which we observe the most conspicuous inhomogeneity in the distribution of galaxies. It would be of particular interest in constraining theories of galaxy formation if we could observe temperature fluctuations due to the same scales which now display clustering. The upper limits to $\Delta T/T$ on the relevant scales of a few arc minutes (cf. (3.1)) are at the level of a few times 10^{-5}.

On angles corresponding to scales $l > l_*$ we are basically just seeing 'metric fluctuations' which have been unaffected by pressure gradients and have evolved acausally throughout the (post-inflationary?) Friedmannian expansion phase of the universe. But if $l < l_*$ various complications arise.

  (i) The Sachs–Wolfe contribution to $\Delta T/T$ may not be the dominant one. Doppler effects due to peculiar motions may be even more important; so also may effects due to the changes in the recombination and decoupling time in perturbed regions.

  (ii) Pressure gradients and damping may have affected the perturbations on scales $<l_*$ at epochs before decoupling. This complicates the relationship between ε and $(\delta\rho/\rho)_{z=z*}$ for these scales.

(iii) On the scales below $(\Delta z_*/z_*)l_*$ the temperature fluctuations are smeared out by the gradual nature of the decoupling – i.e. the fact that the last scattering surface is not sharp compared with the scale being studied. Furthermore, these scales would be attenuated by photon viscosity during recombination; this further complicates the relationship between $\Delta T/T$ and the value of $\delta\rho/\rho$ which survives after decoupling. (In the case of no reheating, these effects due to smearing and viscosity are important only for $l < ct_H/300$.)

There is an extensive literature attempting to quantify the effects (i)–(iii) above: to relate ε, $\Delta T/T$ and the post-decoupling $(\delta\rho/\rho)$ on the scales relevant to cluster formation (Press and Vishniac, 1980; Peebles, 1981, 1982; Silk, 1982; Bond and Efstathiou, 1986, 1987). Most calculations are predicated on the assumption that there is no early reheating, and that the 'cosmic photosphere' is determined by the recombination behaviour of the primeval H–He plasma. This work offers a consistency check on scenarios which preclude sufficiently early reheating; but it is important to stress that the interpretation of $\Delta T/T$ isotropy limits on small angular scales is very sensitive to the value of z_* and Δz_*. Under more general assumptions, smearing effects can be very large for scales $l \lesssim 10^{-2} ct_H$; furthermore, any anisotropies actually observed on small angular scales may be of 'secondary' origin, arising from reheated gas, dust, or discrete sources rather than being directly related to ε. (See Hogan, Kaiser and Rees (1982) and Bond, Carr and Hogan (1986) for further discussion of secondary fluctuations.)

Even 15 years ago, microwave background isotropy measurements seemed tantalisingly close to the sensitivity level where they were bound to reveal some effects. This still seems so. If they could be pushed below the 10^{-5} level, constant curvature fluctuations with any value of ε could be excluded (see Fig. 10.5), unless one adopted a quite different picture for galaxy formation. The microwave fluctuations on very small angular scales, incidentally, which are harder to interpret because they are sensitive to the thermal history of the intergalactic medium after recombination and to discrete (perhaps pregalactic) sources, can for this reason serve as probes for the 'dark age' between 10^6 and 10^9 years, during which the first structures developed.

10.4 Formation of protogalaxies

There are, as described briefly in Section 10.2, at least three serious candidates for the dark matter in galactic halos and clusters: low mass stars; black hole remnants of very massive objects; or non-baryonic matter, perhaps in the form of supersymmetric particles or axions. I would myself lay similar odds on these three options at the moment. However, it is gratifying that we can expect the odds to change quite rapidly, owing either to improved observational and experimental searches, or to progress in particle physics.

Fig. 10.5. This diagram, from Rees (1983), depicts the astrophysical limits on the amplitude ε of adiabatic metric fluctuations, on various scales l. On large scales, the microwave background isotropy offers stringent upper limits. On very small scales, ε must merely be not so large that too much of the universe collapses as the relevant scales enter the horizon; the absence of distortion in the microwave background sets a slightly better limit for mass-scales 10^4–10^{13} M$_\odot$. The requirement that bound systems have 'turned around' by the present epoch gives *lower* limits (also plotted). A spectrum with $\varepsilon \simeq 10^{-4}$ on all scales is acceptable if the universe is dominated by non-baryonic matter which can start clustering before t_{rec}. If ε has a power-law dependence on l and is $\sim 10^{-4}$ on the scales relevant to galaxy formation, then it cannot fall off more steeply than $\propto l^{-0.15}$ without causing excessive production of primordial mini-holes. Also plotted (dotted lines) are the amplitudes ε_g of primordial gravitational waves that can, within the next few years, be probed by: (a) doppler tracking of spacecraft, (b) timing of 'quiet' pulsars, and (c) timing of the orbit of the binary pulsar. This diagram is drawn assuming a 'flat' background universe with $\Omega_0 = 1$. If $\Omega_0 < 1$ the limits on large scales are modified. Indeed the appropriate definition of ε is ambiguous on scales exceeding the Robertson–Walker curvature radius.

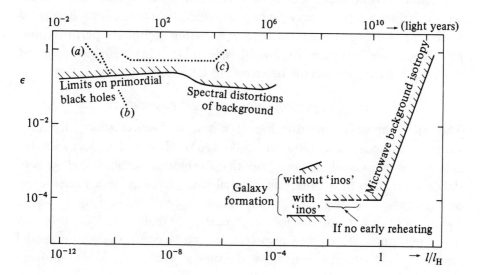

A separate line of attack that might shorten the list of candidates entails exploring the implications of each for galaxy formation – specifically, for processes whereby small primordial perturbations evolve into proto-galaxies and clusters. The key parameter here is the spectrum of the density fluctuations at the era of recombination, which is determined by initial conditions (described, for instance, by (3.2), modified by damping processes during the fireball phase ($t \lesssim 10^6$ years)). The mass-scale of the first bound system (i.e. the scale on which the growing perturbations after recombination first become non-linear) roughly corresponds to the scale at which the fluctuation spectrum peaks. Fig. 10.6 depicts three rival cosmogonic schemes (see caption for further details). The disparity between them is highlighted if we envisage what the universe would have been like, according to each, at the epoch corresponding to $z < 10$. In hierarchical 'bottom-up' schemes, a lot could have happened already. In others, the universe is then still amorphous neutral hydrogen, with all parts expanding in nearly undeviated Hubble flow. The cosmological 'dark age' starts when microwave background photons shift longward of the visible band, when the universe is 10^6 years old: it ends when the first bound systems (or their constituent stars) light up. We know that quasars – and so, presumably, some galaxies – had formed at $z \sim 4$ ($t \sim 10^9 h^{-1}$ years if $\Omega_0 = 1$), but this is merely a lower limit to the redshift of the first non-linearity. We do not know whether the 'dark age' is brief, or lasts for a billion years. We are more confused and ignorant about this phase of cosmic history than many seem to be about the first 10^{-35} s.

A related cosmogonic issue is this. To what extent is the present large-scale structure a rather direct consequence of initial fluctuations, imprinted at 10^{-35} s? Or does it, contrariwise, result from secondary perturbations generated or seeded by the first bound systems? Hogan (1986) has suggested the phrase 'paleogeny versus neogeny', to denote this dichotomy.

10.4.1 The 'cold dark matter' cosmogony

The three schemes depicted in Fig. 10.6 have all been reviewed elsewhere (see, for instance, Efstathiou and Silk, 1983). Here I shall focus on the specific cosmogony which postulates that the hidden mass is so-called 'cold dark matter'. In other words, this material is envisaged as weakly interacting non-baryonic stuff (e.g. GeV supersymmetric particles or axions) whose thermal motions are too low for free streaming to damp out fluctuations on any significant scale. (In this respect, it should be contrasted with a model where light (~ 10 eV) neutrinos dominate: these would be moving

Fig. 10.6. The cosmogonic processes after (re)combination depend on the spectrum of fluctuations which have survived damping processes, etc., at earlier times. After t_{rec}, linear growth proceeds roughly according to the law $(\delta\rho/\rho) \propto t^{2/3}$, and the first gravitationally bound systems to form will have the mass for which the density contrasts at t_{rec} are biggest. In case (1), (super)clusters are the first systems to condense, and they form quite recently. Indeed, models of this kind run into difficulties because we observe quasars with redshifts z as large as 4, whereas if superclusters had collapsed as early as this, they would now be denser, and display a higher density contrast,

(caption continued overleaf)

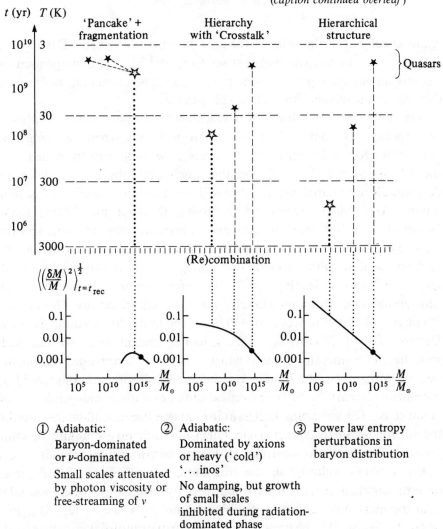

① Adiabatic:

Baryon-dominated or ν-dominated

Small scales attenuated by photon viscosity or free-streaming of ν

② Adiabatic:

Dominated by axions or heavy ('cold') '...inos'

No damping, but growth of small scales inhibited during radiation-dominated phase

③ Power law entropy perturbations in baryon distribution

Fluctuations scale-independent on entering horizon (Harrison/Zeldovich spectrum)

Fig. 10.6 (*continued*)
than is seen. In case (2), baryonic systems of $\sim 10^6\, M_\odot$ condense in potential
wells produced by 'cold dark matter' or 'inos', which are presumed to be slow
moving, so that they are not homogenised on small scales as neutrinos are.
(This specific case is discussed more fully in the text.) The third case shows the
spectrum that might arise from primordial entropy perturbations. In cases
(2) and (3), subgalactic systems would form *before* galaxies; if these
subgalactic systems provided an energy input, they could in principle
generate 'secondary' perturbations on larger scales which could swamp the
genuinely primordial ones.

relativistically until the stage in the cosmic expansion when kT fell below
$\sim 10\,$eV, with the consequence that scales $\lesssim 10^{15}\, M_\odot$ are homogenised by
free-streaming, leading to a cosmogony where superclusters form first and
then have to fragment into individual galaxies.)

It is relatively straightforward to calculate the growth factor for density
perturbations in 'cold dark matter' between the era when they enter the
particle horizon and the present. All scales grow roughly as the scale factor
(i.e. $(\delta\rho/\rho) \propto R \propto T^{-1} \propto t^{2/3}$) after the epoch when the universe becomes
dynamically dominated by dark matter. This result can be derived by simple
arguments based on Newtonian cosmology (McCrea and Milne, 1934).
However, at earlier times the universe is dominated by photons, and the
expansion timescale is proportional to $(G\rho_{\text{rad}})^{-1/2}$; perturbations in the
dark matter alone with a growth time $(G\rho_{\text{CDM}})^{-1/2}$ do not then have time to
grow significantly. See Fig. 10.7. This effect, termed 'stagspansion' by
Blumenthal and Primack (1983), was first calculated by Guyot and
Zel'dovich (1970) and Meszaros (1974). If the initial fluctuations have the
Harrison (1970)–Zel'dovich (1972) scale-independent form, so that each
scale has the same amplitude ε when it is first encompassed within the
particle horizon (i.e. at a time $\propto M$), the present-day fluctuations would be
proportional to $M^{-2/3}$ for large masses, and be essentially independent of M
at low mass. The spectrum flattens at low masses because all scales entering
the horizon before the epoch of equal densities undergo essentially the same
growth. The resultant spectrum of density perturbations actually has a
rather gradual 'rollover' in the transition region (Fig. 10.8). There is
nevertheless a characteristic mass imprinted by cosmology, which is smaller
than the mass now within a 'Hubble volume' by a factor $(\Omega_{\text{rad}}/\Omega_{\text{DM}})^{3/2}$,
where $\Omega_{\text{rad}}(\sim 10^{-4})$ is the present contribution of primordial radiation to Ω.

The amplitude of the fluctuations (i.e. the value of ε in (3.2)) cannot yet be
reliably predicted theoretically. The vertical scaling in Fig. 10.8 can

therefore be normalized to ensure that $\langle(\delta\rho/\rho)^2\rangle^{1/2}$ be now ~ 1 on a scale $\sim 5h^{-1}$ Mpc. (This is inferred from the clustering properties of galaxies; the normalization would need some adjustment if, as discussed further in Section 10.5, the galaxies do not trace the overall mass distribution.)

Fig. 10.7. The growth of adiabatic fluctuations in a universe dominated by 'cold dark matter'. The mass of cold dark matter within the particle horizon is shown as a function of time on a log–log plot. For $t > t_{eq}$ (corresponding to the redshift indicated on the figure), all scales grow at the same rate. Before t_{eq}, when the expansion is radiation-dominated, there is essentially no growth, because the growth timescale much exceeds the expansion timescale. If the fluctuations on all scales enter the horizon with equal amplitude (the 'Harrison/Zel'dovich' hypothesis), then the present-day spectrum would have the approximate form written on the right-hand side. (The accurately calculated spectrum is shown in Fig. 10.8.) The cold dark matter perturbations start to grow at t_{eq}, whereas radiation pressure inhibits growth of baryonic fluctuations on relevant scales until the (later) recombination time t_{rec}. Cold dark matter fluctuations thus have a 'head start' (baryons being able to fall into the resultant potential wells after t_{rec}); this permits an acceptable cosmogonic scheme with lower fluctuation amplitude ε, and smaller microwave background fluctuations, than in a baryon-dominated universe (cf. Fig. 10.5).

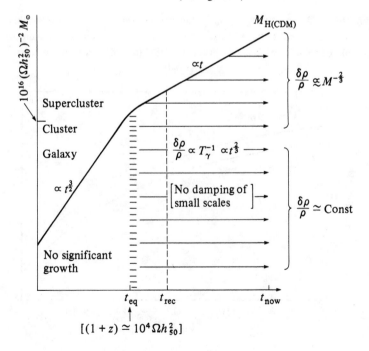

Because the spectrum in Fig. 10.8 is so nearly flat for small M, the typical fluctuation of $\sim 10^6 \, M_\odot$ would collapse no earlier than the epoch corresponding to $z \approx 10$. The build-up of structure is hierarchical, in the sense that smaller scales tend to form earlier. However, because of the flat spectrum, there would be complicated 'crosstalk' between many different scales (cf. middle panel of Fig. 10.6): the 3σ peaks in the density distribution on $10^{11} \, M_\odot$ scales would have the same amplitude as typical peaks of mass $\sim 10^6 \, M_\odot$, and would therefore collapse at the same time. It is consequently hard to analyse (either analytically or numerically) even the purely dynamical and non-dissipative aspects of the clustering. However, those studies that have been done (e.g. Davis *et al.*, 1985; Frenk *et al.*, 1985) are encouraging, in that, when the amplitude of the fluctuations is normalised so as to match the data on galaxy clustering, the finer-scale disposition of the dark matter closely reproduces the sizes and profiles of galactic halos (see Fig. 10.9).

10.4.2 The first stars

Pressure gradients in the baryons would prevent bound systems of $\lesssim 10^5 \, M_\odot$ from collapsing, even after recombination. If there were no primordial perturbations in the baryon–photon ratio (i.e. the initial fluctuations were

Fig. 10.8. The root-mean-square fluctuations within a randomly-placed sphere containing mass M for cold dark matter, for two specific models calculated by Blumenthal and Primack (1983) (adapted from Blumenthal *et al.* (1984).

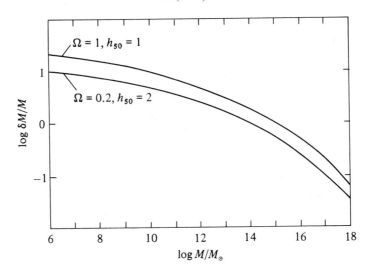

adiabatic), the first bound baryonic systems would be associated with high-σ peaks in the dark matter distribution on mass scales 10^5–10^6 M$_\odot$. These would collapse at a redshift z of order 10ν, where ν is the number of standard deviations above the mean. Everything is straightforward while all perturbations remain linear, but once the first bound baryonic systems have formed – once 'first light' occurs – what happens next could be crucially influenced by feedback from these systems. This influence depends on the mass spectrum of the first-generation stellar objects.

Our poor understanding of early star formation impedes quantitative study of all galaxy formation schemes, particularly the hierarchical ones where the first stars form at pregalactic epochs. We cannot reliably decide

Fig. 10.9. This shows three stages in the evolution, at redshifts $z = 2.5$, 1.0 and 0 respectively, of an N-body computer simulation intended to represent the contents of a cubical volume of present dimensions $14\,h^{-1}$ Mpc in a universe with $\Omega = 1$ (Frenk *et al.*, 1985). The initial fluctuations were chosen to match those expected for cold dark matter (see Fig. 10.8). When the amplitude is chosen so as to match the present-day scale of clustering there is no further freedom in the model; it is therefore gratifying that the matter is aggregated in systems whose properties match those of galactic halos. The baryonic component, which settles dissipatively in these halos to form the luminous content of galaxies, is unimportant for the overall dynamics. The present structure is the outcome of gravitational instability, in an initially almost homogeneous expanding universe, developing into the non-linear regime. In a celebrated letter to Bentley, Newton wrote 'It seems to me, that if the matter of our sun and planets and all the matter of the universe were evenly scattered throughout all heavens, and every particle had an innate gravity toward all the rest, and ... if the matter were evenly disposed throughout an infinite space, it could never convene into one mass; but some of it would convene into one mass and some into another, so as to make an infinite number of great masses, scattered at great distances from one another throughout all that infinite space. And thus might the sun and fixed stars be formed ...'

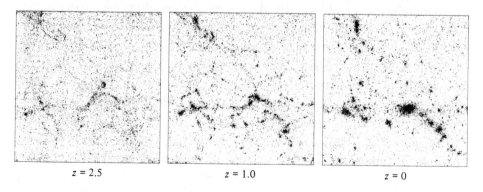

$z = 2.5$ $z = 1.0$ $z = 0$

whether a collapsing baryonic cloud of $\sim 10^5\,M_\odot$ will turn into a single very massive star, or fragment into 10^7 stars of ultralow mass. In the latter case, the first stars would not inject any significant energy into the remaining gas. On the other hand, if the first stars were ordinary massive stars or supermassive objects, their ultraviolet radiation would photoionize the medium. Indeed, only $\sim 10^{-4}$ of the initial mass need turn into such stars in order to generate enough UV radiation to photoionize everything else (Hartquist and Cameron, 1977). An additional possibility is that heavy elements generated by pregalactic supernovae may be spread through the intergalactic medium at an early stage. Details of these feedback processes are discussed by Couchman and Rees (1986). The baryonic Jeans mass would be raised from $\sim 10^5\,M_\odot$ to $\sim 10^8\,M_\odot$, with a consequent pause in the cosmogonic process until systems exceeding this larger mass undergo collapse; moreover, no more than 10^{-4} of the baryons (i.e. the $\gtrsim 3\sigma$ peaks) need condense into massive stars before reheating chokes off any further collapse of $\sim 10^5\,M_\odot$ systems.

Gas would subsequently condense into dark matter potential wells in the mass range 10^8–$10^{12}\,M_\odot$. The lower limit is set by the Jeans mass after reionization, and the upper is the maximum scale where bremsstrahlung and hydrogen and helium recombination cooling is efficient enough (see Fig. 10.10; Compton cooling would allow larger and hotter clouds to collapse at $z \gtrsim 10$). The luminosity function of galaxies depends primarily on two things.

(i) We need to know the mass distribution of isolated virialized systems of dark matter. This is a hard problem when the initial fluctuations have the spectrum shown in Fig. 10.8 because of the cross-talk between various scales, and its solution awaits N-body simulations with larger dynamic range than have so far been carried out. At the moment it is unclear whether well-defined substructure survives, or whether bound systems on scales of, say, $10^8\,M_\odot$ are rapidly engulfed in a larger system before baryons have much chance to condense within them.

(ii) The luminosity of a galaxy resulting from infall into a given potential well depends on what fraction of infalling gas is retained in each virialized clump of dark matter (likely to be larger for deeper potential wells, i.e. those of large mass and/or those which evolve from high-σ peaks) and on the kind of stars it turns into.

The above two issues will need to be settled, before we can reliably estimate the luminosity function of galaxies, even when our starting point is the specific assumption of a cold dark matter spectrum that evolved from Harrison/Zel'dovich initial fluctuations.

The following are among the observations that would help to decide among the three cosmogonic schemes illustrated in Fig. 10.6, and to test the 'cold dark matter' model in particular.

(i) The upper limits on the microwave background fluctuations on small angular scale constrain the amplitude of $\langle(\delta\rho/\rho)^2\rangle^{1/2}$ at the

Fig. 10.10. This diagram, adapted from Rees and Ostriker (1977), delineates the mass–radius relation for a self-gravitating gas cloud whose cooling and dynamical timescales are equal (assuming cooling due only to bremsstrahlung, H and He recombination, and line emission). A cloud of given mass whose radius was initially very large would deflate quasistatically (because $t_{cool} > t_{dyn}$) until it crossed the critical line; it would then collapse in free fall and could fragment into stars. This simple argument (which can readily be modified to allow for non-spherical geometry, a non-baryonic component of mass, etc.) suggests why, irrespective of the cosmological details, no galaxies form with baryonic masses $> 10^{12}\,M_\odot$ and radii $> 10^5$ pc. We would like to understand why galaxies have these observed masses and radii to the same extent that we understand the dimensions of individual stars. The order-of-magnitude considerations summarised in this diagram are probably part of the story, but to fill in the details we need to know more about the initial fluctuations, and also about the efficiency of star formation in protogalaxies.

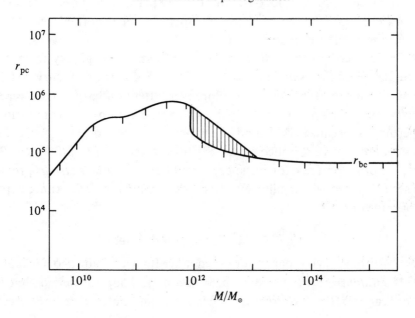

recombination epoch, and are already sensitive enough to exclude some models. Uson and Wilkinson (1984), for instance, find $\langle(\Delta T/T)^2\rangle^{1/2} <$ 2.8×10^{-5} for angles $\sim 3.5'$. A further improvement, even by a factor no more than 2, could be crucial. Measurements on larger angular scales are directly probing the metric fluctuation, and could test the consistency of a Harrison/Zel'dovich spectrum where this amplitude is scale-independent.

(ii) It would be valuable to improve data on clustering correlations (and non-Hubble velocities) on scales > 10 Mpc. Because the cold dark matter fluctuation spectrum falls off steeper than a 'white noise' spectrum at long wavelengths, the correlation function in this model should go negative for scales $> 18h^{-2}$ Mpc, unless (for some astrophysical reason) the galaxy distribution does not trace mass on these scales.

(iii) One would like to know whether there is dark matter in dwarf galaxies. The observational evidence here is uncertain, because the stellar velocities are below 10 km s^{-1} (Fig. 10.2). However, such dark matter certainly could not be light neutrinos, because they would need a higher phase space density than they had in the big bang (Tremaine and Gunn, 1979; Madsen and Epstein, 1984); *cold* non-baryonic matter would be an acceptable possibility, however.

(iv) Most crucial of all would of course be firm evidence for particles that could constitute 'cold dark matter'. Experimental hopes for axion searches, and for the detection of supersymmetric particles, are by no means fanciful (see Section 10.2).

As a final comment, note that most discussions of galaxy formation postulate initial *Gaussian* fluctuations (of the kind that might survive from the quantum or inflationary era). Moreover, attention has focussed on so-called 'adiabatic' perturbations, in which the photon–baryon ratio is unperturbed, because GUT models suggest that this ratio may be fixed by physical processes controlled by the expansion rate at very early times. But there are other options: for instance, it could be that *cosmic strings* provide the seeds around which galaxies and clusters condense (Vilenkin, Chapter 11 of this volume).

10.5 A flat ($\Omega_0 = 1$) universe?

In an influential review paper published more than a decade ago, Gott *et al.* (1974) summarized the evidence bearing on Ω. They concluded that the dynamical evidence favoured a value 0.1–0.2, and noted that if the matter were all baryonic, the lower end of this range was compatible with the value favoured by standard big-bang nucleosynthesis (for a Hubble time $t_H \approx 2 \times 10^{10}$ yr, a value consistent with the ages of globular clusters, etc.). Much new evidence has accumulated since 1974, especially on cluster dynamics and element abundances; and some relevant theoretical issues have been refined and elaborated. But, if one were to update Gott *et al.*'s discussion, their net conclusion would not change much.

There has, however, been a marked change in theorists' attitudes. This is partly because non-baryonic matter is now taken much more seriously, and seems in some ways almost a natural expectation. But the main element in the discussion is the concept of 'inflation': this is so appealing, and resolves some well-known and stubborn cosmological paradoxes in such a natural way, that it instils a strong prejudice in favour of $\Omega_0 = 1$. It is perhaps worth spelling out the basis for this prejudice.

For all the present observable universe to have evolved from a region that was in causal contact at the earliest times, inflation by a factor of at least 10^{30} is required. In most versions of inflation the exponential growth, once started, rapidly continues for many expansion timescales: it is likely to overshoot, stretching any small part of an initial chaotic hypersurface so that it becomes essentially flat over our present horizon scale. This would yield $\Omega_0 = 1$, with a precision of order 1 part in 10^4 (the expected fluctuation amplitude). For inflation to yield the dynamically preferred value $\Omega = 0.1$ or 0.2 the inflation factor would have to be 'just' 10^{30}, making the present Robertson–Walker curvature radius of the order of the Hubble radius. This would demand some coincidence. But there would then be an additional requirement that appears still more contrived: our presently observable part of the universe would have to arise from a segment of the initial hypersurface with the seemingly very special property that its curvature was uniform to a few parts in 10^5; otherwise the curvature fluctuations that could induce quadrupole effects in the microwave background would not be at least 10^4 times smaller than the overall Robertson–Walker curvature (Wilkinson, 1986). Our universe could thus not have inflated from a typical element of an initial chaotic hypersurface: if $\Omega \neq 1$, the required region would have to be special, rather as a sphere would seem specially smooth if its surface irregularities amounted to 10^{-5} of the uniform mean curvature.

10.5.1 *Inferring Ω_0 from the cosmic deceleration*

In Friedmann models with zero cosmological constant, Ω is directly proportional to the *deceleration* of the cosmological expansion.

Programmes attempting to measure this deceleration have been pursued throughout the last 20 years, but there is now not much optimism that they can succeed until we understand better how galaxies evolve.

There is a spread in the properties of galaxies, just as there is among individual stars. One needs, therefore, to find some class of galaxies which serve as 'standard candles', so that their distance can be inferred from their apparent brightness. Otherwise there will be a big scatter in the magnitude–redshift plot. The brightest galaxies in clusters are the best candidates, but even they display a 30 per cent scatter. Another problem is that galaxies become almost undetectably faint, even when observed with the largest telescopes, when they are far enough away that the effects of the deceleration should really show up.

But the worst problems stem from *galactic evolution*. The galaxies seen at large distances are systematically younger than nearby ones. Even if a certain class of galaxy provided precise standard candles at the present, one needs to know how each candle changes as it burns. The galaxies that are crucial for cosmological tests are those whose light has been journeying for at least 5 billion years towards us, which are therefore being seen at only about half their present age.

There are two aspects to the evolutionary correction. In a younger elliptical galaxy many stars would be shining which by now have died, and the present stars would be seen at an earlier stage in their evolution. This alters the brightness and the colour of the galaxy, the trend being that a younger galaxy would appear somewhat brighter than its present-day counterpart. But there is a secondary evolutionary correction stemming from the fact that a galaxy is not a self-contained isolated system. We can see many instances where galaxies seem to be colliding and merging with each other; and in rich clusters the large central galaxies may be cannibalising their smaller neighbours. (In a few billion years this may, incidentally, happen in our own local group. The Andromeda galaxy is falling towards our own Milky Way, and there may be a collision between these two large disc galaxies, the likely remnant being a bloated amorphous 'star pile' resembling an elliptical galaxy.) Many big galaxies, particularly those in the centres of clusters, may result from such mergers, traces of disturbance having by now been erased. This process would obviously result in big galaxies, having been, on average, fainter in the past. There are thus two evolutionary effects, both uncertain (and with opposite signs), either of which could be large enough to mask the geometrical difference between a universe with $\Omega_0 = 1$ and one with $\Omega_0 \approx 0.2$.

A new cosmological test, involving the *number* of galaxies of a given type as a function of redshift, has recently been carried out (Loh and Spillar, 1986). This is still somewhat vulnerable to evolutionary effects but obviously utilises more information than simple number counts of galaxies. It yields results consistent with Ω_0 as high as unity.

Note that the cosmic deceleration $\ddot{R}$ is related to Ω via $-R\ddot{R}/\dot{R}^2=\Omega/2$ only if the Friedmann equations apply, if the cosmical constant is zero, and if the dominant form of mass-energy has an associated pressure $\ll \frac{1}{3}\rho c^2$. Ideally, therefore, one would like to determine Ω and $\ddot{R}$ independently, as a test of these assumptions.

10.5.2 'Biased' galaxy formation

The dynamical evidence from clusters and from galactic halos (surveyed in Section 10.2) does not offer evidence for any value of Ω_0 higher than ~ 0.2. If Ω_0 is indeed unity, then 80 per cent of the mass is unaccounted for even by dynamical considerations: it is not just the light, but evidence for gravitating matter itself, which is 'missing'. This raises the interesting question of whether the dynamical evidence is nonetheless compatible with $\Omega_0 = 1$.

If $\Omega_0 = 1$, then the dominant mass must not participate fully in the observed clustering: galaxies must be more 'clumped' than mass in general, so that their spatial distribution enhances and exaggerates the inhomogeneity on large scales. This requires some kind of 'bias' in the formation of galaxies. There are three ways this would come about (see Dekel and Rees, 1987, for fuller discussion):

(i) The entire universe may be pervaded by a uniform component of 'missing mass' of a different nature from the clustered dark matter, contributing $\Omega \sim 0.8$.

(ii) There may be one important kind of dark matter, but the baryonic component may be segregated from it even on scales $\sim 30h^{-1}$ Mpc, so that galaxy formation occurs only in certain regions. This could result from large-scale fluctuations in the initial baryon/photon ratio, from gas dissipation within superclusters of collisionless dark matter (e.g. neutrinos), or from very energetic winds or blast waves pushing the gas over large distances.

(iii) Less extravagant in energy is the possibility that the large-scale baryon distribution does trace dark matter on scales $\gtrsim 1h^{-1}$ Mpc, but the efficiency with which baryons turn into luminous galaxies is modulated by large-scale environmental effects.

The universe may be dynamically dominated by 'ultrahot' weakly interacting particles which do not cluster because they have velocities 10^3 km s^{-1} (alternative (i) above). But if the 'ultrahot' particles are relics of an early epoch, their mass would have always been dynamically dominant over the baryons, and would have inhibited gravitational clustering altogether. This would also yield an unacceptably fast expansion timescale during nucleosynthesis. A way round this difficulty – though maybe a rather contrived one – involves supposing that these particles arise from *non-radiative decay of heavy particles* with lifetimes only slightly shorter than the age of the universe.

The universe could be flat with $\Omega_0 < 1$ if a non-zero *cosmological constant* contributed to the curvature such that $\Omega_0 + \frac{1}{3}\Lambda t_H^2 = 1$. In some respects this idea resembles 'ultrahot' particles, but it has the advantage that the Λ-term is unimportant at early epochs so it would not have had such a serious inhibiting effect on galaxy formation. However, for the Λ-contribution today to be comparable with the ordinary matter, the required fine-tuning is as *ad hoc* as the one we intended to avoid by adopting $\Omega_0 = 1$.

A wide class of possibilities in category (iii) involve the idea that galaxies formed only from *exceptionally high peaks* of the initial (fluctuating) density distribution. If the local distribution function of $\delta\rho$ has a steeply decreasing tail, like a Gaussian, high peaks occur with enhanced probability in the crests rather than the troughs of a large-scale fluctuation mode, and hence display enhanced clustering (if the distribution function has a flat tail the effect is weaker and it might even be reversed). See Bardeen *et al.* (1986) for a full discussion of the theory.

The high-amplitude peaks would collapse *earlier*, and have higher density at turnaround, than more typical fluctuations on a given mass scale. This could, in principle, by itself account for the biasing. For instance, collapse and fragmentation of gas within protogalaxies is highly sensitive to the ratio of the cooling time ($\propto \rho^{-1}$ for bremsstrahlung and recombination cooling) to the collapse time ($\propto \rho^{-1/2}$): it occurs more rapidly when this ratio is small. Moreover, if the high-amplitude perturbations collapsed at $z \gtrsim 10$, Compton scattering of the microwave background would guarantee efficient cooling; this process, whose efficiency is proportional to $(1+z)^4$, would be unimportant for lower-amplitude systems collapsing at late times. Protogalaxies that formed from high-σ fluctuations would have higher velocity dispersions (and escape velocities) than later-forming systems of similar mass. This makes it easier for them to retain gas that might otherwise be expelled by, for instance, supernova-driven winds (Dekel and Silk, 1986).

The formation of galaxies from lower amplitude perturbations may be further impeded owing to some kind of negative feedback induced by the first galaxies, suppressing the baryonic infall or affecting the stellar initial mass function (Rees, 1985; Silk, 1985).

10.5.3 Biasing in the cold dark matter (CDM) cosmogony

The standard CDM cosmogonic scheme ($\Omega_0 = 1$ and scale-invariant initial curvature fluctuations) cannot be reconciled with observations without some form of biasing. Detailed simulations of such models, hypothesizing that galaxies form only from regions which are $>2.5\sigma$ peaks of the initial mass distribution (smoothed on a galactic scale), show a good fit with the galaxy–galaxy correlation function, the M/L ratio in clusters and the Virgo infall (Davis *et al.*, 1985). Although this is encouraging, we still need a genuine physical justification for the biasing.

The CDM cosmogony will not be able to predict the luminosity function of galaxies until we develop a physical understanding of several distinct phases of galaxy formation:

(i) Baryons are presumed to condense in virialized 'halos' of dark matter in the approximate mass range 10^8–$10^{12} \, M_\odot$. For larger masses, dissipative cooling may be inefficient; below $\sim 10^8 \, M_\odot$ the potential wells may be too shallow to capture primordial gas (especially if this gas has been ionized and heated to $\sim 10^4 \, K$). The mass distribution of isolated virialized systems can in principle be learned from N-body simulations. In practice, progress is impeded by the large dynamic range involved; galactic-mass halos build up hierarchically, via a complicated series of mergers, between $z \approx 5$ and the present epoch (Frenk *et al.*, 1985).

(ii) Even if the dissipationless clustering of the dark matter were accurately known, the fate of the baryonic component – how much gas falls into each potential well, and how much is retained – involves complex gas dynamics.

(iii) The resultant luminosity and brightness profile of the galaxy depends on the IMF of the stars that form from the gas, and on the efficiency of star formation.

The biasing could in principle involve any (or all) of these three aspects of galaxy formation. The N-body simulations suggest that even stage (i) may lead to substantial biasing: the dark halos associated with bright galaxies – those with circular velocities $v_c > 200 \, \mathrm{km \, s^{-1}}$ – are themselves more clustered than the overall mass distribution, essentially because the binding energy of galactic mass perturbations is significantly affected by whether they are in a peak or a trough of the fluctuation spectrum at larger wavelengths. (The growth of the perturbation on galactic scales is boosted or suppressed in surroundings whose overall expansion mimics an $\Omega > 1$ or an $\Omega < 1$ universe respectively.) This effect alone might account for the requisite biasing of bright galaxies, although it would not explain the tendency of galaxies of low luminosity to cluster like brighter galaxies. Nevertheless, any of the other mechanisms involving autonomous biasing or feedback could also be operative, modifying stages (ii) and/or (iii) of the formation process.

Very-large-scale structures. There are indications that the *very-large-scale* superclusters and voids may call for a formation (and bias) mechanism over and above the type of biasing process that could plausibly account for the galaxy correlations on scales $\sim 10h^{-1}$ Mpc. One crucial observation in addition to the presence of big 'voids' is the enhanced superclustering of rich Abell clusters over the galaxies (and the matter) characterized by a correlation function which is positive at least out to $100h^{-1}$ Mpc (Bahcall and Soneira, 1983; Bahcall, 1986), and by associated velocities of ~ 1000 km s^{-1}. Another puzzling finding of potentially great importance is the bulk motion of a region $\sim 100h^{-1}$ Mpc in diameter around us relative to the microwave background frame of reference with a velocity of ~ 600 km s^{-1} (Burstein *et al.*, 1987). This superstructure, on such large scales, is hard to reproduce in any of the standard scenarios that assume initial fluctuations with a scale-invariant spectrum and random-phases; it is difficult also to reconcile with the upper limits on the microwave background fluctuations on arc minute scales.

If the density fluctuations began as quantum fluctuations of a free scalar field during the era of inflation they are indeed expected to be Gaussian, but it is also possible that the fluctuations arose from a different mechanism, in which they would not in general be Gaussian, and have *non-random phases*. If the initial fluctuations had non-random phases, then the universe might be inherently smoother in some places than in others: galactic-scale fluctuations with amplitudes large enough to form bound systems could then abound in some regions, but be entirely absent in large 'calm' patches, the latter corresponding to 'voids'.

Biasing is not just an *ad hoc* idea introduced by theorists to save the philosophically attractive $\Omega_0 = 1$ model when confronted with apparently conflicting evidence. A bias is essential in order to understand even the gross features of the large-scale structure, in particular the big 'voids' and the superclustering of clusters, and to reconcile any of the cosmogonic scenarios under current discussion with the observed universe. Therefore, the search for an appropriate physical bias mechanism, and especially for confirming observational evidence, are of great importance.

On the theoretical side, the bias mechanism is intimately related to the cosmogonic scenario and the nature of the dark matter. Although some of the proposed bias mechanisms may seem somewhat contrived, others are very plausible physically: some or all of them must have affected galaxy formation to some degree and they should be worked out in more detail. It is evident that the notion that galaxies trace mass is an unjustified assumption.

Answers to the following observational questions would help to distinguish between various options outlined in this section.

(i) How much diffuse gas is there in the voids?

(ii) Are voids empty of all types of galaxy, or only those types that are most conspicuous? Any evidence that galaxies of different morphological types display unequal degrees of clustering is relevant here.

(iii) Are there any galactic-mass dark halos with no luminous galaxy within them? Such objects might be expected in the cold dark matter model (assuming fluctuations with random phases) if biasing is indeed important; and could, if their core radius were small enough, account for gravitational lensing of quasars even when no lens is visible.

(iv) Have the large-scale inhomogeneities in the galaxy distribution (superclusters, etc.) given rise to substantial deviations from the Hubble flow on corresponding scales?

I shall conclude by summarising two candidate cosmologies, both of which now seem appealing in their different ways:

(i) The first model has $\Omega_b = \Omega_{total} \approx 0.15$ (and a Hubble time $t_H = 2 \times 10^{10}$ yr). The baryons would be partly in 'luminous' form (stars and gas) and partly in 'dark matter' (low mass stars and/or black holes).

(ii) Alternatively, theoretical predilection may dispose us to favour $\Omega_{total} = 1 \pm 10^{-4}$. (The small uncertainty would arise only from the initial curvature fluctuations, which need to be 10^{-4}–10^{-5} on the scale of galaxies and clusters, and would, in the simplest models, extend with similar amplitude up to the Hubble radius and even beyond.) Conventional big bang nucleosynthesis then suggests $\Omega_b \approx 0.1$, $\Omega_{non-baryon} \approx 0.9$, and introduces a new dimensionless ratio into cosmology. The large-scale structure of the universe then involves some kind of biasing in the formation of bright galaxies.

It may become easier to decide between the options when we have a better knowledge of galactic evolution and star formation. It is still unclear whether the luminous parts of galaxies result from infall into a preformed halo composed of quite different stuff, or whether, contrariwise, galaxies and their halos resulted from a single dissipative collapse process whereby the IMF gradually changed as contraction proceeded. Observations of quasars back to redshifts $z \approx 4$ can now provide direct evidence of what the universe was like at an era when galaxy formation was still going on. Numerical simulations which can follow not only gravitational clustering (cf. Fig. 10.9) but also the complexities of gas dynamics – shock waves, radiative cooling and fragmentation – will soon become feasible.

The problems of large-scale cosmogony are so intermeshed that we will not really solve any until the whole picture comes into better focus. For instance, we cannot test theories of galaxy formation and evolution until we understand star formation (and the possible role of active galactic nuclei) as well as the initial fluctuations. The relevant *physics* is not specially recondite – indeed most of the physics that astrophysicists need is in 'Landau and Lifshitz' – and the most intractable phenomena are complex manifestations of Newtonian gravity and dissipative gas dynamics. But when we ask where the initial fluctuations come from or what the dark halos are made of, we realise that even the most ordinary galaxies pose questions that may transcend the physics we understand.

References

Alcock, C. and Fahri, H. (1985). *Phys. Rev.*, **D32**, 1273.

Applegate, J. and Hogan, C. J. (1985). *Phys. Rev.*, **D31**, 3037.

Applegate, J., Hogan, C. and Scherrer, R. (1986). Preprint.

Bahcall, N. A. (1986). In *Galactic Distances and Deviations from Universal Expansion*, ed. B. F. Madore and R. B. Tully. Reidel: Dordrecht.

Bahcall, N. A. and Soneira, R. M. (1983). *Astrophys. J.*, **270**, 20.

Bardeen, J., Bond, J. R., Kaiser, N. and Szalay, A. (1986). *Astrophys. J.*, **304**, 15.

Blandford, R. D. and Jaroszynski, M. (1981). *Astrophys. J.*, **246**, 1.

Blumenthal, G., Faber, S. M., Primack, J. R. and Rees, M. J. (1984). *Nature*, **311**, 517.

Blumenthal, G. and Primack, J. R. (1983). In *Fourth Workshop on Grand Unification*, ed. H. A. Wilson *et al.*, p. 256.

Bond, J. R., Carr, B. J. and Hogan, C. (1986). *Astrophys. J.*, **306**, 428.

Bond, J. R. and Efstathiou, G. (1986). *Mon. Not. R. Astron. Soc.*, **218**, 103.

Bond, J. R. and Efstathiou, G. (1987). *Rev. Mod. Phys.* (in preparation).

Burstein, D., Davis, R. L., Dressler, A., Faber, S. M., Lynden-Bell, D., Terlevich, R. J. and Wegner, G. (1987). *Astrophys. J.* (in press).

Canizares, C. R. (1982). *Astrophys. J.*, **263**, 50.

Carr, B. J., Bond, J. R. and Arnett, W. D. (1984). *Astrophys. J.*, **277**, 445.

Chang, K. and Refsdal, S. (1979). *Nature*, **282**, 561.

Couchman, H. M. P. and Rees, M. J. (1986). *Mon. Not. R. Astron. Soc.*, **221**, 53.

Cowsik, R. and McLelland, J. (1973). *Astrophys. J.*, **180**, 7.

Davis, M., Efstathiou, G., Frenk, C. S. and White, S. D. M. (1985). *Astrophys. J.*, **292**, 371.

Dekel, A., Einasto, J. and Rees, M. J. (1987). *Rev. Mod. Phys.* (in press).

Dekel, A. and Rees, M. J. (1987). *Nature* (submitted).

Dekel, A. and Silk, J. I. (1986). *Astrophys. J.*, **303**, 39.

De Rejula, A. and Glashow, S. L. (1984). *Nature*, **312**, 734.

Drukier, A. K., Freese, K. and Spergel, D. N. (1986). *Phys. Rev.*, **D33**, 3495.

Dyer, C. (1976). *Mon. Not. R. Astron. Soc.*, **175**, 429.

Efstathiou, G., Schneider, P. and Wagoner, R. V. (1986). Preprint.

Efstathiou, G. and Silk, J. I. (1983). *Fundamentals Cosmic Phys.*, **9**, 1.

Einasto, J., Kaasik, A. and Saar, E. (1974). *Nature*, **250**, 309.

Einstein, A. (1917). *Sitz. Preuss. Acad. Wiss.*, 142.

Faber, S. M. (1984). *Proceedings First ESO/CERN Conference*, ed. G. Setti and L. van Hove.

Fall, S. M. and Rees, M. J. (1985). *Astrophys. J.*, **298**, 18.

Frenk, C. S., White, S. D. M., Efstathiou, G. and Davis, M. (1985). *Nature*, **317**, 595.

Friedman, A. (1922). *Z. Phys.*, **10**, 377.

Goodman, M. W. and Witten, E. (1985). *Phys. Rev.*, **D31**, 3059.

Gott, J. R. (1981). *Astrophys. J.*, **243**, 140.

Gott, J. R., Gunn, J. E., Schramm, D. N. and Tinsley, B. M. (1974). *Astrophys. J.*, **194**, 543.

Grishchuk, L. and Zel'dovich, Y. B. (1978). *Sov. Astron.*, **22**, 125.

Gunn, J. E., Lee, B. W., Lerche, I., Schramm, D. N. and Steigman, G. (1978). *Astrophys. J.*, **223**, 1015.

Guyot, M. and Zel'dovich, Y. B. (1970). *Astron. Astrophys.*, **9**, 227.

Harrison, E. R. (1970). *Phys. Rev.*, **D1**, 2726.

Hartquist, T. W. and Cameron, A. G. W. (1977). *Astrophys. Sp. Sci.*, **48**, 105.

Hawking, S. W. and Ellis, G. F. R. (1968). *Astrophys. J.*, **152**, 25.

Hodge, P. W. (1981). *Ann. Rev. Astr. Astrophys.*, **19**, 357.

Hogan, C. (1986). In *Inner Space/Outer Space*, ed. E. W. Kolb *et al.*, p. 279. Chicago University Press: Chicago.

Hogan, C., Kaiser, N. and Rees, M. J. (1982). *Phil. Trans. Roy. Soc.*, **A307**, 97.

Hoyle, F. and Tayler, R. J. (1964). *Nature*, **203**, 1108.

Hubble, E. (1929). *Proc. Nat. Acad. Sci.*, **15**, 168.

Kaiser, N. (1982). *Mon. Not. R. Astron. Soc.*, **198**, 1033.

Kaiser, N. and Silk, J. I. (1986). *Nature*, **324**, 529.

Knapp, J. and Kormendy, J. (1986). *Dark Matter in the Universe, Proceedings IAU Symposium 117*. Reidel: Dordrecht.

Kolb, E. W. and Wolfram, S. (1980). *Nucl. Phys.*, **B172**, 224.

Loh, E. D. and Spillar, E. J. (1986). *Astrophys. J. Lett.*, **307**, L1.

Lyubimov, V. A., Novikov, E. G., Nozik, V. Z., Tretyakov, E. F. and Kozek, V. S. (1980). *Phys. Lett.*, **394**, 266.

McCrea, W. H. and Milne, E. A. (1934). *Quart. J. Math. (Oxford)*, **5**, 64.

Madsen, J. and Epstein, R. I. (1984). *Astrophys. J.*, **282**, 11.

Marx, G. and Szalay, A. (1972). *Proc. Neutrino 82, Technoinform, Budapest*, p. 191.

Meszaros, P. (1974). *Astron. Astrophys.*, **37**, 225.

Ostriker, J. P., Peebles, P. J. E. and Yahil, A. (1974). *Astrophys. J. Lett.*, **193**, L1.

Paczynski, B. (1986*a*). *Astrophys. J.*, **301**, 503.

Paczynski, B. (1986*b*). *Astrophys. J.*, **304**, 1.

Partridge, R. B. (1986). *Proceedings IAU Symposium on Cosmology* (in press).

Peebles, P. J. E. (1966). *Astrophys. J.*, **146**, 542.

Peebles, P. J. E. (1981). *Astrophys. J.*, **248**, 885.

Peebles, P. J. E. (1982). *Astrophys. J.*, **258**, 415.

Penzias, A. A. and Wilson, R. W. (1965). *Astrophys. J.*, **142**, 419.

Press, W. H. and Gunn, J. E. (1973). *Astrophys. J.*, **185**, 397.

Press, W. H. and Vishniac, E. (1980). *Astrophys. J.*, **236**, 323.

Rees, M. J. (1983). In *The Very Early Universe*, ed. G. W. Gibbons *et al.*, p. 29. Cambridge University Press: Cambridge.

Rees, M. J. (1985). *Mon. Not. R. Astron. Soc.*, **213**, 75P.

Rees, M. J. (1986). *Proceedings 2nd ESO/CERN Conference*, ed. G. Setti and L. van Hove.

Rees, M. J. and Ostriker, J. P. (1977). *Mon. Not. R. Astron. Soc.*, **179**, 541.

Rees, M. J. and Sciama, D. W. (1968). *Nature*, **217**, 511.

Refsdal, S. (1964). *Mon. Not. R. Astron. Soc.*, **128**, 295.

Refsdal, S. (1970). *Astrophys. J.*, **159**, 357.

Rowan-Robinson, M. (1985). *The Cosmic Distance Scale*. Freeman: New York.

Rubin, V. (1986). In *Highlights of Astronomy*, ed. P. Swings. Reidel: Dordrecht.

Ryle, M. (1958). *Proc. Roy. Soc.*, **A248**, 289.

Sachs, R. K. and Wolfe, A. M. (1967). *Astrophys. J.*, **147**, 73.

Sikivie, P. (1985). *Phys. Rev.*, **D32**, 2988.

Silk, J. I. (1982). In *Astrophysical Cosmology*, ed. H. A. Bruck *et al.*, p. 427. Vatican Publications.

Silk, J. I. (1985). *Astrophys. J.*, **297**, 1.

Silk, J. I., Olive, K. I. and Srednicki, M. (1985). *Phys. Rev. Lett.*, **55**, 257.

Smith, P. F. (1986). *Proceedings 2nd ESO/CERN Conference*, ed. G. Setti and L. van Hove.

Spergel, D. N. and Press, W. H. (1985). *Astrophys. J.*, **294**, 663.

Steigman, S., Sarazin, C. L., Quintana, H. and Faulkner, J. (1978). *Astron. J.*, **83**, 1050.

Tremaine, S. and Gunn, J. E. (1979). *Phys. Rev. Lett.*, **42**, 407.

Uson, J. and Wilkinson, D. T. (1984). *Nature*, **312**, 427.

Wagoner, R. V., Fowler, W. A. and Hoyle, F. (1967). *Astrophys. J.*, **148**, 3.

Wilkinson, D. (1986). *Phil. Trans. R. Soc.*, **A320**, 595.

Witten, E. (1984). *Phys. Rev.*, **D30**, 272.

Yang, J., Turner, M. S., Steigman, G., Schramm, D. N. and Olive, K. A. (1984). *Astrophys. J.*, **281**, 493.

Young, P. J. (1981). *Astrophys. J.*, **244**, 756.

Zel'dovich, Y. B. (1972). *Mon. Not. R. Astron. Soc.*, **160**, 1P.

Zwicky, F. (1933). *Helv. Phys. Acta*, **6**, 110.

11

Gravitational interactions of cosmic strings

ALEXANDER VILENKIN

11.1 Introduction

The origin of structure in the universe is one of the great cosmological mysteries. Newton thought that it could not be explained by natural causes and attributed it to God. Now, 300 years after the publication of *Principia*, the problem is still unresolved, and possible ways to approach it have started to emerge only in the last several years. One of the possibilities is related to cosmic strings, which could arise as a random network of line-like defects at a phase transition in the early Universe. In this scenario, massive closed loops of string serve as seeds for the formation of galaxies and clusters of galaxies. While matter is being accreted onto the loops, they oscillate violently, lose their energy by gravitational radiation, shrink and disappear.

Apart from their possible role in galaxy formation, strings are fascinating objects in their own right. The physical properties of strings are very different from those of more familiar systems and can give rise to a rich variety of unusual physical phenomena. In particular, if strings exist, they can produce a number of characteristic observational effects detectable with existing astronomical instruments.

The physical properties, evolution and cosmological consequences of strings have been studied extensively for several years, but the subject is still rapidly expanding. For an up-to-date guide to the literature the reader is referred to Vilenkin (1985) and Preskill (1985). A review of all aspects of cosmic strings would require writing a book, and here I have chosen just one aspect which I think is the most appropriate one for a volume dedicated to Newton's *Principia*.

This article reviews the gravitational properties of cosmic strings. Some of these properties have been used to analyse various cosmological effects of strings, but here my emphasis will be on the basic physics. Cosmological and astrophysical applications will be mentioned only occasionally. The article can be roughly divided into three parts. The first part, including Sections 11.2–11.6, reviews the dynamics of strings in flat spacetime. After a brief summary of the relevant physical properties of strings, I derive the string equations of motion and discuss their solutions describing oscillating closed loops and waves on infinite strings. In the second part (Sections 11.7–11.10) I give some exact solutions of Einstein's equations for static straight strings and analyze their physical properties. In the third part (Sections 11.11–11.14) the gravitational field of oscillating strings is studied in the linear perturbation theory. The string configurations found in the first part of the paper are used as sources of the linearized Einstein equations. The gravitational field produced by oscillating closed loops and the gravitational radiation from loops are discussed in detail.

Throughout the paper I use the system of units in which $\hbar = c = 1$. The metric signature is $(+, -, -, -)$. Greek indices take values from 0 to 3 and Latin indices take values 1 to 3, or, where it is specifically indicated, only the values 1 and 2. Notation (x^0, x^1, x^2, x^3) and (t, x, y, z) for spacetime coordinates is used interchangeably.

11.2 Some properties of strings

Strings arise in gauge theories with spontaneously broken symmetries. If the symmetry groups satisfy certain topological conditions, the field equations of the theory have stable string-like solutions with a unit flux of a massive gauge field running along each string. A close condensed-matter analogue of strings is a quantized tube of magnetic flux in superconductors.

Another type of string, called a global string, is similar to a vortex line in liquid helium. It arises as a result of a global symmetry breaking and does not carry a gauge magnetic flux. Such strings have long-range interactions due to their coupling to a massless Goldstone boson. This coupling is much stronger than their coupling to gravity, and so gravitational interactions do not play a significant role in the dynamics of global strings (Davis, 1985). In this paper we shall concentrate on gauge-symmetry strings, for which gravity is the dominant long-range interaction.

In some gauge models, strings do not have ends: they are either infinite or closed. This is the most interesting type of string for cosmological applications. In other models strings can end at magnetic monopoles. In this case the magnetic flux of the monopoles is confined in the string.

Some properties of strings can be deduced just from dimensionality and Lorentz invariance arguments. Strings are typically characterized by a single dimensional parameter η – the energy scale of symmetry breaking. Then, from dimensionality, the thickness of the string is $\delta \sim \eta^{-1}$ and the mass per unit length of string is

$$\mu \sim \eta^2. \tag{2.1}$$

For a grand unification scale $\eta \sim 10^{16}$ GeV we get $\delta \sim 10^{-30}$ cm and $\mu \sim 10^{22}$ g/cm. Strings of cosmological interest have sizes much greater than their thickness and can be treated as one-dimensional objects.

Consider now a static straight string lying along the z-axis. It is described by a solution of Lorentz-invariant field equations which is independent of z and t, and should therefore be invariant under Lorentz boosts along the z-axis. Hence the rest frame of the string is defined only up to longitudinal Lorentz boosts, and only transverse motion of the string has physical meaning. The physical characteristics of the string, such as the energy–momentum tensor T^ν_μ, should have the same invariance. This implies that

$$T^0_0 = T^3_3 \tag{2.2}$$

with all other components equal to zero, except perhaps T^i_j with $i, j = 1, 2$. Using the conservation law $T^i_{j,i} = 0$ (in flat spacetime) it can be shown that the latter components vanish when averaged over the cross-section of the string

$$0 = \int T^j_{i,j} x^k \, d^2 x = - \int T^j_i \, \partial_j x^k \, d^2 x$$

$$= - \int T^k_i \, d^2 x \quad (i, j, k = 1, 2). \tag{2.3}$$

Then, neglecting the width of the string, we can write

$$T^\nu_\mu = \mu \, \delta(x) \, \delta(y) \, \text{diag}(1, 0, 0, 1). \tag{2.4}$$

We see that the tension along the string is equal to the linear mass density.

11.3 Nambu action

A curved string oscillates under the influence of tension. Its world history corresponds to a two-dimensional surface in spacetime

$$x^\mu = x^\mu(\zeta^0, \zeta^1), \tag{3.1}$$

where ζ^0 and ζ^1 are, respectively, a time-like and a space-like parameter on this surface. (Here again we neglect the string thickness, δ, assuming that it is much smaller than the curvature radius of the string.) The action functional for a string should satisfy the following requirements:

(i) It should be invariant under general coordinate transformations,

$$x^\mu \to x^{\mu'}(x^\nu) \tag{3.2}$$

and under arbitrary reparametrization of the surface,

$$\zeta^a \to \zeta^{a'}(\zeta^b). \tag{3.3}$$

(ii) It should be local, since there is no long-range interaction between different parts of the string.† The general form of the action is then

$$S = \int \mathscr{L}(-\gamma)^{1/2} \, d^2\zeta, \tag{3.4}$$

where the Lagrangian $\mathscr{L}$ is invariant under (3.2) and (3.3), γ is the determinant of the metric tensor of the surface,

$$\gamma_{ab} = g_{\mu\nu} \frac{\partial x^\mu}{\partial \zeta^a} \frac{\partial x^\nu}{\partial \zeta^b}, \tag{3.5}$$

$$\gamma = \dot{x}^2 x'^2 - (\dot{x} \cdot x')^2 \tag{3.6}$$

and $g_{\mu\nu}$ is the four-dimensional metric. Here, dots and primes stand for derivatives with respect to ζ^0 and ζ^1, respectively, $\dot{x}^2 = \dot{x}^\mu \dot{x}_\mu$, etc.

The 'building blocks' for the Lagrangian $\mathscr{L}$ are the string tension μ and the geometric quantities, such as the intrinsic and extrinsic curvature of the surface (3.1) and their covariant derivatives. Note that the 4-velocity u^μ is not among the building blocks; the reason is that the local rest frame of the string, and therefore the 4-velocity, is defined only up to longitudinal Lorentz boosts.

The dimension of $\mathscr{L}$ is mass squared, and we can write

$$\mathscr{L} = -\mu + \alpha K + \beta K^2/\mu + \cdots, \tag{3.7}$$

where K stands for the curvature (with indices suppressed) and α, β are numerical coefficients. For a string with a typical curvature radius R, we have $K \sim R^{-2} \ll \mu$ (since the string thickness is $\delta \sim \mu^{-1/2}$ and we assumed that $R \gg \delta$). Hence, the curvature-dependent terms in (3.7) can be neglected and we obtain

$$S = -\mu \int (-\gamma)^{1/2} \, d^2\zeta. \tag{3.8}$$

This is the Nambu action for a string (see, for example, Scherk, 1975). Up to an overall factor, it is given by the surface area traversed by the string in

† Note that the second requirement is not satisfied for global strings. The long-range force in this case is due to the interaction of strings with Goldstone bosons; the corresponding action functional is derived in Vilenkin and Vachaspati (1986).

spacetime. Note that (3.8) is similar to the action for a relativistic particle,

$$S = -m \int ds, \qquad (3.9)$$

which is proportional to the length of the particle's world line.

Varying the action (3.8) with respect to $x_\mu(\zeta^a)$ we obtain the equations of motion for a string:

$$\frac{\partial}{\partial \zeta^0} \{(-\gamma)^{-1/2} [(\dot{x} \cdot x') x'^\mu - x'^2 \dot{x}^\mu]\}$$

$$+ \frac{\partial}{\partial \zeta^1} \{(-\gamma)^{-1/2} [(\dot{x} \cdot x') \dot{x}^\mu - \dot{x}^2 x'^\mu]\} = 0. \quad (3.10)$$

The energy–momentum tensor can be found by varying S with respect to $g_{\mu\nu}$ (Turok, 1984; Vachaspati, 1986):

$$T^{\mu\nu} = \mu \int d^2\zeta (-\gamma)^{-1/2} \delta^{(4)}(x^\mu - x^\mu(\zeta^a))$$

$$\times \{x'^2 \dot{x}^\mu \dot{x}^\nu + \dot{x}^2 x'^\mu x'^\nu - (\dot{x} \cdot x')(\dot{x}^\mu x'^\nu + x'^\mu \dot{x}^\nu)\}. \quad (3.11)$$

It is easily verified that for a straight string in flat spacetime lying along the z-axis, $t = \zeta^0$, $z = \zeta^1$, $x = y = 0$, and eq. (3.11) reduced to eq. (2.3).

11.4 Gauge conditions

Since the Nambu action is invariant under transformations (3.3), we are free to choose any parametrization of the world surface. This can be done explicitly, for example,

$$\zeta^0 = t, \quad \zeta^1 = z, \qquad (4.1)$$

or by imposing two gauge conditions. A very convenient choice of gauge for a string in flat spacetime is (Scherk, 1975)

$$\dot{x} \cdot x' = 0, \quad \dot{x}^2 + x'^2 = 0. \qquad (4.2)$$

The string equation of motion in this gauge is simply the wave equation,

$$\ddot{x}^\mu - x''^\mu = 0. \qquad (4.3)$$

We note further that $t \equiv x^0 = \zeta^0$ solves eq. (4.3). With this choice the string trajectory can be written as

$$\tilde{x}(\zeta, t), \qquad (4.4)$$

where $\zeta \equiv \zeta^1$ is the space-like parameter on the string, and eqs. (4.2), (4.3) take the form

$$\dot{\tilde{x}} \cdot \tilde{x}' = 0; \quad \dot{\tilde{x}}^2 + \tilde{x}'^2 = 1; \qquad (4.5)$$

$$\ddot{\tilde{x}} - \tilde{x}'' = 0. \qquad (4.6)$$

The energy–momentum tensor is also greatly simplified in this gauge:

$$T^{\mu\nu}(\vec{r}, t) = \mu \int d\zeta (\dot{x}^\mu \dot{x}^\nu - x'^\mu x'^\nu) \, \delta^{(3)}(\vec{r} - \vec{x}(\zeta, t)).$$ (4.7)

The total energy of the string is given by

$$E = \int T^0_0 \, d^3x = \mu \int d\zeta.$$ (4.8)

In the following two sections we shall find some solutions of the string equations of motion in flat spacetime.

11.5 Oscillating loops

The general solution of eq. (4.6) is

$$\vec{x}(\zeta, t) = \tfrac{1}{2}[\vec{a}(\zeta - t) + \vec{b}(\zeta + t)],$$ (5.1)

and eqs. (4.5) give the following constraints for the otherwise arbitrary functions $\vec{a}$ and $\vec{b}$ (Kibble and Turok, 1982):

$$\vec{a}'^2 = \vec{b}'^2 = 1.$$ (5.2)

The motion of a closed loop of string in its center-of-mass frame is described by a solution of the form (5.1) where $\vec{a}(\zeta)$ and $\vec{b}(\zeta)$ are periodic functions,

$$\vec{a}(\zeta + L) = \vec{a}(\zeta); \quad \vec{b}(\zeta + L) = \vec{b}(\zeta).$$ (5.3)

From eq. (4.8) we find

$$L = M/\mu,$$ (5.4)

where M is the mass of the loop. It is clear from eq. (5.1) that the motion of the loop must also be periodic in time. The period is $T = L/2$, rather than $T = L$, since

$$\vec{x}(\zeta + L/2, t + L/2) = \vec{x}(\zeta, t).$$ (5.5)

In fact, the period can be smaller than $L/2$ for some special loop trajectories.

An interesting property of the loop solutions is that the string typically reaches the velocity of light at some points at certain moments during its period (Turok, 1984). From eq. (5.1) we have

$$\dot{\vec{x}}^2(\zeta, t) = \tfrac{1}{4}[\vec{a}'(\zeta - t) - \vec{b}'(\zeta + t)]^2.$$ (5.6)

Now, it follows from (5.2) and (5.3) that the vector functions $\vec{a}'(\zeta)$ and $-\vec{b}'(\zeta)$ describe closed curves on a unit sphere as ζ runs from 0 to L. These functions should satisfy

$$\int_0^L \vec{a}' \, d\zeta = \int_0^L \vec{b}' \, d\zeta = 0$$ (5.7)

and are otherwise arbitrary. If the two curves intersect, $\vec{a}'(\zeta_a) = -\vec{b}'(\zeta_b)$, then eq. (5.6) gives $\dot{x}^2(\zeta, t) = 1$ for $2\zeta = \zeta_a + \zeta_b + nL$, $2t = \zeta_b - \zeta_a + nL$, where n is an integer. It is possible to construct loops which never reach the velocity of light, but that requires a somewhat contrived choice of the functions $\vec{a}(\zeta)$ and $\vec{b}(\zeta)$.

To study the behavior of the string near the point of luminal motion, it is convenient to choose the origin of spacetime coordinates and the parametrization of the string so that the luminal point is at $\zeta = t = 0$ and $\dot{x} = 0$. Expanding the functions $\vec{a}$ and $\vec{b}$ near $\zeta = 0$,

$$\vec{a}(\zeta) = \vec{a}_0'\zeta + \tfrac{1}{2}\vec{a}_0''\zeta^2 + \tfrac{1}{6}\vec{a}_0^{(3)}\zeta^3 + \cdots,$$
$$\vec{b}(\zeta) = \vec{b}_0'\zeta + \tfrac{1}{2}\vec{b}_0''\zeta^2 + \tfrac{1}{6}\vec{b}_0^{(3)}\zeta^3 + \cdots, \tag{5.8}$$

and using eq. (5.2) we obtain

$$\vec{a}_0' = -\vec{b}_0'$$
$$|\vec{a}_0'| = |\vec{b}_0'| = 1,$$
$$\vec{a}_0' \cdot \vec{a}_0'' = \vec{b}_0' \cdot \vec{b}_0'' = 0 \tag{5.9}$$
$$\vec{a}_0''^2 + \vec{a}_0' \cdot \vec{a}_0^{(3)} = \vec{b}_0''^2 + \vec{b}_0' \cdot \vec{b}_0^{(3)} = 0, \ldots.$$

The first of these relations just expresses the fact that the curves $\vec{a}'(\zeta)$ and $-\vec{b}'(\zeta)$ intersect at $\zeta = t = 0$. From eqs. (5.6), (5.8) and (5.9) we find

$$\dot{x}^2(\zeta, t) = 1 - \tfrac{1}{4}[(\zeta - t)\vec{a}_0'' + (\zeta + t)\vec{b}_0'']^2 + \cdots. \tag{5.10}$$

The shape of the string at $t = 0$ is given by

$$\vec{x}(\zeta, 0) = \tfrac{1}{4}(\vec{a}_0'' + \vec{b}_0'')\zeta^2 + \tfrac{1}{12}(\vec{a}_0^{(3)} + \vec{b}_0^{(3)})\zeta^3 + \cdots. \tag{5.11}$$

It is easily seen that for $(\vec{a}_0'' + \vec{b}_0'') \neq 0$ the string momentarily develops a cusp, as shown in Fig. 11.1, and it follows from eq. (5.9) that the direction of the cusp $(\vec{a}_0'' + \vec{b}_0'')$ is orthogonal to that of the luminal velocity $(-\vec{a}_0')$. With an

Fig. 11.1. The string momentarily develops a cusp at the point of luminal motion.

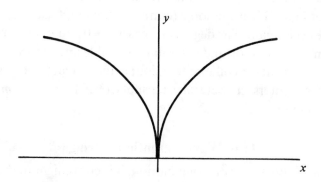

appropriate choice of axes (Fig. 11.1) the shape of the string near a cusp is given by $y \propto x^{2/3}$.

The rms string velocity in a loop can be defined as $v_{\text{rms}} = (\langle v^2 \rangle)^{1/2}$, where

$$\langle v^2 \rangle = \frac{2}{L} \int_0^{L/2} \mathrm{d}t \, \frac{1}{L} \int \mathrm{d}\zeta \dot{x}^2(\zeta, t)$$

$$= \tfrac{1}{2} - \tfrac{1}{2} \langle \vec{a}'(\zeta - t) \cdot \vec{b}'(\zeta + t) \rangle \qquad (5.12)$$

and I have used eq. (5.6). Using the identity

$$\frac{\mathrm{d}}{\mathrm{d}t} \int \vec{a}' \cdot \vec{b} \, \mathrm{d}\zeta = \int (-\vec{a}'' \cdot \vec{b} + \vec{a}' \cdot \vec{b}') \, \mathrm{d}\zeta$$

$$= 2 \int \vec{a}' \cdot \vec{b}' \, \mathrm{d}\zeta \qquad (5.13)$$

and the fact that an average over period of a time derivative is equal to zero, we see that the last term in (5.12) vanishes, and thus

$$\langle v^2 \rangle = 0.5. \qquad (5.14)$$

To give a specific example of loop trajectories, we can choose $\vec{a}$ and $\vec{b}$ to describe circular motion in planes at an angle ϕ to one another:

$$\vec{a}(\zeta) = \alpha^{-1}(\hat{e}_1 \sin \alpha\zeta + \hat{e}_3 \cos \alpha\zeta),$$

$$\vec{b}(\zeta) = \beta^{-1}[(\hat{e}_1 \cos \phi + \hat{e}_2 \sin \phi) \sin \beta\zeta + \hat{e}_3 \cos \beta\zeta]. \qquad (5.15)$$

Here, $\hat{e}_i$ is a unit vector along the x^i-axis, $\alpha = 2\pi m/L$, $\beta = 2\pi n/L$, m and n are relatively prime integers (otherwise the parameter ζ traverses the loop more than once between 0 and L and can be redefined to make m and n relatively prime). It is easily checked that the period of the loop (5.15) is $T = L/2n$. The family of solutions (5.15) was found by Burden (1985). Solutions with $(m, n) = (1, 1)$ as well as other families of solutions had been studied earlier by Kibble and Turok (1982) and Turok (1984).

Eq. (5.15) with $m = n = 1$ describes an elliptical loop which rotates and turns into a double line; at that moment the ends of the line are moving at the speed of light. Then the loop returns to the elliptical shape and goes through the cycle again. The degenerate cases $\phi = 0$ and $\phi = \pi$ correspond to an oscillating circular loop and to a rotating double line, respectively. Loops with $(m, n) \neq (1, 1)$ never collapse to a double line; in fact, loops with $m = 1$, $n \neq 1$ never self-intersect. Several snapshots of a loop with $m = 1$, $n = 2$, $\phi = \pi/2$ are shown in Fig. 11.2.

11.6 Waves on infinite strings

To illustrate another type of string motion, it is convenient to use the gauge

(4.1). Then the string trajectory is described by two functions, $x(z, t)$ and $y(z, t)$, and it is easily verified that

$$x = f(z \pm t), \quad y = g(z \pm t) \tag{6.1}$$

is a solution of eq. (3.10) for arbitrary functions f and g. These solutions describe waves of arbitrary shape propagating along the string with the velocity of light. Note that a superposition of waves travelling in opposite directions is not a solution, since eq. (3.10) is nonlinear.

11.7 Gravitational field of a straight string

In this section we shall find a solution of Einstein's equations describing the gravitational field of a static straight string lying along the z-axis. Such a string is left unchanged by time translations, spatial translations in the z-direction, rotations around the z-axis and Lorentz boosts in the z-direction. Assuming that the metric has the same symmetries, we can write it as

$$ds^2 = A^2(\rho)(dt^2 - dz^2) - d\rho^2 - B^2(\rho)\, d\phi^2. \tag{7.1}$$

The energy–momentum tensor of the string vanishes exponentially at $\rho \gg \delta$, where δ is the string thickness, and so outside the string core the metric is accurately described by the vacuum Einstein equations. Kasner (1921) has shown that the general vacuum solution depending on one spatial variable has the form (up to a coordinate transformation)

$$ds^2 = \rho^{2p_1}\, dt^2 - d\rho^2 - \rho^{2p_2}\, d\phi^2 - \rho^{2p_3}\, dz^2, \tag{7.2}$$

$$p_1 + p_2 + p_3 = p_1^2 + p_2^2 + p_3^2 = 1. \tag{7.3}$$

The metrics (7.1) and (7.2) are consistent with one another only if $p_1 = p_3$. The only solutions of eq. (7.3) with $p_1 = p_3$ are

$$p_1 = p_3 = 0, \quad p_2 = 1; \tag{7.4}$$

$$p_1 = p_3 = \tfrac{2}{3}, \quad p_2 = -\tfrac{1}{3}. \tag{7.5}$$

In the solution (7.5), the length of a circle, $\rho = \text{const}$, decreases as $\rho^{-1/3}$ when

Fig. 11.2. The xz-projections of the loop, e.g. (5.15), with $m = 1$, $n = 2$, $\phi = \pi/2$ at four moments during its period.

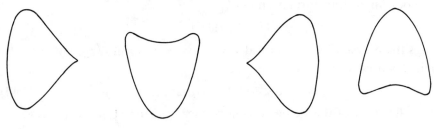

$\rho \to \infty$. It is unlikely that this metric has anything to do with strings which could have formed in our Universe, and we dismiss the case (7.5) as unphysical. The remaining solution (7.4) is just the metric of flat space in cylindrical coordinates,

$$ds^2 = dt^2 - d\rho^2 - \rho^2\, d\phi^2 - dz^2. \tag{7.6}$$

Thus, we have reached a surprising conclusion that the spacetime outside a straight string is flat. We shall see, however, that it is only locally flat and that globally it is not equivalent to Minkowski space (Vilenkin, 1981a).

To represent the interior of the string, we shall first assume a simple energy–momentum tensor of the form (Linet, 1985; Gott, 1985; Hiscock, 1985)

$$T^\nu_\mu = \sigma(\rho)\, \mathrm{diag}(1, 0, 0, 1). \tag{7.7}$$

The general case will be discussed later. The form (7.7) is clearly consistent with all the symmetries of the string. In the limit of negligible string thickness, $\sigma(\rho)$ is a δ-function and (7.7) reduces to (2.4). Substituting (7.1) and (7.7) in Einstein's equations, it is easily shown that $A(\rho) = \mathrm{const}$. By a suitable rescaling of t and z we can set $A = 1$; then

$$ds^2 = dt^2 - dz^2 - d\rho^2 - B^2(\rho)\, d\phi^2 \tag{7.8}$$

and Einstein's equations reduce to a single equation for the function $B(\rho)$:

$$B''/B = -8\pi G\sigma. \tag{7.9}$$

The metric (7.8) is nonsingular on the axis $\rho = 0$ only if

$$B(0) = 0, \quad B'(0) = 1. \tag{7.10}$$

The mass per unit length of string is

$$\mu = \int_0^\infty d\rho \int_0^{2\pi} d\phi\, \sigma(^{(2)}g)^{1/2} = \frac{1}{4G}\left[1 - B'(\infty)\right], \tag{7.11}$$

where $^{(2)}g_{ij}$ is the metric on the surface $(t, z) = \mathrm{const}$ and $^{(2)}g = B^2$ is its determinant. At large distances from the string $(\rho \gg \delta)$, $\sigma \to 0$, $B'(\rho) \to B'(\infty)$ and we obtain

$$ds^2 = dt^2 - dz^2 - d\rho^2 - (1 - 4G\mu)^2\rho^2\, d\phi^2. \tag{7.12}$$

A coordinate transformation

$$(1 - 4G\mu)\phi \to \phi \tag{7.13}$$

brings the metric (7.14) to a locally Minkowskian form (7.6), but then the angle ϕ varies in the range

$$0 < \phi < (1 - 4G\mu)2\pi. \tag{7.14}$$

Thus, the effect of the string is to introduce an azimuthal 'deficit angle'

$$\delta = 8\pi G\mu, \tag{7.15}$$

with the result that a surface of constant t and z has the geometry of a cone rather than that of a plane. The point of the cone is smoothed on a scale $\rho \sim \delta$. In the limit of an infinitely thin string the cone has a sharp point; the corresponding spacetime is called conical.

The dimensionless parameter $G\mu$ plays an important role in the physics of cosmic strings. Its magnitude can be estimated using eq. (2.1),

$$G\mu \sim (\eta/m_p)^2 \ll 1, \tag{7.16}$$

where η is the symmetry breaking scale of strings, m_p is the Planck mass and it is assumed that $\eta \ll m_p$. The string scenario of galaxy formation requires $G\mu \sim 10^{-6}$, which corresponds to $\eta \sim 10^{16}$ GeV.

11.8 Gravitational field of a straight string (continued)

We derived eq. (7.15) for the conical deficit angle assuming a simple form of the energy–momentum tensor (7.7). In the general case, the components T_ρ^ρ and T_ϕ^ϕ do not vanish inside the string and eq. (7.15) has to be modified. We shall see, however, that this equation holds in the cosmologically interesting case of $G\mu \ll 1$ (the reason, basically, is that for $G\mu \ll 1$ the metric inside the string is approximately flat and that transverse tensions average out to zero in flat spacetime (see eq. (2.3))).

We shall first express the deficit angle in terms of the curvature tensor in the interior of the string (Ford and Vilenkin, 1981). Let S denote a surface $(t, z) = \text{const}$ and let $^{(2)}g_{ij}$ be the metric on S. The scalar curvature of S is

$$^{(2)}R = -2B''/B. \tag{8.1}$$

Integrating $^{(2)}R$ over the surface and using eq. (7.12) we obtain the following expression for the deficit angle:

$$\delta = \tfrac{1}{2} \int {}^{(2)}R(^{(2)}g)^{1/2} \, d^2x. \tag{8.2}$$

This is a version of the Gauss–Bonnet theorem.

The curvature of the surface S can be expressed in terms of the Riemann tensor $^{(4)}R_{\nu\sigma\tau}^\mu$ of the four-dimensional spacetime. First consider a 3-surface $t = \text{const}$. Its Riemann tensor $^{(3)}R_{jkl}^i$ is expressible in terms of $^{(4)}R_{\nu\sigma\tau}^\mu$ and the extrinsic curvature K_{ij} by means of the Gauss–Codazzi equations. However, for the 3-surface $t = \text{const}$, the extrinsic curvature vanishes and $^{(3)}R_{jkl}^i = {}^{(4)}R_{jkl}^i$. Similarly, one obtains that the Riemann tensor of the 2-surface S is $^{(2)}R_{jkl}^i = {}^{(4)}R_{jkl}^i$. Hence,

$$^{(2)}R = {}^{(2)}g^{ik(2)}g^{jl(4)}R_{ijkl}. \tag{8.3}$$

This result, combined with eq. (8.2), enables us to find the deficit angle in terms of the Riemann tensor of the four-dimensional spacetime.

In general, it does not seem to be possible to express δ in terms of the four-dimensional Ricci tensor, and hence of the energy–momentum tensor of the string, $T_{\mu\nu}$. However, in the case that the gravitational field is sufficiently weak that the linearized theory may be applied, such an expression can be given. Let the metric be

$$g_{\mu\nu} = \eta_{\mu\nu} + h_{\mu\nu}, \tag{8.4}$$

where $\eta_{\mu\nu} = \mathrm{diag}(1, -1, -1, -1)$ and $h_{\mu\nu} \ll 1$. With the gauge condition

$$\partial_\nu(h_\mu^\nu - \tfrac{1}{2}\delta_\mu^\nu h) = 0 \tag{8.5}$$

the linearized field equations become

$$\Box h_{\mu\nu} = -16\pi G S_{\mu\nu}, \tag{8.6}$$

where $\Box = \partial_t^2 - \nabla^2$, the indices are raised and lowered using the flat metric $\eta_{\mu\nu}$ and

$$S_{\mu\nu} = T_{\mu\nu} - \tfrac{1}{2}\eta_{\mu\nu}T. \tag{8.7}$$

The linearized Riemann tensor is

$$^{(4)}R_{\alpha\mu\beta\nu} = \tfrac{1}{2}(h_{\alpha\nu,\mu\beta} + h_{\mu\beta,\alpha\nu} - h_{\mu\nu,\alpha\beta} - h_{\alpha\beta,\mu\nu}). \tag{8.8}$$

For a metric independent of t and z,

$$^{(2)}R = 2^{(4)}R_{1212} = 2h_{12,12} - h_{11,22} - h_{22,11}. \tag{8.9}$$

Using (8.5) and (8.6), this can be rewritten as

$$^{(2)}R = -16\pi G(T_{11} + T_{22} - \tfrac{1}{2}T). \tag{8.10}$$

Substituting this in (8.2) and using (2.2), (2.3), we obtain

$$\delta = 8\pi G \int T_0^0\, \mathrm{d}^2x = 8\pi G\mu. \tag{8.11}$$

The use of the linear perturbation theory is justified for $G\mu \ll 1$, which is the case for strings of cosmological interest.

A non-perturbative analysis of Einstein's equations with a realistic energy–momentum tensor of a string has been given by Garfinkle (1985). The gravitational field of global strings, for which the energy–momentum tensor falls off only as an inverse square of the distance from the string, has been discussed by Aryal and Everett (1986). I should mention also that spacetimes with conical singularities had been studied long before their relevance to cosmic strings was recognized. See, for example, Bach and Weyl (1922), Sokolov and Starobinsky (1977) and Israel (1977).

11.9 Physics in conical space

We have found that, neglecting the string thickness, the gravitational field of a straight string is given by the metric (7.12). It describes a conical space, which is just a flat space with a wedge of angular size $\delta = 8\pi G\mu$ removed and the two faces of the wedge identified. The geodesics in this space are straight lines, and it is clear that a test particle initially at rest relative to the string will remain at rest and will not experience any gravitational force.

Although the metric (7.12) is locally flat, its global structure is different from that of Minkowski space. The most striking effect of this difference is the formation of double images of objects located behind the string. (Vilenkin, 1981a, 1984; Gott, 1985). This is illustrated in Fig. 11.3. The light rays from a quasar intersect behind the string and the observer sees two images of the quasar. If l and d are the distances from the string to the quasar and to the observer, respectively, and θ is the angle between the string and the line of sight, then it is easily shown that the angular separation between the images is (for $G\mu \ll 1$).

$$\delta\phi_0 = 8\pi G\mu l(d+l)^{-1} \sin\theta. \tag{9.1}$$

Gravitational lensing by a string is a classical analogue of the Aharonov–Bohm effect. Spacetime curvature is confined to the string core, but its effect is 'felt' by the photons propagating in flat spacetime region around it. As in the Aharonov–Bohm case, a Minkowskian coordinate system can be chosen in any region on one side of the string, but such a system does not exist in a region surrounding the string.

The subscript of $\delta\phi_0$ in eq. (9.1) refers to the fact that the string was assumed to be at rest with respect to the observer. Cosmological strings are expected to move at relativistic speeds, and so eq. (9.1) has to be generalized to the case of a moving string. In the general case, let $\hat{s}$ and $\hat{v}$ be the direction and the velocity of the string in the observer's frame S. Since only transverse motion of strings is observable, we can take $\hat{v} \perp \hat{s}$. Let k and k' be the four-

Fig. 11.3. Light rays emitted by the quasar intersect behind the string and the observer sees two images of the same quasar.

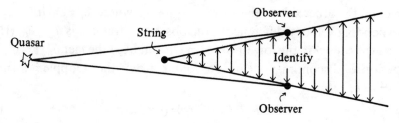

dimensional wave vectors of the light waves corresponding to the two images and consider the invariant

$$k \cdot k' = \omega\omega'[1 - \cos(\delta\phi)] \approx \tfrac{1}{2}\omega\omega'(\delta\phi)^2. \qquad (9.2)$$

In the rest frame of the string, S_0, the two frequencies are equal, $\omega = \omega' = \omega_0$. In observer's frame, S, ω and ω' are slightly different because of the difference in the directions of $\vec{k}$ and $\vec{k}'$. However, the difference is $\sim \delta\phi$ and can be neglected to the lowest order in $G\mu$. Then $\omega\delta\phi = \omega_0\delta\phi_0$. The frequencies ω and ω_0 are related by $\omega = \gamma(1 - \hat{n} \cdot \hat{v})\omega_0$, where n is the direction from the observer to the quasar and $\gamma = (1 - v^2)^{-1/2}$. Hence,

$$\delta\phi = \gamma^{-1}(1 - \hat{n} \cdot \hat{v})^{-1}\delta\phi_0. \qquad (9.3)$$

Depending on the relative direction of $\hat{v}$ and $\hat{n}$, the angular separation takes a value in the range

$$\kappa^{1/2}\,\delta\phi_0 < \delta\phi < \kappa^{-1/2}\,\delta\phi_0, \qquad (9.4)$$

where $\kappa = (1 - v)/(1 + v)$.

For $G\mu \sim 10^{-6}$, the typical angular separation of images is of the order of a few seconds of arc. Several double quasars with such angular separations are known, and it has been argued that some of them may be due to strings (Gott, 1985; Vilenkin, 1984; Paczynski, 1986).

Another interesting effect arises because the relative velocity of two objects in a conical space is double-valued. The answer depends on which way one parallel transports the velocities around the string. If velocities are measured using direct light signals, then the relative velocity experiences a discontinuous change when the string passes between the two objects. In a cosmological setting, this results in a steplike discontinuity of the microwave background temperature along the string. The velocity of the observer relative to the surface of last scattering has different values in front of and behind a moving string; the corresponding Doppler shift is responsible for the temperature discontinuity.

Let us now describe this effect quantitatively (Kaiser and Stebbins, 1984; Vachaspati, 1986; Vilenkin, 1986). In the rest frame of the string, S_0, the observer moves by the string with velocity $-\hat{v}$. If he is at rest with respect to the sources of radiation on one side of the string, then the velocity of the sources on the other side of the string is $-\hat{v} + \hat{u}_0$, where $\hat{u}_0 = 8\pi G\mu\hat{v} \times \hat{s}$. The corresponding velocity difference in the observer's frame S is $\hat{u} = \gamma\hat{u}_0$. (Here, I have expanded $\hat{u}$ to the lowest order in $G\mu$ and used the fact $\hat{u}_0 \cdot \hat{v} = 0$.) The temperature discontinuity can be found from the Doppler formula, $\delta T/T = \hat{u} \cdot \hat{n}$,

$$\delta T/T = 8\pi G\mu\gamma\hat{n} \cdot (\hat{v} \times \hat{s}). \qquad (9.5)$$

The existence of strings is consistent with the observed isotropy of the cosmic microwave background ($\delta T/T < 10^{-4}$ on large scales) only if $G\mu < 10^{-5}$.

Smith (1986) has studied the effects of a conical singularity on classical and quantum fields. He has found, in particular, that although a straight string exerts no gravitational force on a particle outside it, the particle's own gravitational field is distorted by the string and results in a classical attractive self-force proportional to the particle mass squared:

$$F = \tfrac{1}{4}\pi G\mu \, Gm^2/r^2. \tag{9.6}$$

This result was also obtained by Linet (1986).

11.10 Other metrics with straight strings

Solutions of Einstein's equations with conical singularities describing straight strings can easily be constructed (Aryal, Ford and Vilenkin, 1986). One needs only a spacetime with a symmetry axis. If one then cuts a wedge out of this space, which is done by requiring the azimuthal angle around the axis to run over the range (7.14) or by multiplying $g_{\phi\phi}$ by $(1-4G\mu)^2$, one has a space with a string along the axis.

For example, the metrics

$$ds^2 = \left(1 - \frac{2m}{r}\right) dt^2 - \left(1 - \frac{2m}{r}\right)^{-1} dr^2 - r^2 \, d\Omega^2, \tag{10.1}$$

$$ds^2 = dt^2 - a^2(t)[(1-kr^2)^{-1} \, dr^2 + r^2 \, d\Omega^2] \tag{10.2}$$

with

$$d\Omega^2 = d\theta^2 + (1-4G\mu)^2 \sin^2 \theta \, d\phi^2 \tag{10.3}$$

describe a black hole with a string passing through it, and a Robertson–Walker universe with a string, respectively.

Some well-known solutions of Einstein's equations can be re-interpreted in terms of cosmic strings. For example, the solution found many years ago by Bach and Weyl (1922) describes, with an appropriate choice of integration constants, a pair of black holes held apart by cosmic strings extending to infinity in opposite directions. If one of the black holes is removed to a very large, but not infinite, distance, one obtains a solution representing a black hole suspended by a string in a weak uniform gravitational field, g. To linear order in g,

$$ds^2 = (1 - 2m/r)(1 + 2gz) \, dt^2 + (1 - 2gz)[(1 - 2m/r)^{-1} \, dr^2$$
$$+ r^2 \, d\theta^2 + b^2 r^2 \sin^2 \theta \, d\phi^2], \tag{10.4}$$

where $z = (r - m) \cos \theta$, $b = 1$ for $z < -m$ and $b = 1 - 4mg$ for $z > m$. The string tension $\mu = mg$ exactly balances the weight of the black hole.

11.11 Gravitational field of oscillating loops

The most striking feature of the gravitational field of a straight string is that there is no Newtonian force. Its origin can be understood using the linear perturbation theory (8.4)–(8.7). The source of the Newtonian potential, $\phi = \frac{1}{2} h_{00}$, is

$$S_{00} = S_0^0 = \frac{1}{2}(T_0^0 - T_i^i). \tag{11.1}$$

For the energy–momentum tensor of a string, eq. (4.7),

$$S_0^0(\vec{r}, t) = \mu \int d\zeta \dot{\vec{x}}^2 \, \delta^{(3)}(\vec{r} - \vec{x}(\zeta, t)). \tag{11.2}$$

In a rest frame of a straight string, $S_0^0 = 0$ and the gravitational potential vanishes. A string can be static only if it is straight, and so curved strings produce non-zero gravitational potentials. In this section we shall analyze the gravitational field produced by oscillating closed loops.

The retarded solution of the wave equation (8.6) can be written as

$$h^{\mu\nu}(\vec{r}, t) = -4G \int \frac{d^3 r'}{|\vec{r} - \vec{r}'|} S^{\mu\nu}(\vec{r}, t - |\vec{r} - \vec{r}'|). \tag{11.3}$$

Using eqs. (4.7) and (8.7) and integrating over $\vec{r}'$ we obtain (Turok, 1984; Vachaspati, 1986)

$$h^{\mu\nu}(\vec{r}, t) = -4G\mu \int \frac{F^{\mu\nu}(\zeta, \tau) \, d\zeta}{|\vec{r} - \vec{x}(\zeta, \tau)|(1 - \hat{\varepsilon} \cdot \dot{\vec{x}}(\zeta, \tau))}, \tag{11.4}$$

where

$$\hat{\varepsilon} = \frac{\vec{r} - \vec{x}(\zeta, \tau)}{|\vec{r} - \vec{x}(\zeta, \tau)|}, \tag{11.5}$$

$$F^{\mu\nu} = \dot{x}^\mu \dot{x}^\nu - x'^\mu x'^\nu + \eta^{\mu\nu} x'^\sigma x'_\sigma \tag{11.6}$$

and the retarded time τ is defined by

$$\tau = t - |\vec{r} - \vec{x}(\zeta, \tau)|. \tag{11.7}$$

The motion of non-relativistic test particles is determined mainly by the time-averaged field of the loop,

$$\langle h^{\mu\nu}(\vec{r}) \rangle = \frac{2}{L} \int_0^{L/2} dt \, h^{\mu\nu}(\vec{r}, t). \tag{11.8}$$

The period of the oscillating component of the field, $T = L/2$, is much shorter than the time it takes for the particle to traverse a distance $\sim L$. The effect of the oscillating component on the particle trajectory at a distance $\lesssim L$ from the loop is to introduce a small perturbation of the position $\Delta x \sim aT^2 \sim G\mu L$. Here, $a \sim GM/L^2$ is the typical acceleration and $M = \mu L$ is the mass of the loop. The acceleration can be enhanced by the 'beaming factor' $(1 - \dot{\bar{\varepsilon}} \cdot \dot{x})^{-1}$ in eq. (11.4) (Turok, 1984). This factor becomes infinite along the beams extending from the cusps where $|\dot{x}| = 1$ in the direction of $\dot{x}$. Because of the retardation, the divergence occurs at different moments for different points along the beam. If a particle passes near the beam at the right time, the effect can be very large, but it affects only a rather special class of particle trajectories. Here, we shall first discuss the average field (11.8) and then consider what happens at the beams. The gravitational radiation from loops is discussed in the next section.

Using eqs. (11.4), (11.8) and the relation

$$dt = d\tau[1 - \dot{\bar{\varepsilon}} \cdot \dot{x}(\zeta, \tau)], \qquad (11.9)$$

we obtain for the average field (Turok, 1983)

$$\langle h^{\mu\nu}(\vec{r}) \rangle = -\frac{8G\mu}{L} \int_0^{L/2} dt \int_0^L d\zeta \, \frac{F^{\mu\nu}(\zeta, \tau)}{|\vec{r} - \dot{x}(\zeta, \tau)|}. \qquad (11.10)$$

This result is easily interpreted: the time-averaged field of a loop is equal to that of the surface traced out by its motion with a surface energy–momentum tensor

$$\sigma^{\mu\nu} = \frac{2\mu}{L|\dot{x}||\dot{x}'|} (\dot{x}^\mu \dot{x}^\nu - x'^\mu x'^\nu). \qquad (11.11)$$

Here, I have used the expression for the surface element, $dS = |\dot{x}||\dot{x}'| \, d\zeta \, d\tau$. The mass per unit area of the surface is given by $\sigma^{00} = 2\mu/L|\dot{x}||\dot{x}'|$.

We note that $\sigma^{\mu\nu}$ becomes singular at the cusps, where $|\dot{x}| = 1$ and $|\dot{x}'| = 0$. However, the singularity is integrable and the total mass of the shell is equal to the mass of the loop, M:

$$\int \sigma^{00} \, dS = \frac{2\mu}{L} \int_0^{L/2} dt \int_0^L d\zeta = \mu L = M. \qquad (11.12)$$

We also note an interesting possibility that the peaks of $\sigma^{\mu\nu}$ near the cusps can be directly observable as overdensity regions in the distribution of matter accreted by the loops.

At a large distance from the loop $r \gg L$, eq. (11.10) gives

$$\langle h^{\mu\nu}(\vec{r}) \rangle = -A^{\mu\nu}/r, \qquad (11.13)$$

where

$$A^{\mu\nu} = 8G\mu L^{-1} \int_0^{L/2} d\tau \int_0^L d\zeta F^{\mu\nu}(\zeta, \tau). \qquad (11.14)$$

The integration in (11.14) is easily performed. For example,

$$A^{00} = 4G\mu L\langle \dot{x}^2\rangle = 2GM,\tag{11.15}$$

where $\langle \dot{x}^2\rangle$ is defined by eq. (5.12) and I have used eq. (5.14). Similarly, we find

$$A^{0i} = 0; \quad A^{ij} = 2Gm\delta^{ij}.\tag{11.16}$$

Combining eqs. (11.13), (11.15) and (11.16), we see that the long-distance behavior of the average field coincides with that of a Schwarzschild field of the same mass (Turok, 1984). It should be emphasized that the Schwarzschild limit is approached only in the average sense. Since the motion of the loop is relativistic, the oscillating component of $h^{\mu\nu}$ describing gravitational waves has the same order of magnitude as the average field.

We now turn to the discussion of the bursts of gravitational field produced when the beaming factor $(1 - \dot{\varepsilon}\cdot\dot{x})^{-1}$ in eq. (11.4) becomes infinite (T. Vachaspati and A. Vilenkin, 1986, unpublished results). The main contribution to $h^{\mu\nu}$ in the bursts comes from a small vicinity of the cusp, which we take to be at $\dot{x} = t = \zeta = 0$. Taking the slowly varying factors out of the integral, we can rewrite (11.4) as

$$h^{\mu\nu}(\vec{r}, t) \approx -4G\mu F_0^{\mu\nu} r^{-1} \int_{-\infty}^{\infty} d\zeta (1 - \dot{\varepsilon}\cdot\dot{x}(\zeta, \tau))^{-1},\tag{11.17}$$

where $F_0^{\mu\nu} = F^{\mu\nu}(0, 0)$ and τ is a function of $\vec{r}, t$ and ζ implicitly defined by eq. (11.7). The integral in (11.17) can be estimated using the expansion (5.8) near the cusp and the relations (5.9). Here, we shall calculate $h^{\mu\nu}$ only on the beam, where $\dot{\varepsilon} = -\ddot{a}_0' = \vec{b}_0'$. With the notation $\alpha_0 = |\ddot{a}_0''|^2/4$, $\beta_0 = |\vec{b}_0''|^2/4$, we obtain for the points on the beam,

$$3(t - r) + \alpha_0(\zeta - \tau)^3 - \beta_0(\zeta + \tau)^3 \approx 0;\tag{11.18}$$

$$1 - \dot{\varepsilon}\cdot\dot{x} \approx \alpha_0(\zeta - \tau)^2 + \beta_0(\zeta + \tau)^2;\tag{11.19}$$

$$\dot{\varepsilon}\cdot\dot{x}' \approx \alpha_0(\zeta - \tau)^2 - \beta_0(\zeta + \tau)^2.\tag{11.20}$$

Introducing a new variable, $u = \zeta/\tau$, and using eqs. (11.18)–(11.20) and the relation

$$\partial\tau/\partial\zeta = \dot{\varepsilon}\cdot\dot{x}'(1 - \dot{\varepsilon}\cdot\dot{x})^{-1}\tag{11.21}$$

we can rewrite (11.17) as

$$h^{\mu\nu}(z, t) \approx -16G\mu I(\alpha_0/\beta_0)F_0^{\mu\nu} z^{-1}(12\beta_0^2|t - z|)^{-1/3},\tag{11.22}$$

where the z-axis is chosen along the beam and

$$I(x) = \int_{-\infty}^{\infty} du[(u + 1)^3 - x(u - 1)^3]^{-2/3}.\tag{11.23}$$

The integrand of $I(x)$ is singular at $u = u_0 = (x^{1/3} + 1)(x^{1/3} - 1)^{-1}$, but the singularity is integrable and $h^{\mu\nu}$ is finite for $z - t \neq 0$. For a particular value $x = 1$, $I(1) = 2^{-2/3} 3^{-1/2} B(\frac{1}{2}, \frac{1}{6}) = 4.6$, where $B(x, y) = \Gamma(x)\Gamma(y)/\Gamma(x + y)$ is the beta function.

The metric (11.22) is singular at $z = t$, $h^{\mu\nu} \propto |z - t|^{-1/3}$, and the z-component of the force acting on a test particle is

$$F_z \propto (z - t)^{-4/3} \, \text{sign}(z - t). \tag{11.24}$$

The force which is initially repulsive, grows in magnitude, becomes infinite and instantaneously changes into an infinite attractive force, which then decreases. (Note, however, that the total momentum transferred to the particle is finite.) This burst of gravitational field propagates along the beam at the speed of light. Of course, the linear perturbation theory cannot be trusted at points where $h^{\mu\nu}$ diverges, but we expect eq. (11.24) to apply for $|z - t|$ not too small. It is possible that nonlinear effects and the back reaction of the gravitational field on the string near the cusps make $h^{\mu\nu}$ finite at $z = t$.

Eq. (11.17) gives the most singular components of the gravitational field on the beam. The analysis of other components and of the gravitational field slightly off the beam is rather complicated and will not be given here.

11.12 Gravitational radiation from loops

Gravitational radiation is the main energy-loss mechanism for oscillating loops of string. The rate of energy loss can be roughly estimated using the quadrupole radiation formula:

$$\dot{E} \sim GM^2 L^4 \omega^6. \tag{12.1}$$

The typical frequency of oscillation is $\omega \sim L^{-1}$, and we obtain (Vilenkin, 1981b)

$$\dot{E} = \gamma G \mu^2, \tag{12.2}$$

where the numerical coefficient γ is independent of the loop size, but may depend on its shape and trajectory. The lifetime of the loop is

$$\tau \sim M/\dot{E} \sim L/\gamma G \mu. \tag{12.3}$$

The quadrupole radiation formula is reliable only for nonrelativistic sources. The motion of loops is ultrarelativistic near the cusps, and so the quadrupole formula cannot be used to calculate γ in eq. (12.2). The power of gravitational radiation from a periodic source of period $T = L/2$ to the lowest order in G, and without any further assumptions about the source, can be found from the following equations (Weinberg, 1972):

$$P = \dot{E} = \sum_n P_n = \sum_n \int d\Omega \frac{dP_n}{d\Omega}, \qquad (12.4)$$

$$\frac{dP_n}{d\Omega} = \frac{G\omega_n^2}{\pi} \{ T_{\mu\nu}^*(\omega_n, \vec{k}) T^{\mu\nu}(\omega_n, \vec{k}) - \tfrac{1}{2} |T_\nu^\nu(\omega_n, \vec{k})|^2 \}. \qquad (12.5)$$

Here, $dP_n/d\Omega$ is the radiation power at frequency $\omega_n = 2\pi n/T = 4\pi n/L$ per unit solid angle in the direction of $\vec{k}$, $|\vec{k}| = \omega_n$ and

$$T^{\mu\nu}(\omega_n, \vec{k}) = \frac{2}{L} \int_0^{L/2} dt \exp(i\omega_n t) \int d^3x \exp(-i\vec{k}\cdot\vec{x}) T^{\mu\nu}(\vec{x}, t) \qquad (12.6)$$

is the Fourier transform of the energy–momentum tensor.

This formalism has been applied to simple loop configurations by Turok (1984), Vachaspati and Vilenkin (1985) and Burden (1985). For loops described by eq. (5.15) the angular distribution $dP_n/d\Omega$ can be expressed analytically in terms of Bessel functions. The total power, P_n, at frequency ω_n is then calculated by a straightforward numerical integration. An asymptotic analysis at large n shows that

$$P_n \sim \text{const } n^{-4/3}, \qquad (12.7)$$

indicating that the convergence of the series in (12.4) is slow. The best strategy, then, is to keep calculating P_n until the asymptotic regime of eq. (12.7) is reached and then to use that equation to estimate the remainder of the series. The resulting values of γ for several loop configurations are shown in Fig. 11.4. We see that, somewhat surprisingly, γ is typically ~ 100.

The large value of γ is due partly to the large n contributions in eq. (12.4). For 'normal' sources of radiation, P_n decreases exponentially in the large n limit. The power-law dependence (12.7) in the case of loops is due to the singular behavior of string near the cusps. The angular distribution of the radiation is also marked by the presence of cusps. Large amounts of radiation are beamed from the cusps in the direction of luminal velocity. The analysis in Vachaspati and Vilenkin (1985) shows that near the beam

$$\frac{dP}{d\Omega} = \sum_n \frac{dP_n}{d\Omega} \propto \theta^{-1}, \qquad (12.8)$$

where θ is the angle between the beam and the wave vector $\vec{k}$. The radiation intensity diverges on the beam, but the divergence is integrable and the total power is finite. The immediate vicinity of the beam, where the linear perturbation theory breaks down, contributes very little to the total power (12.2).

The gravitational waves emitted by oscillating loops at different epochs add up to a stochastic gravitational wave background with a scale-invariant spectrum (Vilenkin, 1981c; Hogan and Rees, 1984; Vachaspati and Vilenkin, 1985)

$$\Omega_g(\omega) \sim 10^{-4}(G\mu)^{1/2}. \tag{12.9}$$

Here, $\Omega_g(\omega) = (\omega/\rho_c)\,(d\rho_g/d\omega)$, ρ_g is the energy density of gravitational waves and ρ_c is the critical density. Eq. (12.9) is expected to apply in a wide range of frequencies,

$$10^{-2}\,\mathrm{yr}^{-1} \lesssim \omega \lesssim 10^5\,\mathrm{s}^{-1}. \tag{12.10}$$

With $G\mu \sim 10^{-6}$, as required by the cosmic string scenario of galaxy formation, eq. (12.9) gives $\Omega_g(\omega) \sim 10^{-7}$. A gravitational background of such intensity should be observable using the millisecond pulsar (Hogan and Rees, 1984; Witten, 1984).

11.13 Gravitational rocket effect

Gravitational radiation from an oscillating loop carries away not only energy, but also momentum. As a result, the loop can accelerate like a rocket. The rate of momentum radiation is given by

$$\frac{d\vec{P}}{dt} = \sum_n \int \frac{dP_n}{d\Omega}\,\vec{k}\,d\Omega. \tag{13.1}$$

Fig. 11.4. Gravitational radiation power for the family of loops, eq. (5.15), with $m=1$, $n=1,3,5$ for several values of ϕ.

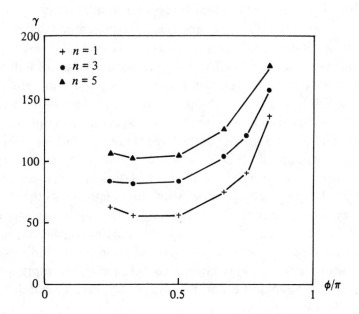

It vanishes for loops described by eq. (5.15) because of their high symmetry. In general, one expects $|\dot{\vec{P}}|$ to be of the same order as $\dot{E}$ in eq. (12.2). A numerical calculation for several asymmetric loops gives (Vachaspati and Vilenkin, 1985)

$$|\dot{\vec{P}}|=\gamma_p G\mu^2 \qquad (13.2)$$

with $\gamma_p \sim 10$.

A loop radiating momentum at the rate (13.2) will move with acceleration

$$\dot{v}=\gamma_p G\mu/L. \qquad (13.3)$$

If it starts from rest, then by the end of its life it will reach the velocity

$$v \sim \dot{v}\tau \sim \gamma_p/\gamma \sim 0.1, \qquad (13.4)$$

where τ is from eq. (12.3).

In deriving the estimate (13.4) we assumed that the direction of $\dot{\vec{P}}$ does not change appreciably during the loop's lifetime. This is far from being clear. For example, one can argue that angular momentum radiation can prevent the loop from accumulating a large velocity. The angular momentum of a loop is $l \sim \mu L^2$ and the angular momentum radiation rate can be written as (on dimensional grounds)

$$|\dot{\vec{l}}|=\gamma_l G\mu^2 L, \qquad (13.5)$$

where γ_l is a numerical coefficient. Let us first consider the component of $\dot{\vec{l}}$ parallel to $\vec{l}$, which changes only the magnitude of $\vec{l}$. If the torque (13.5) causes the loop to rotate as a solid body, then the corresponding angular acceleration is $\ddot{\theta} \sim G\mu L^{-2}$. The time it takes the loop to rotate by about one radian is $\Delta t \sim \ddot{\theta}^{-1/2} \sim (G\mu)^{-1/2}L$, and the velocity accumulated during this time with acceleration (13.3) is $v \sim (G\mu)^{1/2} \ll 1$. Note, however, that the assumption that the loop reacts to a torque like a solid is not well justified. Besides, even if it does, the rotation axis is not, in general, at right angles to the direction of $\dot{\vec{P}}$, and then the loop will accelerate along the direction of $\vec{l}$.

Torques perpendicular to $\vec{l}$ tend to change the direction of angular momentum. However, because of the large value of l, the loop behaves as a relativistic gyroscope, and the direction of its rotation axis is very stable (Hogan, 1986). According to eq. (13.5), this direction changes on a time scale, $l/|\dot{\vec{l}}| \sim L/(\gamma_l G\mu)$, comparable with the lifetime of the loop. To summarize, our qualitative arguments suggest that loops can accelerate to velocities $v \sim 0.1$, at least if $\dot{\vec{P}} \cdot \vec{l} \neq 0$. To reach a reliable conclusion, one has to study the back reaction of gravitational radiation on the loop. Note also that here we discussed the gravitational rocket effect in flat empty space. The effects of the cosmological expansion and of the gravitational drag due to surrounding matter have been studied by Vachaspati and Vilenkin (1985), Hogan and Rees (1984) and Hogan (1986).

11.14 Gravitational field of waves on infinite strings

Waves propagating along infinite strings are described by solutions of the form (6.1) to the string equations of motion. In this section we shall study the gravitational field produced by such waves (Vachaspati, 1986). A wave propagating in the positive z-direction is given by (in the gauge (4.1))

$$x=f(z-t), \quad y=g(z-t), \tag{14.1}$$

where f and g are arbitrary functions. The source of the linearized Einstein's equations (8.6) is found by a straightforward calculation using eq. (3.11):

$$S^{\mu\nu} = \mu F^{\mu\nu}(z-t)\,\delta(x-f)\,\delta(y-g), \tag{14.2}$$

where

$$F^{\mu\nu} = \begin{pmatrix} f'^2+g'^2 & -f' & -g' & f'^2+g'^2 \\ -f' & 1 & 0 & -f' \\ -g' & 0 & 1 & -g' \\ f'^2+g'^2 & -f' & -g' & f'^2+g'^2 \end{pmatrix} \tag{14.3}$$

Now it is not difficult to verify that

$$h^{\mu\nu} = hF^{\mu\nu}(z-t) \tag{14.4}$$

with

$$h = 4G\mu \ln[r_0^{-2}\{(x-f)^2 + (y-g)^2\}] \tag{14.5}$$

is a solution of eq. (8.6). Indeed, $(\partial_t^2 - \partial_z^2)h^{\mu\nu}=0$ because $h^{\mu\nu}$ depends on t and z only in the combination $(t-z)$. On the other hand, $(\partial_x^2 - \partial_y^2)$ acts only on the logarithm and gives the δ-function in $S^{\mu\nu}$. r_0 in eq. (14.5) is an arbitrary constant of integration.

In the case of a static straight string parallel to the z-axis, all components of $F^{\mu\nu}$ vanish except $F^{11}=F^{22}=1$, and the solution can be written as

$$ds^2 = dt^2 - dz^2 - (1-h)(d\rho^2 + \rho^2\,d\phi^2), \tag{14.6}$$

where $\rho^2 = x^2 + y^2$. A coordinate transformation

$$(1-h)\rho^2 = (1-8G\mu)\rho'^2 \tag{14.7}$$

brings it to the form (7.12). The metric (14.6) is locally flat, but for a nontrivial choice of f and g the metric (14.4) has a nonvanishing Riemann tensor.

To elucidate some interesting properties of the gravitational field (14.4), let us consider a pulse of length d propagating on a straight string. We can choose the origin of z so that at any time t the pulse is localized in the range

$$t \leqslant z \leqslant t + d. \tag{14.8}$$

An unusual feature of the gravitational field of the pulse is that for values of z outside this range the metric is the same as for a straight string. Hence, the pulse exerts a gravitational force on a test particle only for a short period $\Delta t \sim d$ when the particle is within the slab (14.8). Suppose the particle is initially at rest at a large distance, ρ, from the string (much greater than the amplitude of the pulse). Then it can be shown that, after the pulse has passed by, the particle has a velocity

$$v = \frac{4G\mu}{\rho} \int (f'^2 + g'^2) \, dt \tag{14.9}$$

directed towards the string. Note that an analogous situation arises in electromagnetism when one considers the electric field of an ultra-relativistic charge. In the limit $v \to 1$, the field is confined to the plane perpendicular to the direction of motion. The impulse received as this plane passes through a test charge is also inversely proportional to the distance from the source.

Acknowledgements

I am grateful to Tanmay Vachaspati for many helpful discussions and to Bruce Allen for his useful comments on the manuscript. This work was supported in part by the National Science Foundation and by the General Electric Company.

References

Aryal, M., Ford, L. H. and Vilenkin, A. (1986). *Phys. Rev.*, **D34**, 2263.
Aryal, M. and Everett, A. E. (1986). *Phys. Rev.*, **D34**, 434.
Bach, R. and Weyl, H. (1922). *Math. Z.*, **13**, 134.
Burden, C. J. (1985). *Phys. Lett.*, **164B**, 277.
Davis, R. L. (1985). *Phys. Rev.*, **D32**, 3172.
Ford, L. H. and Vilenkin, A. (1981). *J. Phys.*, **A14**, 2353.
Garfinkle, D. (1985). *Phys. Rev.*, **D32**, 986.
Gott, J. R. (1985). *Ap. J.*, **288**, 422.
Hiscock, W. A. (1985). *Phys. Rev.*, **D31**, 3288.
Hogan, C. J. (1986). University of Arizona preprint.
Hogan, C. J. and Rees, M. J. (1984). *Nature*, **311**, 109.
Israel, W. (1977). *Phys. Rev.*, **15**, 935.
Kaiser, N. and Stebbins, A. (1984). *Nature*, **310**, 191.
Kasner, E. (1921). *Amer. J. Math.*, **43**, 217.
Kibble, T. W. B. (1976). *J. Phys.*, **A9**, 1387.
Kibble, T. W. B. and Turok, N. (1982). *Phys. Lett.*, **116B**, 141.
Linet, B. (1985). *Gen. Relativ. Gravit.*, **17**, 1109.

Linet, B. (1986). *Phys. Rev.*, **D33**, 183.
Paczynski, B. (1986). *Nature*, **319**, 567.
Preskill, J. (1985). *Les Houches Lectures.*
Scherk, J. (1975). *Rev. Mod. Phys.*, **47**, 123.
Smith, A. G. (1986). *Phys. Rev.*, **D**, in press.
Sokolov, D. D. and Starobinsky, A. A. (1977). *Sov. Phys. Dokl.*, **22**, 312.
Turok, N. (1983). *Phys. Lett.*, **123B**, 387.
Turok, N. (1984). *Nucl. Phys.*, **B242**, 520.
Vachaspati, T. (1986). *Nucl. Phys.*, **B**, in press.
Vachaspati, T. and Vilenkin, A. (1985). *Phys. Rev.*, **D31**, 3052.
Vilenkin, A. (1981a). *Phys. Rev.*, **D23**, 852.
Vilenkin, A. (1981b). *Phys. Rev. Lett.*, **46**, 1169, 1496(E).
Vilenkin, A. (1981c). *Phys. Lett.*, **107B**, 47.
Vilenkin, A. (1984). *Ap. J. Lett.*, **282**, L51.
Vilenkin, A. (1985). *Phys. Rep.*, **121**, 263.
Vilenkin, A. (1986). *Nature*, **322**, 613.
Vilenkin, A. and Vachaspati, T. (1986). To be published.
Weinberg, S. (1972). *Gravitation and Cosmology.* Wiley: New York.
Witten, E. (1984). *Phys. Rev.*, **D30**, 272.

12

Inflationary cosmology

STEVEN K. BLAU† and ALAN H. GUTH‡

12.1 Introduction

During the last decade, experimental particle physics has tended to confirm the notion that the standard $SU(3) \times SU(2) \times U(1)$ model accounts for essentially all the physics that we have seen. In hopes of discovering new physical laws, many particle theorists have turned to speculations on what happens beyond the standard model, and much of this speculation has centered on grand unified theories (GUTs). (For a review, see Langacker, 1981.) GUTs explain the quantization of charge and also make a very good prediction for $\sin^2 \theta_W$ ($\theta_W \equiv$ Weinberg angle), giving grand unification a certain amount of plausibility. But the most dramatic predictions of GUTs occur only at the extraordinary energy scale of 10^{14} GeV. By the standards of the local power company, this is not an extraordinary amount of energy – it is roughly what it takes to light a 100 W light bulb for about a minute. However, the idea of having that much energy on a single elementary particle is extraordinary. If we were to try to build a 10^{14} GeV accelerator with present technology, we could in principle do it, more or less. It would be a linear accelerator with a length of about one light-year. Now such an accelerator is unlikely to be funded, so we must turn to other means to see the 10^{14} GeV consequences of GUTs. According to standard cosmology, the universe had a temperature with $kT = 10^{14}$ GeV at about 10^{-35} s after

† This work is supported in part by funds provided by the Robert A. Welch Foundation and in part by the US National Science Foundation (NSF) under contract PHY 8304629.
‡ This work is supported in part by funds provided by the US Department of Energy (DOE) under contract DE-AC02-76ERO3069, and in part by the National Aeronautics and Space Administration (NASA) under grant NAGW-553.

the big bang, and thus the universe itself becomes the best laboratory for studying the physics of very high energies.

The interface between particle physics and cosmology has become a very active field in recent years, and one of the outcomes has been the development of inflationary cosmology. The inflationary universe is a modification of the standard hot big bang model, motivated by several flaws that emerge when the standard model is extrapolated backward to very early times. The inflationary model agrees precisely with the standard model description of the observed universe for all times later than about 10^{-30} s, and all the successes of the standard model are preserved. For the first fraction of a second, however, the scenario is dramatically different. According to the inflationary model, the universe underwent a brief period of exponential expansion, or inflation, during which its scale factor increased by a factor perhaps 10^{50} times larger than in standard cosmology. In the course of this spectacular growth spurt all the matter, energy, and entropy in the universe could have been created from virtually nothing.

The goal of this article is to explain the basics of inflationary models, and then to summarize some of the more recent research. Emphasis will be placed on research in which we have been involved – we do not mean to imply that these investigations are more important than others, but we want to write about what we understand well. Furthermore, there are already very good reviews which emphasize other lines of development. Linde's review (1984a) gives particular attention to the recent work in which he has been involved, such as chaotic inflation, inflation in the context of supergravity, and also quantum cosmology. There is also an article by Linde (Chapter 13, this volume) which we have not yet seen. Brandenberger's review (1985) emphasizes the underlying quantum field theory, such as the effective potential, the finite temperature effective potential, the decay of the false vacuum, and Hawking radiation in de Sitter space. Steinhardt's article (1986) features a detailed explanation of the properties that an underlying particle theory must have in order to be consistent with inflationary cosmology. Turner's 'Inflationary Paradigm' (1985) emphasizes the role of density fluctuations, the requirements on the underlying particle theory, and some simple examples of particle theories that work. As the title implies, the paper also stresses that inflation is not a specific theory, but rather a mechanism which can be implemented in a number of different ways. We would also like to recommend the reprint volume edited by Abbott and Pi (1986).

Throughout this paper we will set $\hbar = c = k = 1$, and we will take the GeV as the fundamental unit for everything. We will set $G \equiv 1/M_P^2$, where $M_P = 1.22 \times 10^{19}$ GeV is the Planck mass. Note that $1\,\text{GeV} = 1.16 \times 10^{13}\,\text{K} = 1.78 \times 10^{-24}$ g, that $1\,\text{GeV}^{-1} = 1.97 \times 10^{-14}\,\text{cm} = 6.58 \times 10^{-25}$ s, and that 1 megaparsec (Mpc) $= 3.26 \times 10^6$ light-yr $= 3.09 \times 10^{24}\,\text{cm} = 1.56 \times 10^{38}\,\text{GeV}^{-1}$. We will use a metric with signature $(-1, 1, 1, 1)$.

12.2 Summary of the standard cosmological model

There is a rather well-defined standard cosmological model, the basic features of which will be reviewed in this section. The standard model rests on the assumption that the universe is homogeneous and isotropic, which implies that it can be described in comoving coordinates by the Robertson–Walker metric:

$$ds^2 = -dt^2 + R^2(t)\left[\frac{dr^2}{1-kr^2} + r^2(d\theta^2 + \sin^2\theta\,d\phi^2)\right]. \tag{2.1}$$

The universe is closed if $k > 0$, open if $k < 0$, and flat if $k = 0$. By convention, we rescale r and $R(t)$ so that the parameter k assumes one of the three discrete values -1, 0, or 1. We will refer to distances measured by the coordinate r as 'coordinate' distances, and the corresponding physical distances will be given by multiplication with the scale factor $R(t)$. The expansion of the universe is described by the growth of $R(t)$. The Einstein field equations relate the behavior of $R(t)$ to the energy–momentum tensor, which is restricted by the assumptions of homogeneity and isotropy to have the form $Ag^{\mu\nu} + Bu^\mu u^\nu$, where $u^\mu = (1,0,0,0)$ is a unit vector in the time direction. The energy–momentum tensor can therefore be written in the perfect fluid form

$$T^{\mu\nu} = pg^{\mu\nu} + (\rho + p)u^\mu u^\nu. \tag{2.2}$$

In a locally inertial comoving frame this expression reduces to $T^{\mu\nu} = \text{diag}[\rho, p, p, p]$, so one identifies ρ as the energy density, p as the pressure, and u^μ as the four-velocity of the fluid. The field equations then imply that

$$\ddot{R} = -\frac{4\pi}{3}G(\rho + 3p)R, \tag{2.3}$$

$$\left[\frac{\dot{R}}{R}\right]^2 = \frac{8\pi}{3}G\rho - \frac{k}{R^2}, \tag{2.4}$$

where the dot denotes differentiation with respect to time t. By combining eqs. (2.3) and (2.4), one obtains the relation for conservation of energy, which can be written as

$$\frac{d}{dt}(R^3\rho) = -p\frac{d}{dt}(R^3).$$ (2.5)

The quantity $H \equiv \dot{R}/R$ is known as the Hubble 'constant' even though it evolves as the universe evolves. Its value today is denoted by

$$H_0 \equiv h_0 \times (100 \text{ km/s Mpc})$$
$$= h_0 \times (9.78 \times 10^9 \text{ yr})^{-1},$$ (2.6)

where empirically h_0 probably lies between 0.4 and 1 (Aaronson *et al.*, 1980; Branch, 1979; de Vaucouleurs and Bollinger, 1979; Sandage and Tammann, 1976). (We will use the subscript '0' to denote the present value of any quantity.)

From eq. (2.4) one can define a 'critical density'

$$\rho_c = 3H^2/8\pi G,$$ (2.7)

which is the energy density that gives a flat ($k=0$) universe. Its value today is given by

$$\rho_{c0} = 1.87 \times 10^{-29} h_0^2 \text{ g cm}^{-3}.$$ (2.8)

The ratio of the actual energy density to the critical mass density (ρ/ρ_c) is called Ω, and its value today is believed to lie in the range (Schramm and Steigman, 1981; Faber and Gallagher, 1979; Peebles, 1979)

$$0.1 \lesssim \Omega_0 \lesssim 1.$$ (2.9)

The universe is assumed to expand adiabatically, with an energy density which is dominated at early times by the thermal radiation of effectively massless (i.e., $M \ll T$) particles at temperature T:

$$\rho = \frac{\pi^2}{30}N_{\text{eff}}T^4,$$ (2.10)

where

$$N_{\text{eff}} = N_b + \tfrac{7}{8}N_f.$$ (2.11)

Here N_b denotes the number of effectively massless bosonic spin degrees of freedom (e.g., photons contribute two units), and N_f denotes the number of effectively massless fermionic spin degrees of freedom (e.g. electrons and positrons together contribute four units). The entropy density s is given by

$$s = \frac{2\pi^2}{45}N_{\text{eff}}T^3.$$ (2.12)

Since the universe expands adiabatically, the entropy per comoving volume

$$S = R^3 s$$ (2.13)

is conserved. Thus, as long as the number of effectively massless particle species does not change, one has

$$R \propto T^{-1}. \tag{2.14}$$

Using eq. (2.10) one finds

$$\rho \propto R^{-4} \quad \text{(radiation-dominated)}, \tag{2.15}$$

which can be substituted into the Einstein equation (2.4). For the early universe R is small enough so that the curvature term $-k/R^2$ may be ignored, leading to the simple solution

$$R \propto t^{1/2} \quad \text{(radiation-dominated)}. \tag{2.16}$$

It follows that $H(t) = 1/2t$, and eqs. (2.4) and (2.10) can then be used to find

$$T^2 = \left(\frac{45}{\pi^3 N_{\text{eff}}}\right)^{1/2} \frac{M_P}{4t} \quad \text{(radiation-dominated)}, \tag{2.17}$$

where $M_P \equiv 1/G^{1/2} = 1.2 \times 10^{19}$ GeV is the Planck mass.

At the highest temperatures all of the fundamental particles of nature contributed to the thermal radiation. As the temperature fell below the mass of a given particle species, those particles disappeared from the thermal equilibrium gas – they are sometimes said to have 'frozen out'. Modern particle physics also predicts that there must have been a number of phase transitions which took place as the universe cooled, but in the standard model one assumes that these phase transitions were inconsequential. That is, they occurred quickly when the temperature fell to the critical temperature, and the release of latent heat was negligible.

When the temperature reached 1 MeV, the effectively massless degrees of freedom were the photons, electrons, positrons, and neutrinos. If there are three species of neutrinos, then $N_{\text{eff}} = 10\frac{3}{4}$ and $t = 0.74$ s. Calculations of the neutrino interaction rates indicate that at this time the neutrinos began to 'decouple' – i.e., they lost thermal contact with the rest of matter. For all times somewhat later than this, the neutrinos can be described as a collisionless gas. It follows that the neutrinos maintained a thermal distribution, with RT_ν equal to the same value it had before decoupling. When $T \approx \frac{1}{2}$ MeV (at $t \approx 3.0$ s), the electron–positron pairs began to freeze out. Their entropy (given by eq. (2.12) with $N_{\text{eff}} = 2$) was imparted to the photons, and as a result RT_γ was increased by a factor of $(\frac{11}{4})^{1/3}$. The ratio of the neutrino temperature to the photon temperature,

$$T_\nu/T_\gamma = (\tfrac{4}{11})^{1/3}, \tag{2.18}$$

has been maintained until the present. If we assume that there are three species of massless neutrinos, then the radiation energy density after electron–positron annihilation is given by eq. (2.10) with

$$N_{\text{eff}} = 2 + 6 \times \tfrac{7}{8}(\tfrac{4}{11})^{4/3} = 3.363. \tag{2.19}$$

The interactions which convert protons into neutrons and vice versa all involve neutrinos, so the ratio of neutrons to protons became essentially constant as the neutrinos decoupled. After neutrino decoupling, the ratio was changed only by the free decay of the neutron, which has a mean life of 898 ± 16 s. At $t \approx 3.5$ min the temperature fell low enough for deuterium to become relatively stable, and the processes of big bang nucleosynthesis began. It is possible to calculate the resulting abundances of ^{4}He, ^{3}He, ^{2}H, and ^{7}Li, and the comparisons with observation are excellent (Yang *et al.*, 1984; Boesgaard and Steigman, 1985). The success of these calculations, along with the observation of the cosmic background radiation, provides the primary evidence for the standard big bang model of the universe. (Note that the abundances of heavier elements are not predicted, as these elements were primarily produced later in the history of the universe in the interior of stars.)

In addition to thermal radiation, the energy density of the early universe also contained a contribution from 'matter' – i.e., from massive particles, such as protons and neutrons. During the time period when these particles were non-relativistic (i.e., $T \ll M$), their number was conserved and their contribution to the total energy density was dominated by their rest mass. This implies that

$$\rho \propto R^{-3} \quad \text{(matter-dominated)}. \tag{2.20}$$

For simplicity, we will now assume that the universe is described by $\Omega = 1$. If we insert eq. (2.20) into eq. (2.4) and set $k = 0$, we obtain

$$R \propto t^{2/3} \quad \text{(flat, matter-dominated)}. \tag{2.21}$$

Today the energy density appears to be almost entirely in the form of massive particles,† so the density of these massive particles at earlier times would be given by

$$\rho_m(t) = \rho_{c0} \left[\frac{R(t_0)}{R(t)} \right]^3 \approx \rho_{c0} \left[\frac{T_\gamma(t)}{T_{\gamma 0}} \right]^3, \tag{2.22}$$

where t_0 denotes the present time, and $T_{\gamma 0}$ denotes the present value of the photon temperature. This expression will equal the energy density ρ_r of

---

† It is conceivable that the universe today is dominated by some unseen relativistic decay product, but we will not deal with this possibility here. (See Turner, Steigman and Krauss, 1984.)

radiation of massless particles when $T_y = 5.79\,h_0^2\tau_0^{-3}$ eV, where

$$T_{y0} \equiv \tau_0 \times 2.7\ \text{K}. \tag{2.23}$$

According to eq. (2.17) this temperature was reached at time $t = 1.25 \times 10^3\,\tau_0^6 h_0^{-4}$ yr. From that time onward the universe has been matter-dominated.

12.3 Problems of the standard cosmological model

The standard hot big bang model has been accepted by most of the scientific community, and with good reason. The model is very successful in its predictions of the Hubble expansion law, the cosmic microwave background radiation, and the abundances of the light chemical elements. However, these successes do not probe the behavior of the model all the way back to time zero. Of the three predictions, the calculation of the element abundances provides the test of the model which probes it at the earliest time – the relevant processes began at about one second after the big bang. For times earlier than one second, there is no direct experimental evidence which validates the standard model.

When the standard model is extrapolated backward to times much earlier than one second, there are at least four problems which become apparent. It is the existence of these problems which motivates the search for a modification of the model.

The first of these problems is the monopole problem – that is, in the context of GUTs of particle physics, the standard model leads to a tremendous overproduction of magnetic monopoles. To see how this comes about, recall that the monopoles are in fact topologically stable knots in the Higgs field expectation value.

Suppose the Higgs field is characterized by a correlation length ξ. Then for each cubic region of space with edge length ξ, the Higgs field on one face of the cube is essentially independent of the Higgs field on the opposite face. We may estimate that each such cube with volume ξ^3 corresponds to a topological knot in the Higgs field expectation value, so that the monopole density is roughly given by (Kibble, 1976)

$$n_\text{M} \approx \xi^{-3}. \tag{3.1}$$

While this argument is crude, we nonetheless expect the relation to be accurate to within one or two orders of magnitude. This accuracy will be sufficient to establish the conclusion which will follow.

Let $T_c \approx 10^{14}$ GeV denote the temperature of the GUT phase transition to the $SU(3) \times SU(2) \times U(1)$ phase. The corresponding time is obtained from eq. (2.17) – if we take $N_{eff} \approx 10^2$, which is typical of simple GUTs, we find $t_c \approx 10^{-35}$ s. It can be shown (Ellis and Steigman, 1980) that the Higgs field had time to reach thermal equilibrium before the phase transition took place, so that for $T > T_c$ the correlation length must be given by the only relevant length scale, T^{-1}. (Note that even if one were to imagine that the universe started out with the Higgs field uniformly aligned, thermal agitation would destroy this alignment and produce a randomly oriented Higgs field with correlation length T^{-1}.) When the universe cooled below T_c, it suddenly became thermodynamically favorable for the Higgs field to align uniformly over large distances. However, it takes time for these correlations to be established. Causality constrains the Higgs field correlation length to be less than the horizon distance

$$l_H(t) = R(t) \int_0^t \frac{dt'}{R(t')}, \tag{3.2}$$

which is the total distance a light pulse could have traveled since the initial singularity. In the standard cosmology $R(t) \propto t^{1/2}$, so $l_H = 2t$. Since the phase transition is assumed to have occurred quickly at time t_c, eq. (3.1) provides an approximate lower bound on the magnetic monopole density:

$$n_M \approx \xi^{-3} \geq \frac{1}{8t_c^3}. \tag{3.3}$$

Preskill has shown (1979; see also Zeldovich and Khlopov, 1978; Goldman, Kolb and Toussaint, 1981) that at densities near $1/8t_c^3$ monopole–antimonopole annihilation is totally ineffectual. Since the monopole mass $m_M \approx 10^{16}$ GeV, the monopoles behave as non-relativistic particles with a number density which is diluted with the expansion of the universe as $1/R^3(t)$. The energy density is then given by

$$\rho_M = n_M m_M \propto R^{-3}. \tag{3.4}$$

It is useful to evaluate the ratio ρ_M/s; since the numerator and denominator each fall off as $1/R^3(t)$ as the universe expands, the ratio remains constant. The value of the ratio immediately after the phase transition can be bounded by using eqs. (2.12), (2.17), (3.3), and (3.4), which lead to the inequality

$$\rho_M/s \geq c(N_{eff})^{1/2} \frac{m_M T_c^3}{M_P^3}, \tag{3.5}$$

where $c = 4\pi^2(\pi/45)^{1/2} = 10.4$.

The present-day entropy density s_0 can be calculated using eq. (2.12), assuming that it is dominated by the cosmic background radiation of photons and neutrinos. Thus

$$s_0 = \frac{2\pi^2}{45}(2T_\gamma^3 + \tfrac{7}{4}N_v T_v^3), \tag{3.6}$$

where N_v is the number of species of massless neutrinos, each assumed to contribute two spin degrees of freedom, and T_v is the neutrino temperature, given by eq. (2.18). Taking $N_v = 3$ to reflect the observed families of neutrinos, one finds

$$s_0 = 2.8 \times 10^3 \, \text{cm}^{-3} \tag{3.7}$$

and

$$\rho_M \gtrsim 5 \times 10^{-18} \, \text{g cm}^{-3} \tag{3.8}$$

today.

It may not seem so, but ρ_M is an outrageously high energy density. We may compare ρ_M to the critical density ρ_c given by eq. (2.8), and we find (taking $h_0 \approx 1$)

$$\Omega_M \equiv \rho_M/\rho_c \gtrsim 3 \times 10^{11}, \tag{3.9}$$

which is totally incompatible with the observational limits on the present value of Ω, shown in eq. (2.9). In particular, an energy density as high as that described by eq. (3.9) would imply that the expansion rate of the universe is decreasing rapidly – it would have slowed to its present value in only 30 000 years!

The second problem of the standard model is called the flatness problem, and is based on the realization by Dicke and Peebles (1979) that it is rather puzzling for the value of Ω today to be of order unity. $\Omega(t) = 1$ is an unstable equilibrium point in the standard model, and the only time scale which appears in the equations describing the early universe is the Planck time

$$t_P = (M_P)^{-1} = 5.4 \times 10^{-44} \, \text{s}. \tag{3.10}$$

Since the universe is about 10^{60} Planck times old, it is remarkable that Ω is still in the vicinity of one.

Of course for a $k=0$ universe Ω is always exactly equal to one, but we regard this possibility as very unlikely. Recall that the Robertson–Walker metric is defined for any real value of k, and that the discrete choices $(+1, -1, \text{or } 0)$ are obtained by rescaling the coordinates. Thus any positive k is scaled to $+1$, any negative k is scaled to -1, but $k=0$ is obtained only by starting with $k=0$, which is a set of measure zero on the real line. In the following discussion the possibility $k=0$ is ignored.

We can express the flatness problem more quantitatively by considering the behavior in time of $(\Omega-1)/\Omega$, which using eqs. (2.4) and (2.7) can be written as

$$\frac{\Omega-1}{\Omega}=\frac{3k}{8\pi G\rho R^2}.$$ (3.11)

The scale factor R can be eliminated from this equation by using (2.13), leading to

$$\frac{\Omega-1}{\Omega}=\frac{3ks^{2/3}}{8\pi G\rho S^{1/3}}.$$ (3.12)

This equation is useful because the quantities on the right-hand side are easy to evaluate: ρ and s are expressed in terms of the temperature by eqs. (2.10) and (2.12), and S is constant as the universe evolves. To calculate the value of S, note that eqs. (2.4) and (2.7) imply that

$$R^2=\frac{k}{H^2(\Omega-1)},$$ (3.13)

and therefore

$$S=\left[\frac{k}{H^2(\Omega-1)}\right]^{2/3}s.$$ (3.14)

The right-hand side of this equation can be evaluated for the present era, taking $H\approx(10^{10}\,\mathrm{yr})^{-1}$, $|\Omega-1|<1$, and $s=2.8\times10^3\,\mathrm{cm}^{-3}$. The result is

$$S>10^{87},$$ (3.15)

which is an extraordinarily large number. The fact that S is so large is an alternative statement of the flatness problem. In the context of the standard cosmological model S is a parameter whose value is fixed by the initial conditions, and one would expect its value to be of order unity unless there is some reason to believe otherwise. In this language, the flatness problem is the fact that the standard model provides no reason to believe otherwise.

One can calculate, for example, the allowed range of Ω when the temperature was $1\,\mathrm{MeV}$, when the processes of big bang nucleosynthesis were just beginning (at about one second after the big bang). Using eq. (3.12) one finds

$$|\Omega-1|\lesssim10^{-15}.$$ (3.16a)

At the time of the GUT phase transition, when $T\approx10^{14}\,\mathrm{GeV}$, one has

$$|\Omega-1|\lesssim10^{-49}.$$ (3.16b)

Finally, if one extrapolates all the way back to the Planck time, when $T\approx10^{19}\,\mathrm{GeV}$, then

$$|\Omega-1|\lesssim10^{-59}.$$ (3.16c)

The flatness problem is then the difficulty in understanding why Ω was so extraordinarily close to one at these time scales.

Note that the value of Ω will approach one at early times in any Friedmann–Robertson–Walker model which is either radiation or matter-dominated, and so some cosmologists have concluded that there is no flatness problem. However, if one adopts the point of view that the Planck time t_P is the natural time scale of the system, then t_P is not an early time. The fact that Ω is so extraordinarily close to one at t_P is then not explained by saying that t_P is small, and some other explanation must be sought.

Note also that the flatness problem is not an inconsistency in the standard cosmological model. If one assumes that the initial value of Ω was within the required range, then the model will evolve in a way that agrees with our observations. The flatness problem is really just a severe shortcoming in the predictive power of the standard model. The dramatic flatness of the early universe cannot be predicted by the standard model, but must instead be assumed in the initial conditions.

The third problem is the horizon problem, first pointed out by Rindler (1956; see also Weinberg, 1972, pp. 525–6; Misner, Thorne and Wheeler, 1973, pp. 740 and 815). The observational basis for this problem is the large-scale homogeneity of the universe, seen most strikingly in the uniformity of the cosmic background radiation. The effective temperature of the cosmic background radiation is known to be isotropic to about one part in 10^3, and even this anisotropy can be accounted for by the assumption that the earth is moving through the background radiation. If one removes a component which can be attributed to the earth's motion, then the residual anisotropy is known to be less than one part in 10^4 (Wilkinson, 1986). This extreme uniformity is very difficult to understand in the context of the standard cosmological model, in which the horizon distance (i.e., the distance that a light pulse could have traveled since the initial singularity) is rather short.

Consider, for example, two microwave antennae pointing in opposite directions. Each is receiving radiation that is believed to have been emitted (or last scattered) at the time of hydrogen recombination, t_r, about 10^5 years after the big bang, when the temperature T_r was about 4000 K. (At earlier times the plasma which filled the universe was opaque to this radiation.) At the time of emission, these two sources were separated from each other by many horizon distances. To calculate how many, let us assume that $\Omega \approx 1$, and that the integral in eq. (3.2) for the horizon distance is dominated by the matter-dominated period, with $R(t) = bt^{2/3}$ for some constant b. The coordinate distance between the source of the cosmic background radiation and us is then given by

$$r_{\text{coord}} = \int_{t_r}^{t_0} \frac{dt'}{bt'^{2/3}} = 3b^{-1}(t_0^{1/3} - t_r^{1/3}). \tag{3.17}$$

The coordinate value of the horizon distance at t_r is given by

$$l_{\text{H,coord}} = \int_0^{t_r} \frac{dt'}{bt'^{2/3}} = 3b^{-1}t_r^{1/3}. \tag{3.18}$$

Thus, the number of horizon distances separating the two sources in opposite directions is given by

$$N = \frac{2r_{\text{coord}}}{l_{\text{H,coord}}} = 2\left[\left(\frac{t_0}{t_r}\right)^{1/2} - 1\right] = 2\left[\left(\frac{T_r}{T_0}\right)^{1/2} - 1\right] \approx 75. \tag{3.19}$$

The problem is to understand how two regions over 75 horizon distances apart came to be at the same temperature at the same time.† In the standard model this large-scale homogeneity is simply assumed as an initial condition.

Like the flatness problem, the horizon problem is not an inconsistency in the standard model, but represents instead a lack of explanatory power. If the universe is assumed to have begun homogeneously, then it will continue to evolve homogeneously. The problem is that the very striking large-scale homogeneity of the universe is not explained or predicted by the model, but instead must simply be assumed.

The fourth problem is the density fluctuation problem – the difficulty in understanding the origin of structure in the universe. Although the universe appears homogeneous on scales greater than a few times 10^8 light years, on smaller scales one sees galaxies, clusters, voids, etc. To account for this structure, one must assume an initial spectrum of mass density perturbations $\delta\rho/\rho$. In the standard model there is no explanation for these perturbations, and the entire spectrum must be assumed as part of the initial conditions.

The fact that the spectrum of perturbations is unexplained in the standard model is a drawback in itself, but the situation becomes even more puzzling when the starting time is taken as early as $t \approx 10^{-35}$ s. The problem hinges on the fact that mass density perturbations are gravitationally unstable – a

---

† In the above discussion we have assumed for simplicity that $\Omega = 1$ and that the universe was matter-dominated throughout the relevant portion of its history. The calculation has been carried out (Guth, 1983a) without using either of the approximations, and it was found that the problem is even a little worse – the two regions were at least 90 horizon distances apart.

region with excess mass will produce an attractive gravitational field, further increasing the mass density contrast. Thus, if one extrapolates to very early times one must find a much more uniform distribution of mass. At $t \approx 10^{-35}$ s one must assume that these perturbations were present, but it is difficult to understand why they were so incredibly small.

To determine how small the perturbations must have been at $t = 10^{-35}$ s, we must review some of the basic facts concerning the growth of mass density perturbations (Olson, 1976; Bardeen, 1980; Press and Vishniac, 1980). Since we are interested in the early universe, it will be sufficient to deal with perturbations in a flat, radiation-dominated Robertson–Walker universe. One begins by decomposing the mass density function into homogeneous and inhomogeneous pieces:

$$\rho(\check{x}, t) = \rho_0(t) + \delta\rho(\check{x}, t), \tag{3.20}$$

where $\check{x}$ is the comoving coordinate of the Robertson–Walker coordinate system. The next step is to write down the general relativity equations that describe the evolution of $\delta\rho$ and the changes in the metric which are induced. Since $\delta\rho/\rho$ is assumed to be small in the early universe, one can simplify the equations by expanding to first order in this quantity. Finally, the linearized equations can be Fourier transformed, and the resulting equations are then soluble. The results that one finds depend somewhat on how one chooses to define the equal-time hypersurfaces in the perturbed spacetime – one popular choice is the 'comoving gauge', in which the time variable is defined by a set of clocks that move with the matter fluid. One then finds that the behavior of the Fourier transformed quantity $\delta\tilde{\rho}(\vec{k}, t)/\rho_0$ depends on the relative magnitudes of the physical wavelength and the Hubble length $H^{-1} = 2t$. (Since the horizon length $l_{\mathrm{H}} = 2t$ as well, one often hears the words 'Hubble length' and 'horizon length' used interchangeably in this context.) The physical wavelength corresponding to any given comoving wave number $\vec{k}$ grows with the scale factor as $\sqrt{t}$, and so at early times the wavelength is always larger than the Hubble length, and at late times it is smaller. When the wavelength is much larger than the Hubble length, one finds that

$$\frac{\delta\tilde{\rho}}{\rho}(\vec{k}, t) \propto t, \tag{3.21}$$

while the perturbation is found to oscillate with constant amplitude when the wavelength becomes much less than the Hubble length. (The perturbations will start to grow again after the universe becomes matter-dominated, but that era is not relevant for the present discussion.)

Although an elaborate calculation is necessary to verify these results, there is a simple argument which demonstrates their plausibility. For long wavelengths the behavior is the same as for an infinite wavelength – i.e., a homogeneous perturbation from a flat universe. By inserting into eq. (3.12) the behavior of s and ρ for a flat, radiation-dominated universe, one finds that $(\Omega - 1)/\Omega$ grows linearly with t.

Now let us consider in particular the growth of perturbations on the length scale corresponding to a typical galaxy, with a mass $M_g \sim 10^{12} M_\odot$, where $M_\odot \approx 2 \times 10^{33}$ g is the solar mass. In order for galaxies to have formed, it is necessary for the perturbations on this scale to have become non-linear several billion years into the history of the universe, and this implies that the value of $\delta\tilde{\rho}/\rho$ must have been about 10^{-4} when the wavelength was equal to the Hubble length (Zeldovich, 1972; Press and Vishniac, 1980). To estimate when this occurred, one can calculate for any given time the total baryonic mass in a sphere with radius equal to the Hubble length. The baryonic mass density varies as $1/R^3(t)$, and so it can be written as

$$\rho_B(t) = \frac{3H_0^2\Omega_{B0}}{8\pi G} \frac{R^3(t_0)}{R^3(t)}, \tag{3.22}$$

where H_0, Ω_{B0}, and $R(t_0)$ refer to the present values of the Hubble constant, the baryonic contribution to Ω, and the scale factor, respectively. Using the fact that $RT \approx constant$, the last factor can be replaced by $T^3(t)/T^3(t_0)$, and $T(t)$ can be evaluated using eq. (2.17). The baryonic mass within the sphere is then given by

$$M_B = \left(\frac{45}{\pi^3 N_{\text{eff}}}\right)^{3/4} \frac{M_P^{7/2}\Omega_{B0}H_0^2 t^{3/2}}{2T^3(t_0)}$$

$$\approx 1.0(\Omega_{B0}h_0^2)M_\odot \left(\frac{t}{1\,\text{s}}\right)^{3/2}. \tag{3.23}$$

In the second line above we took the value of N_{eff} from eq. (2.19), which applies for times later than a few seconds. Taking $\Omega_{B0}h_0^2 \approx 10^{-2}$ and setting $M_B = M_g$, one finds

$$t(M_g) \approx 2 \times 10^9 \text{ s.} \tag{3.24}$$

The galaxy scale perturbations will grow linearly with t for times earlier than $t(M_g)$, and so they will grow by a factor of about 2×10^{44} between $t = 10^{-35}$ s and $t(M_g)$. Given the desired value of 10^{-4} at $t(M_g)$, one has

$$\frac{\delta\tilde{\rho}}{\rho} \text{ (galaxy scale)} \approx 5 \times 10^{-49} \quad \text{at } t = 10^{-35} \text{ s.} \tag{3.25}$$

This sounds like a very small number, but to decide for sure one should compare it to some reasonable standard. As a comparison, we will consider the $1/N^{1/2}$ Poisson fluctuations which are characteristic of any gas of particles with random locations.

The universe contains about 10^{10} particles per baryon – mostly photons and neutrinos – and so the number of particles associated with a typical galaxy is given by

$$N = 10^{12} M_\odot \times \frac{10^{57} \text{ baryons}}{M_\odot} \times \frac{10^{10} \text{ particles}}{\text{baryon}} = 10^{79} \text{ particles.} \quad (3.26)$$

Thus, the Poisson fluctuations in the mass density are

$$\frac{\delta\rho}{\rho} = \frac{1}{N^{1/2}} \approx 3 \times 10^{-40}, \quad (3.27)$$

These Poisson fluctuations are nine orders of magnitude larger than those permitted by the standard cosmological model. If we extrapolate this model back to the Planck time,

$$\frac{\delta\tilde{\rho}}{\rho} (\text{galaxy}) \approx 5 \times 10^{57} \quad \text{at } t = 10^{-43} \text{ s}, \quad (3.28)$$

and so at this time the fluctuations required in the standard model are seventeen orders of magnitude smaller than Poisson fluctuations.

Thus, if one is to begin the standard cosmological model at times as early as 10^{-35} s, then one must begin with density fluctuations which are non-zero, but which are incredibly small – much smaller than $1/N^{1/2}$. Initial conditions of this sort seem very peculiar, since density fluctuations of order $1/N^{1/2}$ are almost universal in macroscopic systems – they are present (in flat space) for a classical gas in thermal equilibrium, and also for quantum thermal radiation of either bosons or fermions. Density fluctuations are much less than $1/N^{1/2}$ only for highly ordered systems, such as a configuration of particles arranged uniformly on a lattice. Density fluctuations are also much less than $1/N^{1/2}$ for a degenerate Fermi gas at zero temperature – in this case, the fluctuations are suppressed by the long-range correlations imposed by the Pauli exclusion principle.

The density fluctuation problem may be viewed as a local version of the flatness problem, and it was described in general terms in the same paper by Dicke and Peebles (1979). The flatness problem is, in essence, the observation that the universe in the large is unstable against perturbations away from $\Omega = 1$. The density fluctuation problem hinges upon the fact that the universe is unstable against mass density perturbations over a very wide range of mass scales.

12.4 The original inflationary universe

The inflationary universe is a modification of the standard hot big bang model which avoids the problems discussed above. According to the inflationary model, the early universe underwent a brief period during which the matter was in a metastable 'false vacuum' state, driving the evolution of the universe into exponential expansion. During this period the scale factor of the universe increased by a tremendous factor – perhaps 10^{50} times larger than in standard cosmology.

There were a number of precursors to the inflationary model, as has recently been discussed in a very interesting historical article by Lindley (1985). Gliner (1965, 1970) and Gliner and Dymnikova (1975) discussed a state of matter which they called the 'μ-vacuum', which has the same energy–momentum tensor as the false vacuum. The μ-vacuum, however, was motivated solely by symmetry considerations. The theory therefore seemed highly speculative, and a totally *ad hoc* recipe had to be introduced to describe the decay of the μ-vacuum. Zeldovich (1968) was perhaps the first to realize that a vacuum energy is equivalent to a cosmological constant – he suggested that an effective cosmological constant might arise from the renormalization of the QED vacuum. Linde (1974) realized that a spontaneously broken gauge theory would give rise to an effective cosmological constant in the symmetric phase, but he concluded (correctly) that this term could never become dominant at temperatures for which the symmetric phase is stable. Similar but more detailed derivations were carried out later by Bludman and Ruderman (1977) and by Kolb and Wolfram (1980). These papers assumed that the universe remained in thermal equilibrium, and therefore they found no inflation. (Later Guth and Sher (1983) showed on general grounds that a departure from thermal equilibrium is a prerequisite for inflation.) Sato (1981a, b) became interested in strongly supercooled phase transitions, attempting to build a consistent baryon number symmetric universe in which *CP*-violation occurs spontaneously in domains of supercluster proportions. At about the same time Guth and Tye (1980) and Einhorn and Sato (1981) studied supercooled phase transitions as a method of solving the monopole problem. It was this work that led Guth (1981) to the realization that an extremely supercooled phase transition could potentially solve not only the monopole problem, but the horizon and flatness problems as well.

In this section we will describe the original form of the inflationary universe model (Guth, 1981). While this model solves the flatness and horizon problems, and perhaps the monopole problem as well, it nonetheless has a fatal flaw, which is sometimes called the 'graceful exit problem'. This graceful exit problem is avoided by the 'new inflationary universe', which will be discussed in Section 12.7. Here we discuss the original inflationary universe model for its pedagogical value – the model demonstrates how the flatness, horizon, and monopole problems can be solved, and the failure of the model provides the motivation for the special features of the new inflationary universe.

The original inflationary scenario requires a Higgs field effective potential $V(\phi)$ similar to the one shown in Fig. 12.1. While we graph ϕ schematically as if it were one-dimensional, it is in reality a multicomponent field. At zero temperature, the global minimum of the effective potential is called the true vacuum, with $\phi = \phi_{\text{true}}$; the scalar field in the universe today is undergoing small fluctuations about this true vacuum value. The value of $V(\phi_{\text{true}})$ represents the energy density of the vacuum, and is related to the cosmological constant Λ by

$$\Lambda = 8\pi G V(\phi_{\text{true}}).\tag{4.1}$$

The cosmological constant might be large enough to have significant effects in the present universe, but the mere fact that it does not overwhelm the present universe implies that the energy density of the vacuum is utterly negligible compared with the energy densities of relevance to the early universe.† Thus, we can assume that $V(\phi_{\text{true}}) = 0$.

The zero temperature effective potential is also required to have a secondary local minimum, and the value of ϕ at this minimum will be called ϕ_{false}. The state of lowest energy density for which the Higgs field expectation value $\langle \phi \rangle = \phi_{\text{false}}$ is called the false vacuum, and the energy density of this state will be called ρ_{f}. Furthermore, we assume that there is a critical temperature T_{c} above which the finite temperature effective potential has a lower value near the false vacuum than it has near the true vacuum, as shown in Fig. 12.1.‡ Thus, T_{c} is the critical temperature for a first-order phase transition.

† While it is clear observationally that the energy density of the vacuum is negligible compared with that of the early universe, there is no accepted theoretical explanation of this fact. Naively, one would expect the vacuum energy density to be of order of the fourth power of the largest mass scale appearing in the theory, which would presumably mean of order M_{p}^4. The fact that the empirical limit is more than 120 orders of magnitude smaller than this theoretical estimate is regarded by many to be one of the deepest mysteries in physics.

‡ The finite temperature effective potential is actually the free energy density of the system, and it must be minimized in the thermal equilibrium state.

For a wide range of parameters (Guth and Tye, 1980) the Higgs field in the minimal $SU(5)$ GUT has an effective potential with precisely these properties. The false vacuum is a Higgs field configuration which breaks the gauge symmetry to $SU(4) \times U(1)$ rather than $SU(3) \times SU(2) \times U(1)$. The critical temperature is $T_c \approx 10^{14}$ GeV.

Although a false vacuum has never been observed, its essential properties are independent of the details of the particle theory and can therefore be predicted rather unambiguously. The false vacuum has a constant energy density $\rho = \rho_f$, the value of which is determined by the parameters of the particle theory. The energy density is typically of the order of the fourth power of the characteristic mass scale of the theory, which for GUTs means that

$$\rho_f \approx (10^{14}\,\text{GeV})^4 \approx 10^{73}\,\text{g cm}^{-3}. \tag{4.2}$$

This energy density is almost unimaginably large. It is the energy density that a large star would have if it were compressed to the size of a proton.

We may deduce the energy–momentum tensor $T_{\mu\nu}$ of the Higgs field in the false vacuum by observing that $T_{\mu\nu}$ is a covariantly conserved tensor constructed from the Higgs field, the metric, and the first and second derivatives of these fields. The Higgs field in the false vacuum is constant, so the only tensors that may be constructed are $g_{\mu\nu}$ and

$$G_{\mu\nu} \equiv R_{\mu\nu} - \tfrac{1}{2} g_{\mu\nu} R. \tag{4.3}$$

Thus, the energy–momentum tensor must assume the form

$$T_{\mu\nu} = A g_{\mu\nu} + B G_{\mu\nu}, \tag{4.4}$$

where A and B are constants. The Einstein field equations are

$$G_{\mu\nu} = -8\pi G T_{\mu\nu}, \tag{4.5}$$

Fig. 12.1. A schematic Higgs effective potential appropriate for the original inflationary universe.

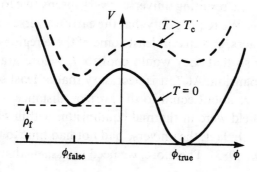

so that the term $BG_{\mu\nu}$ in eq. (4.4) may be absorbed into the left-hand side of (4.5) by redefining the gravitational constant. Having done this we write

$$T_{\mu\nu} = -\rho_f g_{\mu\nu}, \tag{4.6}$$

where eq. (2.2) has been used to identify the constant A with the energy density of the false vacuum. Eq. (4.6) implies that the pressure p of the false vacuum equals $-\rho_f$; it is large and negative. Since the pressure is constant, it has no gradient and therefore does not produce any mechanical forces. However, the pressure does have very dramatic gravitational effects which will be discussed below. The energy–momentum tensor of eq. (4.6) produces a term in the Einstein field equations which is identical in form to a positive cosmological constant – the only difference is that the energy–momentum tensor of the false vacuum is not permanent, but changes form when the false vacuum decays.

The false vacuum is unstable, and decays by the spontaneous nucleation of bubbles of the new phase. For some parameters it is also possible for magnetic monopoles, which could remain as relics from a high temperature past, to serve as nucleation sites for bubbles of the new phase (Steinhardt, 1981a, b; Guth and E. Weinberg, 1981). However, it appears that a phase transition driven by monopoles could never be slow enough to allow for a successful inflationary model, so we will assume that these monopoles are stable against nucleation. (Steinhardt (1981a, b) has shown that this assumption is valid for a wide range of parameters in the context of the minimal $SU(5)$ GUT.) The process of random bubble nucleation will be discussed in more detail in Section 12.6, but for now it will be sufficient to simply assume that the nucleation rate is very low.

We are now ready to describe the chronology of the original inflationary model. As with other cosmological scenarios, the starting point is somewhat a matter of taste and philosophical prejudice. An advantage of the inflationary scenario is that it appears to allow a wide variety of starting configurations – the resulting universe is very insensitive to the details of the initial conditions. We require only that the early universe was hot ($T > T_c$) in at least some places, and that at least some of these regions were expanding rapidly enough so that they would cool to T_c before gravitational effects reversed the expansion. At T_c it is necessary that at least some of these hot regions had a size about equal to the horizon distance.

If the Higgs field were in thermal equilibrium within such a hot region, then $\langle \phi \rangle \approx \phi_{false}$. In fact, the universe had not had time to thermalize at this point (Steigman, 1983). Therefore, we need to *assume* that there were some regions of high energy density with $\langle \phi \rangle \approx \phi_{\text{false}}$. These regions would cool to T_c and would then start to supercool below T_c. At this point the phase transition would begin, occurring through the spontaneous nucleation of bubbles of the new phase. We assume, however, that the nucleation rate is very slow, so that the initially very hot regions would supercool to T near zero while remaining near the false vacuum. (This assumption is known (Guth and E. Weinberg, 1981) to be valid for a wide range of parameters in the minimal $SU(5)$ model.)

We now follow the evolution of those regions which have supercooled and approached the false vacuum state. To see what happens next, it is easiest to begin by assuming that the region is homogeneous, isotropic, and flat. (Later we will describe what happens when these assumptions are dropped.) The region can then be described by the metric of eq. (2.1) with $k = 0$, and eq. (2.4) becomes

$$\left[\frac{\dot{R}}{R} \right]^2 = \frac{8\pi}{3} G \rho_{\text{f}}, \tag{4.7}$$

which has the solution

$$R(t) \propto e^{\chi t} \tag{4.8}$$

where

$$\chi = \left(\frac{8\pi G}{3} \rho_{\text{f}} \right)^{1/2} \approx 10^{10} \, \text{GeV} \approx 10^{34} \, \text{s}^{-1}. \tag{4.9}$$

The remarkable consequence of the huge positive energy density of the false vacuum is the exponential growth of the scale factor $R(t)$, with minuscule time constant χ^{-1}. This exponential expansion is the hallmark of the inflationary model. The space described by a flat Robertson–Walker metric with an exponentially growing scale factor is called de Sitter space.† The dynamics of the inflationary era can also be understood by considering the second-order Einstein equation, eq. (2.3). This equation implies that the deceleration of the cosmic scale factor, $-\ddot{R}(t)$, is proportional to $\rho + 3p$. In the standard cosmology ρ and p are both positive so that gravity acts to slow down the expansion rate of the universe – $\dot{R}$ decreases monotonically with time. During the inflationary era, on the other hand, $p = -\rho = -\rho_{\text{f}}$ so that the pressure contributes a negative term which overwhelms the positive contribution of the energy density – the effect of gravity is reversed, and the expansion rate $\dot{R}(t)$ increases with time. While the inflationary model does not attempt to explain the formation of the initially hot expanding regions

† The properties of de Sitter space are well described by Hawking and Ellis (1973).

which subsequently supercool into the false vacuum state, it does explain the origin of most of the momentum of the cosmic expansion: the big bang gets its big push from the false vacuum. The standard cosmology, by contrast, makes no attempt to explain the expansion of the big bang.

Now let us consider what would happen if the initial region were not homogeneous, isotropic, and flat. In that case, one must examine the behavior of perturbations about the de Sitter metric. These perturbations seem to be governed by a 'cosmological no-hair theorem', which states that whenever the energy–momentum tensor is given by eq. (4.6), any locally measurable perturbation about the de Sitter metric is damped exponentially on the time scale of χ^{-1}. Any initial particle density is diluted to negligibility, and any initial distortion of the metric is stretched (i.e., redshifted) until it is no longer locally detectable.

If other matter is present in addition to that described by eq. (4.6), then the theorem is still applicable provided that the energy–momentum tensor of the additional matter obeys certain conditions, known as the strong and dominant energy conditions. The strong energy condition states that

$$(T_{\mu\nu} - \tfrac{1}{2}g_{\mu\nu}T^{\lambda}{}_{\lambda})W^{\mu}W^{\nu} \geq 0 \qquad (4.10)$$

for any timelike vector W^{μ}, and reduces in the case of a perfect fluid to the condition $\rho + 3p \geq 0$. The dominant energy condition states that for any timelike W^{μ}, $T_{\mu\nu}W^{\mu}W^{\nu} \geq 0$ and $T_{\mu\nu}W^{\nu}$ is non-spacelike. For a perfect fluid this condition reduces to $\rho \geq |p|$.

The no-hair theorem has been demonstrated (Frieman and Will, 1982; Barrow, 1983; Boucher and Gibbons, 1983; Ginsparg and Perry, 1983) in the context of linearized perturbation theory, and it is conjectured to hold even for large perturbations (Gibbons and Hawking, 1977; Hawking and Moss, 1982). Its validity in the non-perturbative regime has also been verified in certain exactly soluble models (Wald, 1983). Recently Jensen and Stein-Schabes (1986) have given a proof which holds whenever a synchronous reference frame exists – i.e., whenever spacetime can be filled with a family of timelike geodesics which do not intersect each other. For non-perturbative situations, the statement of the no-hair conjecture must include some kind of restriction to insure that the false vacuum energy density becomes dominant. In particular, this restriction must exclude the case of a closed universe that has a positive cosmological constant which is too weak to prevent it from collapsing.† Although to our knowledge this extra restriction has never been stated, one would guess that it is not

† We thank David Garfinkle for making us aware of this simple example.

stringent. Thus, it seems reasonable to conclude that a smooth de Sitter metric arises naturally, without any need to fine-tune the initial conditions.‡

Since the smoothing effect of de Sitter expansion occurs primarily through the redshifting of perturbations, it is essential in the arguments for the 'no-hair' conjecture that the initial perturbations have some minimum wavelength. In a world described by quantum theory, this assumption about the initial perturbations is very reasonable. A short wavelength excitation corresponds to a particle of high energy, so as long as there is some maximum energy available, there will also be some minimum wavelength.

As the space continues to supercool and exponentially expand, the energy density is fixed at ρ_f. Thus, the total energy (i.e., all energy other than gravitational) is increasing! If the inflationary model is right, this false vacuum energy is the source of essentially all the energy in the observed universe.

This production of energy seems naively to violate the conservation of energy, but we must remember that the gravitational field can exchange energy with the matter fields. The energy–momentum tensor for matter obeys a covariant conservation equation, which in the Robertson–Walker metric reduces to eq. (2.5):

$$\frac{d}{dt}(R^3\rho) = -p\frac{d}{dt}(R^3).$$

Thus, even in standard cosmology the energy of matter is not by itself conserved. During the radiation-dominated period, for example, the number of particles is conserved as the universe expands, but the energy of each particle is redshifted as $1/R(t)$. From one second after the big bang until the present, the total matter energy in a comoving volume has fallen by a factor of about 10^5. Eq. (2.5) implies that the energy per comoving volume $R^3\rho$ decreases with time when $p > 0$, but it increases with time when $p < 0$.

---

‡ Although the authors continue to believe that the 'no-hair' conjecture is valid under reasonable circumstances, the issues have become controversial. Ford (1985) and Barrow (1986) have discussed instabilities of de Sitter space, but they deal with situations in which there is no false vacuum of the type we have discussed. Traschen and Hill (1986) conclude that de Sitter space can be destabilized by minimally coupled scalar fields, but their method is invalid unless there are N species of such fields, with $NH^2 \gg M_p^2$. For typical GUT parameters, this would require an unrealistic 10^{18} scalar fields. Antoniadis, Iliopoulos and Tomaras (1986) have argued that the graviton propagator in de Sitter space has infrared singularities, but it appears (Allen, 1986) that the pathologies which they discovered were caused by their choice of gauge conditions. Other papers discussing instabilities of de Sitter space include Myhrvold (1983a, b), Mottola (1985, 1986) and Mazur and Mottola (1986).

During the inflationary era one has $p = -\rho = -\rho_f$, and one can see that eq. (2.5) is satisfied identically, with the energy of the expanding gas increasing due to the negative pressure. If the spacetime were asymptotically Minkowskian it would be possible to define a conserved total (i.e., matter plus gravitational) energy (see, for example, S. Weinberg, 1972, pp. 165–72; or Witten, 1981a). The Robertson–Walker metric, however, does not admit a global conservation law of this type.

Perhaps the most startling idea to come out of recent developments in cosmology is the suggestion that the universe may possess no conserved quantities which distinguish it from the vacuum. The total matter energy in the observed universe is of course very large – $\gtrsim 10^{78}$ GeV – but we have just seen that this quantity is *not* conserved. The baryon number of the universe is apparently also huge – $\gtrsim 10^{78}$ – but according to GUTs, baryon number is also not conserved. On the other hand, there are several quantities in nature which we believe are exactly conserved, such as electric charge and angular momentum. The consistency of our theories would break down if these quantities were not conserved. It is therefore very suggestive to note that the observed value for each of these quantities is compatible with zero. Thus, provided that baryon number is not conserved, the universe appears to be devoid of all conserved quantities. In that case, it is tempting to believe that the universe began from nothing, or from almost nothing. The inflationary universe illustrates the latter possibility. The idea that the universe may have emerged from 'absolutely nothing' was suggested by Tryon (1973), and further suggestions along these lines were discussed by Brout, Englert and Spindel (1979); Gott (1982); Atkatz and Pagels (1982); Vilenkin (1982, 1983a, b, 1984, 1985b); and Linde (1983a, 1984b, c).

Let us now return to discuss the fate of those regions which have been undergoing exponential expansion in the false vacuum state. We suppose that inflation continued for a time Δt, during which the regions expanded by a factor

$$Z \equiv e^{\chi \Delta t}. \tag{4.11}$$

The inflationary era then ended with a phase transition from the metastable phase (with $\langle \phi \rangle \approx \phi_{\text{false}}$) to the stable phase (with $\langle \phi \rangle \approx \phi_{\text{true}}$). The original inflationary universe model relies on the assumption that the phase transition took place suddenly, with rapid thermalization of the latent heat. This assumption is now known to be false, and in Section 12.6 we will discuss what would actually happen. For pedagogical purposes, however, we will now follow the logic of the original model. We will therefore suppose that when the energy density ρ_f of the false vacuum was released, it rapidly thermalized to produce a hot gas of particles – precisely the initial state that was postulated in the standard cosmological model. Thus, in the inflationary model the false vacuum energy is also the source of essentially all the entropy in the observed universe.

The temperature to which this gas reheated can be calculated for any particular particle theory by using the conservation of energy. Typically one finds that the reheating temperature is given by

$$T_{\text{reheat}} \sim \tfrac{1}{3}T_c. \tag{4.12}$$

From here on, the inflationary scenario joins the standard cosmology. As the hot gas cooled below the GUT scale, the baryon non-conserving interactions produced a small net excess of quarks over antiquarks. These excess quarks eventually resulted in the baryons which we observe in the universe today (see, for example, Kolb and Turner, 1983; Yoshimura, 1981). Thus, in this model the false vacuum energy is also the source of essentially all the matter in the universe. Note that the baryon number production must occur after the inflationary era so that the baryon number density is not unacceptably diluted.

While the inflationary scenario was certainly motivated by GUTs, it is worth noting that there are really only two properties of the GUTs that are essential to allow inflation to take place. First, the theory must contain some sort of a false vacuum – i.e., a metastable state with $\rho + 3p < 0$ – to drive the inflation. And, second, the theory must allow for baryon production to take place after the decay of the false vacuum.

12.5 Successes of the original inflationary model

Let us now discuss the extent to which the inflationary model provides solutions to the problems of the standard cosmology that were discussed in Section 12.3. The monopole problem centers on the fact that the Higgs field correlation length is bounded from above by the horizon length at the time t_f immediately after the phase transition. In the standard cosmology, the phase transition is instantaneous, so $t_f = t_c$ and one has

$$\xi < l_H(t_f) = R(t_c) \int_0^{t_c} \frac{dt'}{R(t')} = 2t_c. \tag{5.1}$$

In the inflationary model, on the other hand, the value of $R(t)$ increases by a factor of Z during the period of inflation, so

$$l_H(t_f) \approx 2Zt_c. \tag{5.2}$$

The bound (3.3) on the number density of magnetic monopoles is then weakened by a factor of Z^3, so the argument which was used to deduce a vast overproduction of monopoles in the standard model is obviated as long as $Z \gtrsim 10^4$. However, it is not clear in the original inflationary scenario how many magnetic monopoles would be produced in the phase transition. The answer to this question depends on the details of the phase transition, which we will discuss later. Thus, a large discrepancy is replaced by an open question. The new inflationary scenario provides a much more definitive solution to the monopole problem, as we will see in Section 12.7.

The flatness problem is solved rather simply by the original inflationary model. Recall that the problem is solved if we can explain why the dimensionless parameter $S \equiv R^3 s$ has such an incredibly large value. To understand why, note that eq. (4.12) indicates that T_{reheat}, the temperature immediately after the inflationary era, is roughly the same as T_c, the value of the temperature immediately before the inflationary era. The entropy densities are therefore related by

$$s(\text{after}) \sim s(\text{before}). \tag{5.3}$$

The scale factors, however, are related by

$$R(\text{after}) \approx ZR(\text{before}), \tag{5.4}$$

so that

$$S(\text{after}) \sim Z^3 S(\text{before}). \tag{5.5}$$

In order to explain why $S(\text{after}) \gtrsim 10^{87}$, assuming that $S(\text{before})$ has a value of order unity, we require $Z \gtrsim 10^{29}$. This seems like a very large number, but because the expansion is exponential it is very easy to find parameters (in the minimal $SU(5)$ model, for example) which give a nucleation rate low enough to yield this value of Z. In fact, reasonable parameters can lead (Guth and E. Weinberg, 1981) to values of Z as high as $10^{10^{10}}$, which is also quite acceptable; we require only that $S(\text{after})$ be at least 10^{87} – there is no upper limit. If one does not carefully fine-tune the parameters of the underlying particle theory, then typically the expansion factor overshoots the minimum required value, leading to $S(\text{after}) \gg 10^{87}$. Recalling eq. (3.12), we see that this leads to the prediction

$$\Omega = 1 \quad \text{(today)} \tag{5.6}$$

to extremely high precision.

The prediction (5.6) is somewhat problematic, since observations do not indicate a mass density this large. Observations of the motion of galaxies within clusters typically indicate a value for Ω in the range of 0.1–0.3, and big bang nucleosynthesis calculations (Yang et al., 1984) place a limit on the baryonic contribution to Ω given by

$$\Omega_B < 0.035\, h_0^{-2}, \tag{5.7}$$

where h_0 was defined in eq. (2.6). Thus, if inflation is correct then most of the mass in the universe must somehow remain hidden. Note that it is not sufficient for the mass merely to be dark – most dark matter candidates that one might invent would fall into clusters of galaxies and contribute to the mass that would be measured by observing the cluster dynamics. The proposals to account for $\Omega = 1$ include the possibility of dark galaxies associated with 'biased galaxy formation' (Kaiser, 1986; Bardeen *et al.*, 1986), and the possibility of a smooth component of relativistic particles created by the decay of some unstable particle (Turner *et al.*, 1984; Dicus, Kolb and Teplitz, 1977). Another possibility is a positive cosmological constant (Peebles, 1984). Note that the effects of a cosmological constant can be included in the energy–momentum tensor by setting $8\pi G T^{\mu\nu}_{\text{vac}} \equiv -\Lambda g^{\mu\nu}$, giving

$$\rho_{\text{vac}} = \Lambda/8\pi G,$$
$$p_{\text{vac}} = -\Lambda/8\pi G. \tag{5.8}$$

The prediction of inflation can then be written as

$$\Omega = \Omega_{\text{matter}} + \Omega_{\text{vac}} = \Omega_{\text{matter}} + \Lambda/3H^2 = 1. \tag{5.9}$$

A cosmological constant corresponding to $\Omega_{\text{vac}} \approx 1$ is consistent with observations, but most particle theorists consider it to be somewhat implausible. Naively one would expect that ρ_{vac} would be of order M_P^4, which would give $\Omega_{\text{vac}} \approx 10^{120}$. Since some unknown mechanism is apparently suppressing Ω_{vac} by 120 orders of magnitude, one would guess that the mechanism would most likely suppress it by at least a few more orders of magnitude, making the cosmological constant negligible. This argument, however, is obviously not compelling.

Returning to our discussion of the solutions to the problems of standard cosmology, we come next to the horizon problem. This problem hinges on the fact that the standard cosmological model requires that the universe was homogeneous, at the time of hydrogen recombination, on scales exceeding 75 horizon distances. In the inflationary model, on the other hand, the horizon distance is stretched during the inflationary era by a factor of Z, and at all subsequent times the entire observed universe is much smaller than the horizon distance. Furthermore, the inflationary model does more than simply enlarge the horizon distance – it actually provides a mechanism to create the observed large-scale homogeneity. In the inflationary model the size of the observable universe at times before the GUT phase transition was smaller than it would have been in the standard scenario by a factor of Z. if $Z \gtrsim 10^{25}$, then the entire observable universe would have been within its horizon at $t \approx t_c$; it would have become homogeneous at this time by normal thermal processes, and then this very small homogeneous region would have been stretched by inflation to become large enough to encompass the observed universe. The region would then remain homogeneous as it continued to evolve.

The original inflationary model does not in any way help us to understand the density perturbation problem. As we will see in the next section, the key failure of the original inflationary model is the disastrously unrealistic distribution of mass which it predicts. Fortunately the new inflationary universe model offers a possible answer to this problem, which we will discuss in Section 12.7.

12.6 Problems of the original inflationary model

The successes of the original inflationary model depended on the assumption that the phase transition occurred quickly, with rapid thermalization of the energy that was released. It is now known that this assumption is false, and in this section we will discuss what actually happens when a slow first-order phase transition occurs in an exponentially expanding space. We will find that the randomness of the bubble nucleation process leads to gross inhomogeneities, rendering the original inflationary model untenable. The problems discussed here are avoided by the new inflationary scenario, which will be discussed in the next section.

The decay of the false vacuum has been studied by Coleman (1977) and Callan and Coleman (1977; see also Coleman, 1979), and the description has been extended to curved spacetime by Coleman and De Luccia (1980). The false vacuum is classically stable, but it can decay by quantum mechanical tunneling. The tunneling is a local process, with bubbles of the new phase nucleating randomly in the midst of the old phase. The original inflationary model would have worked well if all of the bubbles nucleated at once, but this is not the case. Rather, the bubbles nucleate with a constant probability per physical volume per time, which we will call Γ. (The procedure for calculating Γ is discussed in the references cited above.) These bubbles are approximately spherical; on the inside $\langle \phi \rangle \approx \phi_{\text{true}}$, and on the outside $\langle \phi \rangle$ approaches ϕ_{false} as one moves away from the bubble. The nucleation of the bubble cannot rigorously be described in classical terms, but one can roughly think of the bubble as materializing at rest. The bubble then grows at a speed which rapidly approaches the speed of light. Energy is released as the false vacuum is converted into true, and this energy remains concentrated in the vicinity of the bubble wall. The energy density in any region of space inside the bubble will approach zero as the bubble wall moves outward. If these bubbles were to frequently collide with each other, then the energy of the bubble walls would thermalize and the original inflationary cosmology would be workable. As we will see, however, this does not happen.

To simplify the problem, we will assume that the bubbles materialize with zero size, and then begin to grow immediately at the speed of light. The bubbles actually form with an initial size determined by the microscopic particle physics; our approximation will be valid as long as we are interested in times which are late enough for this initial size to be negligible.

Since the bubble walls expand with the speed of light, the exterior of the bubble is causally disconnected from the interior. Therefore, the spacetime outside the bubble continues to be described by the de Sitter metric

$$ds^2 = -dt^2 + R^2(t)\,d\bar{x}^2, \tag{6.1}$$

with $R(t) = e^{\chi t}$. When viewed from the outside, the bubble walls will travel along lightlike geodesics in this metric. Thus, if a bubble nucleates at time t_N, its coordinate radius at a time $t > t_N$ is given by

$$r(t, t_N) = \int_{t_N}^{t} \frac{dt'}{R(t')} = \chi^{-1}(e^{-\chi t_N} - e^{-\chi t}). \tag{6.2}$$

As $t \to \infty$

$$r(t, t_N) \to \chi^{-1} e^{-\chi t_N}. \tag{6.3}$$

The fact that the limit is finite reflects the existence of event horizons in de Sitter space; if an event takes place at time t_E, a comoving observer whose coordinate distance from the event is greater than $\chi^{-1} e^{-\chi t_E}$ will never be able to detect it. The physical horizon distance is then $R(t_E)\chi^{-1} e^{-\chi t_E} = \chi^{-1}$, independent of time. Note that (6.3) does not imply that the bubble stops growing – it continues to grow at the speed of light. However, the scale of the coordinate system is changing so fast that the *coordinate* velocity of light approaches zero.

Now consider what happens when random bubble formation begins, at a time which we will call t_B. The first bubbles which form will have the largest asymptotic coordinate radii, while the asymptotic coordinate radii of bubbles which form at a later time t will be smaller by a factor $\exp\{-\chi(t - t_B)\}$. The asymptotic coordinate volume of bubbles which form at a later time t will then be smaller by a factor $\exp\{-3\chi(t-t_{\mathrm{B}})\}$. However, the nucleation rate per coordinate volume for the bubbles which form at the later time is larger by a factor $\exp\{3\chi(t-t_{\mathrm{B}})\}$, since the nucleation rate per physical volume remains constant. Thus there are many more of the smaller bubbles, and the larger multiplicity compensates exactly for the smaller size. The total coordinate volume occupied by all bubbles nucleated at time t is therefore reasonably constant over a very wide range of nucleation times. The bubbles which form at later times will fill a slightly smaller coordinate volume because of an exclusion effect – bubbles do not form in regions which are already in the new phase, so the bubbles which form later have a smaller region of comoving coordinate space in which to nucleate. This exclusion effect will be examined quantitatively in the next paragraph. A qualitative picture of this situation is illustrated in Fig. 12.2.

The bubbles will eventually fill the entire space, in the sense that the probability that any given point will not be covered by a bubble will approach zero. To prove this assertion we calculate $p(t)$, the probability that an arbitrary point remains outside of all bubbles at time t. Note that $p(t)$ is the fraction of the comoving coordinate volume available for nucleation at time t, and therefore gives a quantitative statement of the exclusion effect described in the previous paragraph. In calculating $p(t)$ we will use two simplifications. First, we will ignore the fact that bubbles cannot form inside of bubbles; instead we will assume that bubbles nucleate at a constant rate $\Gamma e^{3\chi t}$ per coordinate volume at all points in space. This simplification will produce no error in $p(t)$ whatever, since any bubble which forms inside of

Fig. 12.2. Random bubble nucleation in an exponentially expanding space. The largest circles represent those bubbles which nucleate at the earliest times.

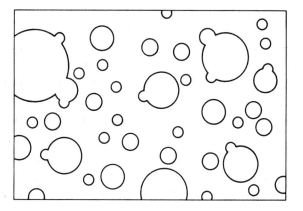

another bubble will remain forever inside it, having no effect on $p(t)$. Second, we will pretend that each bubble appears instantaneously with its asymptotic coordinate radius $\chi^{-1} e^{-\chi t_N}$. Since χ^{-1} is a short time for our purposes, this is a good approximation.

Now consider some arbitrary coordinate volume V and the time interval between t and $t+dt$. The expectation value of the number of bubbles that will nucleate in this spacetime region is given by $\Gamma V e^{3\chi t} dt$. The coordinate volume of each bubble is $(4\pi/3)\chi^{-3} e^{-3\chi t}$, so the expected total coordinate volume of the bubbles formed is $(4\pi/3)\Gamma V \chi^{-3} dt$. Thus, a fraction $(4\pi/3)\Gamma\chi^{-3} dt$ of the volume is covered with new bubbles. Since the new bubbles are uncorrelated with any pre-existing bubbles, it follows that given any region, no matter how chosen, the expectation value for the fraction of the region that will be covered with bubbles formed during this time interval is given by $(4\pi/3)\Gamma\chi^{-3} dt$. In particular, it follows that during the time interval between t and $t+dt$ the fraction $(4\pi/3)\Gamma\chi^{-3} dt$ of the region which was previously uncovered with bubbles will become covered, and this leads to the exponential form for $p(t)$

$$p(t) = \exp\left\{-\frac{4\pi}{3}\,\varepsilon\chi(t-t_B)\right\} \quad \text{(for } t > t_B\text{)}, \tag{6.4}$$

where the dimensionless quantity ε is given by

$$\varepsilon = \Gamma/\chi^4. \tag{6.5}$$

Thus, as $t \to \infty$, the fraction of space which remains in the metastable phase goes exponentially to zero. Note, however, that for $\varepsilon < 9/4\pi$ the physical volume of the region that remains in the old phase increases exponentially with time – the scale factor grows faster than $p(t)$ decreases. For more details about $p(t)$, see Guth and Tye (1980) and Guth and E. Weinberg (1981, 1983).

If inflation is to continue long enough to achieve the successes discussed in the previous section, it is necessary for ε to be small. To estimate how small it must be, we can consider the inflationary era to have ended at a time t_e when $p(t=t_e)$ reaches some suitably small value $p_e \ll 1$; the exact choice of p_e is not important. Then

$$p(t=t_e) = \exp\left\{\frac{4\pi}{3}\,\varepsilon \ln Z\right\} = p_e \tag{6.6}$$

where $Z = \exp\{\chi(t_e - t_c)\}$ is the inflationary scale factor. In order to have $Z \gtrsim 10^{29}$ we require

$$\varepsilon \lesssim 10^{-2}. \tag{6.7}$$

The range of values for ε which can arise for reasonable values of the parameters in the minimal $SU(5)$ GUT has been investigated (Guth and E. Weinberg, 1983). It was found that values anywhere in the range

$$10^{-10\,000} \lesssim \varepsilon \lesssim 10^{17} \qquad (6.8)$$

are quite plausible. There are two main reasons why one finds such a spectacularly large range for ε. It is actually the exponents that appear in eq. (6.8) which are calculated, and these exponents in turn depend on rather high powers of the GUT parameters. Thus, values of ε which satisfy eq. (6.7) are quite plausible, but it would require significant fine-tuning of parameters to have ε near to 10^{-2}. It is much more plausible to have ε many orders of magnitude smaller.

For $\varepsilon \ll 1$, the instantaneous phase transition assumed in the previous section must be replaced by one in which $p(t)$ approaches zero with an exponential time constant which is very long compared with the expansion time χ^{-1}.

To see if such a slow phase transition might work, one must examine the properties of the resulting distribution of bubbles (Guth and E. Weinberg, 1983). It can be proven rigorously that if $\varepsilon < 10^{-6}$, then the bubbles will form finite sized clusters only, no matter how long one waits, even though $p(t) \rightarrow 0$. In mathematical terminology, the bubbles never 'percolate'. It can also be shown that the bubbles *do* percolate if $\varepsilon \gtrsim 0.24$. There is some numerical evidence that the threshold for percolation is at $\varepsilon \approx 0.02$.

Since the regions that are in the new phase exist as isolated clusters only, one must look in more detail at these clusters. It can be shown that a typical cluster is dominated by its largest bubble, with the second largest bubble being smaller by a factor of order $\varepsilon^{1/2}$. For $\varepsilon \ll 1$, this means that a typical cluster would appear as a single dominant bubble, with an insignificant fuzz of smaller bubbles around its boundary. The energy of such a cluster would be concentrated near the wall of the large bubble. Clearly this structure bears no resemblance to the observed universe.

The problems of the original inflationary universe have also been analyzed by Hawking, Moss and Stewart (1982). Taking $\varepsilon \approx \frac{1}{200}$ (which is roughly the maximal value consistent with the inflationary scenario), these authors calculate that about one-third of all bubbles will grow to a galactic size (i.e., a size which will evolve to galactic proportions in the present universe) before undergoing their first collision. They then point out that, in the standard model, a region of galactic size does not come within the horizon until $t \approx 10^8$ s, when $T \approx 10^6$ K. Thus, they argue, the energy in the wall of such a bubble could not thermalize until T fell below 10^6 K, a value too low for baryon production or even nucleosynthesis.

The conclusion is therefore clear that the original inflationary scenario is unworkable. What is needed is a more graceful way to end the period of inflation. The model would work if one could find a mechanism which would allow ε to be much less than one during the inflationary era, and then to simultaneously and quickly change to a large value throughout the space. No mechanism of this type has ever been found. In the next section we will discuss the new inflationary model, which provides an elegant solution to this 'graceful exit' problem.

12.7 The new inflationary universe

We now turn to the new inflationary universe model, proposed by Linde (1982*a*, *b*) and Albrecht and Steinhardt (1982). The crucial observation is that for certain types of scalar field effective potential functions – the Coleman–Weinberg potential (Coleman and E. Weinberg, 1973) was the first example to be studied – the phase transition is expected to be quite different in character from the one which occurs in the original inflationary model. This new type of phase transition has come to be called a 'slow rollover' transition, a name which refers to the slow evolution of a scalar field down a hill in its effective potential diagram. Like the process of Coleman–Callan tunneling discussed earlier, this type of phase transition is driven by random processes – in this case it is the random fluctuations of the scalar field. However, in this model the transition occurs gradually, so that significant inflation can take place after the phase transition is already well underway, and after the random choices have been made. The inflation would then produce huge regions of homogeneous space, and we would be living today deep within one of these regions. The problems of bubble coalescence which plagued the original inflationary model do not arise in this context. The new inflationary model maintains all of the successes of the original model, and does significantly better. The monopole problem is clearly solved in the new inflationary model, and the model even offers an elegant solution to the density fluctuation problem.

We will describe the new inflationary universe model in the context of a hot initial state. This description follows the original proposal by Linde (1982*a*, *b*) and Albrecht and Steinhardt (1982), but we want to emphasize that there are certainly other possibilities. The process of inflation essentially erases all visible evidence of how it started, so a wide variety of starting points are possible – the observable predictions of inflation depend only on how it ends. Some time ago Tryon (1973) proposed the speculative but attractive idea that the universe may have begun from a quantum fluctuation of 'absolutely nothing'. This idea has since been pursued in the context of the inflationary universe by Vilenkin (1982, 1983a, b, 1984, 1985b) and Linde (1983a, 1984b, c). In these scenarios the universe tunnels directly from a state of 'absolute nothingness' into the false vacuum, with no need for an intermediate hot phase. In a similar spirit Hartle and Hawking (1983) have proposed a unique wave function for the universe, incorporating dynamics which leads to an inflationary era. (See also Hawking, 1984a, b, 1986; Moss and Wright, 1984; Hawking and Luttrell, 1984; and Hawking and Wu, 1985.) Linde (1983c, d, 1986a, b, c) has proposed and developed the idea of chaotic inflation, in which inflation is driven by a scalar field which is initially chaotic but far from thermal equilibrium. (See also Goncharov and Linde, 1984a, b, c; Goncharov, Linde and Vysotskii, 1984; and Khlopov and Linde, 1984.)

In order for the new inflationary universe scenario to occur, the underlying particle theory must contain a scalar field ϕ which has the following properties:[†]

(i) The effective potential function $V(\phi)$ must have a minimum at a value of ϕ not equal to zero.

(ii) $V(\phi)$ must be very flat in the vicinity of $\phi=0$. The value $\phi=0$ is usually assumed to be a local maximum of $V(\phi)$.

(iii) At high temperature T, the thermal equilibrium value of ϕ (i.e., the minimum of the finite temperature effective potential $V_T(\phi, T)$) should lie at $\phi=0$.

An effective potential function of this general form is shown in Fig. 12.3. The point $\phi=0$ is an equilibrium point which in the example shown is just barely unstable. We will refer to the field configuration $\phi=0$ at zero temperature as the false vacuum, even though the term is traditionally reserved for configurations that are classically stable. The example shown is in fact the minimal $SU(5)$ Coleman–Weinberg potential (Coleman and E. Weinberg, 1973), which will be discussed in more detail in Section 12.10.

The early papers on the new inflationary universe assumed that the Higgs field responsible for breaking the grand unified symmetry would also play the role of the field ϕ which drives the inflation. However, it was soon found (as we will discuss in Sections 12.8 and 12.10) that the effective potential for

---

[†] The properties required of the scalar field are described in more detail by Steinhardt and Turner (1984).

the Higgs field is not flat enough, due to the radiative corrections arising from the gauge couplings. The quantum fluctuations in the Higgs field then result in mass density fluctuations which are unacceptably large. Thus, in most newer models the ϕ field is a gauge singlet 'inflation' field, which in many cases serves no function other than the driving of inflation. Pi (1984), however, has proposed that the axion field can drive inflation, and Ovrut and Steinhardt (1983, 1984a, b, c, 1986; see also Albrecht and Steinhardt, 1983; and Lindblom, Ovrut and Steinhardt, 1986) have shown that inflation can be driven by the same scalar field that breaks supersymmetry. We will use the term 'new inflation' to refer to any model based on a slow rollover phase transition.

The initial conditions of the universe required by the new inflationary cosmology are almost identical to those assumed in the original inflationary scenario. In either case the initial conditions imposed are much less strict than they are in the standard cosmology. The early universe must have contained some regions with temperature $T > T_c$ which were expanding rapidly enough so that they would cool down to T_c before gravitational effects had a chance to reverse the expansion. Such regions must have had a size of about a horizon length χ^{-1} when $T = T_c$, or else they would have subsequently collapsed.

Fig. 12.3. The form of the scalar field effective potential function required for the new inflationary universe model. The example shown is the potential for the minimal $SU(5)$ theory with Coleman–Weinberg parameters.

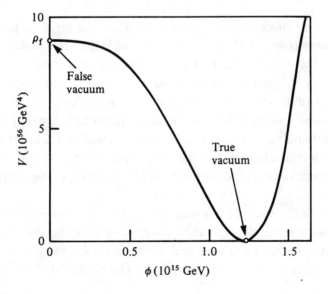

If the scalar field in such a region were in thermal equilibrium, then $\langle\phi\rangle\approx 0$. However, the universe would not have had time to thermalize at this point (Steigman, 1983), so we must assume that there were hot expanding regions within which $\langle\phi\rangle$ was near to its thermal equilibrium value of zero. This assumption, however, is not as stringent as it may at first appear. Thermal equilibrium properties are determined by averaging over microscopic states, so there are necessarily a large number of microscopic states with properties similar to thermal equilibrium. In addition, the Hubble damping (i.e., redshifting) caused by the expansion tends to suppress perturbations in the scalar field, driving it toward a homogeneous state with $\phi\approx 0$. Albrecht, Brandenberger and Matzner (1985) have carried out a classical numerical analysis for a variety of initial conditions, and have found that inflation is quite generic.

As the temperature of a hot expanding region fell below T_c, the process of bubble nucleation in principle began. However, the rate for this nucleation process is typically very low for potential functions with the properties listed above, and we will assume that it is negligible. The temperature therefore continued to fall until it reached the order of χ, when the situation became more complicated for at least two reasons. First, curved space corrections to the quantum field theory became important at this time. Gibbons and Hawking have shown that quantum fluctuations in de Sitter space mimic thermal fluctuations at the Gibbons–Hawking temperature

$$T_{GH} = \frac{\chi}{2\pi}, \tag{7.1}$$

so curved space effects become important when the physical temperature becomes comparable with this value. In addition, the scalar field was perched at the top of the plateau in the effective potential diagram. Depending on the precise choice of parameters, the situation was either classically unstable or at least very nearly so. Under these circumstances the approximations used in bubble nucleation calculations break down, and a more detailed analysis is necessary. These issues will be discussed in Section 12.9, but for now the behavior of the scalar field as it begins to roll down the hill of the effective potential diagram will be described only in qualitative terms.

The scalar field hovered for a while in the vicinity of $\phi=0$, and was then pushed from the top of the plateau by quantum and/or thermal fluctuations. These fluctuations were random, and therefore the field began to roll down the hill at different times in different places. The evolution of the scalar field can be characterized by a coherence length – the distance over which the field is expected to have been approximately uniform. This coherence length must be of order χ^{-1}, since that is the only relevant length scale in the problem. To see this, note that the scalar field dynamics is governed by the effective potential function, and that the behavior of small fluctuations about $\phi=0$ is therefore governed by the effective potential in the vicinity of $\phi=0$. We are assuming, however, that the effective potential is very flat in this vicinity. More specifically, we can assume that at $\phi=0$ one has

$$\frac{\partial V}{\partial \phi}=0, \quad \left|\frac{\partial^2 V}{\partial \phi^2}\right| \lesssim \chi^2, \quad \left|\frac{\partial^3 V}{\partial \phi^3}\right| \lesssim \chi \qquad (7.2)$$

In some models the natural mass scale of the effective potential in the vicinity of $\phi=0$ might be much less than χ, but the Gibbons–Hawking thermal effects mentioned above would in any case prevent the coherence length from being greater than a number of order χ^{-1}. For typical GUT parameters, $\chi^{-1} \approx 10^{-24}$ cm.

The regions over which the scalar field was approximately uniform will be called 'coherence regions'. We avoid the use of the word 'bubble' in describing these regions, for they are non-spherical and they do not have sharp boundaries.

The initial fluctuations must be described quantum mechanically, but the subsequent evolution can be described classically. (This transition from a quantum to a classical description will be discussed in detail in Section 12.9.) The scalar field within a coherence region was essentially homogeneous, so we may ignore the gradient terms in the classical equations of motion. Thermal effects were important in the initial fluctuations, but as inflation continued the temperature rapidly became negligible. One then has $\Box\phi= -\partial V/\partial\phi$, which in the de Sitter metric of eq. (6.1) becomes

$$\ddot{\phi}+3\chi\dot{\phi}= -\frac{\partial V}{\partial \phi}. \qquad (7.3)$$

If one views ϕ as a position coordinate, then this equation describes a particle rolling in the potential V in the presence of a damping force $3\chi\dot{\phi}$. The damping term reflects the expansion of the universe. The presence of this damping term helps to make the new inflationary scenario possible – it is necessary for the scalar field to roll slowly down the potential so that the coherence region can undergo a large amount of inflation.

As long as $\phi \approx 0$, the energy density ρ was very nearly equal to ρ_f. Thus the exponential expansion continued during this slow-rollover process. The typical size of a coherence region was then stretched to $Z\chi^{-1}$, where $Z = e^{\chi \Delta t}$ and Δt denotes the time period of the exponential expansion.

When the scalar field ϕ reached the steep part of the potential, it fell quickly to the minimum and then oscillated about it. These oscillations were damped by terms in the equations of motion which arise due to the coupling of ϕ to the other fields in the theory (Albrecht, Steinhardt, Turner and Wilczek, 1982; Abbott, Farhi and Wise, 1982; Dolgov and Linde, 1982). Note that the oscillations of ϕ describe a coherent state of zero momentum scalar particles – the momentum is zero because the scalar field is homogeneous. The damping of these oscillations may therefore be interpreted as the decay of the scalar particles into lighter species. This decay rate is typically greater than the expansion rate χ of the de Sitter space, in which case there was no significant energy loss to the expansion during the decay process. The decay products collided with each other and perhaps decayed into still lighter species, and we will assume that the energy was rapidly thermalized. The reheating temperature is calculated, just as in the original inflationary universe, by using conservation of energy. A reheating temperature of the order of $\frac{1}{3}T_c$ is again typical.

At this point the scenario joined the standard cosmology, as is the case in the original inflationary universe. In the standard cosmology the observable universe had a size of about 10 cm when T was 10^{14} GeV, and we therefore want to be sure that the new inflationary model provides us with regions that are at least this large. Thus we require $Z\chi^{-1} \gtrsim 10$ cm, which implies at least 25 orders of magnitude (or 58 exponential time constants) of inflation. Thus the inflationary era must have lasted at least about $58\chi^{-1}$, which for typical GUT parameters is about 10^{-32} s. Inflation could, however, have lasted for much longer than 10^{-32} s, in which case the observed universe would be a relatively small speck within a much larger homogeneous region.

We can now describe how the four problems of the standard cosmological scenario discussed in Section 12.3 are avoided in the new inflationary scenario. First, let us consider the monopole problem. Recall that in the standard scenario, the tremendous excess of monopoles was produced by the disorder that the Higgs field retained from the previous high-temperature phase (i.e., by the Kibble mechanism). To see how this can be avoided in the new inflationary model, let us first consider the possibility that ϕ and the Higgs field are distinct (as is the case in most recent versions of the model). In such cases one must arrange for the Higgs field to acquire its non-zero expectation value either before or during the inflationary era. The monopole density would then become negligible as it is diluted by the inflation. Alternatively, we can consider the possibility that inflation was driven by the Higgs field. (We do not know of any successful models of this type, but it is not clear that the possibility is completely ruled out.) In such cases the Higgs field would have been approximately uniform throughout the coherence region, which was stretched by inflation to become much larger than the observed universe – it would therefore be unlikely for the Kibble mechanism to produce even a single monopole in the observed universe. Some monopoles would still have been produced by thermal fluctuations after reheating, but this production is suppressed by a large Boltzmann factor – it would be negligible in the minimal $SU(5)$ model (Lazarides, Shafi and Trower, 1982; Guth, 1983*b*), and presumably in most other models as well. It takes much less energy, however, to cause a fluctuation in the Higgs field as it begins to roll down the hill in the effective potential diagram. Monopole production by this mechanism may be significant in the context of the minimal $SU(5)$ model (Moss, 1983, 1984). However, as we will discuss in Section 12.9, the quantum fluctuations in the minimal $SU(5)$ model are too large to be consistent with the observed density fluctuations of the universe. It seems likely that the monopole production would be totally negligible in any GUT in which the quantum fluctuations are small enough to be consistent with observation.

The flatness and horizon problems were solved by the original inflationary universe, and there is nothing more to be said – the same mechanisms lead to solutions to these problems in the context of the new inflationary cosmology. The fourth problem discussed in Section 12.3 was the density perturbation problem, which will be the topic of the next section.

12.8 Density perturbations in the new inflationary universe

Since inflation stretches microscopic scales into astronomical ones, it suggests the possibility that the density perturbations which provide the seeds for galaxy formation might have originated as microscopic quantum fluctuations. Although it is still not clear if the predictions of the inflationary model are consistent in detail with the properties of the observed universe, we nonetheless feel that this consequence of inflation is very attractive. The question of density fluctuations was not even considered when the model was formulated, so it is impressive that the model yields predictions that are at least reasonable. Furthermore, it must be remembered that the inflationary model is the first cosmological scenario with enough detail to even allow a prediction of the density perturbations from first principles.

The question of density perturbations resulting from the new inflationary scenario was studied in the summer of 1982 by a number of authors at the Nuffield Workshop on the Very Early Universe in Cambridge, England (Starobinsky, 1982; Guth and Pi, 1982; Hawking, 1982; Bardeen, Steinhardt and Turner, 1983; see also Brandenberger, Kahn and Press, 1983). These authors used a variety of methods to treat the problem, but they all agreed on the answer. Since this time several authors have questioned the methods used (Hawking and Moss, 1983; Mazenko, Unruh and Wald, 1985), but we believe that these questions have been answered (Brandenberger, 1984; Albrecht and Brendenberger, 1985; Guth and Pi, 1985, 1986). The controversies will be discussed further in Section 12.10, but here we will describe the results that were derived at the Nuffield meeting, which we believe are correct. In particular we will describe the formalism used by Guth and Pi (1982).

In the new inflationary universe the primordial density perturbations arise solely from the zero-point fluctuations of the quantized fields. Although the region which ultimately expanded to become the observed universe may have contained excitations above the vacuum, these excitations would not have any significant effect on the present state of the universe. To see this, note that since the region contained only a finite amount of energy, quantum theory implies that the excitations must have had some minimum wavelength. As we discussed in Section 12.4 in the context of the 'no-hair' conjecture, a sufficiently large amount of inflation would have redshifted these excitations to immeasurably long wavelengths. The zero-point fluctuations, on the other hand, have arbitrarily small wavelengths – indeed they are most significant at very small length scales. Under ordinary circumstances these zero-point fluctuations are imperceptible because of their short wavelengths, but the process of inflation can stretch these wavelengths to macroscopic, and eventually to astronomical dimensions.

Eq. (7.3) describes the time evolution of a homogeneous scalar field evolving in the potential $V(\phi)$. We would now like to consider the possibility that the scalar field might undergo inhomogeneous quantum fluctuations. The full quantum field $\phi(\check{x}, t)$ can be written as

$$\phi(\check{x}, t) = \phi_0(t) + \delta\phi(\check{x}, t), \tag{8.1}$$

where $\phi_0(t)$ is the classical homogeneous field which obeys eq. (7.3). Including the spatial derivative term in $\Box\phi$, the equation of motion (using the de Sitter metric of eq. (6.1)) becomes

$$\ddot{\phi} + 3\chi\dot{\phi} = -\frac{\partial V}{\partial \phi} + e^{-2\chi t}\,\partial_i^2\phi, \tag{8.2}$$

where ∂_i represents differentiation with respect to the comoving spatial coordinate x^i. Note that the last term denotes the Laplacian with respect to the physical coordinates. Substituting eq. (8.1) into eq. (8.2), we obtain

$$\delta\ddot{\phi} + 3\chi\delta\dot{\phi} = -\frac{\partial^2 V}{\partial \phi^2}(\phi_0)\delta\phi + e^{-2\chi t}\,\partial_i^2\delta\phi. \tag{8.3}$$

To proceed, we introduce a Fourier expansion with respect to the comoving coordinates. For an arbitrary function $f(\vec{x}, t)$ we can write

$$f(\vec{x}, t) = \int d^3k\, e^{i\vec{k}\cdot\vec{x}}\tilde{f}(\vec{k}, t). \tag{8.4}$$

Eq. (8.3) then becomes

$$\delta\ddot{\tilde{\phi}} + 3\chi\delta\dot{\tilde{\phi}} = -\frac{\partial^2 V}{\partial \phi^2}(\phi_0)\delta\tilde{\phi} - e^{-2\chi t}k^2\delta\tilde{\phi}. \tag{8.5}$$

Note that the second term on the right-hand side tends to stabilize the fluctuations of the scalar field. It is important at early times, but becomes unimportant at late times. For the potentials of interest $\partial^2 V/\partial\phi^2 < 0$ for $\phi \approx 0$, so the first term on the right-hand side tends to destabilize scalar field fluctuations. Thus, the fluctuation amplitude $\delta\tilde{\phi}(\vec{k}, t)$ will oscillate at early times, and will begin unstable growth at a time $t^*(k)$ given by

$$e^{-2\chi t^*}k^2 = -\frac{\partial^2 V}{\partial \phi^2}(\phi_0(t^*)). \tag{8.6}$$

In other words, the physical wavelength at the time the fluctuation begins unstable growth is given by

$$\lambda^* = \frac{2\pi R(t^*)}{k} = \frac{2\pi}{\left\{ -\dfrac{\partial^2 V}{\partial \phi^2}(\phi_0(t^*)) \right\}^{1/2}}. \tag{8.7}$$

From eq. (7.2) it follows that $\lambda^* \gtrsim \chi^{-1}$.

The history of a fluctuation with a given comoving wave number k is shown in Fig. 12.4. The physical wavelength λ grows exponentially during the inflationary era, which extends from t_i to t_f, and then grows as $t^{1/2}$ during the radiation-dominated era. The graph also shows the Hubble length, $H^{-1}(t)$, which is constant during the inflationary era and grows linearly with t during the radiation-dominated era. Note that the curves for λ and H^{-1} cross at two points, labelled A and B – the physical wavelength is initially less than the Hubble length, then becomes larger, and then becomes smaller again. The growth of the fluctuation amplitude becomes unstable at about the time of point A (the first crossing of the two curves) or perhaps shortly afterward. Fig. 12.4 is not drawn to scale – for a typical galactic length scale, the physical wavelength is about $10^{21}\chi^{-1}$ at the time of the phase transition; the time at point B is about 10^9 s, much longer than the duration of the inflationary era.

As we stated in Section 12.3, a density fluctuation $\delta\tilde{\rho}(\vec{k}, t)/\rho$ in a radiation-dominated flat universe will oscillate with constant amplitude when $\lambda \ll H^{-1}$, after the second crossing of the two curves at B. We will denote this asymptotic density perturbation amplitude by

$$\left.\frac{\delta\tilde{\rho}(\vec{k})}{\rho}\right|_H.$$

Our goal will therefore be to calculate the root-mean-square expectation value for this constant amplitude. We will be interested in wavelengths such that

$$\chi(t_f - t^*(k)) \gg 1 \tag{8.8a}$$

Fig. 12.4. Wavelengths and the Hubble length as a function of time. One curve shows the evolution of the physical wavelength λ of a typical fluctuation, proportional to $R(t)$, and the other curve shows the evolution of the Hubble length $H^{-1}(t)$.

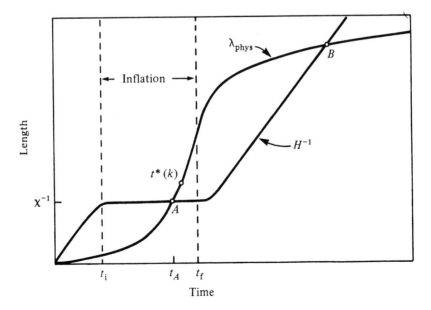

and

$$\chi(t^*(k) - t_i) \gg 1. \tag{8.8b}$$

Note that there are no physical processes which operate at scales larger than the Hubble length, since points at this separation are (at least temporarily) causally disconnected. Thus the amplitude of the fluctuation shown in Fig. 12.4 is actually determined by processes that occur in the vicinity of point A, when the physical wavelength is $\lesssim \chi^{-1}$. The form of Fig. 12.4, therefore, illustrates an important difference between inflationary and standard cosmology. If Fig. 12.4 had been drawn for standard cosmology, the physical wavelength would cross the Hubble length only once, and at all times earlier than the crossing time the physical wavelength would be larger than the Hubble length. For this reason it is hard even to conceive of a mechanism, within the standard cosmology, by which the density perturbations could be determined by physical processes.

Eq. (8.5) is a linear equation of second order. The general solution can therefore be expressed in terms of two linearly independent basis functions, and in principle two initial conditions must be given to uniquely specify a solution. We will be interested in the behavior of the solution, however, only during the period when the scalar field begins to roll off the hill, at $t \approx t_f$. Assuming that the time scale for this rolling is large compared with χ^{-1}, it will be shown in the Appendix that the damping term in eq. (8.5) causes one of the two linearly independent solutions to become negligible. Specifically, we will demonstrate that for $t > t^*(k)$ one of the two basis functions grows monotonically, and the other falls off faster than $e^{-3\chi t}$. (To see that this result is reasonable, the reader may wish to check the simplified case in which the right-hand side of the equation is replaced by $m^2 \delta\tilde{\phi}$.) By the assumption (8.8a) the falling basis function can be neglected, and then the growing basis function becomes the essentially unique solution. Since the neglected basis function falls off so rapidly, the expression '$\gg 1$' in eq. (8.8a) can be interpreted roughly as '$\gtrsim 5$'.

For $\chi(t - t^*) \gg 1$ the second term on the right-hand side of eq. (8.5) can be neglected, and the equation is then identical to the one satisfied by $\phi_0(t)$. Since we have argued that the solution to the equation is essentially unique, it follows that $\delta\tilde{\phi}(\vec{k}, t) \propto \phi_0(t)$. The proportionality constant may depend on $\vec{k}$, and we give it the name $-\delta\tilde{\tau}(\vec{k})$. Thus,

$$\delta\tilde{\phi}(\vec{k}, t) = -\delta\tilde{\tau}(\vec{k})\dot{\phi}_0(t), \tag{8.9}$$

from which it follows that

$$\delta\phi(\vec{x}, t) = -\delta\tau(\vec{x})\dot{\phi}_0(t). \tag{8.10}$$

To linear order in $\delta\phi$, eq. (8.1) can then be rewritten as

$$\phi(\vec{x}, t) = \phi_0(t - \delta\tau(\vec{x})).\qquad(8.11)$$

The inhomogeneities in the scalar field $\phi(\vec{x}, t)$ are therefore characterized by a single function $\delta\tau(\vec{x})$, which can be interpreted as a time delay function: the scalar field rolls off the hill at different times in different places. Since the rolling of the scalar field results in the release of the false vacuum energy density, this inhomogeneous time delay results in inhomogeneities in the mass density of the universe.

The completion of the calculation involves two steps: one must calculate the asymptotic density perturbation amplitude in terms of the time delay function, and one must estimate the magnitude of the time delay function.

The first step involves a calculation in linearized general relativity. We will omit the calculation here, referring the reader to the literature which emerged from the Nuffield Workshop, cited at the beginning of this section. The result is

$$\left.\frac{\delta\tilde{\rho}(\vec{k})}{\rho}\right|_H = 4\chi\delta\tilde{\tau}(\vec{k}).\qquad(8.12)$$

The second step is the estimate of the magnitude of the time delay function $\delta\tilde{\tau}(\vec{k})$. Since $\delta\tilde{\tau}(\vec{k})$ is a stochastic quantity, we want to estimate something like its root-mean-square value. There is a small complication, however, arising from the infinite volume of the space which is used for the calculations. Let us for the moment consider an arbitrary stochastic function $\delta f(\vec{x}, t)$ defined on a translationally invariant space. We assume that $\langle \delta f(\vec{x}, t)\rangle = 0$, and that the correlations have a finite range:

$$\lim_{|\vec{x}|\to\infty} \langle\delta f(\vec{x}, t)\,\delta f(\vec{0}, t)\rangle = 0.\qquad(8.13)$$

One can then see that the expectation value

$$\langle\delta\tilde{f}^*(\vec{k}, t)\,\delta\tilde{f}(\vec{k}, t)\rangle = \int\frac{d^3x}{(2\pi)^3}\frac{d^3x'}{(2\pi)^3}\,e^{i\vec{k}\cdot(\vec{x}'-\vec{x})}\langle\delta f(\vec{x}', t)\,\delta f(\vec{x}, t)\rangle\qquad(8.14)$$

diverges as the volume of space, since the integrand depends only on $\vec{x}' - \vec{x}$. This infinity can be factored out by considering the expectation value $\langle\delta\tilde{f}^*(\vec{k}', t)\,\delta\tilde{f}(\vec{k}, t)\rangle$, which can be written as

$$\langle\delta\tilde{f}^*(\vec{k}', t)\,\delta\tilde{f}(\vec{k}, t)\rangle = k^{-3}\delta^3(\vec{k}' - \vec{k})[\Delta f(\vec{k}, t)]^2,\qquad(8.15)$$

where

$$\Delta f(\vec{k}, t) \equiv \left\{k^3\int\frac{d^3x}{(2\pi)^3}\,e^{i\vec{k}\cdot\vec{x}}\langle\delta f(\vec{x}, t)\,\delta f(\vec{0}, t)\rangle\right\}^{1/2}.\qquad(8.16)$$

We therefore adopt the finite quantity $\Delta f(\vec{k}, t)$ as our root-mean-square measure of fluctuations. The factor k^3 has been inserted in the definition of $\Delta f(\vec{k}, t)$ so that it has the same units at $\delta f(\vec{x}, t)$. The mean-squared fluctuations at a given point can be expressed in terms of $\Delta f(\vec{k})$ by integrating with a dimensionless measure of integration:

$$\langle [\delta f(\vec{x}, t)]^2 \rangle = \int \frac{d^3 k}{k^3} [\Delta f(\vec{k}, t)]^2. \tag{8.17}$$

Note that eq. (8.13) implies that $\Delta f(\vec{k}, t)$ is a local quantity, in the sense that it depends on the behavior of $\delta f(\vec{x}, t)$ only in the vicinity of a given point, taken here to be the origin. We therefore expect that our discussion will be valid even if the physical region is finite, as long as its size is large compared with the wavelengths under consideration.

The fluctuations of the scalar field are most easily determined at very early times, when $t \ll t^*(k)$. (The inequality (8.8b) allows us, however, to restrict our attention to $t \gg t_i$, so that the de Sitter metric applies.) At such times the right-hand side of eq. (8.5) is dominated by the second term. If we ignore the first term on the right-hand side, the equation is precisely that of a massless (minimally coupled) scalar field in de Sitter space. The two-point correlation function is therefore given by the propagator of the quantum field theory, which is well known (Bunch and Davies, 1978; Gibbons and Hawking, 1977):

$$\Delta \phi(\vec{k}, t) = \left\{ \frac{\chi^2}{16\pi^3} \left[1 + \frac{k^2 e^{-2\chi t}}{\chi^2} \right] \right\}^{1/2} \quad (t \ll t^*(k)). \tag{8.18}$$

The derivation of this formula will be sketched in Section 12.9. Notice that the two terms in the square brackets can be given a simple interpretation. Since k is the comoving wave number, $k e^{-\chi t}$ is the physical wave number. Thus the second term in the square brackets of eq. (8.18) is precisely the term that would be calculated in flat space. The first term in the square brackets represents the Hawking radiation and is most important when k is small, i.e., at large wavelengths. One may understand this heuristically by observing that the large zero-point fluctuations which exist at small wavelengths are stretched by the exponentially expanding de Sitter space. Thus at large wavelengths the fluctuations in de Sitter space are significantly larger than they would be in flat space.

In free field theory approximation, the evolution of the propagator is determined by the classical equations of motion. (The normalization of the propagator, however, is proportional to $\hbar$ and is inherently quantum mechanical.) Thus, we could determine the behavior of $\Delta \phi(\vec{k}, t)$ at late times by following the evolution using eq. (8.5). Depending on the form of $\partial^2 V/\partial\phi^2(\phi_0)$, this method may or may not be tractable analytically – it would certainly be tractable numerically. One can get a good approximation to the answer, however, by using a simple matching argument. While eq. (8.18) is accurate only for $t \ll t^*(k)$, it should be a reasonable estimate at $t \approx t^*(k)$. Similarly the equation

$$\Delta\phi(\vec{k}, t) = \Delta\tau(\vec{k})\dot{\phi}_0(t), \tag{8.19}$$

which follows from eq. (8.9), is accurate only for $t \gg t^*(k)$, but it should also be a reasonable estimate for $t \approx t^*(k)$. Combining this equation with (8.12), one finds

$$\left.\frac{\Delta\rho(\vec{k})}{\rho}\right|_H \approx \frac{4\chi\,\Delta\phi(\vec{k}, t^*(k))}{\dot{\phi}_0(t^*(k))}. \tag{8.20}$$

Using eqs. (8.18) and (8.6) one then has

$$\left.\frac{\Delta\rho(\vec{k})}{\rho}\right|_H \approx \frac{\chi^2}{\pi^{3/2}\dot{\phi}_0(t^*(k))}\left[1 + \frac{1}{\chi^2}\left|\frac{\partial^2 V}{\partial\phi^2}(\phi_0(t^*(k)))\right|\right]^{1/2}. \tag{8.21}$$

In Section 12.10 we will apply this result to the specific case of the $SU(5)$ Coleman–Weinberg potential.

Since the rollover process must be slow for the new inflationary scenario to be viable, the function $\dot{\phi}_0(t^*(k))$ must be a slowly varying function of k – it follows that the fluctuation amplitude given by eq. (8.21) is slowly varying. A spectrum for which $[\Delta\rho(\vec{k})/\rho]|_H = constant$ is completely scale invariant, and is known as the Harrison–Zeldovich spectrum (Harrison, 1970; Zeldovich, 1972). Thus, inflationary models yield a nearly Harrison–Zeldovich spectrum.

Note that the perturbations arise from a nearly free quantum field theory, and therefore their probability distribution is Gaussian – each Fourier mode obeys a Gaussian probability distribution, and has a phase and magnitude uncorrelated with those of the other modes.

The perturbations produced in this way are 'adiabatic', in the sense that the baryon number to entropy ratio is unperturbed. The adiabaticity is a consequence of the short range nature of the baryon production processes, which occurred in these models very shortly after the GUT phase transition. The natural length scale for these interactions was necessarily smaller than the Hubble length, but at this time the perturbations of approximately galactic scale had wavelengths of order 10^{21} times the Hubble length. These perturbations are described by an inhomogeneous time delay $\delta\tau(\vec{x})$ of the phase transition, which occurred in a background geometry which was not yet significantly perturbed. Thus the time delay function $\delta\tau(\vec{x})$ was essentially constant within any volume of Hubble size. Any two distinct volumes of Hubble size therefore underwent the same evolution, although slightly out of synchronization. The net baryon number produced per unit of entropy was determined by this evolution, and is therefore the same everywhere.

Thus, the generic prediction of inflation is a nearly Harrison–Zeldovich spectrum of adiabatic, Gaussian fluctuations. A spectrum of this type, with an amplitude of $[\Delta\rho(\vec{k})/\rho]|_H \approx 10^{-4}$ or 10^{-5}, appears to be a viable candidate for the spectrum that produced the observed structure of the universe. It is very difficult, however, to know whether this spectrum is compatible in detail with the observed universe. To answer this question one would have to calculate the evolution of these perturbations up to present time – a calculation which requires an understanding of the dark matter in the universe and the roles of non-linear dynamics and perhaps non-gravitational forces. The difficulties of the calculation can be minimized by focussing attention on the large-scale structure of the microwave background radiation (see, for example, Abbott and Wise, 1984a, b), since in this case a linearized, purely gravitational calculation is expected to be valid. Unfortunately, though, it is very difficult to obtain data with sufficient precision. The problem of comparing theory with observation is further complicated by the need to predict the *observable* consequences of the perturbations; i.e., one must understand how enhancements in the matter density lead eventually to the production of detectable radiation. It has frequently been assumed naively that 'light follows mass', but now the possibility of 'biased galaxy formation' (Kaiser, 1986; Bardeen *et al.*, 1986) is taken quite seriously. The uncertainties about the connection between light and mass can be bypassed by directly observing the peculiar velocity field (see, for example, Kaiser, 1983; Vittorio and Silk, 1985; Vittorio, Juszkiewicz and Davis, 1986; Vittorio and Turner, 1986), but these measurements are very difficult. Research on these questions is very active at present, but it seems too early to expect a definitive answer. There is some evidence, however, that a Harrison–Zeldovich spectrum of adiabatic, Gaussian fluctuations cannot account for the detailed structure of the observed universe. For a sample of the literature, see IAU (1986).

The perturbations discussed above are generated by the non-uniformity of the slow-rollover phase transition, and are expected to be present in any inflationary model. It is therefore very economical to assume that these fluctuations are responsible for the formation of structure in the universe. It is possible, however, that the amplitude $[\Delta\rho(\vec{k})/\rho]|_H$ of these perturbations was smaller than 10^{-5}, and that the perturbations which produced the structure in the universe arose from another source.

In models that contain axions which dominate the mass of the universe (Preskill, Wise and Wilczek, 1983; Abbott and Sikivie, 1983; Dine and Fischler, 1983; Ipser and Sikivie, 1983), quantum fluctuations in the axion field would have produced 'isocurvature perturbations', in which the local axion to photon ratio varied from place to place while the initial mass density and baryon to photon ratio were uniform (Axenides, Brandenberger and Turner, 1983; Steinhardt and Turner, 1983; Turner, Wilczek and Zee, 1983; Linde, 1984*d*, 1985; Seckel and Turner, 1985; Turner, 1985, 1986).

Cosmic strings provide another plausible candidate for the origin of density fluctuations in the universe. (See the review by Vilenkin, 1985*a*, and references therein, and also the more recent articles by: Kibble, 1986; Turok and Brandenberger, 1986; Brandenberger and Turok, 1986; Turok, 1986; Scherrer and Frieman, 1986, Bennett, 1986; Chudnovsky *et al.*, 1986; Aryal *et al.*, 1986; Ostriker, Thompson and Witten, 1986.) Strings would also have an approximately Harrison–Zeldovich spectrum, but the probability distribution would be distinctly non-Gaussian.

Finally, in the event that the simple Harrison–Zeldovich spectrum of adiabatic, Gaussian fluctuations is shown to be unworkable, Silk and Turner (1986) have suggested the possibility of 'double inflation', in which the amplitude of fluctuations on small length scales is independent of the amplitude on large scales. In this scenario there is a primary episode of inflation identical to what we have been discussing, but it is followed by a second episode of inflation which lasts for only about 38–47 exponential time constants.

12.9 Quantum theory of the new inflationary universe phase transition

In the previous section we discussed the quantum fluctuations of the scalar field ϕ which drives inflation, but the analysis rested on one important simplifying idealization: we assumed that the problem could be described in terms of local quantum fluctuations about a globally homogeneous classical solution $\phi_0(t)$. In this section we will discuss a fully quantum mechanical treatment of the phase transition in the new inflationary universe, following the treatment of Guth and Pi (1985, 1986), hereafter called GP. As we will see, the model of GP is also based on some simplifying idealizations, but we believe that it qualitatively describes the correct physics and provides believable answers to the most important questions. We will not present the calculations here, but we will describe the assumptions and the conclusions.

The idealization of a globally homogeneous classical background solution is undoubtedly invalid at some level, and a number of questions can be raised about its use:

(i) Is the picture of a classical slow rollover valid? It has been pointed out by Mazenko *et al.* (1985) that at high temperatures, when one says that $\phi \approx 0$, one really means that the spatial or time average of ϕ is about equal to zero – the field itself is undergoing large fluctuations. As the system cools, they argue, it is possible that these fluctuations, which extend initially out to the minimum of the potential ϕ_c or beyond, will cause the scalar field to settle quickly into small regions with $\phi = \pm \phi_c$ in each of these regions. When one looks at the spatial average one might see what appears to be a rolling motion, but the actual local dynamics could be quite different.

(ii) What is the physical significance of the classical function $\phi_0(t)$? Hawking and Moss (1983) point out that the system begins in a thermal ensemble which possesses an exact symmetry, $\phi \to -\phi$. The dynamics is also consistent with this symmetry, and it therefore follows that $\langle \phi(\vec{x}, t) \rangle$ remains zero for all time – the field presumably does roll down the hill but, since it is equally likely to roll in any direction, the expectation value remains zero. A number of authors have suggested that the quantity $\{\langle \phi^2(\vec{x}, t) \rangle\}^{1/2}$ can play the role of $\phi_0(t)$, but this identification is problematic. At the quantum level the operator ϕ^2 is infinite and therefore requires a subtraction which is more or less arbitrary – it is then not clear how the expectation value of this operator is related to the behavior of ϕ itself. Furthermore, even for a classical probability ensemble the quantity $\{\langle \phi^2(\vec{x}, t) \rangle\}^{1/2}$ does not in general obey the classical equations of motion for ϕ, so one can hardly expect the expectation value of a quantum ensemble to have this property.

(iii) How long will ϕ hover in the vicinity of $\phi \approx 0$ before rolling down the hill in the effective potential diagram? If one treats ϕ_0 purely classically, then this question cannot be addressed – there is a classical homogeneous solution which remains at $\phi = 0$ for all time. This question is crucial, since it determines whether or not sufficient inflation is obtained. The question has been studied by Linde (1982c) and by Vilenkin and Ford (1982; see also Vilenkin, 1983c), who calculated the behavior of $\langle \phi^2(t) \rangle$. As we stated above, the required regularization of this operator makes it somewhat unclear how to interpret the results of this calculation. The methods discussed below will address the question in what we feel is a more transparent way. The results obtained with these newer methods are essentially in agreement with those of the earlier authors.

In this section we will try to provide answers for these questions. To understand the quantum mechanical behavior of unstable systems better, we begin by discussing the relatively simple problem of an upside-down harmonic oscillator. Consider a single particle of mass m moving in one dimension, under the influence of the potential

$$V(x) = -\tfrac{1}{2}kx^2. \tag{9.1}$$

We will track the behavior of a wave packet initially having the Gaussian form

$$\psi(x, t=0) \propto \exp\{-x^2/2h_0^2\}. \tag{9.2}$$

The solution is expressed most simply by introducing the new variables

$$b^2 = h/(mk)^{1/2} \tag{9.3a}$$

and

$$\omega^2 = k/m. \tag{9.3b}$$

Note that b corresponds to the characteristic quantum mechanical length scale of the problem, analogous to the Bohr radius of the hydrogen atom. More precisely, b corresponds to the width of the ground state wave function of the right-side-up harmonic oscillator potential with the same absolute value of k. Similarly, ω describes the natural frequency of the corresponding right-side-up harmonic oscillator. The solution maintains the Gaussian form $\psi(x, t) = A(t) \exp(-B(t)x^2)$ for all time, where $A(t)$ and $B(t)$ are complex functions. $B(t)$ can be expressed as

$$B(t) = \frac{1}{2b^2} \tan(\phi - i\omega t), \tag{9.4}$$

where $\tan \phi \equiv b^2/h_0^2$. For large times this solution has the form

$$\psi(x, t) \propto \exp\left\{ -\frac{x^2}{2h^2(t)} + i\frac{x}{2b^2} \right\} \tag{9.5a}$$

where

$$h(t) = \frac{(b^4 + h_0^4)^{1/2}}{2h_0} e^{\omega t}. \tag{9.5b}$$

The behavior of eq. (9.5b) is easily understood – the width of the wave packet at large times is minimized for a specific value of h_0, which turns out to be b. If h_0 were chosen very small, then the width would become large due to the

h_0 in the denominator, corresponding to the fact that the uncertainty principle would require the initial wave packet to have a large spread in momentum. On the other hand, a large value of h_0 means that the initial wave packet is already part way down the hill, and therefore spreads quickly.

The main point in introducing this model is to show that the quantum mechanical wave function at large times is accurately described by classical physics. Most textbooks speak of the classical limit of quantum mechanics in terms of sharply peaked wave packets, which is certainly not the case for eq. (9.5). However, classical physics also applies whenever the length over which the phase of the quantum mechanical wave function changes by 2π (i.e., the de Broglie wavelength) is much smaller than any other length in the physical problem. In such cases, though, the system cannot be described by a single classical trajectory. It must instead be described by a classical probability distribution, which for this system is given by:

$$f(x, p, t) \propto e^{-x^2/h^2(t)} \delta(p - (mk)^{1/2} x). \tag{9.6}$$

The use of the classical probability distribution is justified because the following two facts can be demonstrated:

(i) $f(x, p, t)$ describes classical physics – i.e., it obeys the classical equations of motion.

(ii) For any dynamical variable (i.e., any function $Q(x, p)$), the expectation value of Q can be computed by using either the quantum mechanical wave function ψ or the classical probability distribution f. It can be shown that

$$\langle Q \rangle_f = \langle Q \rangle_\psi [1 + O(e^{-2\omega t})]. \tag{9.7}$$

The probability distribution f describes an ensemble of particles, each of which rolls from rest at $x=0$. The various classical trajectories $x_i(t)$ contained in the ensemble differ from each other only by an offset of the time variable:

$$x_i(t) = x(t - \delta t_i), \tag{9.8}$$

where the function $x(t)$ is a solution to the classical equations of motion which approaches 0 as $t \to -\infty$. Thus, at large times the particle rolls classically down the hill, but the time at which it started to roll is described by a probability distribution. In this situation an ensemble average (whether it be an average of x or x^2) can obscure the physics, since it averages over systems that are in very different stages of their evolution.

Having discussed this toy problem, we can now describe the quantum field theory model proposed by GP. Any perturbative treatment of quantum field theory (including effective action methods) begins with a free field theory approximation, but the standard methods were developed for the purpose of treating perturbations about stable configurations. The GP model, on the other hand, is a free field theory approximation to a scalar field in an unstable configuration, perched at the top of a hill in the potential energy diagram. The model is somewhat crude, but we believe that it qualitatively describes the correct physics and provides a reasonable approximation to the behavior of the scalar field in the new inflationary universe. It is plausible that this model could serve as a valid zero order approximation to a systematic calculation in which interactions are taken into account perturbatively – so far, however, no such calculation has been carried out.

We consider a real scalar field with a potential energy function

$$V(\phi) = \tfrac{1}{4}a\left(\phi^2 - \frac{\mu^2}{a}\right)^2 = -\tfrac{1}{2}\mu^2\phi^2 + \tfrac{1}{4}a\phi^4 + \frac{\mu^4}{4a}. \qquad (9.9)$$

The dynamics of the model is simplified by two assumptions. First, we assume a perfect background de Sitter space, as described by eq. (6.1). This is a reasonable approximation for most stages of the rollover. Corrections due to the transition from the previous hot phase have not been considered. They are presumably significant, but only for the behavior of the very long wavelength modes. Modes with periods at the time of the transition which are short compared with the Hubble time will behave adiabatically and will will therefore not be excited. Second, we approximate the scalar field as a free scalar quantum field in de Sitter space, governed by the time-dependent potential

$$V_0(\phi) = -\tfrac{1}{2}(\mu^2 - \tfrac{1}{4}aT^2)\phi^2 + \frac{\mu^4}{4a}, \qquad (9.10)$$

where T represents a background temperature obeying the relation

$$T = \bar{T}\,e^{-\chi t}. \qquad (9.11)$$

The value of the parameter $\bar{T}$ depends on the arbitrary origin of t, and hence has no physical significance. The temperature-dependent term in eq. (9.10) describes the leading high-temperature behavior of the finite temperature effective potential. To justify its use in this context, GP use an approach due to S. Weinberg (1974) to argue that the validity of perturbation theory requires that the term be added to the zero-order potential – the negative of the term is then added to the counterterm part of the Lagrangian, and it is used in higher orders to cancel high-temperature contributions which would otherwise cause the perturbation expansion to be badly non-convergent. As we mentioned before, the potential energy function (9.10) describes a free field theory which is only an approximation to the truth. In particular, it is stable for early times and becomes unstable for late times, so ϕ will roll indefinitely to larger and larger values. The model will clearly not describe how ϕ settles into its minimum, but we believe that it will give a good description of the slow rollover.

In order to complete the model we must also assume an initial state. However, in order to describe our assumption it will be necessary first to discuss the behavior of this quantum system for very early times. The discussion will emphasize the fact that, at very short length scales, de Sitter space is essentially indistinguishable from Minkowski space. For example, it is conceivable that our universe has a positive cosmological constant, and we might today be living at the beginning of a de Sitter phase with an exponential time constant of the order of ten billion years – this situation, however, would in no way affect experiments being done at Fermilab today.

To proceed, we express the scalar field ϕ as a Fourier expansion:

$$\phi(\vec{x}, t) = \frac{1}{(2\pi)^{3/2}} \int d^3k [d(\vec{k}) \, e^{i\vec{k}\cdot\vec{x}} \psi(\vec{k}, t) + d^\dagger(\vec{k}) \, e^{-i\vec{k}\cdot\vec{x}} \psi^*(\vec{k}, t)]. \quad (9.12)$$

The mode functions $\psi(\vec{k}, t)$ then obey the equations of motion:

$$\ddot{\psi} + 3\chi\dot{\psi} = [\mu^2 - e^{-2\chi t}(k^2 + \gamma^2)]\psi, \quad (9.13)$$

where

$$\gamma^2 = \tfrac{1}{4}a\bar{T}^2. \quad (9.14)$$

As $t \to -\infty$, each $\psi(\vec{k}, t)$ obeys an equation of motion which approaches that of a simple harmonic oscillator, with angular frequency

$$\omega(t) = e^{-\chi t}(k^2 + \gamma^2)^{1/2}. \quad (9.15)$$

The frequency of the harmonic oscillator shifts as the universe exponentially expands, with a rate of order χ. However, as $t \to -\infty$, the harmonic oscillator frequency $\omega \to \infty$, and thus for early times $\omega \gg \chi$. Although the frequency of the harmonic oscillator is changing, the change is adiabatic; each mode then behaves precisely as a mode of a scalar field in Minkowski space, and we treat it accordingly.

Eq. (9.13) has two linearly independent solutions, each given by the product of an exponential and a Bessel function. The desired linear combination can be chosen to match the standard treatment of free field theory in Minkowski space. Hence we choose

$$\psi(\vec{k}, t) = \tfrac{1}{2}(\pi/\chi)^{1/2}\, e^{-3\chi t/2} H_p^{(1)}(z), \qquad (9.16a)$$

where $H_p^{(1)}(z)$ is the Hankel function (see, for example, Abramowitz and Stegun, 1972),

$$p = \left(\frac{9}{4} + \frac{\mu^2}{\chi^2}\right)^{1/2}, \qquad (9.16b)$$

and

$$z = \frac{(k^2 + \gamma^2)^{1/2}}{\chi}\, e^{-\chi t}. \qquad (9.16c)$$

The linear combination in (9.16) was chosen because at asymptotically early times it has the form

$$\psi(\vec{k}, t) \sim e^{i\theta(t)}\, e^{-3\chi t/2}\, \frac{e^{-i\omega(t)t}}{[2\omega(t)]^{1/2}}, \qquad (9.17)$$

where $e^{i\theta(t)}$ is a slowly varying phase factor. The factor $e^{-3\chi t/2}$ would not be present in Minkowski space – it is a slowly varying shift in the normalization conventions. Using the canonical commutation relations with eqs. (9.16) and (9.12), it can be shown that

$$[d(\vec{k}), d^\dagger(\vec{k}')] = \delta^3(\vec{k}' - \vec{k}). \qquad (9.18)$$

Given the mode function behavior (9.17) and the commutation relations (9.18), it follows that at early times the operators $d^\dagger(\vec{k})$ and $d(\vec{k})$ can be interpreted as creation and annihilation operators of nearly Minkowskian particles.[†]

For the special case $\mu = 0$, the Hankel function reduces to the simple closed form expression

$$H_{3/2}^{(1)}(z) = -\left(\frac{2}{\pi z}\right)^{1/2}\left(\frac{z + i}{z}\right) e^{iz}. \qquad (9.19)$$

Note that the choice of the linear combination of solutions to define the mode function $\psi(\vec{k}, t)$ is merely a choice of convention, and does not constitute a physical assumption. A different choice for $\psi(\vec{k}, t)$ would lead to a different meaning for the operators $d^\dagger(\vec{k})$ and $d(\vec{k})$. The statement that the operators $d(\vec{k})$ annihilate the vacuum, on the other hand, is a physical assumption. For the conventions used here (which are more or less

---

[†] To compare the normalization conventions, however, one must be aware of several differences between this formalism and the standard Minkowski space formalism: in this formalism $\psi(\vec{k}, t)$ contains an explicit factor $e^{-3\chi t/2}$; the measure of integration in (9.12) and the δ-function in (9.18) are expressed in terms of coordinate momenta rather than physical momenta.

standard), the state $|vac\rangle$ defined by

$$d(\vec{k})|vac\rangle=0 \quad \text{for all } \vec{k} \tag{9.20}$$

is the usual Gibbons–Hawking vacuum (Gibbons and Hawking, 1977; Bunch and Davies, 1978). The propagator for this state is used to derive eq. (8.18). Here we are interested, however, in a high-temperature state, so we assume that each of these harmonic oscillator variables is described at asymptotically early times by a thermal equilibrium density matrix at temperature T. Since T is changing with time (eq. (9.11)), one might think that thermal equilibrium can hold at one time only. However, ω is also changing, and the ratio $\omega/T \to constant$ as $t \to -\infty$. This means that thermal equilibrium can be maintained as $t \to -\infty$. Expectation values in this ensemble can be calculated straightforwardly by using the thermal ensemble occupation numbers

$$\langle d^{\dagger}(\vec{k}')\,d(\vec{k})\rangle = \frac{1}{\exp\{(k^2+\gamma^2)^{1/2}/T\}-1}\,\delta^3(\vec{k}'-\vec{k}). \tag{9.21}$$

For late times it can be shown that the Fourier amplitudes of the field obey a classical probability distribution, just as the variable x did in the example of the upside-down harmonic oscillator. In this case the classical approximation becomes valid for a given mode when its frequency (see eq. (9.15)) becomes small compared with the de Sitter expansion rate χ.

Since the GP model is a free field theory, essentially everything about it can be calculated exactly. Here we will summarize the most important results.

(i) *Qualitative behavior of* $\phi(\vec{x}, t)$. One might ask what $\phi(\vec{x}, t)$ would typically look like if one measured it at all points in space at a given time, but one must remember that a quantum field at a single point is invariably described by a probability distribution with an infinite width. To obtain a measurable operator one must 'smear' the quantum field over some finite volume. One therefore defines a smeared field

$$\hat{\phi}_l(\vec{x}, t)=\frac{1}{(2\pi)^{3/2}l^3}\int d^3y\,\exp\{-\tfrac{1}{2}(\vec{x}-\vec{y})^2/l^2\}\phi(\vec{y}, t), \tag{9.22}$$

where l is a smearing length measured in comoving coordinates. It is then possible to construct pictures of typical field configurations by generating a random value for each Fourier amplitude, using the probability distribution that has been computed. (To carry this out explicitly one could discretize the momentum integral by putting the system in a box, with periodic boundary conditions.) One can also calculate $\langle\hat{\phi}_l^2(t)\rangle$; this quantity requires no subtractions, but one must bear in mind the caveat that such an expectation value averages over systems in different stages of evolution. For potentials which are extremely flat (i.e., $\mu^2 \lesssim \chi^2$ and $a \ll 1$), the calculation of $\langle \hat{\phi}_l^2(t) \rangle$ for $l \approx 1/\bar{T}$ shows that $\hat{\phi}_l(\check{x}, t)$ has a high probability of hovering at $\hat{\phi}_l \approx 0$ for a long time before beginning to roll down the hill of $V(\phi)$. Thus, the picture of the classical slow rollover is valid under these circumstances, but it would not be valid for a value of a of order one.

(ii) *The meaning of* $\phi_0(t)$. In the new inflationary universe scenario, the observed universe develops from part of a much larger region which has evolved into the false vacuum. Let ξ denote the coordinate diameter of the region which evolves into the observed universe, and define a 'wave number of the universe' $k_U \equiv \xi^{-1}$. One can then write a Fourier expansion for the smeared field, separating it into two parts:

$$\hat{\phi}_l(\check{x}, t) = \int_{k < k_U} d^3k [\cdots] + \int_{k > k_U} d^3k [\cdots], \qquad (9.23)$$

where the integrand $[\cdots]$ has the same form as in eq. (9.12), but with an additional damping factor $e^{-(1/2)k^2 l^2}$ which arises from the smearing. The first term on the right-hand side describes fluctuations with wavelengths longer than the diameter of the observed universe, and thus can be considered homogeneous for astronomical purposes. It is this term that we identify with $\phi_0(t)$. In this formalism it obeys a classical equation of motion, but the time at which the rolling begins is described by a probability distribution. The second term represents inhomogeneities on scales less than that of the observed universe, and these correspond to the term $\delta\phi(\check{x}, t)$ in eq. (8.1).

(iii) *How long will* $\phi_0(t)$ *hover around* $\phi \approx 0$? In the GP model this question can be answered unambiguously by calculating the probability distribution of $\phi_0(t)$. One finds that the probability distribution is Gaussian, and that $\langle \phi_0^2(t) \rangle \approx \langle \hat{\phi}_l^2(t) \rangle$ for $l = 1/\bar{T}$, so the comments made above in paragraph (i) apply. The GP formalism can also be used to discuss the case of a Coleman–Weinberg potential. The results are similar to those obtained by Linde (1982c) and Vilenkin and Ford (1982), but the picture is somewhat different. We will return to this discussion in the next section, where we will discuss the Coleman–Weinberg potential in detail.

(iv) *Calculation of density perturbations.* The probability distribution for density perturbations $\delta\rho/\rho$ in the GP model can be calculated exactly. The results are in excellent agreement with those obtained by applying the approximate methods described in Section 12.8 to a scalar field with the potential energy function (9.10).

12.10 Inflation in the minimal $SU(5)$ GUT

The new inflationary universe was originally proposed (Linde, 1982*a*, *b*; Albrecht and Steinhardt, 1982) in the context of the minimal $SU(5)$ GUT (Georgi and Glashow, 1974), with the Higgs field playing the role of the scalar field ϕ which drives the inflation. The Higgs field in this theory necessarily has two of the three properties which were listed in Section 12.7 as requirements for inflation – the minimum of the effective potential function $V(\phi)$ is not equal to zero, and at high temperatures the thermal equilibrium value of ϕ is equal to zero. The remaining requirement, that $V(\phi)$ be very flat in the vicinity of $\phi=0$, can be obtained by a special choice of parameters, known as the Coleman–Weinberg values (Coleman and E. Weinberg, 1973). The effective potential function which was shown in Fig. 12.3 is in fact the minimal $SU(5)$ Coleman–Weinberg potential.

It is now known, however, that the new inflationary scenario does not work in its original context – the density fluctuations produced by the quantum fluctuations of the scalar field have an amplitude which is much too large. The problem is associated with the fact that although the Coleman–Weinberg potential is flat in the vicinity of $\phi=0$, it is not flat enough. When this inconsistency between the new inflationary cosmology and the minimal $SU(5)$ GUT was first discovered in 1982, it was a serious disappointment for the people working on inflationary models – at that time the minimal $SU(5)$ model was considered by many physicists to be by far the most attractive of the GUTs. Since that time, however, the minimal $SU(5)$ model has been essentially ruled out by the proton decay experiments, and there is now no longer any clear favorite among the variety of GUTs that are consistent with observations. Here we discuss the minimal $SU(5)$ GUT because it provides a simple context in which to carry out detailed calculations, and because an understanding of its failure is necessary to motivate the subsequent developments.

In this section we will make use of a number of particle theory results which we will not derive. We will attempt, however, to state clearly the facts that we are using, so that the reader can follow our line of reasoning whether or not he is familiar with the derivations. For the reader who would like more information about the $SU(5)$ model, we suggest the articles by Georgi and Glashow (1974), Georgi, Quinn and S. Weinberg (1974), or Buras, Ellis, Gaillard and Nanopoulos (1978), or the review article by Langacker (1981).

For details about the effective potential function, the reader can refer to the articles by Coleman and E. Weinberg (1973), S. Weinberg (1973), Jackiw (1974), and Dolan and Jackiw (1974a). For more details about the finite temperature effective potential, there are articles by Kirzhnits and Linde (1972), Dolan and Jackiw (1974b), S. Weinberg (1974), and Bernard (1974). There is a review article on effective potentials by Coleman (1975), and a review article on finite temperature effects by Linde (1979). The review article by Brandenberger (1985) contains a thorough treatment of both the zero temperature and finite temperature effective potentials.

In the minimal $SU(5)$ theory, the full gauge symmetry is broken to the subgroup $SU(3) \times SU(2) \times U(1)$ by a set of Higgs fields Φ which transform according to the adjoint representation of $SU(5)$. That is, Φ represents a traceless, hermition, 5×5 matrix of fields (with 24 independent components) which transforms under the gauge group as

$$\Phi'(x) = g^{-1}(x)\Phi(x)g(x), \qquad (10.1)$$

where $g(x)$ denotes an $SU(5)$ matrix. The symmetry breaking is accomplished by the fields Φ acquiring a vacuum expectation value of the form

$$\Phi = (\tfrac{2}{15})^{1/2}\phi \, \text{diag}[1, 1, 1, -\tfrac{3}{2}, -\tfrac{3}{2}], \qquad (10.2)$$

or a form which is gauge-equivalent to this. The prefactor $(\tfrac{2}{15})^{1/2}$ is chosen so that

$$\text{Tr} \, \Phi^2 = \phi^2. \qquad (10.3)$$

The space of generators of the unbroken subgroup are the hermitian traceless matrices which commute with the matrix displayed in (10.2): the matrices with non-zero entries in the upper left 3×3 block generate an $SU(3)$ subgroup, those with non-zero entries in the lower right 2×2 block generate an $SU(2)$ subgroup, and the matrix in (10.2) generates a $U(1)$ subgroup.

By restricting Φ to the form of eq. (10.2), the effective potential becomes a function of ϕ. The Coleman–Weinberg condition is then

$$\left. \frac{\partial^2 V}{\partial \phi^2} \right|_{\phi=0} = 0. \qquad (10.4)$$

Computing the effective potential including the one-loop radiative corrections for the gauge interactions, one finds at zero temperature

$$V(\phi) = \tfrac{25}{16}\alpha^2[\phi^4 \ln(\phi^2/\sigma^2) + \tfrac{1}{2}(\sigma^4 - \phi^4)], \qquad (10.5)$$

where $\alpha = g^2/4\pi \approx \tfrac{1}{45}$, where g is the gauge field coupling constant, and $\sigma \approx 1.2 \times 10^{15}$ GeV. The form of this potential was shown in Fig. 12.3. The minimum lies at $\phi = \sigma$, corresponding to the true vacuum. The point $\phi = 0$ is an equilibrium point which is just barely unstable. The configuration with $\phi = 0$ at zero temperature will be called the false vacuum.

The finite temperature effective potential is given by

$$V_T(\phi, T) = V(\phi) + \frac{18T}{\pi^2} \int_0^\infty q^2 \, dq \ln\left(1 - \exp\left\{-\left(q^2 + \frac{5\pi}{3}\alpha\phi^2\right)^{1/2}\Big/T\right\}\right).$$

$$(10.6)$$

The behaviour of $V(\phi, T)$ is shown in Fig. 12.5. The curves show a first-order phase transition at $T = T_c = 1.25 \times 10^{14}$ GeV. For $T < T_c$ the minimum of the effective potential is in the vicinity of $\phi = \sigma$, while for $T > T_c$ the minimum lies at $\phi = 0$. For all temperatures $0 < T < T_c$ there is a metastable phase with $\phi = 0$, corresponding to a local minimum of the effective

Fig. 12.5. The Coleman–Weinberg finite temperature effective potential for the minimal $SU(5)$ Higgs field.

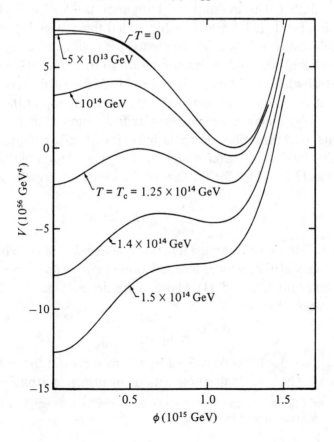

potential. For small T the dip which creates the minimum at $\phi = 0$ has a depth of order T^4, and a width of order T.

Having summarized the particle physics background, we are now prepared to discuss the details of the inflationary scenario in the context of the minimal $SU(5)$ GUT. Following the general scenario laid out in Section 12.7, we assume that the early universe contained hot expanding regions within which $\langle \phi \rangle$ was near to its thermal equilibrium value of zero. As the temperature of such a region fell below T_c, it became possible for bubbles of the new phase to nucleate. However, as long as $T \gg \chi \approx 10^{10}$ GeV one can rely on flat space calculations (Sher, 1981; see also Abbott, 1981; Billoire and Tamvakis, 1982; and Cook and Mahanthappa, 1982) that show that the nucleation rate is completely negligible.

Once the temperature fell to the order of χ, the scalar field was agitated by quantum and thermal fluctuations which tended to start it rolling down the hill of the potential energy diagram. It is important to know whether the scalar field would remain on the top of the hill for long enough to allow for sufficient inflation. This question has been addressed by Linde (1982c) and Vilenkin and Ford (1982), who concluded that this model does not lead to enough inflation. It will be easier to discuss the issues, however, after more groundwork has been laid, so we will return to this topic at the end of the section. Meanwhile, in order to continue our pedagogical discussion, we will assume that the scalar field remained perched at the top of the hill of the potential energy diagram for some unspecified length of time.

The scalar field then rolled down the hill of the effective potential diagram, obeying the classical equation of motion. The general form of this equation was given as (7.3), which for the case of the Coleman–Weinberg potential becomes

$$\ddot{\phi} + 3\chi\dot{\phi} = -\frac{\partial V}{\partial \phi} = \frac{25}{4}\alpha^2\phi^3\left|\ln(\phi^2\sigma^2)\right|. \tag{10.7}$$

The logarithm is a slowly varying function compared with ϕ^3. The coupling constant α may also be viewed as a function of ϕ, because it is a 'running' coupling constant (Sher, 1981) whose value depends on the choice of a renormalization scale μ:

$$\alpha(\mu^2) = \frac{4\pi}{\frac{40}{3}\ln(\mu^2/\Lambda_{SU(5)}^2)}, \tag{10.8}$$

where $\Lambda_{SU(5)} = 2.5 \times 10^5$ GeV is fixed by the requirement that $\alpha \approx \frac{1}{45}$ at the GUT scale.† The renormalization scale is in principle arbitrary, but the

† We have used here the renormalization group equations for the unbroken $SU(5)$ gauge group, since we are interested in the case $\phi \approx 0$.

choice $\mu = \phi$ is preferred – it helps to keep the contributions of higher-order perturbation theory small by avoiding the occurrence of large logarithmic factors. This dependence on ϕ, however, is also slowly varying compared with ϕ^3. Thus we may approximate $\alpha^2 |\ln(\phi^2/\sigma^2)|$ as a constant and write

$$-\frac{\partial V}{\partial \phi} = a\phi^3, \qquad (10.9)$$

where $a \approx \frac{1}{2}$ for $\phi \approx \chi$, the only scale relevant to the problem. As we will see, it is also safe to ignore the $\ddot{\phi}$ term in eq. (10.7), which is then solved by

$$\phi_0^2(t) = \frac{3\chi}{2a(t_f - t)}. \qquad (10.10)$$

We have chosen the constant of integration so that $\phi_0 \to \infty$ at a time which we call t_f. Of course the approximation (10.9) breaks down when ϕ_0 becomes comparable with σ, but t_f can still be thought of as the time when ϕ_0 becomes large. The neglect of the $\ddot{\phi}$ term can now be justified self-consistently, provided that $\chi(t_f - t) \gg 1$. Using (10.10) one has

$$\frac{\ddot{\phi}_0}{3\chi\dot{\phi}_0} = \frac{1}{2\chi(t_f - t)} \ll 1. \qquad (10.11)$$

A plot of eq. (10.10) is shown in Fig. 12.6. The field ϕ_0 changes almost imperceptibly until $\chi(t_f - t) \approx 10$ or so, and then approaches infinity very sharply. If the ϕ-axis had been drawn with a linear scale, the motion of ϕ_0 would have been imperceptible until $\chi(t_f - t) \approx 1$, at which time the approximations become invalid.

As long as $\phi_0 \ll \sigma$, the energy density ρ was very nearly equal to ρ_f. Thus

Fig. 12.6. The evolution of the scalar field during a slow-rollover phase transition in the minimal $SU(5)$ theory.

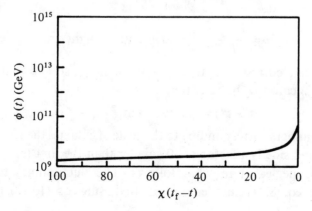

the exponential expansion continued during this slow-rollover process. For the scenario to work, it is necessary for the length scale of homogeneity to have stretched from χ^{-1} to at least about 10 cm before the scalar field ϕ_0 rolled off the plateau in Fig. 12.3. This corresponds to an expansion factor of about 10^{25}, which requires about $\ln 10^{25} = 58$ time constants (χ^{-1}) of expansion. Thus, the scenario will work if the classical rolling described by eq. (10.10) can be considered a valid approximation for times as early as $\chi(t_f - t) \approx 58$, which according to (10.10) implies a value of ϕ_0 as small as 0.23χ. On dimensional grounds one expects the initial fluctuations to have been of order χ, so detailed calculations are necessary to decide if they were large or small compared with 0.23χ.

We are now prepared to use the formalism of Section 12.8 to determine the spectrum of primordial density fluctuations $[\Delta\rho(\vec{k})/\rho]|_H$. The quantity $t^*(k)$ is given by eq. (8.6), which can be evaluated for the Coleman–Weinberg potential by using eqs. (10.9) and (10.10). One then has

$$e^{2\chi(t_f - t^*)} = \frac{9(\chi k^{-1} e^{\chi t_f})^2}{2\chi(t_f - t^*)}. \tag{10.12}$$

In order to use this relation we will need to express the comoving coordinate wave number k in terms of the physical distance scale measured at the present time, l_0. At the time t_f of the phase transition the universe was reheated to some temperature $T_{\text{reheat}} \approx 10^{14}$ GeV, so taking $TR \approx constant$ and using $T_0 \approx 3\,K \approx 3 \times 10^{-13}$ GeV to denote the current temperature, one has

$$l_0 = R(t_0)k^{-1} = \frac{R(t_0)}{R(t_f)} k^{-1} e^{\chi t_f} \approx \left(\frac{T_{\text{reheat}}}{T_0} \chi^{-1}\right)(\chi k^{-1} e^{\chi t_f}). \tag{10.13}$$

Thus

$$\chi k^{-1} e^{\chi t_f} \approx l_0/b, \tag{10.14}$$

where

$$b = \frac{T_{\text{reheat}}}{T_0} \chi^{-1} \approx 10\,m \approx 10^{-15}\text{ light-yr.} \tag{10.15}$$

A galactic scale corresponds to about 10^6 light-yr, so $l_0/b \approx 10^{21}$.

Eq. (10.12) can now be rewritten as

$$\chi(t_f - t^*) = \ln(l_0/b) - \tfrac{1}{2}\ln[\tfrac{2}{9}\chi(t_f - t^*)]. \tag{10.16}$$

For wave numbers corresponding to the scale of galaxies the second term on the right-hand side is a factor of 40 smaller than the first term, and it will therefore be neglected in what follows. The fluctuation amplitude as expressed by eq. (8.21) can then be evaluated using eqs. (10.10), (10.16), and

(10.9), with the result†

$$\frac{\Delta\rho(\vec{k})}{\rho}\bigg|_H \approx \left(\frac{8a}{3\pi^3}\right)^{1/2} \ln^{3/2}(l_0/b). \tag{10.17}$$

In deriving (10.17) the second term inside the square brackets on the right-hand side of eq. (8.21) was neglected – on the scale of galaxies this term has a value of 0.09, which is small compared with one.

Inserting numbers into (10.17), one finds that on the scale of galaxies

$$\frac{\Delta\rho}{\rho}\bigg|_H \approx 70 \quad (l_0 \approx 10^6 \text{ light-yr}). \tag{10.18}$$

The density fluctuation amplitude varies slowly with wavelength – for the scale of the observed universe one finds

$$\frac{\Delta\rho}{\rho}\bigg|_H \approx 91 \quad (l_0 \approx 10^{10} \text{ light-yr}). \tag{10.19}$$

These values of $(\Delta\rho/\rho)|_H$ are much too large to be acceptable – the desired number would be about 10^{-4}. In fact, the numbers obtained here are so large that the predictions of the theory are uncertain. The linearized calculation which we have carried out requires that $(\Delta\rho/\rho)|_H \ll 1$, so the calculation is not valid. The calculation suggests, however, that the density perturbations are strongly non-linear at the time when the wavelength becomes comparable with the Hubble length, so it seems reasonable to speculate that the matter would collapse to black holes on approximately this scale. Note that in a flat universe any sphere with a radius r equal to the Hubble length encloses a mass $M = \frac{4}{3}\pi r^3\rho$ such that $r = r_{\text{Schwarzschild}} = 2GM$. In an unperturbed flat universe the momentum of the expansion prevents the collapse of such a sphere to a black hole, but one would expect that any increase in mass density by a factor of 2 or more would result in black hole formation. Since $(\Delta\rho/\rho)|_H \gg 1$ on all relevant scales, we expect that the black holes would continually coalesce to produce larger black holes on the scale of the ever-increasing Hubble length.

In any case it seems clear that the theory is inconsistent with the observed universe. If the true prediction of the theory were $(\Delta\rho/\rho)|_H \approx 10^{-4}$, then the linearized calculation would have been valid and could not have led to the results (10.18) and (10.19).

Note that the desired answer of about 10^{-4} could be obtained if a were

† This formula is a factor of $\sqrt{2}$ larger than that in Guth and Pi (1982) because here we derive the asymptotic *amplitude* of the wave, while the earlier reference derived the asymptotic *root-mean-square* value.

smaller, but it would have to be very much smaller – the required value is about 10^{-12}. Thus, it seems that an *extremely* weakly interacting scalar field is needed to drive inflation.

We can now return to the question of whether or not the Higgs field in this model would remain perched at the top of the hill for long enough to provide an adequate amount of inflation. As mentioned above, this question has been studied by Linde (1982c) and Vilenkin and Ford (1982; see also Vilenkin, 1983c), who found a negative answer. Their argument begins with the observation that for $\phi \approx 0$, the Coleman–Weinberg potential can be approximated by $V(\phi) = constant$, describing a free, massless scalar field in de Sitter space (minimally coupled to gravity). This quantum field theory has the peculiar property that there exists no stable vacuum state – the scalar field never settles near any particular value, but instead undergoes an endless random walk up and down the real number line. Since there is no stable vacuum the renormalized expectation value $\langle \phi^2 \rangle$ is time-dependent, and is found to behave as

$$\langle \phi^2(t) \rangle = \frac{1}{4\pi^2} \chi^3 (t - t_0), \tag{10.20}$$

where t_0 depends on the arbitrary infinite subtraction. If one interprets $\langle \phi^2(t) \rangle$ as the quantum version of the classical $\phi_0^2(t)$ (an interpretation which we do not recommend), then one can compare the time derivative of (10.20) with the time derivative of (10.10). If one insists that $d\langle \phi^2(t) \rangle / dt \lesssim d\phi_0^2/dt$, one finds that

$$\chi(t_f - t) \lesssim \left(\frac{6\pi^2}{a} \right)^{1/2} \approx 11. \tag{10.21}$$

Thus, the growth in ϕ^2 due to quantum fluctuations (10.20) will be smaller than the growth due to classical motion (10.10) only during the last 11 exponential time constants of the slow rollover. Since adequate inflation requires the classical solution to be valid for at least 58 time constants, it seems clear that the scalar field does not remain long enough near the top of the hill.

The formalism of GP described in Section 12.9 leads to a conclusion similar to the pioneering results of Linde, Vilenkin, and Ford, but the picture appears slightly different. To approximate the minimal $SU(5)$ Coleman–Weinberg situation within the GP formalism, we take $\mu^2 = 0$ – this gives the zero temperature potential $V(\phi) = constant$. To determine a, we examine the high-temperature behavior (Dolan and Jackiw, 1974b) of the minimal $SU(5)$ finite temperature effective potential (eq. (10.6)):

$$V_T(\phi, T) \approx V(\phi) - \frac{2\pi^2}{5} T^4 + \frac{5\pi}{2} \alpha\phi^2 T^2 + \cdots. \tag{10.22}$$

Comparing the coefficient of T^2 with eqs. (9.10) and (9.14), one sets

$$a = 20\pi\alpha \approx 1.4 \tag{10.23a}$$

$$\gamma^2 = \tfrac{1}{4}a\bar{T}^2 \approx 0.35\bar{T}^2. \tag{10.23b}$$

(Note that the term in eq. (10.22) proportional to T^4 is independent of ϕ, and therefore has no dynamical effect. The term is not included in the GP calculation.)

As explained in the previous section, GP define the smeared field $\hat{\phi}_l(\hat{x}, t)$, and then calculate the finite quantity $\langle \hat{\phi}_l^2(\hat{x}, t) \rangle$. This quantity averages over systems in different stages of evolution, and therefore the interpretation has to be done carefully – nonetheless, a small value for the quantity implies unambiguously that quantum fluctuations are small. The calculation is carried out using eqs. (9.12), (9.22), (9.16), and (9.21), and then eq. (9.19) is used to reduce the answer to the special case $\mu = 0$. The result is given by

$$\langle \hat{\phi}_l^2(\hat{x}, t) \rangle = \frac{\chi^2}{4\pi^2} \int_0^\infty \frac{k^2 \, dk \, e^{-k^2 l^2}}{(k^2 + \gamma^2)^{3/2}} \left\{ 1 + \frac{k^2 + \gamma^2}{\chi^2} e^{-2\chi t} \right\}$$

$$\times \coth\left\{ \left(\frac{k^2 + \gamma^2}{2\bar{T}} \right)^{1/2} \right\}. \tag{10.24}$$

As an example, we can take the comoving smearing length as $l = 1/\bar{T}$, so the physical smearing length at any time is $1/T$, where $T = \bar{T} e^{-\chi t}$ is the background temperature. On dimensional grounds, one expects this smearing length to correspond roughly to a coherence length. The integral can then be carried out numerically, yielding

$$\langle \hat{\phi}_l^2(t) \rangle = 0.0219 \chi^2 + 0.0198 \, T^2. \tag{10.25}$$

Thus, the mean-squared value of the smeared field consists of a thermal contribution which redshifts to zero, and a residual contribution. Note that $\langle \hat{\phi}_l^2(t) \rangle$ does *not* contain a contribution that grows linearly with t, as in eq. (10.20). The difference is in the nature of the short distance cutoff. Eq. (10.24) is expressed in terms of a fixed *coordinate* smearing length l – this type of smearing is appropriate to discuss the behavior of the mean value of ϕ throughout the region which will evolve to become the observed universe. The renormalized expectation value $\langle \phi^2 \rangle$, on the other hand, is calculated with a fixed *physical* cutoff momentum – this type of cutoff is appropriate to discuss local measurements of ϕ, where the short wavelength fluctuations are operationally regulated by the resolution of the measuring device. If eq. (10.24) is evaluated for a fixed physical smearing length, so that $l = l_{phys} e^{-\chi t}$, then one can show that at large times the right-hand side grows linearly with time, with the same coefficient as in eq. (10.20).

Eq. (10.25) shows that $\hat{\phi}_l(t)$ typically remains rather small. Note that the residual contribution corresponds to a root-mean-square value of $\sqrt{(0.0219)}\chi = 0.15\chi$, which is smaller than the value of 0.23χ at which the classical equations of motion are required to be valid.

Similarly we can study the quantity $\phi_0(t)$, as described at the end of Section 12.9. Taking $\xi = 1/\bar{T}$ as the coordinate diameter of the region hypothesized to evolve into the observed universe, and taking $l \ll \xi$, we can calculate $\langle \phi_0^2(t) \rangle$. The result is an integral of the same form as (10.24), except that $l \approx 0$ and the upper limit of integration is $\bar{T}$. Again the integration can be carried out numerically, with the result

$$\langle \phi_0^2(t) \rangle = 0.0272\chi^2 + 0.0210 T^2. \tag{10.26}$$

Here the residual root-mean-square value is 0.16χ – again acceptably small.

Thus, quantum and thermal fluctuations do *not* cause the mean value of ϕ on large scales to fluctuate to a value large enough to upset the validity of the classical equations of motion. However, the fluctuations $\delta\phi(\dot{x}, t)$ on smaller scales are very large, completely destroying the validity of the slow-rollover picture. The problem can be seen by calculating $\langle [\delta\phi(\dot{x}, t)]^2 \rangle$, using eqs. (8.17) and (8.18). If we assume that the second term in the square brackets of eq. (8.18) is negligible (as is the case when the physical wavelength is large compared with the Hubble length), and if we assume that the integral in (8.17) receives contributions from wave numbers in a finite range $k_{min} < k < k_{max}$, then we find

$$\langle [\delta\phi(\dot{x}, t)]^2 \rangle = \frac{\chi^2}{4\pi^2} \ln\left(\frac{k_{max}}{k_{min}}\right). \tag{10.27}$$

Using the decomposition $\phi = \phi_0 + \delta\phi$ of eq. (9.23), one has a time-independent lower limit of integration, $k_{min} = k_U = \xi^{-1}$. The integral of eq. (8.17) should be cut off at large k by the smearing function, which gives a factor $e^{-k^2 l^2}$ (as in eq. (10.24)). For present purposes, however, we will approximate the smearing by a sharp cutoff at some physical wave number $k_{phys, max}$. Then $k_{max} = e^{\chi t} k_{phys, max}$, and the right-hand side of eq. (10.27) is seen to have the same behavior as the right-hand side of eq. (10.20). The significance of the numerical result obtained in eq. (10.21) can now be restated: the growth of the classical solution $\phi_0(t)$ will exceed the growth in the root-mean-square value of $\delta\phi(\dot{x}, t)$ only during the last 11 exponential time constants of the slow rollover. The assumption that $\delta\phi(\check{x}, t)$ is a small perturbation that can be treated linearly is therefore badly violated, and the entire picture of the slow rollover breaks down. Thus, this model seems to exhibit the pathological behavior suggested by Mazenko, Unruh and Wald (1985).

We conclude that the new inflationary model with a minimal $SU(5)$ GUT has two flaws: the density fluctuations are unacceptably large, and the amount of inflation is insufficient. Both of these flaws, however, are symptoms of one underlying problem: the quantum fluctuations in $\delta\phi(\check{x}, t)$ are too large. As we saw in eq. (10.27), however, the size of these quantum fluctuations is controlled by general principles of quantum field theory, and does not depend on the detailed shape of the minimal $SU(5)$ Coleman–Weinberg potential. What is needed, then, is a potential for which the classical solution gives a value of $\phi(t)$ large enough to suppress the effect of these fluctuations. The example studied here suggests that the requirement of obtaining $[\Delta\rho(\vec{k})/\rho]|_H \lesssim 10^{-4}$ is more stringent than the requirement of obtaining sufficient inflation. Thus, it seems plausible that in models for which $[\Delta\rho(\vec{k})/\rho]|_H$ has an acceptable value, the expected amount of inflation will greatly exceed the minimal requirement of 58 time constants.

Once the failure of the minimal $SU(5)$ theory was discovered, efforts turned toward the construction of particle theories that have the properties necessary for the new inflationary cosmology. The first model to be studied (Dimopoulos and Raby, 1983; Albrecht *et al.*, 1982, 1983) was the 'geometric hierarchy model' (proposed by Witten (1981*b*)), a supersymmetric model which makes use of a fundamental energy scale of about 10^{12} GeV. This model provided acceptable density fluctuations and much more than adequate inflation, but the reheating temperature of the model was too low to allow for baryogenesis. Subsequent work has led to a number of models that are believed to be completely successful. In a series of papers (Ellis *et al.*, 1982, 1983*a, b*; Nanopoulos *et al.*, 1983*a, b*; Nanopoulos and Srednicki, 1983; Nanopoulos, Olive and Srednicki, 1983; Gelmini, Nanopoulos and Olive, 1983; Olive, 1983; Linde, 1983*e, f*; Ellis *et al.*, 1984; Kounnas and Quiros, 1985; Ellis *et al.*, 1985; Gelmini, Kounnas and Nanopoulos, 1985; Enqvist *et al.*, 1985*a, b*; Jensen and Olive, 1985, 1986), a group loosely centered at CERN has developed a model called 'primordial inflation', in which inflation is driven by a scalar field which breaks supersymmetry near the Planck scale. Alternative approaches using supersymmetry and/or supergravity have been developed by Ovrut and Steinhardt (1983, 1984*a, b, c*, 1986; see also Albrecht and Steinhardt, 1983;

and Lindblom, Ovrut and Steinhardt, 1986) and by Holman, Ramond and Ross (1984). Shafi and Vilenkin (1984) have shown that supersymmetry is not necessary for inflation – they constructed a model based on a non-minimal $SU(5)$ theory that includes a weakly coupled gauge singlet 'inflation' field. Pi (1984) developed a variant of this model in which the role of the inflaton is played by the axion field. Other particle physics models that are believed to lead to successful inflationary cosmologies have been constructed by Hung (1984), Gupta and Quinn (1984), and Shafi and Stecker (1984).

At the least, the models mentioned above demonstrate that the problems associated with inflation in the context of the minimal $SU(5)$ GUT can be overcome. However, we think it is fair to say that all of these theories appear somewhat contrived. The ultimate goal of constructing a theory which is consistent with inflation and which also provides an elegant solution to the problems of particle physics has yet to be achieved.

12.11 False vacuum bubbles and child universes

In this section we will briefly discuss the behavior of false vacuum bubbles – regions of inflation that are surrounded by regions that do not inflate. This issue is relevant to all theories of the early universe in which the conditions before the inflationary phase transition were highly inhomogeneous. It is also relevant to the intriguing question of whether it is possible in principle to create an inflationary universe 'in the laboratory' (i.e., by man-made processes). Furthermore, the geometry of false vacuum bubbles provides a very interesting example of a highly non-Euclidean situation.

The mathematics of inhomogeneous spacetimes can be very complicated, but a detailed calculation can be carried out for one idealized system: a spherical region of false vacuum surrounded by an infinite region of true vacuum, with a thin domain wall separating the two regions. This situation has been studied by a number of authors (Sato *et al.*, 1981, 1982; Kodama *et al.*, 1981, 1982; Sato, 1981c; Maeda *et al.*, 1982; Berezin, Kuzmin and Tkachev, 1983, 1985; Aurilia *et al.*, 1984, 1985; Blau, Guendelman and Guth, 1986).

Although this idealized system is highly simplified, it nonetheless raises a significant paradox. If the false vacuum region is sufficiently large, then an observer deep within it would unambiguously expect to see inflation. An observer watching the domain wall, however, would note that the false vacuum region has negative pressure and is surrounded by the zero pressure true vacuum, so the pressure force is inward. The spherical symmetry

implies that the metric in the true vacuum region has the usual
Schwarzschild form, so the gravitational field is not expected to oppose the
force of the pressure gradient. Thus, the second observer would not expect to
see inflation. The resolution of this paradox relies on a highly non-Euclidean
geometry, allowing the false vacuum region to inflate without moving
outward into the true vacuum region.

An example of a false vacuum bubble solution is illustrated in Fig. 12.7.
The figure shows a diagram of spacetime, with the angular coordinates θ and
ϕ suppressed – light-like lines travel at 45°. To the right of the bubble wall
(shown as a heavy line with an arrow on it) the diagram represents a region
of Schwarzschild space, shown in Kruskal–Szekeres coordinates. The
behavior of the solution can be illustrated more intuitively by the diagrams
in Fig. 12.8, which show the spatial hypersurfaces at successive values of the
time coordinate used in Fig. 12.7. The diagrams are drawn by suppressing
one dimension of the hypersurface and embedding the resulting two-
dimensional surface in a three-dimensional space so that the curvature can
be displayed. The labels (*a*)–(*d*) in Fig. 12.8 correspond to the horizontal
slices indicated in Fig. 12.7.

The resolution of the paradox is now apparent: the shape is distorted so
that a force driving the bubble wall from the true vacuum to the false
vacuum pushes the wall to *larger* values of the radius. Meanwhile the false

Fig. 12.7. A spacetime diagram of a false vacuum bubble solution. Angular
coordinates are suppressed, and lightlike lines travel at 45°. The bubble wall
is shown as a heavy line with an arrow on it. The true vacuum region is to the
right of the wall, where the metric describes part of Schwarzschild space,
shown in Kruskal–Szekeres coordinates. The false vacuum region is to the
left, where the metric describes part of de Sitter space. The horizontal lettered
lines indicate spacelike hypersurfaces to be illustrated in Fig. 12.8.

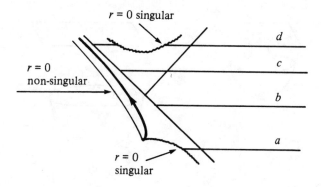

vacuum region expands into a bulge which soon disconnects completely from the original space, forming an isolated closed universe. Sato *et al.* (1982) have called this disconnecting region a 'child universe'. A black hole singularity is left in the original spacetime, but for typical GUT parameters it would evaporate in about 10^{-14} s.

Figs. 12.7 and 12.8 illustrate one type of solution to the false vacuum bubble equations of motion, but there are also solutions describing small bubbles which are crushed by the pressure forces. There are also 'bounce' solutions, describing bubbles which approach infinite size in the asymptotic past, fall to a minimum size, and then grow without bound. At late times these bounce solutions resemble the solution shown in Figs. 12.7 and 12.8.

Although the problem that has been solved is very idealized, we believe that it contains the essential physics of more complicated inhomogeneous spacetimes. The paradox discussed above will exist whenever an inflating region is surrounded by a non-inflating region, and the qualitative behavior of the system seems to be determined by the way in which this paradox is

Fig. 12.8. The evolution of a false vacuum bubble solution. Each lettered diagram illustrates a spacelike hypersurface indicated in Fig. 12.7. The diagrams are drawn by suppressing one dimension of the hypersurface and embedding the resulting two-dimensional surface in a three-dimensional space so that the curvature can be displayed. The false vacuum region is indicated by shading. Note that diagram (*d*) shows a child universe detaching from the original spacetime.

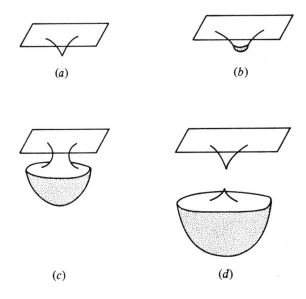

(*a*)   (*b*)

(*c*)   (*d*)

resolved. Thus, we believe that if inflation occurred in an inhomogeneous universe, then many isolated 'child universes' would have been ejected.

Furthermore, the analysis of Section 12.9 indicates that even if inflation began homogeneously, child universes are likely to have been produced. To see this, consider the field $\hat{\phi}_l(\dot{x}, t)$, smeared over a physical size χ^{-1}. In the approximations of Section 12.9, this quantity obeyed a Gaussian probability distribution, with a root-mean-square value which (for $\mu > 0$) increased exponentially with time. The probability density for $\hat{\phi}_l \approx 0$ therefore decreased exponentially with time, but for $\mu^2 \lesssim \chi^2$ this decrease was considerably slower than the exponential increase in the volume, proportional to $e^{3\chi t}$. Thus, the expected number of horizon-sized volumes with $\hat{\phi}_l \approx 0$ *increased* with time.† Although these regions were presumably non-spherical, it seems likely that they would have evolved into child universes.

The question of whether a false vacuum bubble, and hence an inflationary universe, can in principle be produced in the laboratory has been investigated by Farhi and Guth (1986). The issue concerns the initial singularity shown in Fig. 12.7. Although an initial singularity is often hypothesized to have been present at the big bang, there do not appear to be any initial singularities available today. Thus, to produce an inflationary universe in the laboratory, one must find some mechanism to modify the early stages of this picture, so that the initial matter distribution is smooth. It can be shown, however, that the singularity theorems of general relativity require the initial singularity, provided that

(i) the false vacuum region inflates without bound;

(ii) the energy–momentum tensor obeys the condition that

$$T_{\mu\nu} W^\mu W^\nu \geq 0 \tag{11.}$$

for any null vector W^μ; and

(iii) the false vacuum bubble is spherically symmetric.

The proof does not require that the machinery which produces the bubble be spherically symmetric. Intuitively, the theorem means that the outward velocity required for the false vacuum bubble to grow without bound is so large that it can emerge only from an initial singularity. Although a rigorous argument exists only for the limited case described above, it seems reasonable to conjecture that the theorem applies even in the absence of spherical symmetry (condition (iii)). Note, however, that this argument does not exclude the possiblity of the creation of a false vacuum bubble from

---

† We thank Alex Vilenkin for calling our attention to this fact.

smooth initial conditions by a quantum tunneling process, since the quantum mechanical energy–momentum tensor violates condition (ii) above.

12.12 Conclusion

As emphasized by Turner (1985), inflation is really a paradigm, and not a specific theory. In particular, there are a number of variations of inflation that appear to be capable of achieving the same objectives. Of these, probably the best known is Linde's proposal of chaotic inflation (Linde, 1983b, c, d, 1986a, b, c; see also Goncharov and Linde, 1984a, b, c; Goncharov et al., 1984; and Khlopov and Linde, 1984). Our primary reason for omitting a discussion of chaotic inflation was the simple fact that Linde is also contributing an article to this volume (Chapter 13). The key idea of chaotic inflation is the observation that inflation can work for a scalar field potential as simple as $V(\phi) = \lambda\phi^4$, provided that one makes some assumptions about the initial conditions. Linde proposes that the scalar field begins in a chaotic state, so that there are some regions in which the value of ϕ is a few times larger than M_P. These regions must exceed some minimal size, which is estimated to be somewhere between $2H^{-1}$ (Linde, 1984a) and $9H^{-1}$ (Turner, 1985), where H denotes the Hubble constant. Then ϕ rolls down the hill of the potential energy diagram, and a straightforward calculation indicates that there is an adequate amount of inflation. The Hubble 'constant' is not a constant in this case, but it is slowly varying, so the expansion can be called 'quasi-exponential'. In order for the density fluctuations to have the desired magnitude of $(\Delta\rho/\rho)|_H \approx 10^{-4}$, the coupling constant λ must be very small: $\lambda \approx 10^{-13}$. This number is typical of working inflationary models. Thus, chaotic inflation can greatly enlarge the class of particle theories consistent with inflationary cosmology – the plateau of the effective potential diagram is no longer required. It seems difficult to assess the plausibility of the initial conditions required for chaotic inflation, but the same can be said for the standard new inflationary scenario. Either model appears to be much more plausible than the standard cosmology (Section 12.2), and probably both models could benefit from more work concerning this question.

Another variation of inflation is the Starobinsky model (Starobinsky, 1979, 1980), a model proposed in its original form several years before inflation itself. In this model the energy density of the false vacuum is supplied by the curved space quantum corrections to the renormalized expectation value of the energy–momentum tensor of one or more scalar fields in the theory (Dowker and Critchley, 1976). While the initial conditions required for this model do not appear plausible at the level of classical physics, the desired conditions can perhaps be achieved in models in which the universe is created by a quantum fluctuation from 'nothing'. For a summary of the model and its literature, see Linde (1984*a*) and Vilenkin (1985*c*).

Other alternative inflationary models include induced gravity inflation (Accetta, Zoller and Turner, 1985; Spokoiny, 1984), and inflation driven by a scalar field which represents the radius of a compactified dimension (Shafi and Wetterich, 1983, 1985). Both of these possibilities, along with chaotic inflation, are summarized in Turner's review (1985).

In conclusion, we want to say that the basic idea of inflation – the idea that the universe went through a period during which it expanded exponentially while trapped in a false vacuum – appears to us to be probably correct. It is a very simple and natural idea in the context of spontaneously broken gauge theories, and it seems to solve some very fundamental cosmological problems. On the other hand, we clearly do not yet have the details straight. In order to understand the density fluctuations, for example, we must at the same time understand the details of particle physics at GUT energy scales. We are presumably some distance from that goal.

Finally, we wish to emphasize that probably the most dramatic recent development in cosmology is the realization that the universe may be completely devoid of conserved quantum numbers. If so, then, even if we do not understand the precise scenario, it becomes very plausible that our observed universe emerged from nothing or from almost nothing. The universe may indeed be the ultimate free lunch.

Acknowledgements

We would like to thank the many people with whom we have interacted over the years, exchanging information, insights, opinions, and suggestions concerning inflationary cosmology. A complete list is impossible to construct, but the list should certainly include Larry Abbott, Jim Bardeen, Marc Davis, Lawrence Ford, Stephen Hawking, Lawrence Krauss, Andrei Linde, Ian Moss, So-Young Pi, Bill Press, David Schramm, Marc Sher, Gary Steigman, Paul Steinhardt, Michael Turner, Henry Tye, Alex Vilenkin, and Erick Weinberg. In addition, our understanding of density fluctuations in the new inflationary universe was aided by conversations with Aleksei Starobinsky and Jemal Guven, and our understanding of false vacuum bubbles was aided by the input of Edward Farhi, José Figueroa-

O'Farrill, Eduardo Guendelman, Werner Israel, Richard Matzner, Malcolm Perry, Ian Redmount, Bill Unruh, Robert Wald, and Neil Woodward. We also benefited from discussions about the stability of de Sitter space with Bruce Allen, Joshua Frieman, David Garfinkle, Emil Mottola, and Jennie Transchen, but we do not mean to imply that all of these physicists agree with our point of view.

This paper is based in part on lectures delivered by A.H.G. at the First Jerusalem Winter School of Theoretical Physics, and we would like to acknowledge the warm hospitality of that meeting.

The work of S.K.B. is supported in part by funds provided by the Robert A. Welch Foundation and in part by the US National Science Foundation (NSF) under contract PHY 8304629. The work of A.H.G. is supported in part by funds provided by the US Department of Energy (DOE) under contract DE-AC02-76ERO3069, and in part by the National Aeronautics and Space Administration (NASA) under grant NAGW-553.

Appendix

In this appendix we will prove a mathematical property of the solutions to a differential equation which was used in Section 12.8. The equation has the form

$$\ddot{u} + 3\chi\dot{u} = \omega^2(t)u, \tag{A.1}$$

where the dot denotes differentiation with respect to t, and where there exists a t^* such that $\omega^2(t) > 0$ for $t > t^*$. We will show that one can find two linearly independent solutions $u_1(t)$ and $u_2(t)$ with the property that for $t > t^*$, $u_1(t)$ is monotonically increasing and $|u_2(t)| < constant \times e^{-3\chi t}$. Thus, at large times there is an essentially unique solution $u_1(t)$.

To prove the theorem, we first construct $u_1(t)$. This can be done by choosing initial data at $t = t^*$, with $u_1(t^*) > 0$ and $\dot{u}_1(t^*) > 0$. By writing eq. (A.1) as

$$\frac{d}{dt}(e^{3\chi t}\dot{u}) = e^{3\chi t}\omega^2(t)u, \tag{A.2}$$

one can see immediately that $\dot{u} > 0$ for all $t > t^*$.

To construct $u_2(t)$, consider an arbitrary solution $v(t)$ and construct the Wronksian

$$W(t) = u_1\dot{v} - \dot{u}_1 v. \tag{A.3}$$

One can easily verify that $\dot{W} = -3\chi W$, so one has $W = W_0 e^{-3\chi t}$. It follows that

$$u_1^2 \frac{\mathrm{d}}{\mathrm{d}t}\left(\frac{v}{u_1}\right) = W_0\, e^{-3\chi t}, \tag{A.4}$$

which can be integrated to give

$$v(t) = c_1 u_1(t) - W_0 u_1(t) \int_t^{\infty} \frac{e^{-3\chi t'}}{u_1^2(t')}\, \mathrm{d}t'. \tag{A.5}$$

Thus we can take

$$u_2(t) = u_1(t) \int_t^{\infty} \frac{e^{-3\chi t'}}{u_1^2(t')}\, \mathrm{d}t'. \tag{A.6}$$

The monotonicity of u_1 implies that for $t > t^*$,

$$0 < u_2(t) < \frac{e^{-3\chi t}}{3\chi u_1(t)} < \frac{e^{-3\chi t}}{3\chi u_1(t^*)}. \tag{A.7}$$

References

Aaronson, M., Mould, J., Huchra, J., Sullivan, W. T., Schommer, R. A. and Bothun, G. D. (1980). *Astrophys. J.*, **239**, 12, 66-B1.

Abbott, L. F. (1981). *Nucl. Phys.*, **B185**, 233.

Abbott, L. F., Farhi, E. and Wise, M. B. (1982). *Phys. Lett.*, **117B**, 29.

Abbott, L. and Pi, S.-Y. (1986). *Inflationary Cosmology*. World Scientific: Singapore.

Abbott, L. F. and Sikivie, P. (1983). *Phys. Lett.*, **120B**, 133.

Abbott, L. F. and Wise, M. B. (1984a). *Phys. Lett.*, **135B**, 279.

Abbott, L. F. and Wise, M. B. (1984b). *Astrophys. J.*, **282**, L47.

Abramowitz, M. and Stegun, I. A. (1972). *Handbook of Mathematical Functions with Formulas, Graphs, and Mathematical Methods*. Dover: New York.

Accetta, F., Zoller, D. and Turner, M. S. (1985). *Phys. Rev.*, **D31**, 3046.

Albrecht, A. and Brandenberger, R. (1985). *Phys. Rev.*, **D31**, 1225.

Albrecht, A., Brandenberger, R. and Matzner, R. (1985). *Phys. Rev.*, **D32**, 1280.

Albrecht, A., Dimopoulos, S., Fischler, W., Kolb, E. W., Raby, S. and Steinhardt, P. J. (1982). *Proceedings of the 3rd Marcel Grossman Meeting on the Recent Developments in General Relativity*, ed. Hu Ning, p. 511. North-Holland: Amsterdam.

Albrecht, A., Dimopoulos, S., Fischler, W., Kolb, E. W., Raby, S. and Steinhardt, P. J. (1983). *Nucl. Phys.*, **B229**, 528.

Albrecht, A. and Steinhardt, P. J. (1982). *Phys. Rev. Lett.*, **48**, 1220.

Albrecht, A. and Steinhardt, P. J. (1983). *Phys. Lett.*, **131B**, 45.

Albrecht, A., Steinhardt, P. J., Turner, M. S. and Wilczek, F. (1982). *Phys. Rev. Lett.*, **48**, 1437.

Allen, B. (1986). The graviton propagator in de Sitter space, Tufts preprint TUTP 86-9.

Antoniadis, I., Iliopoulos, J. and Tomaras, T. N. (1986). *Phys. Rev. Lett.*, **56**, 1319.

Aryal, M., Everett, A. E., Vilenkin, A. and Vachaspati, T. (1986). *Phys. Rev.*, **D34**, 434.

Atkatz, D. and Pagels, H. (1982). *Phys. Rev.*, **D25**, 2065.

Aurilia, A., Denardo, G., Legovini, F. and Spallucci, E. (1984). *Phys. Lett.*, **147B**, 258.

Aurilia, A., Denardo, G., Legovini, F. and Spallucci, E. (1985). *Nucl. Phys.*, **B252**, 523.

Axenides, M., Brandenberger, R. and Turner, M. S. (1983). *Phys. Lett.*, **126B**, 178.

Bardeen, J. M. (1980). *Phys. Rev.*, **D22**, 1882.

Bardeen, J. M., Bond, J. R., Kaiser, N. and Szalay, A. S. (1986). The statistics of peaks of Gaussian random fields. To be published in *Astrophys. J.*

Bardeen, J. M., Steinhardt, P. J. and Turner, M. S. (1983). *Phys. Rev.*, **D28**, 679.

Barrow, J. D. (1983). In *The Very Early Universe* (Proceedings of the Nuffield Workshop), ed. G. W. Gibbons, S. W. Hawking and S. T. C. Siklos, p. 267. Cambridge University Press: Cambridge.

Barrow, J. D. (1986). The deflationary universe: an instability of the de Sitter universe. Sussex preprint.

Bennett, D. P. (1986). *Phys. Rev.*, **D33**, 872.

Berezin, V. A., Kuzmin, V. A. and Tkachev, I. I. (1983). *Phys. Lett.*, **120B**, 91.

Berezin, V. A., Kuzmin, V. A. and Tkachev, I. I. (1985). In *Proceedings of 3rd Seminar on Quantum Gravity, 1984*, ed. M. A. Markov, V. A. Berezin and V. P. Frolov, p. 605. World Scientific: Singapore.

Bernard, C. (1974). *Phys. Rev.*, **D9**, 3313.

Billoire, A. and Tamvakis, K. (1982). *Nucl. Phys.*, **B200**, 329.

Blau, S. K., Guendelman, E. I. and Guth, A. H. (1986). The dynamics of false vacuum bubbles. Submitted to *Phys. Rev.*, **D**.

Bludman, S. A. and Ruderman, M. A. (1977). *Phys. Rev. Lett.*, **38**, 255.

Boesgaard, A. and Steigman, G. (1985). To be published in *Ann. Rev. Astron. Astrophys.*

Boucher, W. and Gibbons, G. W. (1983). In *The Very Early Universe* (Proceedings of the Nuffield Workshop)¡ ed. G. W. Gibbons, S. W. Hawking and S. T. C. Siklos, p. 273. Cambridge University Press: Cambridge.

Branch, D. (1979). *Mon. Not. R. Astron. Soc.*, **186**, 609.

Brandenberger, R. H. (1984). *Nucl. Phys.*, **B245**, 328.

Brandenberger, R. H. (1985). *Rev. Mod. Phys.*, **57**, 1.

Brandenberger, R., Kahn, R. and Press, W. H. (1983). *Phys. Rev.*, **D28**, 1809.

Brandenberger, R. H. and Turok, N. (1986). *Phys. Rev.*, **D33**, 2182.

Brout, R., Englert, F. and Spindel, P. (1979). *Phys. Rev. Lett.*, **43**, 417.

Bunch, T. S. and Davies, P. C. W. (1978). *Proc. Roy. Soc. London, Ser.*, **A360**, 117.

Buras, A. J., Ellis, J., Gaillard, M. K. and Nanopoulos, D. V. (1978). *Nucl. Phys.*, **B135**, 66.

Callan, C. G. and Coleman, S. (1977). *Phys. Rev.*, **D16**, 1762.

Chudnovsky, E. M., Field, G. B., Spergel, D. N. and Vilenkin, A. (1986). Superconducting cosmic strings. Submitted to *Phys. Rev.*, **D**.

Coleman, S. (1975). In *Laws of Hadronic Matter* (Erice, 1973), ed. A. Zichichi, p. 139. Academic Press: New York.

Coleman, S. (1977). *Phys. Rev.*, **D15**, 2929.

Coleman, S. (1979). In *The Whys of Subnuclear Physics* (Proceedings of the International School of Subnuclear Physics, Ettore Majorana, Erice, 1977), ed. A. Zichichi. Plenum: New York.

Coleman, S. and De Luccia, F. (1980). *Phys. Rev.*, **D21**, 3305.

Coleman, S. and Weinberg, E. J. (1973). *Phys. Rev.*, **D7**, 1888.

Cook, G. P. and Mahanthappa, K. T. (1982). *Phys. Rev.*, **D25**, 1154.

Dicke, R. H. and Peebles, P. J. E. (1979). In *General Relativity: An Einstein Centenary Survey*, ed. S. W. Hawking and W. Israel. Cambridge University Press: Cambridge.

Dicus, D., Kolb, E. W. and Teplitz, V. (1977). *Phys. Rev. Lett.*, **39**, 168.

Dimopoulos, S. and Raby, S. (1983). *Nucl. Phys.*, **B219**, 479.

Dine, M. and Fischler, W. (1983). *Phys. Lett.*, **120B**, 137.

Dolan, L. and Jackiw, R. (1974a). *Phys. Rev.*, **D9**, 2904.

Dolan, L. and Jackiw, R. (1974b). *Phys. Rev.*, **D9**, 3320.

Dolgov, A. D. and Linde, A. D. (1982). *Phys. Lett.*, **116B**, 329.

Dowker, J. S. and Critchley, R. (1976). *Phys. Rev.*, **D13**, 3224.

Einhorn, M. B. and Sato, K. (1981). *Nucl. Phys.*, **B180** [FS2], 385.

Ellis, J., Enqvist, K., Gelmini, G., Kounnas, C., Masiero, A., Nanopoulos, D. V. and Smirnov, A. Yu. (1984). *Phys. Lett.*, **147B**, 27.

Ellis, J., Enqvist, K., Nanopoulos, D. V., Olive, K. A. and Srednicki, M. (1985). *Phys. Lett.*, **152B**, 175 [*Erratum: Phys. Lett.*, **156B**, 452 (1985)].

Ellis, J., Nanopoulos, D. V., Olive, K. A. and Tamvakis, K. (1982). *Phys. Lett.*, **118B**, 335.

Ellis, J., Nanopoulos, D. V., Olive, K. A. and Tamvakis, K. (1983a). *Nucl. Phys.*, **B221**, 524.

Ellis, J., Nanopoulos, D. V., Olive, K. A. and Tamvakis, K. (1983b). *Phys. Lett.*, **120B**, 331.

Ellis, J. and Steigman, G. (1980). *Phys. Lett.*, **B89**, 186.

Enqvist, K., Nanopoulos, D. V., Quiros, M. and Kounnas, C. (1985a). *Nucl. Phys.*, **B262**, 538.

Enqvist, K., Nanopoulos, D. V., Quiros, M. and Kounnas, C. (1985b). *Nucl. Phys.*, **B262**, 556.

Faber, S. M. and Gallagher, J. S. (1979). *Ann. Rev. Astron. Astrophys.*, **17**, 135.

Farhi, E. and Guth, A. H. (1986). An obstacle to creating a universe in the laboratory, unpublished.

Ford, L. H. (1985). *Phys. Rev.*, **D31**, 710.

Frieman, J. A. and Will, C. M. (1982). *Astrophys. J.*, **259**, 437.

Gelmini, G. B., Kounnas, C. and Nanopoulos, D. V. (1985). *Nucl. Phys.*, **B250**, 177.

Gelmini, G. B., Nanopoulos, D. V. and Olive, K. A. (1983). *Phys. Lett.*, **131B**, 53.

Georgi, H. and Glashow, S. L. (1974). *Phys. Rev. Lett.*, **32**, 438.

Georgi, H., Quinn, H. R. and Weinberg, S. (1974). *Phys. Rev. Lett.*, **33**, 451.

Gibbons, G. W. and Hawking, S. W. (1977). *Phys. Rev.*, **D15**, 2738.

Ginsparg, P. & Perry, M. J. (1983). *Nucl. Phys.*, **B222**, 245.

Gliner, E. B. (1965). *Zh. Eksp. Teor. Fiz.*, **49**, 542 (*JETP Lett.*, **22**, 378 (1966)).

Gliner, E. B. (1970). *Dokl. Akad. Nauk SSSR*, **192**, 771 (*Sov. Phys. Doklady*, **15**, 559 (1970)).

Gliner, E. B. and Dymnikova, I. G. (1975). *Pis'ma Astron. Zh.*, **1**, 7 (*Sov. Astr. Lett.*, **1**, 93 (1975)).

Goldman, T., Kolb., E. W. and Toussaint, D. (1981). *Phys. Rev.*, **D23**, 867.

Goncharov, A. S. and Linde, A. D. (1984a). *Zh. Eksp. Teor. Fiz.*, **86**, 1594 (*JETP*, **59**, 930 (1984)).

Goncharov, A. S. and Linde, A. D. (1984b). *Phys. Lett.*, **139B**, 27.

Goncharov, A. S. and Linde, A. D. (1984c). *Class. Quantum Grav.*, **1**, L75.

Goncharov, A. S., Linde, A. D. and Vysotskii, M. I. (1984). *Phys. Lett.*, **147B**, 279.

Gott, J. R. (1982). *Nature*, **295**, 304.

Gupta, S. and Quinn, H. R. (1984). *Phys. Rev.*, **D29**, 2791.

Guth, A. H. (1981). *Phys. Rev.*, **D23**, 347.

Guth, A. H. (1983a). In *Asymptotic Realms of Physics: Essays in Honor of Francis E.*

Low, ed. A. H. Guth, K. Huang and R. L. Jaffe. MIT Press: Cambridge, Massachusetts.

Guth, A. H. (1983*b*). In *Magnetic Monopoles* (Proceedings of the NATO Advanced Study Institute on Magnetic Monopoles, Wingspread, Wisconsin, 1982), ed. R. A. Carrigan and W. P. Trower. Plenum: New York.

Guth, A. H. and Pi, S.-Y. (1982). *Phys. Rev. Lett.*, **49**, 1110.

Guth, A. H. and Pi, S.-Y. (1985). *Phys. Rev.*, **D32**, 1899.

Guth, A. H. and Pi, S.-Y. (1986). In *Inner Space/Outer Space: The Interface between Cosmology and Particle Physics*, ed. E. W. Kolb, M. S. Turner, D. Lindley, K. Olive and D. Seckel, p. 345. Chicago University Press: Chicago.

Guth, A. H. and Sher, M. (1983). *Nature*, **302**, 505.

Guth, A. H. and Tye, S.-H. (1980). *Phys. Rev. Lett.*, **44**, 631 [*Erratum: Phys. Rev. Lett.*, **44**, 963 (1980)].

Guth, A. H. and Weinberg, E. J. (1981). *Phys. Rev.*, **D23**, 876.

Guth, A. H. and Weinberg, E. J. (1983). *Nucl. Phys.*, **B212**, 321.

Harrison, E. R. (1970). *Phys. Rev.*, **D1**, 2726.

Hartle, J. B. and Hawking, S. W. (1983). *Phys. Rev.*, **D28**, 2960.

Hawking, S. W. (1982). *Phys. Lett.*, **115B**, 295.

Hawking, S. W. (1984*a*). In *Relativity, Groups and Topology II* (Proceedings of the Les Houches Summer School in Theoretical Physics, June 27–4 Aug., 1983), ed. B. S. DeWitt and R. Stora, p. 333. North Holland: Amsterdam.

Hawking, S. W. (1984*b*). *Nucl. Phys.*, **B239**, 257.

Hawking, S. W. (1986). The density matrix of the universe. Cambridge University preprint 86-0918.

Hawking, S. W. and Ellis, G. F. R. (1973). *The Large Scale Structure of Space–Time*. Cambridge University Press: Cambridge.

Hawking, S. W. and Luttrell, J. C. (1984). *Phys. Lett.*, **143B**, 83.

Hawking, S. W. and Moss, I. G. (1982). *Phys. Lett.*, **110B**, 35.

Hawking, S. W. and Moss, I. G. (1983). *Nucl. Phys.*, **B224**, 180.

Hawking, S. W., Moss, I. G. and Stewart, J. M. (1982). *Phys. Rev.*, **D26**, 2681.

Hawking, S. W. and Wu, Z. C. (1985). *Phys. Lett.*, **151B**, 15.

Holman, R., Ramond, P. and Ross, G. G. (1984). *Phys. Lett.*, **137B**, 343.

Hung, P. Q. (1984). *Phys. Rev.*, **D30**, 1637.

IAU (1986). *Dark Matter in the Universe* (Proceedings of IAU Symposium 117, Princeton, New Jersey, June 1985). Reidel: Dordrecht.

Ipser, J. and Sikivie, P. (1983). *Phys. Rev. Lett.*, **50**, 925.

Jackiw, R. (1974). *Phys. Rev.*, **D9**, 1686.

Jensen, L. G. and Olive, K. A. (1985). *Phys. Lett.*, **159B**, 99.

Jensen, L. G. and Olive, K. A. (1986). *Nucl. Phys.*, **B263**, 731.

Jensen, L. G. and Stein-Schabes, J. A. (1986). A no-hair theorem for inhomogeneous cosmologies, Fermilab-Pub-86/51-A, unpublished.

Kaiser, N. (1983). *Astrophys. J.*, **273**, L17.

Kaiser, N. (1986). In *Inner Space/Outer Space: The Interface between Cosmology and Particle Physics*, ed. E. W. Kolb, M. S. Turner, D. Lindley, K. Olive and D. Seckel. Chicago University Press: Chicago.

Khlopov, M. Yu. and Linde, A. D. (1984). *Phys. Lett.*, **138B**, 265.

Kibble, T. W. B. (1976). *J. Phys.*, **A9**, 1387.

Kibble, T. W. B. (1986). *Phys. Rev.*, **D33**, 328.

Kirzhnits, D. A. and Linde, A. D. (1972). *Phys. Lett.*, **42B**, 471.

Kodama, H., Sasaki, M. and Sato, K. (1982). *Prog. Theor. Phys.*, **68**, 1979.
Kodama, H., Sasaki, M., Sato, K. and Maeda, K. (1981). *Prog. Theor. Phys.*, **66**, 2052.
Kolb, E. W. and Turner, M. S. (1983). *Ann. Rev. Nucl. Part. Sci.*, **33**, 645.
Kolb, E. W. and Wolfram, S. (1980). *Astrophys. J.*, **239**, 428.
Kounnas, C. and Quiros, M. (1985). *Phys. Lett.*, **151B**, 189.
Langacker, P. (1981). *Phys. Rep.*, **72C**, 185.
Lazarides, G., Shafi, Q. and Trower, W. P. (1982). *Phys. Rev. Lett.*, **49**, 1756.
Lindblom, P. R., Ovrut, B. A. and Steinhardt, P. J. (1986). *Phys. Lett.*, **B172**, 309.
Linde, A. D. (1974). *Zh. Eksp. Teor. Fiz.*, **19**, 320 (*JETP Lett.*, **19**, 183 (1974)).
Linde, A. D. (1979). *Rep. Prog. Phys.*, **42**, 389.
Linde, A. D. (1982*a*). *Phys. Lett.*, **108B**, 389.
Linde, A. D. (1982*b*). *Phys. Lett.*, **114B**, 431.
Linde, A. D. (1982*c*). *Phys. Lett.*, **116B**, 335.
Linde, A. D. (1983*a*). In *The Very Early Universe* (Proceedings of the Nuffield Workshop), ed. G. W. Gibbons, S. W. Hawking and S. T. C. Siklos, p. 205. Cambridge University Press: Cambridge.
Linde, A. D. (1983*b*). *Zh. Eksp. Teor. Fiz.*, **38**, 149 (*JETP Lett.*, **38**, 176).
Linde, A. D. (1983*c*). *Phys. Lett.*, **129B**, 177.
Linde, A. D. (1983*d*). *Phys. Lett.*, **131B**, 330.
Linde, A. D. (1983*e*). *Phys. Lett.*, **132B**, 317.
Linde, A. D. (1983*f*). *Zh. Eksp. Teor. Fiz.*, **37**, 606 (*JETP Lett.*, **37**, 724 (1983)).
Linde, A. D. (1984*a*). *Rep. Prog. Phys.*, **47**, 925.
Linde, A. D. (1984*b*). *Nuovo Cim. Lett.*, **39**, 401.
Linde, A. D. (1984*c*). *Zh. Eksp. Teor. Fiz.*, **87**, 369 (*JETP Lett.*, **60**(2), 211 (1984)).
Linde, A. D. (1984*d*). *Zh. Eksp. Teor. Fiz.*, **40**, 496 (*JETP Lett.*, **40**, 1333 (1984)).
Linde, A. D. (1985). *Phys. Lett.*, **158B**, 375.
Linde, A. D. (1986*a*). *Mod. Phys. Lett.*, **A1**, 81.
Linde, A. D. (1986*b*). Eternally existing, self-reproducing inflationary universe. To be published in Proceedings of Nobel Symposium on Unification of Fundamental Interactions.
Linde, A. D. (1986*c*). Eternally existing self-reproducing chaotic inflationary universe. *Phys. Lett.*, **175B**, 395.
Lindley, D. (1985). The inflationary universe: a brief history. Unpublished.
Maeda, K., Sato, K., Sasaki, M. and Kodama, H. (1982). *Phys. Lett.*, **108B**, 98.
Mazenko, G. F., Unruh, W. G. and Wald, R. M. (1985). *Phys. Rev.*, **D31**, 273.
Mazur, P. and Mottola, E. (1986). Spontaneous breaking of de Sitter symmetry by radiative effects. Santa Barbara Inst. Theor. Phys. preprint NSF-ITP-85-153.
Misner, C. W., Thorne, K. S. and Wheeler, J. A. (1973). *Gravitation*. Freeman: San Francisco.
Moss, I. G. (1983). *Phys. Lett.*, **128B**, 385.
Moss, I. G. (1984). *Nucl. Phys.*, **B238**, 436.
Moss, I. G. and Wright, W. A. (1984). *Phys. Rev.*, **D29**, 1067.
Mottola, E. (1985). *Phys. Rev.*, **D31**, 754.
Mottola, E. (1986). *Phys. Rev.*, **D33**, 1616.
Myhrvold, N. P. (1983*a*). *Phys. Rev.*, **D28**, 2439.
Myhrvold, N. P. (1983*b*). *Phys. Lett.*, **132B**, 308.
Nanopoulos, D. V., Olive, K. A. and Srednicki, M. (1983). *Phys. Lett.*, **127B**, 30.
Nanopoulos, D. V., Olive, K. A., Srednicki, M. and Tamvakis, K. (1983*a*). *Phys. Lett.*, **123B**, 41.

Nanopoulos, D. V., Olive, K. A., Srednicki, M. and Tamvakis, K. (1983*b*). *Phys. Lett.*, **124B**, 171.

Nanopoulos, D. V. and Srednicki, M. (1983). *Phys. Lett.*, **133B**, 287.

Olive, K. A. (1983). In *Galaxies and the Early Universe* (Proceedings of the 18th Rencontre de Moriond, Astrophysics Meeting), ed. J. Audouze and J. Tran Thanh Van, p. 3. Reidel: Dordrecht.

Olson, D. W. (1976). *Phys. Rev.*, **D14**, 327.

Ostriker, J. P., Thompson, C. and Witten, E. (1986). Cosmological effects of superconducting strings. Submitted to *Phys. Lett.*

Ovrut, B. A. and Steinhardt, P. J. (1983). *Phys. Lett.*, **133B**, 161.

Ovrut, B. A. and Steinhardt, P. J. (1984*a*). *Phys. Rev. Lett.*, **53**, 732.

Ovrut, B. A. and Steinhardt, P. J. (1984*b*). *Phys. Rev.*, **D30**, 2061.

Ovrut, B. A. and Steinhardt, P. J. (1984*c*). *Phys. Lett.*, **147B**, 263.

Ovrut, B. A. and Steinhardt, P. J. (1986). In *Inner Space/Outer Space: The Interface between Cosmology and Particle Physics*, ed. E. W. Kolb, M. S. Turner, D. Lindley, K. Olive and D. Seckel. Chicago University Press: Chicago.

Peebles, P. J. E. (1979). *Astron. J.*, **84**, 730.

Peebles, P. J. E. (1984). *Astrophys. J.*, **284**, 439.

Pi, S.-Y. (1984). *Phys. Rev. Lett.*, **52**, 1725.

Preskill, J. P. (1979). *Phys. Rev. Lett.*, **43**, 1365.

Preskill, J., Wise, M. B. and Wilczek, F. (1983). *Phys. Lett.*, **120B**, 127.

Press, W. H. and Vishniac, E. T. (1980). *Astrophys. J.*, **239**, 1.

Rindler, W. (1956). *Mon. Not. R. Astron. Soc.*, **116**, 663.

Sandage, A. and Tammann, G. A. (1976). *Astrophys. J.*, **210**, 7.

Sasaki, M. and Sato, K. (1982). *Prog. Theor. Phys.*, **68**, 1979.

Sato, K. (1981*a*). *Phys. Lett.*, **99B**, 66.

Sato, K. (1981*b*). *Mon. Not. R. Astron. Soc.*, **195**, 467.

Sato, K. (1981*c*). *Prog. Theor. Phys.*, **66**, 2287.

Sato, K., Kodama, H., Sasaki, M. and Maeda, K. (1982). *Phys. Lett.*, **108B**, 103.

Sato, K., Sasaki, M., Kodama, H. and Maeda, K. (1981). *Prog. Theor. Phys.*, **65**, 1443.

Scherrer, R. J. and Frieman, J. A. (1986). *Phys. Rev.*, **D33**, 3556.

Schramm, D. N. and Steigman, G. (1981). *Astrophys. J.*, **243**, 1.

Seckel, D. and Turner, M. S. (1985). *Phys. Rev.*, **D32**, 3178.

Shafi, Q. and Stecker, F. W. (1984). *Phys. Rev. Lett.*, **53**, 1292.

Shafi, Q. and Vilenkin, A. (1984). *Phys. Rev. Lett.*, **52**, 691.

Shafi, Q. and Wetterich, C. (1983). *Phys. Lett.*, **129B**, 387.

Shafi, Q. and Wetterich, C. (1985). *Phys. Lett.*, **152B**, 51.

Sher, M. (1981). *Phys. Rev.*, **D24**, 1699.

Silk, J. and Turner, M. S. (1986). Double inflation. Submitted to *Phys. Rev.*, **D**.

Spokoiny, B. L. (1984). *Phys. Lett.*, **147B**, 39.

Starobinsky, A. A. (1979). *Zh. Eksp. Teor. Fiz.*, **30**, 719 (*JETP Lett.*, **30**, 682 (1979)).

Starobinsky, A. A. (1980). *Phys. Lett.*, **91B**, 99.

Starobinsky, A. A. (1982). *Phys. Lett.*, **117B**, 175.

Steigman, G. (1983). In *Unification of the Fundamental Particle Interactions II* (Proceedings of the Europhysics Study Conference, Erice, Italy, Oct. 6–14, 1981), ed. J. Ellis and S. Ferrara. Plenum: New York.

Steinhardt, P. J. (1981*a*). *Phys. Rev.*, **D24**, 842.

Steinhardt, P. J. (1981*b*). *Nucl. Phys.*, **B190**, 583.

Steinhardt, P. J. (1986). In *High Energy Physics, 1985* (Proceedings of the Yale

Theoretical Advanced Study Institute), ed. M. J. Bowick and F. Gürsey, Vol. 2, p. 567. World Scientific: Singapore.

Steinhardt, P. J. and Turner, M. S. (1983). *Phys. Lett.*, **129B**, 51.

Steinhardt, P. J. and Turner, M. S. (1984). *Phys. Rev.*, **D29**, 2162.

Traschen, J. and Hill, C. T. (1986). *Phys. Rev.*, **D33**, 3519.

Tryon, E. P. (1973). *Nature*, **246**, 396.

Turner, M. S. (1985). In *Proceedings of the Cargese School on Fundamental Physics and Cosmology*, ed. J. Audouze and J. Tran Thanh Van. Editions Frontiers: Gif-Sur-Yvette.

Turner, M. S. (1986). *Phys. Rev.*, **D33**, 889.

Turner, M. S., Steigman, G. and Krauss, L. M. (1984). *Phys. Rev. Lett.*, **52**, 2090.

Turner, M. S., Wilczek, F. and Zee, A. (1983). *Phys. Lett.*, **125B**, 35 [*Erratum: Phys. Lett.*, **125B**, 519 (1983)].

Turok, N. (1986). Recent developments in the cosmic string theory of galaxy formation. Lecture presented at the International Symposium 'Particles and the Universe' at the Aristotle Univ. of Thessaloniki, June 1985.

Turok, N. and Brandenberger, R. H. (1986). *Phys. Rev.*, **D33**, 2175.

Vaucouleurs, G. de and Bollinger, G. (1979). *Astrophys. J.*, **233**, 433.

Vilenkin, A. (1982). *Phys. Lett.*, **117B**, 25.

Vilenkin, A. (1983a). *Phys. Rev.*, **D27**, 2848.

Vilenkin, A. (1983b). *Nucl. Phys.*, **B226**, 527.

Vilenkin, A. (1983c). *Nucl. Phys.*, **B226**, 504.

Vilenkin, A. (1984). *Phys. Rev.*, **D30**, 509.

Vilenkin, A. (1985a). *Phys. Rep.*, **121**, 263.

Vilenkin, A. (1985b). *Nucl. Phys.*, **B252**, 141.

Vilenkin, A. (1985c). *Phys. Rev.*, **D32**, 2511.

Vilenkin, A. and Ford, L. H. (1982). *Phys. Rev.*, **D26**, 1231.

Vittorio, N., Juszkiewica, R. and Davis, M. (1986). Large scale velocity fields as a test of cosmological models. Unpublished.

Vittorio, N. and Silk, J. (1985). *Astrophys. J.*, **293**, L1.

Vittorio, N. and Turner, M. S. (1986). The large-scale peculiar velocity field in flat models of the universe. Submitted to *Astrophys. J.*

Wald, R. M. (1983). *Phys. Rev.*, **D28**, 2118.

Weinberg, S. (1972). *Gravitation and Cosmology*. Wiley: New York.

Weinberg, S. (1973). *Phys. Rev.*, **D7**, 2887.

Weinberg, S. (1974). *Phys. Rev.*, **D9**, 3357.

Wilkinson, D. T. (1986). In *Inner Space/Outer Space: The Interface between Cosmology and Particle Physics*, ed. E. W. Kolb, M. S. Turner, D. Lindley, K. Olive and D. Seckel, p. 126. Chicago University Press: Chicago.

Witten, E. (1981a). *Commun. Math. Phys.*, **80**, 381.

Witten, E. (1981b). *Phys. Lett.*, **105B**, 267.

Yang, J., Turner, M. S., Steigman, G., Schramm, D. N. and Olive, K. (1984). *Astrophys. J.*, **281**, 493.

Yoshimura, M. (1981). In *Grand Unified Theories and Related Topics* (Proceedings of the 4th Kyoto Summer Institute), ed. M. Konuma and T. Maskawa. World Scientific: Singapore.

Zeldovich, Ya. B. (1968). *Usp. Fiz. Nauk.*, **95**, 209 (*Sov. Phys. Uspekhi*, **11**, 381 (1968)).

Zeldovich, Ya. B. (1972). *Mon. Not. R. Astron. Soc.*, **160**, 1P.

Zeldovich, Ya. B. and Khlopov, M. Y. (1978). *Phys. Lett.*, **B79**, 239.

13

Inflation and quantum cosmology

ANDREI LINDE

13.1 Introduction

One of the most popular models for the evolution of the universe developed at the beginning of the 1980s is the inflationary universe scenario. This scenario in its present form (see e.g. Linde, 1984c, 1986c; Blau and Guth, this volume, Ch. 12) makes it possible to give answers to about ten different problems relating both to cosmology and to elementary particle physics. It then becomes possible to understand why the observable part of the universe, with a size $\sim 10^{28}$ cm, is so flat, homogeneous and isotropic, why we do not see any primordial monopoles, what is the origin of galaxies, etc. No alternative theory that can solve all these problems has been suggested so far. Therefore it seems plausible that something like inflation actually did occur in the very early stages of the evolution of the universe.

Historically, there were many different versions of the inflationary universe scenario. For example, about 20 years ago it was argued (Gliner, 1965) that the superdense baryonic matter should have the vacuum-like equation of state, $p = -\rho$, and its energy–momentum tensor should have the form $T_{\mu\nu} \sim g_{\mu\nu}\Lambda$. This would lead to exponential expansion of the universe in the very early stages of its evolution (Gliner, 1970; Gliner and Dymnicova, 1975; Gurevich, 1975). At present, however, it seems that the equation of state of superdense baryonic matter should be not $p = -\rho$ but $p = \frac{1}{3}\rho$.

The next stage in the development of the inflationary universe scenario is associated with the Starobinsky model (Starobinsky, 1979, 1980). He noted that the exponentially expanding Friedmann universe (de Sitter space) is an unstable solution of the Einstein equations with quantum corrections (Dowker and Critchley, 1976), which after the development of the instability transforms into the hot Friedmann universe. The main purpose of this model was to solve the singularity problem, which proved to be too complicated. Moreover, density perturbations, generated during the exponential expansion of the universe in the original version of this model, typically are much greater than the desirable value $\delta\rho/\rho \sim 10^{-4}$ (Mukhanov and Chibisov, 1981, 1982).

Another line of attack was related to the unified models of weak, strong and electromagnetic interactions. As shown in Linde (1974), Veltman (1974, 1975) and Dreitlein (1974), the energy–momentum tensor of a constant scalar field ϕ, which plays an important role in these theories, is given by $T_{\mu\nu}(\phi) \sim g_{\mu\nu} V(\phi)$, where $V(\phi)$ is the effective potential of this field. The value of the field ϕ at the minimum of $V(\phi)$ depends on the value of the temperature T in the very early universe. Typically, the minimum of $V(\phi)$ at large T is displaced to $\phi=0$, and with decrease in T, a phase transition to some state $\phi_0 \neq 0$ occurs (Kirzhnits, 1972; Linde, 1979). If this phase transition proceeds from a strongly supercooled vacuum state $\phi=0$, the total entropy of the universe after the phase transition may increase considerably (Linde, 1979). As pointed out by Guth (1981), this effect can help to solve the horizon problem, the flatness problem and the primordial monopole problem. His model (now called the old inflationary universe scenario) was very simple and attractive. It was based on four important assumptions.

(1) Initially the universe was in a symmetric state $\phi=0$ due to high-temperature effects.

(2) After the universe expands the field ϕ becomes trapped at a local minimum of $V(\phi)$ near $\phi=0$. With a decrease of temperature the total energy–momentum tensor of matter becomes equal to $T_{\mu\nu}=g_{\mu\nu}V(0)$. The universe in such a state expands exponentially.

(3) The stage of exponential expansion (inflation) finishes at the moment of the phase transition to the stable state $\phi=\phi_0 \neq 0$.

(4) The phase transition occurs due to formation of bubbles with $\phi=\phi_0$. The process of reheating the universe occurs due to bubble wall collisions.

Unfortunately, in this scenario the universe becomes extremely inhomogeneous after reheating (Guth, 1981; Hawking and Moss, 1982; Guth and Weinberg, 1983). This problem was solved in the context of the new inflationary universe scenario (Linde, 1982a–d; Albrecht and Steinhardt, 1982). In the new scenario the last two assumptions mentioned above were abandoned. This scenario is still very popular. In my opinion, however, this scenario also is not perfect. It can be realised only in some theories with rather unnatural potentials $V(\phi)$ that are extremely flat near $\phi=0$ and are sufficiently curved near the global minimum of $V(\phi)$. Density perturbations produced in this scenario are sufficiently small only if $V(0)/M_P^4 \sim 10^{-10}$ and if the field ϕ interacts extremely weakly with all other fields (e.g. $\lambda \sim 10^{-12}$ for $V(\phi) \sim V(0) - \frac{1}{4}\lambda\phi^4$). The first condition means that inflation in this scenario starts very late, which makes it impossible to solve the horizon problem in this scenario if the universe is closed: the universe typically collapses before the beginning of inflation (Linde, 1984c). The second condition implies that thermal effects typically cannot raise the field ϕ to the top of the effective potential at $\phi=0$, and in such a case inflation does not occur at all (Linde, 1984c, 1985a, b). Consequently, despite many efforts, no realistic versions of the new inflationary universe scenario have been suggested so far.

The only scenario in which these problems do not arise is the chaotic inflationary scenario (Linde, 1983b, 1984a, b, 1985a, b). The main idea of this scenario is based on the investigation of the possibility of inflation in a universe filled with some non-equilibrium initial distribution of the field ϕ, without making any *ad hoc* assumptions (1)–(4) concerning thermal equilibrium, supercooling, etc. This scenario is a most general one, and, surprisingly enough, it works perfectly well in a wide class of theories with fairly natural effective potentials $V(\phi)$. Therefore in the present paper we shall consider only this version of the inflationary universe scenario.

The concrete models of inflation are modified with each new development of the underlying elementary particle theory. However, some basic features of these models remain intact. Many conceptual problems of the inflationary cosmology have now been solved. But two problems are still widely discussed at present.

The first problem is connected with the initial conditions in the early universe. This problem is also related to the singularity problem. The main part of the problem is not the existence of singularities in the universe, but the statement (or the common belief) that the universe does not exist eternally and that there exists 'some time at which there is no spacetime at all'. What is the origin of the universe? Was it created in a singular state or has it appeared due to a quantum jump 'from nothing'? Which initial conditions in the new-born universe are most natural?

Different cosmologists give different answers to these questions. Some of these answers are based on a phenomenological description of the universe soon after its formation (Linde, 1983b, 1985a, b). Another approach is based on the investigation of the wave function of the universe (DeWitt, 1967; Wheeler, 1968). The choice of a particular wave function in this approach is crucially author-dependent. One of the most interesting suggestions was given by Hartle and Hawking (1983). Later Hawking (1984) and Hawking and Page (1986) have studied a possibility of realisation of the chaotic inflation scenario in quantum cosmology with the help of the Hartle–Hawking wave function of the universe. An alternative choice of the wave function was suggested by some other authors (Linde, 1984*a, b*; Zeldovich and Starobinsky, 1984; Rubakov, 1984; Vilenkin, 1983). These two approaches lead to different answers to the question of which initial conditions were realised in the very early stages of the evolution of the universe. This is one of the main problems to be discussed in the present paper.

Another problem is related to the uniqueness of the universe. The essence of this problem was formulated by Einstein in his talk with E. Straus: 'What I am really interested in is whether God could create the world differently.' The answer to this question in the context of the inflationary cosmology appears to be rather unexpected. Namely, the *local* structure of the universe is determined by inflation, which occurs at the *classical* level. The universe after inflation becomes locally flat, homogeneous and isotropic. However, its *global* structure is determined by *quantum* effects. It proves that the large-scale quantum fluctuations of the scalar field ϕ generated in the chaotic inflation scenario lead to an infinite process of creation and self-reproduction of inflationary parts of the universe (Linde, 1986*a, b, c*). In this scenario the evolution of the inflationary universe has no end and may have no beginning. As a result, the universe becomes divided into many different domains (mini-universes) of exponentially large size, inside which all possible (metastable) vacuum states are realised. One may say therefore that not only could God create the universe differently, but in His wisdom He created a universe which has been unceasingly producing different universes of all possible types.

We shall start our presentation with a discussion of the simplest version of the chaotic inflationary scenario. However, most of the qualitative results that will be obtained are essentially model-independent.

13.2 Chaotic inflation

Let us consider the simplest model of inflation based on the theory of a massive non-interacting scalar field ϕ with the Lagrangian

$$\mathcal{L} = -\frac{M_P^2}{16\pi} R + \frac{1}{2} \partial_\mu \phi \, \partial^\mu \phi - \frac{m^2}{2} \phi^2. \tag{2.1}$$

Here $M_P^{-2} = G$ is the gravitational constant, $M_P \sim 10^{19}$ GeV is the Planck mass, R is the curvature scalar, m is the mass of the scalar field ϕ, $m \ll M_P$. If the classical field ϕ is sufficiently homogeneous in some domain of the universe (see below), then its behaviour inside this domain is governed by equations

$$\ddot{\phi} + 3H\dot{\phi} = -dV/d\phi, \tag{2.2}$$

$$H^2 + \frac{k}{a^2} = \frac{8\pi}{3M_P^2}\left(\tfrac{1}{2}\dot{\phi}^2 + V(\phi)\right). \tag{2.3}$$

Here $V(\phi)$ is the effective potential of the field ϕ (in our case $V(\phi) = \tfrac{1}{2}m^2\phi^2$), $H = \dot{a}/a$, $a(t)$ is the scale factor of the locally Friedmannian universe (inside the domain under consideration), $k = +1, -1$, or 0 for a closed, open or flat universe, respectively. If the field ϕ initially is sufficiently large ($\phi \gtrsim M_P$), then the functions $\phi(t)$ and $a(t)$ rapidly approach the asymptotic regime

$$\phi(t) = \phi_0 - \frac{mM_P}{2(3\pi)^{1/2}}\, t, \tag{2.4}$$

$$a(t) = a_0 \exp\left(\frac{2\pi}{M_P^2}\left(\phi_0^2 - \phi^2(t)\right)\right). \tag{2.5}$$

According to (2.4) and (2.5), during a time $\tau \sim \phi/mM_P$ the value of the field ϕ remains almost unchanged and the universe expands quasi-exponentially:

$$a(t + \Delta t) \sim a(t) \exp(H\,\Delta t) \tag{2.6}$$

for $\Delta t \lesssim \tau = \phi/mM_P$. Here

$$H = \frac{2\pi^{1/2}}{\sqrt{3}} \cdot \frac{m\phi}{M_P}, \tag{2.7}$$

$H \gg \tau^{-1}$ for $\phi \gg M_P$.

The regime of quasi-exponential expansion (inflation) occurs for $\phi \gtrsim \tfrac{1}{5}M_P$. For $\phi \lesssim \tfrac{1}{5}M_P$ the field ϕ oscillates rapidly, and if this field interacts with other matter fields (which are not written explicitly in (2.1)), its potential energy $V(\phi \sim \tfrac{1}{5}M_P) \sim m^2 M_P^2$ is transformed into heat. The reheating temperature T_R may be of the order $(mM_P)^{1/2}$ or somewhat smaller, depending on the strength of the interaction of the field ϕ with other fields. It is important that T_R does not depend on the initial value ϕ_0 of the field ϕ. The only parameter which depends on ϕ_0 is the scale factor $a(t)$, which grows $\exp((2\pi/M_P^2)\phi_0^2)$ times during inflation.

If, as is usually assumed, a classical description of the universe becomes possible only when the energy–momentum tensor of matter becomes smaller than M_P^4, then at this moment $\partial_\mu\phi\,\partial^\mu\phi \lesssim M_P^4$ and $V(\phi) \lesssim M_P^4$.

Therefore the only constraint on the initial amplitude of the field ϕ is given by $\frac{1}{2}m^2/\phi^2 \lesssim M_P^4$. This gives a typical initial value of the field ϕ:

$$\phi_0 \sim \frac{M_P^2}{m}. \tag{2.8}$$

Let us consider for definiteness a closed universe of a typical initial size $O(M_P^{-1})$. It can be shown that if initially $\partial_\mu \phi \, \partial^\mu \phi \lesssim V(\phi) \sim M_P^4$ inside this universe, then very soon $\partial_\mu \phi \, \partial^\mu \phi$ becomes much smaller than $V(\phi)$, the evolution of the universe becomes describable by eqs. (2.2)–(2.7), and after inflation the total size of the universe becomes larger than

$$l \sim M_P^{-1} \exp\left(\frac{2\pi}{M_P^2}\phi_0^2\right) \sim M_P^{-1} \exp\left(\frac{2\pi M_P^2}{m^2}\right). \tag{2.9}$$

For $m \sim 10^{-5}M_P$ (see the next section),

$$l \sim M_P^{-1} \exp(10^{11}) \gtrsim 10^{10^{10}} \text{ cm}, \tag{2.10}$$

which is much greater than the size of the observable part of the universe $\sim 10^{28}$ cm.

After such a large inflation the term k/a^2 in (2.3) becomes negligibly small compared with H^2, which means that the universe becomes flat and its geometry locally Euclidean. For similar reasons the universe becomes locally homogeneous and isotropic. The density of all 'undesirable' objects (monopoles, domain walls, gravitinos) created before or during inflation becomes exponentially small, and they never appear again if the reheating temperature T_R is not too large.

We should like to emphasise that for a realisation of this scenario it is sufficient that initially $\partial_\mu \phi \, \partial^\mu \phi \lesssim V(\phi) \sim M_P^4$ in a domain of a smallest possible size $O(M_P^{-1})$. Since $\partial_\mu \phi \, \partial^\mu \phi \lesssim M_P^4$, $V(\phi) \lesssim M_P^4$ in any classical spacetime, the above-mentioned condition is quite natural. Note that, despite some recent claims (see, for instance, Turner, 1985; Enquist *et al.*, 1986), there is no need for the field ϕ to have a very small kinetic energy $(\partial_0 \phi \, \partial^0 \phi \ll V(\phi))$ and to be uniform on scales much larger than M_P^{-1}. For a more detailed discussion of initial conditions which are necessary for inflation see Linde (1985*a, b*).

This is the general scheme of inflation as it can be understood at the classical level. One may wonder whether one can reach similar conclusions concerning initial conditions in the context of quantum cosmology, and whether quantum fluctuations may lead to some modification of the scenario discussed in this section.

13.3 Inflation and the wave function of the universe

One of the most ambitious approaches to cosmology is based on the investigation of the Wheeler–DeWitt equation for the wave function Ψ of the universe (Wheeler, 1968; DeWitt, 1967). However, this equation has many different solutions, and *a priori* it is not quite clear which of these solutions describes our universe.

A very interesting idea was suggested by Hartle and Hawking (1983). According to their work (Hartle and Hawking, 1983), the wave function of the ground state of the universe with a scale factor a filled with a scalar field ϕ in the semi-classical approximation is given by

$$\Psi_0(a, \phi) \sim \exp(-S_E(a, \phi)). \tag{3.1}$$

Here $S_E(a, \phi)$ is the Euclidean action corresponding to the Euclidean solutions of the Lagrange equation for $a(\tau)$ and $\phi(\tau)$ with the boundary conditions $a(0)=a$, $\phi(0)=\phi$. The reason for choosing this particular solution of the Wheeler–DeWitt equation was explained as follows. Let us consider the Green's function of a particle which moves from the point $(0, t')$ to the point $(\mathbf{x}, t)$:

$$\langle \mathbf{x}, 0 | 0, t' \rangle = \sum_n \Psi_n(\mathbf{x}) \Psi_n(0) \exp(i\, E_n t)$$

$$= \int d\mathbf{x}(t) \exp(i\, S(\mathbf{x}(t))), \tag{3.2}$$

where Ψ_n is a complete set of energy eigenstates corresponding to the energies $E_n \geqslant 0$. To obtain an expression for the ground-state wave function $\Psi_0(\mathbf{x})$, one should make a rotation $t \to -i\tau$ and take the limit as $\tau' \to -\infty$. In the summation (3.2) only the term $n=0$ with $E_0=0$ survives, and the integral transforms into $\int d\mathbf{x}(\tau) \exp(-S_E(\mathbf{x}(\tau)))$. Hartle and Hawking have argued that the generalisation of this result to the case of interest in the semi-classical approximation would yield eq. (3.1).

The gravitational action corresponding to the Euclidean section S_4 of de Sitter space dS_4 with $a(\tau)=H^{-1}(\phi)\cos H\tau$ is negative,

$$S_E(a, \phi) = -\frac{1}{2}\int d\eta \left[\left(\frac{da}{d\eta}\right)^2 - a^2 + \frac{\Lambda}{3}a^4\right]\cdot\frac{3\pi M_P^2}{2} = -\frac{3M_P^4}{16V(\phi)}. \tag{3.3}$$

Here η is the conformal time, $\eta=\int dt/a(t)$, $\Lambda=8\pi V/M_P^2$. Therefore, according to (3.1),

$$\Psi_0(a, \phi) \sim \exp(-S_E(a, \phi)) \sim \exp\left(\frac{3M_P^4}{16V(\phi)}\right). \tag{3.4}$$

This means that the probability P of finding the universe in the state with $\phi = \text{const}$, $a = H^{-1}(\phi) = (3M_P^2/8\pi V(\phi))^{1/2}$ is given by

$$P(\phi) \sim |\Psi_0|^2 \sim \exp\left(\frac{3M_P^4}{8V(\phi)}\right). \tag{3.5}$$

This expression has a very sharp maximum as $V(\phi) \to 0$. Therefore the probability of finding the universe in a state with a large field ϕ and having a long stage of inflation becomes strongly diminished. One can argue, of course, that at $V(\phi) \gg M_P^4$ the function $P(\phi)$ becomes constant, and if one introduces some cut-off at $V(\phi) \to 0$ and integrates over all values of the field ϕ from $-\infty$ to $+\infty$, then the probability of finding the universe with large ϕ (with $V(\phi) \gg M_P^4$) becomes large (Hawking and Page, 1986). However, this would break the rule according to which one should try not to appeal to the processes which occur at densities much greater than the Planck density M_P^4.

There exists an alternative choice of the wave function of the universe. It can be argued that the analogy between the standard theory (3.2) and the gravitational theory (3.3) is incomplete. Indeed, there is an overall minus sign in the expression for $S_E(a, \phi)$ (3.3), which indicates that the gravitational 'energy' associated with the scale factor a is negative. (This is related to the well-known fact that the total energy of a closed universe is zero, being a sum of the positive energy of matter and the negative energy of the scale factor a.) In such a case, to obtain Ψ_0 from (3.2) one should rotate t not to $-i\tau$, but to $+i\tau$,† which leads to (Linde, 1984)

$$P \sim |\Psi_0(a, \phi)|^2 \sim \exp(-2|S_E(a, \phi)|) \sim \exp\left(-\frac{3M_P^4}{8V(\phi)}\right). \tag{3.6}$$

Actually, this result is valid only if the evolution of the field ϕ is very slow, so that this field acts only as a cosmological constant $\Lambda(\phi) = 8\pi V(\phi)/M_P^2$ in (3.3). Fortunately, this is indeed the case during inflation.

Later the same result was obtained by another method, devised by Zeldovich and Starobinsky (1984), Rubakov (1984) and Vilenkin (1983). This result can be interpreted as a probability of quantum tunnelling of the universe from $a=0$ (from 'nothing') to $a=H^{-1}(\phi)$. In complete agreement with the results of the previous section, eq. (3.6) predicts that a typical initial value of the field ϕ is given by $V(\phi) \sim M_P^4$ (if one does not speculate about the possibility that $V(\phi) \gg M_P^4$), which leads to a very long stage of inflation.

It must be said that there is no *rigorous* proof of either eq. (3.1) or eq. (3.6),

---

† In our opinion this does not lead to a change of sign of the gravitational constant, as claimed by Hawking (this volume, Ch. 14). In any case, the usual rule of rotation does not give Ψ_0 if $E_n < 0$.

and the physical meaning of creation of everything from 'nothing' is far from clear. Therefore a deeper understanding of the physical processes in the inflationary universe is necessary in order to investigate the wave function of the universe $\Psi_0(a, \phi)$ and to suggest a correct interpretation of this wave function. With this purpose we shall try to investigate the global structure of the inflationary universe, and go beyond the minisuperspace approach used in the derivation of eqs. (3.1) and (3.6).

13.4 Quantum fluctuations in the inflationary universe

As we have already mentioned, inflation makes the universe locally homogeneous and isotropic. However, inflation also leads to the creation of long-wave fluctuations $\delta\rho(\mathbf{x})$ of the field ϕ (Vilenkin and Ford, 1982; Linde, 1982c; Starobinsky, 1982). Fluctuations generated during a time $\Delta t = H^{-1}$ in our theory form an inhomogeneous distribution of the classical field $\delta\phi(x)$ with a time-independent amplitude

$$|\delta\phi(\mathbf{x})| \sim \frac{H(\phi)}{2\pi} \tag{4.1}$$

and with initial wavelength $\Delta l \sim H^{-1}$. Later, their wavelength grows exponentially as $a(t)$, eq. (2.5), and the field $\phi + \delta\phi(\mathbf{x})$ becomes almost exactly homogeneous, slowly decreasing according to eq. (2.4). However, at the same time new perturbations of the field ϕ are generated, and so on. This process looks like a Brownian motion of the field ϕ. Inhomogeneities of the resulting distribution of the field ϕ lead to density perturbations $\delta\rho(\mathbf{x})$, which grow very slowly (logarithmically) as their wavelengths grow. On a galactic scale $\delta\rho/\rho \sim 10(m/M_P)$, which at $m \sim 10^{-5}M_P \sim 10^{14}$ GeV, gives the desirable value $\delta\rho/\rho \sim 10^{-4}$ necessary for galaxy formation (Mukhanov and Chibisov, 1981, 1982; Hawking, 1982, 1985; Starobinsky, 1982; Guth and Pi, 1982; Bardeen et al., 1983). However, on a much greater scale perturbations $\delta\rho/\rho$ become very large. The estimates of $\delta\rho(\mathbf{x})/\rho$ in the chaotic inflation scenario show (Linde, 1984c) that density perturbations formed at the moment at which the classical scalar field was equal to ϕ have the amplitude

$$\frac{\delta\rho(\phi)}{\rho} \sim C \frac{\phi[V(\phi)]^{1/2}}{M_P^3} \sim \frac{m}{M_P}\left(\frac{\phi}{M_P}\right)^2, \tag{4.2}$$

where $C = O(1)$. This means that $\delta\rho/\rho \sim 1$ for

$$\phi \gtrsim \frac{M_P^{1/2}}{m^{1/2}} \cdot M_P. \tag{4.3}$$

Perturbations which are formed at that moment have at present the wavelength

$$l \sim M_{\mathrm{P}}^{-1} \exp\left(\frac{2\pi M_{\mathrm{P}}}{m}\right) \sim 10^{10^{5'}} \mathrm{cm}, \tag{4.4}$$

for $m/M_{\mathrm{P}} \sim 10^{-5}$. Eq. (4.4) gives the size of a locally-Friedmannian part of our universe after inflation, which is much smaller than the classical estimate $l \sim 10^{10^{10}}$ cm, but is still many orders of magnitude greater than the size of the observable part of the universe, $\sim 10^{28}$ cm. Let us try to understand the origin of large inhomogeneities formed at $\phi \gtrsim (M_{\mathrm{P}}/m)^{1/2} \cdot M_{\mathrm{P}}$, eq. (4.3), and the global structure of the universe at scales greater than $M_{\mathrm{P}}^{-1} \exp[(2\pi M_{\mathrm{P}}/m)]$, eq. (4.4).

Evolution of the fluctuating field ϕ in any given domain can be described with the help of its distribution function $P(\phi)$, or in terms of its average value $\bar{\phi}$ in this domain and its dispersion $\Delta = (\langle \delta\phi^2 \rangle)^{1/2}$. However, one will obtain different results depending on the method of averaging: one can consider the distribution $P_{\mathrm{c}}(\phi)$ over the non-growing *coordinate* volume of the domain (i.e. over its physical volume at some initial moment of inflation), or the distribution $P_{\mathrm{p}}(\phi)$ over its *physical* (proper) volume, which grows exponentially at a different rate in different parts of the domain. It can be shown that the dispersion of the field ϕ in the coordinate volume Δ_{c} is always much smaller than $\bar{\phi}_{\mathrm{c}}$ for $V(\bar{\phi}_{\mathrm{c}}) \ll M_{\mathrm{P}}^4$. Therefore the evolution of the averaged field $\bar{\phi}_{\mathrm{c}}$ can be described approximately by eqs. (2.2)–(2.5). However, if one wishes to know the resulting spacetime structure and the distribution of the field ϕ after (or during) inflation, it is more appropriate to take an average $\bar{\phi}_{\mathrm{p}}$ over the physical volume, and in some cases the behaviour of $\bar{\phi}_{\mathrm{p}}$ and Δ_{p} differs considerably from the behaviour of $\bar{\phi}_{\mathrm{c}}$ and Δ_{c}.

To illustrate this let us note that perturbations $\delta\phi(x)$ generated during inflation always have a wavelength $\Delta l \gtrsim H^{-1}(\phi)$. Each domain of the universe of a size $\Delta l \gtrsim H^{-1}(\phi)$, which is the size of the event horizon in de Sitter space, evolves independently of what occurs outside it (Gibbons and Hawking, 1977; Hawking and Moss, 1982). Therefore, as a result of generation of long-wave perturbations of the field ϕ, during inflation the universe becomes effectively divided into many mini-universes of initial size $\Delta l \gtrsim H^{-1}(\phi)$. Each of these domains may be considered as a separate part of a Friedmann universe with a coordinate volume which effectively does not change during inflation. The field ϕ inside a domain of a size $\Delta l \sim H^{-1}$ looks as if it were almost exactly homogeneous $[\partial_{\mu}(\delta\phi) \cdot \partial^{\mu}(\delta\phi)] \sim H^4 \ll V(\phi)$ for $V \ll M_{\mathrm{P}}^4$ and $\delta\phi$ given by eq. (4.1). During a typical time $\Delta t = H^{-1}$ this field decreases by

$$\Delta\phi = \frac{mM_{\rm P}}{2\sqrt{3}\pi^{1/2}H} = \frac{M_{\rm P}^2}{4\pi\phi}, \tag{4.5}$$

in accordance with eq. (2.4). The physical size of this domain during the time $\Delta t = H^{-1}$ increases e times, and its physical volume increases e^3 times. As a result of generation of perturbations of the field ϕ, eq. (4.1), the value of the field ϕ in this domain becomes $\phi - \Delta\phi + \delta\phi(\mathbf{x})$. Note that for $\phi \gg M_{\rm P}(M_{\rm P}/m)^{1/2}$, $|\delta\phi(\mathbf{x})|$ is much greater than $\Delta\phi$, see Fig. 13.1. Since a typical wavelength of the perturbations $\delta\phi(\mathbf{x})$ generated during the time $\Delta t = H^{-1}$ is $O(H^{-1})$, the domain whose initial size is $O(H^{-1})$, after expanding e times, becomes divided into $O(e^3/2)$ domains of size $O(H^{-1})$, in which the field has a value $\phi - O(H)$, and $O(e^3/2)$ domains of size

Fig. 13.1. The region of possible values of the field ϕ is divided into four parts: (1) $\phi > M_{\rm P}^2/m$. Fluctuations of metric and of the scalar field ϕ are extremely large, and any classical description of this field and of spacetime seems impossible. (2) $M_{\rm P}(M_{\rm P}/m)^{1/2} \lesssim \phi \lesssim M_{\rm P}^2/m$. In this region fluctuations of the field ϕ are large, but dispersion of these fluctuations averaged over the coordinate volume is small, $\Delta_{\rm c} \ll \bar\phi_{\rm c}$. In this region $\bar\phi_{\rm c}$ rolls down according to eq. (2.4), but the field ϕ averaged over the physical volume, $\bar\phi_{\rm p}$, diffuses up to $\phi \sim M_{\rm P}^2/m$. (3) $M_{\rm P} \lesssim \phi \lesssim M_{\rm P}(M_{\rm P}/m)^{1/2}$. In this region $\bar\phi_{\rm c} \approx \bar\phi_{\rm p}$ decreases according to eq. (2.4). Inflation occurs until the field ϕ enters the region $\phi \lesssim M_{\rm P}$. (4) $\phi \lesssim M_{\rm P}$. The field ϕ rolls down, oscillates near the minimum of $V(\phi)$ and its energy transforms into heat.

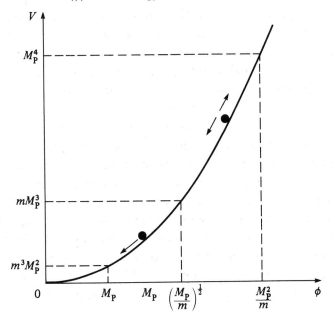

$O(H^{-1})$ with field value $\phi + O(H)$. In other words, the original domain (mini-universe) of size $O(H^{-1})$ after the time $\Delta t = H^{-1}$, expands and separates into $O(e^3)$ mini-universes of a size $O(H^{-1})$, and in (almost) half of these mini-universes the field ϕ *grows* rather than decreases. During the next interval $\Delta t = H^{-1}$ the total number of domains with a growing field ϕ increases again, and so on. This means that the total physical volume of domains containing permanently growing field ϕ increases as $\exp[(3 - \ln 2)Ht] \sim \exp(2.3H(\phi)t)$ for $\phi \gg M_P(M_P/m)^{1/2}$, whereas the total physical volume of domains in which the field ϕ does not decrease grows approximately as $\frac{1}{2}\exp(3Ht)$. Since the value of $H(\phi)$ increases with a growth of ϕ, the main part of the physical volume of the universe emerges as a result of expansion of domains with a maximal possible field ϕ, i.e. with $\phi \sim M_P^2/m$, at which $V(\phi) \sim M_P^4$. (Note that for $\phi \gg M_P^2/m$, if it is possible to consider such domains at a classical level, the process of self-reproduction of inflationary mini-universes with a growing field ϕ becomes suppressed, since at $V(\phi) \gg M_P^4$ a typical value of $\partial_\mu(\delta\phi) \partial^\mu(\delta\phi) \sim H^4$ is greater than $V(\phi)$, which does not lead to creation of *inflationary* mini-universes with $\phi \gg M_P^2/m$.) Therefore, whereas the field ϕ averaged over the coordinate volume of any given domain (i.e. $\bar{\phi}_c$), gradually decreases in accordance with eq. (2.4), the field ϕ averaged over the physical volume of a domain that initially contains the field $\phi \gtrsim M_P(M_P/m)^{1/2}$ grows to $\bar{\phi}_P \sim M_P^2/m$ (Linde, 1986a).

It may be useful to look at the same problem from another point of view. The Brownian motion of the field ϕ at $M_P \lesssim \phi \lesssim M_P^2/m$ can be described (for changes of the field ϕ that are not too large) by the diffusion equation

$$\frac{\partial P_c}{\partial t} = \frac{\partial}{\partial \phi}\left(\frac{\partial(\mathcal{D}P_c)}{\partial \phi} + \frac{P_c}{3H}\frac{\partial V}{\partial \phi}\right), \tag{4.6}$$

where the coefficient of diffusion $\mathcal{D} = H^3/8\pi^2$. This equation for the case $H(\phi) = \text{const.}$ was first derived by Starobinsky (1984, 1986); for a more detailed derivation see Goncharov and Linde (1986, 1987). For the special case $\partial V/\partial \phi = 0$ this equation was obtained by Vilenkin (1983).

The stationary solution ($\partial P_c/\partial t = 0$) would be

$$P_c \sim \exp(3M_P^4/8V(\phi)), \tag{4.7}$$

which is equal to the square of the Hartle–Hawking wave function of the universe (3.5) (Starobinsky, 1986; Linde, 1986b, c). A more general stationary solution would also contain a term

$$3\pi j_0 \frac{M_P^2}{V(\phi)} \cdot \exp\left(\frac{3M_P^4}{8V(\phi)}\right) \int_0^\phi d\phi \exp\left(-\frac{3M_P^4}{V(\phi)}\right) \tag{4.8}$$

(Starobinsky, 1986), which would correspond to a stationary flux of probability j_0 from $\phi = +\infty$ to $\phi = -\infty$. However, eq. (4.6) actually has *no* normalisable stationary solutions, since it is not valid at $|\phi| \lesssim M_P$ (and at $|\phi| \gtrsim M_P^2/m$).† The field ϕ at $|\phi| \lesssim M_P$ is rapidly oscillating rather than slowly rolling down, there is no diffusion of the field ϕ from the region $\phi \lesssim M_P$ to the region $\phi \gtrsim M_P$, and the averaged field ϕ decreases according to eq. (2.4). A detailed discussion of the solutions of eq. (4.6) is included in a separate publication (Goncharov and Linde, 1986, 1987; Linde, 1986c). Here we would like to discuss some qualitative features of these solutions and their physical interpretation.

Let us consider a domain of initial size $l \sim H^{-1}(\phi)$. The field $\phi = \phi_0$ inside such a domain is distributed homogeneously, since its classical part becomes homogeneous due to inflation and perturbations $\delta\phi(\mathbf{x})$ become essential only at a scale $l \gg H^{-1}$. There are two main stages of evolution of the field ϕ inside this domain. The first stage has a duration $\Delta t \sim 2\sqrt{3\pi} \, \phi_0/mM_P$. During this time the average field $\bar\phi_c$ inside this domain remains practically unchanged, $\bar\phi_c = \phi_0$, eq. (2.4), whereas the dispersion $\Delta_c^2 = \langle \delta\phi^2 \rangle$ grows as $(H^3/4\pi^2)\Delta t$ (Vilenkin and Ford, 1982; Linde, 1982c; Starobinsky, 1982) up to the value $\Delta_c^2 \sim (cm^2\phi_0^4/M_P^4)$, $c = O(1)$. Note that, at $m^2\phi_0^2/2 \ll M_P^4$, Δ_c is much smaller than ϕ_0. Therefore the behaviour of the field $\bar\phi_c$ at the next stage of the universe's expansion can be described by eq. (2.4), and the field $\bar\phi_c$ decreases linearly. Fluctuations $\delta\phi$ are also generated at that stage, but they have much smaller amplitude and dispersion, whereas the dispersion of $\delta\phi(\mathbf{x})$ produced at the first stage behaves as $\dot{\bar\phi}_c$ (Guth and Pi, 1982), and it therefore remains constant during inflation, $|\delta\phi(\mathbf{x})| \sim |\dot{\bar\phi}_c| = mM_P/2\sqrt{3\pi} =$ const. As a result, the function $P_c(\phi)$ is mainly determined by the fluctuations produced at the first stage, and is given by

$$P_c(\phi) \sim \exp\left(-\frac{(\phi - \bar\phi_c(t))^2}{2\Delta_c^2}\right) \sim \exp\left(-\frac{(\phi - \bar\phi_c)^2 M_P^4}{2cm^2\phi_0^4}\right). \qquad (4.9)$$

Eq. (4.9) shows that the coordinate volume occupied by a large field ϕ is exponentially small, and it rapidly decreases for $t > 2\sqrt{3\pi} \, \phi_0/mM_P$ (non-stationary regime!). On the other hand, during the time $t \sim 2\sqrt{3\pi} \, \phi_0/mM_P$, at which the distribution (4.9) remains unchanged, domains having a large field ϕ expand $\exp[B(\phi^2/M_P^2)]$ times, $B = O(1)$. This gives the

---

† Eq. (4.8) (without the term (4.7)) may represent such a solution, but only if there exists a constant probability flux from the region $\phi \gg M_P^2/m$, where eq. (4.6) is not valid.

distribution $P_p(\phi)$ over the physical volume of the domain under consideration after the time $t \sim 2\sqrt{3\pi}\, \phi_0/mM_P$:

$$P_p(\phi) \sim \exp\left(-\frac{(\phi - \phi_0)^2 M_P^4}{2cm^2\phi_0^4} + \frac{3B\phi^2}{M_P^2} \right). \tag{4.10}$$

For $\phi_0 \gtrsim M_P(M_P/m)^{1/2}$ the distribution $P_p(\phi)$ grows with increase of ϕ. This means that after a time $t \sim \phi_0/mM_P$ the field ϕ inside the main part of the physical volume of a domain with $\phi_0 \gtrsim M_P(M_P/m)^{1/2}$ *grows* and becomes larger than ϕ_0, in agreement with our previous results.

For completeness we shall mention here another solution of eq. (4.6). If the initial value of the field ϕ is very large, $\phi_0 \gtrsim M_P^2/m$, i.e. if one starts with the spacetime foam with $V(\phi_0) \gtrsim M_P^4$, then the evolution of the field ϕ in the first stage (rapid diffusion) becomes more complicated (the naively estimated dispersion $\Delta_c^2 \sim H^3\,\Delta t$ soon becomes greater than ϕ_0^2). In this case the distribution of the field ϕ is not Gaussian. The solution of eq. (4.6) at the stage of diffusion from ϕ_0 to some field ϕ with $V(\phi) \ll M_P^4$ is given by

$$P_c(\phi) \sim \exp\left(-\frac{3\sqrt{3\pi}\, M_P^3}{m^3\phi t} \right). \tag{4.11}$$

This solution describes quantum creation of domains of a size $l \gtrsim H^{-1}(\phi)$, which occurs due to the diffusion of the field ϕ from $\phi_0 \gtrsim M_P^2/m$ to $\phi \ll \phi_0$. Direct diffusion with formation of a domain filled with the field ϕ is possible only during the time $t = c(2\sqrt{3\pi}\, \phi/mM_P)$, $c = O(1)$. At larger times a more rapid process is a diffusion to some field $\tilde{\phi} > \phi$ and a subsequent classical rolling down from $\tilde{\phi}$ to ϕ. Therefore one may interpret a distribution $P_c(\phi)$ formed after a time $t = c(2\sqrt{3\pi}\, \phi/mM_P)$ as a probability of a quantum creation of a mini-universe filled with a field ϕ (Linde, 1986b, c; Goncharov and Linde, 1987)

$$P_c(\phi) \sim \exp\left(-c\,\frac{3M_P^4}{2m^2\phi^2} \right) \sim \exp\left(-2c\,\frac{3M_P^4}{8V(\phi)} \right), \tag{4.12}$$

which is in agreement with the estimate (3.6) of the probability of a quantum creation of the inflationary universe (Linde, 1984a, b; Zeldovich and Starobinsky, 1984; Rubakov, 1984; Vilenkin, 1983). The same result is valid for all sufficiently steep potentials $V(\phi)$, in particular for all $V(\phi) \sim \phi^n$ (Linde, 1986b, c; Goncharov and Linde, 1987), and, as discussed by Rubakov (1984), no quantum particle production occurs during the process of the mini-universe creation. Presumably, the process considered above is complementary to the process of quantum creation 'from nothing' (see also the next section). In our case, creation of an inflationary (mini-) universe occurs as a diffusion (=tunnelling) not from 'nothing' (whatever that means), but from the spacetime foam with $V(\phi) \gtrsim M_P^4$, which in our case fills most of the physical volume of the universe and plays the role of the unstable (but regenerating) gravitational vacuum.

13.5 Eternal chaotic inflation

From the results obtained in the previous section it follows that the inflationary universe, which contains at least one domain of a size $l \gtrsim H^{-1}(\phi)$, with $\phi \gtrsim M_P(M_P/m)^{1/2}$, endlessly reproduces inflationary mini-universes with $\phi \gtrsim M_P(M_P/m)^{1/2}$. The global geometry of the inflationary universe has nothing in common either with the geometry of an open or flat homogeneous universe with a gradually decreasing energy density or with the geometry of a closed universe, which is created at some initial moment $t=0$ and which disappears as a whole at another moment $t=t_{\max}$.† In our case the universe endlessly regenerates itself, and there is no global 'end of time'. Moreover, it is not necessary to assume that the universe as a whole was created at some initial moment $t=0$. The process of creation of each new mini-universe with $\phi \gtrsim M_P(M_P/m)^{1/2}$ occurs independently of the pre-history of the universe, it depends only on the value of the scalar field inside a domain of a size $O(H^{-1}(\phi))$ and not on the moment of creation of the mini-universe. Therefore the whole process can be considered as an infinite chain reaction of creation and self-reproduction which has no end and which may have no beginning.

A similar situation may occur in the old inflationary universe scenario (Guth, 1981; Guth and Weinberg, 1983) and in the new inflationary universe scenario (Steinhardt, 1983; Linde, 1982e; Vilenkin, 1983). If the probability of bubble production with $\phi \neq 0$ is sufficiently small, or if the average time for the field ϕ to roll down from the extremum of $V(\phi)$ at $\phi=0$ is sufficiently large, then the phase transition to the global minimum of $V(\phi)$ occurs in the major part of the coordinate volume of the universe, but the major part of the *physical* volume of the universe remains in the unstable state $\phi=0$ for ever. This may give us a solution to the problem of the initial cosmological singularity: the major part of the inflationary universe may always exist in the unstable but self-reproducing state $\phi=0$, and it is not necessary to

---

† This means in particular that the *global* geometry of the inflationary universe cannot be studied in the context of the minisuperspace approach. However, the geometry of a locally Friedmannian mini-universe and the process of its formation can be investigated with the help of the minisuperspace approach, but in this approach formation of a mini-universe looks like creation of the universe 'from nothing'.

assume that the universe was created as a whole at some initial moment $t=0$ (Linde, 1982*e*).

There are some problems associated with this suggestion. The parts of the universe with $\phi=0$ locally have the same geometrical properties as de Sitter space. A geodesically complete de Sitter space is closed and its scale factor is $a(t)=H^{-1}\cosh Ht$. For $-\infty<t<0$, $a(t)$ decreases, and the lifetime of such a space in the unstable vacuum state $\phi=0$ is finite. Therefore such a universe cannot remain in the unstable state $\phi=0$ at $t=0$, when inflation starts. One may therefore wonder whether the eternally existing universe without initial singularity and with the major part of its volume in the unstable state $\phi=0$, is geodesically complete, or whether the initial singularity in this scenario is unavoidable (Linde, 1983*a*).

One should note, that the global geometry of such a universe differs considerably from the geometry of de Sitter space, since the part of the coordinate volume which remains in the de Sitter phase $\phi=0$ becomes infinitesimally small in the course of time. Therefore it is not excluded that such a universe is geodesically complete, which gives us a possible solution of the problem of the initial cosmological singularity (Linde, 1982*e*). However, there is no need to enter into a detailed discussion of this possibility here, since no realistic versions of the old and of the new inflationary universe scenario have been elaborated so far, and the probability of quantum creation of the universe with $\phi=0$, $V(0)\lesssim 10^{-10}M_{\rm P}^4$ (see the beginning of this chapter) is vanishingly small, see eq. (3.6).

On the other hand, in the eternally existing chaotic self-reproducing inflationary universe discussed above the major part of the physical volume of the universe emerges from the regions with $V(\phi)\sim M_{\rm P}^4$. In this sense the universe is effectively singular. Thus we do not claim that the nonsingular part of the universe in this scenario is geodesically complete. What is important, however, is that these 'singularities' do not form a global space-like singular hypersurface, which would mean the existence of a 'beginning of time' for the whole universe. The existence of such a hypersurface does not follow from general topological theorems concerning singularities in general relativity (Hawking and Ellis, 1973), but it is usually assumed that our universe looks like a slightly inhomogeneous Friedmann universe, in which such a hypersurface does exist. However, as we have seen, the global geometry of the inflationary universe has nothing in common with the geometry of the Friedmann universe; a global space-like singular hypersurface in the inflationary universe does not exist in the future and well may not exist in the past.

Thus, there exist two main versions of the chaotic inflationary scenario.

(i) There may exist an initial global singular space-like hypersurface. In this case the universe as a whole emerges from a state with a Planck density $\rho \sim M_P^4$ at some moment $t = t_P$, at which it becomes possible to speak about the universe in terms of classical spacetime. A natural initial value of the field ϕ at the Planck time is $\phi \sim M_P^2/m \sim 10^5 M_P$ for $m \sim 10^{-5} M_P$. Then the universe endlessly reproduces itself, due to generation of long-wave fluctuations of the field ϕ. This process occurs for $M_P(M_P/m)^{1/2} \lesssim \phi \lesssim M_P^2/m$ ($300 M_P \lesssim \phi \lesssim M_P^2/m$), i.e. for $m M_P^3 \lesssim V(\phi) \lesssim M_P^4$. In the domains with $\phi \lesssim M_P(M_P/m)^{1/2}$ this process becomes inefficient, and each such domain after inflation looks like a Friedmann mini-universe of a size $l \sim M_P^{-1} \exp[2\pi(M_P/m)] \sim 10^{10^5}$ cm. In this model the universe has a beginning but has no end.

(ii) The possibility that the universe has a global singular space-like hypersurface seems rather improbable unless the universe is compact and its initial size is $O(M_P^{-1})$. There is no reason for different, causally disconnected, regions of the universe to start their expansion simultaneously. If the universe is not compact, there should be no global beginning for its evolution. A model which illustrates this possibility was suggested above. The inflationary universe may infinitely reproduce itself, and it may have no beginning and no end. Any two points of such a universe in a sufficiently distant past could be causally connected and corresponding observers could synchronise their clocks even though later they may live in mini-universes which have become causally disconnected due to the exponential expansion of the universe.

Note that the 'energy' of the scale factor $a(t)$ is negative, and inflation may be considered as a result of instability arising from the pumping of energy from $a(t)$ to the field $\phi(t)$ (Linde, 1984c). For example, the energy of the scalar field ϕ in a closed inflationary universe grows exponentially, whereas the energy of the scale factor $a(t)$ becomes exponentially large and negative, the sum of their energies being equal to zero. It was unclear why this instability, being potentially possible, does not develop *after* inflation. We now know a possible answer. The gravitational vacuum in the major part of its physical volume always remains in the unstable inflationary state with energy density of the order of M_P^4. However, during the evolution of the universe many islands of stability are formed, one of which is the mini-universe in which we now live.

13.6 Global structure of the inflationary universe and the anthropic principle

13.6.1 The domain structure of the universe

The process of mini-universe formation and self-reproduction occurs at densities $mM_P^3 \lesssim V(\phi) \lesssim M_P^4$, i.e. at $10^{-5}M_P^4 \lesssim V(\phi) \lesssim M_P^4$ for $m \sim 10^{-5}M_P$. Note that this process may occur at densities many orders of magnitude smaller than M_P^4. Therefore to prove the very existence of the process of self-reproduction of the inflationary universe in our scenario there is no need to appeal to unknown physical processes at densities greater than the Planck density $\rho_P \sim M_P^4$.

On the other hand, it is very important that independently of the origin of the universe in our scenario (whether the universe was created as a whole at $t \sim t_P$, or whether it exists eternally), it now contains an exponentially large (or even infinite) number of causally disconnected mini-universes, and a considerable proportion of these mini-universes were created when the field ϕ was $O(M_P^2/m)$ and its energy density was (almost) as large as M_P^4. (Note that this is true in the chaotic inflationary scenario only, in which inflation may occur even at $V(\phi) \sim M_P^4$.)

The local structure of spacetime inside a new-born mini-universe formed at $V(\phi) \ll M_P^4$ remains unchanged. However, in realistic theories, in which there exist many types of scalar fields Φ with masses $m_\Phi \ll H(\phi)$, large-scale fluctuations of all these fields similar to the fluctuations of the field ϕ are formed. Since the Hubble parameter $H(\phi)$ during mini-universe formation is very large, $H = (8\pi V(\phi)/3M_P^2)^{1/2} \gtrsim 10^{-2}M_P \sim 10^{17}$ GeV, fluctuations of the scalar fields Φ are strong enough to transfer the classical fields Φ in the new-born mini-universe from one local minimum of the effective potential $V(\Phi, \phi)$ to another (Linde, 1983c, 1984c). This changes the low-energy elementary particle physics inside the new mini-universe. A particular example which is relevant to this effect is the supersymmetric $SU(5)$ model. The effective potential $V(\Phi)$ in this model has several minima of approximately equal depth, and only one of them corresponds to the desirable symmetry breaking $SU(3) \times U(1)$. Even if the universe initially was in one particular vacuum state corresponding to one of these minima of $V(\Phi)$, after inflation it becomes divided into many mini-universes corresponding to *all* possible minima of $V(\Phi)$. A typical time which is necessary for a quantum tunneling from one local minimum of $V(\Phi)$ to another is many orders greater than 10^{10} years, which is the age of the observable part of the universe. In one of these mini-universes the vacuum state corresponds to the $SU(3) \times U(1)$ minimum of $V(\Phi)$. We live in this mini-universe, not for the reason that the whole universe is in the $SU(3) \times U(1)$ symmetric state and not for the reason that the corresponding minimum of $V(\Phi)$ is deeper than other minima, but for the reason that life of our type is impossible in the mini-universes with other types of symmetry breaking. This solves the problem of symmetry breaking in the supersymmetric $SU(5)$ model (Linde, 1983c).

In order to illustrate new possibilities which appear in the context of the scenario discussed above, let us consider now a toy model which may explain the small value of the vacuum energy density ρ_v (of the cosmological constant), in the observable part of the universe ($|\rho_v| \lesssim 10^{-29}$ g cm^{-3}). This model is probably unrealistic, but nevertheless it may be rather instructive. The model describes an inflaton field ϕ, which drives inflation, and a field Φ with an extremely flat effective potential, $V(\Phi) = \alpha M_P^3(\Phi - C)$, where $\alpha \lesssim 10^{-120}$. It can be shown that during the time interval $t \sim 10^{10}$ years after inflation such a field Φ remains essentially unchanged due to the very small slope of $V(\Phi)$. However, the Brownian motion of this field during inflation is very rapid, and it divides the universe into an exponentially large number of mini-universes containing *all* possible values of the field Φ for which $|V(\phi) + V(\Phi)| \lesssim M_P^4$. After inflation the vacuum energy ρ_v inside these mini-universes is given by $V(\Phi, \phi) = \alpha M_P^3(\Phi - C) + V(\phi_0)$, where ϕ_0 corresponds to the minimum of $V(\phi)$. This quantity in different mini-universes changes continuously from $-M_P^4$ to $+M_P^4$, but life of our type is possible only in those mini-universes in which $|\rho_v| \lesssim 10^{-29}$ g cm^{-3}. Indeed, domains with $V(\Phi, \phi) < 0$, $|V(\Phi, \phi)| \gg 10^{-29}$ g cm^{-3} correspond to anti-de Sitter mini-universes with a lifetime $\tau \ll 10^{10}$ years. The domains with $V(\Phi, \phi) \gg 10^{-29}$ g cm^{-3} remain inflationary for $t \gtrsim 10^{10}$ years, and the present density of matter in such domains is negligibly small. For this reason we see ourselves inside a domain with $|\rho_v| \lesssim 10^{-29}$ g cm^{-3}.

To make this model realistic it would be desirable to explain why $V(\Phi)$ is so flat (though similar potentials sometimes appear in realistic models describing unified theories of elementary particles). In any case, the main idea suggested above may be used in other models as well. In our scenario it is not necessary to insist (as is usually done) that the vacuum energy must disappear in a 'true' vacuum state. It is quite sufficient if there exists some relatively stable vacuum-like state with $|\rho_v| \lesssim 10^{-29}$ g cm^{-3}. This requirement is still very restrictive, but it can be satisfied much more easily than the previous one. For a discussion of a similar approach to the cosmological constant problem, see also Sakharov (1984) and Linde (1984*c*).

13.6.2 *A scenario of inflationary compactification in the Kaluza–Klein cosmology*

In the mini-universes of initial size $O(H^{-1}) \simeq O(M_P^{-1})$, which are formed at $V(\phi) \sim M_P^4$, fluctuations of the metric on a scale $O(M_P^{-1})$ are of the order of unity. Therefore, if a theory of the type of eq. (2.1) can be considered as a part of a Kaluza–Klein (or superstring) theory, then at the moment of mini-universe formation the type of compactification and the number of compactified dimensions inside the new-born mini-universe can be changed (almost) independently of what occurs in causally disconnected regions outside it. (The only possibly constraint on the local changes of spacetime structure stems from topological considerations and may lead, for example, to the formation of a pair of mini-universes with opposite topological numbers.) As a result, our universe becomes divided into an exponentially large number of domains, and all possible types of compactification and all possible (metastable) vacuum states should exist in different domains of the universe. All these domains (mini-universes), in which inflation remains possible after their formation, later become exponentially large (Linde, 1986*a, b, c*).

For a detailed investigation of this possibility it would be desirable to have a realistic model of inflation in Kaluza–Klein or superstring theories. Unfortunately, the models of Kaluza–Klein inflation that now exist are not completely satisfactory (Maeda and Pollock, 1986; Linde, 1986*c*). On the other hand, we do not know any alternative possibility for solving all the problems that are solved by the inflationary universe scenario. Therefore one may consider the difficulties with the Kaluza–Klein (and superstring) inflation as difficulties of these theories (Maeda and Pollock, 1986). Actually, however, our understanding of these theories is incomplete, and hopefully the corresponding problem will be resolved in future. At present we are just trying to understand some possible features of the Kaluza–Klein inflationary cosmology, which are rather unusual. To illustrate some of the ideas discussed above, let us again consider inflation in the toy model, eq. (2.1).

In our investigation of inflation in this model we have studied only four-dimensional de Sitter-like solutions dS_4 with metric

$$ds^2 \approx dt^2 - H^{-2}\cosh^2 Ht(d\chi^2 + \sin^2\chi(d\theta^2 + \sin^2\theta \, d\phi^2)), \qquad (6.1)$$

where $H^2 = 8\pi V(\phi)/3M_P^2$. However, there exist some other interesting solutions, such as

$$ds^2 = dt^2 - \tfrac{1}{3}H^{-2}((\cosh^2(\sqrt{3}\,Ht)\,d\chi^2 + d\theta^2 + \sin^2\theta\,d\phi^2). \qquad (6.2)$$

This solution describes a universe that is a product of a two-dimensional de Sitter space dS_2 and a compact sphere S_2 with a very small radius $R_2 = (\sqrt{3}H)^{-1}$. This is a particular case of a Kantowski–Sachs universe (Kantowski and Sachs, 1966; Kofman et al., 1983; Paul et al., 1986). This solution is unstable; it (locally) transforms into dS_4. However, it is possible to stabilize it, for example by an appropriate addition of R^n terms to the Lagrangian, eq. (2.1) (Deruelle and Madore, 1986; Linde and Zelnikov, 1987).

Recently it was shown that long-wave quantum perturbations of scalar fields ϕ similar to those discussed in Section 13.4 are generated in the $dS_2 \times S_2$ universe (6.2) as well (Kofman and Starobinsky, 1987).[†] This result is directly related to the main conclusions of the present chapter (Linde, 1986c). Namely, if inflation is initially three dimensional as in ordinary de Sitter space dS_4 (6.1), then fluctuations of the scalar field ϕ create some inflationary domains (mini-universe), in which $V(\phi) \sim M_P^4$. Due to large fluctuations of ϕ and of metric $g_{\mu\nu}$ inside domains of a size $O(H^{-1}) \sim M_P^{-1}$, expansion inside some of these domains may become one dimensional (6.2), *independently of what occurs in the nearby domains*. This leads to formation of Kantowski–Sachs 'branches', which spread out of de Sitter space. However, fluctuations in the Kantowski–Sachs mini-universes lead again to creation of domains with $V(\phi) \sim M_P^4$ where de Sitter mini-universes (6.1) can be created. With account taken of this effect, the inflationary universe in our scenario may appear as a system of huge inflationary bubbles dS_4 connected with each other by thin inflationary tubes $dS_2 \times S_2$ of exponentially large length (Fig. 13.2).

The possibility of existence of such a complicated spacetime structure is directly related to the 'no-hair' theorem for de Sitter space, which is valid also for the one-dimensional exponential expansion (6.2). The processes which occur in a part of a tube (6.2) of initial size $\Delta l \gtrsim H^{-1}$ proceed independently of what occurs in other parts of the universe. Actually, the Kantowski–Sachs universe (6.2) is not the only solution of the Einstein equations with $V(\phi) > 0$ which leads to spontaneous compactification of a

---

† This is actually a general property of all inflationary models related to the existence of d zero modes of a scalar field ϕ in the Euclidean section S_d of d-dimensional de Sitter space dS_d.

four-dimensional space during inflation. We hope to return to the discussion of this question in a separate publication. Here we would only like to mention that the effect considered above is fairly general. For example, in d-dimensional space $(d>4)$ inflation and fluctuations of scalar fields lead to formation of a complicated structure consisting of inflationary mini-universes of all possible types, including mini-universes dS_d, $dS_{d-2} \times S_2$, $dS_{d-3} \times S_3$, etc., connected with each other.

During inflation the radius R_n of each compactified sphere S_n remains of

Fig. 13.2. This picture gives some idea of the global structure of the chaotic inflationary universe. Child; universes created where $V(\phi) \ll M_P^4$ have the same 'genetic code' as their mother universes: they have the same number of dimensions and the same (or almost the same) vacuum structure. However, the universes created where $V(\phi)$ is not much smaller than M_P^4 are 'mutants', which may have different dimensionality and different low-energy elementary particle physics inside them. Each mini-universe may die, but the universe as a whole has no end and may have no beginning. A typical thickness of 'tubes' connecting mini-universes after inflation may become very large. However, during inflation some of the tubes have a very small thickness. For example, some of the de Sitter mini-universes dS_d may be connected by the Kantowski–Sachs tubes $dS_{d-n} \times S_n$, where the radius of the sphere S_n is of the order $O(H^{-1}(\phi))$. The mini-universes compactified during inflation may 'serve as seeds for the next stage of the process of compactification, which occurs after inflation.

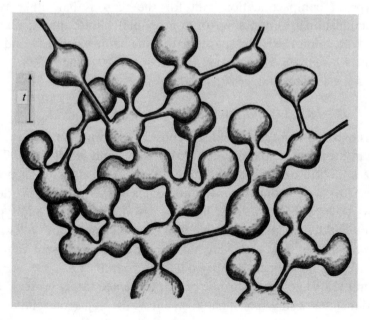

the order $H^{-1}(\phi)$, which initially is as small as M_P^{-1}. The main part of the physical volume of the universes $dS_{d-n} \times S_n$ remains in the state with $V(\phi) \sim M_P^4$ for ever due to the effects connected with the long-wave fluctuations of the field ϕ, see Section 13.4. In those domains of the universe in which the field ϕ slowly decreases, the radius of compactification $R_n(\phi)$ slowly grows as $H^{-1}(\phi)$ unless there exist some dynamical mechanism which fixes the radius of compactification at some value $R_0 \gtrsim M_P^{-1}$. In this sense the domains of the universe compactified during inflation may serve as seeds for different possible types of compactified spaces in the Kaluza–Klein theories.

This provides us with a new scenario of compactification as compared with the scenario of a classical power-law Kasner-like anisotropic expansion discussed by several authors (see, e.g. Freund, 1982; Freund and Oh, 1985). Namely, during inflation at $mM_P^3 \lesssim V(\phi) \lesssim M_P^4$ mini-universes with all possible types of compactification are formed in different causally disconnected regions of the universe, and those mini-universes in which inflation (in the uncompactified directions) remains possible after their formation later become exponentially large. The process of formation of new mini-universes has no end, in the major part of the physical volume of the universe this process occurs even now, and therefore even if the probability of formation of a mini-universe of some particular type is strongly suppressed, many such mini-universes should exist at present. According to this scenario, we live in a four-dimensional spacetime with our type of compactification, not for the reason that other types of compactification are impossible (or improbable), but for the reason that life of *our* type cannot exist in spaces with other dimensionality and with a different low-energy elementary particle physics (Linde, 1983a, 1984c, 1986c; see also Rozental, 1980, 1984).

13.6.3 Inflation and the anthropic principle

Discussion of some of the problems considered above is based on the so-called anthropic principle (Dicke, 1961; Collins and Hawking, 1973; Carr and Rees, 1979; Rozental, 1980; Linde, 1984c). According to this principle, we live in a four-dimensional, homogeneous, isotropic world with $e^2/4\pi = \frac{1}{137}$, $m_e = 0.5$ MeV, etc., for the reason that life of our type in a different world would be impossible. For example, life of our type would be impossible if e, m_e, M_P, etc., were half (or less) of their present values, or if our spacetime had dimensionality different from $d = 4$.

Several years ago the anthropic principle seemed rather esoteric, since it implied that many universes might exist, but we live in just one of them, which is sufficiently suitable for us. It was not clear in what sense one could speak of many universes if our universe is unique, and whether such fundamental constants of nature as the dimensionality of spacetime, the vacuum energy, the value of electric charge, etc., can change when one 'travels' from one universe to another.

A possible answer to the first part of this question was suggested by the many-worlds interpretation of quantum mechanics (Everett, 1957). Further investigation of this possibility may be of profound importance. However, it seems doubtful that it is possible to use this interpretation correctly without a proper understanding of the nature of consciousness. Does an observer just observe the universe, or does he 'create' it? (Wheeler, 1979). What is actually split: the universe or consciousness? Can consciousness exist 'by itself' (like spacetime without matter) or is it merely an arena for the manifestation of spacetime and matter? In what sense can consciousness 'choose' a universe to live in?

Thus the development of physics reveals problems which traditionally were beyond the scope of physics. It seems that to go further we must investigate these problems without prejudice, rather than wait until philosophers try to do it for us. However, such a path is not easy to follow, and it seems encouraging that inflation makes it possible to circumvent some of the above-mentioned problems that precluded the justification of the anthropic principle. Namely, from the scenario discussed in this chapter it follows that even if the universe at some time contains only one domain (mini-universe) of initial size $O(H_{(\phi)}^{-1}) \gtrsim O(M_P^{-1})$, later it splits into many causally disconnected mini-universes of exponentially large size, in which all possible types of compactification and all possible vacuum states are realised. In this sense our universe actually consists of many universes of all possible types. Whereas several years ago the dimensionality of spacetime, the vacuum energy density, the value of electric charge, the Yukawa couplings, etc., were regarded as true constants, it now becomes clear that these 'constants' actually depend on the type of compactification and on the mechanism of symmetry breaking, which may be different in different domains of our universe.

One of the main objections to the anthropic principle was the assertion that for the existence of life of our type there is no need for our universe to be so uniform on scales much greater than the scale of the Solar System. We now know the answer to this question. The size of the homogeneous locally Friedmannian mini-universes in which $\delta\rho/\rho \sim 10^{-4}$ (which is necessary for the formation of galaxies of our type) becomes after inflation much greater than the size of the Solar System. For example, in the model (2.1) $\delta\rho/\rho$ on a galactic scale is given by $O(10)(m/M_P)$, which gives $m \sim 10^{-5} M_P$. On the other hand, the size of a uniform part of our universe is greater than $M_P^{-1} \exp[2\pi(M_P/m)] \gtrsim 10^{10^5}$ cm, which is much greater than the observable part of the universe ($\sim 10^{28}$ cm).

This means that actually it is possible to justify some kind of weak anthropic principle in the inflationary cosmology. The line of thought advocated here is an alternative to the old assumption that in a 'true' theory it must be possible to compute unambiguously all masses, coupling constants, etc. From our point of view, it is rather improbable that a 'true' theory must have only one 'true' ground state. In the context of the unified theories of all fundamental interactions, the validity of this assumption becomes unlikely. According to the scenario discussed in this chapter, this assumption is probably incorrect; in any case it is not necessary. The old question of why our universe is the only possible one now is replaced by a new question: in which theories is the existence of mini-universes of our type possible? This question is still very difficult, but it is much easier than the old one.

13.7 Conclusions

The inflationary universe scenario continues to develop rapidly. It appears that the global structure of the universe according to this scenario is determined by quantum effects, that the major part of the physical volume of the universe should even now be in an inflationary state with $V(\phi) \sim M_P^4$, that the universe eternally reproduces itself, and that it consists of exponentially many mini-universes of different types. Some of these results are model-dependent, some are not. We do not know what the fate of the ideas discussed above will be, nor how they will be modified by future developments of elementary particle physics and of the theory of superstrings. In any case, the inflationary universe scenario may serve as a good example, showing how many exciting surprises the theory of gravity developed three hundred years ago can still give us.

References

Albrecht, A. and Steinhardt, P. J. (1982). *Phys. Rev. Lett.*, **48**, 1220.
Bardeen, J., Steinhardt, P. J. and Turner, M. (1983). *Phys. Rev.*, **D28**, 679.

Carr, B. J. and Rees, M. J. (1979). *Nature*, **278**, 605.

Collins, C. B. and Hawking, S. W. (1973). *Astrophys. J.*, **180**, 317.

Deruelle, N. and Madore, T. (1986). *Mod. Phys. Lett.*, **A1**, 237.

DeWitt, B. S. (1967). *Phys. Rev.*, **160**, 1113.

Dicke, R. H. (1961). *Nature*, **192**, 440.

Dowker, J. S. and Critchley, R. (1976). *Phys. Rev.*, **D13**, 3224.

Dreitlein, J. (1974). *Phys. Rev. Lett.*, **33**, 1243.

Enquist, K., Nanopoulos, D. V., Papantonopoulos, E. and Tamvakis, K. (1986). *Phys. Lett.*, **166B**, 41.

Everett, H. (1957). *Rev. Mod. Phys.*, **29**, 454.

Freund, P. G. O. (1982). *Nucl. Phys.*, **B209**, 146.

Freund, P. G. O. and Oh, P. (1985). *Nucl. Phys.*, **B255**, 688.

Gibbons, G. W. and Hawking, S. W. (1977). *Phys. Rev.*, **D15**, 2738.

Gliner, E. B. (1965). *Sov. Phys.-JETP*, **22**, 378.

Gliner, E. B. (1970). *Dokl. Akad. Nauk SSSR*, **192**, 771.

Gliner, E. B. and Dymnicova, V. (1975). *Pis'ma V. Astron. Zh.*, **1**, 5, 7.

Goncharov, A. S. and Linde, A. D. (1986). *Fiz. Elem. Chast. Atom Yad.*, **17**, 837. *Sov. J. Particles and Nuclei*, to be translated.

Goncharov, A. S. and Linde, A. D. (1987). *ZhETF*, **92**, 1137.

Gurevich, L. E. (1975). *Astrophys. Space Sci.*, **38**, 67.

Guth, A. H. (1981). *Phys. Rev.*, **D23**, 347.

Guth, A. H. and Pi, S.-Y. (1982). *Phys. Rev. Lett.*, **49**, 1110.

Guth, A. H. and Weinberg, E. (1983). *Nucl. Phys.*, **B212**, 321.

Hartle, J. B. and Hawking, S. W. (1983). *Phys. Rev.*, **D28**, 2960.

Hawking, S. W. (1982). *Phys. Lett.*, **115B**, 295.

Hawking, S. W. (1984). *Nucl. Phys.*, **B239**, 257.

Hawking, S. W. (1985). *Phys. Lett.*, **150B**, 339.

Hawking, S. W. and Ellis, G. F. R. (1973). *The Large Scale Structure of the Universe*. Cambridge University Press: Cambridge.

Hawking, S. W. and Moss, I. G. (1982). *Phys. Lett.*, **110B**, 35.

Hawking, S. W. and Page, D. N. (1986). *Nucl. Phys.*, **B264**, 185.

Kantowski, R. and Sachs, R. K. (1966). *J. Math. Phys.*, **7**, 443.

Kirzhnits, D. A. (1972). *JETP Lett.*, **15**, 529.

Kofman, L. A., Sahni, V. and Starobinsky, A. A. (1983). *ZhETF*, **85**, 1876.

Kofman, L. A. and Starobinsky, A. A. (1987). Submitted to *Phys. Lett.*

Linde, A. D. (1974). *JETP Lett.*, **19**, 183.

Linde, A. D. (1979). *Rep. Prog. Phys.*, **42**, 389.

Linde, A. D. (1982a). *Phys. Lett.*, **108B**, 389.

Linde, A. D. (1982b). *Phys. Lett.*, **114B**, 431.

Linde, A. D. (1982c). *Phys. Lett.*, **116B**, 335.

Linde, A. D. (1982d). *Phys. Lett.*, **116B**, 340.

Linde, A. D. (1982e). Non-singular regenerating inflationary universe. Cambridge University preprint.

Linde, A. D. (1983a). In *The Very Early Universe*, ed. G. W. Gibbons, S. W. Hawking and S. Siklos, p. 205. Cambridge University Press: Cambridge.

Linde, A. D. (1983b). *Phys. Lett.*, **123B**, 177.

Linde, A. D. (1983c). *Phys. Lett.*, **131B**, 330.

Linde, A. D. (1984a). *Sov. Phys.-JETP*, **60**, 211.

Linde, A. D. (1984b). *Lett. Nuovo Cimento*, **39**, 401.

Linde, A. D. (1984c). *Rep. Prog. Phys.*, **47**, 925.

Linde, A. D. (1985a). *Phys. Lett.*, **162B**, 281.

Linde, A. D. (1985b). *Prog. Theor. Phys. Suppl.*, **85**, 279.

Linde, A. D. (1986a). *Mod. Phys. Lett.*, **A1**, 81.

Linde, A. D. (1986b). *Phys. Lett.*, **175B**, 395.

Linde, A. D. (1986c). Eternally existing self-reproducing inflationary universe. Trieste preprint IC/86/75, to be published in the *Proceedings of the Nobel Symposium on Unification of Fundamental Interactions*, Physica Scripta, **34**.

Linde, A. D. and Zelnikov, M. (1987). To be published.

Maeda, K. and Pollock, M. D. (1986). *Phys. Lett.*, **173B**, 251.

Mukhanov, V. F. and Chibisov, G. V. (1981). *JETP Lett.*, **33**, 532.

Mukhanov, V. F. and Chibisov, G. V. (1982). *Sov. Phys.-JETP*, **56**, 258.

Paul, B. G., Datta, D. P. and Mikherjee, S. (1986). *Mod. Phys. Lett.*, **A1**, 149.

Rozental, I. L. (1980). *Sov. Phys. Uspekhi*, **23**, 296.

Rozental, I. L. (1984). *Elementary Particles and Structure of the Universe*. Nauka: Moscow. To be translated into English by Springer: Heidelberg.

Rubakov, V. A. (1984). *Phys. Lett.*, **148B**, 280.

Sakharov, A. D. (1984). *Sov. Phys.-JETP*, **60**, 214.

Shafi, Q. and Wetterich, C. (1985). **152B**, 51.

Starobinsky, A. A. (1979). *JETP Lett.*, **30**, 682.

Starobinsky, A. A. (1980). *Phys. Lett.*, **91B**, 99.

Starobinsky, A. A. (1982). *Phys. Lett.*, **117B**, 175.

Starobinsky, A. A. (1984). In *Fundamental Interactions*, p. 55. MGPI Press: Moscow.

Starobinsky, A. A. (1986). In *Current Topics in Field Theory, Quantum Gravity and Strings*, Lect. Notes in Physics, ed. H. J. de Vega and N. Sanchez, **246**, 107. Springer: Heidelberg.

Steinhardt, P. J. (1983). In *The Very Early Universe*, ed. G. W. Gibbons, S. W. Hawking and S. Siklos, p. 251. Cambridge University Press: Cambridge.

Turner, M. S. (1985). In *Proceedings of the Cargese School on Fundamental Physics and Cosmology*, ed. J. Audouze and J. Tran Thanh Van. Editions Frontiers: Gif-Sur-Yvette.

Veltman, M. (1974). Rockefeller University preprint.

Veltman, M. (1975). *Phys. Rev. Lett.*, **34**, 77.

Vilenkin, A. (1983). *Phys. Rev.*, **D27**, 2848.

Vilenkin, A. and Ford, L. H. (1982). *Phys. Rev.*, **D26**, 1231.

Wheeler, J. A. (1968). In *Battelle Recontres*, ed. C. DeWitt and J. A. Wheeler. Benjamin: New York.

Wheeler, J. A. (1979). In *Astrofizica e Cosmologia Gravitazione Quanti e Relativita*. Guinti Barbera: Firenze.

Zeldovich, Ya. B. and Starobinsky, A. A. (1984). *Sov. Astron. Lett.*, **10**, 135.

14

Quantum cosmology

S. W. HAWKING

14.1 Introduction

A few years ago I received a reprint request from an Institute of Quantum Oceanography somewhere in the Soviet Far East. I thought: What could be more ridiculous? Oceanography is a subject that is pre-eminently classical because it describes the behaviour of very large systems. Moreover, oceanography is based on the Navier–Stokes equation, which is a classical effective theory describing how large numbers of particles interact according to a more basic theory, quantum electrodynamics. Presumably, any quantum effects would have to be calculated in the underlying theory.

Why is quantum cosmology any less ridiculous than quantum oceanography? After all, the universe is an even bigger and more classical system than the oceans. Further, general relativity, which we use to describe the universe, may be only a low energy effective theory which approximates some more basic theory, such as string theory.

The answer to the first objection is that the spacetime structure of the universe is certainly classical today, to a very good approximation. However, there are problems with a large or infinite universe, as Newton realised. One would expect the gravitational attraction between all the different bodies in the universe to cause them to accelerate towards each other. Newton argued that this would indeed happen in a large but finite universe. However, he claimed that in an infinite universe the bodies would not all come together because there would not be a central point for them to fall to. This is a fallacious argument because in an infinite universe any point can be regarded as the centre. A correct treatment shows that an infinite universe can not remain in a stationary state if gravity is attractive. Yet so firmly held was the belief in an unchanging universe that when Einstein first proposed general relativity he added a cosmological constant in order to obtain a static solution for the universe, thus missing a golden opportunity to predict that the universe should be expanding or contracting. I shall discuss later why it should be that we observe it to be expanding and not contracting.

If one traces the expansion back in time, one finds that all the galaxies would have been on top of each other about 15 thousand million years ago. At first it was thought that there was an earlier contracting phase and that the particles in the universe would come very close to each other but would miss each other. The universe would reach a high but finite density and would then re-expand (Lifshitz and Khalatnikov, 1963). However, a series of theorems (Hawking and Penrose, 1970; Hawking and Ellis, 1973) showed that if classical general relativity were correct, there would inevitably be a singularity at which all physical laws would break down. Thus classical cosmology predicts its own downfall. In order to determine how the classical evolution of the universe began one has to appeal to quantum cosmology and study the early quantum era.

But what about the second objection? Is general relativity the fundamental underlying theory of gravity or is it just a low energy approximation to some more basic theory? The fact that pure general relativity is not finite at two loops (Goroff and Sagnotti, 1985) suggests it is not the ultimate theory. It is an open question whether supergravity, the supersymmetric extention of general relativity, is finite at three loops and beyond but no-one is prepared to do the calculation. Recently, however, people have begun to consider seriously the possibility that general relativity may be just a low energy approximation to some theory such as superstrings, although the evidence that superstrings are finite is not, at the moment, any better than that for supergravity.

Even if general relativity is only a low energy effective theory it may yet be sufficient to answer the key question in cosmology: Why did the classical evolution phase of the universe start off the way it did? An indication that this is indeed the case is provided by the fact that many of the features of the universe that we observe can be explained by supposing that there was a phase of exponential 'inflationary' expansion in the early universe. This is described in more detail in the articles by Linde, and Blau and Guth (Chapters 13, 12, this volume). In order not to generate fluctuations in the microwave background bigger than the observational upper limit of 10^{-4}, the energy density in the inflationary era cannot have been greater than about $10^{-10}m_p^4$ (Rubakov et al., 1982; Hawking, 1984a). This would put the inflationary era well inside the regime in which general relativity should be a good approximation. It would also be well inside the region in which any possible extra dimensions were compactified. Thus it might be reasonable to hope that the saddle point or semi-classical approximation to the quantum mechanical path integral for general relativity in four dimensions would give a reasonable indication of how the universe began. In what follows I shall assume that the lowest-order term in the action for a spacetime metric is the Einstein one, as it must be for agreement with ordinary, low energy, observations. However, I shall bear in mind the possibilities of higher-order terms and extra dimensions.

14.2 The quantum state of the universe

I shall use the Euclidean path integral approach. The basic assumption of this is that the 'probability' in some sense of a positive definite spacetime metric $g_{\mu\nu}$ and matter fields ϕ on a manifold M is proportional to $\exp(-\tilde{I})$ where $\tilde{I}$ is the Euclidean action. In general relativity

$$\tilde{I} = -\frac{1}{16\pi} \int_M R(g)^{1/2} \, \mathrm{d}^d x + \frac{1}{8\pi} \int_{\partial M} K(h)^{1/2} \, \mathrm{d}^{d-1} x - \int L_m[\phi](g)^{1/2} \, \mathrm{d}^d x,$$

where h and K are respectively the determinant of the first fundamental form and the trace of the second fundamental form of the boundary ∂M of M. In string theory the action $\tilde{I}$ of a metric $g_{\mu\nu}$, antisymmetric tensor field $B_{\mu\nu}$ and dilaton field ϕ is given by the log of the path integral of the string action over all maps of string world sheets into the given space. For most fields the path integral will not be conformally invariant. This will mean that the path integral diverges and $\tilde{I}$ will be infinite. Such fields will be suppressed by an infinite factor. However, the path integral over maps into certain background fields will be conformally invariant. The action for these fields will be that of general relativity plus higher-order terms.

The probability of an observable O having the value A can be found by summing the projection operator Π_A over the basic probability over all Euclidean metrics and fields belonging to some class C.

$$P_O(A) = \int_C \mathrm{d}[g_{\mu\nu}] \, \mathrm{d}[\phi] \Pi_A \exp(-\tilde{I}),$$

where $\Pi_A = 1$ if the value of O is A and zero otherwise. From such probabilities and the conditional probability, the probability of A given B,

$$P(A|B) = \frac{P(A, B)}{P(B)},$$

where $P(A, B)$ is the joint probability of A and B, one can calculate the outcome of all allowable measurements.

The choice of the class C of metrics and fields on which one considers the probability measure $\exp(-\tilde{I})$ determines the quantum state of the universe. C is usually specified by the asymptotic behaviour of the metric and matter fields, just as the state of the universe in classical general relativity can be specified by the asymptotic behaviour of these fields. For instance, one could demand that C consist of all metrics that approach the metric of Euclidean flat space outside some compact region and all matter fields that go to zero at infinity. The quantum state so defined is the vacuum state used in S matrix calculations. In these one considers incoming and outgoing states that differ from Euclidean flat space and zero matter fields at infinity in certain ways. The path integral over all such fields gives the amplitude to go from the initial to the final state.

In these S matrix calculations one considers only measurements at infinity and does not ask questions about what happens in the middle of the spacetime. However, this is not much help for cosmology: it is unlikely that the universe is asymptotically flat, and, even if it were, we are not really interested in what happens at infinity but in events in some finite region surrounding us. Suppose we took the class C of metrics and matter fields that defines the quantum state of the universe to be the class described above of asymptotically Euclidean metrics and fields. Then the path integral to calculate the probability of a value of an observable O would receive contributions from two kinds of metrics. There would be connected asymptotically Euclidean metrics and there would be a disconnected metric which consisted of a compact component that contained the observable O and a separate asymptotically Euclidean component. One can not exclude disconnected metrics from the class C because any disconnected metric can be approximated arbitrarily closely by a connected metric in which the different components are joined by thin tubes with negligible action. It turns out that for observables that depend only on a compact region the dominant contribution to the path integral comes from the compact regions of disconnected metrics. Thus, as far as cosmology is concerned, the probabilities of observables would be almost the same if one took the class C to consist of compact metrics and matter fields that are regular on them.

In fact, this seems a much more natural choice for the class C that defines the quantum state of the universe. It does not refer to any unobserved asymptotic region and it does not involve any boundary or edge to spacetime at infinity or a singularity where one would have to appeal to some outside agency to set the boundary conditions. It would mean that spacetime would be completely self contained and would be determined completely by the laws of physics: there would not be any points where the laws broke down and there would not be any edge of spacetime at which unpredictable influences could enter the universe. This choice of boundary conditions for the class C can be paraphrased as: 'The boundary condition of the universe is that it has no boundary' (Hawking, 1982; Hartle and Hawking, 1983; Hawking, 1984b).

This choice of the quantum state of the universe is very analogous to the vacuum state in string theory which is defined by all maps of closed string world sheets without boundary into Euclidean flat space. More generally, one can define a 'ground' state of no string excitations about any set of background fields that satisfy certain conditions by all maps of closed string world sheets into the background. Thus one can regard the 'no boundary' quantum state for the universe as a 'ground' state (Hartle and Hawking, 1983). It is, however, different from other ground states. In other quantum theories non-trivial field configurations have positive energy. They therefore cannot appear in the zero energy ground state except as quantum fluctuations. In the case of gravity it is also true that any asymptotically flat metric has positive energy, except flat space, which has zero energy. However, in a closed, non-asymptotically flat universe there is no infinity at which to define the energy of the field configuration. In a sense the total energy of a closed universe is zero: the positive energy of the matter fields and gravitational waves is exactly balanced by the negative potential energy which arises because gravity is attractive. It is this negative potential energy that allows non-trivial gravitational fields to appear in the 'ground' state of the universe.

Unfortunately, this negative energy also causes the Euclidean action $\tilde{I}$ for general relativity to be unbounded below (Gibbons *et al.*, 1978), thus causing $\exp(-\tilde{I})$ not to be a good probability measure on the space C of field configurations. In certain cases it may be possible to deal with this difficulty by rotating the contour of integration of the conformal factor in the path integral from real values to be parallel to the imaginary axis. However, there does not seem to be a general prescription that will guarantee that the path integral converges. This difficulty might be overcome in string theory where the string action is positive in Euclidean backgrounds. It may be, however, that the difficulty in making the path integral converge is fundamental to the fact that the 'ground' state of the universe seems to be highly non-trivial. In any event it would seem reasonable to expect that the main contribution to the path integral would come from fields that are near stationary points of the action $\tilde{I}$, that is, near solutions of the field equations.

It should be emphasised that the 'no boundary' condition on the metrics in the class C that defines the quantum state of the universe is just a proposal: it cannot be proved from something else. It is quite possible that the universe is in some different quantum state though it would be difficult to think of one that was defined in a natural manner. The 'no boundary' proposal does have the great advantage, however, that it provides a definite basis on which to calculate the probabilities of observable quantities and compare them with what we see. This basis seems to be lacking in many other approaches to quantum cosmology in which the assumptions on the quantum state of the universe are not clearly stated. For instance, Vilenkin (1986) defines the quantum state in a toy minisuperspace model by requiring that a certain current on minisuperspace be ingoing at one point of the boundary of minisuperspace (corresponding to 'creation from nothing') and outgoing elsewhere on the boundary (annihilation into nothing?). However, he does not seem to have a general prescription that would define the quantum state except in simple minisuperspace cases. Moreover, his state is not CPT invariant, which is a property that one might think the quantum state of the universe should have. Similarly, Linde (1985; Chapter 13, this volume) does not give a definition of the quantum state of the universe. He also suggests that the Wick rotation for the Euclidean action of the gravitational field should be in the opposite direction to that for other fields. This would be equivalent to changing the sign of the gravitational constant and making gravity repulsive instead of attractive.

14.3 The density matrix

One thinks of a quantum system as being described by its state at one time. In the case of cosmology, 'at one time' can be interpreted as on a spacelike surface S. One can therefore ask for the probability that the metric and matter fields have given values on a $d-1$ surface S. In fact, it is meaningful to ask questions only about the $d-1$ metric h_{ij} induced on S by the d metric $g_{\mu\nu}$ on M because the components $n^{\mu}g_{\mu\nu}$ of $g_{\mu\nu}$ that lie out of S can be given any values by a diffeomorphism of M that leaves S fixed. Thus the probability that the surface S has the induced metric h_{ij} and matter fields ϕ_0 is

$$P(h_{ij}, \phi_0) = \int_C d[g_{\mu\nu}] d[\phi_0] \exp(-\tilde{I}) \Pi_{(h_{ij}, \phi_0)},$$

where $\Pi_{(h_{ij}, \phi_0)}$ is the projection operator which has value 1 if the induced metric and matter fields on S have the given values and is zero otherwise. One can cut the manifold M at the surface S to obtain a new manifold $\tilde{M}$

bounded by two copies $\tilde{S}$ and $\tilde{S}'$ of S. One can then define $\rho(h_{ij}, \phi_0; h'_{ij}, \phi'_0)$ to be the path integral over all metrics and matter fields on $\tilde{M}$ which agree with the given values h_{ij}, ϕ_0 on $\tilde{S}$ and h'_{ij}, ϕ'_0 on $\tilde{S}$. The quantity ρ can be regarded as a density matrix describing the quantum state of the universe as seen from a single spacelike surface for the following reasons:

(i) The diagonal elements of ρ, that is, when $h_{ij} = h'_{ij}$ and $\phi_0 = \phi'_0$, give the probability of finding a surface S with the metric h_{ij} and matter fields ϕ_0.

(ii) If S divides M into two parts, the manifold $\tilde{M}$ will consist of two disconnected parts, $\tilde{M}_+$ and $\tilde{M}_-$. The path integral for ρ will factorise:

$$\rho(h_{ij}, \phi_0; h'_{ij}, \phi'_0) = \Psi_+(h_{ij}, \phi_0)\Psi_-(h'_{ij}, \phi'_0),$$

where the wave functions Ψ_+ and Ψ_- are given by the path integral over all metrics and matter fields on $\tilde{M}_+$ and $\tilde{M}_-$ respectively which have the given values on S and S'. If the matter fields ϕ are CP invariant, $\Psi_+ = \Psi_-$ and both are real (Hawking, 1985). Ψ is known as 'The Wave Function Of The Universe'. A density matrix which factorises can be interpreted as corresponding to a pure quantum state.

(iii) If the surface S does not divide M into two parts, the manifold $\tilde{M}$ will be connected. In this case the path integral for ρ will not factorise into the product of two wave functions. This means that ρ will correspond to the density matrix of a mixed quantum state, rather than a pure state for which the density matrix would factorise (Page, 1986; Hawking, 1987).

One can think of the density matrices which do not factorise in the following way: Imagine a set of surfaces T_i which, together with S, divide the spacetime manifold M into two parts. One can take the disjoint union of the T_i and S as the surface which is used to define ρ (there is no reason why this surface has to be connected). In this case the manifold $\tilde{M}$ will be disconnected and the path integral for ρ will factorise into the product of two wave functions which will depend on the metrics and matter fields on two sets of surfaces, S, T_i and S', T'_i. The quantity ρ will therefore be the density matrix for a pure quantum state. However, an observer will be able to measure the metric and matter fields only on one connected component of the surface (say, S) and will not know anything about their values on the other components, T_i, or even if any other components are required to divide the spacetime manifold into two parts. The observer will therefore have to sum over all possible metrics and matter fields on the surfaces $\tilde{T}_i$. This summation or trace over the fields on the $\tilde{T}_i$ will reduce ρ to a density matrix corresponding to a mixed state in the fields on the remaining surfaces $\tilde{S}$ and

$\tilde{S}'$. It is like when you have a system consisting of two parts A and B. Suppose the system is in a pure quantum state but that you can observe only part A. Then, as you have no knowledge about B, you have to sum over all possibilities for B, with equal weight. This reduces the density matrix for the system from a pure state to a mixed state.

The summation over all fields on the surfaces $\tilde{T}_i$ is equivalent to joining the surfaces $\tilde{T}_i$ to $\tilde{T}'_i$ and doing the path integral over all metrics and matter fields on a manifold $\tilde{M}$ whose only boundaries are the surfaces $\tilde{S}$ and $\tilde{S}'$. There is an overcounting because, as well as summing over all metrics and matter fields, one is summing over all positions of the surfaces T_i in $\tilde{M}$. However, the path integral over these extra degrees of freedom can be factored out by introducing ghosts. The reduced path integral is then the same as that for the density matrix ρ for a single pair of surfaces $\tilde{S}$ and $\tilde{S}'$. Thus one can see that the reason that the density matrix for S corresponds to a mixed state is that one is observing the state of the universe on a single spacelike surface and ignoring the possibility that spacetime may be not simply connected and so require other surfaces T_i as well as S to divide M into two parts.

14.4 The Wheeler–DeWitt equation

In a neighbourhood of the boundary surface $\tilde{S}$ of the manifold $\tilde{M}$, one can write the metric $g_{\mu\nu}$ in the $(d-1)+1$ form:

$$ds^2 = (N^2 + N^i N_i)\, dt^2 + 2N_i\, dx^i\, dt + h_{ij}\, dx^i\, dx^j,$$

where $\tilde{S}$ is the surface $t=0$. The Euclidean action can then be written in the Hamiltonian form:

$$\tilde{I} = -\int dt\, d^{d-1}x(\pi^{ij}\dot{h}_{ij} + \pi_\phi\dot{\phi} - NH^0 - N_iH^i),$$

where $\pi^{ij} = -(h^{1/2}/16\pi)(K^{ij} - h^{ij}K)$ is the Euclidean momentum conjugate to h_{ij}, K_{ij} is the second fundamental form of $\tilde{S}$,

$$H_0 = 16\pi G_{ijkl}\pi^{ij}\pi^{kl} - \frac{1}{16\pi}h^{1/2\,d-1}R + T^{00}$$

$$H^i = -2\pi^{ij}_{\;;j} + T^{0i}$$

$$G_{ijkl} = \tfrac{1}{2}h^{-1/2}(h_{ik}h_{jl} + h_{il}h_{jk} - h_{ij}h_{kl}).$$

As was stated above, the components of $g_{\mu\nu}$ that lie out of the surface $\tilde{S}$ can be given any values by a diffeomorphism of $\tilde{M}$ that leaves $\tilde{S}$ fixed. This means that the variational derivative of the path integral for ρ with respect to N and N_i on $\tilde{S}$ must be zero:

$$\frac{\delta\rho}{\delta N_i} = -\int d[g_{\mu\nu}]\, d[\phi_0]\, \frac{\delta\tilde{I}}{\delta N_i}\exp(-\tilde{I}) = \tilde{H}_i\rho = 0$$

$$\frac{\delta\rho}{\delta N} = -\int d[g_{\mu\nu}]\, d[\phi_0]\, \frac{\delta\tilde{I}}{\delta N}\exp(-\tilde{I}) = \tilde{H}\rho = 0,$$

where the operators $\tilde{H}$ and $\tilde{H}_i$ are obtained from the corresponding classical expressions by replacing the Euclidean momentum π^{ij} by $-\delta/\delta h_{ij}$ and π_ϕ by $-\delta/\delta\phi$.

The first equation is called the momentum constraint. It is a first-order equation for ρ on superspace, the space W of all metrics h_{ij} and matter fields ϕ on a surface S. It implies that ρ is the same for metrics and matter fields which can be obtained from each other by coordinate transformations in S. The second equation is called the Wheeler–DeWitt equation. It holds at each point of superspace, except where $h_{ij}=h'_{ij}$ and $\phi_0=\phi'_0$. When this is true, the separation between $\tilde{S}$ and $\tilde{S}'$ in the metric $g_{\mu\nu}$ on the manifold $\tilde{M}$ may be zero. In this case, it is no longer true that the variation of ρ with respect to N is zero. There is an infinite dimensional delta function on the right-hand side of the Wheeler–DeWitt equation. Thus, the Wheeler–DeWitt equation is like the equation for the propagator, $G(x,x')=\langle\phi(x)\phi(x')\rangle$:

$$(-\Box+m^2)G(x,x')=\delta(x,x').$$

As the point x tends towards x', the propagator diverges like r^{2-d}, where r is the distance between x and x'. Thus $G(x,x')$ will be infinite. Similarly, $\rho(h_{ij},\phi_0;h_{ij},\phi_0)$, the diagonal elements of the density matrix, will be infinite. This infinity arises from Euclidean geometries of the form $S\times S^1$, where the S^1 is of very short radius. However, we are interested really only in the probabilities for Lorentzian geometries, because we live in a Lorentzian universe, not a Euclidean one. One can recognise the part of the density matrix ρ that corresponds to Lorentzian geometries by the fact that it will oscillate rapidly as a function of the scale factor of the metrics h_{ij} and h'_{ij} (Hawking, 1984b). One therefore wants to subtract out the infinite, Euclidean, component and leave a finite, Lorentzian, component. One way of doing this is to consider only spacetime manifolds M which the surface S divides into two parts. The density matrix from such geometries will be of the factorised form:

$$\rho(h_{ij},\phi_0;h'_{ij},\phi'_0)=\Psi(h_{ij},\phi_0)\Psi(h'_{ij},\phi'_0),$$

where the wave function Ψ obeys the Wheeler–DeWitt equation with no delta function on the right-hand side. This part of the density matrix will therefore remain finite when $h_{ij} = h'_{ij}$ and $\phi_0 = \phi'_0$. In a supersymmetric theory, such as supergravity or superstrings, the infinity at the diagonal in the density matrix would probably be cancelled by the fermions.

14.5 Minisuperspace

The Wheeler–DeWitt equation can be regarded as a second-order differential equation for ρ or Ψ on superspace, the infinite-dimensional space of all metrics and matter fields on S. It is hard to solve such an equation. Instead, progress has been made by using finite dimensional approximations to superspace, called minisuperspaces, first introduced by Misner (1970). In other words, one reduces the infinite number of degrees of freedom of the gravitational and matter fields and of the gauge to a finite number and solves the Wheeler–DeWitt equation on a finite-dimensional space.

14.5.1 de Sitter model

The simplest example is a homogeneous isotropic four-dimensional universe with a cosmological constant and metric
$$ds^2 = \sigma^2 [N^2\,dt^2 + a^2\,d\Omega_3^2].$$
The action is
$$\tilde{I} = -\frac{1}{2} \int dt N a \left[\frac{1}{N^2} \left[\frac{da}{dt} \right]^2 + 1 - \lambda a^2 \right],$$
where $\sigma^2 = \frac{2}{3}\pi m_p^2$, a is the radius of the 3-sphere space-like surfaces and $\lambda = \frac{1}{3}\sigma^2\Lambda$. One can choose $N = a$. The first two terms in the Euclidean action are negative definite. This means that the path integral over a does not converge. However, one can make the path integral converge by taking a to be imaginary. This corresponds to integrating the conformal factor over a contour parallel to the imaginary axis (Gibbons et al., 1978).

With a imaginary, the action is the same as that of the anharmonic oscillator. The density matrix $\rho(a, a')$ is given by a path integral over all values of a on a manifold $\tilde{M}$ bounded by surfaces with radii a and a'. There are two kinds of such manifold: ones that have two disconnected components, which correspond to spacetimes that are divided in two by S, and connected ones, which correspond to non-simply connected spacetimes that S does not divide.

Consider first the case in which S divides M in two. The density matrix from these geometries that S divides into two is the product of wave functions:

$$\rho(a, a') = \Psi(a)\Psi(a'),$$

where the wave function Ψ is given by a path integral over compact 4-geometries bounded by a 3-sphere of radius a or a'. One would expect this path integral to be approximately $A \exp(-B)$, where B is the action of a solution of the classical Euclidean field equations with the given boundary conditions and the prefactor A is given by a path integral over small fluctuations about the solution of the classical field equations. The compact homogeneous isotropic solution of the Euclidean field equations is a 4-sphere of radius $\lambda^{-1/2}$. A 3-sphere of radius $a < \lambda^{-1/2}$ can fit into such a 4-sphere in two positions: it can bound more or less than half the 4-sphere. The action B of both these solutions of the classical equations is negative, with the action of more than half the 4-sphere being the more negative. One might therefore expect that this solution would provide the dominant contribution to the path integral. However, if one takes the scale factor a to be imaginary, in order to make the path integral converge, and then analytically continues back to real a, one finds that the dominant contribution comes from the solution that corresponds to less than half the 4-sphere, rather than the other solution which corresponds to more than half the 4-sphere, as one might have expected. This conclusion also follows from an analysis of the path integral in the K representation (Hartle and Hawking, 1983).

In terms of the gauge choice $N = a$, used above, the path integral is over a with a the given value at $t = 0$ and $a = 0$ at $t = \pm \infty$. This path integral is the same as that for the propagator for the anharmonic oscillator from ia at $t = 0$ to 0 at $t = \infty$. But this gives the ground state wave function. Thus

$$\Psi(a) \propto \operatorname{Re}(A_0(ia)),$$

where $A_0(x)$ is the ground state wave function of the anharmonic oscillator.

For small x, $A_0(x)$ behaves like $\exp(-\frac{1}{2}x^2)$. Thus $\Psi(a)$ behaves exponentially like $\exp(\frac{1}{2}x^2)$. This agrees with the estimates from the action of less than half the 4-sphere, as above. However, for $a > \lambda^{-1/2}$, there is no Euclidean solution of the classical field equations for a compact homogeneous isotropic 4-space bounded by a 3-sphere of radius a. Instead there are complex metrics which are solutions of the field equations with the required properties. Near the 3-sphere of radius a, one can take a section through the complexified spacetime manifold on which the metric is real and Lorentzian. This is reflected in the fact that $A_0(ia)$ will oscillate for $a > \lambda^{-1/2}$: exponential wave functions correspond to Euclidean 4-geometries and oscillating wave functions correspond to Lorentzian 4-geometries (Hawking, 1984b).

For large a, $\Psi(a)$ behaves like $a^{-1}\cos(\lambda^{1/2}a^3)$. One can interpret this by writing the wave function in the WKB form: $C(\exp(\mathrm{i}\,S)+\exp(-\mathrm{i}\,S))$, where S is a rapidly varying phase factor and C is a slowly varying amplitude. The wave function will satisfy the Wheeler–DeWitt equation to leading order if the phase factor S obeys the classical Hamilton–Jacobi equation. Thus, an oscillating wave function will correspond in the classical limit to an $(n-1)$-dimensional family of solutions of the classical Lorentzian field equations, where n is the dimension of the minisuperspace.

In the example above, $n=1$. The oscillating part of the wave function corresponds to the classical de Sitter solution which collapses from infinite radius to a minimum radius $a=\lambda^{-1/2}$ and then expands again exponentially to infinite radius. The classical Lorentzian solution does not go below a radius of $\lambda^{-1/2}$, so one can interpret the exponentially damped wave function below that radius as corresponding to a Euclidean geometry in the classically forbidden region. Note that, for this explanation to make sense, the wave function has to decrease with decreasing a, and not increase as authors such as Linde and Vilenkin have argued on the analogy of tunnelling 'from nothing'. Anyway, if one believes that the quantum state of the universe is determined by a path integral over compact geometries, one has no freedom of choice of the solution of the Wheeler–DeWitt equation: it has to be the one that increases exponentially with increasing a.

Another feature of the wave function that is worth remarking on is that it is real. This means that, in the oscillating region, the WKB ansatz is $C(\exp(\mathrm{i}\,S)+\exp(-\mathrm{i}\,S))$. One can regard the first term as representing an expanding universe and the second a contracting universe. More generally, if the wave function represents some history of the universe, it also represents the CPT image of that history (Hawking, 1985). This should be contrasted with the approach of Vilenkin and others, who try to choose a solution of the Wheeler–DeWitt equation which corresponds only to expanding universes. The fallacy of this attempt is that the direction of the time coordinate has no intrinsic meaning: it can be changed by a coordinate transformation. The physically meaningful question is: how does the entropy or degree of disorder behave during the histories of the universe that are described by the wave function? The minisuperspace models considered here are too simple to answer this but it will be discussed for models with the full number of degrees of freedom in Section 14.7.

The contribution to the density matrix from geometries that S does not divide into two parts is given by a path integral with a fixed at the given values at $t=0$ and $t=t_1$ for some Euclidean time interval t_1. But this is equal to the real part of the propagator $K(ia,0; ia',t_1)$ for the anharmonic oscillator from ia at $t=0$ to ia' at $t=t_1$.

$$K(ia,0;ia',t_1)=\sum_n A_n(ia)A_n(ia')\exp(-E_nt_1),$$

where $A_n(x)$ are the wave functions of the excited states of the anharmonic oscillator and E_n are the energy levels. To obtain the density matrix one has to integrate over all values of t_1 because the two surfaces can have any time separation:

$$\rho(a,a')=\mathrm{Re}\int_0^\infty K(ia,0;ia',t_1)\,\mathrm{d}t_1=\mathrm{Re}\sum_n\frac{A_n(ia)A_n(ia')}{E_n}.$$

One can interpret this as saying that the universe is in the state specified by the wave function $\mathrm{Re}(A_n(ia))$ with the relative probability $(E_n)^{-1}$. Note that the universe need not be 'on shell' in the sense that the Wheeler–DeWitt operator acting on A_n is not 0, but E_n. This term in the Wheeler–DeWitt equation acts as if the universe contained a certain amount of negative energy radiation. It will cause the classical solution corresponding to A_n by the WKB approximation to bounce at a larger radius than $\lambda^{-1/2}$. Thus, the effect of the universe being in a mixed quantum state might be observable. However, at large values of a, the effect of the negative energy radiation would be very small and the universe would expand exponentially, like the de Sitter solution.

14.5.2 The massive scalar field model

The de Sitter model was interesting because it showed that the 'no boundary' proposal for the quantum state of the universe leads to inflation if there is some process which gives rise to an effective cosmological constant in the early universe. However, the universe is not expanding exponentially at the present time, so there has to be some way in which the cosmological 'constant' can reduce to zero at late times. One mechanism, and possibly the only one, for generating such a decaying effective cosmological constant is a scalar field with a potential which has a minimum at zero and which is exponentially bounded. I shall consider the simplest example, a massive scalar field.

The action of a homogeneous isotropic universe of radius a with a massive scalar field ϕ that is constant on the surfaces of homogeneity is

$$\tilde{I} = -\frac{1}{2} \int dt N a \left[\frac{1}{N^2} \left[\left[\frac{da}{dt} \right]^2 - a^2 \left[\frac{d\phi}{dt} \right]^2 \right] + 1 - a^2 m^2 \phi^2 \right].$$

Unfortunately, in this case, there does not seem to be any simple prescription for making the Euclidean action positive definite. Taking a imaginary leaves the kinetic term for ϕ negative, while taking ϕ imaginary would cure this problem but would make the mass term negative. One could, however, make the action positive in this manner if the potential was pure ϕ^4. On physical grounds, one would not expect that there would be a qualitative difference between the behaviour of a universe in which the scalar potential was ϕ^2 and one in which it was ϕ^4.

In the case that the surface S divides the spacetime into two parts, the wave function will obey the Wheeler–DeWitt equation

$$\frac{1}{2} \left[\frac{1}{a^p} \frac{\partial}{\partial a} a^p \frac{\partial}{\partial a} - \frac{1}{a^2} \frac{\partial^2}{\partial \phi^2} - a^2 + a^4 m^2 \phi^2 \right] \Psi[a, \phi] = 0,$$

where p reflects some of the uncertainty in the factor ordering of the operators in the Wheeler–DeWitt equation. It is thought that the value of p does not have much effect, so it is usual to take $p = 1$, because this simplifies the equation. One can introduce new coordinates:

$$x = a \sinh \phi, \quad y = a \cosh \phi.$$

In these coordinates, the Wheeler–DeWitt equation becomes

$$\left[\frac{\partial^2}{\partial y^2} - \frac{\partial^2}{\partial x^2} + V \right] \Psi(x, y) = 0,$$

where $V = (y^2 - x^2)[-1 + (y^2 - x^2)m^2 (\text{arctanh } x/y)^2]$.

For small values of a, one can expect that Ψ is approximately $A \exp(-B)$, where B is the action of a solution of the Euclidean field equations. If $\phi \gg 1$ and $a < 1/m\phi$, the value of ϕ will not vary much over the solution and the $m^2\phi^2$ term in the action will act as an effective cosmological constant. One would therefore expect B to be the action of the smaller part of a 4-sphere of radius $1/m\phi$, bounded by a 3-sphere of radius a. From the de Sitter model, one would expect the wave function to oscillate for $a > 1/m\phi$ and the phase factor S to be $\frac{1}{3}m\phi a^3$, the analytic continuation of B. Such a wave function is a solution to the Wheeler–DeWitt equation to leading order.

One can interpret the oscillating part of the wave function as corresponding to a complex compact metric which is a solution of the field equations and which is bounded by the surface S. In a neighbourhood of S one can take a section through the complexified spacetime manifold on which the metric is nearly real and Lorentzian. This solution will have a minimum radius of order $1/m\phi$ and will expand exponentially with ϕ slowly decreasing. It will be a quantum realisation of the 'chaotic inflation' model proposed by Linde (1983).

After an exponential expansion of the universe by a factor of order $\exp(\frac{1}{2}\phi^2)$, the scalar field will start to oscillate with frequency m. The energy momentum tensor of the scalar field will change from that of an effective cosmological constant to that of pressure-free matter. The universe will change from an exponential expansion to a matter-dominated one. In a model with other matter fields, one would expect the energy in the massive scalar field oscillations to be converted into zero rest mass particles. The universe would then expand as a radiation-dominated model.

The universe would expand to a maximum radius and then recollapse. One would expect that if such complex, almost Lorentzian, geometries contributed to the wave function in their expanding phase, they would also contribute in their contracting phase. However, although a few solutions will bounce at small radius and expand again (Hawking, 1984b; Page, 1985a, b), most solutions will collapse to a singularity. They will give an oscillating contribution to the wave function, even in the region $a < 1/m\phi$ of superspace where the dominant contribution is exponential. It will also mean that the boundary condition for the Wheeler–DeWitt equation on the light cone $x = \pm y$ is not exactly $\Psi = 1$, as was assumed in some earlier papers (Hawking and Wu, 1985; Moss and Wright, 1983).

The density matrix from geometries that S does not divide into two parts has not been calculated yet. By analogy with the de Sitter model, one might expect that the part which corresponds to Lorentzian geometries would behave like solutions with a massive scalar field and negative energy radiation. One would not expect the negative energy to prevent collapse to a singularity.

To summarise, in this model, the universe begins its expansion from a non-singular state. It expands in an inflationary manner, goes over to a matter or radiation-dominated expansion, reaches a maximum radius and recollapses to a singularity. This will be discussed further in Sections 14.7 and 14.8.

14.6 Beyond minisuperspace

The minisuperspace models were useful because they showed that the 'no boundary' proposal for the quantum state of the universe can lead to a universe like the one that we observe, at least in its large scale features.

However, ultimately one would like to know the density matrix or wave function on the whole of superspace, not just a finite-dimensional subspace. This is a bit of a tall order but one can use a 'midisuperspace' approximation in which one takes the action to all orders in a finite number of degrees of freedom and to second order in the remaining degrees of freedom.

A treatment of the massive scalar field model on these lines has been given by Halliwell and Hawking (1985). The two degrees of freedom of the model described above are treated exactly, and the rest as perturbations on the background determined by the two-dimensional minisuperspace model. As in the model above, the oscillating part of the background wave function corresponds by the WKB approximation to a universe which starts at a minimum radius, expands in an inflationary and then a matter-dominated manner, reaches a maximum radius and recollapses to a singularity.

From the 'no boundary' condition the behaviour of the perturbations is determined by a path integral of the perturbation modes over the compact geometries represented by the background wave function. In the case of Euclidean geometries that are part of a 4-sphere or of complex geometries that are near such a Euclidean geometry, one can use an adiabatic approximation to show that the perturbation modes are in their ground state, with the minimum excitation compatible ·with the uncertainty principle. This means that the Lorentzian geometries that correspond to the oscillating part of the wave function start off at the minimum radius with all the perturbation modes in the ground state. As the universe inflates, the adiabatic approximation remains good and the perturbation modes remain in their ground states until their wavelength becomes longer than the horizon size or, in other words, their frequency is red shifted to less than the expansion time scale. After this, the wave functions of the perturbation modes freeze and do not relax adiabatically to remain in the ground state as the frequency of the modes changes.

The perturbation modes remain frozen until the wavelength of the modes becomes less than the horizon size again during the matter- or radiation-dominated expansion. Because they have not been able to relax adiabatically, they will then be in a highly excited state. After this, they will evolve like classical perturbations of a Friedmann universe. They will have a 'scale free' spectrum, that is, their rms amplitude at the time the wavelength equals the horizon size will be independent of the wavelength. The amplitude will be roughly $10(m/m_P)$, where m is the mass of the scalar field. Thus they would have the right amplitude of about 10^{-4} to account for galaxy formation if m is about 10^{14} GeV.

In order to generate sufficient inflation, the initial value of the scalar field ϕ has to be greater than about 8. However, with $m = 10^{-5}m_P$, the energy density of the scalar field will still be a lot less than the Planck density. Thus it may be reasonable in quantum cosmology to ignore higher-order terms and extra dimensions.

In the recollapse phase the perturbations will continue to grow classically. They will not return to their ground state when the universe becomes small again, as I suggested (Hawking, 1985). The reason is that when they start expanding, the background compact geometry bounded by the surface S is near to the Euclidean geometry of half a 4-sphere. On such a background the adiabatic approximation will hold for the perturbation modes, so they will be in their ground state. However, when the universe recollapses, the background geometry will be near a Lorentzian solution which expands and recontracts. The adiabatic approximation will not hold on such a background. Thus the perturbation modes will not be in their ground state when the universe recollapses, but will be highly excited.

14.7 The direction of time

The quantum state defined by the 'no boundary' proposal is CPT invariant (Hawking, 1985), though this is not true of other quantum states, such as that proposed by Vilenkin (1986). Yet the observed universe shows a pronounced asymmetry between the future and the past. We remember events in the past but we have to predict events in the future. Imagine a tall building which is destroyed by an explosion and collapses to a pile of rubble and dust. If one took a film of this and ran it backwards, one would see the rubble and dust gather themselves together and jump back into their places in the building. One would easily recognise that the film was being shown backwards because this kind of behaviour is never observed: we do not see tower blocks jumping up. Yet it is not forbidden by the laws of physics. These are CPT invariant. In fact, the laws that are important for the structure of buildings are invariant under C and P separately. Thus, they must be invariant under T alone. In other words, if a building can collapse, it can also resurrect itself.

The explanation that is usually given as to why we do not see buildings jumping up is that the second law of thermodynamics says that entropy or disorder must always increase with time, and that an erect building is in a much more ordered state than a pile of rubble and dust. However, this law has a rather different status from other laws, such as Newton's law of gravity. First, it is not an absolute law that is always obeyed: rather it is a statistical law that says what will probably happen. Second, it is not a local law like other laws of physics: it is a statement about boundary conditions. It says that if a system starts off in a state of high order, it is likely to be found in a disordered state at a later time, simply because there are many more disordered states than ordered ones.

The reason that entropy and disorder increase with time and buildings fall down rather than jump up is that the universe seems to have started out in a state of high order in the past. On the other hand, if, for some reason, the universe obeyed the boundary condition that it was in a state of high order at late times, then at earlier times it would be likely to be in a disordered state and disorder would decrease with time. However, human beings are governed by the second law and the boundary conditions, just like everything else in the universe. Our subjective sense of the direction of time is determined by the direction in which disorder increases because to record information in our memories requires the expenditure of free energy and increases the entropy and disorder of the universe. Thus, if disorder decreased with time, our subjective sense of time would also be reversed and we would still say that entropy and disorder increased with time. The second law is almost a tautology: entropy and disorder increase with time because we measure time in the direction in which disorder increases.

However, there remains the question of why should the universe have been in a state of high order at one end of time? Why was it not in a state of complete disorder or thermal equilibrium at all times? After all, that might seem more probable as there are many more disorder states than order ones. And why does the direction of time in which disorder increases coincide with that in which the universe expands? Put it another way: why do we say that the universe is expanding, and not contracting?

These questions can be answered only by some assumption on the boundary conditions of the universe or, equivalently, on the class of spacetime geometries in the path integral. As we have seen, the 'no boundary' condition implies that the universe would have started off in a smooth and ordered state with all the inhomogeneous perturbations in their ground state of minimum excitation. As the universe expanded, the perturbations would have grown and the universe would have become more inhomogeneous and disordered. This would answer the questions above.

But what would happen if the universe, or some region of it, stopped expanding and began to collapse? At first I thought (Hawking, 1985) that entropy and disorder would have to decrease in the contracting phase so that the universe would get back to a smooth state when it was small again. This was because I thought that at small values of the radius a, the wave function would be given just by a path integral over small Euclidean geometries. This would imply $\Psi = 1$ on the light cone $x = \pm y$ in the model described above and that the adiabatic approximation would hold for the perturbation modes, which would therefore be in their ground state. However, Page (1985b) pointed out that there would also be a contribution to the wave function from compact, complex, almost Lorentzian geometries that represented universes that started at a minimum radius, expanded to a maximum and recollapsed, as described above. This was supported by work by Laflamme (1987), who investigated a minisuperspace model in which the surfaces S had topology $S^1 \times S^2$. He also found almost Lorentzian solutions which started in a non-singular manner but recollapsed to a singularity. The adiabatic approximation for the perturbation modes would not hold in the recollapse. Thus they would not return to their ground states, but would get even more excited as the collapse continued. The universe would get more and more inhomogeneous and disorder would continue to increase with time.

There remains the question of why we observe that the direction of time in which disorder increases is also the direction in which the universe is expanding. Because the 'no boundary' quantum state is CPT invariant, there will also be histories of the universe that are the CPT reverses of that described above. However, intelligent beings in these histories would have the opposite subjective sense of time. They would therefore describe the universe in the same way as above: it would start in a smooth state, expand and collapse to a very inhomogeneous state. The question therefore becomes: why do we live in the expanding phase? If we lived in the contracting phase, we would observe entropy to increase in the opposite direction of time to that in which the universe was expanding. To answer this, I think one has to appeal to the weak anthropic principle. The probability is that the universe will not recollapse for a very long time (Hawking and Page, 1986). By that time, the stars would all have burnt out and the baryons would have decayed. The conditions would therefore not be suitable for the existence of beings like us. It is only in the expanding phase that intelligent beings can exist to ask the question: why is entropy increasing in the same direction of time as that in which the universe is expanding?

14.8 The origin and fate of the universe

Does the universe have a beginning and/or end?

If the 'no boundary' proposal for the quantum state is correct, spacetime is compact. On a compact space, any time coordinate will have a minimum and a maximum. Thus, in this sense, the universe will have a beginning and an end.

Will the beginning and end be singularities?

Here one must distinguish between two different questions: whether there are singularities in the geometries over which the path integral is taken, and whether there are singularities in the Lorentzian geometries that correspond to the density matrix by the WKB approximation. A singularity cannot really be regarded as belonging to spacetime because the laws of physics would not hold there. Thus, the requirement of the 'no boundary' proposal that the path integral is over compact geometries only rules out the existence of any singularities in this sense. Of course, one will have to allow compact metrics that are not smooth in the path integral, just as in the integral over particle histories one has to allow particle paths that are not smooth but satisfy a Hölder continuity condition. However, one can approximate such paths by smooth paths. Similarly, in the path integral for the universe, it must be possible to approximate the non-smooth metrics in a suitable topology by sequences of smooth metrics because otherwise one could not define the action of such metrics. Thus, in this sense, the geometries in the path integral are non-singular.

On the other hand, the Lorentzian geometries that correspond to the density matrix by the WKB approximation can and do have singularities. In the minisuperspace model described above, the Lorentzian geometries began at a non-singular minimum radius or 'bounce' and evolve to a singularity in general, in the direction of time defined by entropy increase. I would conjecture that this is a general feature: oscillating wave functions and Lorentzian geometries arise only when one has a massive scalar field which gives rise to an effective cosmological constant and Euclidean solutions which are like the 4-sphere. The Lorentzian solutions will be the analytic continuation of the Euclidean solutions. They will start in a smooth non-singular state at a minimum radius equal to the radius of the 4-sphere and will expand and become more irregular. When and if they collapse, it will be to a singularity.

One could say that the universe was 'created from nothing' at the minimum radius (Vilenkin, 1982). However, the use of the word 'create' would seem to imply that there was some concept of time in which the universe did not exist before a certain instant and then came into being. But time is defined only within the universe, and does not exist outside it, as was pointed out by Saint Augustine (400): 'What did God do before He made Heaven and Earth? I do not answer as one did merrily: He was preparing Hell for those that ask such questions. For at no time had God not made anything because time itself was made by God.'

The modern view is very similar. In general relativity, time is just a coordinate that labels events in the universe. It does not have any meaning outside the spacetime manifold. To ask what happened before the universe began is like asking for a point on the Earth at 91° north latitude; it just is not defined. Instead of talking about the universe being created, and maybe coming to an end, one should just say: The universe is.

References

Gibbons, G. W., Hawking, S. W. and Perry, M. J. (1978). *Nucl. Phys.*, **B138**, 141.

Goroff, M. H. and Sagnotti, A. (1985). *Phys. Lett.*, **160B**, 81.

Halliwell, J. J. and Hawking, S. W. (1985). *Phys. Rev.*, **D31**, 1777.

Hartle, J. B. and Hawking, S. W. (1983). *Phys. Rev.*, **D28**, 2960.

Hawking, S. W. (1982). In *Astrophysical Cosmology*. Proceedings of the Study Week on Cosmology and Fundamental Physics, ed. H. A. Bruck, G. V. Coyne and M. S. Longair. Pontificia Academiae Scientarium: Vatican City.

Hawking, S. W. and Penrose, R. (1970). *Proc. Roy. Soc. Lon.*, **A314**, 529.

Hawking, S. W. and Ellis, G. F. R. (1973). *The Large Scale Structure of Space-Time*. Cambridge University Press: Cambridge.

Hawking, S. W. (1984a). *Phys. Lett.*, **150B**, 339.

Hawking, S. W. (1984b). *Nucl. Phys.*, **B239**, 257.

Hawking, S. W. and Wu, Z. C. (1985). *Phys. Lett.*, **151B**, 15.

Hawking, S. W. (1985). *Phys. Rev.*, **D32**, 2489.

Hawking, S. W. and Page, D. N. (1986). *Nucl. Phys.*, **B264**, 185.

Hawking, S. W. (1987). *Physica Scripta* (in press).

Laflamme, R. (1987). The wave function of a $S^1 \times S^2$ universe. Preprint, to be published.

Lifshitz, E. M. and Khalatnikov, I. M. (1963). *Adv. Phys.*, **12**, 185.

Linde, A. D. (1983). *Phys. Lett.*, **129B**, 177.

Linde, A. D. (1985). *Phys. Lett.*, **162B**, 281.

Misner, C. W. (1970). In *Magic without Magic*, ed. J. R. Klauder. Freeman: San Francisco.

Moss, I. and Wright, W. (1983). *Phys. Rev.*, **D29**, 1067.

Page, D. N. (1985a). *Class. & Q.G.*, **1**, 417.

Page, D. N. (1985b). *Phys. Rev.*, **D32**, 2496.

Page, D. N. (1986). *Phys. Rev.*, **D34**, 2267.

Rubakov, V. A., Sazhin, M. V. and Veryaskin, A. V. (1982). *Phys. Lett.*, **115B**, 189.

Saint Augustine (400). *Confessions*. Re-edited in *Encyclopedia Britannica* (1952).

Vilenkin, A. (1982). *Phys. Lett.*, **117B**, 25.

Vilenkin, A. (1986). *Phys. Rev.*, **D33**, 3560.

15

Superstring unification†

15.1 Introduction

Superstring theory is an approach to the unification of fundamental
particles and forces that has attracted a great deal of attention in the last two
years [1]. It offers the possibility of overcoming many of the shortcomings of
the standard model and providing a unified theory free from much of the
arbitrariness that is inherent in conventional point–particle theories.

The standard model of electroweak and strong forces has enjoyed an
enormous amount of success. Indeed, it appears to be consistent with all
established particle physics experiments. This being the case, the first
question one should ask is why one should even be looking for something
better. Most criticisms of the standard model are based on the fact that it
requires a number of arbitrary choices and fine-tuning adjustments of
parameters. These features do not prove that it is wrong or even incomplete.
However, given the history of successes in elementary particle physics, it is
natural to seek a deeper underlying theory that can account for many of the
arbitrary choices and parameters. These include the choice of gauge groups
and representations, the number of families of quarks and leptons, the
origins of the Higgs symmetry-breaking mechanism, and the specific values
of various parameters.

The aesthetic shortcomings of the standard model or a grand unified
model do not prove that it is wrong or incomplete. The thing that does this
most convincingly is the absence of gravity. There is a straightforward way
to couple Einsteinian gravity to any relativistic field theory by following the

† Work supported in part by the US Department of Energy under contract DEAC 03-
81-ER40050.
‡ Based on a lecture presented at the Twenty-third International Conference on High
Energy Physics in Berkeley, California, July 1986.

dictates of general coordinate invariance, and at the classical level it poses no difficulties. However, the quantum theory, which is renormalizable before gravity is appended, inevitably acquires non-renormalizable ultraviolet divergences. Many years of study strongly suggest that this is a generic feature of all quantum field theories based on point particles. Even Einsteinian gravity by itself has been shown to be singular at two loops [2]. Pure supergravity theories in four dimensions are finite at two loops, but are very likely to be singular at three loops. The addition of extra spatial dimensions (such as in eleven-dimensional supergravity) makes the divergences even more severe, and therefore represents a step in the wrong direction. This is where superstring theory comes to the rescue.

String theory was developed in the late 1960s and early 1970s in an attempt to describe the strong nuclear force [3]. The first string theory was plagued by a number of unrealistic features: the absence of fermions, the presence of a tachyon and massless modes, and the necessity of 26-dimensional spacetime. In an attempt to do better a second string theory was introduced in 1971 [4]. It incorporates a spectrum of fermions that was proposed by Pierre Ramond [5]. In the form first discussed this theory also contained a tachyon. It possesses two-dimensional superconformal symmetry on the string world sheet but not spacetime supersymmetry. Five years later, Gliozzi, Scherk, and Olive proposed a truncation of the spectrum that eliminated the tachyon [6]. They noted that then there are an equal number of bosons and fermions at each mass level and conjectured that this version of the theory possesses spacetime supersymmetry. This result was proved in 1980 by Michael Green and myself [7]. This theory is called type I superstring theory in the modern classification (see Section 15.2). It was discovered in 1972 that superstring theories required ten-dimensional spacetime for their consistency [8].

Neither the bosonic string theory nor superstring theory succeeded in giving a realistic description of hadron physics. The extra dimensions were an embarrassment as were the massless modes, which had no counterparts in the hadronic spectrum. In 1974 Joël Scherk and I demonstrated that the massless spin-2 mode that is present in both theories interacts in the standard Einsteinian way at low energies [9]. (This was also shown by Yoneya [10].) We therefore suggested that it be interpreted as a graviton rather than a hadron and that string theory be used as a basis for constructing a unified description of gravity and all the other forces rather than as a theory of hadrons. In other words, it was proposed that elementary particles are strings rather than points as in conventional quantum field theory.

The idea of using strings for unification offered a number of appealing features. First of all, the existence of gravity would be understood as necessary since it is a generic feature of all consistent string theories. String theories do not possess ultraviolet divergences of the conventional type, and therefore it appeared plausible that this type of theory could overcome the divergence problem discussed above. To state the case as forcefully as possible, string theory requires the existence of gravity, whereas the point-particle theories require that it does not exist. This conviction sustained my enthusiasm for the subject throughout the ten-year period before its widespread acceptance.

The use of string theory for unification rather than hadronic physics has a number of other advantages. The extra dimensions can acquire a sensible interpretation, because spacetime geometry is dynamical in a gravity theory and there is a possibility that the unobserved dimensions could be spontaneously compactified. Also, string theories do not possess adjustable dimensionless parameters, nor much freedom to choose groups and representations. So, if a scheme of this type is successful, it should be much more predictive than the standard model. To put it differently, the consistent introduction of gravity in a unified theory is such a severe constraint that it could have dramatic predictive consequences for low-energy physics.

String theory contains a fundamental length parameter, which is the characteristic size of a string. In order for the strength of gravity to emerge with the usual Newtonian value, it is necessary to associate this length with the Planck length: a few times 10^{-33} cm. Thus in the unification context the strings are taken to be some twenty orders of magnitude smaller than they were in the hadronic context.

15.2 Classification of string theories

Strings have two possible topologies: open (with free ends) or closed (a loop without ends). The most interesting theories consist of oriented closed strings only. Type I superstring theory, which was the first one to be understood, consists of unoriented open and closed strings.

How many string theories are there? The answer to this question is not yet known. Ideally it will turn out that there is just one. This could happen if the ones that are presently known are either inconsistent or equivalent to one another. In order for a theory to be consistent it must be free from tachyons (states of negative mass squared) and ghosts (states of negative norm). It is also necessary that loop amplitudes be free from anomalies and possess modular invariance: properties to which we shall return. Additional criteria, perhaps of non-perturbative origin, may eventually be found to be necessary also.

Six string theories are known that satisfy the consistency properties mentioned above. Each of them requires that spacetime has ten dimensions (nine space and one time) and that the two-dimensional world-sheet action has superconformal symmetry. These theories are listed in Table 15.1. There is some evidence that the three heterotic theories are actually different phases of the same theory, which would reduce the number of different theories to be considered.

Superstring theories possess not only two-dimensional superconformal symmetry but local ten-dimensional super-Poincaré symmetry, a possibility first pointed out by Gliozzi, Scherk and Olive [6]. There are three possibilities for $D = 10$ supersymmetry, each of which can be realized in a superstring theory. The minimal irreducible spinor in $D = 10$ satisfies simultaneous Majorana and Weyl properties and has 16 independent real components. A theory with a single conserved Majorana–Weyl supercharge has $N = 1$ supersymmetry, a possibility realized by three of the six theories listed in the table. There are two distinct possibilities for theories with two conserved Majorana–Weyl supercharges. Either they have opposite chirality (type IIA) or they have the same chirality (type IIB). This is the maximum amount of supersymmetry that is possible in an interacting theory (corresponding to $N = 8$ supersymmetry in four dimensions). The type II theories are not promising for phenomenology because they do not accommodate elementary Yang–Mills fields. However, the type IIB theory is in some respects the most beautiful of all the theories, and it is not totally out of the question that it could form the basis of a realistic phenomenology.

Table 15.1. *Six string theories*

Name	$D = 10$ supersymmetry	Yang–Mills symmetry
Type I [11]	$N = 1$	$SO(32)$
Type IIA [12]	$N = 2$	—
Type IIB [12]	$N = 2$	—
Heterotic [13]	$N = 1$	$SO(32)$
Heterotic [13]	$N = 1$	$E_8 \times E_8$
Heterotic [14]	$N = 0$	$SO(16) \times SO(16)$

An important fact about the standard model is the left–right asymmetry in the classification of quark and lepton multiplets (chirality). The most natural way to achieve this chiral asymmetry, starting from a theory in more than four dimensions, is for the higher-dimensional theory to be left–right asymmetric itself. This property is shared by all the theories in the table except for the type IIA.

Another important consideration in choosing among the theories is a desire to understand the origin of the hierarchy of mass scales. In particular, one wants to understand why radiative corrections to the mass of Higgs scalars do not destroy the enormous ratio between the electroweak scale and the unification scale. The most promising possibility seems to be 'low-energy' supersymmetry spontaneously broken around the electroweak scale. This can be realized in superstring theories with spacetime supersymmetry.

In addition to the string theories listed in the table, there are some others that are consistent in all respects except for the presence of a tachyon in the spectrum. The original $D = 26$ bosonic string is an example of such a theory, and a number of others have been discovered recently [15]. It is an open question whether any of these theories can be given a consistent interpretation by identifying a stable ground state that is free from tachyonic modes. Even if this is possible, one might have to pay an unacceptable price in terms of a large cosmological constant. But that is a threat to all theories once supersymmetry is broken.

15.3 Feynman diagrams

In conventional field theory, Feynman diagrams correspond to the distinct topological possibilities for connecting the world lines of elementary particles. The situation is analogous in string theory. Since strings are one dimensional they sweep out a two-dimensional surface in spacetime. Thus Feynman diagrams correspond to the topologically distinct world sheets representing the possible spacetime histories of interacting strings. In thinking about these surfaces it is extremely convenient to imagine carrying out a Wick rotation and regarding the surfaces as being embedded in a space of Euclidean metric. This makes the corresponding path integrals well defined. This rotation can probably be rigorously justified by arguing that it gives a unique unitary S matrix.

The classification of string Feynman diagrams is especially simple for the theories that consist of oriented closed strings only. In these theories there is a single fundamental string interaction, depicted in Fig. 15.1, that describes one string breaking into two, or two strings joining to give a single one. The surface shown is topologically like a pair of pants. Since the surface is smooth the particular spacetime point at which the interaction takes place is a frame-dependent question. If the time slices are taken using planes that are tilted relative to the ones shown, a different interaction point would be identified. Thus there is no preferred point on the surface and the existence of interaction is an inevitable feature of geometric origin, not something that needs to be appended to the theory in an *ad hoc* way.

In calculating *S*-matrix elements one should consider amplitudes for scattering closed strings that propagate to time $\pm \infty$, i.e., with tubes emerging from the surface that extend to infinity. However, the conformal symmetry of the underlying two-dimensional field theory [16, 17], implies that world sheets that differ by a conformal transformation are equivalent. In particular, one may map the asymptotic strings to finite coordinates. When this is done, they are represented by points on the world sheet. Altogether, the amplitude is then represented by a closed oriented surface with 'punctures' representing the initial and final string states.

The topological classification of closed oriented two-dimensional surfaces is characterized by a single integer, the genus *g*, which can be thought of as the number of handles that is attached to a sphere. Thus, as shown in Fig. 15.2, the sphere has $g = 0$, the torus has $g = 1$, and so forth.

In the type I superstring theory the topological classification of diagrams is a good deal more complicated. Since the strings themselves can be open or closed and have no intrinsic orientation, the Feynman diagrams can have boundaries and need not be orientable. This makes the analysis of this theory more difficult. An important advantage of orientability is that an

Fig. 15.1. A piece of world sheet in the shape of pants. The slice at time t_1 shows two closed strings while the one at time t_2 shows one closed string.

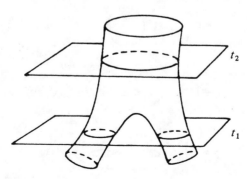

orientable surface can be regarded as a Riemann surface and powerful techniques of complex analysis can be utilized.

In the type II and heterotic string theories, the topology of a diagram is characterized by the genus g, which can be thought of as the number of loops in the diagram. Thus there is just a single diagram at each order of the perturbation expansion. This is a remarkable simplification compared with conventional field theories in which the number of diagrams at n loops is roughly of order $n!$. It has been shown by explicit computation that the one-loop amplitudes are finite [12, 13]. Arguments have been made to suggest that the loop amplitudes should be finite at every order, but this still requires a careful analysis to be definitively established. This is likely to be settled within a year or so. Assuming that the expected result emerges, we shall then be in the happy position of having the first examples of perturbatively finite quantum theories of gravity.

The analysis of multiloop (genus $g > 1$) string amplitudes is a very rapidly developing subject in which much beautiful work has been done in the last year [18]. The details require state-of-the-art methods in the theory of Riemann surfaces and algebraic geometry. All I can realistically hope to do here is to sketch some of the basic ideas.

The loop amplitudes are given by integrals that represent a sum over all conformally inequivalent geometries of the given topologies. Fortunately, these are finite-dimensional integrals. $3g - 3 + n$ complex parameters describe the integration space M_g for a genus g amplitude with n external particles. (This result is due to Riemann.) The structure of an amplitude is

Fig. 15.2. The topology of closed orientable two-dimensional surfaces is given by the genus g. The cases $g = 0, 1, 2$ are illustrated.

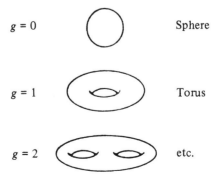

then given by an expression of the form

$$\int_{M_g} \mathrm{d}\mu \prod_{i<j} (F_{ij})^{k_i \cdot k_j}. \tag{1}$$

In this expression the indices i and j refer to the external particles and the k_i are the momenta (assumed to be on-shell). The function $\log F_{ij}$ is proportional to the Green's function on the world sheet between the coordinates of particles i and j. Also, $\mathrm{d}\mu$ represents a suitable integration measure on the 'moduli space' M_g.

Not only the two-dimensional world sheet, but also the $2(3g-3+n)$-dimensional moduli space has a complex structure. As a result it is possible to express the measure $\mathrm{d}\mu$ and functions F_{ij} in terms of holomorphic functions on M_g. The space M_g is naturally expressed as a quotient of a space T_g, known as Teichmüller space, divided by an infinite discrete group of transformations. The expressions $\mathrm{d}\mu$ and F_{ij} can be expressed as products of various terms that are well defined on T_g but are multivalued ('line bundles') on M_g. In order for the loop amplitude to be well defined it is necessary that when the relevant products are formed, expressions that are single-valued on M_g ('modular invariant') result. The fact that this works for each of the various string theories is highly non-trivial. In fact, it almost completely determines the various functions!

Moduli space M_g has a very rich and subtle topology. In fact, it even has boundary components that correspond to singular limits of the geometry in which the world sheet 'degenerates'. The relevant issue for finiteness is whether or not the measure $\mathrm{d}\mu$ diverges on any of these boundary components so fast as to give a singular integral. There are two distinct ways in which the surface can degenerate. One, depicted in Fig. 15.3(a), involves the formation of a long thin tube that separates the diagram into two pieces. The second, shown in Fig. 15.3(b), involves the formation of a long thin tube that does not separate the surface.

15.4 String field theory

The history of string theory has many strange aspects. One of them is that the Feynman diagrams that I have been discussing were formulated before a field theory that gives rise to them. The development of string field theory was carried out in the light-cone gauge for bosonic strings in the 1970s [19] and for superstrings in the early 1980s [20]. However, what one really wants is a covariant gauge-invariant action that makes manifest all of the beautiful underlying symmetries of the theory in a way that is not tied to a particular choice of background fields. In short, one would like to know the string analog of the Hilbert–Einstein action of general relativity. This has not yet been achieved, but an enormous effort by many workers over the past year has brought us much closer to this goal. In particular, a beautiful field theory for open strings, correct at least at tree level, has been achieved. It is almost impossible to give adequate credit to all the workers who have made important contributions. Some of them are Siegel and Zwiebach [21], Banks and Peskin [22], Neveu and West [23], Hata, Itoh, Kugo, Kunitomo and Ogawa [24]. I will give a brief description of the results in the form developed by Witten [25].

One important remark is that in field theory an individual Feynman diagram is given by an integral whose integration region is topologically trivial: some sort of hypercube. Thus, since moduli space M_g has a complicated topology, it cannot arise from a single diagram. What happens is that many different field theory diagrams give different topologically trivial pieces of the integration region M_g, all with the same integrand, and the sum gives the complete genus g amplitude. Thus the individual diagrams provide a 'triangulation' of moduli space [26].

Witten's description of the field theory of open bosonic strings has many analogies with Yang–Mills theory. This is not really surprising in as much as open strings are an infinite-component generalization of Yang–Mills fields.

Fig. 15.3. The boundary of moduli space corresponds to singular limits in which the surface 'degenerates'. This can occur by the formation of a long thin tube that splits the diagram as in (*a*) or that does not split it as in (*b*).

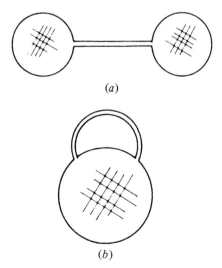

(*a*)

(*b*)

It is pedagogically useful to emphasize these analogies in describing the theory. The basic object in Yang–Mills theory is the vector potential $A_\mu^a(x^\rho)$, where μ is a Lorentz index and a runs over the generators of the symmetry algebra. By contracting with matrices $(\lambda^a)_{\alpha\beta}$ that represent the algebra and differentials dx^μ we can define

$$A_{\alpha\beta} = \sum_{a,\mu} A_\mu^a (\lambda^a)_{\alpha\beta} \, dx^\mu, \qquad (2a)$$

a matrix of one form. This is a natural quantity from a geometric point of view. The analogous object in open-string field theory is the string field

$$A[x^\rho(\sigma), c(\sigma)]. \qquad (2b)$$

This is a functional field that creates or destroys an entire string with coordinates $x^\rho(\sigma), c(\sigma)$, where the parameter σ is taken to have the range $0 \leqslant \sigma \leqslant \pi$. The coordinates $c(\sigma)$ are anticommuting ghost degrees of freedom that arise in the first quantization of the action. They are essential so that, when A is expanded in an infinite sequence of point fields, there is an appropriate set of auxiliary Stückelberg fields at each mass level. The details of such an expansion are quite complicated, but the mathematics of the complete string field is not so bad. The Yang–Mills field is one of the infinity of terms in such an expansion.

The string field A can be regarded as a matrix (in analogy to $A_{\alpha\beta}$) by regarding the coordinates with $0 \leqslant \sigma \leqslant \pi/2$ as providing the left matrix index and those with $\pi/2 \leqslant \sigma \leqslant \pi$ as providing the right matrix index as shown in Fig. 15.4(a). One could also associate quark-like charges with the ends of the

Fig. 15.4. An open string has a left-hand ($\sigma < \pi/2$) and right-hand ($\sigma > \pi/2$) segment, depicted in (a), which can be treated as matrix indices. The multiplication $A*B = C$ is depicted in (b).

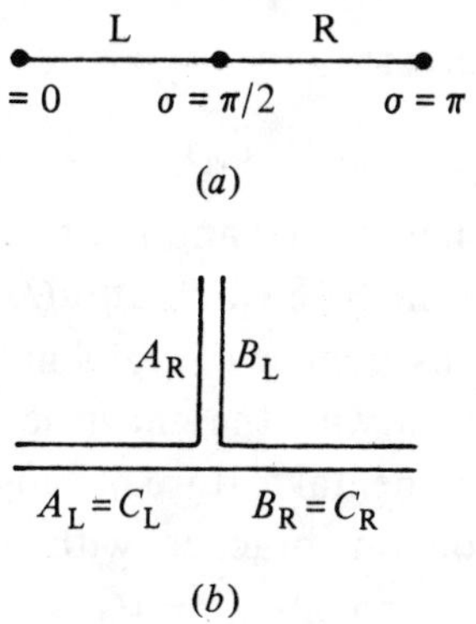

$$L \qquad R$$
$$\sigma = 0 \qquad \sigma = \pi/2 \qquad \sigma = \pi$$

(a)

$$A_R \quad B_L$$
$$A_L = C_L \qquad B_R = C_R$$

(b)

strings which would then be included in the matrix labels as well. This is a minor and inessential complication, which we shall suppress. By not including such charges we are constructing the string generalization of $U(1)$ gauge theory. $U(1)$ gauge theory (without matter fields) is a free theory, but the string extension has non-trivial interactions, as we shall see.

In the case of Yang–Mills theory, we can multiply two fields by the rule

$$\sum_\gamma A_{\alpha\gamma} \wedge B_{\gamma\beta} = C_{\alpha\beta}. \tag{3a}$$

This is a combination of matrix multiplication and antisymmetrization of the tensor indices (the wedge product of differential geometry). This multiplication is associative but non-commutative. A corresponding rule for string fields is given by a $*$ product,

$$A * B = C. \tag{3b}$$

This infinite-dimensional matrix multiplication is depicted in Fig. 15.4(*b*). One identifies the coordinates of the right half of string A with those of the left half of string B and functionally integrates over them. This leaves string C consisting of the left half of string A and the right half of string B. It is also necessary to include a suitable factor involving the ghost coordinates at the midpoint $\sigma = \pi/2$.

A fundamental operation in gauge theory is exterior differentiation $A \to dA$. In terms of components

$$dA = \tfrac{1}{2}(\partial_\mu A_\nu - \partial_\nu A_\mu)\, dx^\mu \wedge dx^\nu,$$

which contains the abelian field strengths as coefficients. Exterior differentiation is a nilpotent operation, $d^2 = 0$, since partial derivatives commute and vanish under antisymmetrization. The non-abelian field strength is given by the matrix-valued two-form

$$F = dA + A \wedge A, \tag{4a}$$

or in terms of tensor indices,

$$F_{\mu\nu} = \partial_\mu A_\nu - \partial_\nu A_\mu + [A_\mu, A_\nu].$$

Let us now construct analogs of d and F for the string field. The operator that plays the roles of d is the BRST operator Q. Q is a conserved fermionic charge that arises as a consequence of a global fermionic symmetry of the gauge-fixed quantum action with the ghost fields. The occurrence of this BRST symmetry is a fundamental feature of gauge theories. Classically Q is nilpotent (vanishing Poisson bracket with itself) for any spacetime dimension. Quantum mechanically, $Q^2 = 0$ is satisfied only for $D = 26$ in the case of bosonic strings [27]. Since this is an essential requirement, we must make this choice. Q can be written explicitly as a differential operator involving the coordinates $X(\sigma)$, $c(\sigma)$, but I shall not write out the formula here. Given the operator Q, there is an obvious formula for the string theory field strength, analogous to the Yang–Mills formula, namely

$$F = QA + A * A. \tag{4b}$$

An essential feature of Yang–Mills theory is gauge invariance. Infinitesimal gauge transformation can be described by a matrix of infinitesimal parameters $\Lambda(x^\rho)$, which are functions of the spacetime coordinates x^ρ. The transformation rules for the potential and the field strength are then

$$\delta A = d\Lambda + [A, \Lambda] \tag{5a}$$

and

$$\delta F = [F, \Lambda]. \tag{6a}$$

We can write down completely analogous formulas for the string theory, namely

$$\delta A = Q\Lambda + [A, \Lambda] \tag{5b}$$

and

$$\delta F = [F, \Lambda]. \tag{6b}$$

In this case $[A, \Lambda]$ means $A * \Lambda - \Lambda * A$, of course. Since the infinitesimal parameter $\Lambda[x^\rho(\sigma), c(\sigma)]$ is a functional, it can be expanded in terms of an infinite number of ordinary functions. Thus the gauge symmetry of string theory is infinitely richer than that of Yang–Mills theory, as required for the consistency of the infinite spectrum of high spin fields contained in the theory.

The next step is to formulate a gauge-invariant action. The key ingredient in doing this is to have a suitably defined integral. In the case of Yang–Mills theory we must integrate over spacetime and take a trace over the matrix indices. Thus it is convenient to define $\int X$ as $\int d^4x \, \mathrm{Tr}(X)$. In this notation the usual Yang–Mills action is

$$I \sim \int g^{\mu\rho} g^{\nu\lambda} F_{\mu\nu} F_{\rho\lambda}, \tag{7}$$

which is easily seen to be gauge invariant. Requiring I to be stationary gives the classical equations of motion $D^\mu F_{\mu\nu} = 0$, where D^μ is a covariant derivative. Since F is quadratic in A, I describes cubic and quartic interactions.

The definition of integration appropriate to string theory is a 'trace' that identifies the left and right segments of the string field Lagrangian. Thus in the case of string theory we define

$$\int X = \int \sum_{\sigma < \pi/2} \delta(x(\sigma) - x(\pi - \sigma)) \cdots e^{-(3i/2)\phi(\pi/2)} X. \tag{8}$$

As indicated in Fig. 15.5(a), this identifies the left and right segments of X. A ghost factor has been inserted at the midpoint. (ϕ is a bosonized form of the ghost coordinates.) The $\cdots$ signifies that analogous integrations should also be performed for the ghost coordinates. We now have the necessary ingredients to write a string action. If we try to emulate the Yang–Mills formula we run into a problem, namely no analog of the metric $g^{\mu\rho}$ has been defined. Rather than trying to find one, it proves more fruitful to look for a gauge-invariant action that does not require one. The simplest possibility is given by the Chern–Simons form

$$I \sim \int A * QA + \tfrac{2}{3}A * A * A. \tag{9}$$

In the context of ordinary Yang–Mills theory the integrand is a three-form and therefore such a term can only be introduced in three dimensions, where it is interpreted as giving mass to the gauge field. In string theory the interpretation is different and the formula makes perfectly good sense. In fact, it gives rise to the deceptively simple field equation $F = 0$.

The fact that the string equation of motion is $F = 0$ does not mean the theory is trivial. If we drop the interaction term the equation of motion for

Fig. 15.5. Integration of a string functional requires identifying the left-hand and right-hand halves as depicted in (a). The three-string vertex, shown in (b), is based on two multiplications and one integration and treats the three strings symmetrically.

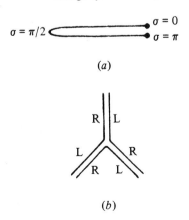

(a)

(b)

the free theory is $QA=0$, which is invariant under the abelian gauge transformation $\delta A = Q\Lambda$ since $Q^2 = 0$. Once one imposes suitable restrictions on ghost number, one can show this precisely reproduces the known spectrum of the bosonic string.

The cubic string interaction is depicted in Fig. 15.5(b). Two of the segment identifications are consequences of the $*$ products in $A*A*A$ and the third is a consequence of the integration resulting in a symmetric expression. As an alternative way of thinking about the interaction we can attach some flat world sheet corresponding to free propagation of each of the three strings in the interaction. This is depicted in Fig. 15.6. The folds in the drawing do not imply intrinsic curvature of the surface and therefore have no physical significance. (They are introduced for ease of depiction only.) There is curvature at the interaction point, however. To see this, imagine drawing a small circle of radius r around it. It has circumference $3\pi r$ (a contribution of πr coming from each side of the surface). Thus the surface is flat everywhere except at this one point where the curvature has a δ function singularity.

A standard theorem implies that a closed orientable two-dimensional surface of genus g has an integrated curvature proportional to $\chi = 2(1-g)$. Thus, except for genus 1, the surface does not admit a flat metric. In string theory one is only interested in equivalence classes of metrics that are related by conformal mappings. It is always possible to find representatives of each equivalence class in which the metric is flat everywhere except at isolated points where the curvature is infinite. Such a metric describes a surface with conical singularities, which is not a manifold in the usual sense. In fact, it is an example of a class of surfaces called orbifolds. The string field theory construction of the amplitude automatically chooses a particular metric which, as we have indicated, is of this type. It is also possible to choose constant curvature metrics, but they do not arise from the string field theory.

Fig. 15.6. The three-string vertex with external string propagators attached is given by a world sheet that is flat everywhere except at the interaction point where a small circle of radius r has circumference $3\pi r$.

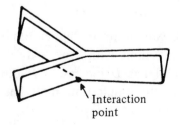

Interaction
point

The formation of string field theory described above is certainly very beautiful. It is also deceptively simple. To really understand it in detail one must define very carefully the various functional integrations that appear. One way of doing this is to expand the fields in a Fock-space basis using an infinite number of harmonic oscillators corresponding to the normal modes of $x^\rho(\sigma)$ and $c(\sigma)$. The functional integrations can then be carried out explicitly leaving a Fock-space expression for the interaction vertex that can be evaluated for any three states of the string spectrum. This calculation has been carried out [28], demonstrating that the formula is in fact well defined and possesses the properties that it should. Evidence has also been found that it reproduces the standard scattering amplitudes [29].

15.5 Anomalies

The gauge invariance of classical gauge theories implies the existence of conserved gauge currents, whose form can be deduced by the standard Noether procedure. In Yang–Mills theories the associated conserved charges are the symmetry generators. In general relativity the conserved current is the energy–momentum tensor and the associated charges are the energy and momentum. The classical conservation of these currents can, under certain circumstances, be destroyed by quantum effects called anomalies. When this happens unphysical modes of the gauge fields become coupled in the S matrix leading to a breakdown of unitarity and causality. Thus, in general, all anomalies in gauge currents must cancel or else the theory must be rejected as inconsistent.

The Feynman diagrams that give rise to anomalies typically arise at one-loop order and involve a loop of chiral fermions. In the well-known case of a gauge theory in four dimensions the simplest diagram that can give rise to anomalies is a triangle diagram, as depicted in Fig. 15.7(a). At one vertex we have attached the current whose conservation we wish to study. Gauge fields are attached to the other two. The anomaly that occurs typically has the form

$$\partial^\mu J_\mu \sim \varepsilon^{\mu\nu\rho\lambda} F_{\mu\nu} F_{\rho\lambda}.$$

Analogous phenomena can occur in higher dimensions, but then it is necessary to consider diagrams with more external lines. The reason is simply that the ε symbol has D indices in D dimensions. Thus in ten dimensions, for example, the simplest anomaly has the structure

$$\partial^\mu J_\mu \sim \varepsilon^{\mu_1\mu_2\cdots\mu_{10}} F_{\mu_1\mu_2} \cdots F_{\mu_9\mu_{10}}.$$

The corresponding Feynman diagram must have at least six legs (hexagon graph) as shown in Fig. 15.7(*b*).

In the case of string theories the situation is similar, but one should consider a Feynman diagram appropriate to string theory, i.e. a two-dimensional surface. For example, in the case of a theory involving closed oriented strings the only relevant diagram is a torus diagram with external strings (one to represent the gauge current and five to represent gauge fields), as shown in Fig. 15.8. In the case of type IIA superstrings all anomalies trivially cancel because of the left–right symmetry of the theory. In the case of the type IIB theory and the heterotic string theories the spectrum is not

Fig. 15.7. Gauge-current anomalies in $D=4$ can occur in triangle graphs such as the one in (*a*). In $D=10$ the simplest anomalous graph is the hexagon diagram shown in (*b*).

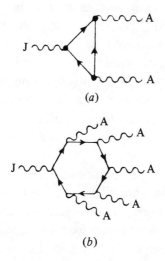

(*a*)

(*b*)

Fig. 15.8. The simplest potentially anomalous diagram in a $D=10$ theory of closed oriented strings is the torus with six external massless strings.

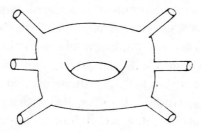

symmetric and the cancellation of anomalies is very non-trivial. In fact, the full-fledged string calculations have not yet been carried out. However, there is evidence (based on a low-energy expansion described below) that the cancellation does in fact take place for the theories listed in Table 15.1 [11, 14, 30]. In the case of type I superstrings surfaces of various different topologies need to be studied. The two that are relevant to pure gauge anomalies were evaluated by Green and me in 1984 [31]. We found that they individually give anomalies but that the anomaly cancels from the sum for the group choice $SO(32)$ only.

At low energies, string theories can be approximated by point-particle theories. This is physically reasonable since, when the wavelengths that occur are much longer than the strings, the spatial extension of the string becomes irrelevant. In practice one associates a quantum field with each massless mode of the string theory and writes an effective Lagrangian in terms of these fields only. The effects of massive string modes, and stringiness in general, are expanded out in a series of terms involving higher powers of derivatives and fields. The correction terms in such an expansion are typically suppressed by powers of E/M_P, where M_P is about 10^{19} GeV. Since the anomaly cancellation must take place at each order in this expansion, such a formulation is sufficient for investigating the cancellations in the leading orders of the low-energy expansion. This already gives very severe constraints. It is believed that if the leading-order constraints are satisfied then all the higher ones will also be, but this is not yet completely established.

For theories with $N=1$ supersymmetry in ten dimensions the contributions to hexagon anomalies arise entirely from chiral fermions. There is one Majorana–Weyl gravitino that makes a certain contribution, an additional Majorana–Weyl spinor from the supergravity multiplet, and n Majorana–Weyl spinors from the Yang–Mills supermultiplet for a gauge group with n generators. By considering the pure gravitational anomaly arising from hexagon diagrams with six external graviton lines one learns that a necessary condition for anomaly cancellation is that the gauge group have 496 generators. This requirement is satisfied, in particular, by $SO(32)$ which has $\frac{1}{2} \cdot 32 \cdot 31 = 496$ generators and by $E_8 \times E_8$ which has $248 + 248 = 496$ generators [11]. Thus one prediction of superstring theory is that there are 484 new forces beyond the 12 that are already known. (This counting does not include gravity, which is not associated with a Yang–Mills generator.) Of course, if this is to be reconciled with nature, almost all these symmetries must be broken at a very high mass scale.

There are many groups in addition to the two that have been mentioned that have 496 generators. To obtain additional restrictions we consider anomaly diagrams involving Yang–Mills fields as well as gravitons. The cancellation of all such anomalies leads to the additional requirement that an arbitrary generator F of the algebra, expressed in the adjoint representation, should satisfy the equation [11]

$$\text{Tr } F^6 = \tfrac{1}{48} \text{Tr } F^2 \text{Tr } F^4 - \tfrac{1}{14400} (\text{Tr } F^2)^3.$$

Remarkably this equation, with exactly the right coefficients, is satisfied for both $SO(32)$ and $E_8 \times E_8$. The only other possible solutions are $E_8 \times [U(1)]^{248}$ and $[U(1)]^{496}$, neither of which seems to correspond to a string theory. In the anomaly analysis an antisymmetric tensor field $B_{\mu\nu}$ that is part of the supergravity multiplet plays an important role. The details are given in the references.

The $SO(16) \times SO(16)$ heterotic string theory, listed in Table 15.1, has only 240 generators. This is possible because it is not supersymmetric. It is chiral, however, and the cancellation of anomalies involves 'miracles' analogous to those of the other theories [14].

The groups $SO(32)$ and $E_8 \times E_8$, singled out by the anomaly cancellation analysis in superstring theories, had previously arisen in another context [32]. Specifically, mathematicians have investigated lattices, like those of solid-state physics, in higher dimensions. They were led to consider in particular lattices that are self-dual (i.e. are coincident with the dual lattice) for which the distance squared of each lattice site from the origin is an even integer. (A cubic lattice generated by orthogonal unit vectors is self-dual but not even.) It turns out that even self-dual lattices only occur in dimensions that are multiples of eight. In eight dimensions there is just one and it is generated by the root vectors of the Lie algebra of E_8. In 16 dimensions there are two of them. One is obvious from the eight-dimensional construction, namely the root lattice generated by $E_8 \times E_8$. The second possibility is closely associated with $SO(32)$. It is generated by root vectors of the algebra $SO(32)$ and certain spinorial weights of its covering group spin(32). More precisely, it is the weight lattice of the group spin(32)/Z_2. This coincidence between self-dual lattices and the algebras singled out by the anomaly analysis was skillfully exploited in the construction of the heterotic string theories. The basic idea is that modes that travel clockwise (right-moving) on the string are described by the mathematics of a ten-dimensional superstring whereas those that travel counterclockwise (left-moving) are described by the mathematics of a 26-dimensional bosonic string. Ten of the

26 are paired with the right-moving coordinates to describe 'ordinary' ten-dimensional spacetime, whereas the remaining 16 are required to form a 16-dimensional torus that is conjugate to one of the two even self-dual lattices. These degrees of freedom, which are the origin of the Yang–Mills symmetry structure of the theory, can equivalently be described using fermionic coordinates instead [13].

15.6 Compactification

It may be that our theoretical understanding of superstring theory is not yet sufficiently developed to be able to do correct phenomenology. If we nevertheless choose to plunge in with reckless abandon, then it is clear that a crucial question is what to do with the six extra spatial dimensions. The natural guess is that they curl up to form a six-dimensional space K that is sufficiently small not to have been observed. In fact, the only fundamental scale in the theory is the Planck length, and it is natural to suppose that the internal space K is roughly of this size. This means that it is about the same size that is characteristic of strings themselves.

Since string theory contains gravity it should determine spacetime geometry dynamically, and so we must require that the 'background geometry' $M_4 \times K$ corresponds to a solution of the classical equations of motion. That is the classical statement sometimes called 'spontaneous compactification'. Quantum mechanically, it should be determined as part of the characterization of the vacuum: the quantum state of lowest energy. It is generally assumed, since that is all we can do at this point, that a classical solution is a good approximation to a quantum ground state. This might not be true. Another potential pitfall arises from describing K in terms of classical differential geometry. In doing this it is implicit that that geometry is determined entirely in terms of the gravitational field (metric tensor). However, this is just one of an infinity of modes of the string. Of course, it is singled out by being massless. The massive modes could play an equally important role, however, in characterizing the geometry and topology of K if K is not much larger than the characteristic string length scale. In this case a whole new type of geometry, let us call it 'string geometry', may be required for a suitable description of K. Since this does not yet exist, one assumes that it is not necessary. I would not be surprised, however, if the prevailing opinion changes when these matters are better understood.

Whatever the correct language for describing it may turn out to be, once the vacuum of the theory is identified correctly we shall be in a strong position for calculating many quantities of physical interest. The particle spectrum would be determined by a small oscillation analysis, i.e. by studying low-lying excitation modes. Also, as it turns out, many interesting quantities are controlled by the topology of K. Thus one can go a long way with qualitative topological information rather than quantitative geometric information. For example, in a large class of models that has been considered the number of generations of quarks and leptons is controlled by the Euler characteristic of K

$$N_{gen} = \tfrac{1}{2}|\chi(K)|.$$

A rather specific compactification scenario was proposed in the paper of Candelas, Horowitz, Strominger and Witten [33]. They begin with the $E_8 \times E_8$ theory and argue that a classical solution is obtained if K is a Calabi–Yau space, a Kähler manifold of $SU(3)$ holonomy (vanishing first Chern class). Such a manifold always admits a Ricci-flat metric. It has recently been realized that, beginning at fourth order in an expansion in the string length scale, this is not the metric that solves the string equations of motion [34]. However, it has been argued convincingly that there is always another 'nearby' metric that does [35]. In formulating the solution one identifies the connection of the space K with an $SU(3)$ subgroup of $E_8 \times E_8$. In fact it is embedded entirely in one E_8 factor which thereby breaks down to E_6, from which the usual gauge group should emerge.

The E_6 group described above can be broken further if K is not simply connected. In this case there are non-contractible loops in K around which gauge fields can wrap to give non-zero Wilson-loop integrals even though the corresponding field strengths vanish. (This is analogous to the Bohm–Aharanov effect.) In this way it is possible to break E_6 down to a group close to the standard model but containing at least an extra $U(1)$. This would imply the existence of a second Z boson at the electroweak scale. There are several alternative scenarios, however.

One alternative 'superstring-inspired model' [36] starts with a specific Calabi–Yau space K with $\chi = 6$ and fundamental group $\pi_1 = Z_3$. Wilson loops are used to break E_6 to $SU(3) \times SU(3) \times SU(3)$. Then, by introducing some optimistic assumptions about a resulting potential, it is suggested that this could break to $SU(3) \times SU(2) \times U(1)$ at an intermediate scale by a standard Higgs mechanism. I am sure Ross and his collaborators would agree that this is unlikely to be the final correct model. Still, it is remarkable how close they can come to accounting for a large number of desiderata in this way. There are indications that the axions may not have the required properties, however [37].

15.7 Remaining problems and conclusions

Superstring theory is an enormously ambitious attempt to account for all properties of fundamental forces and particles. It starts by describing the physics at the Planck scale. From this all the physics at ordinary energies should be derivable. However, the gap is so large that ordinary physics is, in a sense, all hyperfine structure. Working out these connections in detail is surely going to be a very long struggle requiring the efforts of many clever people. Whether it will ultimately succeed is hard to foresee, even if we assume that the theory is correct. The experimental information that will be learned from the next generation of accelerators, especially as concerns supersymmetry and the Higgs sector, should provide very valuable clues. Without it we are unlikely to find our way.

There are a number of theoretical issues, less directly connected with phenomenology, that must also be addressed. We would like to know how many consistent superstring theories there are. Ideally, there is just one and it explains everything, but there is certainly no guarantee that this is the case. Further work is required in formulating the theories. It seems likely that a deep and beautiful principle underlies string theory, but it remains to be elucidated. The work on string field theory is one approach that is being pursued. Some other approach may, however, be required. One alternative that looks interesting is based on the analytic geometric of conformal field theory [38]. Whatever the answer may be, it will serve to define the theory non-perturbatively and undoubtedly lead to new insights. It could happen that by the time the dust settles the subject will look radically different than it does today.

Superstring theory was developed primarily because it seemed promising for overcoming the divergence problems of quantum gravity. However, even classical general relativity has its problems. It has been proved that generic initial data lead to singularities. This is bad because the structure of spacetime subsequent to the formation of the singularity is not determined. The theory is therefore incomplete. It is a plausible conjecture that classical superstring theory is not subject to the same problems (these remarks are mostly due to D. Gross). They are short-distance effects, and it is at short distances that the theory is modified. The 1986 result [34] that a general solution of $R_{\mu\nu} = 0$ is not a solution of string theory is encouraging, since it implies that the Schwarzschild solution, with its singularity, is averted. The actual proof that string theory is a complete classical theory will undoubtedly require a lot of work. The discovery of the optimal formulation and underlying principles discussed above is probably a necessary preliminary.

Other questions that need to be studied are why a particular ground state of the form $M_4 \times K$ should be singled out, if that is the case. It would be very sad if there were thousands of theoretically acceptable vacua, since then much predictivity would be lost despite the uniqueness of the underlying theory. We need to understand the origin of the mass hierarchy $m_W/m_P \sim 10^{-17}$ and the origin of the supersymmetry breaking. These are tough problems, but the one that really worries us the most is to understand why the cosmological constant $|\Lambda|$ is less than 10^{-120} in natural units. Every little effect one can think of makes a much larger contribution to Λ than this. Why they should all precisely cancel is, at the moment, totally baffling.

In conclusion, the theoretical understanding of superstrings is progressing rapidly. The long-term prospects for the subject appear very bright. Developing this subject is an exhilarating intellectual experience, but it requires a major commitment of time and effort that is not appropriate for everyone. In any case, there is plenty of work to keep busy those who choose to make this commitment while we await the next round of experimental results.

References

1. Schwarz, J. H. (1985). *Superstrings. The First Fifteen Years of Superstring Theory.* World Scientific;
   Green, M. B., Schwarz, J. H. and Witten, E. (1987). *Superstring Theory.* Cambridge University Press: Cambridge.
2. Goroff, M. and Sagnotti, A. (1986). *Nucl. Phys.*, **B266**, 709.
3. Jacob, M. (ed.) (1974). *Dual Theory.* Physics Reports Reprint Volume I. North Holland: Amsterdam;
   Scherk, J. (1975). An introduction to the theory of dual models and strings. *Rev. Mod. Phys.*, **47**, 123.
4. Neveu, A. and Schwarz, J. H. (1971). *Nucl. Phys.*, **B31**, 86; *Phys. Rev.*, **D4**, 1108;
   Neveu, A., Schwarz, J. H. and Thorn, C. B. (1971). *Phys. Lett.*, **35B**, 529;
   Thorn, C. B. (1971). *Phys. Rev.*, **D4**, 1112.
5. Ramond, P. (1971). *Phys. Rev.*, **D3**, 2415.
6. Gliozzi, F., Scherk, J. and Olive, D. (1976). *Phys. Lett.*, **65B**, 282; (1977) *Nucl. Phys.*, **B122**, 253.
7. Green, M. B. and Schwarz, J. H. (1981). *Nucl. Phys.*, **B181**, 502.
8. Goddard, P. and Thorn, C. B. (1972). *Phys. Lett.*, **40B**, 235;
   Schwarz, J. H. (1972). *Nucl. Phys.*, **B46**, 61;
   Brower, R. C. and Friedman, K. A. (1973). *Phys. Rev.*, **D7**, 535.
9. Scherk, J. and Schwarz, J. H. (1974). *Nucl. Phys.*, **B81**, 118.

10. Yoneya, T. (1974). *Prog. Theor. Phys.*, **51**, 1907.

11. Green, M. B. and Schwarz, J. H. (1984). *Phys. Lett.*, **149B**, 117.

12. Green, M. B. and Schwarz, J. H. (1982). *Phys. Lett.*, **109B**, 444.

13. Gross, D. J., Harvey, J. A., Martinec, E. and Rohm, R. (1985). *Phys. Rev. Lett.*, **54**, 502; (1985) *Nucl. Phys.*, **B256**, 253; (1986) *Nucl. Phys.*, **B267**, 75.

14. Dixon, L. and Harvey, J. A. (1986). *Nucl. Phys.*, **B274**, 93;
    Alvarez-Gaumé, L., Ginsparg, P., Moore, G. and Vafa, C. (1986). *Phys. Lett.*, **171B**, 155.

15. Seiberg, N. and Witten, E. (1986). *Nucl. Phys.*, **B276**, 272.

16. Brink, L., Di Vecchia, P. and Howe, P. (1976). *Phys. Lett.*, **65B**, 471;
    Deser, S. and Zumino, B. (1976). *Phys. Lett.*, **65B**, 369.

17. Polyakov, A. M. (1981). *Phys. Lett.*, **103B**, 207, 211.

18. Alvarez, O. (1983). *Nucl. Phys.*, **B216**, 125;
    D'Hoker, E. and Phong, D. (1986). *Nucl. Phys.*, **B269**, 205;
    Belavin, A. and Knizhnik, V. (1986). *Phys. Lett.*, **168B**, 201.
    Alvarez-Gaumé, L., Moore, G. and Vafa, C. (1986). *Commun. Math. Phys.*, **106**, 1;
    Mannin, Yu. I. (1986). *Pisma ZETP*, **43**, 161; (1986) *Phys. Lett.*, **172B**, 184;
    Bost, J. B. and Nelson, P. Harvard preprint HUTP-86-A014;
    Bost, J. B. and Jolicoeur, T. (1986). *Phys. Lett.*, **174B**, 273;
    Catenacci, R., Cornalba, M., Martellini, M. and Reina, C. (1986). *Phys. Lett.*, **172B**, 328;
    Gomez, C. (1986). *Phys. Lett.*, **175B**, 32;
    Alvarez-Gaumé, L., Moore, G., Nelson, P., Vafa, C. and Bost, J. B. Harvard preprint HUTP-86/A039.
    Moore, G., Harris, J., Nelson, P. and Singer, I. Harvard preprint HUTP-86/A051;
    Morozov, A. Preprint ITEP 86-88;
    Belavin, A., Knizhnik, V., Morozov, A. and Perelomov. Preprint ITEP 86-59;
    Beilinson, A. A. and Manin, Yu. I. (1986). Moscow preprints.
    Restuccia, A. and Taylor, J. G. (1986). *Phys. Lett.*, **174B**, 56.

19. Cremmer, E. and Gervais, J. L. (1974). *Nucl. Phys.*, **B76**, 209; (1975) *Nucl. Phys.*, **B90**, 410;
    Kaku, M. and Kikkawa, K. (1974). *Phys. Rev.*, **D10**, 1110, 1823.

20. Green, M. B. and Schwarz, J. H. (1984). *Nucl. Phys.*, **B218**, 43; (1984) *Nucl. Phys.*, **B243**, 475;
    Green, M. B., Schwarz, J. H. and Brink, L. (1983). *Nucl. Phys.*, **B219**, 437.

21. Siegel, W. (1985). *Phys. Lett.*, **151B**, 391, 396;
    Siegel, W. and Zwiebach, B. (1986). *Nucl. Phys.*, **B263**, 105;

22. Banks, T. and Peskin, M. (1986). *Nucl. Phys.*, **B264**, 513;
    Peskin, M. E. and Thorn, C. B. (1986). *Nucl. Phys.*, **B269**, 509.

23. Neveu, A. and West, P. C. (1985). *Phys. Lett.*, **165B**, 63, and CERN preprints.

24. Hata, H., Itoh, K., Kugo, T., Kunitomo, H. and Ogawa, K. (1986). *Phys. Lett.*, **172B**, 186, 195, and Kyoto University preprints.

25. Witten, E. (1986). *Nucl. Phys.*, **B268**, 253; **B276**, 291.

26. Giddings, S. B., Martinec, E. and Witten, E. (1986). *Phys. Lett.*, **176B**, 362.

27. Kato, M. and Ogawa, K. (1983). *Nucl. Phys.*, **B212**, 443;
    Hwang, S. (1983). *Phys. Rev.*, **D28**, 2614.

28. Gross, D. and Jevicki, A. (1986). Princeton preprint;
    Cremmer, E., Schwimmer, A. and Thorn, C. B. Ecole Normale Supérieure preprint LPTENS-86-14.

29. Giddings, S. B. (1986). *Nucl. Phys.*, **B278**, 242;
    Giddings, S. B. and Martinec, E. (1986). *Nucl. Phys.*, **B278**, 91.
30. Alvarez-Gaumé, L. and Witten, E. (1983). *Nucl. Phys.*, **B234**, 269.
31. Green, M. B. and Schwarz, J. H. (1985). *Nucl. Phys.*, **B255**, 93.
32. Goddard, P. and Olive, D. (1985). In *Vertex Operators in Mathematics and Physics*,
    ed. J. Lepowsky *et al.*, p. 51. Springer: Berlin.
33. Candelas, P., Horowitz, G., Strominger, A. and Witten, E. (1985). *Nucl. Phys.*, **B258**,
    75.
34. Grisaru, M. T., Van de Van, A. E. M. and Zanon, D. (1986). *Phys. Lett.*, **173B**, 423;
    Freeman, M. D. and Pole, C. N. (1986). *Phys. Lett.*, **174B**, 48;
    Gross, D. and Witten, E. (1986). *Nucl. Phys.*, **B277**, 1.
35. Witten, E. (1986). *Nucl. Phys.*, **B268**, 79.
    Nemeschansky, D. and Sen, A. (1986). SLAC preprint.
36. Greene, B. R., Kirklin, K. H., Miron, P. J. and Ross, G. G. (1986). Oxford
    University preprints.
37. Kim, J. E. (1986). Seoul National University preprint;
    Dine, M. and Seiberg, N. (1986). *Nucl. Phys.*, **B273**, 109.
38. Friedan, D. and Shenker, S. (1987). *Nucl. Phys.*, **B281**, 509.

16

Covariant description of canonical formalism in geometrical theories

CEDOMIR CRNKOVIC and EDWARD WITTEN†

Quantum field theories are usually studied either by means of path integrals or by means of canonical quantization. Path integral quantization has the great virtue of explicitly maintaining all relevant symmetries, such as Poincaré invariance. The canonical approach is usually interpreted as an approach that ruins Poincaré invariance from the beginning through an explicit choice of a 'time' coordinate. This is not necessarily so, however. The essence of the canonical formalism can be developed in a way that manifestly preserves all relevant symmetries, including Poincaré invariance (Witten, 1986; Zuckerman, 1986). The purpose of the present paper is to carry this out in the case of non-abelian gauge theories and general relativity.

In the canonical formalism of a theory with N degrees of freedom, one usually introduces coordinates and momenta p^i and q^j, $i, j = 1, \ldots, N$. One then defines the two-form

$$\omega = \mathrm{d}p^i \wedge \mathrm{d}q^i. \tag{1}$$

It is convenient to combine the p^i and q^j in a variable Q^I, $I = 1, \ldots, 2N$, with $Q^i = p^i$ for $i \leqslant N$ and $Q^i = q^{i-N}$ for $i > N$. One can think of ω as an antisymmetric $2N \times 2N$ matrix ω_{IJ} whose non-zero matrix elements are $\omega_{i,i+N} = -\omega_{i+N,i} = 1$. This matrix is invertible; we will denote the inverse matrix as ω^{IJ}. One defines the Poisson bracket of any two functions $A(Q^I)$ and $B(Q^I)$ by

$$[A, B] = \omega^{IJ} \frac{\partial A}{\partial Q^I} \frac{\partial B}{\partial Q^J}. \tag{2}$$

† Research supported in part by the National Science Foundation under grants PHY80-19754 and PHY86-16129.

Given the form of (1), this is easily seen to coincide with the usual definitions of Poisson brackets. The advantage of the definition in (2) is that, as is well known (see e.g. Abraham and Marsden, 1967), the essential features of ω can be described in an invariant way. Let Z be the phase space of the theory under discussion, that is, the space on which the ps and qs are coordinates. If one interprets ω as a two-form on Z, then it is clearly a closed two-form,

$$d\omega = 0, \tag{3}$$

since its components are constant in the coordinate system used in (1). What is more, we have already noted that the matrix ω is invertible. The converse to this is as follows. Let ω be any two-form on a manifold Z (which for us will be the phase space of a physical theory). Suppose that ω is closed (obeys (3)), and is non-degenerate in the sense that at each point $z \in Z$, the matrix $\omega_{ij}(z)$ is invertible. Then it is a classical theorem that locally one can introduce coordinates on Z to put ω in the standard form (1). (This is not true globally in theories with interesting geometrical content.) A non-degenerate closed two-form is called a symplectic structure. Thus, to describe the canonical formalism of a theory it is not at all necessary to find or choose ps and qs; the essence of the matter is to describe a symplectic structure on the classical phase space.

Clearly, the notion of a 'symplectic structure on phase space' is a more intrinsic concept than the idea of choosing ps and qs. However, at first sight it might appear that the very concept of phase space is a non-covariant concept, tied to a non-covariant, Hamiltonian description. This is not really so. The whole idea, classically, of picking ps and qs is that the initial values of the ps and qs determine a solution of the classical equations. More precisely, classical solutions of any given physical theory, in any given coordinate system, are in one-to-one correspondence with the values of the ps and qs at time zero. This simple consideration leads us to a manifestly covariant definition of what we mean by classical phase space: *in a given physical theory, classical phase space is the space of solutions of the classical equations.* We can always, if we wish, pick a coordinate system and identify the classical solutions with the initial data in that coordinate system, but there is no necessity to make such a non-covariant choice.

Given a relativistic field theory, such as the scalar field theory with Lagrangian

$$L = \int_M \left(\frac{1}{2} \partial_\alpha \phi \, \partial^\alpha \phi - V(\phi) \right), \tag{4}$$

(here M is spacetime, and ϕ is a scalar field), how do we go about finding a covariant description of a symplectic structure on the space, Z, of classical solutions? A point in Z is a solution of the classical equations

$$0 = \Delta\phi - V'(\phi), \tag{5}$$

with $\Delta = -\partial_\alpha \partial^\alpha$ being the standard Laplacian. In order to construct a symplectic structure on Z, we will need to discuss functions, tangent vectors, differential forms, and exterior derivatives on Z.

Functions on Z are, of course, the easiest to discuss. Among the most important are the following. Let $x \in M$ be a spacetime point. If ϕ is a solution of (5), its value $\phi(x)$ at the point x is a real number. The mapping from the *function* ϕ to the *number* $\phi(x)$ is a real valued function on Z. We will denote this function by $\phi(x)$.

Now we will consider tangent vectors. At a point $p \in Z$, corresponding to a solution of (5), a tangent vector would be a small displacement in ϕ which preserves (5). Thus, a tangent vector is the same as a solution of the linearized equations which we obtain by expanding around a solution of (5). Requiring that $\tilde{\phi} = \phi + \delta\phi$ should obey (5) to lowest order in $\delta\phi$, we find

$$0 = \Delta\,\delta\phi - V''(\phi)\,\delta\phi. \tag{6}$$

A solution of (6) is a tangent vector at the point in phase space corresponding to the solution ϕ. The tangent space is the vector space T of solutions of (6).

Now, how do we describe one-forms on Z? The space of one-forms is, of course, the dual of the tangent space which we have just described; it is the space of linear functionals on T. Most important for our purposes are certain one-forms on Z which we will now describe. Let $x \in M$ be a spacetime point. For every solution $\delta\phi$ of (6), its value $\delta\phi(x)$ at the point x is, of course, a number. The transformation from the function $\delta\phi$ to the number $\delta\phi(x)$ is a one-form on phase space which we will call $\delta\phi(x)$. More generally, we can make k-forms as wedge functions of the one-forms $\delta\phi(x)$:

$$A = \int dx_1 \ldots dx_n\, \alpha_{x_1 \ldots x_n}(\phi)\, \delta\phi(x_1)\, \delta\phi(x_2) \ldots \delta\phi(x_n). \tag{7}$$

Here, for each n-tuple of spacetime points $x_1 \ldots x_n$, $\alpha_{x_1 \ldots x_n}(\phi)$ is an arbitrary zero form – an arbitrary real valued function on Z. On the right-hand side of (7), the product of the $\delta\phi(x_i)$ is understood to be a wedge product. In particular, as the wedge product of one-forms is anticommuting, we interpret the $\delta\phi(x)$ as anticommuting objects:

$$\delta\phi(x)\,\delta\phi(y) = -\delta\phi(y)\,\delta\phi(x). \tag{8}$$

The general n-form on Z can be expanded as in (7); this expansion, however, is not unique, since the $\delta\phi(x)$ are not linearly independent, being subject to (6).

Finally, we need an exterior derivative on Z, which we will call δ and which must map k forms to $(k+1)$-forms. It should obey

$$\delta^2 = 0 \tag{9}$$

and the Leibniz rule

$$\delta(AB) = \delta A \cdot B + (-1)^A A \cdot \delta B. \tag{10}$$

We define δ by saying that acting on the zero-form (function) $\phi(x)$, δ gives the one-form that we have called $\delta\phi(x)$. The Leibniz rule then determines the action of δ on an arbitrary zero-form Γ:

$$\delta(\Gamma(\phi)) = \int dx \frac{\delta\Gamma}{\delta\phi(x)} \delta\phi(x), \tag{11}$$

where $\delta\Gamma/\delta\phi(x)$ is just the variational derivative of Γ with respect to $\phi(x)$. Equation (11) is a familiar formula to which we are giving, perhaps, a slightly novel meaning. To act with δ on k-forms of $k > 0$, one must bear in mind, first of all, that as the one-form $\delta\phi(x)$ is the exterior derivative of the zero-form $\phi(x)$, it must be closed:

$$\delta(\delta\phi(x)) = 0. \tag{12}$$

Also, using the Leibniz rule, we then have the exterior derivative of a general k form (7):

$$\delta A = \int dx_0 \ldots dx_n \frac{\delta\alpha_{x_1\ldots x_n}(\phi)}{\delta\phi(x_0)} \delta\phi(x_0)\,\delta\phi(x_1) \ldots \delta\phi(x_n). \tag{13}$$

Although our definition of δ has been rather formal, one can readily see that it possesses the standard properties of the exterior derivative. Thus, if V is a vector field, Λ a zero-form, and i_V the operation of contraction with V, we have $V(\Lambda) = i_V \delta\Lambda$.

Having defined the relevant concepts, how are we to find a symplectic structure in, say, the scalar field theory (4)? The idea (Witten, 1986; Zuckerman, 1986) is to consider the 'symplectic current'

$$J_\alpha(x) = \delta\phi(x)\,\partial_\alpha\,\delta\phi(x). \tag{14}$$

At each spacetime point, (14) is a two-form on Z; but in its dependence on x, J_α is a conserved current:

$$\partial_\alpha J^\alpha(x) = 0. \tag{15}$$

To verify (15), one needs the equation of motion (6) and the fact that $\delta\phi$ is anticommuting. As J_α is conserved, its integral over an initial value hypersurface Σ,

$$\omega = \int_\Sigma d\Sigma_\alpha J^\alpha, \tag{16}$$

is independent of the choice of Σ and so in particular is Poincaré invariant. The two-form ω is our desired symplectic structure on Z. It is evidently closed, in view of (12), and it is easy to see that upon picking Σ to be the standard initial value surface $t = 0$, (16) reduces to

$$\omega = \int_\Sigma \delta\phi \, \delta\dot\phi, \tag{17}$$

which (as in (1)) is the standard formula.

Our goal in the present paper is to implement this procedure in the case of Yang–Mills theory and general relativity. The main novelty that arises is the need to establish gauge invariance as well as Poincaré invariance of the symplectic structure. We will consider the two cases in turn.

16.1 Yang–Mills theory

Considering first Yang–Mills theory, let A be the gauge connection and F the Yang–Mills curvature or field strength. The covariant derivative of a charged field Λ is $D_\alpha \Lambda = \partial_\alpha \Lambda + [A_\alpha, \Lambda]$. The Yang–Mills equation of motion is

$$0 = [D^\mu, F_{\mu\nu}] = \partial^\mu F_{\mu\nu} + [A^\mu, F_{\mu\nu}]. \tag{18}$$

The variation of (18) is

$$D^\mu \, \delta F_{\mu\nu} + [\delta A^\mu, F_{\mu\nu}] = 0. \tag{19}$$

We define the symplectic current $J_\alpha(x)$ as

$$J_\alpha = \mathrm{Tr}[\delta A^\mu \, \delta F_{\mu\alpha}], \tag{20}$$

where

$$\delta F_{\mu\alpha} = D_\mu \, \delta A_\alpha - D_\alpha \, \delta A_\mu. \tag{21}$$

Let us show that J_α is conserved. We have

$$\begin{aligned}
\partial^\alpha J_\alpha &= \mathrm{Tr}[D^\alpha \, \delta A^\mu \, \delta F_{\alpha\mu}] + \mathrm{Tr}[\delta A^\mu D^\alpha \, \delta F_{\alpha\mu}] \\
&= \tfrac{1}{2}\mathrm{Tr}[\delta F^{\alpha\mu} \, \delta F_{\alpha\mu}] - \mathrm{Tr}[\delta A^\mu [\delta A^\alpha, \delta F_{\alpha\mu}]] = 0.
\end{aligned} \tag{22}$$

On the last line of (22), the second term vanishes because δA is anticommuting, and the first because δF is anticommuting. So the two-form

$$\omega = \int_\Sigma \mathrm{Tr}[\delta A^\mu \, \delta F_{\alpha\mu}] \, d\Sigma^\alpha \tag{23}$$

is Poincaré invariant.

It is easy to see that ω is also closed; δA^μ, being the exterior derivative of the zero-form A^μ, is closed, while, in view of (21) and the anticommutativity of δ, we have

$$\delta(\delta F_{\mu\alpha}) = \delta(D_\mu \delta A_\alpha - D_\alpha \delta A_\mu) = \delta([A_\mu, \delta A_\alpha] - [A_\alpha, \delta A_\mu])$$
$$= \{\delta A_\mu, \delta A_\alpha\} - \{\delta A_\alpha, \delta A_\mu\} = 0. \tag{24}$$

Therefore, ω is closed.

It remains to discuss the behavior of ω under gauge transformations. First of all, the gauge transformation law for the gauge field is

$$A_\mu \to A_\mu + [D_\mu, \varepsilon] = A_\mu + \partial_\mu \varepsilon + [A_\mu, \varepsilon]. \tag{25}$$

Varying (25), we find that under gauge transformations, δA transforms as

$$\delta A_\mu \to \delta A_\mu + [\delta A_\mu, \varepsilon]. \tag{26}$$

In particular, δA transforms homogeneously under gauge transformations. And δF transforms in the same way:

$$\delta F_{\mu\alpha} \to \delta F_{\mu\alpha} + [\delta F_{\mu\alpha}, \varepsilon]. \tag{27}$$

Consequently, J^α and ω are gauge invariant.

This is an important step in the right direction, but it is not the end of the story. Let $\hat{Z}$ be the space of solutions of (18), and let Z be the space of solutions of (18) modulo gauge transformations. Thus, $Z = \hat{Z}/G$, with G being the group of gauge transformations. So far we have defined a gauge-invariant closed two-form ω on $\hat{Z}$. What we want is a gauge-invariant closed two-form on Z. We would like to show that the differential form ω that we have defined on $\hat{Z}$ is the pullback from Z to $\hat{Z}$ of a differential form on Z, which we will also call ω. This will be so if the following condition is obeyed. If V is any vector field tangent to the G orbits on $\hat{Z}$, and i_V is the operation of contraction with V, the requirement is $i_V \omega = 0$. This is a fancy way of saying that the components of ω in the gauge directions are zero; one must require this since the gauge directions are eliminated in passing from $\hat{Z}$ to Z, so a differential form on Z cannot have non-zero components in those directions.

In Yang–Mills theory, the gauge directions in field space are simply

$$\delta A_\mu = D_\mu \varepsilon. \tag{28}$$

More generally, we can consider a field variation

$$\delta A_\mu = \delta' A_\mu + D_\mu \varepsilon, \tag{29}$$

which has a gauge component $D_\mu \varepsilon$, and another component $\delta' A_\mu$ which is not pure gauge. To verify that ω has vanishing components in the gauge directions, we must show that if we insert (29) in the definition of ω, the term proportional to ε drops out. The expression which must vanish is

$$\Delta\omega = \int d\Sigma_\alpha \, \mathrm{Tr}[D_\mu \varepsilon \, \delta F^{\alpha\mu} + \delta A_\mu [F^{\alpha\mu}, \varepsilon]] = \int d\Sigma_\alpha \, \partial_\mu \, \mathrm{Tr}(\varepsilon F^{\alpha\mu}). \qquad (30)$$

In the last step, we have used anticommutativity of ε and $\delta F^{\alpha\mu}$ and the equation of motion (19). Indeed, (30) vanishes, being the integral of a total derivative, and this completes the construction of a symplectic structure on the gauge-invariant space Z.

16.2 General relativity

We now turn our attention to general relativity. The discussion is similar, although somewhat more complicated. Let $g_{\mu\nu}$ be the metric tensor of spacetime, and $R_{\mu\nu\alpha\beta}$ the corresponding Riemann tensor. The equation of motion of pure general relativity is the Einstein equation $R_{\mu\nu} = 0$. To find its variation, remember that for any vector field V, one has

$$[D_\mu, D_\nu] V^\alpha = R^\alpha{}_{\lambda\mu\nu} V^\lambda. \qquad (31)$$

One finds for the variation of the Einstein equation

$$0 = D_\alpha \, \delta\Gamma^\alpha_{\mu\nu} - D_\mu \, \delta\Gamma^\alpha_{\nu\alpha}. \qquad (32)$$

Here

$$\delta\Gamma^\alpha_{\mu\nu} = \tfrac{1}{2} g^{\alpha\beta} (D_\mu \, \delta g_{\nu\beta} + D_\nu \, \delta g_{\mu\beta} - D_\beta \, \delta g_{\mu\nu}) \qquad (33)$$

is the variation of the Levi-Civita connection Γ. Note that $\delta\Gamma$ transforms as a tensor. We define the symplectic current

$$J^\alpha = \delta\Gamma^\alpha_{\mu\nu}[\delta g^{\mu\nu} + \tfrac{1}{2} g^{\mu\nu} \, \delta \ln g] - \delta\Gamma^\nu_{\mu\nu}[\delta g^{\alpha\mu} + \tfrac{1}{2} g^{\alpha\mu} \, \delta \ln g], \qquad (34)$$

where $\delta \ln g \equiv \delta \ln \det(g_{\mu\nu}) = g^{\mu\nu} \, \delta g_{\mu\nu} = -g_{\mu\nu} \, \delta g^{\mu\nu}$. Also keep in mind that $\delta g^{\mu\nu} = -g^{\mu\alpha} g^{\nu\beta} \, \delta g_{\alpha\beta}$.

We have to show that $D_\alpha J^\alpha = 0$. Terms with $\delta \ln g$ give

$$\tfrac{1}{2}(D_\alpha \, \delta\Gamma^\alpha_{\mu\nu}) g^{\mu\nu} \, \delta \ln g - \tfrac{1}{2}(D_\alpha \, \delta\Gamma^\nu_{\mu\nu}) g^{\alpha\mu} \, \delta \ln g + \tfrac{1}{2} \delta\Gamma^\alpha_{\mu\nu} g^{\mu\nu} D_\alpha \, \delta \ln g$$
$$-\tfrac{1}{2} \delta\Gamma^\nu_{\mu\nu} g^{\alpha\mu} D_\alpha \, \delta \ln g = \delta\Gamma^\alpha_{\mu\nu} g^{\mu\nu} \, \delta\Gamma^\lambda_{\alpha\lambda}, \qquad (35)$$

where we have used (33), (32), the fact that the metric tensor is covariantly constant, and the anticommutativity of $\delta\Gamma$. The remaining terms are

$$(D_\alpha \, \delta\Gamma^\alpha_{\mu\nu}) \, \delta g^{\mu\nu} + \delta\Gamma^\alpha_{\mu\nu} D_\alpha \, \delta g^{\mu\nu} - (D_\alpha \, \delta\Gamma^\nu_{\mu\nu}) \, \delta g^{\alpha\mu} - \delta\Gamma^\nu_{\mu\nu} D_\alpha \, \delta g^{\alpha\mu}$$
$$= -\delta\Gamma^\nu_{\mu\nu} D_\alpha \, \delta g^{\alpha\mu} = \delta\Gamma^\nu_{\mu\nu} \, \delta\Gamma^\mu_{\alpha\beta} g^{\alpha\beta},$$

which, together with (35), gives $D_\alpha J^\alpha = 0$. This makes

$$\omega = \int_\Sigma d\Sigma_\alpha \sqrt{g} \, J^\alpha$$

Poincaré invariant.

Let us show that ω is closed, $\delta\omega = 0$. We have

$$\delta\omega = \int_\Sigma d\Sigma_\alpha (\delta \sqrt{g}\, J^\alpha + \sqrt{g}\, \delta J^\alpha)$$

$$\delta J^\alpha = -\tfrac{1}{2}\delta\Gamma^\alpha_{\mu\nu}\,\delta g^{\mu\nu}\,\delta\ln g + \tfrac{1}{2}\delta\Gamma^\nu_{\mu\nu}\,\delta g^{\alpha\mu}\,\delta\ln g.$$

Remembering that $\delta\ln g$ is an anticommuting one-form whose square is zero, we have

$$\delta J^\alpha = -\tfrac{1}{2}J^\alpha\,\delta\ln g.$$

As $\delta\sqrt{g} = \tfrac{1}{2}\sqrt{g}\,\delta\ln g$, $\delta\omega = 0$. Thus, ω is closed.

It now remains to investigate gauge invariance of ω. The fact that ω is invariant under diffeomorphisms is relatively trivial; it follows from the fact that all ingredients in the definition of ω, including $\delta\Gamma$, transform homogeneously, like tensors. As in the Yang–Mills case, the more delicate point is to show that we obtain a closed two-form not just on the space $\hat{Z}$ of solutions of the Einstein equations, but also on the subtler space $Z = \hat{Z}/G$, with G being the group of diffeomorphisms. We must show that components of ω tangent to the G orbits vanish. Under a diffeomorphism $x_\mu \to x_u + \varepsilon_\mu$, the metric changes by

$$g_{\mu\nu} \to g_{\mu\nu} + D_\mu\varepsilon_\nu + D_\nu\varepsilon_\mu.$$

We assume ε has compact support or, more generally, is asymptotic at infinity to a Killing vector field. It we write

$$\delta g_{\mu\nu} = \delta' g_{\mu\nu} + D_\mu\varepsilon_\nu + D_\nu\varepsilon_\mu, \tag{36}$$

where the $D\varepsilon$ terms are pure gauge but $\delta' g$ is not, then our task is to show that, with (36) inserted in ω, the $D\varepsilon$ terms do not contribute. It is useful first to rewrite J^α as

$$J^\alpha = \tfrac{1}{2}g^{\alpha\beta}(D_\mu\,\delta g_{\nu\beta} + D_\nu\,\delta g_{\mu\beta} - D_\beta\,\delta g_{\mu\nu})\,\delta g^{\mu\nu}$$
$$- \tfrac{1}{2}D_\mu\,\delta g^{\mu\alpha}\,\delta\ln g + \tfrac{1}{2}\delta g^{\mu\alpha}D_\mu\,\delta\ln g - \tfrac{1}{2}(D^\alpha\,\delta\ln g)\,\delta\ln g. \tag{37}$$

Inserting (36) in (37), the term linear in ε is (after dropping the ' from $\delta' g$)

$$\Delta J^\alpha = [D_\mu, D^\alpha]\varepsilon_\nu\,\delta g^{\mu\nu} + [D_\nu, D_\mu]\,\delta g^{\alpha\mu}\varepsilon^\nu$$
$$+ D_\nu[(D_\mu\,\delta g^{\mu\nu} + D^\nu\,\delta\ln g)\varepsilon^\alpha + D^\nu\,\delta g^{\mu\alpha}\varepsilon_\mu$$
$$+ D_\mu\varepsilon^\alpha\,\delta g^{\mu\nu} + \tfrac{1}{2}D^\nu\varepsilon^\alpha\,\delta\ln g - (\alpha \leftrightarrow \nu)], \tag{38}$$

where we have used (33), (32), and the identity $[D_\mu, D_\nu]\varepsilon^\mu = 0$, which follows from $R_{\mu\nu} = 0$. We are entitled to discard from ΔJ^μ terms of the form $D_\sigma X^{\mu\sigma}$, with $X^{\mu\sigma}$ being an antisymmetric tensor; such terms will vanish when inserted in the integral for ω. Discarding such total derivatives from (38), we are left with

$$R_{v\lambda\mu}{}^{\alpha}\varepsilon^{\lambda}\,\delta g^{\mu\nu} + (R^{\mu}{}_{\lambda\nu\mu}\,\delta g^{\lambda\alpha} + R^{\alpha}{}_{\lambda\nu\mu}\,\delta g^{\mu\lambda})\varepsilon^{\nu}$$

$$= (R_{v\lambda\mu}{}^{\alpha} + R^{\alpha}{}_{v\mu\lambda})\varepsilon^{\lambda}\,\delta g^{\mu\nu} = g^{\alpha\rho}R_{\rho\lambda\mu\nu}\varepsilon^{\lambda}\,\delta g^{\mu\nu} = 0.$$

Therefore, components of ω tangent to the action of the diffeomorphism group are zero.

In conclusion, we have described in a manifestly covariant way the foundations of the canonical formalism of Yang–Mills theory and general relativity. Since a similar treatment of string theory has been given elsewhere, it seems that such an approach is possible for all of the geometrical theories in physics.

References

Abraham, R. and Marsden, J. E. (1967). *Foundations of Mechanics*. Benjamin: New York.

Witten, E. (1986). Interacting field theory of open superstrings. *Nucl. Phys.*, **B276**, 291.

Zuckerman, G. (1986). Action functionals and global geometry. Yale University preprint, to appear in the proceedings of the San Diego workshop, ed. S.-T. Yau *et al.*